Laboratory Manual for Anatomy & Physiology

featuring MARTINI ART

FIFTH EDITION   MAIN VERSION

Laboratory Manual for Anatomy & Physiology

featuring MARTINI ART

FIFTH EDITION   MAIN VERSION

Michael G. Wood

with

William C. Ober, M.D.
Art Coordinator and Illustrator

Claire W. Garrison, R.N.
Illustrator

Ralph T. Hutchings
Biomedical Photographer

Shawn Miller
Organ and Animal Dissector

Mark Nielsen
Organ and Animal Dissection Photographer

PEARSON

Boston Columbus Indianapolis New York San Francisco Upper Saddle River
Amsterdam Cape Town Dubai London Madrid Milan Munich Paris Montreal Toronto
Delhi Mexico City Sao Paulo Sydney Hong Kong Seoul Singapore Taipei Tokyo

Executive Editor: *Leslie Berriman*
Associate Editor: *Katie Seibel*
Director of Development: *Barbara Yien*
Assistant Editor: *Nicole McFadden*
Senior Managing Editor: *Deborah Cogan*
Production Project Manager: *Caroline Ayres*
Copyeditor: *Carey Lange*
Production Management and Composition:
  S4Carlisle Publishing Services
Director, Media Development: *Lauren Fogel*

Media Producer: *Aimee Pavy*
Design Manager: *Marilyn Perry*
Interior Designer: *tani hasegawa*
Cover Designer: *tani hasegawa*
Senior Photo Editor: *Donna Kalal*
Photo Researcher: *Caroline Commins*
Senior Manufacturing Buyer: *Stacey Weinberger*
Marketing Manager: *Derek Perrigo*
Cover Image Credit: *Bryan Christie*

Credits and acknowledgments borrowed from other sources and reproduced, with permission, in this textbook appear on the appropriate page within the text or on p. CR-1.

The Author and Publisher believe that the lab experiments described in this publication, when conducted in conformity with the safety precautions described herein and according to the school's laboratory safety procedures, are reasonably safe for the student to whom this manual is directed. Nonetheless, many of the described experiments are accompanied by some degree of risk, including human error, the failure or misuses of laboratory or electrical equipment, mismeasurement, chemical spills, and exposure to sharp objects, heat, bodily fluids, blood, or other biologics. The Author and Publisher disclaim any liability arising from such risks in connection with any of the experiments contained in this manual. If students have any questions or problems with materials, procedures, or instructions on any experiment, they should always ask their instructor for help before proceeding.

Library of Congress Cataloging-in-Publication Data
Wood, Michael G.
   Laboratory manual for anatomy & physiology / Michael G. Wood. — 5th ed.
        p. ; cm.
   Laboratory manual for anatomy and physiology
   ISBN-13: 978-0-321-79437-6
   ISBN-10: 0-321-79437-0
   ISBN-13: 978-0-321-80365-8
   ISBN-10: 0-321-80365-5
   [etc.]
 I. Title. II. Title: Laboratory manual for anatomy and physiology.
 [DNLM: 1. Anatomy—Laboratory Manuals. 2. Physiology—Laboratory Manuals. QS 25]
 LC Classification not assigned
   612.0078—dc23

                                          2011041340

  5 6 7 8 9 10—WBC—16 15 14

www.pearsonhighered.com

ISBN 10: **0-321-79437-0** (Student Edition, Main Version)
ISBN 13: **978-0-321-79437-6** (Student Edition, Main Version)

Preface

This laboratory manual is designed to serve the lab course that accompanies the two-semester anatomy and physiology lecture course. It provides students with comprehensive coverage of anatomy and physiology, beautiful full-color art and photographs, and an intuitive pedagogical framework. The primary goals of this manual are to provide students with hands-on experiences that reinforce the information they learn in the lecture course and to help them understand three-dimensional relationships, visualize complex structures, and comprehend intricate physiological processes.

The manual is written to correspond to all current two-semester anatomy and physiology textbooks, although those students and instructors using *Fundamentals of Anatomy & Physiology*, Ninth Edition, by Frederic H. Martini, Judi L. Nath, and Edwin F. Bartholomew will recognize here some of the superb art from that text by William Ober and Claire Garrison, Martini's renowned biomedical illustrators.

This fifth edition manual is available in three separate versions. The Main Version covers the full two-semester A&P curriculum, including dissections of the cow eye and of the sheep heart, brain, and kidney. The Cat Version includes all of the same material plus an additional section of nine cat dissection exercises encompassing the major body systems. The Pig Version, similarly, includes all of the material from the Main Version with a separate section of nine fetal pig dissection exercises. The Cat and Pig Versions make the manual more useful to instructors whose students perform animal dissections in the lab.

Organization

The lab manual contains 47 exercises, plus the 9 additional dissection exercises in each of the Cat and Pig Versions. Large systems, such as the skeletal, muscular, and nervous systems, appear across several exercises, the first serving as an overview exercise that introduces the major anatomical organization of the system. Programs with limited lab time might choose the overview exercises for a hands-on summary of these organ systems that can be completed during a short lab period.

Exercise Organization

Each exercise is organized into a series of Lab Activities that divide the material into natural sets of information to focus students on related concepts. Every exercise begins with a list of the Lab Activities and a set of Learning Outcomes for student learning. A general introduction to the exercise gives students a preview of what they are about to learn; then individual activities focus on more specific study. The activities are self-contained, and instructors may easily assign only certain activities within an exercise.

Each Lab Activity section first introduces the activity and reviews the concepts necessary for understanding it. These are followed by two or three QuickCheck Questions that students can use to gauge their comprehension of the material before proceeding. The activity itself begins with a clearly marked list of Materials and the Procedures for carrying it out. Features such as Clinical Application boxes, Study Tip boxes, Draw It! activities, Sketch to Learn! lessons, and Make a Prediction questions provide students with meaningful context and additional practice and review as they complete each activity. The exercise concludes with a Review & Practice Sheet, which includes data reporting, review questions, and labeling and drawing activities to assess and reinforce student learning.

Cat and Pig Dissection Exercises

Dissection gives students perspective on the texture, scale, and relationships of anatomy. For those instructors who choose to teach dissection in their laboratories, this manual is available in two dissection versions, the Cat Version and the Pig Version, featuring sections at the back of the manual detailing the dissection of the cat or fetal pig. Included are nine exercises that progress through the major body systems, with the goal of relating these exercises to students' study of the human body. Safety guidelines and disposal methods are incorporated into each dissection exercise.

BIOPAC® Activities

Beginning with the second edition, this manual has featured exercises using the BIOPAC Student Lab System, an integrated suite of hardware and software that provides students with powerful tools for studies in physiology. BIOPAC is used in Exercises 22, 23, 30, 37, and 40, and can be easily identified by the BIOPAC logo to the left of the activity title. All of these BIOPAC activities feature step-by-step instructions, full-color art, and instructive screen shots to walk students through the procedures. The instructions in this lab manual are for use with the BIOPAC MP36 (or MP35/30) data acquisition unit, and Biopac Student Lab (BSL) Software version 3.7.5 or better. Instructions for use of the new two-channel data acquisition unit, the MP45, can be found in the Instructor Resources at MasteringA&P (masteringaandp.com).

New to This Edition

In addition to the many technical changes in this edition, such as updated terminology and internal reorganization of exercises in response to reviewer feedback, this revision focuses on improving the visual presentation throughout and provides students with more opportunities for practice and review. These are the key changes in this new edition:

- **Larger, more visually effective art from Martini/Nath/Bartholomew *Fundamentals of Anatomy & Physiology*, Ninth Edition** appears throughout the manual. Improved text–art integration in the figure layouts enhances the readability of the art. Part captions are now integrated into the figures so that relevant text is located immediately next to each part of the figure. A new two-column design better showcases the Martini art.

- **Over 25 new and visually stunning histology images** appear throughout the manual.

- **Silhouetted treatment of the bone photos in the bone exercises (Exercises 13–15)** features a cleaner and more contemporary look and makes bone markings easier to see.

- **More labeling activities** are offered within the tear-out Review & Practice Sheets (formerly titled Lab Reports) throughout the manual, including the dissection exercises.

- **Improved "Draw It!" activities,** complete with blank drawing boxes for the student, are now signaled by a repeating color treatment to call out the hands-on learning opportunity for students.

- **New "Make a Prediction" questions** challenge students to think critically by asking conceptual and/or analytical questions. Students are asked to make predictions and propose hypotheses. This feature appears only where relevant—for example, in exercises that require data interpretation and analysis.

- **BIOPAC activities** have been extensively rewritten with an emphasis on streamlining the instructions to enhance usability of the manual in concert with the BIOPAC software. The number of BIOPAC data graphs has been reduced to prompt students to evaluate their own data during these physiological investigations. In addition, a new BIOPAC activity investigates Respiratory Rate and Depth (Exercise 40).

- **This Laboratory Manual comes with MasteringA&P® for the first time.** MasteringA&P is an online learning and assessment system designed to help instructors teach more efficiently and proven to help students learn. Instructors can assign homework from proven media programs such as Practice Anatomy Lab™ (PAL™) 3.0, PhysioEx™ 9.0 laboratory simulations, and A&P Flix™—all organized by exercise—and have assignments automatically graded. There are also abundant assessments from each exercise's content, including pre- and post-lab quizzes. Items are tagged with the exercise Learning Outcomes so that instructors can teach, test, and assess to Learning Outcomes. In the MasteringA&P Study Area, students can access a variety of study tools. See page xx for more information.

- **NEW! Exercise-opening MasteringA&P® banner** includes a detailed list of student media resources in the MasteringA&P Study Area, individually tailored to each exercise. The list showcases Practice Anatomy Lab™ (PAL)™ 3.0 navigation pathways, applicable A&P Flix™, and relevant PhysioEx™ 9.0 activities.

- **Updated terminology** throughout follows the nomenclature of *Terminologia Anatomica*, the standard of anatomical terminology published by the Federative Committee on Anatomical Terminology. Eponyms are frequently included in the narrative to expose students to both scientific and clinical usage of the language.

Exercise-By-Exercise Changes

The following detailed outline summarizes the major changes by exercise:

Exercise 2

- Expanded "Study Tips" get students on the right track early in their lab studies.
- Two new labeling activities are included in the Review & Practice Sheet.

Exercise 3

- New labeling activity in the Review & Practice Sheet covers internal organs of the cat.

Exercise 4

- New photographs of the microscope provide an update of this important lab tool.

Exercise 5

- New labeling activity in the Review & Practice Sheet includes micrographs of cells in various stages of mitosis.

Exercise 6

- Clinical Application on Dialysis now features a new photograph of a dialysis cartridge.

Exercise 7

- New micrographs guide students in viewing epithelia at various magnifications.
- A new presentation of simple squamous epithelium now includes a micrograph of this epithelium in the lungs to clearly show its thin organization.

Exercise 8

- Many new micrographs support this exercise of connective tissues. New low magnification micrograph of hyaline cartilage introduces supporting connective tissues in preparation for more detailed observations.

Exercise 11

- New micrograph of the scalp with hair follicles better matches the integument slides typical in A&P laboratories.

Exercise 13

- New labeling activity in Review & Practice Sheet helps students start their detailed skeletal studies with a stronger background of axial and appendicular bones.

Exercise 14

- This exercise features an improved sequencing of the activities for the skull.
- New labeling activities have been added to the Review & Practice Sheets.

Exercise 15

- New photographs isolating the ulna and radius and new views of the bones articulated at the elbow enrich study of the upper limb.
- Additional labeling activities have been added to the Review & Practice Sheets.

Exercise 16

- Narrative of the classification of joints has been rewritten for clarity and a more focused lab approach to joints.

Exercise 17

- New micrograph of the neuromuscular junction is more similar to prepared slides used in the A&P lab.

Exercises 18–21

- The text narrative of the muscle exercises has been revised to emphasize descriptions and location of muscles.
- New illustrations feature better label sequences for fostering accurate identification of muscles.
- Many new "Study Tips" are incorporated into the muscle exercises to assist in learning, and remembering, muscles.
- Labeling activities to practice spelling and muscle identification have been added to the Review & Practice Sheets.

Exercise 22

- The entire narrative has been condensed to focus on laboratory activities.
- The narrative for both of the BIOPAC activities has been completely revised for better use in conjunction with BIOPAC software.

Exercise 23

- New micrograph of a peripheral nerve has been added.
- The narrative for the BIOPAC activity "Reaction Time and Learning" has been completely revised for better use in conjunction with BIOPAC software.

Exercise 25

- A new illustration detailing major regions of the brain has been added.
- An improved photograph of the sheep brain in midsagittal view provides better guidance during dissection.

Exercises 26–28

- Labeling activities have been added to the Review & Practice Sheets.

Exercise 29

- Improved art with better label sequence details the internal anatomy of the eye.
- A new figure highlights the cavities within the eye.
- A new labeling activity of the cellular organization of the retina, featuring a new micrograph, has been added to the Review & Practice Sheet.

Exercise 30

- The narrative for the BIOPAC activity "Electrooculogram" has been completely revised for better use in conjunction with BIOPAC software.

Exercise 31

- Two new master figures detail, from macro to micro, the anatomy of the inner ear and the structure of each receptor.
- A new labeling activity of the cochlea, featuring a new micrograph, has been added to the Review & Practice Sheet.

Exercise 33

- New micrographs of endocrine glands enhance the microscopic study of endocrine organs and the cells that produce hormones.

Exercise 34

- New micrographs of blood cells allow easier in-lab identification of formed elements during microscopic observations.

Exercise 36

- Coverage of arteries and veins is expanded in both art and narrative.

Exercise 37

- The narrative in both of the BIOPAC activities has been extensively revised for better use in conjunction with BIOPAC software. Fewer BIOPAC graphs are needed, as computer screen-shots cultivate data interpretation skills in students.
- Redesign of data tables guides students through the analysis.

Exercise 38

- Updated text uses the more commonly used term "lymphatic" rather than "lymphoid."

Exercise 40

- New photograph of a wet spirometer provides a visual reference for equipment used in most A&P labs.
- The narrative for the BIOPAC activity "Volumes and Capacities" has been completely revised for better use in conjunction with BIOPAC software.
- The new BIOPAC activity "Respiratory Rate and Depth" has been added.

Exercise 41

- Expanded histological coverage of digestive organs—including salivary glands, stomach, small intestine, pancreas, liver, and gallbladder—is supported with new micrographs and corresponding narrative.

Exercise 42

- Narrative is reworked for stronger association between the process of chemical digestion and the lab activities using various enzymes and substrates.
- Protein digestion activity is redesigned to use albumin for a protein source for more consistent results.

Exercise 43

- Expanded histological coverage of urinary organs, including ureters, bladder, and urethra, is supported with new micrographs and revised narrative.

Exercise 45

- Reorganization offers better pedagogical sequence of male and female anatomy with the study of gametogenesis after the anatomical studies.
- Expanded histological coverage of male and female organs is supported with new micrographs and revised narrative.

Acknowledgments

I am grateful to a number of people for this fifth edition's excellent illustrations and photographs. Frederic H. Martini, main author of the outstanding and widely acclaimed Martini/Nath/Bartholomew, *Fundamentals of Anatomy & Physiology*, Ninth Edition, deserves credit for his insight and creativity in visualizing anatomical and physiological concepts with the talented biomedical illustrators William Ober and Claire Garrison. This lab manual benefits from their work through the inclusion of many illustrations from that book. I have also worked closely with them over the years to create specific illustrations for the manual. Shawn Miller and Mark Nielsen of the University of Utah are a gifted dissector/photographer team whose meticulous work is coupled with the Ober and Garrison illustrations in the cat and pig dissection versions of the manual. The award-winning human photographs in the manual are by biomedical photographer Ralph Hutchings. I also thank Judi L. Nath and Edwin F. Bartholomew, coauthors on *Fundamentals of Anatomy & Physiology*, for their continued support and encouragement.

In addition to the many micrographs that I prepared for the manual, I was fortunate to have Robert B. Tallitsch, an outstanding histologist/microphotographer and one of Ric Martini's coauthors on *Human Anatomy*, Seventh Edition, graciously provide many critical histological images.

Special thanks are extended to my colleagues at Del Mar College: Albert Drumright III, Lillian Bass, Billy Bob Long, Megan McKee, Joel McKinney, Zaldy Doyungan, Angelica Chapa, Joyce Germany, Reba Jones, and Thomas Klepach for their suggestions, encouragement, and support of the manual over the years and editions.

I thank the many students at Del Mar College whom I have had the privilege to be with in the classroom and laboratory. Teachers are life-long learners and I have gained much insight from my students, many of whom are employed in health care and often interject real-life experiences with patients that directly relate to the laboratory topic.

I thank all of the talented and creative individuals at Pearson. I was especially pleased to have the project managed by Katie Seibel, project editor. Katie's expertise in organizing and coordinating the enormous number of details and managing a challenging schedule were essential in all phases of this edition's development and I am sincerely grateful for her contributions. I am thankful for the ongoing support and encouragement of Leslie Berriman, executive editor, whose creative vision shines in the fifth editon. Leslie has a gift for coalescing and managing a resourceful team of editors, dissectors, photographers, and illustrators whose outstanding work is the foundation of this fifth editon. Caroline Ayres, production project manager, masterfully coordinated all aspects of the production process. Thanks also to Deborah Cogan, senior managing editor; Donna Kalal, senior photo editor; and Caroline Commins, photo researcher, for their roles in the production of this text. I thank Lynn Steines and her fine team at S4Carlisle Publishing Services for their creative layout and attention to detail. I also thank tani hasegawa for her outstanding design—of both the cover and the interior—which gives this complex assemblage of text, illustrations, photographs, and procedures a user-friendly look. Marilyn Perry, design manager, oversaw the design process and provided crucial insight into our design complexities. The entire Pearson Science sales team deserves thanks for their fine efforts in presenting this manual to A&P instructors.

I offer thanks to the people who developed the stellar media available with this lab manual. Aimee Pavy, media producer, managed the development of MasteringA&P for the manual. Sarah Young-Dualan was the media producer for Practice Anatomy Lab™ (PAL™) 3.0.

I thank Biopac Systems, Inc., for their continued support and partnership with Pearson and assistance in incorporating activities for their state-of-the-art instrumentation into the fifth edition. I especially thank Jocelyn Kremer at Biopac for her review of the manuscript and for her much appreciated involvement in the revision of the manual to match the latest Biopac software.

Reviewers helped guide the revision of this fifth edition and I thank them for their time and devotion to the manual.

Ronny K. Bridges
Pellissippi State Community College

James Davis
University of Southern Maine

Kurt J. Elliott
Northwest Vista College

Lorraine A. Findlay
Nassau Community College

Roy A. Hyle, II
Thomas Nelson Community College

Corrie Kezer
Rogue Community College

Hui-Yun Li
Oregon Institute of Technology

Sarah E. Matarese
Salve Regina University

Justin W. Merry
Saint Francis University

Karen E. Plucinski
Missouri Southern State University

Debra A. Rajaniemi
Goodwin College

John M. Ripper
Butler County Community College

Will Robison
Northwest Nazarene University

Teaching anatomy and physiology full time at Del Mar College is challenging and rewarding; add to that the rigors of writing academic books and one's lifestyle approaches frenetic. I remain deeply grateful to my wife, Laurie, and daughters, Abi and Beth, for enduring months of my late-night writing and the pressures of never-ending deadlines. I am thankful for my mother, Janis G. Wood, the strong matriarch of the family who cared for Dad during his time of need. I am fortunate to have life-long friends and I sincerely thank Bruce and Marilyn Gambill, Stephen Gambill, Douglas Townsend, Yvonne Newman, Wally Allen and Connie Vallie, and Fred Saenz for always being there when I needed to unwind. I also thank Kit Semtner and Jess Alford for all they have done for the Wood family.

The person who has had the biggest impact on my professional development is Dr. Loyd Poplin. Loyd hired me right out of graduate school and was my mentor as I learned the ropes of teaching in higher education at Del Mar College. He has been my supervisor, mentor, and close friend. He and his wife, Beverly, directed Celtic Menagerie, a Scottish Folk dance group that my wife Laurie and I danced with for over 8 years. I am deeply grateful for the support, encouragement, and friendship the Poplins have extended to me and my family over the many years.

Any errors or omissions in this edition are exclusively my responsibility and are not a reflection of the dedicated editorial and review team. Comments from faculty and students are welcomed and may be directed to me at the addresses below. I will consider each submission in the preparation of the next edition.

Michael G. Wood
Del Mar College
Department of Natural Sciences
101 Baldwin Blvd.
Corpus Christi, TX 78404

About the Author

Michael G. Wood received his Master's of Science in Biology in 1986 at Pan American University, now the University of Texas at Pan American in Edinburg, Texas. His graduate studies included vertebrate physiology and freshwater ecology. Presently he is a tenured Professor of Biology at Del Mar College in Corpus Christi, Texas, where he has taught over 10,000 students in anatomy and physiology and biology over the past 25 years. His excellence in teaching has been recognized by the Del Mar College community, and he is the recipient of numerous honors, including the "Educator of the Year," "Teacher of the Year," and "Master Teacher" awards. Wood is a member of the Human Anatomy and Physiology Society (HAPS) and enjoys attending their annual meeting when not involved in a writing project. He has a passion for science, reading, and playing guitar in the *Doc Gambill Band* with his brother-in-law Bruce Gambill and close friend Douglas Townsend. Mike and his wife, Laurie, also enjoy travel, dance, gardening, exploring the great outdoors, and a life full of cats and dogs.

Dedication

With love to my daughter Abi whose strength and talent inspire.

Outstanding Illustration and Photo Program

Side-by-side figures show multiple views of the same structure or tissue,
allowing students to compare an illustrator's rendering with a photo of the actual structure or tissue as it would seen in the laboratory or operating room.

NEW! Text-art integration in the figure layouts.
Part labels are fully incorporated into the art, making viewing the art and reading the text easier.

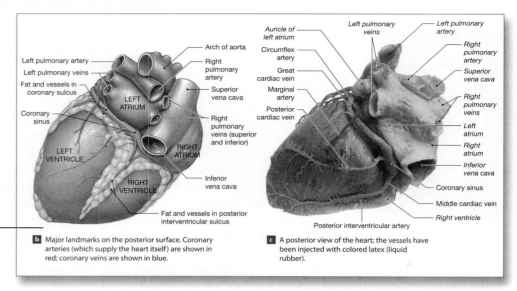

b Major landmarks on the posterior surface. Coronary arteries (which supply the heart itself) are shown in red; coronary veins are shown in blue.

c A posterior view of the heart; the vessels have been injected with colored latex (liquid rubber).

Award-winning bone and cadaver photographs
by biomedical photographer Ralph Hutchings appear throughout the manual.

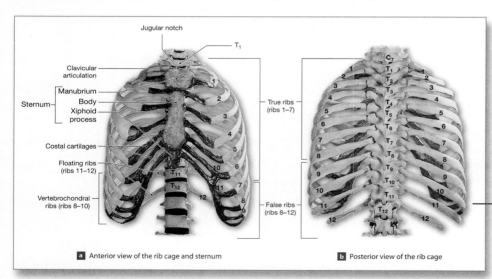

a Anterior view of the rib cage and sternum

b Posterior view of the rib cage

NEW! Silhouetted treatment of the bone photographs
feature a cleaner and more contemporary look and make bone markings easier to see (Exercises 13-15).

NEW! Over 25 brand-new visually stunning histology photomicrographs
often with paired illustrations match the types of slides that students will encounter in their anatomy and physiology labs.

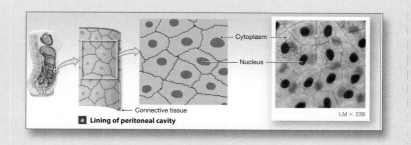

a Lining of peritoneal cavity

Illustration-over-photo figures bring depth, dimensionality, and visual interest to the page and show that the illustrated structures are proportional in size to the human body.

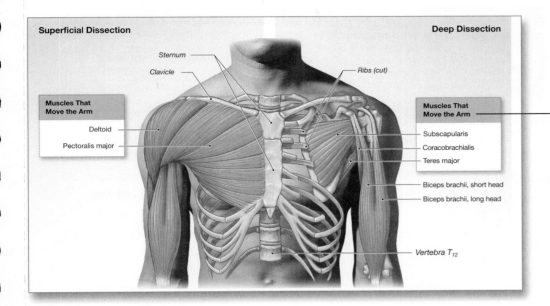

Superficial Dissection

Deep Dissection

Sternum

Clavicle

Ribs (cut)

Muscles That Move the Arm
Deltoid
Pectoralis major

Muscles That Move the Arm
Subscapularis
Coracobrachialis
Teres major
Biceps brachii, short head
Biceps brachii, long head

Vertebra T_{12}

NEW! Reorganized labels chunk information into clear categories for study and reference.

Cat and pig dissection photos and illustrations

were created by the expert dissector/ photographer team of Shawn Miller and Mark Nielsen and the award-winning artists William Ober and Claire Garrison. These photo-illustration combinations provide superb presentations of what students will see on the dissection table. See the Cat Version and Pig Version of this lab manual for the dissection exercises.

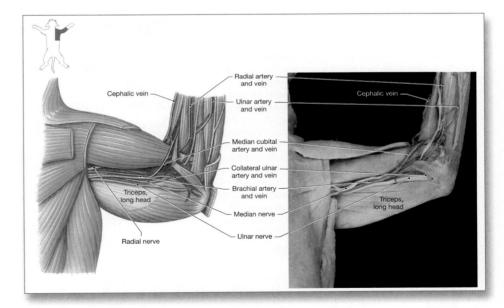

Cephalic vein

Radial artery and vein

Ulnar artery and vein

Cephalic vein

Median cubital artery and vein

Collateral ulnar artery and vein

Brachial artery and vein

Triceps, long head

Median nerve

Ulnar nerve

Radial nerve

Triceps, long head

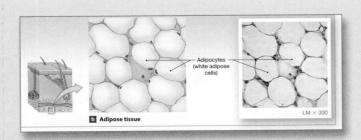

Adipocytes (white adipose cells)

b Adipose tissue

LM × 300

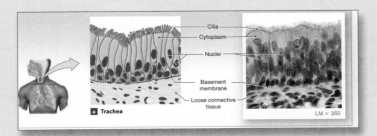

Cilia

Cytoplasm

Nuclei

Basement membrane

Loose connective tissue

a Trachea

LM × 350

Tools For Student Review and Practice

REVIEW IT!

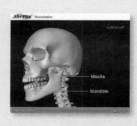 

NEW! Exercise-opening MasteringA&P® banner alerts students to the MasteringA&P Study Area resources that can help them review key exercise material on their own.

QuickCheck Questions

5.1 Where does the sternocleidomastoid muscle attach?

5.2 What is the suffix in the names of muscles that insert on the hyoid bone?

5.3 Where is the digastric muscle located?

QuickCheck questions help students assess their understanding of the material before beginning the actual lab work. Even more review questions can be found in the Review & Practice Sheet at the end of each exercise.

APPLY IT!

Clinical Application Dialysis

Dialysis is a passive process similar to osmosis except that, besides water, small solute particles can pass through a selectively permeable membrane. Large particles are unable to cross the membrane, and thus particles can be separated by size during dialysis. Dialysis does not occur in the body, but it is used in the medical procedure called **kidney dialysis** to remove wastes from the blood of a patient whose kidneys are not functioning properly. Blood from an artery passes into thousands of minute selectively permeable tubules in a dialysis cartridge (Figure 6.4). A dialyzing solution having the same concentration of materials to remain in the blood (nutrients and certain electrolytes) is pumped into the cartridge to flow over the tubules. As blood flows through the tubules, wastes diffuse from the blood, through the selectively permeable tubules, and into the dialyzing solution. Once waste levels in the blood have been reduced to a safe level, the patient is disconnected from the dialysis apparatus. ■

Figure 6.4 Dialysis Cartridge

Clinical Application boxes present clinical information useful to students who plan a career in allied health. Students can apply this information by answering the Clinical Challenge questions in the end-of-exercise Review & Practice Sheet.

Make a Prediction

What kind of muscle contraction occurs when you try to pick up something that is too heavy to lift?

NEW! Make a Prediction questions challenge students to think critically, make predictions, and propose hypotheses within an activity.

PRACTICE IT!

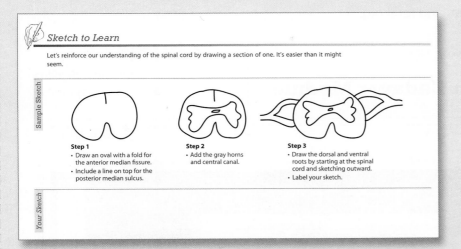

Sketch to Learn activities
guide students through creating a drawing that helps them to make connections and better understand the lab activity.

IMPROVED! Draw It! activities
include blank drawing boxes for the student and are now signaled by a repeating color treatment to call out the hands-on learning opportunity for students.

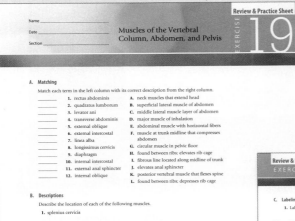

Review & Practice Sheets
follow each exercise and encourage students to answer a variety of questions, draw structures, label illustrations, complete tables, record data, interpret results, apply clinical content, and think critically. These sheets can be handed in for grading.

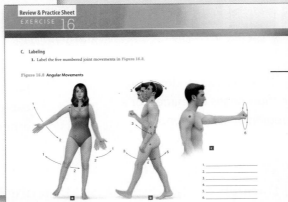

MORE! Labeling activities
give students extra practice.

Practice Anatomy Lab™ (PAL™) 3.0

***NEW!* PAL 3.0** is a virtual anatomy study and practice tool that gives students 24/7 access to the most widely used lab specimens, including human cadaver, anatomical models, histology, cat, and fetal pig.

NEW! Interactive Human Cadaver Module

Carefully prepared dissections show nerves, veins, and arteries across body systems.

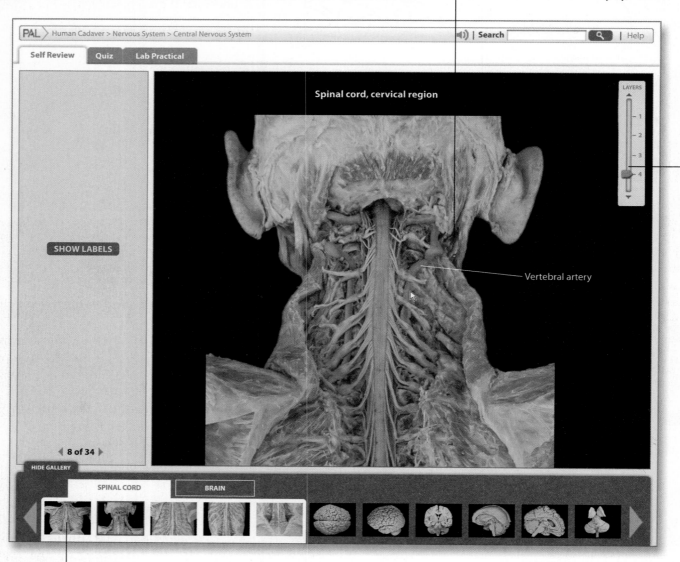

PAL › Human Cadaver > Nervous System > Central Nervous System | Search | Help

Self Review   Quiz   Lab Practical

Spinal cord, cervical region

LAYERS — 1 — 2 — 3 — 4

SHOW LABELS

Vertebral artery

◀ 8 of 34 ▶

HIDE GALLERY

SPINAL CORD   BRAIN

Photo gallery allows students to quickly see thumbnails of images for a particular region or sub-region.

Layering slider allows students to peel back layers of the human cadaver and view and explore hundreds of brand-new dissections especially commissioned for version 3.0.

PAL 3.0 is delivered through the Study Area of MasteringA&P (**www.masteringaandp.com**) and is also available on a DVD.

NEW! Interactive Histology Module

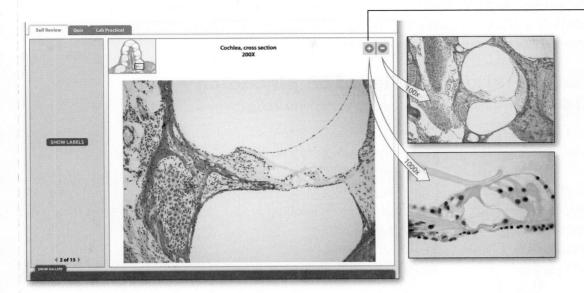

Magnification buttons allow students to view the same tissue slide at varying magnifications, thereby helping them identify structures and their characteristics.

Anatomy animations of origins, insertions, actions, and innervations of over 60 muscles are now viewable in two modules: Human Cadaver and Anatomical Models. Under the Animations tab, over 50 anatomy animations of group muscle actions and joints are also viewable. A new closed-captioning option provides textual presentation of narration to help students retain information and supports ADA compliance.

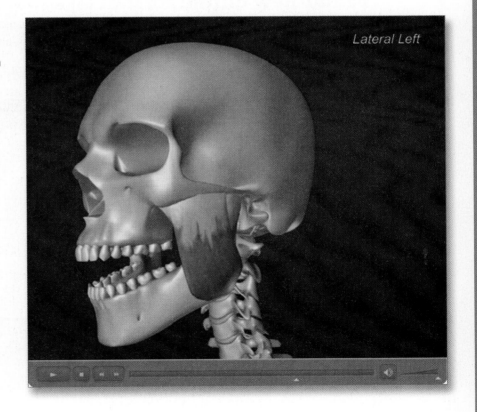

PAL 3.0 also includes:

- *NEW!* Question randomization feature
- *NEW!* Hundreds of new images and views
- *NEW!* Turn-off highlight feature
- *NEW!* IRDVD with Test Bank for PAL 3.0
- Built-in audio pronunciations
- Rotatable bones
- Simulated fill-in-the-blank lab practical exams

SEE FOR YOURSELF! Check out the new PAL 3.0 at www.masteringaandp.com.

PhysioEx™ 9.0

***NEW!* PhysioEx 9.0** is easy-to-use laboratory simulation software that can be used to supplement or substitute for wet labs. PhysioEx allows students to repeat labs as often as they like, perform experiments without harming live animals, and conduct experiments that are difficult to perform in a wet lab environment because of time, cost, or safety concerns.

12 exercises containing a total of 63 physiology lab activities offer a comprehensive approach to the virtual physiology lab.

Exercises:

1: Cell Transport Mechanisms and Permeability
2: Skeletal Muscle Physiology
3: Neurophysiology of Nerve Impulses
4: Endocrine System Physiology
5: Cardiovascular Dynamics
6: Cardiovascular Physiology

7: Respiratory System Mechanics
8: Chemical and Physical Processes of Digestion
9: Renal System Physiology
10: Acid-Base Balance
11: Blood Analysis
12: Serological Testing

Brand-new online format with easy step-by-step instructions includes everything students need in one convenient place. Students read introductory material, gather data, analyze results, and check their understanding—all on screen.

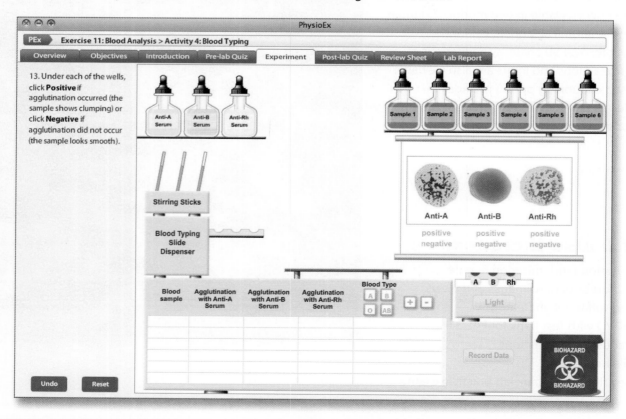

PhysioEx 9.0 is delivered through the Study Area of MasteringA&P (**www.masteringaandp.com**) and is also available on a CD.

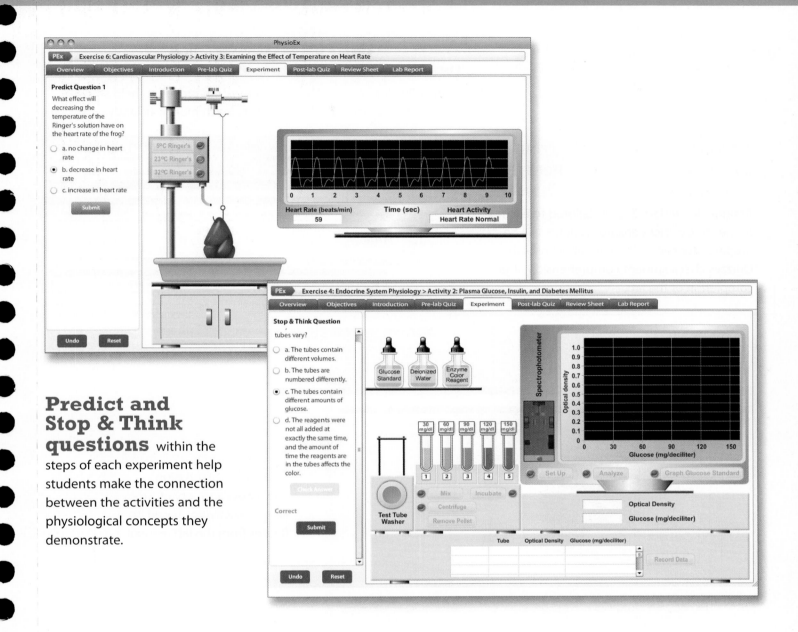

Predict and Stop & Think questions

within the steps of each experiment help students make the connection between the activities and the physiological concepts they demonstrate.

PhysioEx 9.0 also features:

- *NEW!* **Pre- and Post-lab quizzes, as well as Review Sheets,** that give students many opportunities to assess their understanding.

- *NEW!* **Greater data variability** in the results that reflect the more realistic results students would encounter in real wet-lab experiments.

- *NEW!* **A Lab Report** that includes students' answers to all of the questions and their results from the experiment. Students can save their Lab Report as a PDF, which they can print and hand in or email to their instructor.

SEE FOR YOURSELF! Check out the new PhysioEx 9.0 at www.masteringaandp.com.

A Learning and Assessment System

NEW! MasteringA&P®
(www.masteringaandp.com)

is an online learning and assessment system proven to help students learn and designed to help instructors teach more efficiently.

Motivate your students to come to lab prepared, and then check their comprehension of their lab work.

Assignable Pre-lab Quizzes tailored to each of the lab exercises ensure students will be prepared for each lab. Assignable Post-lab Quizzes check student comprehension after completion of the lab.

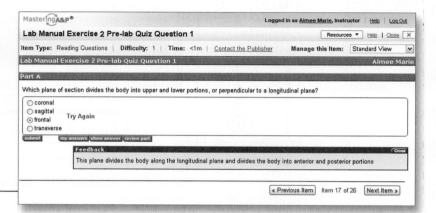

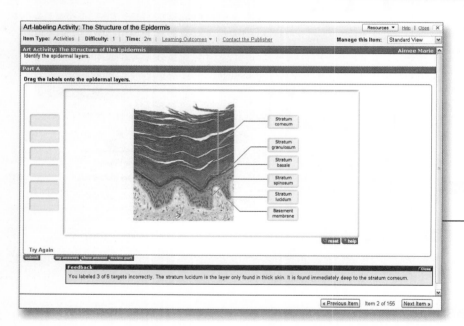

Assign art from the lab manual. Assign and assess a variety of Art-labeling Activities based on figures from the lab manual.

Identify struggling students before it's too late.

MasteringA&P has a color-coded gradebook that helps you identify vulnerable students at a glance. Assignments in MasteringA&P are automatically graded, and grades can be easily exported to course management systems or spreadsheets.

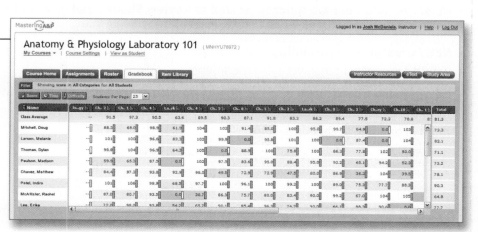

Give your students 24/7 anatomy lab practice. Assign quizzes and lab practicals from the Testbank for Practice Anatomy Lab™ (PAL™) 3.0. PAL 3.0 is a virtual anatomy study and practice tool that gives students 24/7 access to the most widely used lab specimens.

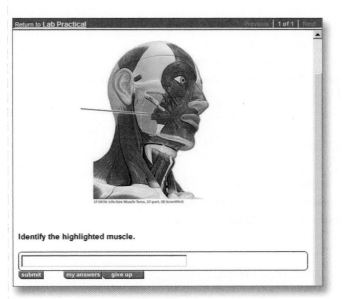

To learn more about PAL 3.0, see pages xvi–xvii.

Give your students 24/7 physiology lab practice. Assign Pre- and Post-lab Quizzes and Review Sheets from PhysioEx™ 9.0. PhysioEx 9.0 is an easy-to-use lab simulation program used to supplement or substitute for wet labs.

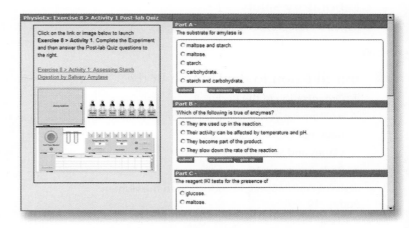

To learn more about PhysioEx 9.0, see pages xviii–xix.

Mastering A&P® Study Area

Mastering A&P® includes a **Study Area** that helps students get ready for tests on their own with its simple three-step approach. Students can:

1. **Take a Pre-lab Quiz** and obtain a personalized study plan.
2. **Learn and practice** with animations and interactive tutorials.
3. **Self-test with a Post-lab Quiz** after completing the lab to check their understanding of the topics.

The Study Area provides 24/7 access to:

- **Pre- and Post-lab Quizzes** (different from those in the assignment area)
- **Practice Anatomy Lab™ (PAL™) 3.0**
- **PhysioEx™ 9.0**
- **A&P Flix™** 3-D movie-quality animations with Quizzes for these anatomy topics:
    Origins, Insertions, Actions, Innervations (over 60 animations on this topic)
    Group Muscle Actions & Joints (over 50 animations on this topic)
- **eText** with functionality to create notes, highlight text, create bookmarks, zoom in and out, view in single-page or two-page view, click terms to view definitions, and search for specific content

Contents

Laboratory Safety

Learning Outcomes

On completion of this exercise, you should be able to:

1. Locate all safety equipment in the laboratory.

2. Show how to handle glassware safely, including insertion and removal of glass rods used with stoppers.

3. Demonstrate how to clean up and dispose of broken glass safely.

4. Demonstrate how to plug in and unplug electrical devices safely.

5. Explain how to protect yourself from and dispose of body fluids.

6. Demonstrate how to mix solutions and measure chemicals safely.

7. Describe the potential dangers of each laboratory instrument.

8. Discuss disposal techniques for glass, chemicals, body fluids, and other hazardous materials.

E xperiments and exercises in the anatomy and physiology laboratory are, by design, safe. Some of the hazards are identical to those found in your home, such as broken glass and the risk of electrical shock. The major hazards can be grouped into six categories: glassware, electrical, body fluids, chemical, laboratory instruments, and preservatives. The following is a discussion of the hazards each category poses and a listing of safety guidelines you should follow to prevent injury to yourself and others while in the laboratory. Proper disposal of biological and chemical wastes ensures that these contaminants will not be released into your local environment.

Need More Practice and Review?

Build your knowledge—and confidence!—in the Study Area of MasteringA&P® at www.masteringaandp.com with Pre-lab Quizzes, Post-lab Quizzes, Practice Anatomy Lab™ (PAL™) 3.0 virtual anatomy practice tool, PhysioEx™ 9.0 laboratory simulations, and A&P Flix™ with Quizzes.

Laboratory Safety Rules

The following guidelines are necessary to ensure that the laboratory is a safe environment for students and faculty alike.

1. No unauthorized persons are allowed in the laboratory. Only students enrolled in the course and faculty are to enter the laboratory.

2. Never perform an unauthorized experiment; unless you have your instructor's permission, never make changes to any experiment that appears either in this manual or in a class handout.

3. Do not smoke, eat, chew gum, or drink in the laboratory.

4. Always wash your hands before and after each laboratory exercise involving chemicals, preserved materials, or body fluids, and immediately after cleaning up spills.

5. Wear shoes at all times while in the laboratory.

6. Be alert to unsafe conditions and to unsafe actions by other individuals in the laboratory. Call attention to those conditions or activities. Someone else's accident can be as dangerous to you as one that you cause.

7. Glass tubes called *pipettes* are commonly used to measure and transfer solutions. Never pipette a solution by mouth. Always use a pipette bulb. Your instructor will demonstrate how to use the particular type of bulb available in your laboratory.

8. Immediately report all spills and injuries to the laboratory faculty.

9. Inform the laboratory faculty of any medical condition that may limit your activities in the laboratory.

Location of Safety Equipment

Write here the location of each piece of safety equipment as your instructor explains how and when to use it:

nearest telephone _____

first aid kit _____

fire exits _____

fire extinguisher _____

eye wash station _____

chemical spill kit _____

fan switches _____

biohazard container _____

Glassware

Glassware is perhaps the most dangerous item in the laboratory. Broken glass must be cleaned up and disposed of safely. Other glassware-related accidents can occur when a glass rod or tube breaks while you are attempting to insert it into a cork or rubber stopper.

Broken Glass

- Sweep up broken glass immediately. Never use your hands to pick up broken glass. Instead, use a whisk broom and dustpan to sweep the area clear of all glass shards.

- The laboratory most likely has a "sharps box" similar to the plastic receptacle that a nurse uses to dispose of used needles. Discard all broken glass in this container. Otherwise place broken glass in a box, tape the box shut, and write "BROKEN GLASS INSIDE" in large letters across it. This will alert custodians and other waste collectors to the hazard inside. Your laboratory instructor will arrange for disposal of the sealed box.

Inserting Glass into a Stopper

- Never force a dry glass rod or tube into the hole cut in a cork or rubber stopper. Use a lubricant such as glycerin or soapy water to ease the glass through the stopper.

- When inserting a glass rod or tube into a stopper, always push on the rod/tube near the stopper. Doing so reduces the length of glass between the stopper and your hand and greatly decreases the chance of your breaking the rod and jamming glass into your hand.

Electrical Equipment

Electrical hazards in the laboratory are similar to those in your home. A few commonsense guidelines will almost eliminate the risk of electrical shock.

- Uncoil an electrical cord completely before plugging it into an electrical outlet. Electrical cords are often wrapped tightly around the base of a microscope, and users often unwrap just enough cord to plug in the microscope. Moving the focusing mechanism of the microscope may pinch the part of the cord still wrapped around the base and shock the user. Inspect the cord for fraying and the plug for secure connections.

- Do not force an electrical plug into an outlet. If the plug does not easily fit into the outlet, inform your laboratory instructor.

- Unplug all electrical cords by pulling on the plug, not the cord. Pulling on the cord may loosen wires inside the cord, which can cause an electrical short and possibly an electrical shock to anyone touching the cord.

- Never plug in or unplug an electrical device in a wet area.

Body Fluids

The three body fluids most frequently encountered in the laboratory are saliva, urine, and blood. Because body fluids can harbor infectious organisms, safe handling and disposal procedures must be followed to prevent infecting yourself and others.

- Work only with your own body fluids. It is beyond the scope of this manual to explain proper protocol for collecting and experimenting on body fluids from another individual.

- Never allow a body fluid to touch your unprotected skin. Always wear gloves and safety glasses when working with body fluids—even though you are using your own fluids.

- Always assume that a body fluid can infect you with a disease. Putting this safeguard into practice will prepare you for working in a clinical setting where you may be responsible for handling body fluids from the general population.

- Clean up all body-fluid spills with either a 10 percent bleach solution or a commercially prepared disinfectant labeled for this purpose. Always wear gloves during the cleanup, and dispose of contaminated wipes in a biohazard container.

Chemicals

Most chemicals used in laboratories are safe. Following a few simple guidelines will protect you from chemical hazards.

- Read the label describing the chemical at hand. Be aware of chemicals that may irritate skin or stain clothing. Chemical containers are usually labeled to show contents and potential hazards. Handling and disposal of all chemicals should follow OSHA guidelines and regulations. Most laboratories and chemical stockrooms keep copies of technical chemical specifications, called *Material Safety Data Sheets (MSDS)*. These publications from chemical manufacturers detail the proper use of the chemicals and the known adverse effects they may cause. All individuals have a federal right to inspect these documents. Ask your laboratory instructor for more information on MSDS.

- Never touch a chemical with unprotected hands. Wear gloves and safety glasses when weighing and measuring chemicals and during all experimental procedures involving chemicals.

- Always use a spoon or spatula to take a dry chemical from a large storage container. Do not shake a dry chemical out of its jar; doing so may result in your dumping the entire container of chemical onto yourself and your workstation.

- When pouring out some volume of a solution kept in a large container, always pour the approximate amount required into a smaller beaker first and then pour from this beaker to fill your glassware with the solution. Attempting to pour from a large storage container directly into any glassware other than a beaker may result in spilled solution coming into contact with your skin and clothing.

- To keep from contaminating a storage container, do not return the unused portion of a chemical to its original container. Dispose of the excess chemical as directed by your instructor. Do not pour any chemicals—unused or used—down the sink unless directed to do so by your instructor.

- When mixing solutions, always add a chemical to water; never add water to the chemical. By adding the chemical to the water, you reduce the chance of a strong chemical reaction occurring.

Laboratory Instruments

You will use a variety of scientific instruments in the anatomy laboratory. Safety guidelines for specific instruments are included in the appropriate exercises. This discussion concerns the instruments most frequently used in laboratory exercises.

- *Microscope:* The microscope is the main instrument you will use in the study of anatomy. (Exercise 4 of this manual is devoted to the use and care of this instrument.) A few simple safety rules will prevent injury to yourself and damage to the microscope.

  1. Always carry a microscope with two hands, and do not swing the instrument as you carry it.

  2. Use only the special lens paper and cleaning solution provided by your laboratory instructor to clean the microscope lenses. Other papers and cloths may scratch the optical coatings on the lenses. An unapproved cleaning agent may dissolve the adhesives used in the lenses.

  3. Always unwrap the electrical cord completely before plugging in the microscope.

  4. Unplug the microscope by pulling on the plug, not by tugging on the electrical cord.

- *Dissection Tools:* Working with sharp blades and points always presents the possibility of injury. Always cut away from yourself, and never force a blade through a tissue. Use small knife strokes for increased blade control rather than large cutting motions. Always use a sharp blade, and dispose of used blades in a specially designated "sharps" container. Carefully wash and dry all instruments upon completion of each dissection.

  Special care is necessary while changing disposable scalpel blades. Your instructor may demonstrate the proper technique for blade replacement. Always wash the used blade before removing it from the handle. Examine the handle and blade, and determine how the blade fits onto the handle. Do not force the blade off the handle. If you have difficulty changing blades, ask your instructor for assistance.

- *Water Bath:* A water bath is used to incubate laboratory samples at a specific temperature. Potential hazards involving water baths include electrical shock due to contact with water and burn-related injuries caused by touching hot surfaces or spilling hot solutions. Electrical hazards are minimized by following the safety rules concerning plugging and unplugging electrical devices. Avoid burns by using tongs to immerse or remove samples from a water bath. Point the open end of all glassware containing a sample away from yourself and others. If the sample boils, it could splatter out and burn your skin. Use a water-bath rack to support all glassware, and place hot samples removed from a water bath in a cooling rack. Monitor the temperature and water level of all water baths. Excessively high temperatures increase the chance of burns and usually ruin an experiment. When using boiling water baths, add water frequently, and do not allow all the water to evaporate.

- *Microcentrifuge:* A microcentrifuge is used for blood and urine analyses. The instrument spins at thousands of revolutions per minute. Although the moving parts are housed in a protective casing, it is important to keep all loose hair, clothing, and jewelry away from the instrument. Never open the safety lid while the centrifuge is on or spinning. Do not attempt to stop a spinning centrifuge with your hand. The instrument has an internal braking mechanism that stops it safely.

Preservatives

Most animal and tissue specimens used in the laboratory have been treated with chemicals to prevent decay. These preservatives are irritants and should not contact your skin or your mucous membranes (linings of the eyes, nose, and mouth, and urinary, digestive, and reproductive openings). The following guidelines will protect you from these hazards.

- If you are pregnant, limit your exposure to all preservatives. Discuss the laboratory exercise with your instructor. Perhaps you can observe rather than perform the dissection.
- Always wear gloves and safety glasses when working with preserved material.
- Your laboratory may be equipped with exhaust fans to ventilate preservative fumes during dissections. Do not hesitate to ask your instructor to turn on the fans if the preservative odor becomes bothersome.

- Many preservatives are either toxic or carcinogenic, and all require special handling. Drain as much preservative as possible from a specimen before beginning a dissection. Pour the drained preservative into either the specimen storage container or a dedicated container provided by your instructor. Never pour preservative down the drain.
- Promptly wipe up all spills and clean your work area when you have completed a dissection. Keep your gloves on during the cleanup, and dispose of gloves and paper towels in the proper biohazard container.

Disposal of Chemical and Biological Wastes

To safeguard the environment and individuals employed in waste collection, it is important to dispose of all potentially hazardous wastes in specially designed containers. State and federal guidelines detail the storage and handling procedures for chemical and biological wastes. Your laboratory instructor will manage the wastes produced in this course.

- *Body Fluids:* Objects contaminated with body fluids are considered a high-risk biohazard and must be disposed of properly. Special biohazard containers will be available during exercises that involve body fluids. A special biohazard sharps container may be provided for glass, needles, and lancets.
- *Chemical Wastes:* Most chemicals used in undergraduate laboratories are relatively harmless and may be diluted in water and poured down the drain. Your instructor will indicate during each laboratory session which chemicals can be discarded in this manner. Other chemicals should be disposed of in a dedicated waste container.
- *Preservatives and Preserved Specimens:* Dispose of preservatives in a central storage container maintained for that purpose. As noted earlier, never pour preservative solutions down the drain. Dispose of all preserved specimens by wrapping them in a plastic bag filled with an absorbent material such as cat litter and placing the bag in a designated area for pickup by a hazardous-waste company.

Name _____

Date _____

Section _____

Laboratory Safety

A. Short-Answer Questions

1. Discuss how to protect yourself from body fluids, such as saliva and blood.

2. Why should you consider a body fluid capable of infecting you with a disease?

3. Describe how to dispose of materials contaminated with body fluids.

4. Explain how to safely plug and unplug an electrical device.

5. Discuss how to protect yourself from preservatives used on biological specimens.

6. Why are special biohazard containers used for biological wastes?

7. Explain how to clean up broken glass.

8. List the location of the following safety items in the laboratory.

first aid kit _____

nearest telephone _____

eye wash station _____

fire exits _____

fire extinguisher _____

chemical spill kit _____

fan switches _____

biohazard container _____

9. Your instructor informs you that a chemical is not dangerous. How should you dispose of the chemical?

10. What precautions should you take while using a centrifuge?

11. How are preservatives correctly discarded?

12. Discuss how to safely measure and mix chemicals.

Introduction to the Human Body

EXERCISE 2

Learning Outcomes

On completion of this exercise, you should be able to:

1. Define *anatomy* and *physiology* and discuss the specializations of each.

2. Describe each level of organization in the body.

3. Describe anatomical position and its importance in anatomical studies.

4. Use directional terminology to describe the relationships of the surface anatomy of the body.

5. Describe and identify the major planes and sections of the body.

6. Locate all abdominopelvic quadrants and regions on laboratory models.

7. Locate the major organs of each organ system and briefly describe each organ's function.

8. Identify the location of the cranial, spinal, and ventral body cavities.

9. Describe the two main divisions of the ventral cavity.

10. Describe and identify the serous membranes of the body.

Knowledge about what lies beneath the skin and how the body works has been slowly amassed over a span of nearly 3000 years. It may be obvious to us now that any logical practice of medicine depends on an accurate knowledge of human anatomy, yet people have not always realized this. Through most of human history, corpses were viewed with superstitious awe and dread. Observations of anatomy

Need More Practice and Review?

Build your knowledge—and confidence!—in the Study Area of MasteringA&P® at www.masteringaandp.com with Pre-lab Quizzes, Post-lab Quizzes, Practice Anatomy Lab™ (PAL™) 3.0 virtual anatomy practice tool, PhysioEx™ 9.0 laboratory simulations, and A&P Flix™ with Quizzes.

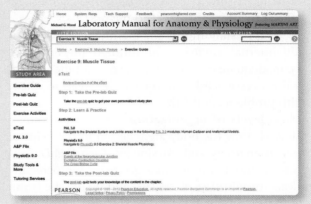

MasteringA&P®

by dissection were illegal, and medicine therefore remained an elusive practice that often harmed rather than helped the unfortunate patient. Despite these superstitions and prohibitions, however, there have always been scientists who wanted to know the human body as it really is rather than how it was imagined to be.

The founder of anatomy was the Flemish anatomist and physician Andreas Vesalius (1514–1564). Vesalius set about to describe human structure accurately. In 1543, he published his monumental work, *De Humani Corporis Faberica (On the Structure of the Human Body)*, the first meaningful text on human anatomy. In this work he corrected more than 200 errors of earlier anatomists and produced drawings that are still useful today. The work done by Vesalius laid the foundation for all future knowledge of the human body. Merely imagining the body's internal structure at last became unacceptable in medical literature.

Many brilliant anatomists and physiologists since the time of Vesalius have contributed significantly to the understanding of human form and function. Advances in medicine and in the understanding of the human body continue at an accelerated pace. For accuracy and consistency, this manual follows the terminology of the publication *Terminologica Anatomica* as endorsed by the International Federation of Associations of Anatomists.

Lab Activity 1 Organization of the Body

Anatomy is the study of body structures. Early anatomists described the body's **gross anatomy,** which includes the large parts such as muscles and bones. As knowledge of the body advanced and scientific tools permitted more detailed observations, the field of anatomy began to diversify into such areas as **microanatomy,** the study of microscopic structures; **cytology,** the study of cells; and **histology,** the study of **tissues,** which are groups of cells coordinating their effort toward a common function.

Physiology is the study of how the body functions and of the work that cells must do to keep the body stable and operating efficiently. **Homeostasis** (hō-mē-ō-STĀ-sis; *homeo-*, unchanging + *stasis*, standing) is the maintenance of a relatively steady internal environment through physiological work. Stress, inadequate diet, and disease disrupt the normal physiological processes and may, as a result, lead to either serious health problems or death.

The various **levels of organization** at which anatomists and physiologists study the body are reflected in the fields of specialization in anatomy and physiology. Each higher level increases in structural and functional complexity, progressing from chemicals to cells, tissues, organs, and finally the organ systems that function to maintain the organism.

Figure 2.1 uses the cardiovascular system to illustrate these levels of organization. The simplest is the **molecular level,** sometimes called the *chemical level*, shown at the bottom of the figure. Atoms such as carbon and hydrogen bond together and form molecules. The heart, for instance, contains protein molecules that are involved in contraction of the cardiac muscle. Molecules are organized into cellular structures called *organelles*, which have distinct shapes and functions. The organelles collectively constitute the next level of organization, the **cellular level.** Cells are the fundamental level of biological organization because it is cells, not molecules, that are alive. Different types of cells working together constitute the **tissue level.** Although tissues lack a distinct shape, they are distinguishable by cell type, such as the various cells that comprise the pancreas. Tissues function together at the **organ level;** at this level, each organ has a distinct three-dimensional shape and a range of functions that is broader than the range of functions for individual cells or tissues. The **organ system level** includes all the organs of a system interacting to accomplish a common goal. The heart and blood vessels, for example, constitute the cardiovascular organ system and physiologically work to move blood through the body. All organ systems make up the individual, which is referred to as the **organism level.**

QuickCheck Questions

1.1 What is the lowest living level of organization in the body?

1.2 What is homeostasis?

In the Lab 1

Materials

☐ Variety of objects and object sets, each representing a level of organization

☐ Torso models

☐ Articulated skeleton

☐ Charts

Procedures

1. Classify each object or object set as to the level of organization it represents. Write your answers in the spaces provided.

   ▪ Molecular level _____

   ▪ Cellular level _____

   ▪ Tissue level _____

   ▪ Organ level _____

   ▪ Organ system level _____

   ▪ Organism level _____ ▪

Figure 2.1 Levels of Organization

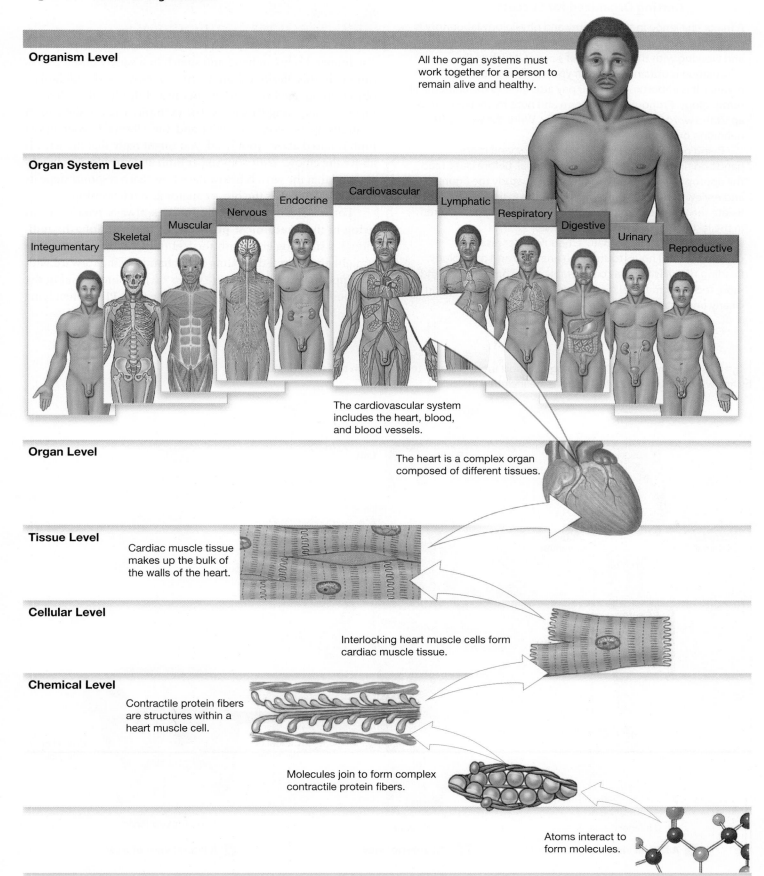

Organism Level

All the organ systems must work together for a person to remain alive and healthy.

Organ System Level

Integumentary Skeletal Muscular Nervous Endocrine Cardiovascular Lymphatic Respiratory Digestive Urinary Reproductive

The cardiovascular system includes the heart, blood, and blood vessels.

Organ Level

The heart is a complex organ composed of different tissues.

Tissue Level

Cardiac muscle tissue makes up the bulk of the walls of the heart.

Cellular Level

Interlocking heart muscle cells form cardiac muscle tissue.

Chemical Level

Contractile protein fibers are structures within a heart muscle cell.

Molecules join to form complex contractile protein fibers.

Atoms interact to form molecules.

Study Tip Getting Organized for Success

A major challenge in the anatomy and physiology laboratory is organizing and processing a substantial volume of information and working with the language of science. Much of the information is obtained through your reading of the lab manual. It is important that you pay attention to the anatomical terminology. Pronounce each term and note its spelling. Break apart the word into its prefix and suffix. Write the word with a definition or example.

Being prepared for lab enables you to spend more hands-on time with the laboratory material. Before class, read the appropriate exercise(s) in this manual, study the figures, and review the Laboratory Activities in the assigned sections. Relate the laboratory material to the theory concepts covered in the lecture component of the course.

Management of your daily schedule is necessary to dedicate several hours to studying anatomy and physiology. Reading typically takes a considerable time commitment, and more technical material may require several readings for you to clearly understand the concepts. ■

Lab Activity 2 Anatomical Position and Directional Terminology

The human body can bend and stretch in a variety of directions. Although this flexibility allows us to move and manipulate objects in our environment, it can cause difficulty when describing and comparing structures. For example, what is the correct relationship between the wrist and the elbow? If your upper limb is raised above your head, you might reply that the wrist is above the elbow. With your upper limb at your side, you would respond that the wrist is below the elbow. Each response appears correct, but which is the proper anatomical relationship?

For anatomical study, the body is always referred to as being in the **anatomical position.** In this position, the individual is standing erect with the feet pointed forward, the eyes straight ahead, and the palms of the hands facing forward with the upper limbs at the sides (**Figure 2.2**). An individual in the anatomical position is said to be **supine** (soo-PĪN) when lying on the back and **prone** when lying face down.

Figure 2.2 **Directional Terminology** Important directional terms used in this text are indicated by arrows; definitions and descriptions are included in Table 2.1.

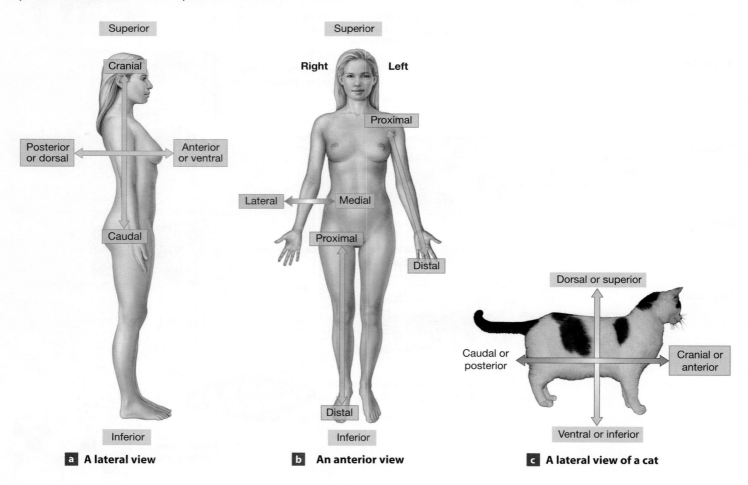

a A lateral view

b An anterior view

c A lateral view of a cat

Imagine attempting to give someone directions if you could not use terms such as *north* and *south* or *left* and *right*. These words have a unique meaning and guide the traveler toward a destination. Describing the body also requires specific terminology. Expressions such as *near*, *close to*, or *on top of* are too vague for anatomical descriptions. To prevent misunderstandings, precise terms are used to describe the locations and spatial relationships of anatomy. These terms have their roots in the Greek and Latin languages. **Table 2.1** and Figure 2.2 display the most frequently used directional terms. Notice that most of them can be grouped into opposing pairs, or antonyms.

- **Superior** and **inferior** describe vertical positions. *Superior* means above, *inferior* means below. For example, on a person in the anatomical position, the head is superior to the shoulders and the knee is inferior to the hip.
- **Anterior** and **posterior** refer to front and back. *Anterior* means in front of or forward, and *posterior* means in back of or toward the back. The anterior surface of the body comprises all front surfaces, including the palms of the hand, and the posterior surface includes all the back surfaces. In addition to describing locations, these directional terms describe position *relationships*, which means that one body part can be described using both terms. The heart, for example, is posterior to the breastbone and anterior to the spine.
- In four-legged animals, the anatomical position is with all four limbs on the ground, and therefore the meanings of some directional terms change (Figure 2.2c). *Superior* now refers to the back, or **dorsal**, surface, and *inferior* refers to the belly, or **ventral**, surface. **Cranial** and *anterior* mean toward the head in four-legged animals, and **caudal** and *posterior* mean toward the tail in four-legged animals and toward the coccyx in humans.

- **Medial** and **lateral** describe positions relative to the body's *midline*, the vertical middle of the body or any structure in the body. *Medial* has two meanings. It describes one structure as being closer to the body's midline than some other structure; for instance, the ring finger is medial to the middle finger when the hand is held in the anatomical position. *Medial* also describes a structure that is permanently between others, as the nose is medial to the eyes. *Lateral* means either farther from the body's midline or permanently to the side of some other structure; the eyes are lateral to the nose, and, in the anatomical position, the middle finger is lateral to the ring finger.
- **Proximal** refers to parts near another structure. **Distal** describes structures that are distant from other structures. These terms are frequently used to describe the proximity of a structure to its point of attachment on the body. For example, the thigh bone (femur) has a proximal region where it attaches to the hip and a distal region toward the knee.
- **Superficial** and **deep** describe layered structures. *Superficial* refers to parts on or close to the surface. Underneath an upper layer are *deep*, or *bottom*, structures. The skin is superficial to the muscular system, and bones are usually deep to the muscles.

Table 2.1	Directional Terms (See Figure 2.2)	
Term	Region or Reference	Example
Anterior	The front; before	The navel is on the *anterior* surface of the trunk.
Ventral	The belly side (equivalent to anterior when referring to human body)	In humans, the navel is on the *ventral* surface.
Posterior	The back; behind	The shoulder blade is located *posterior* to the rib cage.
Dorsal	The back (equivalent to posterior when referring to human body)	The *dorsal* body cavity encloses the brain and spinal cord.
Cranial or cephalic	The head	The *cranial*, or *cephalic*, border of the pelvis is on the side toward the head rather than toward the thigh.
Superior	Above; at a higher level (in human body, toward the head)	In humans, the cranial border of the pelvis is *superior to* the thigh.
Caudal	The tail (coccyx in humans)	The hips are *caudal* to the waist.
Inferior	Below; at a lower level	The knees are *inferior* to the hips.
Medial	Toward the body's longitudinal axis; toward the midsagittal plane	The *medial* surfaces of the thighs may be in contact; moving medially from the arm across the chest surface brings you to the sternum.
Lateral	Away from the body's longitudinal axis; away from the midsagittal plane	The thigh articulates with the *lateral* surface of the pelvis; moving laterally from the nose brings you to the eyes.
Proximal	Toward an attached base	The thigh is *proximal* to the foot; moving proximally from the wrist brings you to the elbow.
Distal	Away from an attached base	The fingers are *distal* to the wrist; moving distally from the elbow brings you to the wrist.
Superficial	At, near, or relatively close to the body surface	The skin is *superficial* to underlying structures.
Deep	Farther from the body surface	The bone of the thigh is *deep* to the surrounding skeletal muscles.

Some directional terms seem to be interchangeable, but there is usually a precise term for each description. For example, *superior* and *proximal* both describe the upper region of limb bones. When discussing the point of attachment of a bone, *proximal* is the more descriptive term. When describing the location of a bone relative to an inferior bone, the term *superior* is used.

QuickCheck Questions

2.1 Why is having a precisely defined anatomical position important in anatomical studies?

2.2 What is the relationship of the shoulder joint to the elbow joint?

2.3 Which directional term describes the relationship of muscles to the skin?

In the Lab 2

Materials

☐ Yourself or a laboratory partner
☐ Torso models
☐ Anatomical charts
☐ Anatomical models

Procedures

1. Assume the anatomical position. Consider how this orientation differs from your normal stance.

2. Review each directional term presented in Figure 2.2.

3. Use the laboratory models and charts and your own body (or your partner's) to practice using directional terms while comparing anatomy. The Review & Practice Sheet at the end of this exercise may be used as a guide for comparisons. ■

Lab Activity 3   Regional Terminology

Approaching the body from a regional perspective simplifies the learning of anatomy. Body surface features are used as anatomical landmarks to assist in locating internal structures, and as a result many internal structures are named after an overlying surface structure. For example, the back of the knee is called the popliteal (pop-LIT-ē-al) region, and the major artery in the knee is the popliteal artery. Table 2.2 and Figure 2.3 present the major regions of the body.

The head is referred to as the **cephalon** and consists of the **cranium,** or skull, and the **face.** The neck is the **cervical** region. The main part of the body is the **trunk,** which attaches the neck, upper limbs, and lower limbs. The thorax is the chest, or **pectoral,** region. Below the chest is the **abdominal** region,

which narrows at the **pelvis.** The back surface of the trunk, the **dorsum,** includes the **loin,** or lower back, and the **gluteal** region of the buttock. The side of the trunk below the ribs is the **flank.**

The shoulder, or **scapular,** region attaches the **upper limb,** which is the arm and forearm, to the trunk and forms the **axilla,** the armpit. The proximal part of the upper limb is the **brachium;** the **antebrachium** is the forearm. Between the brachium and antebrachium is the **antecubitis** region, the elbow. The wrist is called the **carpus,** and the inside surface of the hand is the **palm.**

The pelvis attaches the **lower limb** to the trunk at the **inguinal** area, or **groin.** The proximal part of the lower limb is the **thigh,** the back of the knee is the **popliteal** region, and the leg is the calf, or **sura. Tarsus** refers to the ankle, and the sole of the foot is the **plantar** surface.

Reference to the position of internal abdominal organs is simplified by partitioning the trunk into four equal **quadrants,** the right and left upper quadrants and the right and left lower quadrants (Figure 2.4). Observe in Figure 2.4a the vertical and horizontal planes used to delineate the quadrants. Quadrants are used to describe the positions of organs.

Table 2.2	Regions of the Human Body (See Figure 2.3)
Structure	**Region**
Cephalon (head)	Cephalic region
Cervicis (neck)	Cervical region
Thoracis (thorax, or chest)	Thoracic region
Axilla (armpit)	Axillary region
Brachium (arm)	Brachial region
Antecubitis (elbow)	Antecubital region
Antebrachium (forearm)	Antebrachial region
Carpus (wrist)	Carpal region
Manus (hand)	Manual region
Abdomen (belly)	Abdominal region
Lumbus (loin)	Lumbar region
Gluteus (buttock)	Gluteal region
Pelvis (hip)	Pelvic region
Pubis (anterior pelvis)	Pubic region
Inguen (groin)	Inguinal region
Femur (thigh)	Femoral region
Popliteus (back of knee)	Popliteal region
Crus (anterior leg)	Crural region
Sura (calf)	Sural region
Tarsus (ankle)	Tarsal region
Pes (foot)	Pedal region
Planta (sole)	Plantar region

Figure 2.3 Regional Terminology Anatomical terms are shown in boldface type, common names in plain type, and anatomical adjectives in parentheses.

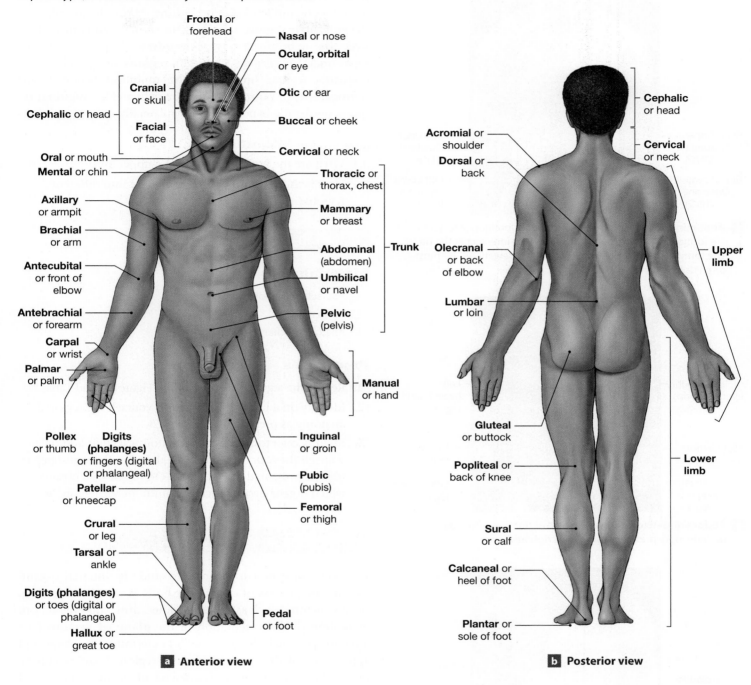

a Anterior view

b Posterior view

The stomach, for example, is mostly located in the left upper quadrant.

For more detailed descriptions, the abdominal surface is divided into nine **abdominopelvic regions,** shown in Figure 2.4b. Four planes are used to define the regions: two vertical and two transverse planes arranged in the familiar tic-tac-toe pattern. The vertical planes, called the right and left **lateral planes,** are positioned slightly medial to the

nipples, each plane on the side of the nipple that is closer to the body center. The lateral planes divide the trunk into three nearly equal vertical regions. A pair of transverse planes crosses the vertical planes to isolate the nine regions. The **transpyloric plane** is superior to the umbilicus (navel) at the level of the pylorus, the lower region of the stomach. The **transtubercular plane** is inferior to the umbilicus and crosses the abdomen at the level of the superior hips.

Figure 2.4 Abdominopelvic Quadrants and Regions

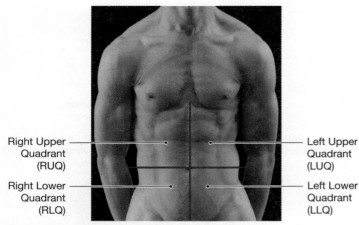

Right Upper
Quadrant
(RUQ)

Right Lower
Quadrant
(RLQ)

Left Upper
Quadrant
(LUQ)

Left Lower
Quadrant
(LLQ)

a **Abdominopelvic quadrants.** The four abdominopelvic quadrants are formed by two perpendicular lines that intersect at the navel. The terms for these quadrants, or their abbreviations, are most often used in clinical discussions.

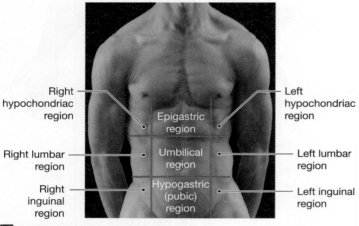

Right
hypochondriac
region

Epigastric
region

Left
hypochondriac
region

Right lumbar
region

Umbilical
region

Left lumbar
region

Right
inguinal
region

Hypogastric
(pubic)
region

Left inguinal
region

b **Abdominopelvic regions.** The nine abdominopelvic regions provide more precise regional descriptions.

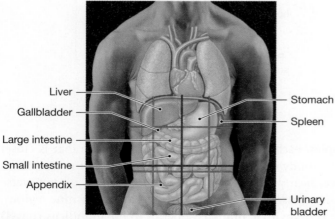

Liver

Gallbladder

Large intestine

Small intestine

Appendix

Stomach

Spleen

Urinary
bladder

c **Anatomical relationships.** The relationship between the abdominopelvic quadrants and regions and the locations of the internal organs are shown here.

The nine abdominopelvic regions are as follows: The **umbilical region** surrounds the umbilicus. Lateral to this region are the right and left **lumbar regions.** Above the umbilicus is the **epigastric region** containing the stomach and much of the liver. The right and left **hypochondriac** (hī-pō-KON-drē-ak; *hypo,* under + *chondro,* cartilage) **regions** are lateral to the epigastric region. Inferior to the umbilical region is the **hypogastric,** or **pubic, region.** The right and left **inguinal regions** border the hypogastric region laterally.

QuickCheck Questions

3.1 What are the major regions of the upper limb?

3.2 How is the abdominal surface divided into different regions?

In the Lab   3

Materials

☐ Yourself or a laboratory partner

☐ Torso models

☐ Anatomical charts

Procedures

1. Review the regional terminology in Figure 2.3 and Table 2.2.

2. Identify on a laboratory model or yourself the regional anatomy as presented in Figure 2.3.

3. Identify on a torso model or anatomical chart and on yourself the four quadrants and the nine abdominopelvic regions presented in Figure 2.4. On the model, observe which organs occupy each abdominopelvic region. ■

Lab Activity   4   **Planes and Sections**

The body must be cut in order to study its internal organization. The process of cutting the body is called **sectioning.** Most structures, such as the trunk, knee, arm, and eyeball, can be sectioned. The orientation of the **plane of section** (the direction in which the cut is made) determines the shape and appearance of the exposed internal region. Imagine cutting one soda straw crosswise (crosswise plane of section) and another straw lengthwise (lengthwise plane of section). The former produces a circle, and the latter produces a concave U-shaped tube.

Three major types of sections are used in the study of anatomy: two vertical and one transverse (**Figure 2.5**). **Transverse** sections are perpendicular to the vertical orientation of the body. (The crosswise cut you made on the imaginary straw yielded a transverse section.) Transverse sections are often called **cross sections** because they go across the body axis.

Figure 2.5 Planes of Section The three primary planes of section. The photographs of sectional images were derived from the Visible Human data set.

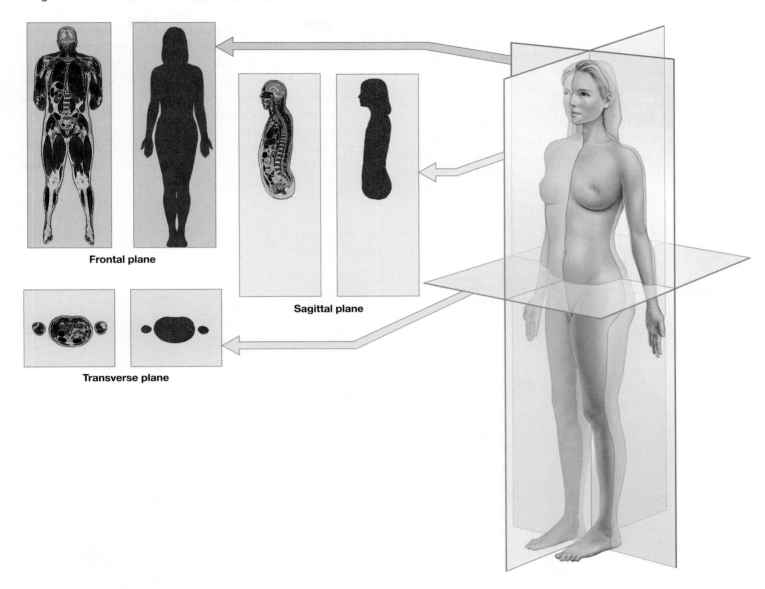

Frontal plane

Sagittal plane

Transverse plane

A transverse section divides superior and inferior structures. **Vertical** sections are parallel to the vertical axis of the body and include sagittal and frontal sections. A **sagittal** vertical section divides a body or organ into right and left portions. A **midsagittal** vertical section equally divides structures, and a **parasagittal** vertical section produces nearly equal divisions. A **frontal,** or **coronal,** vertical section separates anterior and posterior structures.

QuickCheck Questions

4.1 Which type of section separates the kneecap from the lower limb?

4.2 Amputation of the forearm is performed by which type of section?

In the Lab   4

Materials

☐ Anatomical models with various sections
☐ Knife and objects for sectioning

Procedures

1. Review each plane of section shown in Figure 2.5.

2. Identify the sections on models and other materials presented by your instructor.

3. Cut several common objects, such as an apple and a hot dog, along their sagittal and transverse planes. Compare the exposed arrangement of the interior. ∎

Lab Activity 5 | Body Cavities

Body cavities are internal spaces that house internal organs, such as the brain in the cranium and the digestive organs in the abdomen. The walls of a body cavity support and protect the soft organs contained in the cavity. In the trunk, large cavities are subdivided into smaller cavities that contain individual organs. The smaller cavities are enclosed by thin sacs, such as those around the heart, lungs, and intestines. Most of the space in a cavity is occupied by the enclosed organ and by a thin film of liquid.

The **cranial cavity** is the space within the oval *cranium* of the skull that encases and protects the delicate brain. The **spinal cavity** is a long, slender canal that passes through the vertebral column. The cranial and spinal cavities are continuous with each other and join at the base of the skull, where the spinal cord meets the brain. The brain and spinal cord are contained within the **meninges,** a protective three-layered membrane. Some anatomists group the cranial and spinal cavities into a larger *dorsal body cavity;* however, this term is not recognized by the international reference *Terminologica Anatomica* and is therefore not used in this manual.

The **ventral body cavity,** also called the **coelom** (SĒ-luhm; *koila,* cavity), is the entire space of the body trunk anterior to the vertebral column and posterior to the sternum (breastbone) and the abdominal muscle wall (**Figure 2.6**). This large cavity is divided into two major cavities, the **thoracic** (*thorax,* chest)

Figure 2.6 Body Cavities

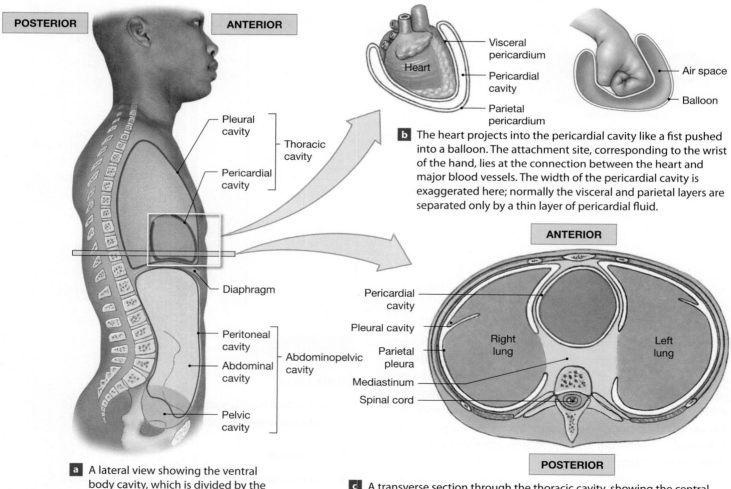

b The heart projects into the pericardial cavity like a fist pushed into a balloon. The attachment site, corresponding to the wrist of the hand, lies at the connection between the heart and major blood vessels. The width of the pericardial cavity is exaggerated here; normally the visceral and parietal layers are separated only by a thin layer of pericardial fluid.

a A lateral view showing the ventral body cavity, which is divided by the muscular diaphragm into a superior thoracic (chest) cavity and an inferior abdominopelvic cavity. Three of the four adult body cavities are shown and outlined in red; only one of the two pleural cavities can be shown in a sagittal section.

c A transverse section through the thoracic cavity, showing the central location of the pericardial cavity. Notice how the mediastinum divides the thoracic cavity into two pleural cavities. Note that this transverse or cross-sectional view is oriented as though the observer were standing at the subject's feet and looking toward the subject's head. This is the standard presentation for clinical images, and unless otherwise noted, sectional views in this text use this same orientation.

cavity and the **abdominopelvic cavity.** These cavities, in turn, are further divided into the specific cavities that surround individual organs. The heart, lungs, stomach, and intestines are covered with a double-layered **serous** (SĒR-us; *seri-*, watery) **membrane.** Each serous membrane isolates one organ and reduces friction and abrasion on the organ surface.

The walls of the thoracic cavity are muscle and bone. The main subdivisions of this cavity are the **mediastinum** (mē-dē-as-TI-num *or* mē-dē-AS-ti-num; *media-*, middle) and two **pleural cavities.** The mediastinum is the mass of organs and tissues separating the pleural cavities. Each pleural cavity contains one lung. Inside the mediastinum is a smaller cavity, the **pericardial** (*peri-*, around + *kardia*, heart) **cavity,** and the heart is most often described as being contained inside this cavity rather than simply inside the mediastinum.

The abdominopelvic cavity is separated from the thoracic cavity by a dome-shaped muscle, the diaphragm. The abdominopelvic cavity is the space between the diaphragm and the floor of the pelvis. This cavity is subdivided into the abdominal cavity and the pelvic cavity. The **abdominal cavity** contains most of the digestive organs, such as the liver, gallbladder, stomach, pancreas, kidneys, and small and large intestines. The **pelvic cavity** is the small cavity enclosed by the pelvic girdle of the hips. This cavity contains the internal reproductive organs, parts of the large intestine, the rectum, and the urinary bladder.

The heart, lungs, stomach, and intestines are encased in double-layered serous membranes that have a minuscule fluid-filled cavity between the two layers. Directly attached to the exposed surface of an internal organ is the **visceral** (VIS-er-al; *viscera*, internal organ) **layer** of the serous membrane. The **parietal** (pah-RĪ-e-tal; *pariet-*, wall) **layer** is superficial to the visceral layer and lines the wall of the body cavity. The **serous fluid** between these layers is a lubricant that reduces friction and abrasion between the layers as the enclosed organ moves.

Figure 2.6b highlights the anatomy of the serous membrane of the heart, the **pericardium.** This membrane consists of an outer **parietal pericardium** and an inner **visceral pericardium.** The parietal pericardium is a fibrous sac attached to the diaphragm and supportive tissues of the thoracic cavity. The visceral pericardium is attached to the surface of the heart. The space between these two serous layers is the pericardial cavity.

The serous membrane of the lungs is called the **pleura** (PLOO-rah). The **parietal pleura** lines the thoracic wall, and the **visceral pleura** is attached to the surface of the lung. Because each lung is contained inside a separate pleural cavity, a puncture wound on one side of the chest usually collapses only the corresponding lung.

Most of the digestive organs are encased in the **peritoneum** (per-i-ton-Ē-um), the serous membrane of the abdomen. The **parietal peritoneum** has numerous folds that wrap around and attach the abdominal organs to the posterior abdominal wall. The **visceral peritoneum** lines the organ surfaces. The **peritoneal cavity** is the space between the parietal and visceral peritoneal layers. The peritoneum has many blood vessels, lymphatic vessels, and nerves that support the digestive organs. The kidneys are **retroperitoneal** (*retro-*, behind) and are located outside the peritoneum.

Clinical Application Problems with Serous Membranes

Serous membranes may become inflamed and infected as a result of bacterial invasion or damage to the underlying organ. Liquids often build up in the cavity of the serous membrane, causing additional complications. **Peritonitis** is an infection of the peritoneum that occurs when the digestive tract is damaged—often by ulceration, rupture, or a puncture wound—in a way that permits intestinal bacteria to contaminate the peritoneum. **Pleuritis,** or **pleurisy,** is an inflammation of the pleura often caused by tuberculosis, pneumonia, or thoracic abscess. Breathing is made painful as the inflamed membranes move when a person inhales and exhales. **Pericarditis** is an inflammation of the pericardium resulting from infection, injury, heart attack, or other causes. In advanced stages, a buildup of liquid causes the heart to compress, a condition resulting in decreased cardiac function. ■

QuickCheck Questions

5.1 What structures form the walls of the cranial and spinal cavities?

5.2 Name the various subdivisions of the ventral body cavity.

5.3 Describe the two layers of a serous membrane.

5.4 Name the three serous membranes of the body.

In the Lab   5

Materials

☐ Torso models
☐ Articulated skeleton
☐ Anatomical charts

Procedures

1. Review each cavity and serous membrane illustrated in Figure 2.6.

2. Locate each body cavity on the torso models, anatomical charts, and articulated skeleton.

3. Identify the organ(s) in the various cavities of the ventral body cavity on the torso models.

4. Identify the pericardium, pleura, and peritoneum on the torso models and charts.

For additional practice, complete the *Sketch to Learn* activity (on p. 18). ■

Sketch to Learn

Drawing is an excellent study technique to review material. Let's sketch the thoracic cavity and its major organs. "Hey, I'm not an artist," you may say. Just follow the simple examples here and throughout the manual, and you just might learn to draw nearly anything. Most sketching we will do in this manual requires little more than the skill to draw simple shapes with lines, uncomplicated curves, and circles.

First, a few hints are in order:

- Use a pencil while sketching. I personally need more eraser than lead!

- Take your time and plan your sketch. Look at the example and be sure you understand what you need to draw.

- Consider the size of the sketch and its components so your drawing is easy to label and study.

- Organize your labels into anatomical groups as shown in the example.

- Most rewarding of all, relax and enjoy the creative process. Further develop your drawing by using color pencils and adding shade and texture.

Sample Sketch

Step 1
- Draw thorax with the heart and lungs.

Step 2
- Add another layer around organs to show serous membranes.

Step 3
- Label your sketch.

Your Sketch

Name _____

Date _____

Section _____

Introduction
to the Human Body

A. Definitions

Define each directional term.

1. anterior

2. lateral

3. proximal

4. ventral

5. posterior

6. medial

7. distal

8. superficial

9. superior

10. dorsal

11. inferior

12. deep

B. Fill in the Blanks

Use the correct term(s) to complete each sentence.

1. The heart is surrounded by a small cavity called the _____, which is inside a larger cavity, the _____.

2. The _____ cavity surrounds the digestive organs in the abdominal cavity.

3. The kidneys are _____ because they are located superficial to the _____.

4. The inner membrane layer surrounding a lung is the _____.

5. The brain is contained in a cavity called the _____.

6. A lubricating substance in body cavities is called _____.

7. The large medial area of the chest is called the _____.

8. The muscle that divides the ventral body cavity horizontally is the _____.

9. The outer layer of a serous membrane is the _____ layer.

10. In the anatomical position, the palms of the hands are _____.

11. The index finger is _____ to the ring finger.

12. The trunk is _____ to the pubis.

13. Where it attaches to the elbow, the brachium is _____ to the elbow.

14. The buttock is _____ to the pubis.

15. The shoulders are _____ to the hips.

C. Short-Answer Questions

1. Describe the six main levels of organization in the body.

2. List the nine abdominopelvic regions and the location of each.

3. Compare the study of anatomy with that of physiology.

4. Define the term *homeostasis*.

5. In which quadrant is the liver located?

6. Name the abdominopelvic region that contains the urinary bladder.

7. Describe a parasagittal plane of section.

8. What do the brachium, antecubitis, and antebrachium constitute?

9. Where is the dorsal surface of a four-legged animal?

D. Labeling

1. Label the regions of the body in **Figure 2.7**.

Figure 2.7 **Regional Terminology**

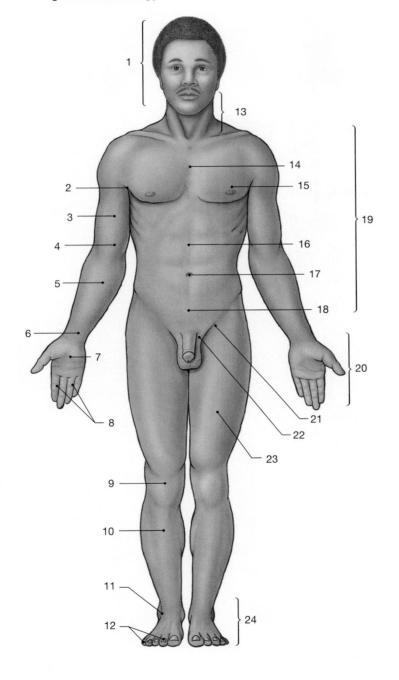

1. _____
2. _____
3. _____
4. _____
5. _____
6. _____
7. _____
8. _____
9. _____
10. _____
11. _____
12. _____
13. _____
14. _____
15. _____
16. _____
17. _____
18. _____
19. _____
20. _____
21. _____
22. _____
23. _____
24. _____

2. Label the directional terms in **Figure 2.8**.

Figure 2.8 Directional References

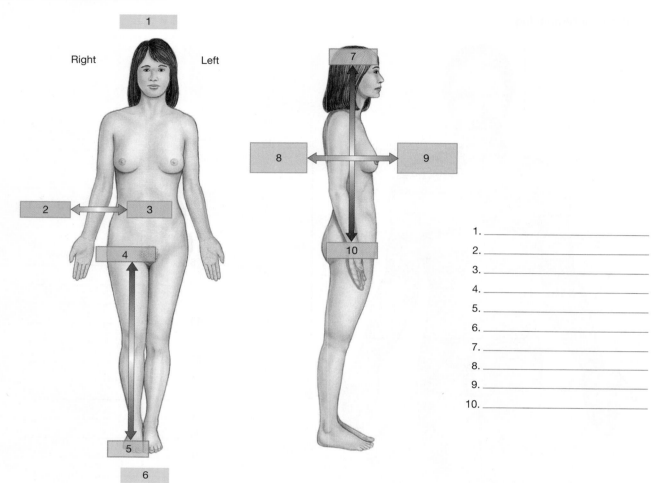

Right   Left

1. _____
2. _____
3. _____
4. _____
5. _____
6. _____
7. _____
8. _____
9. _____
10. _____

3. Label the structures in **Figure 2.9**.

Figure 2.9 Body Cavity

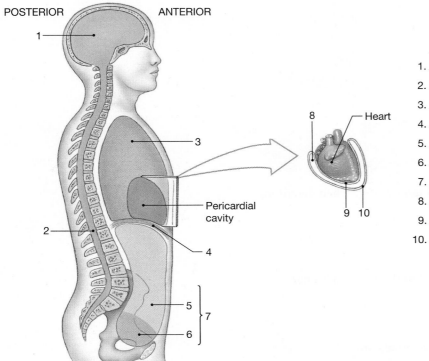

POSTERIOR

ANTERIOR

1

3

2

Pericardial cavity

4

5

6

7

8

Heart

9   10

1. _____

2. _____

3. _____

4. _____

5. _____

6. _____

7. _____

8. _____

9. _____

10. _____

E. Drawing

1. ***Draw It!*** Draw two pictures of a bagel sectioned by a plane. In one drawing, make the sectioning plane parallel to the circular surface of the bagel; in the other drawing, make the sectioning plane perpendicular to that surface.

2. ***Draw It!*** Sketch the body trunk and the planes that designate the nine abdomino-pelvic regions.

F. Analysis and Application

1. Explain why it is important to use anatomical terminology when describing body parts.

2. Compare the body axis of a four-legged animal to the axis of a human.

3. Describe the cavities that protect the brain and spinal cord.

4. What is the benefit for organs in the ventral body to be surrounded by double-layered membranes instead of single-layered membranes?

G. Clinical Challenge

1. Nicole has a respiratory infection that has caused her right pleura to dry out. Describe the symptoms that could be related to this condition.

2. Doug has a skateboard accident and scrapes his knees, left hip, and left elbow. Using the appropriate regional terminology, describe his injuries as if you were writing them in his medical chart.

Organic Systems Overview

Learning Outcomes

On completion of this exercise, you should be able to:

1. Describe the main functions of each organ system.

2. Identify the major organs of each organ system on charts and lab models.

3. Identify the major organs in the ventral body cavity of the cat.

Lab Activity 1 — Introduction to Organ Systems

The human body consists of 11 **organ systems,** each responsible for specific functions (**Figure 3.1**). The organs of a system coordinate their activities to accomplish that system's role in maintaining homeostasis. Organ systems adjust and regulate their activities to meet the ever-changing demands of the organism. For example, to clean the blood, the urinary system relies on the cardiovascular system to deliver blood within specific volume and pressure requirements.

Most anatomy and physiology courses are designed to progress through the lower levels of organization first and then examine each organ system. Because organ systems work together to maintain the organism, it is important that you have a basic understanding of the function of each one.

QuickCheck Questions

1.1 What is a function of the endocrine system?

1.2 Name the major organs of the integumentary system.

Need More Practice and Review?

Build your knowledge—and confidence!—in the Study Area of MasteringA&P® at www.masteringaandp.com with Pre-lab Quizzes, Post-lab Quizzes, Practice Anatomy Lab™ (PAL™) 3.0 virtual anatomy practice tool, PhysioEx™ 9.0 laboratory simulations, and A&P Flix™ with Quizzes.

PAL | practice anatomy lab   For this lab exercise, follow these navigation paths:
- PAL>Cat>Respiratory System
- PAL>Cat>Digestive System
- PAL>Cat>Reproductive System

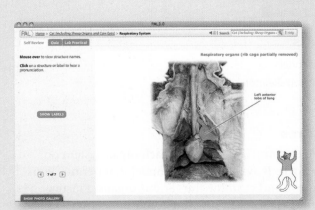

MasteringA&P®

Figure 3.1 **An Introduction to Organ Systems**

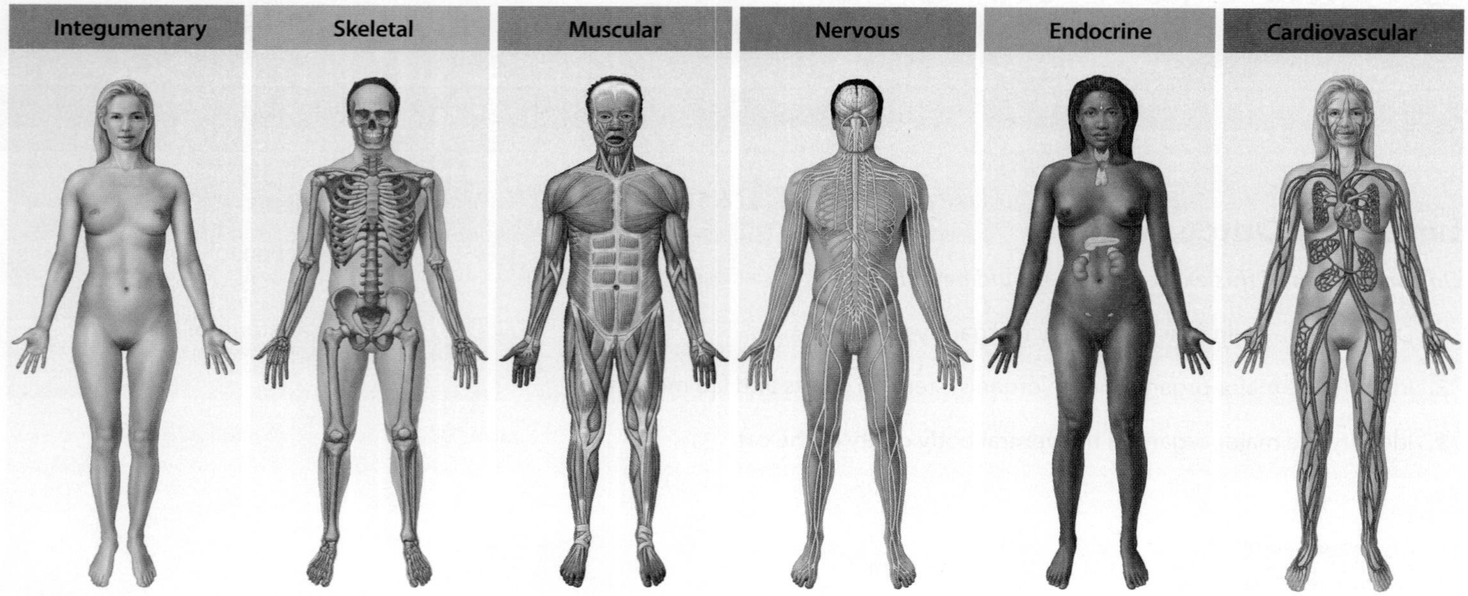

Integumentary	Skeletal	Muscular	Nervous	Endocrine	Cardiovascular

Major Organs
- Skin
- Hair
- Sweat glands
- Nails

Functions
- Protects against environmental hazards
- Helps regulate body temperature
- Provides sensory information

Major Organs
- Bones
- Cartilages
- Associated ligaments
- Bone marrow

Functions
- Provides support and protection for other tissues
- Stores calcium and other minerals
- Forms blood cells

Major Organs
- Skeletal muscles and associated tendons

Functions
- Provides movement
- Provides protection and support for other tissues
- Generates heat that maintains body temperature

Major Organs
- Brain
- Spinal cord
- Peripheral nerves
- Sense organs

Functions
- Directs immediate responses to stimuli
- Coordinates or moderates activities of other organ systems
- Provides and interprets sensory information about external conditions

Major Organs
- Pituitary gland
- Thyroid gland
- Pancreas
- Adrenal glands
- Gonads
- Endocrine tissues in other systems

Functions
- Directs long-term changes in the activities of other organ systems
- Adjusts metabolic activity and energy use by the body
- Controls many structural and functional changes during development

Major Organs
- Heart
- Blood
- Blood vessels

Functions
- Distributes blood cells, water and dissolved materials including nutrients, waste products, oxygen, and carbon dioxide
- Distributes heat and assists in control of body temperature

In the Lab   1

Materials

☐ Torso models
☐ Articulated skeleton
☐ Charts

Procedures

1. Locate the principal organs of each organ system on the models. If available, use a model that permits you to remove and examine the various organs. Practice returning each organ to its anatomical location.

2. On your own body, identify the general location of as many organs as possible. ■

Lab Activity   2   Gross Anatomy of the Cat

The terminology used to describe the position and location of body parts in four-legged animals differs slightly from that used for the human body, because four-legged animals move forward head first, with the abdominal surface parallel to the ground. Anatomical position for a four-legged animal is all four limbs on the ground. *Superior* refers to the back (dorsal) surface, and *inferior* relates to the belly (ventral) surface. *Cephalic* means toward the front or anterior, and *caudal* refers to posterior structures.

Figure 3.1 An Introduction to Organ Systems *(continued)*

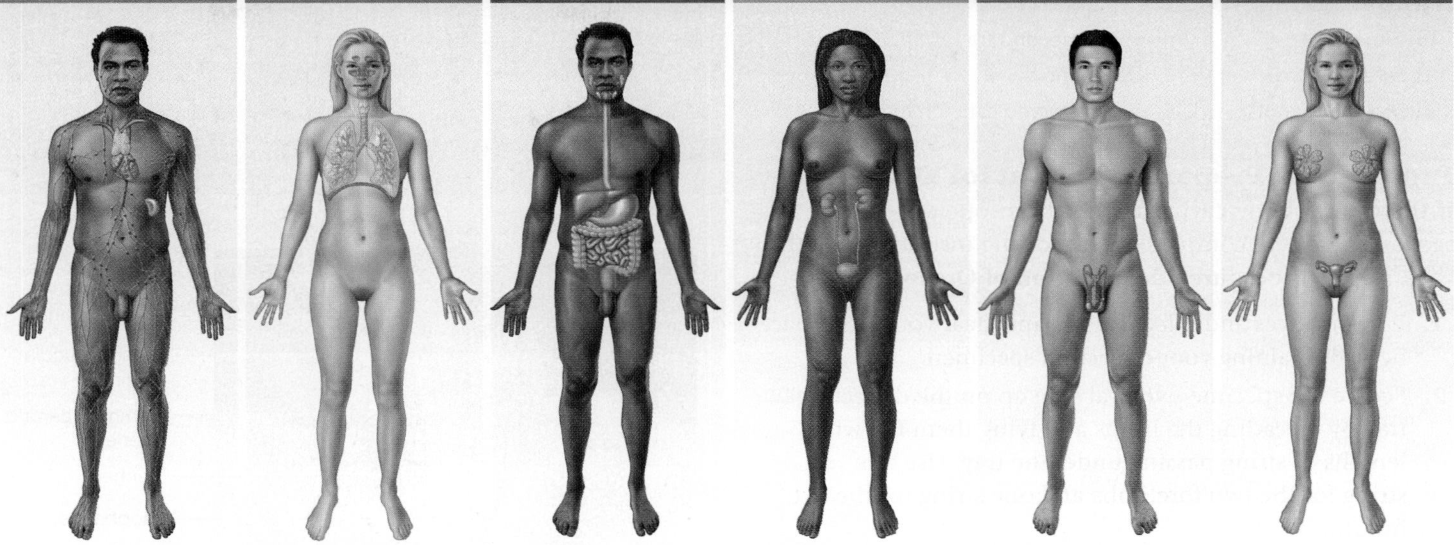

Lymphatic	Respiratory	Digestive	Urinary	Male Reproductive	Female Reproductive

Major Organs
- Spleen
- Thymus
- Lymphatic vessels
- Lymph nodes
- Tonsils

Functions
- Defends against infection and disease
- Returns tissue fluids to the bloodstream

Major Organs
- Nasal cavities
- Sinuses
- Larynx
- Trachea
- Bronchi
- Lungs
- Alveoli

Functions
- Delivers air to alveoli (sites in lungs where gas exchange occurs)
- Provides oxygen to bloodstream
- Removes carbon dioxide from bloodstream
- Produces sounds for communication

Major Organs
- Teeth
- Tongue
- Pharynx
- Esophagus
- Stomach
- Small intestine
- Large intestine
- Liver
- Gallbladder
- Pancreas

Functions
- Processes and digests food
- Absorbs and conserves water
- Absorbs nutrients
- Stores energy reserves

Major Organs
- Kidneys
- Ureters
- Urinary bladder
- Urethra

Functions
- Excretes waste products from the blood
- Controls water balance by regulating volume of urine produced
- Stores urine prior to voluntary elimination
- Regulates blood ion concentrations and pH

Major Organs
- Testes
- Epididymides
- Ductus deferentia
- Seminal vesicles
- Prostate gland
- Penis
- Scrotum

Functions
- Produces male sex cells (sperm), suspending fluids, and hormones
- Sexual intercourse

Major Organs
- Ovaries
- Uterine tubes
- Uterus
- Vagina
- Labia
- Clitoris
- Mammary glands

Functions
- Produces female sex cells (oocytes) and hormones
- Supports developing embryo from conception to delivery
- Provides milk to nourish newborn infant
- Sexual intercourse

QuickCheck Questions

2.1 How is anatomical position different for a four-legged animal compared to a human?

2.2 Which term refers to the head of a four-legged animal?

 Safety Alert: Cat Dissection Basics

You *must* practice the highest level of laboratory safety while handling and dissecting the cat. Keep the following guidelines in mind during the dissection.

1. Wear gloves and safety glasses to protect yourself from the fixatives used to preserve the specimen.

2. Do not dispose of the fixative from your specimen. You will later store the specimen in the fixative to keep the specimen moist and to keep it from decaying.

3. Be extremely careful when using a scalpel or other sharp instrument. Always direct cutting and scissor motion away from yourself to prevent an accident if the instrument slips on moist tissue.

4. Before cutting a given tissue, make sure it is free from underlying and adjacent tissues so that they will not be accidentally severed.

5. Never discard tissue in the sink or trash. Your instructor will inform you of the proper disposal procedure. ▲

In the Lab 2

Materials

- ☐ Gloves
- ☐ Safety glasses
- ☐ Dissecting tools
- ☐ Dissecting tray
- ☐ String
- ☐ Preserved cat, skin removed

Procedures: Preparing the Cat for Dissection

If the ventral body cavity has not been opened on your dissection specimen, complete the following instructions. Otherwise, skip to **Procedure: Identification of Organs**.

1. Put on gloves and safety glasses and clear your workspace before obtaining your dissection specimen.

2. Secure the specimen ventral side up on the dissecting tray by spreading the limbs and tying them flat with lengths of string passing under the tray. Use one string for the two forelimbs and one string for the two hindlimbs.

3. If the ventral body cavity has not been opened, use scissors to cut a midsagittal section through the muscles of the abdomen to the sternum.

4. To avoid cutting through the bony sternum, angle your incision laterally approximately 0.5 inch and cut the costal cartilages. Continue the parasagittal section to the base of the neck.

5. Make a lateral incision on each side of the diaphragm. Use care not to damage the diaphragm or the internal organs. Spread the thoracic walls to observe the internal organs.

6. Make a lateral section across the pubic region and angle toward the hips. Spread the abdominal walls to expose the abdominal organs.

Procedures: Identification of Organs

Organs in the Neck

1. The **larynx** is the cartilaginous structure on the anterior neck (Figure 3.2). The airway through the larynx is called the **glottis.** The **trachea** is the windpipe that passes through the midline of the neck. The trachea is kept open by C-shaped pieces of hyaline cartilage called the **tracheal rings.**

2. On the dorsal side of the trachea is the food tube, the **esophagus,** which passes through the thoracic cavity and into the abdomen.

3. Spanning the trachea on both sides is the **thyroid gland.**

4. If the cat's blood vessels have been injected with latex, the red **common carotid arteries** and the blue **external jugular veins** will be clearly visible.

Figure 3.2 Organs of the Neck and Thoracic Cavity
Ventral dissection of the cat neck and thoracic cavity. The vascular system has been injected with red latex in the arteries and blue latex in the veins.

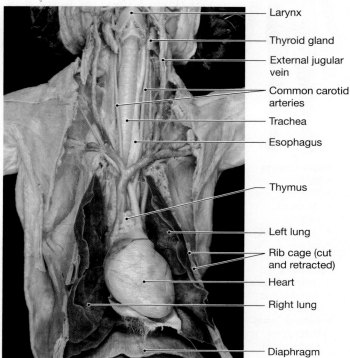

- Larynx
- Thyroid gland
- External jugular vein
- Common carotid arteries
- Trachea
- Esophagus
- Thymus
- Left lung
- Rib cage (cut and retracted)
- Heart
- Right lung
- Diaphragm

Organs in the Thoracic Cavity

1. Follow the trachea to where it divides into left and right **primary bronchi** that enter the **lungs** (Figure 3.2). Observe the many lobes of the cat lungs and the glossy **pleura** surrounding each lung.

2. The **heart** is positioned medial to the lungs. The **aorta** is the main artery that curves posteriorly and passes into the abdominal cavity.

3. Examine the superior surface of the heart and identify the **thymus.**

4. The **diaphragm** is the sheet of muscle that divides the thoracic cavity from the abdominal cavity and is one of the major muscles involved in respiration.

Organs in the Abdominal Cavity

1. The brown **liver** is the largest organ in the abdominal cavity and is located posterior to the diaphragm (Figure 3.3). The cat liver is divided into more lobes than the human

Figure 3.3 Organs of the Abdominopelvic Cavity Ventral dissection of the cat abdominopelvic cavity.

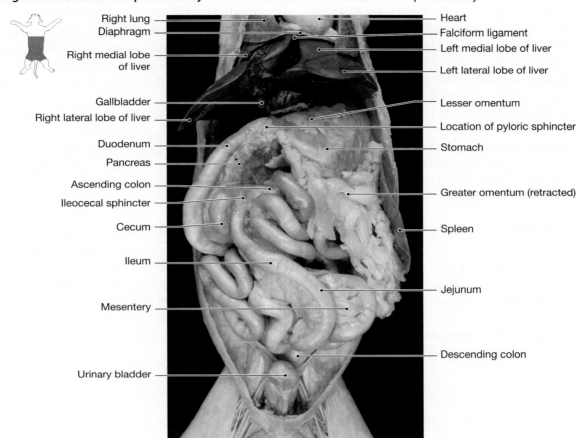

Right lung
Diaphragm
Right medial lobe of liver
Gallbladder
Right lateral lobe of liver
Duodenum
Pancreas
Ascending colon
Ileocecal sphincter
Cecum
Ileum
Mesentery
Urinary bladder

Heart
Falciform ligament
Left medial lobe of liver
Left lateral lobe of liver
Lesser omentum
Location of pyloric sphincter
Stomach
Greater omentum (retracted)
Spleen
Jejunum
Descending colon

liver. The **gallbladder** is a dark green sac immediately posterior to the liver. The liver produces bile, a substance that emulsifies (breaks down) lipids into small drops for digestion.

2. The abdominal organs are protected by a fatty extension of the peritoneum from the lateral margin of the stomach called the **greater omentum.** The **lesser omentum** is a peritoneal sheet of tissue that suspends the stomach from the liver.

3. The esophagus empties into the bag-shaped **stomach** located posterior to the liver. Posterior and to the left of the stomach is the dark brown **spleen.**

4. The **small intestine** receives the stomach contents and secretions from the gallbladder and pancreas. The small intestine has three regions. The first 6 inches is the C-shaped **duodenum.** The **jejunum** comprises the bulk of the remaining length of the small intestine. The **ileum** is the last region of the small intestine and joins with the large intestine.

5. Locate the **pancreas** lying between the stomach and small intestine. The pancreas is a "double gland" because it has both exocrine and endocrine functions.

6. To view the **large intestine,** gently pull the loops of the small intestine to the cat's left and let them drape out of the body cavity. The large intestine is divided into three regions: the cecum, colon, and rectum. The first, following the terminus of the small intestine, is the **cecum,** which is wider than the rest of the large intestine and noticeably pouch shaped. In humans, the appendix is attached to the cecum, but cats do not have an appendix. The greatest portion of the large intestine is the **colon,** which runs anterior from the cecum, across the abdominal cavity, and then posterior, terminating in the third region of the large intestine, the **rectum.** The rectum opens at the **anus** where fecal material is eliminated. Sheets of peritoneum, called **mesentery,** extend between the loops of intestines. The **mesocolon** is the mesentery of the large intestine.

7. Reflect the abdominal viscera to one side of the abdominal cavity, and locate the large, bean-shaped **kidneys.** As in humans, the kidneys are **retroperitoneal** (outside the peritoneal cavity). Identify the **renal artery** (injected red) and the **renal vein** (injected blue), and the cream-colored tube known as the **ureter.**

8. Follow the ureter as it descends posteriorly along the dorsal body wall to drain urine into the **urinary bladder.** Gently pull the bladder anteriorly and observe how the bladder narrows into the **urethra,** the tube through which urine passes to the exterior of the body. Note where the urethra terminates. If your specimen is male, follow the urethra as it passes into the penis. If your specimen is female, notice how the urethra and the vagina empty into a common **urogenital sinus.**

9. Locate the **adrenal glands,** superior to the kidneys and close to the aorta. Identify the **suprarenal arteries** (in red if your specimen has been injected with latex paint) and the **suprarenal veins** (injected blue).

Reproductive Organs

Male Cat

1. The male cat reproductive tract (**Figure 3.4**) is very similar to its counterpart in human males. As in all other mammals, the feline **testes,** the gonads that produce

spermatozoa, are outside the pelvic cavity and housed inside a covering called the **scrotum.**

2. Ventral to the scrotum is the **penis,** the tubular shaft through which the urethra passes. The expanded tip of the penis is the **glans.** The **ductus deferens** carries spermatozoa from testes to the urethra for transport out of the body.

3. Locate the **prostate gland** at the base of the urinary bladder. It is a large, hard mass surrounding the urethra. The **urethra** drains urine from the bladder and transports semen, the sperm-rich fluid expelled during ejaculation, to the tip of the penis.

Female Cat

1. The female cat reproductive system is an excellent example of the interplay of structure and function (**Figure 3.5**). Cats give birth to litters of offspring and the uterus is structured to accommodate multiple gestations. Move (reflect) the abdominal viscera to one side and

Figure 3.4 Male Organs of the Urinary and Reproductive Systems Ventral dissection of the male cat showing the urinary and reproductive systems.

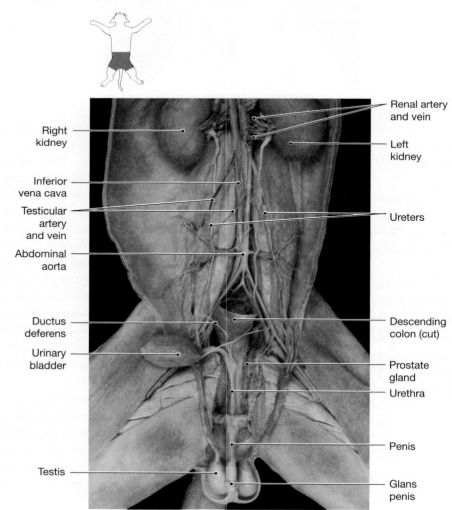

Figure 3.5   Female Organs of the Urinary and Reproductive Systems Ventral dissection of the female cat showing the urinary and reproductive systems.

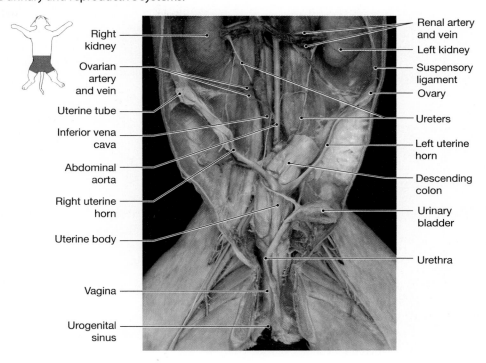

Right kidney

Ovarian artery and vein

Uterine tube

Inferior vena cava

Abdominal aorta

Right uterine horn

Uterine body

Vagina

Urogenital sinus

Renal artery and vein

Left kidney

Suspensory ligament

Ovary

Ureters

Left uterine horn

Descending colon

Urinary bladder

Urethra

locate the paired, oval **ovaries,** lying on the dorsal body wall lateral to the kidneys. On the surface of the ovaries, find the small, coiled **uterine tubes.** Note that, unlike the pear-shaped uterus of the human, the uterus of the cat is Y-shaped (bicornate) and consists of two large **uterine horns** joining a single **uterine body.** Each uterine tube leads into a uterine horn. The horns are where the fertilized eggs implant and develop into the litter of offspring.

2. Trace the uterine body caudally into the pelvic cavity, where it is continuous with the **vagina.**

3. Locate the **urethra** that emerges from the urinary bladder. The vagina is dorsal to the urethra. At the posterior end of the urethra, the vagina and urethra unite to form a common passage called the **urogenital sinus** for the urinary and reproductive systems. In humans, females have separate urethral and vaginal openings.

Procedures: Storing the Cat and Cleaning Up

To store your specimen, wrap it in the skin and moisten it with fixative. Use paper towels if necessary to cover the entire specimen. Return it to the storage bag and seal the bag securely. Label the bag with your name, and place it in the storage area as indicated by your instructor. Wash all dissection tools and the tray, and set them aside to dry. Dispose of your gloves and any tissues from the dissection as indicated by your laboratory instructor. ■

Name _____

Date _____

Section _____

Organ Systems Overview

A. Matching

Match each term on the left with the correct description on the right.

_____	**1.** greater omentum	**A.**	site where bile empties into small intestine
_____	**2.** esophagus	**B.**	organ that directs the airway into lungs
_____	**3.** pylorus	**C.**	organ superior to kidney
_____	**4.** aorta	**D.**	passes swallowed food to stomach
_____	**5.** cecum	**E.**	main artery of body
_____	**6.** diaphragm	**F.**	fatty sheet protecting abdominal organs
_____	**7.** bronchi	**G.**	pouch region of large intestine
_____	**8.** liver	**H.**	empties to duodenum
_____	**9.** duodenum	**I.**	muscle that divides ventral body cavity
_____	**10.** pancreas	**J.**	tubular organ that transports urine
_____	**11.** adrenal gland	**K.**	glandular organ near duodenum
_____	**12.** urethra	**L.**	largest organ in abdomen

B. Fill in the Blanks

Complete the following statements.

1. The heart is located in a small cavity called the _____, which is inside a larger cavity, the _____.
2. The _____ surrounds the digestive organs in the abdominal cavity.
3. The kidneys are _____ because they are located outside the _____.
4. The first section of the small intestine is the _____.
5. Urine is transported from the kidneys to the bladder by the _____.

C. Short-Answer Questions

1. Which organ systems protect the body from infection?

2. Long-term coordination of body function is regulated by which organ system?

3. Which organ system stores minerals for the body?

D. Labeling

Label the organs of the cat in Figure 3.6.

Figure 3.6 **Organs of the Cat Abdominopelvic Cavity**

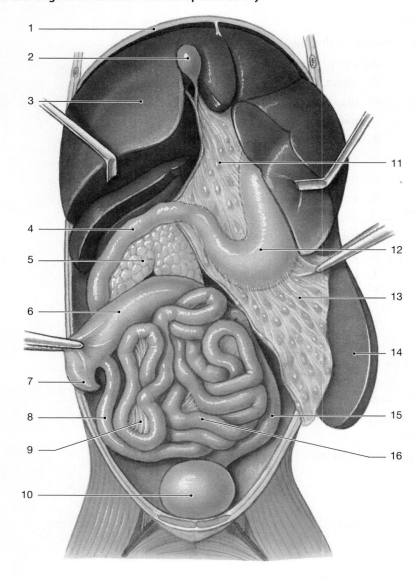

1. _____
2. _____
3. _____
4. _____
5. _____
6. _____
7. _____
8. _____
9. _____
10. _____
11. _____
12. _____
13. _____
14. _____
15. _____
16. _____

E. Analysis and Application

1. How does the uterus of a cat differ from the uterus of a human?

2. Trace a bite of food as it passes through the digestive tract from the mouth to the anus.

Use of the Microscope

Learning Outcomes

On completion of this exercise, you should be able to:

1. Describe how to properly carry, clean, and store a microscope.

2. Identify the parts of a microscope.

3. Focus a microscope on a specimen and adjust the illumination and magnification.

4. Calculate the total magnification for each objective lens.

5. Measure the field diameter at each magnification and estimate the size of cells on a slide.

6. Make and view a wet-mount slide of newspaper print.

7. Observe a slide using the oil-immersion lens and correctly clean the oil off the lens and slide.

Lab Activities

As a student of anatomy and physiology, you will explore the organization and structure of cells, tissues, and organs. The basic research tool for your observations is the microscope. The instrument is easy to use once you learn its parts and how to adjust them to produce a clear image of a specimen. Therefore, it is important that you complete each activity in this exercise and that you are able to use a microscope to observe a specimen on a slide by the end of the laboratory period.

The **compound microscope** uses several lenses to direct a narrow beam of light through a thin specimen mounted on a glass slide. Focusing knobs move either the lenses or the slide to bring the specimen into focus within the round viewing area

Need More Practice and Review?

Build your knowledge—and confidence!—in the Study Area of MasteringA&P® at www.masteringaandp.com with Pre-lab Quizzes, Post-lab Quizzes, Practice Anatomy Lab™ (PAL™) 3.0 virtual anatomy practice tool, PhysioEx™ 9.0 laboratory simulations, and A&P Flix™ with Quizzes.

MasteringA&P®

of the lenses, an area called the **field of view.** Lenses magnify objects so that the objects appear larger than they actually are. As magnification increases, the viewer can more easily see details that are close together. It is this increase in **resolution**— the ability to distinguish between two objects located close to each other—that makes the microscope a powerful observational tool.

Lab Activity 1 Parts and Care of the Compound Microscope

Because the microscope is a precision scientific instrument with delicate optical components, you should always observe the following guidelines when using one. Your laboratory instructor will provide you with specific infor-mation regarding the use and care of microscopes in your laboratory.

General Care of the Microscope

1. Carry the microscope with two hands, one hand on the arm and the other hand supporting the base. (See Figure 4.1 for the parts of the microscope.) Do not swing the microscope as you carry it to your laboratory bench, because such movement could cause a lens to fall out. Avoid bumping the microscope as you set it on the laboratory bench.

2. If the microscope has a built-in light source, completely unwind the electrical cord before plugging it in.

3. To clean the lenses, use only the lens-cleaning fluid and lens paper provided by your instructor. Facial tissue is unsuitable for cleaning because it is made of small wood

Figure 4.1 Parts of the Compound Microscope

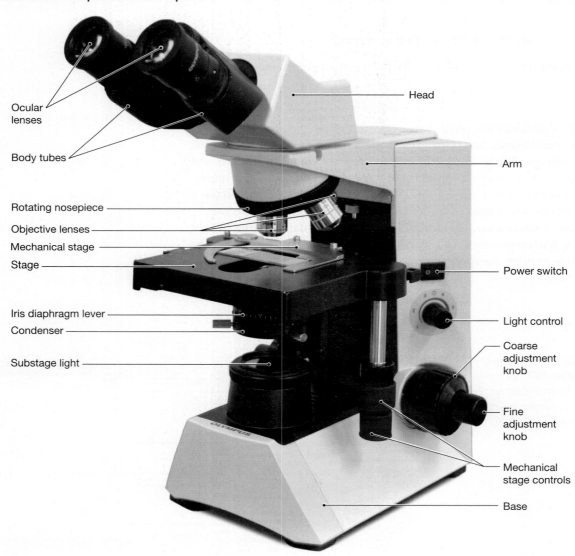

- Ocular lenses
- Body tubes
- Rotating nosepiece
- Objective lenses
- Mechanical stage
- Stage
- Iris diaphragm lever
- Condenser
- Substage light
- Head
- Arm
- Power switch
- Light control
- Coarse adjustment knob
- Fine adjustment knob
- Mechanical stage controls
- Base

fibers that will damage the special optical coating on the lenses.

4. When you are finished using it, store the microscope with the low-power objective lens in position and the stage in the uppermost position. Either return the microscope to the storage cabinet or cover it with a dust cover. The cord may be wrapped neatly around the base; some cords are removable for separate storage.

The parts of a typical compound microscope are presented in Figure 4.1 and **Table 4.1**. Your laboratory may be equipped with a different type; if so, your instructor will discuss the type of microscope you will use.

Microscopes use a **compound lens system,** with each lens consisting of many pieces of optical glass. The magnification is stamped on the barrel of each objective lens, as is the magnifying power of the ocular lens. To calculate the **total magnification** of the microscope at a particular lens setting, you multiply the ocular lens magnification by the objective lens magnification. For example, a $10\times$ ocular lens used with a $10\times$ objective lens produces a total magnification of $100\times$.

QuickCheck Questions

1.1 What is the proper way to hold a microscope while carrying it?

1.2 Why is a facial tissue not appropriate for cleaning microscope lenses?

1.3 How do you change the magnification of a microscope?

1.4 What is the function of the iris diaphragm on a microscope?

Table 4.1	Parts of the Compound Microscope
Microscope Part	**Description and Function**
Arm	The arm is the supportive frame of the microscope that joins the body tube to the base. The microscope is correctly carried with one hand on the arm and the other on the base.
Base	The base is the broad, flat, lower support of the microscope.
Body tube	The body tube is the cylindrical tube that supports the ocular lens and extends down to the nosepiece. A microscope has one body tube if it has one ocular lens, and two body tubes if it has two ocular lenses.
Ocular lens	The ocular lens is the eyepiece where you place your eye(s) to observe the specimen. The magnification of most ocular lenses is $10\times$. This results in an image 10 times larger than the actual size of the specimen. **Monocular** microscopes have a single ocular lens; **binocular** microscopes have two ocular lenses, one for each eye. The ocular lenses may be moved closer together or farther apart by adjusting the body tubes.
Nosepiece	The nosepiece is a rotating disk at the base of the body tube where several objective lenses of different lengths are attached. Turning the nosepiece moves an objective lens into place over the specimen being viewed.
Objective lenses	Mounted on the nosepiece are objective lenses. Magnification of the viewed image is determined by the choice of objective lens. The longer the objective lens, the greater is its magnifying power. The **working distance** is the distance between the tip of the lens and the top surface of the microscope slide. Your microscope may also have an **oil-immersion objective lens,** which is usually $100\times$. With this lens, a small drop of immersion oil is used on the slide to eliminate the air between the lens and the slide, thereby improving the resolution of the microscope. It is important to carefully clean the lens and slide to completely remove the oil.
Stage	The **stage** is a flat, horizontal shelf under the objective lenses that supports the microscope slide. The center of the stage has an **aperture,** or hole, through which light passes to illuminate the specimen on the slide. Most microscopes have a **mechanical stage** that holds and moves the slide with more precision than is possible manually. The mechanical stage has two **mechanical stage controls** on the side that move the slide around on the stage in horizontal and vertical planes.
Focus knobs	The **coarse adjustment knob** is the large dial on the side of the microscope that is used only at low magnification to find the initial focus on a specimen. The small dial on the side of the microscope is the **fine adjustment knob**. This knob moves the objective lens for precision focusing after coarse focus has been achieved. The fine adjustment knob is used at all magnifications and is the only focusing knob used at magnifications greater than low.
Condenser	The **condenser** is a small lens under the stage that narrows the beam of light and directs it through the specimen on the slide. A **condenser adjustment knob** moves the condenser vertically. For most microscope techniques, the condenser should be in the uppermost position, near the stage aperture.
Iris diaphragm	The **iris diaphragm** is a series of flat metal plates at the base of the condenser that slide together and create an aperture in the condenser to regulate the amount of light passing through the condenser. Most microscopes have a small **diaphragm lever** extending from the iris diaphragm; this lever is used to open or close the diaphragm to adjust the light for optimal contrast and minimal glare.
Lamp	A **lamp** provides the light that passes through the specimen, through the lenses, and finally into your eyes. Most microscopes have a built-in light source underneath the stage. The **light control knob,** a rheostat dial located on either the base or the arm, controls the brightness of the light. Microscopes without a light source use a mirror to reflect ambient light into the condenser.

In the Lab   1

Material

☐ Compound microscope

Procedures

1. Identify and describe the function of each part of the microscope.

2. Determine the magnification of the ocular lens and each objective lens on the microscope. Enter this information in the second and third columns of Table 4.2, and then fill in the fourth column by calculating the total magnification for each ocular/objective combination.

3. Use a ruler to measure the working distance between objective lens and slide for each magnification. Record your data in column 5 of Table 4.2. ■

Lab Activity 2   Using the Microscope

Four basic steps are involved in successfully viewing a specimen under the microscope: setup, focusing, magnification control, and light intensity control.

Setup

- Plug in the electrical cord and turn the microscope lamp on. If the microscope does not have a built-in light source, adjust the mirror to reflect light into the condenser.

- Check the position of the condenser; it should be in the uppermost position, near the stage aperture.

- Rotate the nosepiece to swing the low-magnification objective lens into position over the aperture.

- Place the slide on the stage and use the stage clips or mechanical slide mechanism to secure the slide. Move the slide so that the specimen is over the stage aperture.

- After you have finished your observations, reset the microscope to low magnification, remove the slide from the stage, and store the microscope.

Focusing and Ocular Lens Adjustment

- To focus on a specimen, first move the low-power objective lens, or the stage on older microscopes, to its lowest position. Next, look into both ocular lenses and raise the low-power objective lens by slowly turning the coarse adjustment knob. The image should come into focus.

- Once the image is clear, use the fine adjustment knob to examine the detailed structure of the specimen.

- When you are ready to change magnification, do not move the adjustment knobs before changing the objective lens. Most microscopes are **parfocal,** which means they are designed to stay in focus when you change from one objective lens to another. After changing magnification, use only the fine adjustment knob to adjust the objective lens.

- The distance between the ocular lenses is adjustable for your *interpupillary distance,* the distance between the two pupils of your eyes, so that a single image is seen in the microscope. Move the body tubes apart and look into the microscope. If two images are visible, slowly move the body tubes closer together until you see, with *both* eyes open, a single circular field of view.

Magnification Control

- Always use low magnification during your initial observation of a slide. You will see more of the specimen and can quickly select areas on the slide for detailed studies at higher magnification.

- To examine part of the specimen at higher magnification, move that part of the specimen to the center over the aperture before changing to a higher-magnification objective lens. This repositioning keeps the specimen in the field of view at the higher magnification. Because a higher-magnification lens is closer to the slide, less of the slide is visible in the field of view. The image of the specimen enlarges to fill the field of view.

- Highest magnification is achieved on most microscopes by using an oil-immersion objective lens that utilizes a drop of immersion oil between the slide and the lens.

Table 4.2	Microscope Data				
	Ocular Lens	**Objective Lens**	**Total Magnification**	**Working Distance**	**Field Diameter**
Low power	_____	_____	_____	_____	_____
Medium power	_____	_____	_____	_____	_____ *
High power	_____	_____	_____	_____	_____ *
*Calculated field diameter					

Light Intensity Control

- Use the light control knob to regulate the intensity of light from the bulb. Adjust the brightness so that the image has good contrast and no glare.
- Adjust the iris diaphragm by moving the diaphragm lever side to side. Notice how the field illumination is changed by different settings of the iris.
- At higher magnifications, increase the light intensity and open the iris diaphragm.

QuickCheck Questions

2.1 When is the coarse adjustment knob used on a microscope?

2.2 What is the typical view position for the condenser?

In the Lab 2

Materials

- ☐ Compound microscope, slide, and coverslip
- ☐ Newspaper cut into small pieces
- ☐ Dropper bottle containing water
- ☐ Prepared slide: simple cuboidal epithelium (kidney slide)

Procedures

Preparing and Observing a Wet-Mount Slide

1. Make a wet-mount slide of a small piece of newspaper as follows:

    a. Obtain a slide, a coverslip, and a small piece of newspaper that has printing on it.

    b. Place the paper on the slide and add a small drop of water to it.

    c. Put the coverslip over the paper as shown in Figure 4.2. The coverslip will keep the lenses dry.

2. Move the low-magnification objective lens into position (if it is not already there), and place the slide on the stage.

3. Use the coarse adjustment knob to move the objective lens as close to the specimen as possible without touching the slide.

Figure 4.2 **Preparing a Wet Mount** Using tweezers or your fingers, touch the water or stain with the edge of the coverslip.

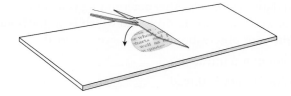

4. Move the slide until the printing is directly over the stage aperture. Look into the ocular lens and slowly turn the coarse adjustment knob until you see the fibers of the newspaper. Once they are in focus, adjust the light source for optimal contrast and resolution.

5. Use the fine adjustment knob to bring the image into crisp focus. Remember, the microscope you are using is a precise instrument and produces a clear image when adjusted correctly. Be patient and keep at it until you get a perfectly clear image.

6. Once the image is correctly focused, do the following and record your observations in the spaces provided.

    a. Locate a letter "a" or "e." Describe the ink and the paper fibers. _____.

    b. Slowly move the slide forward with the mechanical stage knob. In which direction does the image move? _____.

    c. Move the slide horizontally to the left using the mechanical stage knob. In which direction does the image move? _____.

    d. Is the image of the letter oriented in the same direction as the real letter on the slide? _____.

Observing Cells and Tissues on a Dry Slide

1. Obtain a slide provided by your lab instructor. A slide of simple cuboidal epithelium will have cells organized into rings and small pipes called tubules.

2. Focus on the stained tissue using low magnification and the same steps outlined in the previous section.

3. Increase magnification to medium power and scan the slide for rings of cells. Select an area with several tubules and view them at high magnification.

For additional practice, complete the *Sketch to Learn* activity (on p. 40). ■

Lab Activity 3   Depth of Field Observation

Depth of field, or **focal depth,** is a measure of how much depth (thickness) of a specimen is in focus; the in-focus thickness is called a **focal plane** (Figure 4.3). Depth of field is greatest at low power and decreases as magnification increases. In other words, the focal plane is thicker at low power and thinner at higher powers (Figure 4.3a). Because depth of field is reduced at higher power, you use the fine adjustment knob to move the focal plane up and down through the thickness of the specimen and in this way scan the specimen layers. As you turn the fine adjustment knob, the objective lens moves either closer to or farther from the slide surface (Figure 4.3b). This lens movement causes the focal plane to move through the layers of the

Sketch to Learn

Let's practice drawing what we see with the microscope. We will sketch the cells that form a ring of tissue.

Sample Sketch

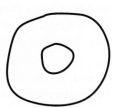

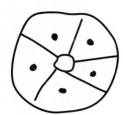

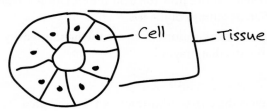

Step 1
• Draw a doughnut shape.

Step 2
• Draw lines to make cells.

Step 3
• Label your sketch.

Your Sketch

Figure 4.3 Focal Plane and Magnification

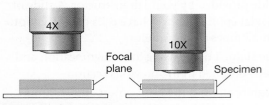

a Focal-plane thickness decreases as magnification increases.

b Using the fine adjustment knob changes the distance between the objective and the stage, thereby moving the focal plane up and down through the specimen.

specimen. Most specimens are many cell layers thick. By slowly rotating the fine adjustment knob back and forth, you will see different layers of the specimen come into or go out of focus.

QuickCheck Questions

3.1 What is depth of field in a microscope?

In the Lab   3

Materials

☐ Compound microscope
☐ Slide of colored threads (or slide of hairs from different students if thread slides are unavailable)

Procedures

To see how depth of field works, you will examine a slide of overlapping colored threads (or hairs). In examining your slide, notice how the threads are layered and how much of each thread is in focus at each magnification.

1. Move the low-power objective lens into position, and place the slide on the stage with the threads over the aperture.

2. Use the coarse adjustment knob to bring the threads into focus. Find the area where the threads overlap.

3. Rotate the nosepiece to select the medium-power objective lens.

4. Use the fine adjustment knob to focus through the overlapping threads. After determining which thread is on top, which is in the middle, and which is on the bottom, write your observations in the space provided.

   Color of top thread _____

   Color of middle thread _____

   Color of bottom thread _____ ▪

Lab Activity 4 Relationship Between Magnification and Field Diameter

At low magnification, the diameter of the field of view is large and most of the slide specimen is visible. As magnification increases, the field diameter decreases, because at higher power the objective lens is closer to the slide and magnifies a smaller area. **Figure 4.4** reviews the relationship between magnification and field diameter.

Field diameter at low and medium magnifications can be measured using millimeter graph paper glued to a microscope slide. By aligning a vertical marking on the paper with the edge of the field and then counting the number of millimeter (mm) squares across the field, you can determine the diameter (**Figure 4.5**). Knowing the diameter of the field of view enables you to estimate the actual size of an object. For

Study Tip Field Diameter

You can demonstrate the relationship between magnification and field diameter by curling your fingers until the thumb of each hand overlaps the index and middle fingers of the same hand. The space enclosed by the curled fingers of each hand forms the barrel of a "lens." Place these two "lenses" to your eyes, and, while sitting up straight in your chair, look at this page. Notice that you can see the entire page at this "low magnification." Now slowly bend forward until the "lenses" are just a few inches away from the page. In this "high-magnification" view, the field of view is much smaller, and you can see only part of the page. ■

Figure 4.4 Magnification and Field Diameter Each circle on the slide illustrates the field diameter for a particular magnification; the corresponding circle outside the slide represents that magnification.

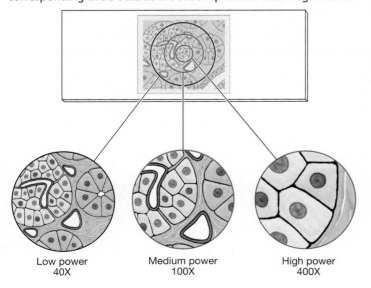

Low power 40X
Medium power 100X
High power 400X

Figure 4.5 Calculation of Field Diameter Using Millimeter Graph Paper In this sample, the field is approximately 3.5 mm in diameter.

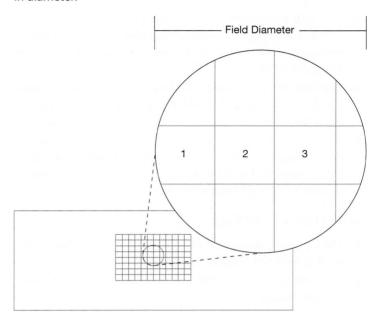

example, if the field diameter is 4 mm and an object occupies one-half of the field, the object is approximately 2 mm wide.

Once you know the field diameter for one magnification—we call this lens A in the following formula—you can calculate the field diameter for other magnifications (lens B) using the formula

Field diameter of lens B

$$= \frac{\text{Field diameter of lens A} \times \text{total magnification of lens A}}{\text{Total magnification of lens B}}$$

QuickCheck Questions

4.1 What happens to the field diameter as magnification increases?

In the Lab 4

Materials

☐ Compound microscope
☐ Graph-paper slides
☐ Practice slides (epithelium or cartilage recommended)

Procedures

Measurement of Field Diameter

1. Place the graph-paper slide on the microscope stage and focus at low magnification. Position the slide so that a vertical line on the paper lines up with the edge of the field.

2. Count the number of millimeters across the field to measure the field diameter. Record this value in Table 4.2 (p. 38).

3. Use your low-power measured field diameter in the formula provided to calculate the field diameter for the microscope set at medium power, and record this value in Table 4.2 (p. 38).

4. Use your low-power measured field diameter in the formula provided to calculate the field diameter for the microscope set at high power, and record this value in Table 4.2 (p. 38).

5. If your microscope has an oil-immersion objective lens, use the formula provided and any of the three field-diameter values you listed in Table 4.2 (p. 38) to calculate the field diameter of the oil-immersion lens.

Estimation of Cell Size

1. On a practice slide selected by your instructor, observe the cells at medium magnification.

2. Estimate the size of some cells using your field-diameter measurements. ∎

Lab Activity 5 Using the Oil-Immersion Objective Lens

The oil-immersion objective lens is used with a drop of special oil applied between the lens and the specimen. The immersion oil eliminates air between the slide and the lens and, because the oil has the same optical properties as the glass slide, it improves image resoultion. The immersion objective is typically a $100\times$ lens and in conjunction with the ocular produces an image at $1,000\times$. Take care to ensure that all of the immersion oil is correctly removed from the immersion lens and slide. Never use immersion oil on the other nonimmersion objectives because the oil will seep into and damage these lenses.

QuickCheck Questions

5.1 Where is immersion oil applied?

In the Lab 5

Materials

☐ Compound microscope
☐ Immersion oil
☐ Lens-cleaning fluid and paper
☐ Prepared slide (blood smear recommended)

Procedures

Using the Oil-Immersion Lens

1. Place the blood smear slide on the stage and focus on the slide at low power. The blood cells are just visible as small ovals.

2. Focus on the blood cells with the medium and the high-dry magnification objectives.

3. Move the oil-immersion objective into position, then swing it away from the slide and add a small drop of the immersion oil on the coverslip where the light is passing through the specimen. Slowly move the oil lens into place and ensure the oil is between the lens and the slide.

4. Carefully focus the lens with the fine focus knob. The nuclei of the stained white blood cells should be clearly discernible at this high magnification.

Cleaning the Oil-Immersion Lens

1. Use the coarse focus knob and move the objective lens away from the slide to create a working space for cleaning.

2. Use a lens tissue and gently wipe the oil off the lens. Repeatedly clean the lens with clean areas of the tissue.

3. Place a drop of lens-cleaning solution on the paper and completely remove any remaining oil. Do not saturate the paper or the lens with cleaner.

4. After the lens is clean, dry it with a new tissue.

5. Repeat the cleaning procedure for the microscope slide. ∎

Name _____

Date _____

Section _____

Use of the Microscope

A. Matching

Match the part of the microscope on the left with the correct description on the right.

_____	1. ocular lens	A.	used for precise focusing
_____	2. aperture	B.	lower support of microscope
_____	3. body tube	C.	narrows beam of light
_____	4. mechanical stage	D.	hole in stage
_____	5. fine adjustment knob	E.	used only at low power
_____	6. base	F.	has knobs to move slide
_____	7. objective lens	G.	special paper for cleaning
_____	8. coarse adjustment knob	H.	eyepiece
_____	9. condenser	I.	holds ocular lens
_____	10. lens paper	J.	lens attached to nosepiece

B. Short-Answer Questions

1. Which parts of a microscope are used to regulate the intensity and contrast of light? What is the function of each of these parts?

2. How is magnification controlled in a microscope?

3. Why should you always view a slide at low power first?

4. Briefly explain how to care for a microscope.

5. Describe when to use the coarse adjustment knob and when to use the fine adjustment knob.

C. Labeling

Label the parts of the microscope in **Figure 4.6**.

Figure 4.6 Parts of the Compound Microscope

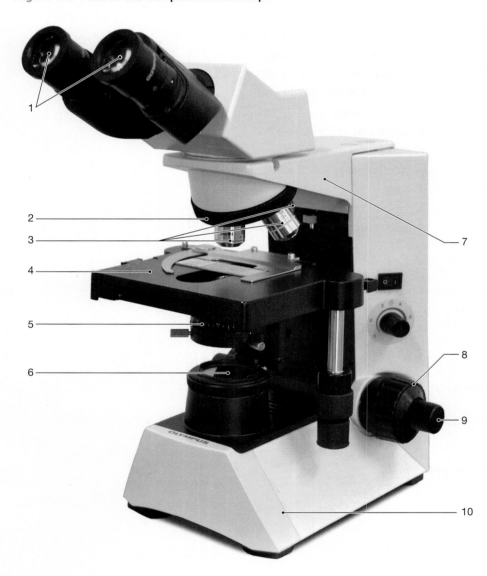

1. _____
2. _____
3. _____
4. _____
5. _____
6. _____
7. _____
8. _____
9. _____
10. _____

D. Analysis and Application

1. You are looking at a slide in the laboratory and observe a cell that occupies one-quarter of the field of view at high magnification. Use your field-diameter calculation from Lab Activity 4 to estimate the size of this cell.

2. Describe how the field diameter changes when magnification is increased.

Anatomy of the Cell and Cell Division

Learning Outcomes

On completion of this exercise, you should be able to:

1. Identify cell organelles on charts, models, and other laboratory material.

2. Use the microscope to identify the nucleus and plasma membrane of cells.

3. State a function of each organelle.

4. Discuss a cell's life cycle, including the stages of interphase and mitosis.

5. Identify the stages of mitosis using the whitefish blastula slide.

Cells were first described in 1665 by a British scientist named Robert Hooke. Hooke examined a thin slice of tree cork with a microscope and observed that it contained many small open spaces, which he called *cells.* Over the next two centuries, scientists examined cells from plants and animals and formulated the **cell theory,** which states that (1) all plants and animals are composed of cells, (2) all cells come from preexisting cells, (3) cells are the smallest living units that perform physiological functions, (4) each cell works to maintain itself at the cellular level, and (5) homeostasis is the result of the coordinated activities of all the cells in an organism.

Your cells are descendants of your parents' sperm and egg cells that combined to create your first cell, the zygote. You are now composed of approximately 75 trillion cells, more cells than you could count in your lifetime. These cells must coordinate their activities to maintain homeostasis for your entire body. If a population of cells becomes dysfunctional, disease may result. Some organisms, like amoebas, are composed of a single cell that

Need More Practice and Review?

Build your knowledge—and confidence!—in the Study Area of MasteringA&P® at **www.masteringaandp.com** with Pre-lab Quizzes, Post-lab Quizzes, Practice Anatomy Lab™ (PAL™) 3.0 virtual anatomy practice tool, PhysioEx™ 9.0 laboratory simulations, and A&P Flix™ with Quizzes.

PAL   For this lab exercise, follow this navigation path:
- PAL>Histology>Cytology (Cell Division)

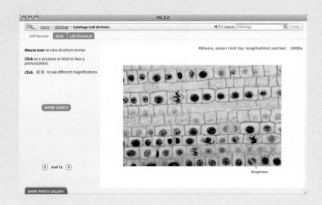

Mastering**A&P**®

performs all functions necessary to keep the organism alive. In humans and other multicellular organisms, cells are diversified, which means that different cells have different specific functions. This specialization leads to dependency among cells. For example, muscle cells are responsible for movement of the body. Because movement requires a large amount of energy, muscle cells rely on the cells of the cardiovascular system to distribute blood rich with oxygen and nutrients to them.

In this exercise you will examine the structure of the cell and how cells reproduce to create new cells that can be used for growth and repair of the body.

Lab Activity 1   Anatomy of the Cell

Although the body is made of a variety of cell types, a generalized composite cell, as illustrated in Figure 5.1, is used to describe cell structure. All cells have an outer boundary, the **plasma membrane,** also called the *cell membrane.* This physical boundary separates the **extracellular fluid** surrounding the cell from the cell interior. It regulates the movement of ions, molecules, and other substances into and out of the cell.

Inside the volume defined by the plasma membrane are a central structure called the **nucleus** of the cell and other internal

Figure 5.1 The Anatomy of a Composite Cell

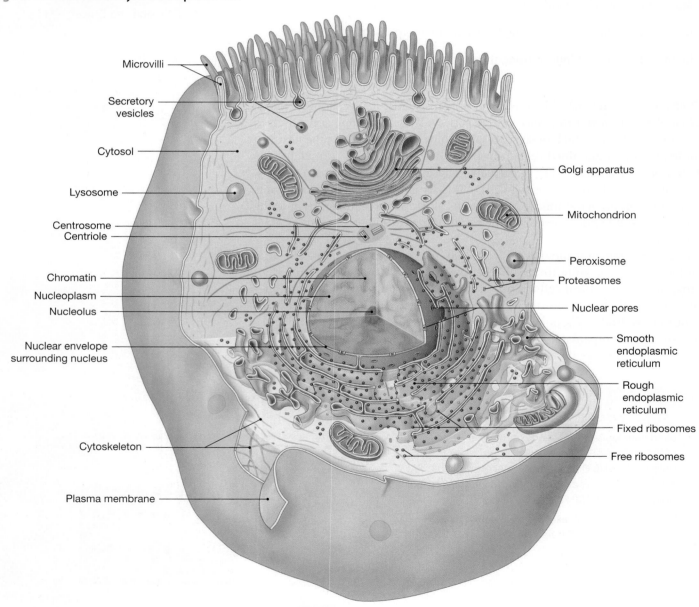

structures. Collectively, these internal structures are called **organelles** (or-gan-ELZ). All the volume inside the plasma membrane but outside the nucleus is referred to as the **cytoplasm.** This region is made up of solid components (all the cell's organelles except the nucleus) suspended in a liquid called the **cytosol.**

Each organelle has a distinct anatomical organization and is specialized for a specific function. Organelles are grouped into two broad classes: nonmembranous and membranous. **Nonmembranous organelles** lack an outer membrane and are directly exposed to the cytosol. Ribosomes, microvilli, centrioles, the cytoskeleton, cilia, and flagella are nonmembranous organelles. **Membranous organelles** are enclosed in a phospholipid membrane that isolates them from the cytosol. The nucleus, endoplasmic reticulum, Golgi apparatus, lysosomes, peroxisomes, and mitochondria are membranous organelles.

Keep in mind while studying cell models that most organelles are not visible with a light microscope. The nucleus typically is visible as a dark-stained oval. It encases and protects the **chromosomes,** which store genetic instructions for protein production by the cell.

Nonmembranous Organelles

- **Microvilli** are small folds in the plasma membrane that increase the surface area of the cell. With more membrane surface, the cell can absorb extracellular materials, such as nutrients, at a greater rate.

- **Centrioles** are paired organelles composed of **microtubules,** which are small hollow tubes made of the protein **tubulin.** The **centrosome** is the area surrounding the pair of centrioles in a cell. When a cell is not dividing, it contains one pair of centrioles. When it comes time for the cell to divide, one of the first steps is replication of the centriole pair, so that the cell contains two pairs. The two centrioles in one pair migrate to one pole of the nucleus, and the two centrioles in the other pair migrate to the opposite pole of the nucleus. As the two pairs migrate, a series of **spindle fibers** radiate from them. The spindle fibers pull the chromosomes of the nucleus apart to give each of the forming daughter cells a full complement of genetic instructions.

- Cells have a **cytoskeleton** for structural support and anchorage of organelles. Like the centrioles, the cytoskeleton is made of microtubules.

- Many cells of the respiratory and reproductive systems have nonmembranous organelles called **cilia,** which are short, hairlike projections that extend from the plasma membrane. One type of human cell, the spermatozoon, has a single, long **flagellum** (fla-JEL-um) for locomotion.

- **Ribosomes** direct protein synthesis. Instructions for making a protein are stored in deoxyribonucleic acid (DNA) molecules in the cell nucleus. The "recipe" for a protein is called a gene and is copied from a segment of DNA onto a molecule of messenger RNA that then carries the instructions out of the nucleus and to the ribosome. Each ribosome consists of one large subunit and one small subunit. Both subunits clamp around the messenger RNA molecule to coordinate protein synthesis. Ribosomes occur either as **free ribosomes** in the cytoplasm or as **fixed ribosomes** attached to the endoplasmic reticulum (ER).

Membranous Organelles

- The **nucleus** controls the activities of the cell, such as protein synthesis, gene action, cell division, and metabolic rate. The material responsible for the dark appearance of the nucleus in a stained specimen is **chromatin,** uncoiled chromosomes consisting of DNA and protein molecules. A **nuclear envelope** surrounds the nuclear material and contains pores through which instruction molecules from the nucleus pass into the cytosol. A darker-stained region inside the nucleus, the **nucleolus,** produces ribosomal RNA molecules for the creation of ribosomes.

- Surrounding the nucleus is the **endoplasmic reticulum** (en-dō-PLAZ-mik re-TIK-yoo-lum). Two types of ER occur: **rough ER,** which has ribosomes attached to its surface; and **smooth ER,** which lacks ribosomes. Generally, the ER functions in the synthesis of organic molecules, transport of materials within the cell, and storage of molecules. Materials in the ER may pass into the Golgi apparatus for eventual transport out of the cell. Proteins produced by ribosomes on the rough ER surface enter the ER and assume the complex folded shape characteristic of the ER. Smooth ER is involved in the synthesis of many organic molecules, such as cholesterol and phospholipids. In reproductive cells, smooth ER produces sex hormones. In liver cells, it synthesizes and stores glycogen, while in muscle and nerve cells it stores

---

Study Tip Information Linking

Practice connecting information together rather than memorizing facts and terms. An effective and fun approach is to compare the cell to a mass-production factory. Each organelle in the cell, like each station in the factory, has a specific task that integrates into the overall function of the cell. As you identify organelles on cell models, consider their function. Once you are familiar with all the organelles, begin to associate them with one another as functional teams. For example, molecules made in the organelles called the endoplasmic reticulum are transported to a neighboring organelle known as the Golgi apparatus, and so you should associate these two organelles with each other. Assimilating information in this way improves your ability to apply knowledge in a working context. ■

calcium ions. Intracellular calcium ions are stored in the smooth ER in muscle, nerve, and other types of cells.

■ The **Golgi** (GŌL-jē) **apparatus** is a series of flattened saccules adjoining the ER. The ER can pass protein molecules in transport vesicles to the Golgi apparatus for modification and secretion. Cell products such as mucus are synthesized, packaged, and secreted by the Golgi apparatus. In a process called **exocytosis,** small **secretory vesicles** pinch off the saccules, fuse with the plasma membrane, and then rupture to release their contents into the extracellular fluid. The phospholipid membranes of the empty vesicles contribute to the renewal of the plasma membrane.

■ **Lysosomes** (LĪ-sō-sōms; *lyso-*, dissolution + *soma*, body) are vesicles produced by the Golgi apparatus. They are filled with powerful enzymes that digest worn-out cell components and destroy microbes. As certain organelles become worn out, lysosomes dissolve them, and some of the materials are used to rebuild the organelles. White blood cells trap bacteria with plasma membrane extensions and pinch the membrane inward to release a vesicle inside the cell. Lysosomes fuse with the vesicle and release enzymes to digest the bacteria. Injury to a cell may result in the rupture of lysosomes, followed by destruction or autolysis of the cell. Autolysis is implicated in the aging of cells owing to the accumulation of lysosomal enzymes in the cytosol.

■ **Peroxisomes** are vesicles filled with enzymes that break down fatty acids and other organic molecules. Metabolism of organic molecules can produce free-radical molecules, such as hydrogen peroxide (H_2O_2), that damage the cell. Peroxisomes protect cell structure by metabolizing hydrogen peroxide to oxygen and water.

■ **Mitochondria** (mī-tō-KON-drē-uh) produce useful energy for the cell. Each mitochondrion is wrapped in a double-layered phospholipid membrane. The inner membrane is folded into fingerlike projections called **cristae** (the singular is *crista*). The region of the inner membrane between cristae is the **matrix.** To provide the cell with energy, molecules from nutrients are passed along a series of **metabolic enzymes** in the cristae to produce a molecule called *adenosine triphosphate (ATP)*, the energy currency of the cell. The abundance of mitochondria varies greatly among cell types. Muscle and nerve cells have large numbers of mitochondria that supply energy for contraction and generate nerve impulses, respectively. Mature red blood cells lack mitochondria and subsequently have a low metabolic rate.

QuickCheck Questions

1.1 What are the two major categories of organelles?

1.2 Which organelles are involved in the production of protein molecules?

In the Lab   1

Materials

☐ Cell models and charts
☐ Toothpicks
☐ Microscope slide and coverslip
☐ Physiological saline in dropper bottle
☐ Iodine stain or methylene blue stain
☐ Compound microscope

Procedures

1. Review the nonmembranous and membranous organelles in Figure 5.1.

2. Identify each organelle on a cell model.

3. Prepare a wet-mount slide from cells of the inner lining of your cheek.

   a. Place a drop of saline on a microscope slide.

   b. Gently scrape the inside of your cheek with the blunt end of a toothpick.

   c. Stir the scraping into the drop of saline on the slide.

   d. Add 1 drop of stain, carefully stir again with the same toothpick, and add a coverslip.

   e. Dispose of your used toothpick in a biohazard bag as indicated by your instructor.

4. Examine your cheek cell slide at low power and note the many flattened epithelial cells. These cells are thin and often become folded by the coverslip.

5. Observe individual cells at medium and high magnifications (**Figure 5.2**). Identify the nucleus, cytoplasm, and plasma membrane of a cell.

For additional practice, complete the *Sketch to Learn* activity (on p. 49). ■

Lab Activity 2   Cell Division

Cells must reproduce if an organism is to grow and repair damaged tissue. During cell reproduction, a cell divides its genes equally and then splits into two identical cells. The division involves two major events: mitosis and cytokinesis. During **mitosis** (mī-TŌ-sis), the chromatin in the nucleus condenses into chromosomes and is equally divided between the two forming cells. Toward the end of mitosis, **cytokinesis** (sī-tō-ki-NĒ-sis; *cyto-*, cell + *kinesis*, motion) separates the cytoplasm to produce the two daughter cells. The daughter cells have the same number of chromosomes as the parent cell. Human cells have 23 pairs of chromosomes that carry the genetic code of approximately 10,000 genes.

Sketch to Learn

Let's draw our cheek cells. This sketch will be simple: A few circles with dots inside and we have cells!

It is important to develop the ability to sketch the cells and tissues that you observe with a microscope. Remembering your simple sketch will enhance your recall of the observation and help you form a mental link to the details of the material.

Sample Sketch 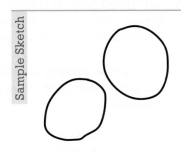    *Your Sketch*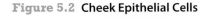

Step 1
- Draw several irregularly shaped ovals.

Step 2
- Add a nucleus and smaller dots to represent other organelles.

Figure 5.2 Cheek Epithelial Cells

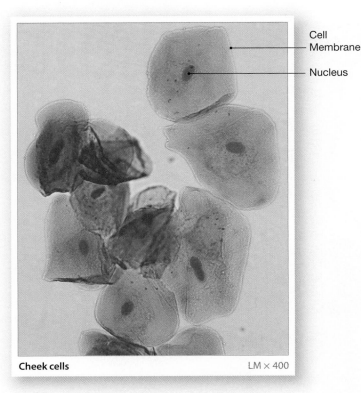

Cell Membrane

Nucleus

Cheek cells   LM × 400

Interphase

Examine the cell life cycle in **Figure 5.3**. Most of the time, a cell is not dividing and is in **interphase.** This is not a resting period for the cell, however, because during this phase the cell carries out various functions and prepares for the next cell division. Distinct phases occur during interphase, each related to cell activity. At this time, the nucleus is visible, as is the darker nucleolus. During the G_0 **phase** of interphase, the cell performs its specialized functions and is not preparing to divide. The G_1 **phase** is a time for protein synthesis, growth, and replication of organelles, including the centriole pair. Replication of DNA occurs during the **S phase.** After DNA replication, each chromosome is double stranded and consists of two **chromatids;** one chromatid is the original strand and the other is an identical copy. The chromatids are held together by a **centromere.** The G_2 **phase** is another time for protein synthesis; at this time, replication of the centriole pair is completed.

Make a Prediction

Which cells would you expect to spend most of their time in the G_0 phase: cells lining the inside of your mouth or nerve cells in the brain?

Figure 5.3 The Cell Life Cycle

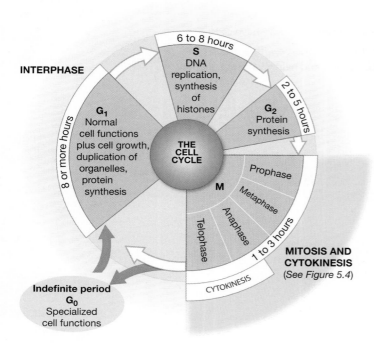

Mitosis

The **M phase** of the cell cycle is the time of mitosis, during which the nuclear material divides (**Figure 5.4**). After chromosomes are duplicated in the S phase of interphase, the double-stranded chromosomes migrate to the middle of the cell, and spindle fibers attach to each chromatid. Chromosomes are divided when the spindle fibers drag sister chromatids to opposite ends of the cell. The division is complete when the cell undergoes cytokinesis and pinches inward to distribute the cytosol and chromosomes into two new daughter cells.

The four stages of mitosis are prophase, metaphase, anaphase, and telophase. Telophase and the latter part of anaphase are together referred to as cytokinesis.

■ ***Prophase:*** Mitosis starts with prophase (PRO-fō-z; *pro-*, before), when chromosomes become visible in the nucleus (Figure 5.4). In early prophase, the chromosomes are long and disorganized, but as prophase continues the nuclear envelope breaks down, and the chromosomes shorten and move toward the middle of the cell. In the cytosol, the two centriole pairs begin moving to opposite sides of the cell. Between the centrioles, microtubules fan out as spindle fibers and extend across the cell.

Figure 5.4 Interphase, Mitosis, and Cytokinesis Diagrammatic and microscopic views of representative cells undergoing cell division.

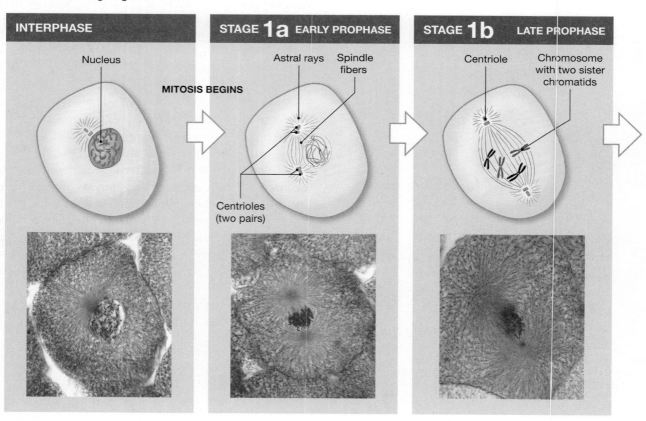

- *Metaphase:* Metaphase (MET-a-fēz; *meta-,* change) occurs when the chromosomes line up in the middle of the cell at the **metaphase plate.** Spindle fibers extend across the cell from one pole to the other and attach to the centromeres of the chromosomes. The cell is now prepared to partition the genetic material and give rise to two new cells.

- *Anaphase:* Separation of the chromosomes is the event that defines anaphase (AN-a-fāz; *ana-,* apart). Spindle fibers pull apart the chromatids of a chromosome and drag them toward opposite poles of the cell. Once apart, individual chromatids are considered chromosomes. Cytokinesis marks the end of anaphase as a **cleavage furrow** develops along the metaphase plate and the plasma membrane pinches. Cytokinesis continues into the next stage of mitosis, telophase.

- *Telophase:* In telophase (TEL-ō-fāz; *telo-,* end), cytokinesis partitions the cytoplasm of the cell and mitosis nears completion as each batch of chromosomes unwinds inside a newly formed nuclear envelope. Each daughter cell has a set of organelles and a nucleus containing a complete set of genes. Telophase ends as the cleavage furrow deepens along the metaphase plate and separates the cell into two identical daughter cells. These daughter cells are in interphase and, depending on their cell type, may divide again.

Clinical Application **Cell Division and Cancer**

A tumor is a mass of cells produced by uncontrolled cell division. The mass replaces normal cells, and cellular and tissue functions are compromised. If **metastasis** (me-TAS-tā-sis), which means spreading of the abnormal cells, occurs, secondary tumors may develop. Cells that metastasize are often cancerous. ■

QuickCheck Questions

2.1 What must the cell do with the genetic material in the nucleus before mitosis?

2.2 Name the four stages of mitosis and list what happens during each stage.

In the Lab **2**

Materials

☐ Compound microscope
☐ Whitefish blastula slide

Figure 5.4 Interphase, Mitosis, and Cytokinesis *(continued)*

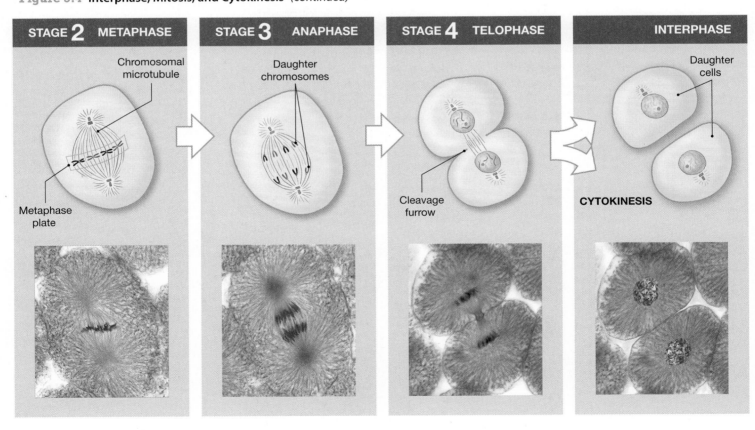

STAGE **2** METAPHASE — Chromosomal microtubule, Metaphase plate

STAGE **3** ANAPHASE — Daughter chromosomes

STAGE **4** TELOPHASE — Cleavage furrow

INTERPHASE — Daughter cells, CYTOKINESIS

Procedures

1. Obtain a slide of the whitefish blastula. A **blastula** is a stage in early development when the embryo is a rapidly dividing mass of cells that is growing in size and, eventually, in complexity. For microscopic observation of the cells, the whitefish embryo is sectioned and stained. A typical slide preparation usually has several sections, each showing cells in various stages of mitosis.

2. Scan the slide at low power, and observe the numerous cells of the blastula.

3. Slowly scan a group of cells at medium power to locate a nucleus, centrioles, and spindle fibers. The chromosomes appear as dark, thick structures.

4. Using Figure 5.4 as a reference, locate cells in the following phases:

   - Interphase with a distinct nucleus
   - Prophase with disorganized chromosomes
   - Metaphase with equatorial chromosomes attached to spindle fibers
   - Anaphase with chromosomes separating toward opposite poles
   - Telophase with nuclear envelope forming around each set of genetic material
   - Cytokinesis in late anaphase and telophase.

5. ***Draw It!*** Draw and label cells in each stage of mitosis in the space provided. ■

Interphase

Prophase

Metaphase

Anaphase

Telophase

Daughter Cells

Name _____

Date _____

Section _____

Anatomy of the Cell and Cell Division

A. Definitions

Describe or state a function of each of the following cellular structures.

1. plasma membrane

2. centrioles

3. ribosome

4. smooth ER

5. chromatid

6. lysosome

7. cilia

8. cytoplasm

9. cytosol

10. nucleus

B. Fill in the Blanks

1. Replication of genetic material results in chromosomes consisting of two _____.

2. A cell in metaphase has chromosomes located in the _____ of the cell.

3. Division of the cytoplasm to produce two daughter cells is called _____.

4. Double-stranded chromosomes separate during the _____ stage of mitosis.

5. During interphase, DNA replication occurs in the _____ phase.

6. Microtubules called _____ attach to chromatids and pull them apart.

7. Chromosomes become visible during the _____ stage of mitosis.

8. The last stage of mitosis is _____.

9. Division of the nuclear material is called _____.

10. Matching chromatids are held together by a _____.

C. Labeling

1. Label the organelles in Figure 5.5.

Figure 5.5 **The Anatomy of a Cell**

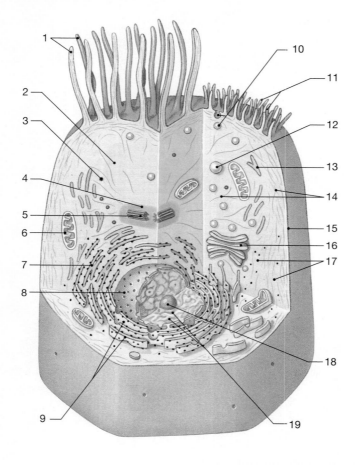

1. _____
2. _____
3. _____
4. _____
5. _____
6. _____
7. _____
8. _____
9. _____
10. _____
11. _____
12. _____
13. _____
14. _____
15. _____
16. _____
17. _____
18. _____
19. _____

2. Label the cell division photos in **Figure 5.6**.

Figure 5.6 Cell Division

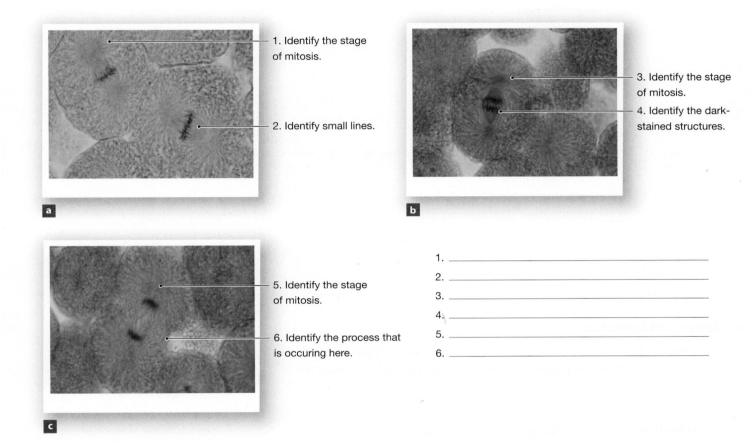

1. Identify the stage of mitosis.

2. Identify small lines.

3. Identify the stage of mitosis.

4. Identify the dark-stained structures.

a

b

5. Identify the stage of mitosis.

6. Identify the process that is occuring here.

c

1. _____

2. _____

3. _____

4. _____

5. _____

6. _____

D. Short-Answer Questions

1. What is the purpose of cell division?

2. Describe a phospholipid molecule and its interaction with water.

3. What is the function of the spindle fibers during mitosis?

4. What structures in the plasma membrane regulate ion passage?

E. Drawing

1. ***Draw It!*** Draw and label a cell with the following organelles: nucleus, RER, Golgi apparatus, mitochondria, and centrioles.

2. ***Draw It!*** Draw and label a cell with four chromosomes in interphase and each stage of mitosis.

F. Analysis and Application

1. Describe how the nucleus, ribosomes, rough ER, Golgi apparatus, and plasma membrane interact to produce and release a protein molecule from the cell.

2. What happens in a cell during the S portion of interphase?

3. Describe how chromosomes are evenly divided during mitosis.

4. Identify where in a cell the production of protein, carbohydrate, and lipid molecules occurs.

G. Clinical Challenge

Lysosomes are sometimes referred to as "suicide bags." Describe what would happen to a cell if its lysosomes ruptured.

Movement Across Plasma Membranes

Learning Outcomes

On completion of this exercise, you should be able to:

1. Describe the two main processes by which substances move into and out of cells.

2. Explain what Brownian movement is and how it can be shown.

3. Discuss osmosis, diffusion, concentration gradients, and equilibrium in a solution.

4. Describe the effect on cells of isotonic, hypertonic, and hypotonic solutions.

5. Discuss the effects of solute concentration on the rate of diffusion and osmosis.

Cells are the functional living units of the body. In order for them to survive, materials must be transported across the plasma membrane. Cells import nutrients, oxygen, hormones, and other regulatory molecules from the extracellular fluid and export wastes and cellular products to the extracellular fluid. Cells rely on the **selectively permeable plasma membrane** to regulate the passage of these materials. Small molecules, such as water and many ions, cross the membrane without assistance from the cell. This movement is called **passive transport** and requires no energy expenditure by the cell. Diffusion and osmosis are the primary passive processes in the body and will be studied in this laboratory exercise.

The plasma membrane consists of a **phospholipid bilayer,** which is a double layer of phospholipid molecules, plus several other structural components, such as cholesterol molecules and glycolipid molecules (**Figure 6.1**). Each phospholipid molecule consists of a **hydrophilic** (*hydro-*, water + *philic*, loving) **head** and two **hydrophobic** (*phobic*, fearing) **tails.** In a plasma membrane, the phospholipids are

Need More Practice and Review?

Build your knowledge—and confidence!—in the Study Area of MasteringA&P® at **www.masteringaandp.com** with Pre-lab Quizzes, Post-lab Quizzes, Practice Anatomy Lab™ (PAL™) 3.0 virtual anatomy practice tool, PhysioEx™ 9.0 laboratory simulations, and A&P Flix™ with Quizzes.

PhysioEx 9.0 For this lab exercise, go to this topic in PhysioEx:
- PhysioEx Exercise 1: Cell Transport Mechanisms and Permeability

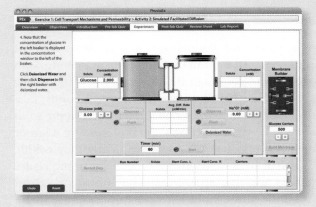

Figure 6.1 Diffusion Across the Plasma Membrane

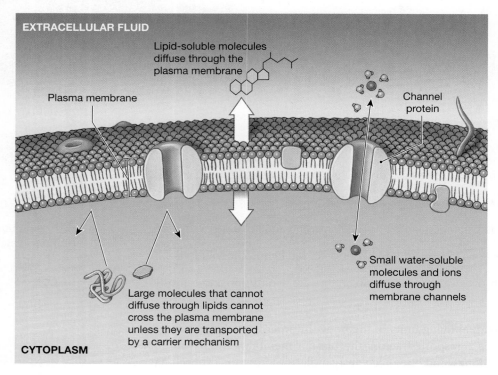

arranged in a double sheet of molecules with the hydrophilic heads facing the watery internal and external environments of the cell. The hydrophobic tails are sandwiched between the phospholipid heads.

Floating like icebergs in the phospholipid bilayer are a variety of **integral proteins.** These proteins have membrane **channels** that regulate the passage of specific ions through the membrane. Lipid-soluble molecules move through the fatty phospholipid bilayer. Other materials, such as proteins and other macromolecules, are too big to pass through channels in the plasma membrane. Movement of these larger molecules requires the use of carrier molecules in a process called **active transport,** a cell function that consumes a cell's energy.

Lab Activity 1   Brownian Movement

Molecules in gases and liquids are in a constant state of **Brownian movement,** motion that causes them to bump into adjacent molecules. (Although the molecules in a solid have this motion, they are held in place by chemical bonds; as a result, the bulk of the molecules remain in place and the solid retains its shape.) The more closely the molecules in a gas or liquid are packed together, the more frequently they collide with one another. Because of Brownian collisions, molecules initially packed together spread out and move toward an equal distribution throughout the container holding them.

Brownian movement supplies the **kinetic energy** for passive transport mechanisms.

Make a Prediction

How would temperature affect the rate of Brownian movement?

QuickCheck Questions

1.1 How does the kinetic energy of Brownian movement cause molecules to spread out in a container?

1.2 Why do solids retain their shape?

In the Lab   1

Materials

☐ Compound microscope
☐ Microscope slide and coverslip
☐ Small dropper bottle with eyedropper
☐ Tap water
☐ Waterproof ink
☐ Powdered kitchen cleanser

Procedures

1. Fill the dropper bottle three-fourths full of tap water and add a small amount of cleanser powder and waterproof ink. Add 10 to 15 mL of additional tap water.

2. Shake the bottle gently to mix the contents. Place a drop of the mixture on a microscope slide and place a coverslip over the drop.

3. Focus on the slide and locate the small granules of cleanser. Observe how the particles move and occasionally collide with one another.

4. Describe the movement observed under the microscope in section B of the Review & Practice Sheet. ■

Lab Activity 2   Diffusion of a Liquid

Diffusion is the net movement of substances from a region of greater concentration to a region of lesser concentration. Simply put, diffusion is the spreading out of substances owing to collisions between moving molecules. Cells cannot directly control diffusion; it is a passive transport process much like a ball rolling downhill. If a substance is unequally distributed, a **concentration gradient** exists, and one region will have a greater concentration of the substance than other regions. The substance will diffuse until an equal distribution occurs, at a point called **equilibrium.**

Figure 6.2 illustrates diffusion with a cube of colored sugar. Before the cube is placed in the water, the sugar molecules are concentrated in it. Once submerged, the sugar dissolves, and the molecules disperse as they bump into other sugar molecules and water molecules. Eventually, the colored molecules become evenly distributed, and the solution is in equilibrium. Even at equilibrium, the molecules are in motion. When one molecule bumps another out of position, the shift forces some other molecule into the vacated space. One movement cancels the other, and no net movement occurs.

Diffusion occurs throughout the body—in extracellular fluid, across plasma membranes, and in the cytosol of cells. Examples of diffusion include oxygen moving from the lungs into pulmonary capillaries, odor molecules moving through the nasal lining to reach olfactory cells, and ions moving in and out of nerve cells to produce electrical impulses. Molecules diffuse through cells by two basic mechanisms: lipid-soluble molecules diffuse through the phospholipid bilayer of the plasma membrane; and small, water-soluble ions and molecules pass through the channels of integral proteins. Molecules like proteins are too large to enter the membrane channels and therefore do not diffuse across the membrane.

Temperature, pressure, and concentration gradient are some factors that influence the rate of diffusion. Brownian movement slows as temperature decreases. Therefore, diffusion is also slower at lower temperatures, because the more slowly a molecule is moving, the longer it takes that molecule to travel far enough to collide with another molecule.

Figure 6.2 **Diffusion** Placing a colored sugar cube in a glass of water establishes a steep concentration gradient. As the cube dissolves, many sugar and dye molecules are in one location, and none are elsewhere. As diffusion occurs, the molecules spread through the solution. Eventually, diffusion eliminates the concentration gradient. The sugar cube has dissolved completely, and the molecules are distributed evenly. Molecular motion continues, but there is no net directional movement.

QuickCheck Questions

2.1 Is diffusion a passive process or an active process?

2.2 Describe a solution that is at equilibrium.

In the Lab   2

Materials

☐ Two beakers, 250 mL or larger

☐ Tap water

☐ Ice

☐ Hotplate or microwave oven

☐ Food coloring dye

Procedures

1. Fill one beaker three-fourths full with tap water and small ice chips.

2. Fill the other beaker three-fourths full with tap water and warm the water in a microwave oven or on a hotplate. Do not boil the water.

3. Leave both beakers undisturbed for several minutes to let the water settle. Remove any remaining ice chips from the chilled water.

4. Carefully add 1 or 2 drops of food coloring dye to each beaker.

5. Observe for several minutes as the dye diffuses. Continue observing the beakers every three to four minutes until equilibrium is reached. Record your observations in section B of the Review & Practice Sheet. ■

Lab Activity   3   Diffusion of a Solid in a Gel

This experiment demonstrates the diffusion of a solid chemical in a thick gelatinous material. As the solid slowly dissolves, it diffuses into the gel. Two chemicals with different molecular masses are used to illustrate the relationship between diffusion rate and molecular mass.

QuickCheck Questions

3.1. Make a prediction: Which would you expect to diffuse at a faster rate, a molecule that has a high molecular mass or one that has a low molecular mass?

In the Lab   3

Materials

☐ Petri dish with plain agar

☐ Cork bore or soda straw

☐ Potassium permanganate crystals

☐ Iodine crystals

☐ Ruler

Procedures

1. Use the bore or straw to punch two small holes in the agar, approximately equidistant from the center of the petri dish.

2. Place a small amount of potassium permanganate crystals in one hole and an equal amount of iodine crystals in the other hole. Do not spill crystals on any other part of the petri dish.

3. After 30 to 45 minutes, measure the distance each chemical has diffused and record your data in the provided blanks. Measure from the edge of the hole farthest from the center of the dish to the outer boundary of the diffusion area.

   potassium permanganate _____

   iodine _____

4. Dispose of the petri dish as instructed by your laboratory instructor. ■

Lab Activity   4   Osmosis

Osmosis (oz-MŌ-sis; *osmos*, thrust) is the net movement of water through a selectively permeable membrane, from a region of greater water concentration to a region of lesser water concentration. We can define osmosis as the *diffusion* of water through a selectively permeable membrane. It occurs when two solutions of different solute concentrations are separated by a selectively permeable membrane. A **solution** is the result of dissolving a **solute** in a **solvent.** In a 1 percent aqueous solution of some salt, for example, the salt is the solute and occupies 1 percent of the solution volume; the solvent—water—makes up the remaining 99 percent of the solution volume. As solute concentration increases, the space available for water molecules decreases, and we can think of this as the water concentration decreasing. For osmosis to occur in a cell, there must be a difference in water concentrations on the two sides of the plasma membrane. This difference in concentration establishes the concentration gradient for osmosis.

Figure 6.3 shows a U-shaped pipe with a selectively permeable membrane located at the bottom where the two arms of the U meet. There are identical molecules on either side of the

> **Study Tip** **Water, Ions, and Membranes**
>
> Only water moves across the membrane during osmosis. If the membrane were permeable to solute molecules, those molecules would move across the membrane until solute equilibrium was reached. Once the solute molecules were in equilibrium, the water molecules would also be in equilibrium. The water concentration gradient would be eliminated, and with no concentration gradient, there can be no osmosis. ■

Figure 6.3 Osmosis The osmotic pressure of solution B is equal to the amount of hydrostatic pressure required to stop the osmotic flow.

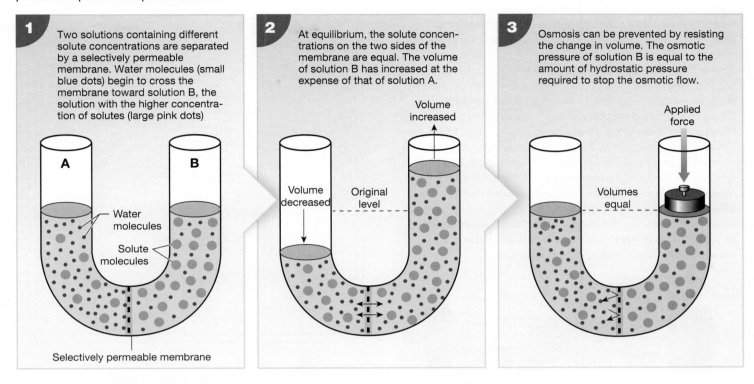

1 Two solutions containing different solute concentrations are separated by a selectively permeable membrane. Water molecules (small blue dots) begin to cross the membrane toward solution B, the solution with the higher concentration of solutes (large pink dots)

2 At equilibrium, the solute concentrations on the two sides of the membrane are equal. The volume of solution B has increased at the expense of that of solution A.

3 Osmosis can be prevented by resisting the change in volume. The osmotic pressure of solution B is equal to the amount of hydrostatic pressure required to stop the osmotic flow.

membrane, but in different concentrations. The small blue dots represent water molecules, and the large pink dots are solute molecules that cannot cross the membrane. (The pale blue background also represents water molecules, but you should concentrate just on the ones represented by the dots.) The numbers of blue dots in **1** tell you that, before our experiment begins, arm A has more water molecules and fewer solute molecules than arm B. Note that in **1**, the water level is the same in the two arms. As water moves from arm A to arm B through the selectively permeable membrane, the volume in arm B increases until equilibrium is reached in **2**. Water and solute concentrations are now equal on the two sides of the membrane.

Solutions have an **osmotic pressure** because of the presence of solute. The greater the solute concentration, the greater the osmotic pressure of the solution. During osmosis, the solution with the greater osmotic pressure causes water to move toward it. In effect, "water follows solute," and osmotic pressure is a "pulling" pressure that draws water toward the higher solute concentration. Notice in **3** of Figure 6.3 that a force applied to arm B will stop osmosis if the pressure resulting from that force is equal to the osmotic pressure causing the osmosis.

Drinking water is often purified by **reverse osmosis,** a process in which the pressure applied to arm B is greater than the osmotic pressure. This increased external pressure forces water molecules across the membrane from right to left in Figure 6.3. As more and more water is forced into arm A, the

concentration of solute molecules in that arm gets lower and lower until at some point the solute concentration is so low that we consider the water to be pure.

The following experiment demonstrates the movement of materials through dialysis tubing. Like a plasma membrane, dialysis tubing is selectively permeable. Small pores in the tubing allow the passage of small particles but not large ones.

QuickCheck Questions

4.1 How is osmosis different from diffusion?

4.2 Which has a greater water concentration, a 1 percent solute solution or a 2 percent solute solution?

4.3 What is osmotic pressure?

In the Lab   4

Materials

☐ Dialysis tubing

☐ Two dialysis tubing clips or two pieces of thread

☐ Gram scale

☐ 500-mL beaker

☐ Distilled water

☐ 5 percent starch solution

☐ Lugol's iodine solution

Clinical Application Dialysis

Dialysis is a passive process similar to osmosis except that, besides water, small solute particles can pass through a selectively permeable membrane. Large particles are unable to cross the membrane, and thus particles can be separated by size during dialysis. Dialysis does not occur in the body, but it is used in the medical procedure called **kidney dialysis** to remove wastes from the blood of a patient whose kidneys are not functioning properly. Blood from an artery passes into thousands of minute selectively permeable tubules in a dialysis cartridge (Figure 6.4). A dialyzing solution having the same concentration of materials to remain in the blood (nutrients and certain electrolytes) is pumped into the cartridge to flow over the tubules. As blood flows through the tubules, wastes diffuse from the blood, through the selectively permeable tubules, and into the dialyzing solution. Once waste levels in the blood have been reduced to a safe level, the patient is disconnected from the dialysis apparatus. ∎

Figure 6.4 Dialysis Cartridge

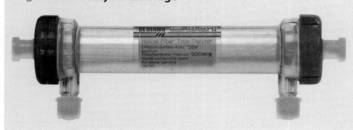

Procedures

1. Cut a strip of dialysis tubing 15 cm (6 in.) long.

2. Add approximately 100 mL of distilled water to the beaker. Soak the dialysis tubing in the water for three to four minutes and then remove it from the beaker.

3. Fold one end of the tubing over and seal it securely with a tubing clip or a piece of thread, forming what is called a dialysis bag. Rub the unclipped end of the bag between your fingers to open the tubing.

4. Fill the bag approximately three-quarters full with starch solution, then fold the end of the tubing over. Clip or tie this end closed without trapping too much air inside.

5. Rinse the bag to remove traces of starch solution from its outside surface. Dry the outside of the filled bag and weigh it to determine its mass. Record your mass measurement in the Initial Observations column of Table 6.1.

6. Submerge the bag completely in the beaker of water, as shown in Figure 6.5. Add enough Lugol's solution to discolor the water in the beaker, and then complete the Initial Observations column in Table 6.1.

Figure 6.5 Osmosis Setup Using Dialysis Membrane
Note: The tubing may be tied with thread if clips are unavailable.

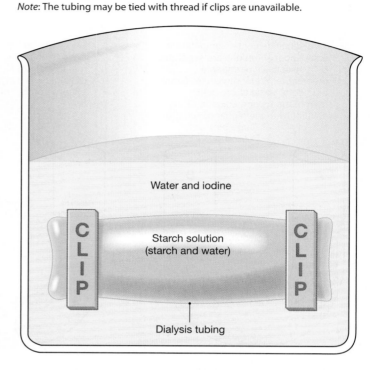

Table 6.1	Dialysis Experiment Observations	
Dialysis Bag	**Initial Observations**	**Final Observations**
Mass of bag plus starch solution		
Shape of filled bag		
Color of starch solution		
Color of beaker water		

Table 6.2	Dialysis Experimental Results	
	Movement (in, out, none)	**Process Substance (diffusion, osmosis)**
Water		
Starch		
Iodine		

7. After 60 minutes:

    a. Without disturbing the setup, examine the beaker and bag and decide if starch, iodine, or water moved either way across the tubing membrane. Record your observations in Table 6.2 and in the second, third, and fourth rows of the Final Observations column of Table 6.1.

    b. Remove the bag from the beaker, dry the outer surface, and determine the mass of the bag plus contents. Record your measurement in the Final Observations column of Table 6.1. ∎

Lab Activity 5 Concentration Gradients and Osmotic Rate

This experiment demonstrates the relationship between concentration gradient and rate of osmosis. Molecules at greater concentrations are packed closer together and have a higher incidence of collisions with neighboring molecules. By comparing changes in mass in a series of dialysis bags, you will measure the osmotic rate at different solute concentrations.

In the Lab 5

Materials

- ☐ Three 15-cm (6-in.) strips of dialysis tubing
- ☐ Six dialysis tubing clips or six pieces of thread
- ☐ Three 500-mL beakers
- ☐ Distilled water
- ☐ 1 percent, 5 percent, and 10 percent sugar solutions

Procedures

1. Add approximately 100 mL of distilled water to each beaker. Place one tubing strip in each beaker, soak for three to four minutes to loosen the tubing, and then remove the strips from the beakers.

2. Fold one end of one piece of tubing over and seal it securely with a tubing clip or a piece of thread, forming a dialysis bag. Rub the unclipped end of the bag between your fingers to open the tubing.

3. Fill the bag approximately three-quarters full with the 1 percent sugar solution, then fold the end of the tubing over. Clip or tie this end closed without trapping too much air inside.

4. Prepare two other bags with the remaining two pieces of tubing. Fill one with the 5 percent sugar solution and the other with the 10 percent solution.

5. Rinse each bag to remove any sugar solution from the outside surface. Dry the outside of each bag, determine its mass, and then submerge it completely in one of the beakers of water, one bag to a beaker. Record your mass measurements in the Initial Mass column of Table 6.3.

Table 6.3	Osmosis Experimental Data	
Dialysis Bag	**Initial Mass**	**Final Mass**
1% sugar	_____	_____
5% sugar	_____	_____
10% sugar	_____	_____

6. After 60 minutes, remove each bag from its beaker, dry the outer surface, and determine the mass of the bag plus contents. Record the final masses in Table 6.3. Use the graph paper in the Review & Practice Sheet and plot the change in mass for each bag. ■

Lab Activity 6 Observation of Osmosis in Cells

A solution that has the same solute concentrations as a cell is an **isotonic solution** (Figure 6.6). If the solute concentrations are the same, the solvent concentrations are also the same. A solution containing more solute (and therefore less solvent) than a cell is a **hypertonic solution,** and a solution containing less solute than a cell is a **hypotonic solution.** The cell is the reference point; solute concentrations in solutions are compared with solute concentrations in the cell. Sitting in a hypertonic solution, a cell will lose water as a result of osmotic movement and will shrink, or **crenate.** Sitting in a hypotonic solution, a cell will gain water and perhaps burst, or **lyse. Hemolysis** is the process of a blood cell rupturing in hypotonic solution.

Your laboratory instructor may choose to use plant cells rather than blood cells to study **tonicity,** the effect of solutions on cells. Plant cells have a thick outer cell wall that provides structural support for the plant. Pushed against the inner surface of the cell wall is the plasma membrane. To study osmosis in plant cells, observe the distribution of the cell's organelles and attempt to locate the plasma membrane. In a hypertonic solution, the plant cell loses water and the plasma membrane shrinks away from the cell wall.

In the Lab 6

Materials

- ☐ Blood (supplied by instructor) or live aquatic plant (Elodea)
- ☐ Microscope slides and coverslips, microscope, eyedroppers, wax pencil
- ☐ 0.90 percent saline solution (isotonic)
- ☐ 2.0 percent saline solution (hypertonic)
- ☐ Distilled water (hypotonic)
- ☐ Gloves and safety glasses

Procedures: Blood Cells

1. With the wax pencil, write along one of the shorter slide edges. Label one slide "Iso," one "Hypo," and one "Hyper."

2. Put on safety glasses and disposable gloves before handling any blood.

Figure 6.6 Osmotic Flow Across Plasma Membranes

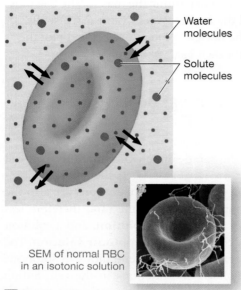
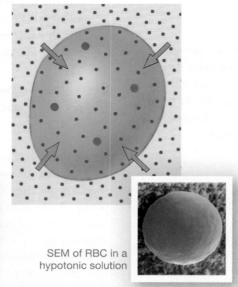
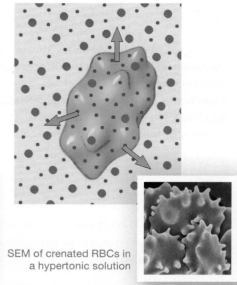

Water molecules

Solute molecules

SEM of normal RBC in an isotonic solution

SEM of RBC in a hypotonic solution

SEM of crenated RBCs in a hypertonic solution

a In an isotonic saline solution, no osmotic flow occurs, and these red blood cells appear normal.

b Immersion in a hypotonic saline solution results in the osmotic flow of water into the cells. The swelling may continue until the plasma membrane ruptures, or lyses.

c Exposure to a hypertonic solution results in the movement of water out of the cell. The red blood cells shrivel and become crenated.

3. Add a small drop of blood to each slide. Do not touch the blood. Place a coverslip over each slide.

4. Add a drop of isotonic solution to the outer edge of the coverslip of the "Iso" slide. Repeat with the other slides and solutions.

5. Use the microscope to observe changes in cell shape as osmosis occurs. Compare your results with the cells in Figure 6.6. Record your observations in section B of the Review & Practice Sheet.

6. Dispose of materials contaminated with blood in a biohazard waste container.

Procedures: Elodea Leaf

1. With the wax pencil, write along one of the shorter slide edges. Label one slide "Iso," one "Hypo," and one "Hyper."

2. Place one Elodea leaf flat on each slide. Place a coverslip over each leaf.

3. Add a drop of isotonic solution to the outer edge of the coverslip of the "Iso" slide. Repeat with the other slides and solutions.

4. Use the microscope to observe changes in cell shape as osmosis occurs. Compare your results with the cells in Figure 6.5.

5. Rinse and clean the slides or dispose of them in a sharps box for glass. ■

Lab Activity 7   Active Transport Processes

Cells use carrier molecules to move nondiffusible materials through the plasma membrane. Unlike passive processes, this carrier-assisted movement may occur against a concentration gradient. Movement of this type is called *active transport* and requires the cell to use energy. Whereas the passive processes of diffusion and osmosis may occur in both living and dead cells, only living cells can supply the energy necessary for active transport.

Endocytosis is the active transport of materials into a cell. **Figure 6.7** presents **phagocytosis,** the movement of large particles into the cell. The figure shows a cell ingesting a bacterium. The cell forms extensions of its plasma membrane, called *pseudopodia*, to capture the bacterium. When the pseudopodia touch one another, they fuse and trap the bacterium in a membrane vesicle. Inside the cell, lysosomes surround and empty their powerful enzymes into the vesicle and destroy the bacterium. During the process called **pinocytosis,** the cell invaginates a small area of the plasma membrane and traps not the large particles of phagocytosis but small particles and fluid. The forming vesicle continues to pinch inward.

Exocytosis is the active transport of materials out of the cell. An intracellular vesicle fills up with materials, fuses with the plasma membrane, and releases its contents into the extracellular fluid. The Golgi apparatus secretes cell products

Figure 6.7 Pinocytosis and Phagocytosis

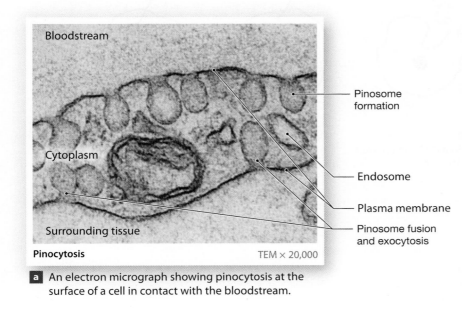

Bloodstream

Cytoplasm

Pinosome formation

Endosome

Plasma membrane

Pinosome fusion and exocytosis

Surrounding tissue

Pinocytosis   TEM × 20,000

a An electron micrograph showing pinocytosis at the surface of a cell in contact with the bloodstream.

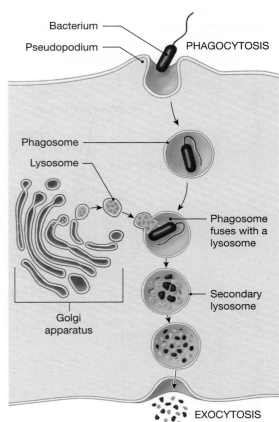

Bacterium

Pseudopodium

PHAGOCYTOSIS

Phagosome

Lysosome

Phagosome fuses with a lysosome

Secondary lysosome

Golgi apparatus

EXOCYTOSIS

b In phagocytosis, material is brought into the cell enclosed in a phagosome that is subsequently exposed to lysosomal enzymes. After nutrients are absorbed from the vesicle, the residue is discharged by exocytosis.

by pinching off small secretory vesicles that fuse with the plasma membrane for exocytosis. Cells also eliminate debris and excess fluids by exocytosis.

QuickCheck Questions

7.1 How do active and passive transports differ?

7.2 Give an example of endocytosis.

In the Lab   7

Materials

☐ *Amoeba proteus* culture, starved for 48 hours
☐ *Tetrahymena* culture
☐ Eyedropper
☐ Microscope depression slide and coverslip
☐ Compound microscope

Procedures

1. Add a drop of the *Amoeba proteus* culture in the well of the depression slide. Place a coverslip over the well to protect the objective lenses.

2. Examine the slide at low magnification and verify that the culture has living amoeba. Reduce the intensity of the microscope's light to keep the slide from heating up.

3. Lift the coverslip and introduce a drop of the *Tetrahymena* culture. *Tetrahymena* are small, ciliated protists that the amoeba will ingest.

4. Observe the activity of the amoeba feeding on the *Tetrahymena*. Select an amoeba ingesting a *Tetrahymena* and observe the process of pseudopodia formation and endocytosis at low and medium powers. Record your observations in section B of the Review & Practice Sheet.

5. Upon completion of your observations, rinse and dry the depression slide and coverslip.

For additional practice, complete the **Sketch to Learn** activity (on p. 66). ∎

Sketch to Learn

Let's draw phagocytosis as observed in the Lab Activity 7 procedures. Redraw the cell in each step to illustrate the sequence of events during phagocytosis.

Sample Sketch

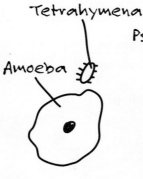

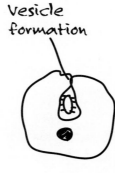

Step 1
• Draw the amoeba with a small ciliated tetrahymena cell nearby.

Step 2
• Add a U-shaped pocket in the cell.

Step 3
• As the pseudopodia touch a vesicle forms.

Step 4
• Vesicle is moved into cell.

Your Sketch

Name _____

Date _____

Section _____

Movement Across
Plasma Membranes

A. Definitions

Define the following terms.

1. Brownian movement

2. osmosis

3. diffusion

4. concentration gradient

5. solute

6. solvent

7. osmotic pressure

8. dialysis

9. hypertonic solution

10. hemolysis

11. crenation

12. equilibrium

B. Results

Lab Activity 1

1. Describe the movement you observed under the microscope.

2. How does this motion occur?

Lab Activity 2

1. How does the dye diffuse in the water?

2. Was there a difference in diffusion rates between the chilled and warmed water?

3. How did temperature influence the diffusion rate?

Lab Activity 3

1. Which crystal diffused farther?

2. Which crystal has the larger molecular mass? How does molecular mass affect diffusion rate?

Lab Activity 4

1. Which had the greater osmotic pressure, the starch solution in the dialysis bag or the iodine solution in the beaker?

2. Use osmotic pressure to explain what happened in this experiment.

3. How did you detect whether it was the starch solution that moved or the iodine solution?

Lab Activity 5

1. Which solution had the greatest osmotic pressure? Because of this highest pressure, what happened more with this solution than with the other two solutions?

2. Describe the relationship between solution concentration and osmosis rate.

3. On the graph paper provided, plot the increase in mass for each bag on the vertical axis and the solution concentration on the horizontal axis. This graph shows how solute concentration affects the rate of osmosis.

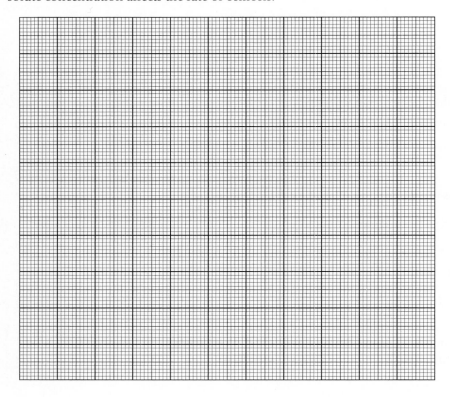

Lab Activity 6

1. Describe the appearance of the blood (or plant) cells in the hypotonic solution. What has happened to these cells?

2. Describe the appearance of the blood (or plant) cells in the hypertonic solution. What has happened to these cells?

3. What is the importance of blood plasma being isotonic to the cytosol of a cell?

Lab Activity 7

1. Describe the movement of the *Amoeba* and the *Tetrahymena*.

2. How are the *Amoeba* feeding on the *Tetrahymena*?

C. Short-Answer Questions

1. Describe the components of a 2 percent sugar solution.

2. After a long soak in the tub, you notice that your skin has become wrinkled and that your fingers and toes feel bloated. Describe why this occurs.

D. Analysis and Application

1. Describe how molecules of a gaseous substance can diffuse through the air.

2. A blood cell is placed in a 1.5 percent salt solution. Will osmosis, diffusion, or both occur across the plasma membrane? Why?

3. Why is a concentration gradient necessary for passive transport?

E. Clinical Challenge

A patient is scheduled for hemodialysis and has a blood sample taken for lab work. Examine **Table 6.4** and discuss how the dialysis procedure will adjust this patient's blood.

Table 6.4 The Composition of Dialysis Fluid		
Component	Patient's Blood Plasma	Dialysis Fluid
ELECTROLYTES (mEq/L)		
Potassium	4	3
Bicarbonate	27	36
Phosphate	3	0
NUTRIENTS (mg/dL)		
Glucose	155	125
NITROGENOUS WASTES (mg/dL)		
Urea	94	0
Creatinine	7	0

Epithelial Tissue

Learning Outcomes

On completion of this exercise, you should be able to:

1. List the characteristics used to classify epithelia.

2. Describe how epithelia are attached to the body.

3. Describe the microscopic appearance of each type of epithelia.

4. List the location and function of each type of epithelia.

5. Identify each type of epithelia under the microscope.

Histology is the study of tissues. A **tissue** is a group of similar cells working together to accomplish a specific function. It may be difficult for us to appreciate how individual cells contribute to the life of the entire organism, but we readily see the effect of tissues in our bodies. Consider how much effort is focused on reducing fat tissue and exercising muscle tissue. An understanding of histology is required for the study of organ function. The stomach, for example, plays major digestive roles, as it secretes digestive juices and is involved in the mixing and movement of food. Each of these functions is performed by specialized tissue.

Figure 7.1 is an overview of tissues of the body. Molecules and atoms combine to form cells, which secrete materials into the surrounding extracellular fluid. The cells and their secretions compose the various tissues of the body. There are four major categories of tissues in the body: **epithelial, connective, muscle,** and **neural.** Each category includes specialized tissues that have specific locations and functions. Many tissues form organs, such as the stomach, a muscle, or a bone.

Need More Practice and Review?

Build your knowledge—and confidence!—in the Study Area of MasteringA&P® at www.masteringaandp.com with Pre-lab Quizzes, Post-lab Quizzes, Practice Anatomy Lab™ (PAL™) 3.0 virtual anatomy practice tool, PhysioEx™ 9.0 laboratory simulations, and A&P Flix™ with Quizzes.

PAL For this lab exercise, follow this navigation path:
• PAL>Histology>Epithelial Tissue

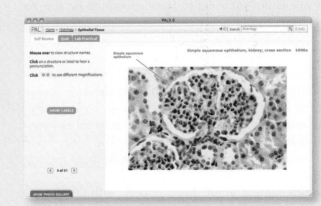

MasteringA&P®

Figure 7.1   An Orientation to the Tissues of the Body

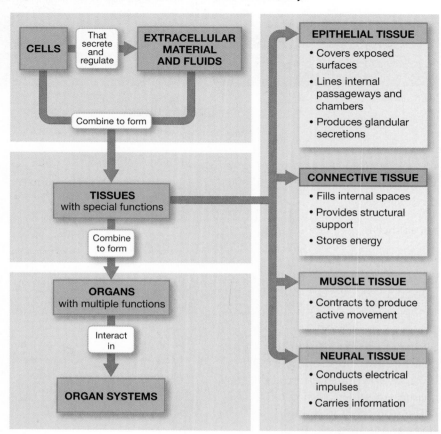

Organs working together to accomplish major processes (such as digestion, movement, or protection) constitute an organ system.

During your microscopic observations of tissues in the following laboratory activities, it is important to scan the entire slide to examine the tissue at low power. A slide may have several tissues, and you must survey the specimen to locate a particular tissue. Once you have located the tissue, increase the magnification and observe the individual cells of the tissue. Take your time when studying a tissue; a quick glance through the microscope is not sufficient to learn enough to be able to identify a tissue on a laboratory examination.

Introduction to Epithelia

Epithelia (e-pi-THĒ-lē-a; singular *epithelium*), or epithelial tissues, are lining and covering tissues. They are the only tissues visible on the body. The respiratory, digestive, reproductive, and urinary systems all have openings to the external environment, and each is lined with an epithelium. The entire body surface is covered with an epithelium in the form of the top layer of the skin.

Epithelium is made up of sheets of cells, with the cells in a given sheet tightly joined together, like ceramic floor tiles,

by a variety of strong intercellular connections. An epithelium always has one surface where the cells are exposed either to the external environment or to an internal passageway or cavity; this surface is called the **free surface** of the epithelium. Because epithelia are surface and lining tissues, they are **avascular** and do not contain blood vessels. The cells obtain nutrients and other necessary materials by diffusion of substances from connective tissue underlying the epithelia. Each epithelium is attached to the body by a **basal lamina** (LA-mi-nah; *lamina*, plate) located between the epithelium and its connective tissue layer.

Epithelia have a wide range of functions, each dependent on the type of cells in the tissue. On exposed surfaces, thick layers of stratified epithelium protect against excessive friction, prevent dehydration, and keep microbes and chemicals from invading the body. Thin, one-layered simple epithelium provides a surface for exchange of materials, such as the exchange of gases between the lungs and blood. Absorption, secretion, and diffusion all occur across simple epithelia. The epithelial tissue that covers the body surface contains many of the body's sensory organs. In some epithelial tissue, such as that associated with glands, the cells are short lived, and in these cases **stem cells** must constantly produce new cells to replenish the tissue.

Classification of Epithelia

Epithelia are classified and identified by the number of cell layers and by the general shape of the tissue cells (**Figure 7.2**). A **simple epithelium** has a single layer of cells and provides a thin surface for exchange of materials, such as the exchange of gases between the lungs and blood. At body surfaces exposed to the external environment, multiple layers of cells in **stratified epithelium** protect against friction, prevent dehydration, and keep microbes and chemicals from invading the body. A **transitional epithelium** is a special kind of stratified epithelium with cells of many shapes that permit the tissue to stretch and recoil. In a **pseudostratified epithelium,** all the cells touch the basal lamina, but the cells grow to different heights. Taller cells grow over shorter ones and cover them, thereby preventing them from reaching the free surface.

Make a Prediction

The esophagus, commonly called the food tube, connects the pharynx to the stomach. What type of layering does the epithelium have in this organ?

Figure 7.2 Classifying Epithelia All epithelia are classified by the number of cell layers and by the shape of the cells.

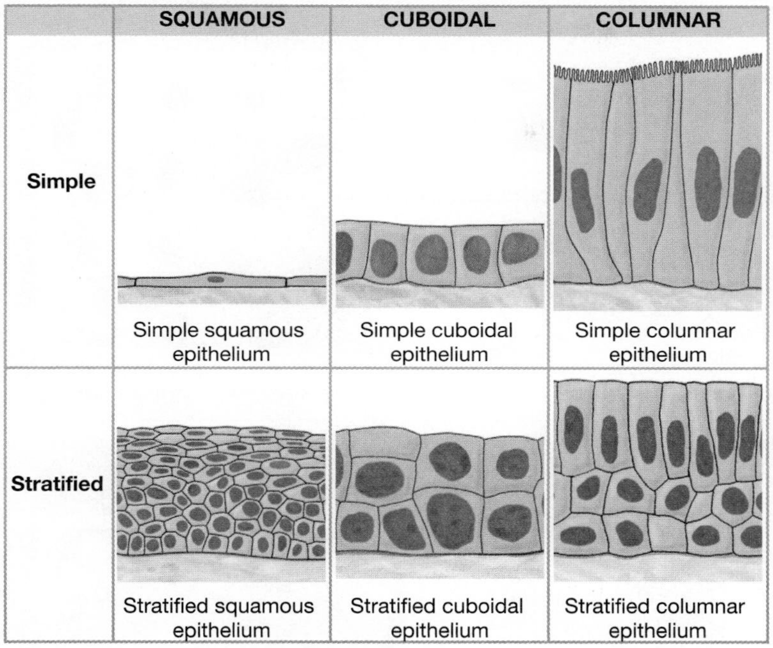

	SQUAMOUS	CUBOIDAL	COLUMNAR
Simple	Simple squamous epithelium	Simple cuboidal epithelium	Simple columnar epithelium
Stratified	Stratified squamous epithelium	Stratified cuboidal epithelium	Stratified columnar epithelium

Cell shape is a second way of classifying epithelium. **Squamous** (SKWA-mus; *squama,* scale) epithelial cells are irregularly shaped, flat, and scalelike. These cells, depending on how they are organized, function either in protection or in secretion and diffusion. **Cuboidal** epithelial cells are cubic (that is, their cross section is approximately square) and have a large nucleus. They are found in the tubules of the kidneys and in many glands, and can secrete and absorb materials across the tubular/glandular wall. **Columnar** epithelial cells are taller than they are wide.

Lab Activity 1 Simple Epithelia

The main functions of simple epithelium are diffusion, absorption, and secretion (Figure 7.3). Many glands consist of simple cuboidal epithelium. The epithelium lining the small intestine has scattered **goblet cells** that secrete mucus to coat and protect the epithelia.

- **Simple squamous epithelium** (Figure 7.3a) is a thin tissue that in a superficial preparation appears as a sheet of cells that looks like ceramic floor tiles. In serous membranes, this tissue is called **mesothelium.** In locations where it lines blood vessels and the heart chambers, it is **endothelium.** Simple squamous epithelium also constructs the thin walls of air sacs in the lungs, where gas exchange occurs.

- **Simple cuboidal epithelium** (Figure 7.3b) lines kidney tubules, the thyroid and other glands, and ducts. On slides of cuboidal epithelium from the kidney, the tubules sectioned longitudinally appear as two rows of square cells; in transverse sections, the cuboidal cells are arranged in a ring to form the round wall of the tubule. Typically the basal lamina is conspicuous in simple cuboidal epithelium.

- **Simple columnar epithelium** (Figure 7.3c) lines most of the digestive tract, the uterine tubes, and the renal collecting ducts. In the small intestine, the wall is folded and covered with simple columnar epithelium to increase the surface area available for digestion and absorption of nutrients. In the uterine tubes, the cilia transport released eggs to the uterus.

Figure 7.4 shows simple epithelia at low and higher magnification.

QuickCheck Questions

1.1 How are epithelia organized and classified?

1.2 What are the functions of simple epithelia?

In the Lab 1

Materials

- ☐ Compound microscope
- ☐ Prepared microscope slides of
  - Simple squamous epithelium
  - Simple cuboidal epithelium
  - Simple columnar epithelium

Procedures

1. Examine each simple epithelium under the microscope at low, medium, and high magnification. Refer to the photomicrographs in Figures 7.3 and 7.4 and locate the featured structures.

   a. Simple squamous epithelium is often observed from a superficial view on a microscope slide and the cells appear much like a ceramic tile floor. Observe how the cells are closely fitted together, with little space between cells available for extracellular material.

Study Tip Looking at Epithelia

When observing epithelia microscopically, find the free surface of the tissue and then look on the opposite edge of the cells. The basal lamina is located right under this edge. It appears as a dark line between the epithelial cells and the connective tissue. ■

Figure 7.3 Simple Epithelia

Simple Squamous Epithelium

LOCATIONS: Epithelia lining ventral body cavities; lining heart and blood vessels; portions of kidney tubules (thin sections of nephron loops); inner lining of cornea; alveoli (air sacs) of lungs

FUNCTIONS: Reduces friction; controls vessel permeability; performs absorption and secretion

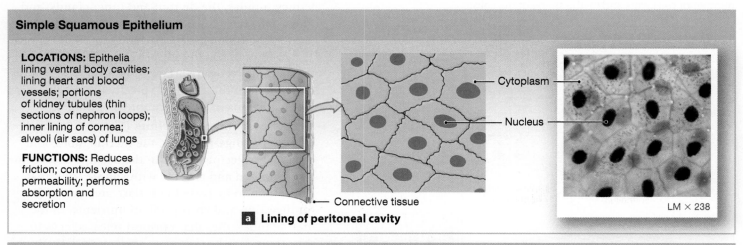

Cytoplasm

Nucleus

Connective tissue

a **Lining of peritoneal cavity**

LM × 238

Simple Cuboidal Epithelium

LOCATIONS: Glands; ducts; portions of kidney tubules; thyroid gland

FUNCTIONS: Limited protection, secretion, absorption

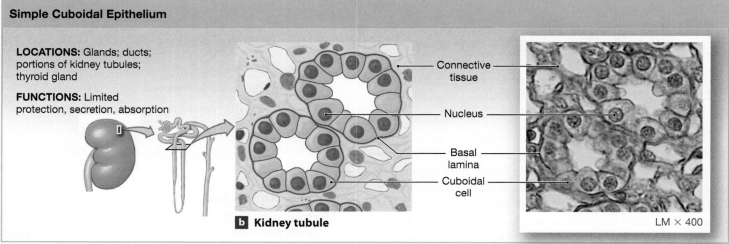

Connective tissue

Nucleus

Basal lamina

Cuboidal cell

b **Kidney tubule**

LM × 400

Simple Columnar Epithelium

LOCATIONS: Lining of stomach, intestine, gallbladder, uterine tubes, and collecting ducts of kidneys

FUNCTIONS: Protection, secretion, absorption

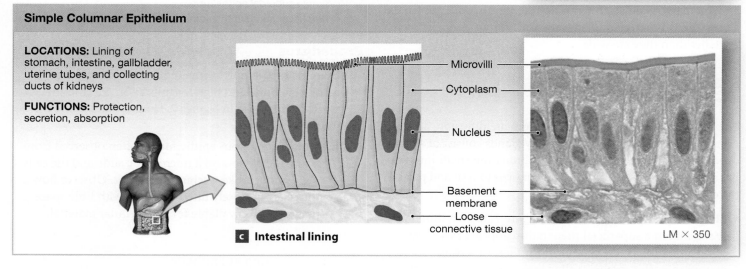

Microvilli

Cytoplasm

Nucleus

Basement membrane

Loose connective tissue

c **Intestinal lining**

LM × 350

Figure 7.4 Observing Epithelia Epithelial cells are close together and cover surfaces such as air sacs in the lungs, tubules in the kidneys, and the passageway of the digestive tract.

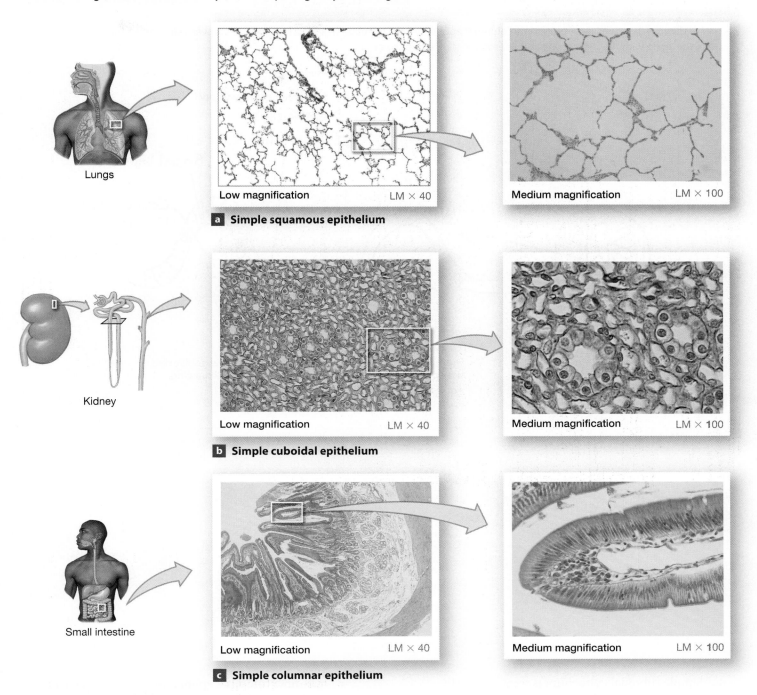

Lungs

Low magnification LM × 40

Medium magnification LM × 100

a Simple squamous epithelium

Kidney

Low magnification LM × 40

Medium magnification LM × 100

b Simple cuboidal epithelium

Small intestine

Low magnification LM × 40

Medium magnification LM × 100

c Simple columnar epithelium

Sketch to Learn

For this sketch we will illustrate simple columnar epithelium at two magnifications: low and high powers. Observe the tissue again with the microscope and refer to the slide as we sketch.

Sample Sketch

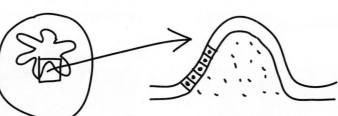

Simple columnar
epithelium
Goblet cell
Basal lamina
Connective
tissue

Sketch 1
Step 1
• Draw the intestine with folds on the inner wall.

Sketch 2
Step 2
• Draw a close-up of the boxed area of sketch 1.
• Add tall, skinny cells across the entire sketch.

Sketch 2
Step 3
• Add a few goblet cells and label drawing.

Your Sketch

b. Simple cuboidal epithelium is recognizable by its cube-shaped cells organized into rings to form ducts and tubules. The base of the tissue is the outer periphery of the tubules and, on most slides, the basal lamina is clearly visible. The free surface of this epithelium is the inner wall of the tubules where ions and molecules are exchanged across the plasma membranes.

c. In simple columnar epithelium, the cell nuclei are uniformly located at the base of the cells. If your slide is of the intestine, interspersed between the columnar cells are oval and light-stained goblet cells. Notice that the wall of the intestine is folded to increase the surface area available for digestion and absorption of nutrients. Simple columnar epithelium covers the folded wall and is in direct contact with the contents of the intestine. For additional practice, complete the *Sketch to Learn* activity.

2. ***Draw It!*** Draw each tissue, as viewed at high magnification, in the space provided.

Simple squamos epithelium

Simple cuboidal epithelium

Simple columnar epithelium

Lab Activity 2 Stratified Epithelia

Stratified epithelia are multilayered tissues with only the bottom layer of cells in contact with the basal lamina and only the upper cells exposed to the free surface (**Figure 7.5**). Stratified epithelia are found in areas exposed to abrasion and friction, such as the body surface and upper digestive tract. When a stratified epithelium contains more than one type of epithelial cell, the type at the free surface determines the classification of the tissue.

■ **Stratified squamous epithelium** (Figure 7.5a) forms the superficial region of the skin, called the **epidermis.** Stem

cells produce new cells at the basal lamina and are pushed toward the free surface by the next group of new cells. The cells manufacture the protein **keratin** (KER-a-tin; *keros,* horn), which toughens the cells but also kills them. The cells then dehydrate and interlock into a broad sheet, forming a dry protective barrier against abrasion, friction, chemical exposure, and even infection. Stratified squamous epithelium of the skin is thus said to be **keratinized** and has a dry surface.

Stratified squamous epithelium also lines the tongue, mouth, pharynx, esophagus, anus, and vagina. The epithelium in these regions is kept moist by lining cells on the tissue surface. This moist tissue is described as being **nonkeratinized** (mucosal type) stratified squamous epithelium.

■ **Stratified cuboidal epithelium** (Figure 7.5b) is uncommon. It is found in the ducts of certain sweat glands.

■ **Stratified columnar epithelium** (Figure 7.5c) is found in parts of the mammary glands, in salivary-gland ducts, and in small regions of the pharynx, epiglottis, anus, and urethra.

QuickCheck Questions

2.1 How is stratified epithelia organized and classified?

2.2 What is the difference between keratinized and nonkeratinized epithelia?

In the Lab 2

Materials

☐ Compound microscope

☐ Prepared microscope slides of
  • Stratified squamous epithelium
  • Stratified cuboidal epithelium
  • Stratified columnar epithelium

Procedures

1. Examine each stratified epithelium under the microscope at low, medium, and high magnification. Refer to the photomicrographs in Figure 7.5 and locate the featured structures.

   a. Stratified squamous epithelium is usually stained red or purple on a microscope slide. Observe that the cells at the free surface are squamous while some of the cells in the middle layers are cuboidal and columnar cells. Remember, it is the cells at the free surface that determine epithelium type.

   b. Stratified cuboidal epithelium is normally only two cell layers thick. Locate a small duct of a sweat gland. With its thick wall, the sectioned duct will look like a donut. Increase magnification and locate the basal lamina.

   c. Stratified columnar epithelium is typically only two to three cell layers thick.

Figure 7.5 Stratified Epithelia

Stratified Squamous Epithelium

LOCATIONS: Surface of skin; lining of mouth, throat, esophagus, rectum, anus, and vagina

FUNCTIONS: Provides physical protection against abrasion, pathogens, and chemical attack

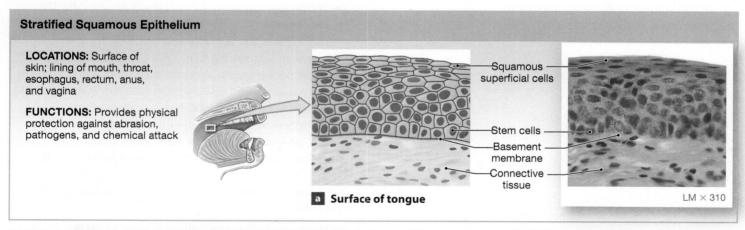

Squamous superficial cells

Stem cells

Basement membrane

Connective tissue

a Surface of tongue

LM × 310

Stratified Cuboidal Epithelium

LOCATIONS: Lining of some ducts (rare)

FUNCTIONS: Protection, secretion, absorption

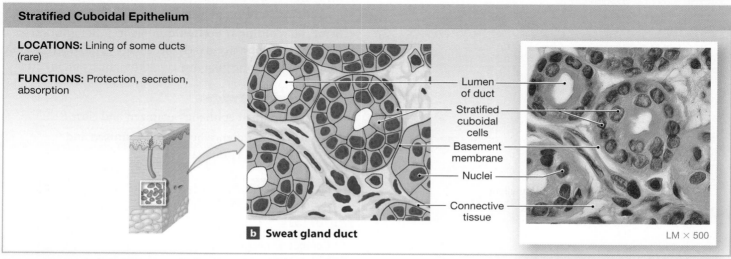

Lumen of duct

Stratified cuboidal cells

Basement membrane

Nuclei

Connective tissue

b Sweat gland duct

LM × 500

Stratified Columnar Epithelium

LOCATIONS: Small areas of the pharynx, epiglottis, anus, mammary glands, salivary gland ducts, and urethra

FUNCTION: Protection

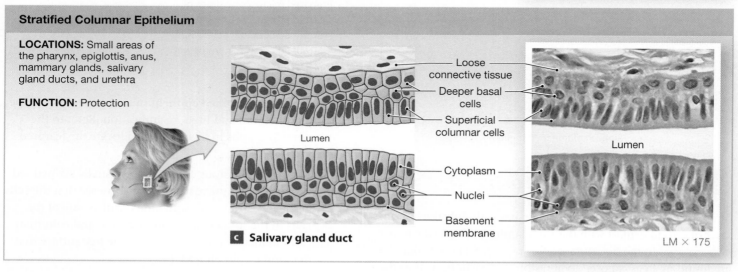

Loose connective tissue

Deeper basal cells

Superficial columnar cells

Lumen

Cytoplasm

Nuclei

Basement membrane

Lumen

c Salivary gland duct

LM × 175

2. *Draw It!* Draw each tissue, as viewed at high magnification, in the space provided. ■

Stratified squamos epithelium

Stratified cuboidal epithelium

Stratified columnar epithelium

Lab Activity 3 Pseudostratified and Transitional Epithelia

Pseudostratified epithelium appears to be a stratified tissue because the nuclei are scattered in the cells, giving the appearance of a stratified tissue. All of the cells adhere to the basal lamina, but, as noted earlier, some cells grow taller and cover other cells at the free surface. The specialized epithelium known as transitional epithelium occurs in the urinary bladder,

among other places in the body. The tissue allows the bladder to fill and empty (Figure 7.6).

■ **Pseudostratified columnar epithelium** (Figure 7.6a) lines the nasal cavity, the trachea, bronchi, and parts of the male reproductive tract. The tissue has columnar cells and smaller stem cells, which replenish the tissue. It appears stratified but is not (hence the *pseudo-* prefix), because every cell touches the basal lamina. Typically the columnar cells are ciliated, and the tissue is called *pseudostratified ciliated columnar epithelium*. Large goblet cells interspersed among the columnar cells secrete mucus onto the epithelial free surface. The mucus traps dust and other particles in the inhaled air. Cilia at the free surface sweep the mucus to the throat, where it is swallowed and disposed of in the digestive tract.

■ **Transitional epithelium** lines organs, such as the urinary bladder, that must stretch and shrink (Figure 7.6b). The cells have a variety of shapes and sizes, and not all of them touch the basal lamina. Most transitional tissue slides are prepared from relaxed transitional tissue, and the tissue appears thick, with many cells stacked one upon another. If the organ is stretched, the transitional epithelium gets thinner.

QuickCheck Questions

3.1 How is pseudostratified epithelia different from simple epithelia?

3.2 What is the function of transitional epithelia and where does it occur in the body?

In the Lab 3

Materials

☐ Compound microscope
☐ Prepared microscope slides of
  • Pseudostratified columnar epithelium
  • Transitional epithelium

Procedures

1. Examine each epithelium under the microscope at low, medium, and high magnification. Refer to the photomicrographs in Figure 7.6 and locate the featured structures.

  a. Notice how the nuclei in the pseudostratified columnar epithelium are unevenly distributed, creating a stratified appearance. At high magnification, slowly turn the fine focus knob back and forth approximately one-quarter turn and examine the tissue surface for cilia. Identify the goblet cells interspersed between the columnar cells. Be sure to include a few goblet cells in your sketch.

  b. From the thickness of the transitional epithelium, determine which one your specimen was made from—an empty, relaxed bladder or a full, stretched bladder.

Figure 7.6 Pseudostratified and Transitional Epithelia

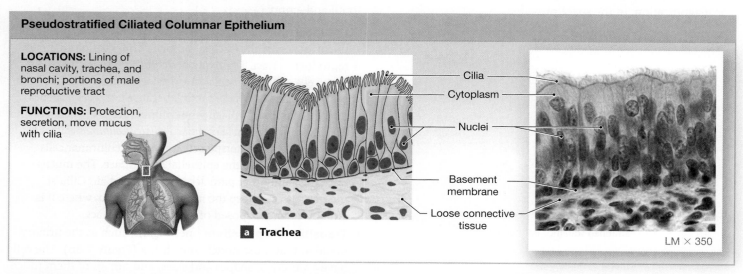

Pseudostratified Ciliated Columnar Epithelium

LOCATIONS: Lining of nasal cavity, trachea, and bronchi; portions of male reproductive tract

FUNCTIONS: Protection, secretion, move mucus with cilia

Cilia
Cytoplasm
Nuclei
Basement membrane
Loose connective tissue

a Trachea

LM × 350

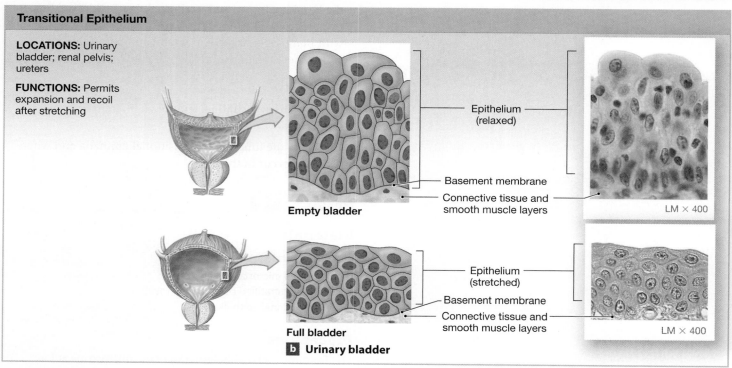

Transitional Epithelium

LOCATIONS: Urinary bladder; renal pelvis; ureters

FUNCTIONS: Permits expansion and recoil after stretching

Epithelium (relaxed)
Basement membrane
Connective tissue and smooth muscle layers

Empty bladder

LM × 400

Epithelium (stretched)
Basement membrane
Connective tissue and smooth muscle layers

Full bladder

b Urinary bladder

LM × 400

2. ***Draw It!*** Draw each tissue, as viewed at high magnification, in the space provided.

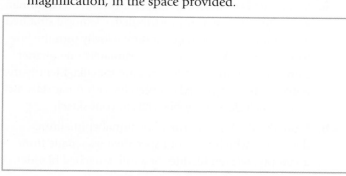

Pseudostratified columnar epithelium

3. ***Draw It!*** Draw transitional epithelium, as viewed at high magnification, in the space provided. ■

Transitional epithelium

Name _____

Date _____

Section _____

Epithelial Tissue

A. Descriptions

Write a brief description of the following:

1. free surface

2. simple cuboidal epithelium

3. stratified squamous epithelium

4. transitional epithelium

5. simple columnar epithelium

6. simple squamous epithelium

7. pseudostratified ciliated columnar epithelium

8. basal lamina

9. goblet cell

10. endothelium

B. Fill in the Blanks

Complete the following statements.

1. Epithelium that occurs in a single layer of flat cells is _____ epithelium.

2. The tissue deep to epithelium is _____.

3. Epithelium that stretches and relaxes is _____ epithelium.

4. Epithelia are attached to a deep layer of connective tissues by a membrane called the

_____.

5. Cells that secrete mucus are called _____.

6. The epithelium that lines the stomach and small intestine is _____ epithelium.

7. Epithelium that occurs in the facing layers of serous membranes is _____ epithelium.

8. Epithelium that occurs in a thick layer of cells is _____ epithelium.

C. Drawing

1. ***Draw It!*** Draw and label a high magnification view of stratified squamous epithelium, including the basal lamina and the underlying connective tissue.

Stratified squamous epithelium

2. ***Draw It!*** Draw and label a high magnification view of pseudostratified ciliated columnar epithelium.

Pseudostratified ciliated columnar epithelium

D. Analysis and Application

1. Suppose you are examining the inner lining of the stomach. Describe the tissue you are observing and relate its structure to its function in that location.

2. Describe the function of stem cells in glands and in the epithelium of the skin.

E. Labeling

Label each structure in Figure 7.7.

Figure 7.7 **Epithelial Tissues**

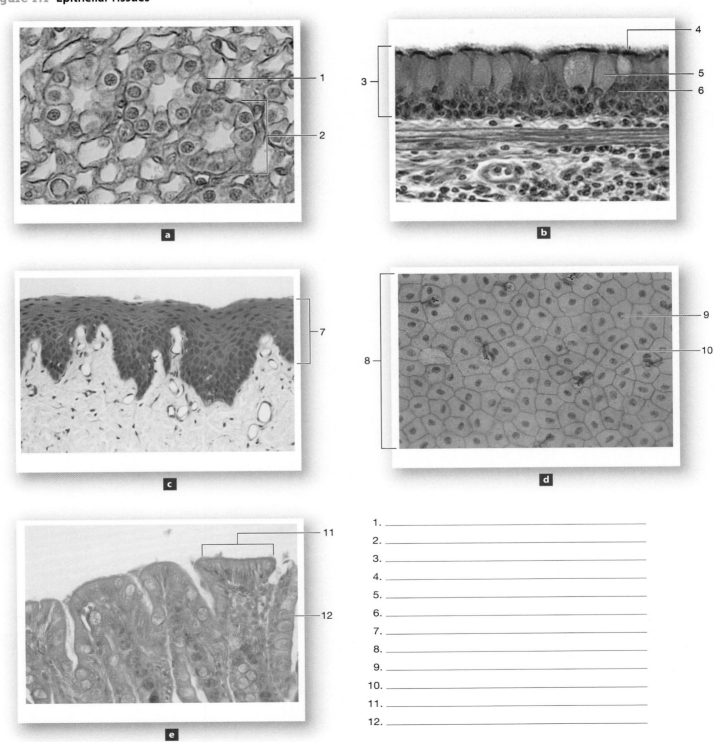

1. _____
2. _____
3. _____
4. _____
5. _____
6. _____
7. _____
8. _____
9. _____
10. _____
11. _____
12. _____

F. Clinical Challenge

Both the facial and oral surfaces of the cheeks are covered with stratified squamous epithelium, yet infection can be transmitted more easily in the mouth. Consider the characteristics of the epithelium on each side of the cheek and write a brief explanation in support of the previous statement.

Connective Tissue

Learning Outcomes

On completion of this exercise, you should be able to:

1. List the major types of connective tissue and the characteristics of each.

2. Describe the structure and function of embryonic connective tissue.

3. Describe the location and function of each type of connective tissue.

4. Identify each type of connective tissue and its cell and matrix structure under the microscope.

Connective tissue provides the body with structural support and with a means of joining various structural components to one another. Unlike the cells in epithelia, cells in connective tissue are widely scattered throughout the tissue. These cells produce and secrete protein **fibers** and a **ground substance** that together form an extracellular **matrix.** The ground substance is composed mainly of glycoprotein and polysaccharide molecules that surround the cells as either a thick, syrupy liquid; a gelatinous layer; or a solid, crystalline material. Suspended in the ground substance are **collagen fibers,** which give tissues strength, and **elastic fibers,** which provide flexibility. **Reticular fibers** are interwoven proteins found in reticular connective tissue; they provide a framework for support of internal soft organs, such as the liver and spleen. As we age, cells secrete fewer protein fibers into the matrix, resulting in brittle bones and wrinkled skin. Leather is mostly collagen fibers from the dermis of animal skins that have been tanned and preserved.

The matrix of a connective tissue determines the physical nature of the tissue. Blood, for example, has a liquid matrix called *blood plasma* that allows the blood to flow freely through vessels. Adipose tissue has a thick liquid matrix that is syrupy,

Need More Practice and Review?

Build your knowledge—and confidence!—in the Study Area of MasteringA&P® at www.masteringaandp.com with Pre-lab Quizzes, Post-lab Quizzes, Practice Anatomy Lab™ (PAL™) 3.0 virtual anatomy practice tool, PhysioEx™ 9.0 laboratory simulations, and A&P Flix™ with Quizzes.

PAL | practice anatomy lab   For this lab exercise, follow this navigation path:

• PAL>Histology>Connective Tissues

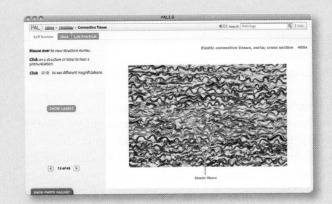

MasteringA&P®

Figure 8.1 **Classification of Connective Tissues**

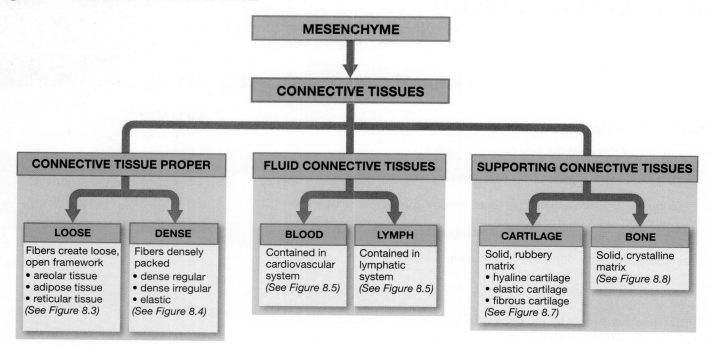

like honey. Cartilage has a thick, gelatinous matrix that allows this connective tissue to slide easily over other structures. Bone has a solid matrix and provides the structural framework for the body.

Connective tissues are classified into three broad groups, distinguished primarily by cellular composition and the characteristics of the extracellular matrix (**Figure 8.1**). **Connective tissue proper** has a thick liquid matrix and a variety of cell types. **Fluid connective tissues** are liquid tissues that flow through blood vessels and lymphatic vessels. **Supporting connective tissues** have a strong gelatinous or solid matrix that acts as support for other tissues.

All connective tissue is produced in the embryo from an unspecialized tissue called **mesenchyme.** Cells in this embryonic connective tissue differentiate into cartilage, bone, and other types of connective tissue. Each tissue group is discussed in the following activities.

Lab Activity 1 **Embryonic Connective Tissue**

All connective tissues are produced in the embryo (**Figure 8.2**) from an unspecialized tissue called **mesenchyme** (mez-en-kīm) (Figure 8.2a). Early in the third week of embryonic development, mesenchyme appears and produces the specialized cells needed to construct mature connective tissues such as bone and cartilage. Mesenchyme is a loose meshwork of star-shaped

cells. Unlike adult connective tissue, mesenchyme has no visible protein fibers in its ground substance. **Mucous connective tissue,** also called **Wharton's jelly,** is an embryonic connective tissue found only in the umbilical cord (Figure 8.2b).

QuickCheck Questions

1.1 Which embryonic tissue produces the various connective tissues?

1.2 What type of embryonic connective tissue is found in the umbilical cord?

In the Lab 1

Materials

☐ Compound microscope

☐ Prepared microscope slide of mesenchyme

Procedures

1. Scan the mesenchyme slide at low magnification and note the concentration of cells.

2. Increase the magnification first to medium and then to high. Observe the shape and distribution of the mesenchyme cells and a matrix lacking protein fibers. Compare the microscope image of the tissue with Figure 8.2.

Figure 8.2 Connective Tissues in Embryos

Mesenchymal cells

Mesenchyme   LM × 140

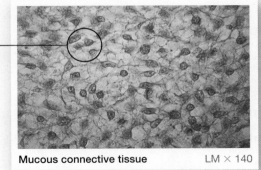

Mesenchymal cells

Mucous connective tissue   LM × 140

3. ***Draw It!*** Draw mesenchyme as you viewed it at high magnification in the space provided. ▪

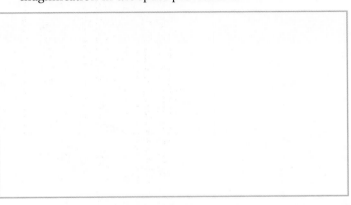

Mesenchyme

Lab Activity 2 Connective Tissue Proper

Connective tissue proper includes two groups of tissues: loose and dense. **Loose connective tissue** has an open network of protein fibers in a thick, syrupy ground substance. *Areolar, adipose,* and *reticular* tissues are the three main types of loose connective tissue (**Figure 8.3**). **Dense connective tissue** is made up of two types of fibers: *protein fibers* assembled into thick bundles of collagen, and *elastic fibers* with widely scattered cells. There are two types of dense connective tissue: *dense regular,* in which the protein fibers in the matrix are arranged in parallel bands, and *dense irregular,* in which the fibers are interwoven (**Figure 8.4**).

Study Tip Cell Names Are Meaningful

The suffix of a cell's name typically indicates the function of the cell: The *blast* cells are tissue *builders,* so remember blasts are builders. These cells produce more cells and tissue matrix. The *cyte* cells are like the maintenance *crew* of a tissue and *clasts* are a demolition squad that tears down a tissue. Bone tissue has all three types of cells; osteoblasts build bone matrix, osteocytes maintain the tissue, and osteoclasts dissolve the matrix to release calcium into the blood for important contributions to blood clotting and muscle contraction. ▪

Connective tissue proper contains a variety of cell types in addition to the collagen fibers and elastic fibers just described (Figure 8.3). **Fibroblasts** (FĪ-brō-blasts) are fixed (stationary) cells that secrete proteins that join other molecules in the matrix to form the collagen and elastic fibers. Phagocytic **macrophage** (MAK-rō-fā-jez; *phagein,* to eat) cells patrol these tissues, ingesting microbes and dead cells. Macrophages are mobilized during an infection or injury, migrate to the site of disturbance, and phagocytize damaged tissue cells and microbes. **Mast cells** release histamines that cause an inflammatory response in damaged tissues. **Adipocytes** (AD-i-Pō-sīts) are fat cells and contain vacuoles for the storage of lipids.

▪ **Areolar tissue** (Figure 8.3a) is distributed throughout the body. This tissue fills spaces between structures for support and protection. It is very flexible and permits muscles to move freely without pulling on the skin. Most of the cells in areolar tissue are oval-shaped fibroblasts that usually stain light.

Figure 8.3 Loose Connective Tissues The body's "packing material" fills spaces between other structures.

Areolar Tissue

LOCATIONS: Within and deep to the dermis of skin, and covered by the epithelial lining of the digestive, respiratory, and urinary tracts; between muscles; around blood vessels, nerves, and around joints

FUNCTIONS: Cushions organs; provides support but permits independent movement; phagocytic cells provide defense against pathogens

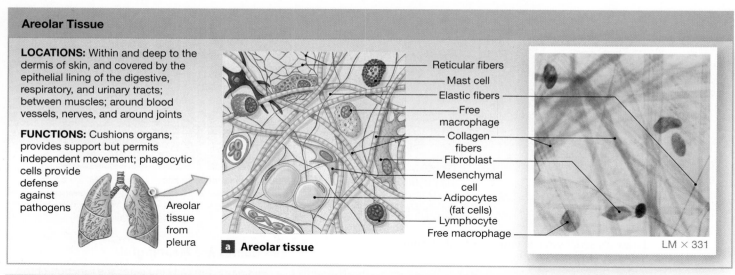

Areolar tissue from pleura

Reticular fibers
Mast cell
Elastic fibers
Free macrophage
Collagen fibers
Fibroblast
Mesenchymal cell
Adipocytes (fat cells)
Lymphocyte
Free macrophage

a Areolar tissue

LM × 331

Adipose Tissue

LOCATIONS: Deep to the skin, especially at sides, buttocks, breasts; padding around eyes and kidneys

FUNCTIONS: Provides padding and cushions shocks; insulates (reduces heat loss); stores energy

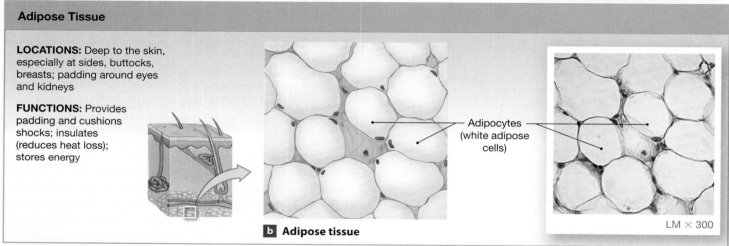

Adipocytes (white adipose cells)

b Adipose tissue

LM × 300

Reticular Tissue

LOCATIONS: Liver, kidney, spleen, lymph nodes, and bone marrow

FUNCTIONS: Provides supporting framework

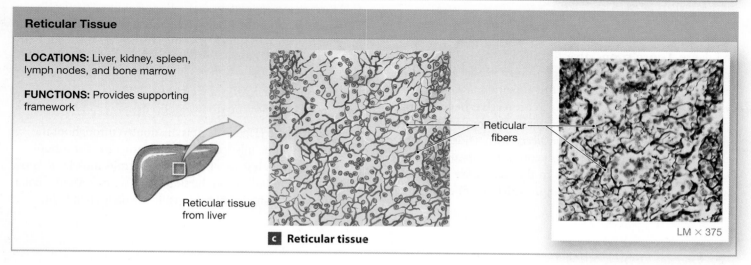

Reticular tissue from liver

Reticular fibers

c Reticular tissue

LM × 375

Figure 8.4 **Dense Connective Tissues**

Dense Regular Connective Tissue

LOCATIONS: Between skeletal muscles and skeleton (tendons and aponeuroses); between bones or stabilizing positions of internal organs (ligaments); covering skeletal muscles; deep fasciae

FUNCTIONS: Provides firm attachment; conducts pull of muscles; reduces friction between muscles; stabilizes relative positions of bones

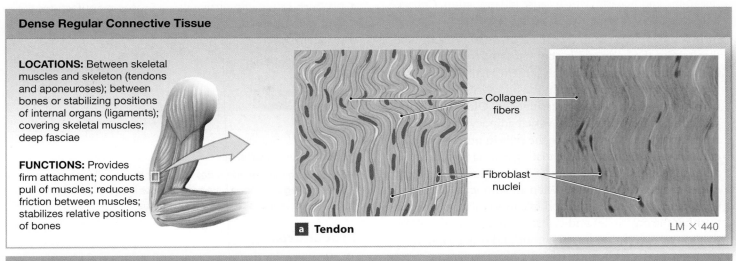

Collagen fibers

Fibroblast nuclei

a **Tendon**

LM × 440

Dense Irregular Connective Tissue

LOCATIONS: Capsules of visceral organs; periostea and perichondria; nerve and muscle sheaths; dermis

FUNCTIONS: Provides strength to resist forces applied from many directions; helps prevent overexpansion of organs such as the urinary bladder

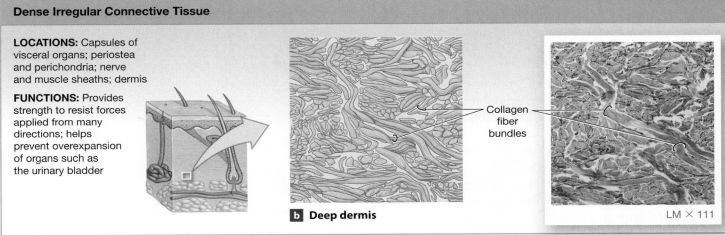

Collagen fiber bundles

b **Deep dermis**

LM × 111

Elastic Tissue

LOCATIONS: Between vertebrae of the spinal column (ligamentum flavum and ligamentum nuchae); ligaments supporting penis; ligaments supporting transitional epithelia; in blood vessel walls

FUNCTIONS: Stabilizes positions of vertebrae and penis; cushions shocks; permits expansion and contraction of organs

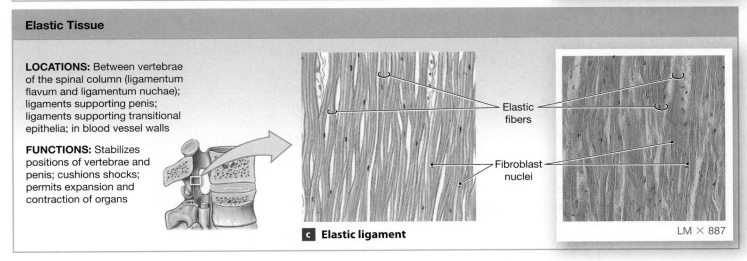

Elastic fibers

Fibroblast nuclei

c **Elastic ligament**

LM × 887

Mast cells are small and filled with dark-stained granules of histamine and heparin, which cause inflammation. Collagen and elastic fibers are clearly visible in the matrix.

- **Adipose tissue** (commonly called *fat tissue*) is distributed throughout the body and is abundant under the skin and in the buttocks, breasts, and abdomen (Figure 8.3b). Two types of adipose tissue occur in the body. Infants have **brown fat,** which is highly vascularized. Older children and adults have **white fat,** in which adipocytes are packed more closely together than are the cells in other types of connective tissue proper. The distinguishing feature of adipose tissue is displacement of the nucleus and cytoplasm due to the storage of lipids. When an adipocyte stores fat, its vacuole expands with lipid and fills most of the cell while pushing the organelles and cytosol to the periphery. When the body needs lipids, for metabolic and other uses, the adipocytes release the lipid into the bloodstream.

- **Reticular tissue** forms the internal supporting framework for soft organs, such as the spleen, liver, and lymphatic organs. The tissue is composed of an extensive network of **reticular fibers** interspersed with small, oval **reticulocytes** (Figure 8.3c).

- **Dense regular connective tissue** consists mostly of collagen or elastic fibers organized into thick bands, with fibroblasts widely interspersed in the fibrous matrix. This strong tissue forms tendons, which attach muscles to bones, and ligaments, which support articulating bones. Because tendons and ligaments conduct pulling forces mainly from one direction, the protein fibers in dense regular tissues are parallel. Tendons transfer strong pulling forces from muscle to bone and have an abundance of strong bands of collagen fibers in the matrix (Figure 8.4a). Flat layers of dense regular connective tissue called *fascia* (FASH-ē-uh) protect and isolate muscles from surrounding structures and allow muscle movement.

- **Dense irregular connective tissue** (Figure 8.4b) is a mesh of collagen fibers with interspersed fibroblasts. Dense irregular connective tissue is located in the **dermis,** which is the skin layer just deep to the epidermis, and in the layers surrounding cartilage and bone. The kidneys, liver, and spleen are protected inside a capsule of dense irregular connective tissue. With its meshwork of collagen fibers, this connective tissue supports areas that receive stress from many directions.

- **Elastic tissue** (Figure 8.4c) is a dense regular connective tissue with elastic fibers in the matrix rather than collagen fibers. The elastic fibers are thicker than collagen fibers and are in large bundles. **Elastic ligaments** have more elasticity than tendons and have a large quantity of elastic fibers in the matrix. Elastic tissue supports the vertebrae of the spine as elastic ligaments and occurs in the blood chambers in the penis.

QuickCheck Questions

2.1 Which cell in connective tissue proper manufactures the protein fibers for the matrix?

2.2 What fiber types are common in the matrix of connective tissue proper?

In the Lab 2

Materials

- ☐ Compound microscope
- ☐ Prepared microscope slides of areolar tissue, adipose tissue, dense regular connective tissue, dense irregular connective tissue, elastic tissue, and reticular tissue

Procedures

1. Scan areolar tissue at all magnifications and identify the prominently stained fibroblasts, mast cells, and macrophages. Note the thick collagen fibers and thin elastic fibers in the matrix.

2. *Draw It!* Draw tissue in the space provided.

Areolar tissue

3. Scan the adipose tissue slide at all magnifications and observe individual adipocytes with their cytoplasm displaced to the cell's edge by fat vacuoles, giving the cell a "signet" appearance similar to a graduation ring.

4. *Draw It!* Draw the tissue in the space provided.

Adipose tissue

Make a Prediction

What is happening to adipocytes in a person who is exercising and losing weight?

5. Scan the reticular tissue slide at all magnifications and locate the reticular fibers and reticulocytes.

6. *Draw It!* Draw the tissue in the space provided.

Reticular tissue

7. Scan the dense regular (tendon) tissue slide at low and medium magnifications. Note the abundance of collagen fibers organized into parallel bands, with few fibroblasts scattered in between. On slides with limited stain, the profusion of collagen fibers makes dense regular connective tissue of tendons appear yellow under the microscope.

8. *Draw It!* Draw the tissue in the space provided.

Dense regular tissue

9. Scan the dense irregular tissue slide at low and medium magnifications. Note the interwoven network of collagen that distinguishes this tissue from dense regular.

10. *Draw It!* Draw the tissue in the space provided.

Dense irregular tissue

11. Scan elastic tissue slide at low and medium magnifications. Observe the branched organization of the elastic fibers and the arrangement of fibroblasts.

12. *Draw It!* Draw the tissue in the space provided. ∎

Elastic Tissue

Lab Activity 3 Fluid Connective Tissue

Fluid connective tissue includes **blood** and **lymph** tissues. These tissues have a liquid matrix and circulate in blood vessels or lymphatic vessels. Blood is composed of cells collectively called the *formed elements* (Figure 8.5), which are supported in a liquid ground substance called blood **plasma.** Protein fibers are dissolved in the matrix of both blood and lymph tissues. During blood clotting, in a process called *coagulation*, fibers in blood produce a fibrin net to trap cells as they pass through the wound. Fibers in blood also regulate the viscosity, or thickness, of the blood.

The formed elements are grouped into three general categories: red blood cells, white blood cells, and platelets. Red blood cells, called **erythrocytes,** transport blood gases. The cells are biconcave discs, with a center so thin that it often looks hollow when viewed with a microscope. White blood cells, called **leukocytes,** are the cells of the immune

Figure 8.5 Formed Elements of the Blood

Red blood cells	White blood cells	Platelets
Red blood cells, or erythrocytes (e-RITH-rō-sīts), are responsible for the transport of oxygen (and, to a lesser degree, of carbon dioxide) in the blood.	**White blood cells**, or leukocytes (LOO-kō-sīts; *leuko-*, white), help defend the body from infection and disease.	**Platelets** are membrane-enclosed packets of cytoplasm that function in blood clotting.

Neutrophil

Eosinophil   Basophil

Red blood cells account for roughly half the volume of whole blood and give blood its color.	**Monocytes** are phagocytes similar to the free macrophages in other tissues. **Lymphocytes** are uncommon in the blood but they are the dominant cell type in lymph, the second type of fluid connective tissue. **Eosinophils** and **neutrophils** are phagocytes. **Basophils** promote inflammation much like mast cells in other connective tissues.	These cell fragments are involved in the clotting response that seals leaks in damaged or broken blood vessels.

system and protect the body from infection. Upon injury to a blood vessel, **platelets** become sticky and form a plug to reduce bleeding.

The most common cells of the lymphatic system are lymphocytes, which are white blood cells produced in lymphoid tissues.

QuickCheck Questions

3.1 Which tissues comprise the fluid connective tissues?

3.2 Describe the ground substance and fibers of blood.

In the Lab   3

Materials

☐ Compound microscope

☐ Prepared microscope slide of blood

Procedures

1. Scan the blood slide at low power, then increase magnification to medium power and identify the three types of blood cells.

   a. Erythrocytes—The majority of the cells on the slide are these red blood cells. How do these cells differ from most other cells in the body?

   b. Leukocytes—These are the dark-stained cells. Note the variation in the morphology, or shape, of the nucleus in the different types of leukocytes.

   c. Platelets—Look closely between the erythrocytes and leukocytes and observe these fragile formed elements.

2. ***Draw It!*** Sketch several erythrocytes and leukocytes in the space provided.

Erythrocytes   Leukocytes

3. ***Draw It!*** Sketch platelets in the space provided. ■

Platelets

Lab Activity 4 Supporting Connective Tissue

Cartilage and bone, the two types of supporting connective tissues, contain a strong matrix of fibers capable of supporting body weight and stress. **Cartilage** is a rubbery, avascular tissue with a gelatinous matrix and many fibers for structural support. **Bone** has a solid matrix of calcium phosphate and calcium carbonate. These salts crystallize on collagen fibers and form a hard material called **hydroxyapatite.**

A membrane surrounds all supporting connective tissues to protect the tissue and supply new tissue-producing cells. The **perichondrium** (per-i-KON-drē-um) is the membrane surrounding cartilage (**Figure 8.6**) and produces **chondroblasts** (KON-drō-blasts; *chondros*, cartilage), which secrete the fibers and ground substance of the cartilage matrix. Eventually, chondroblasts become trapped in the matrix in small spaces called **lacunae** (la-KOO-nē, *lacus*, pool) and lose the ability to produce additional matrix. These cells are then called **chondrocytes** and function in maintenance of the mature tissue. **Figure 8.7** shows the three types of cartilage.

Figure 8.6   Structure of Cartilage An outer membrane called the perichondrium separates cartilage from other tissues. Mitosis by chondroblasts in the perichondrium pushes older cells into the middle of the tissue where they become chondrocytes surrounded by a lacuna in the gel matrix.

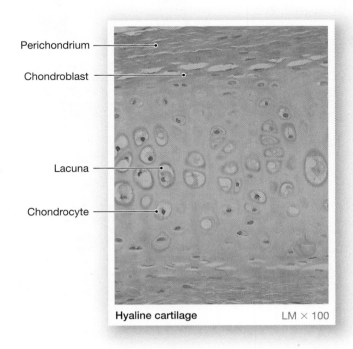

Perichondrium

Chondroblast

Lacuna

Chondrocyte

Hyaline cartilage   LM × 100

- **Hyaline** (HĪ-uh-lin; *hyalus,* glass) **cartilage** (Figure 8.7a) is the most common cartilage in the body. The tissue is distinguishable from other cartilages by the apparent lack of fibers in the matrix. Hyaline cartilage contains elastic and collagen fibers, but it does not stain and therefore is not visible.
- **Elastic cartilage** (Figure 8.7b) has many elastic fibers in the matrix and is therefore easily distinguished from hyaline cartilage. The elastic fibers permit considerable binding and twisting of the tissue.
- **Fibrocartilage** contains irregular collagen fibers that are visible in the matrix (Figure 8.7c). This cartilage is very strong and durable, and its function is to cushion joints and limit bone movement.

Bones for the most part are made of bone tissue. A bone is surrounded by a membrane called the **periosteum** (per-ē-OS-tē-um), which contains cells called **osteoblasts** (OS-tē-ō-blasts) for bone growth and repair (**Figure 8.8**). Like chondroblasts, osteoblasts secrete the organic components of the matrix, become trapped in lacunae, and mature into **osteocytes.** Rings of matrix called **concentric lamellae** (lah-MEL-lē; *lamella,* thin plate) surround a **central canal** that contains blood vessels. **Canaliculi** (kan-a-LIK-ū-lē; little canals) are small channels in the lamellae that provide passageways through the solid matrix for diffusion of nutrients and wastes. Other bone cells, called **osteoclasts,** secrete small quantities of carbonic acid to dissolve portions of the bone matrix and release calcium ions into the blood for various physiological processes. The functions of bone are body support, attachment of skeletal muscles, and protection of internal organs.

QuickCheck Questions

4.1 Describe the matrix of cartilage.

4.2 How are cartilage and bone tissues similar to each other?

In the Lab 4

Materials

☐ Compound microscope

☐ Prepared microscope slides of hyaline cartilage, elastic cartilage, fibrocartilage, and bone

Procedures

1. Scan the hyaline cartilage slide and locate the perichondrium and chondroblasts on the hyaline cartilage slide. Examine the deeper middle region of the cartilage, where the chondroblasts have migrated and become chondrocytes inside lacunae.

Figure 8.7 Types of Cartilage These three types of cartilage provide flexible support of the body.

Hyaline Cartilage

LOCATIONS: Between tips of ribs and bones of sternum; covering bone surfaces at synovial joints; supporting larynx (voice box), trachea, and bronchi; forming part of nasal septum

FUNCTIONS: Provides stiff but somewhat flexible support; reduces friction between bony surfaces

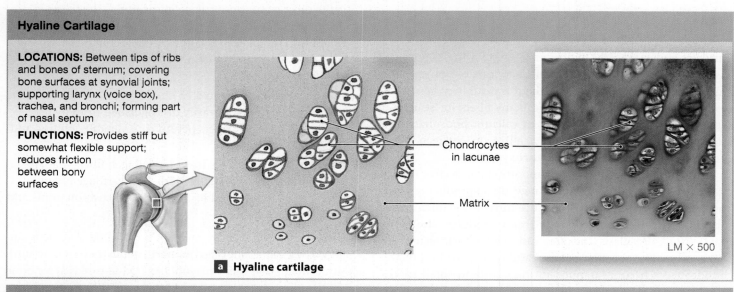

Chondrocytes in lacunae

Matrix

LM × 500

a Hyaline cartilage

Elastic Cartilage

LOCATIONS: Auricle of external ear; epiglottis; auditory canal; cuneiform cartilages of larynx

FUNCTIONS: Provides support, but tolerates distortion without damage and returns to original shape

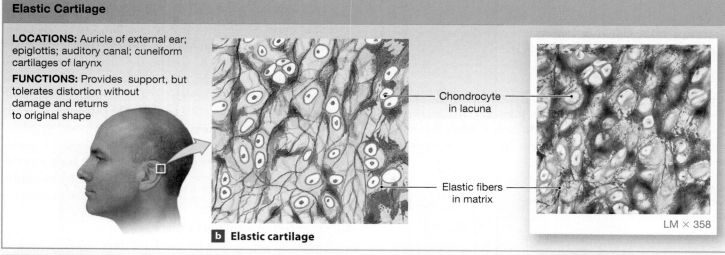

Chondrocyte in lacuna

Elastic fibers in matrix

LM × 358

b Elastic cartilage

Fibrocartilage

LOCATIONS: Pads within knee joint; between pubic bones of pelvis; intervertebral discs

FUNCTIONS: Resists compression; prevents bone-to-bone contact; limits movement

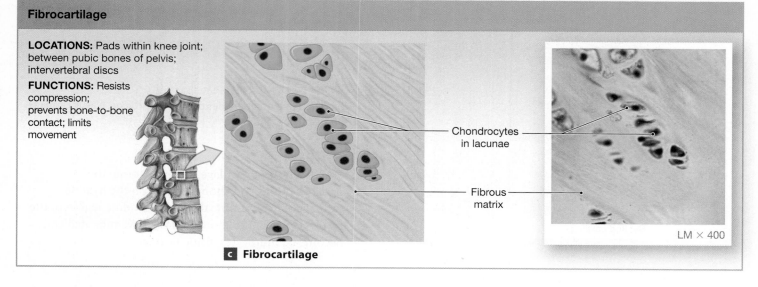

Chondrocytes in lacunae

Fibrous matrix

LM × 400

c Fibrocartilage

Figure 8.8 Bone Compact bone tissue is organized into bony columns called osteons. Each osteon consists of many concentric lamellae that encase a blood vessel passageway called the central canal.

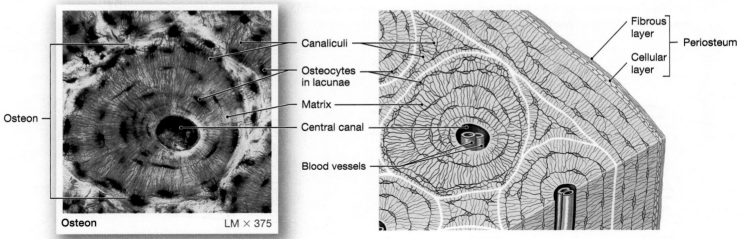

2. ***Draw It!*** Sketch tissue in the space provided.

Hyaline cartilage

3. Scan the elastic tissue slide and observe the many elastic fibers in the matrix of the elastic cartilage. Identify the perichondrium with its many small chondroblasts and the chondrocytes trapped in lacunae deeper in the tissue.

4. ***Draw It!*** Draw the tissue in the space provided.

Elastic cartilage

5. Scan the fibrocartilage slide and observe the thick bundles of collagen and the absence of a perichondrium. Chondrocytes are in lacunae and may be either scattered or stacked in small groups. For additional practice, complete the ***Sketch to Learn*** activity (on p. 98).

6. ***Draw It!*** Draw the tissue in the space provided.

Fibrocartilage

7. Scan the bone slide at each magnification to observe the organization of an osteon with concentric lamellae around a central canal.

8. ***Draw It!*** Draw the tissue in the space provided. ■

Bone

Sketch to Learn

Let's draw elastic cartilage. Observe the tissue with the microscope at each magnification but use a magnification that includes as many structures as possible as shown in the sketch provided.

Sample Sketch

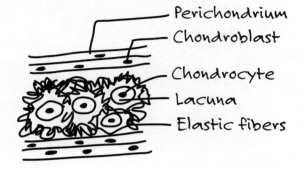

Step 1
• Draw some wavy lines for perichondrium.

Step 2
• Add flat dots for chondroblasts.

Step 3
• In the middle of the sketch, draw circles with dots and add a bunch of scribbly lines. Add labels, too.

Your Sketch

Name _____

Date _____

Section _____

Connective Tissue

A. Definitions

Define each term.

1. collagen fiber

2. perichondrium

3. osteon

4. lacuna

5. fibroblast

6. matrix

7. elastic fiber

8. ground substance

9. mast cell

10. canaliculi

B. Drawing

Draw It! Draw and label areolar tissue and hyaline cartilage as viewed with a microscope at medium magnification.

Areolar tissue   Hyaline cartilage

C. Labeling

Label the following in **Figure 8.9**.

1. Identify the tissue. _____

2. Identify the thick line. _____

3. Identify the thin line. _____

4. Identify the cell. _____

5. Identify the tissue. _____

6. Identify the cell. _____

7. Identify the clear cell structure. _____

8. Identify the pink substance. _____

9. Identify the tissue. _____

10. Identify the membrane. _____

11. Identify the cell. _____

12. Identify the cell and space. _____

13. Identify the tissue. _____

14. Identify the thin lines. _____

15. Identify the tissue. _____

16. Identify the hole. _____

17. Identify the ring of bone. _____

18. Identify the entire bony column. _____

19. Identify the tissue. _____

20. Identify the cell. _____

21. Identify the thick bands. _____

Figure 8.9 **Connective Tissues** Label each connective tissue.

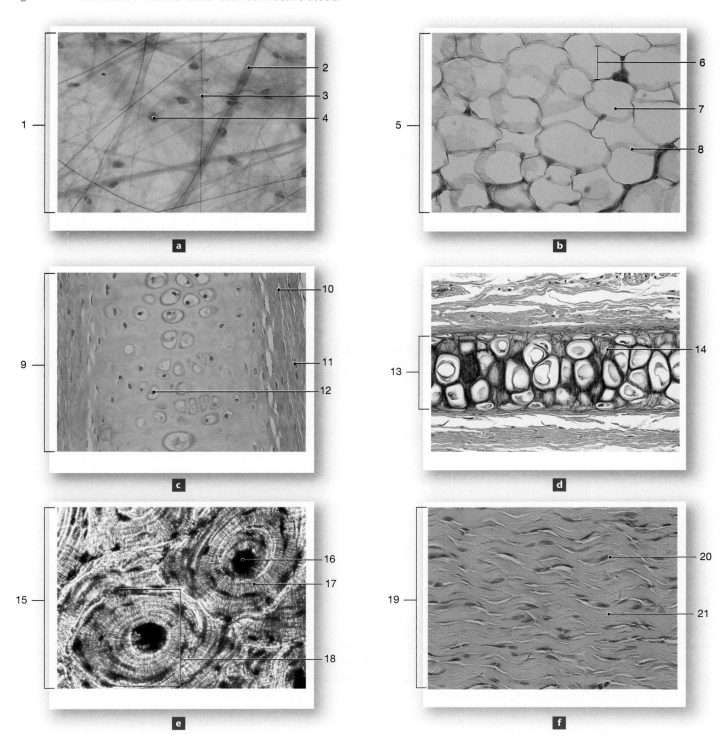

D. Short-Answer Questions

1. Which tissue in the embryo is a precursor of adult connective tissue?

2. What type of fiber is embedded in loose connective tissue?

3. What is the matrix composed of in elastic cartilage?

4. List the three major groups of connective tissue, and give an example of each.

E. Application and Analysis

1. What are the structural and functional differences between dense regular and dense irregular connective tissues?

2. How does connective tissue differ from epithelial tissue?

3. Describe the ground substance and fibers found in each type of connective tissue.

F. Clinical Challenge

The surgical procedure called *liposuction* removes unwanted adipose tissue with a suction wand. The treatment is dangerous and may damage blood vessels or nerves near the site of fat removal. Overlying skin may appear pocketed and marbled after the procedure. Considering that adult connective tissues contain some mesenchyme cells, what would most likely occur if a liposuction patient continued an unhealthy lifestyle of little exercise and poor diet choices?

Muscle Tissue

Learning Outcomes

On completion of this exercise, you should be able to:

1. List the three types of muscle tissue and describe a function of each.

2. Describe the histological appearance of each type of muscle tissue.

3. Identify each type of muscle tissue in microscope preparations.

There are three types of muscle tissue, each named for its location in the body (Figure 9.1). **Skeletal muscle** is attached to bone and provides the means by which the body skeleton moves, as in walking or moving the head. **Cardiac muscle** forms the walls of the heart and pumps blood through the vascular system. **Smooth,** or **visceral, muscle** is found inside hollow organs, such as the stomach, intestines, blood vessels, and uterus; this muscle type controls such functions as the movement of material through the digestive system, the diameter of blood vessels, and uterine contraction during labor.

Muscle tissue specializes in contraction. Muscle cells shorten during contraction, and this shortening produces a force, or tension, that causes movement. During the contraction phase of a heartbeat, for example, blood is pumped into blood vessels. The pressure generated by cardiac muscle contraction forces blood to flow through the vascular system to supply cells with oxygen, nutrients, and other essential materials.

Make a Prediction

Muscle tissue contraction is regulated by either voluntary or involuntary control. Which type of nerve control regulates skeletal muscles?

Need More Practice and Review?

Build your knowledge—and confidence!—in the Study Area of MasteringA&P® at www.masteringaandp.com with Pre-lab Quizzes, Post-lab Quizzes, Practice Anatomy Lab™ (PAL™) 3.0 virtual anatomy practice tool, PhysioEx™ 9.0 laboratory simulations, and A&P Flix™ with Quizzes.

PAL ⟨practice anatomy lab⟩ For this lab exercise, follow this navigation path:

• PAL>Histology>Muscle Tissue

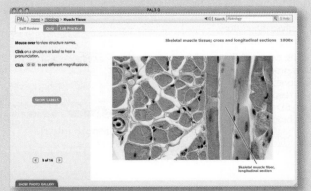

Figure 9.1 **Muscle Tissue**

Skeletal Muscle Tissue

Cells are long, cylindrical, striated, and multinucleate.

LOCATIONS: Combined with connective tissues and neural tissue in skeletal muscles

FUNCTIONS: Moves or stabilizes the position of the skeleton; guards entrances and exits to the digestive, respiratory, and urinary tracts; generates heat; protects internal organs

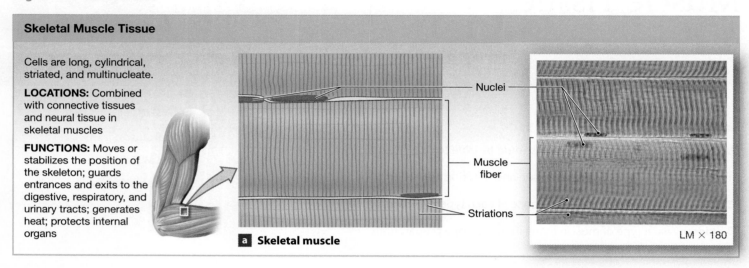

a Skeletal muscle

Nuclei

Muscle fiber

Striations

LM × 180

Cardiac Muscle Tissue

Cells are short, branched, and striated, usually with a single nucleus; cells are interconnected by intercalated discs.

LOCATION: Heart

FUNCTIONS: Circulates blood; maintains blood (hydrostatic) pressure

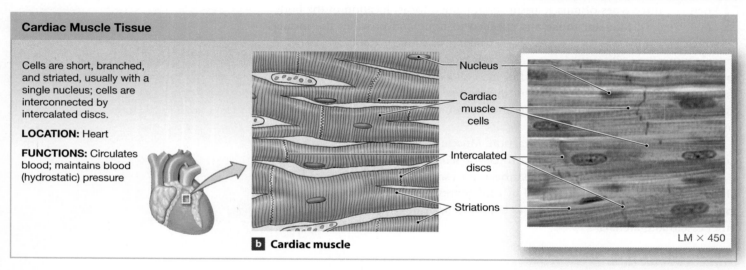

b Cardiac muscle

Nucleus

Cardiac muscle cells

Intercalated discs

Striations

LM × 450

Smooth Muscle Tissue

Cells are short, spindle-shaped, and nonstriated, with a single, central nucleus.

LOCATIONS: Found in the walls of blood vessels and in digestive, respiratory, urinary, and reproductive organs

FUNCTIONS: Moves food, urine, and reproductive tract secretions; controls diameter of respiratory passageways; regulates diameter of blood vessels

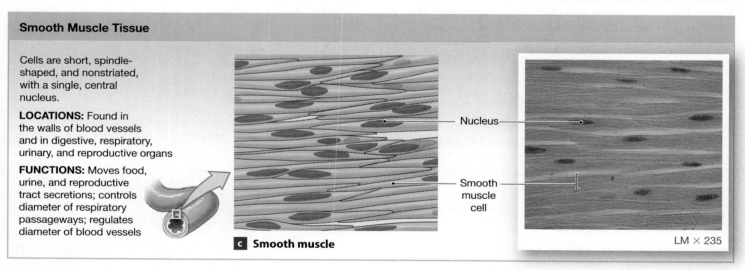

c Smooth muscle

Nucleus

Smooth muscle cell

LM × 235

Lab Activity 1 Skeletal Muscle

Skeletal muscle tissue (**Figure 9.2**) is attached to bones of the skeleton by **tendons** made of dense regular connective tissue proper. When skeletal muscle tissue contracts, it pulls on a tendon that, in turn, pulls and moves a bone. The functions of skeletal muscle tissue include movement for locomotion, facial expressions, and speech; maintenance of body posture and tone; and heat production during shivering.

Skeletal muscle tissue is composed of long cells called **muscle fibers.** During development, a number of embryonic cells called **myoblasts** (*myo-*, muscle + *-blast*, precursor) fuse into one large cellular structure that is the muscle fiber; because each fiber forms from numerous myoblasts, it has many nuclei and is said to be **multinucleated.** The nuclei are clustered under the **sarcolemma** (sar-cō-LEM-uh; *sarco*, flesh), which is the muscle fiber's cell membrane. Muscle fibers are **striated** with a distinct banded pattern resulting from the repeating organization of internal contractile proteins called **filaments.** Skeletal muscle tissue may be consciously stimulated to contract and is therefore under **voluntary control.**

Figure 9.2 Skeletal Muscle Tissue Skeletal muscle tissue with several striated muscle fibers. Note the nuclei at the edges of the fibers.

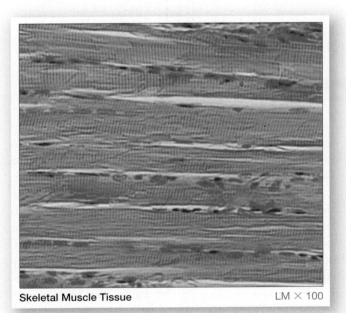

Skeletal Muscle Tissue   LM × 100

QuickCheck Questions

1.1 Where is skeletal muscle tissue located in the body?

1.2 What are the functions of skeletal muscle tissue?

In the Lab 1

Materials

☐ Compound microscope

☐ Skeletal muscle slide (striated muscle or voluntary muscle)

Procedures

1. Place the slide of skeletal muscle tissue on the microscope stage and move the low-power objective lens into position. Rotate the coarse adjustment knob to bring the image into focus.

2. Using Figures 9.1 and 9.2 for reference, examine the tissue at low magnification.

   ▪ Identify an individual skeletal muscle fiber.

   ▪ How many nuclei does it have?

3. Change to medium magnification and observe the skeletal muscle fibers again.

   ▪ Can you see striations across the fibers?

   ▪ How are the nuclei positioned in the fibers?

4. If both transverse and longitudinal sections are on your slide, compare the appearance of the skeletal muscle fibers in the two sections.

   ▪ How do the muscle fibers appear in transverse section?

   ▪ How are the nuclei positioned in the fibers?

Sketch to Learn

It is time to sketch a few skeletal muscle fibers as seen in Figure 9.2. Take a look at skeletal muscle tissue with a microscope; we'll draw the tissue at approximately medium power. Take your time and plan your sketch, and importantly, enjoy the creative process of drawing and self-expression. Let's draw!

Sample Sketch

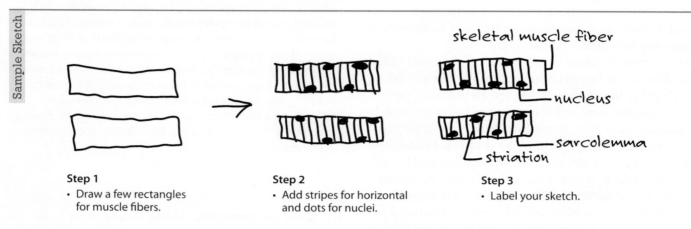

Step 1
- Draw a few rectangles for muscle fibers.

Step 2
- Add stripes for horizontal and dots for nuclei.

Step 3
- Label your sketch.

Your Sketch

5. **Draw It!** Draw and label the microscopic structure of skeletal muscle tissue in the space provided. For additional practice, complete the *Sketch to Learn* activity above. ■

Skeletal muscle tissue

Lab Activity 2   Cardiac Muscle

Cardiac muscle tissue (**Figure 9.3**) occurs only in the walls of the heart. Compare the cardiac and skeletal muscle tissues in Figures 9.1, 9.2, and 9.3. Cardiac muscle tissue is striated like skeletal muscle tissue. Each **cardiac muscle cell,** also called a **cardiocyte,** has a single nucleus (the cell is said to be **uninucleated**) and is branched. Cardiocytes are connected to one another by **intercalated discs,** special gap junctions that conduct contraction stimuli from one cardiocyte to the next. Unlike skeletal muscles, cardiac muscle is under **involuntary control.** For example, when you exercise, nerves of the autonomic nervous system cause an increase in your heart rate in order to deliver more blood to the active tissues. When you relax or sleep, the autonomic nervous system lowers your heart rate.

QuickCheck Questions

2.1 What is the function of cardiac muscle tissue?

2.2 How are cardiocytes connected to one another?

Figure 9.3 **Cardiac Muscle Tissue** The main feature of cardiac muscle tissue is the intercalated discs that connect the cardiocytes. Each cardiocyte is striated, is uninucleated, and branches to join with other cells.

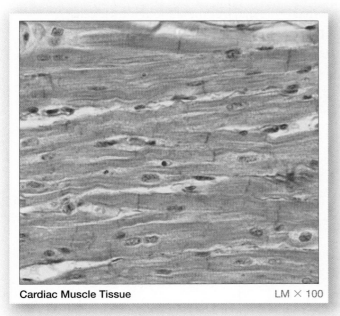

Cardiac Muscle Tissue   LM × 100

Materials

☐ Compound microscope
☐ Cardiac muscle slide

Procedures

1. Place the slide of cardiac muscle tissue on the microscope stage and swing the low-power objective lens into position. Rotate the coarse adjustment knob to bring the image into focus.

2. Using Figures 9.1 and 9.3 for reference, examine the heart muscle at low and medium magnifications. If your cardiac muscle slide has different sections, observe the longitudinal section first.

   ■ How many nuclei are in each cardiac muscle cell?

   ■ How do cardiac muscle cells compare in size with skeletal muscle fibers?

3. Increase the magnification to high and observe several cardiocytes.

   ■ Do you see striations and branching?

   ■ What structure connects adjacent cells?

4. ***Draw It!*** Draw and label the microscopic structure of cardiac muscle tissue in the space provided. ■

Cardiac muscle tissue

Lab Activity 3   **Smooth Muscle**

Figure 9.1c and Figure 9.4 show smooth muscle tissue. The muscle cells are **nonstriated** and lack the bands found in skeletal and cardiac muscle tissue. Each smooth muscle cell is **uninucleated** and spindle shaped, thick in the middle and tapered at the ends like a toothpick. The tissue usually occurs

Figure 9.4 **Smooth Muscle Tissue Teased** The smooth muscle tissue in this micrograph has been teased apart to highlight individual cells. Smooth muscle cells are spindle shaped (pointed on both ends like a toothpick) and have a central nucleus. They do not branch or have striations.

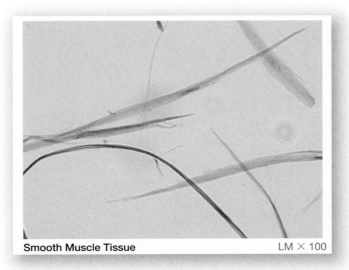

Smooth Muscle Tissue   LM × 100

in double sheets of muscle with one sheet positioned at a right angle to the other. This arrangement permits the tissue to shorten structures and decrease the diameter of vessels and passageways. Smooth muscle is under involuntary control.

QuickCheck Questions

3.1 Where is smooth muscle tissue located in the body?

3.2 How are smooth muscle cells different from skeletal muscle fibers and cardiac muscle cells?

In the Lab 3

Materials

☐ Compound microscope

☐ Smooth muscle slide (visceral muscle; tissue may be teased apart)

Procedures

1. Place the slide of smooth muscle tissue on the microscope stage and swing the low-power objective lens into position. Rotate the coarse adjustment knob to bring the image into focus.

2. Using Figures 9.1 and 9.4 for reference, examine the smooth muscle tissue at low magnification. Figure 9.4 shows smooth muscle that has been teased apart to separate the cells.

3. Locate the smooth muscle cells and center them in the microscope field. Increase the magnification to medium power.

   ▪ Where is the nucleus located in a typical cell?

   ▪ What is the shape of the smooth muscle cells?

   ▪ Do you see any striations?

4. ***Draw It!*** Draw and label the microscopic structure of smooth muscle tissue in the space provided. ▪

Smooth muscle tissue

Name _____

Date _____

Muscle Tissue

Section _____

A. Matching

Match each structure in the left column with its correct description from the right column.
Each term in the right column may be used more than once.

_____ **1.** muscle fiber membrane

_____ **2.** cellular connections between cardiocytes

_____ **3.** striated, uninucleated cells

_____ **4.** muscle tissue in tip of tongue

_____ **5.** muscle tissue in artery

_____ **6.** voluntary muscle

_____ **7.** nonstriated cells

_____ **8.** involuntary, striated cells

A. sarcolemma

B. intercalated disc

C. cardiac muscle

D. skeletal muscle

E. smooth muscle

B. Drawing

1. *Draw It!* Draw and label cardiac muscle tissue as you observed it at medium magnification.

2. *Draw It!* Draw and label skeletal muscle tissue at medium magnification.

C. Short-Answer Questions

    1. Which types of muscle tissue are striated?

    2. Where in the body does smooth muscle occur?

    3. What is the function of intercalated discs in cardiac muscle?

    4. Which muscle tissues are controlled involuntarily?

D. Analysis and Application

    1. Describe how skeletal muscle fibers become multinucleated.

    2. Give an example that illustrates the involuntary control of cardiac muscle tissue.

    3. How are smooth muscle cells similar to skeletal muscle fibers? How are they different from skeletal muscle fibers?

    4. How are skeletal and cardiac muscle tissues similar to each other? How do these two types of muscle tissue differ from each other?

Neural Tissue

Learning Outcomes

On completion of this exercise, you should be able to:

1. List the basic functions of neural tissue.

2. Describe the two basic types of cells found in neural tissue.

3. Identify a neuron and its basic structure under the microscope.

4. Describe how neurons communicate with other cells across a synapse.

5. List several functions of glial cells.

To maintain homeostasis, the body must constantly evaluate internal and external conditions and respond quickly and appropriately to environmental changes. **Sensory receptors** detect changes inside and around the body and send a constant stream of information to the central nervous system for processing and initiating motor adjustments to muscles and glands, the responders which are collectively called the body's **effectors.**

The nervous system processes information from sensory organs and responds with motor instructions. Cells responsible for receiving, interpreting, and sending the electrical signals of the nervous system are called **neurons.** Neurons are excitable, which means they can respond to environmental changes by processing stimuli into electrical impulses called **action potentials.** Sensory neurons detect changes in the environment and communicate these changes to the central nervous system (CNS), which consists of the brain and spinal cord. The CNS responds to the sensory input with motor commands to glands and muscle tissues. This constant monitoring and adjustment play a vital role in homeostasis.

Need More Practice and Review?

Build your knowledge—and confidence!—in the Study Area of MasteringA&P® at www.masteringaandp.com with Pre-lab Quizzes, Post-lab Quizzes, Practice Anatomy Lab™ (PAL™) 3.0 virtual anatomy practice tool, PhysioEx™ 9.0 laboratory simulations, and A&P Flix™ with Quizzes.

PAL | practice anatomy lab   For this lab exercise, follow this navigation path:
• PAL>Histology>Neural Tissue

MasteringA&P®

The most numerous cells in the nervous system are **glial cells** (*glia,* glue), which make up a network of cells and fibers called the **neuroglia.** Large populations of one type of glial cell will support and anchor neurons. Other types of glial cells wrap around neurons, creating a myelin sheath that greatly increases the communication speed of the neuron. Neurons and glial cells are collectively referred to as either *nerve tissue* or *neural tissue.* (The two terms are synonyms, and you will see both in textbooks and in scientific literature.)

Lab Activity 1 Neuron Structure

A typical neuron has distinct cellular regions. Examine **Figure 10.1** and locate the **cell body** surrounding the **nucleus.** This area, also known as the **soma,** contains most of the

neuron's organelles. Many fine extensions, called **dendrites** (DEN-drīts; *dendron,* a tree), receive information from other cells and send impulses toward the soma. The signal is then conducted into a single **axon** that carries information away from the soma, either to other neurons or to effector cells. At the end of the axon is the **synaptic terminal** that contains membranous **synaptic vesicles.** These vesicles contain **neurotransmitter molecules,** which are chemical messengers used either to excite or to inhibit other cells. The axon releases neurotransmitters onto an adjacent neuron across a specialized junction called the **synapse** (SIN-aps; *synap,* union). Notice in the bottom drawing of Figure 10.1 the three axons on the left synapsing on the dendrites of the larger neuron. At the synapse, cells do not touch; they are separated by a small **synaptic cleft.** When an action potential reaches the end of a *presynaptic* axon, the neurotransmitter molecules released by

Figure 10.1 Neural Tissue Neural tissue consists of two major types of cells: neurons and neuroglia cells. Neurons are the communicative cells of the nervous system that send electrical signals to other cells in the body. Neuroglia is a group of different kinds of supportive cells in the nervous system.

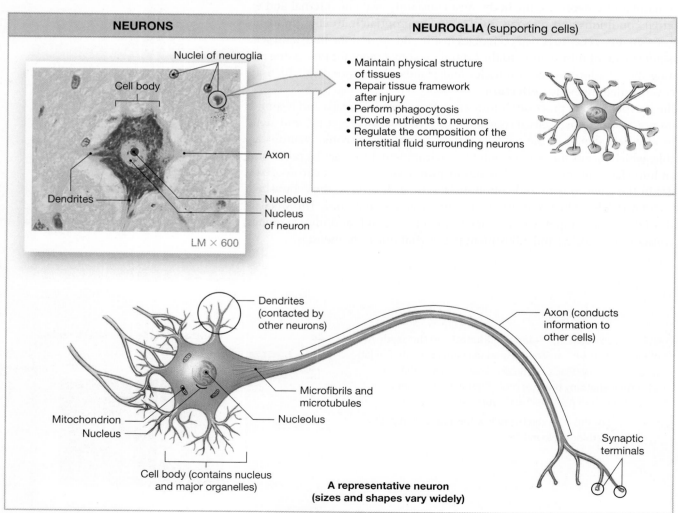

NEURONS	NEUROGLIA (supporting cells)

Nuclei of neuroglia
Cell body
Axon
Dendrites
Nucleolus
Nucleus of neuron
LM × 600

- Maintain physical structure of tissues
- Repair tissue framework after injury
- Perform phagocytosis
- Provide nutrients to neurons
- Regulate the composition of the interstitial fluid surrounding neurons

Dendrites (contacted by other neurons)
Axon (conducts information to other cells)
Microfibrils and microtubules
Mitochondrion
Nucleus
Nucleolus
Synaptic terminals
Cell body (contains nucleus and major organelles)
A representative neuron (sizes and shapes vary widely)

the synaptic vesicles diffuse across the synaptic cleft and either excite or inhibit the *postsynaptic* cell. A presynaptic neuron may communicate across a synapse to either a neuron, a muscle cell, or a glandular cell.

Study Tip Identifying Axons

On most neuron slides, it is difficult to distinguish axons from dendrites. Locate one neuron that is isolated from the others and examine the soma for a large extension. This is most likely the axon. ■

QuickCheck Questions

1.1 What part of a neuron sends information to another neuron?

1.2 In general, how do neurons communicate with other cells?

In the Lab 1

Materials

☐ Compound microscope

☐ Prepared microscope slide of neurons

Procedures

1. Scan the slide at low magnification to locate the neurons. Select one neuron to examine more closely. Center this neuron in the field of view and increase the magnification. Adjust the light setting of the microscope if necessary.

2. On the neuron you have chosen, identify the soma, nucleus, dendrites (thin extensions), and axon (thicker extension).

3. ***Draw It!*** Draw and label several neurons in the space provided. Refer to **Figure 10.2**. ■

Neurons

Figure 10.2 Neurons Neurons have three main structures: dendrites that receive information from other cells, a cell body with the nucleus, and an axon that sends signals to other cells.

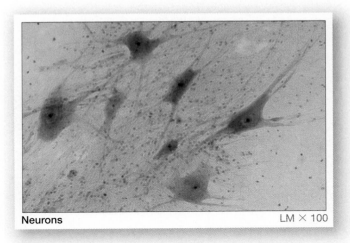

Neurons   LM × 100

Lab Activity 2   Neuroglia

The glial cells of the neuroglia are the supportive cells of the nervous system. There are six types of glial cells, each with a specific function. The glial cells of the CNS are astrocytes, microglia, ependymal cells, and oligodendrocytes. **Astrocytes** attach blood vessels to neurons or anchor neurons in place. Phagocytic **microglia** are responsible for housekeeping chores in the nervous system. **Ependymal cells** line the spaces of the brain and spinal cord; they assist in the production and circulation of cerebrospinal fluid. **Oligodendrocytes** protect neurons by wrapping around them to isolate them from chemicals present in the interstitial fluid.

The part of the nervous system outside the CNS is called the *peripheral nervous system*, and the glial cells here are Schwann cells and satellite cells. **Schwann cells** wrap around peripheral neurons to increase the speed at which they transmit action potentials. Repair of peripheral neural tissue, which is any neural tissue outside the brain and spinal cord, is made possible by Schwann cells that build a "repair tube" to reconnect the severed axons. **Satellite cells** help regulate the environment around peripheral neural tissue.

In this exercise you will examine the most common glial cell in the nervous system, the astrocyte, shown in **Figure 10.3**. Astrocytes are major structural cells of the brain and spinal cord and serve a variety of functions. They provide a framework to support neurons. Cytoplasmic extensions of astrocytes, called *feet*, wrap around capillaries and form the blood-brain barrier that protects the brain and regulates the composition of the extracellular fluid.

Figure 10.3 Astrocytes Astrocytes are neuroglia cells that hold neurons in position and help to regulate the extracellular fluid surrounding neurons.

Astrocytes LM × 100

QuickCheck Questions

2.1 What are the two major types of cells in neural tissue?

2.2 What are the major functions of astrocytes?

In the Lab 2

Materials

☐ Compound microscope
☐ Prepared microscope slide of astrocytes

Procedures

1. Move the low-power objective lens into position. Set the astrocyte slide on the stage and slowly turn the coarse adjustment knob until you can clearly see the specimen. Now use the fine focus adjustment as you examine the tissue.

2. Scan the slide and locate a star-shaped astrocyte. Center the cell in the field of view, and then increase the magnification to medium. Notice the numerous feet extending from the cell.

3. **Draw It!** Draw and label several astrocytes in the space provided. ■

Astrocytes

Name _____

Date _____

Section _____

Neural Tissue

A. Definitions

Define each term.

1. soma

2. synaptic vesicle

3. dendrite

4. neurotransmitter

5. axon

6. synaptic cleft

B. Short-Answer Questions

1. What are the basic functions of neural tissue?

2. How do neurons communicate with other cells?

3. Which part of a neuron conducts an impulse toward the soma?

4. In which direction does an action potential travel in an axon?

C. Labeling

Label the anatomy of the neuron in **Figure 10.4**.

Figure 10.4 Neural Tissue

What are the major functions of glial cells?

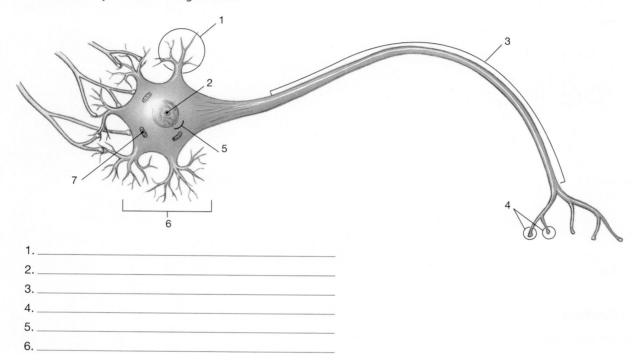

1. _____
2. _____
3. _____
4. _____
5. _____
6. _____
7. _____

D. Analysis and Application

1. List the cellular structures over which an action potential travels, starting at the dendrites and traveling to where neurotransmitter molecules are released.

2. What are the major functions of neuroglial cells?

Integumentary System

Learning Outcomes

On completion of this exercise, you should be able to:

1. Identify the two layers of the skin.

2. Identify the layers of the epidermis.

3. Distinguish between the papillary and reticular layers of the dermis.

4. Identify the accessory structures of the skin.

5. Identify a hair follicle, the parts of a hair, and an arrector pili muscle.

6. Distinguish between sebaceous and sweat glands.

7. Describe three sensory organs of the integument.

The **integumentary** (in-TEG-ū-MEN-ta-ree) **system** is the most visible organ system of the human body. The integument (in-TEG-ū-ment), or skin, is classified as an organ system because it is composed of many different types of tissues and organs. Organs of the skin include oil-, wax-, and sweat-producing glands; sensory organs for touch; muscles attached to hair follicles; and blood and lymphatic vessels.

The integument seals the body in a protective barrier that is flexible yet resistant to abrasion and evaporative water loss. People interact with the external environment with the skin. Caressing a baby's head, feeling the texture of granite, and testing the temperature of bath water all involve sensory organs of the integumentary system. Sweat glands in the skin cool the body to regulate body temperature. When exposed to sunlight, the integument manufactures vitamin D_3, a vitamin essential in calcium and phosphorus balance.

Need More Practice and Review?

Build your knowledge—and confidence!—in the Study Area of MasteringA&P® at www.masteringaandp.com with Pre-lab Quizzes, Post-lab Quizzes, Practice Anatomy Lab™ (PAL™) 3.0 virtual anatomy practice tool, PhysioEx™ 9.0 laboratory simulations, and A&P Flix™ with Quizzes.

PAL | practice anatomy lab   For this lab exercise, follow these navigation paths:
- PAL>Anatomical Models>Integumentary System
- PAL>Histology>Integumentary System

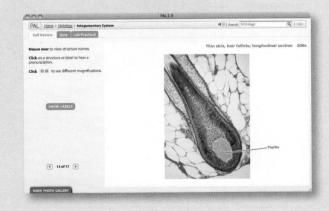

There are two principal tissue layers in the integument: a superficial layer of epithelium called the *epidermis* and a deeper layer of connective tissue, the *dermis* (Figure 11.1). The **epidermis** consists of a stratified squamous epithelium organized into many distinct layers, or *strata*, of cells, as shown in Figure 11.2. Thick-skinned areas, such as the palms of the hands and soles of the feet, have five layers; thin-skinned areas have only four. Cells called **keratinocytes** are produced deep in the epidermis and pushed superficially toward the surface of the skin. It takes from 15 to 30 days for a cell to migrate from the basal region to the surface of the epidermis. During this migration, the keratinocytes synthesize and accumulate the protein keratin, the internal organization of the cell is disrupted, and the cells die. These dry, scalelike keratinized cells on the surface of the stratified squamous epidermis are resistant to dehydration and friction. Because of these characteristics the integument is also called the **cutaneous membrane.**

Moving superficially from the basal lamina, the five layers of the epidermis are as follows:

- The **stratum basale** (STRA-tum bah-SA-le), or the **stratum germinativum** (STRA-tum jer-mi-na-TĒ-vum), is a layer just one cell thick that joins the basal lamina of the epidermis to the upper surface of the dermis. The cells in this stratum are stem cells and so are in a constant state of mitosis, replacing cells that have rubbed off the epidermal surface. Other cells in this layer, called **melanocytes,** produce the pigment **melanin** (MEL-ā-nin), which protects deeper cells from the harmful effects of ultraviolet (UV) radiation from the sun. Prolonged exposure to UV light causes an increase in melanin synthesis, resulting in a darkening, or tanning, of the integument.

- Superficial to the stratum germinativum is the **stratum spinosum,** which consists of five to seven layers of cells, interconnected by strong protein molecules between cell membranes, forming cell attachments called **desmosomes.** When a slide of epidermal tissue is being prepared, cells in

Figure 11.1 Components of the Integumentary System This diagrammatic view of skin illustrates the relationships among the epidermis, dermis, and accessory structures of the integumentary system (with the exception of nails, shown in Figure 11.7).

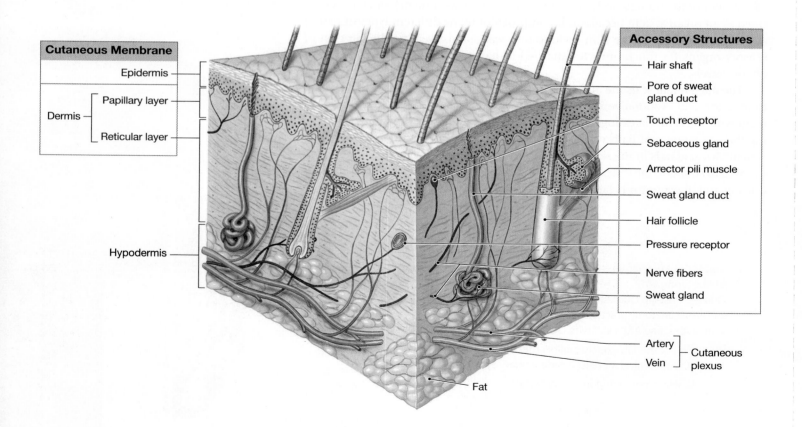

Cutaneous Membrane
- Epidermis
- Dermis
  - Papillary layer
  - Reticular layer
- Hypodermis

Accessory Structures
- Hair shaft
- Pore of sweat gland duct
- Touch receptor
- Sebaceous gland
- Arrector pili muscle
- Sweat gland duct
- Hair follicle
- Pressure receptor
- Nerve fibers
- Sweat gland
- Artery ⎤ Cutaneous
- Vein ⎦ plexus
- Fat

Figure 11.2 **Organization of the Epidermis**

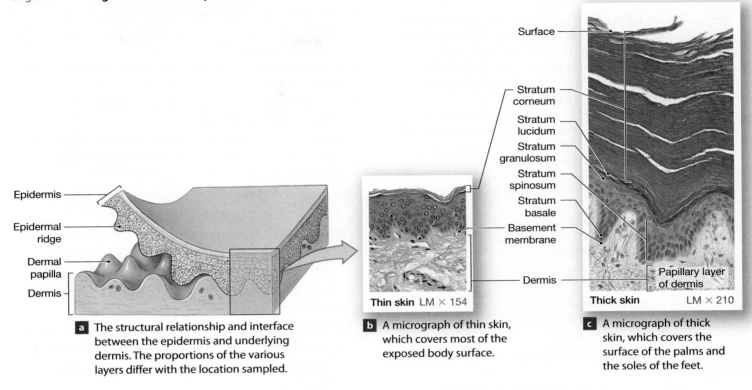

a The structural relationship and interface between the epidermis and underlying dermis. The proportions of the various layers differ with the location sampled.

b A micrograph of thin skin, which covers most of the exposed body surface.

c A micrograph of thick skin, which covers the surface of the palms and the soles of the feet.

this layer often shrink, but the desmosome bridges between cells remain intact. This results in cells with a spiny outline; hence the name "spinosum."

- Superficial to the stratum spinosum is a layer of darker cells that make up the **stratum granulosum.** As cells from the stratum germinativum are pushed superficially, they synthesize the protein **keratohyalin** (ker-a-tō-HĪ-a-lin), which increases durability and reduces water loss from the integument surface. Keratohyalin granules stain dark and give this layer its color.

- In thick skin, a thin, transparent layer of cells called the **stratum lucidum** lies superficial to the stratum granulosum. Only the thick skin of the palms and the soles of the feet have the stratum lucidum; the rest of the skin is considered thin and lacks this layer.

- The **stratum corneum** (KOR-nē-um; *cornu,* horn) is the most superficial layer of the epidermis and contains many layers of flattened, dead cells. As cells from the stratum granulosum migrate superficially, keratohyalin granules are converted to the fibrous protein keratin. Cells in the stratum corneum also accumulate the yellow-orange pigment **carotene,** which is common in light-skinned individuals.

Deep to the epidermis is the second of the two layers of the integument, the **dermis,** a thick layer of irregularly arranged connective tissue that supports and nourishes the epidermis

and secures the integument to the underlying structures (Figure 11.1). The dermis is divided into two layers: *papillary* and *reticular.* Although there is no distinct boundary between these layers, the superficial portion of the dermis is designated the **papillary layer.** It consists of areolar tissue containing numerous collagen and elastic fibers. Folds in the tissue are called **dermal papillae** (pa-PIL-la; *papilla,* a small cone) and project into the epidermis as the swirls of fingerprints. Within the dermal papillae are small sensory receptors for light touch, movement, and vibration, termed **tactile corpuscles** (also called *Meissner's corpuscles*).

Deep to the papillary layer is the **reticular layer** of the dermis. This layer is distinguished by a meshwork of thick bands of collagen fibers in dense irregular connective tissue. Hair follicles and glands from the epidermis penetrate deep into the reticular layer. Sensory receptors in this layer, called **lamallated corpuscles** (*Pacinian corpuscles*), detect deep pressure.

Attaching the dermis to underlying structures is the **hypodermis,** or **subcutaneous layer,** which is composed primarily of adipose tissue and areolar tissue. *Cutaneous membrane* is yet another name for the skin, thus the name *subcutaneous* for this layer. The hypodermis is not part of the integumentary system.

QuickCheck Questions

1.1 Describe the two layers of the skin.

1.2 Why does the epidermis constantly replace its cells?

Clinical Application Skin Cancer

Skin cancer can be deadly. Protect yourself and your loved ones' skin with sunscreen and use common sense when out in the sun. Sunburns greatly increase the chances of getting cancer. During summer, some people can start to burn in just 20 minutes! Know the warning signs for skin cancer and self-examine your skin on a regular basis.

Basal cell carcinoma (Figure 11.3) is a tumor starting in stem cells in the stratum germinativum. Approximately 65% of the tumors occur in areas of skin exposed to excessive UV light (too much sun). Basal cell carcinoma is the most common form of skin cancer. It rarely spreads and there is a very high survivor rate. Squamous cell carcinoma only occurs in areas with high UV exposure.

Malignant melanoma is an extremely dangerous malignant tumor of melanocytes. Cancer cells grow rapidly and metastasize through the lymphatic system which drains into the bloodstream. This type has only a 14% survival rate if widespread in the lymph.

The ABCDs of Skin Cancer—Warning Signs

A = *asymmetry,* the skin tumor has an uneven shape and may bleed.

B = *border,* the edge is irregular instead of round and smooth.

C = *color,* many colors in a spot may indicate skin cancer.

D = *diameter,* 5 mm or larger is dangerous. ■

Figure 11.3 **Skin Cancers**

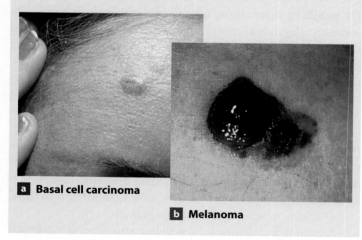

a **Basal cell carcinoma**

b **Melanoma**

In the Lab 1

Materials

☐ Skin model

☐ Compound microscope

☐ Prepared slide of the scalp (cross section)

Procedures

1. Examine a skin model and identify the epidermis, dermis, and hypodermis. Identify the specific layers of the epidermis and dermis.

2. Place the scalp slide on the microscope and focus on the specimen at low magnification.

3. Scan the slide vertically and identify the epidermis, dermis, and hypodermis.

4. Increase the magnification to medium and examine the epidermis. Locate the epidermal layers, beginning with the deepest layer, the stratum germinativum.

   ■ What is the shape of cells in the stratum spinosum?

   ■ What color is the stratum granulosum?

   ■ Does the scalp specimen have a stratum lucidum?

   ■ What is the top layer of cells called? Are these cells alive?

5. Study the dermis at low, medium, and high magnifications.

   ■ Distinguish between the papillary and reticular layers.

   ■ Are Meissner's corpuscles visible at the papillary folds?

   ■ What type of connective tissue is in the reticular layer? ■

Lab Activity 2 Accessory Structures of the Skin

During embryonic development, the epidermis produces accessory integumentary structures called **epidermal derivatives,** which include oil and sweat glands, hair, and nails. These structures are exposed on the surface of the skin and project deep into the dermis.

■ **Sebaceous** (se-BĀ-shus) **glands** are associated with hair follicles and secrete the oily substance **sebum,** which coats the hair shafts and the epidermal surface to reduce brittleness and prevents excessive drying of the integument (Figure 11.4). **Sebaceous follicles** secrete sebum onto the surface of the skin to lubricate the skin and provide limited antibacterial action. These follicles are not associated with hair and are distributed on the face, most of the trunk, and the male reproductive organs.

■ **Sweat glands,** or **sudoriferous** (sū-dor-IF-er-us) **glands,** are scattered throughout the dermis of most of the integument. They are exocrine glands that secrete their liquid either into sweat ducts leading to the skin surface or into sweat ducts leading to hair follicles (Figure 11.5).

Figure 11.4 The Structure of Sebaceous Glands and Sebaceous Follicles Sebaceous glands empty their oil (sebum) into hair follicles; sebaceous follicles secrete sebum onto the surface of the skin.

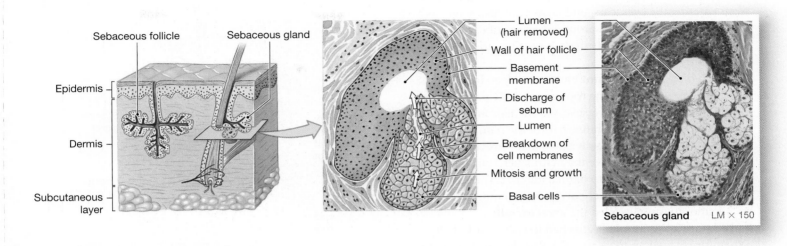

Figure 11.5 Sweat Glands

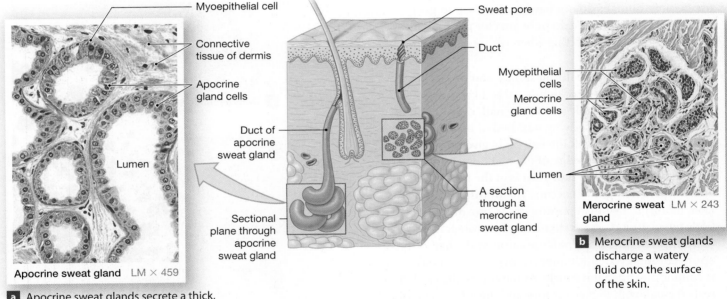

a Apocrine sweat glands secrete a thick, odorous fluid into hair follicles.

b Merocrine sweat glands discharge a watery fluid onto the surface of the skin.

The liquid we call **sweat** can be a thick or a thin substance. To cool the body, **merocrine** (MER-ō-krin) **sweat glands** secrete onto the body surface a thin sweat containing electrolytes, proteins, urea, and other compounds. The sweat absorbs body heat and evaporates from the skin, cooling the body. It also contributes to body odor because of the presence of urea and other wastes. Merocrine glands, also called *eccrine* (EK-rin) *glands*, are not associated with hair follicles and are distributed throughout most of the skin. **Apocrine sweat glands** are found in the groin, nipples, and axillae. These glands secrete a thick sweat into ducts associated with hair follicles. Bacteria on the hair metabolize the sweat and produce the characteristic body odor of, for example, axillary sweat.

- **Hair** covers most of the skin, with only the lips, nipples, portions of the external genitalia, soles, palms, fingers, and toes being without hair. Three major types of hair are found in humans. **Terminal hairs** are the thick, heavy hairs on the scalp, eyebrows, and eyelashes. **Vellus hairs** are lightly pigmented and distributed over much of the skin as fine "peach fuzz." **Intermediate hairs** are the hairs on the arms and legs. Hair generally serves a protective function. It cushions the scalp and prevents foreign objects from entering the eyes, ears, and nose. Hair also serves as a sensory receptor. Wrapped around the base of each hair is a **root hair plexus,** a sensory neuron sensitive to movement of the hair.

Each hair has a **hair root** embedded deep in a hair follicle (**Figure 11.6**). At the root tip is a **hair papilla** containing nerves, blood vessels, and the hair **matrix,** which is the living, proliferative part of the hair. Cells in the matrix undergo mitotic divisions that cause the hair to elongate (it "grows"). Above the matrix, keratinization of the hair cells causes them to harden and die. The resulting **hair shaft** contains an outer **cortex** and an inner **medulla.**

A smooth muscle called the **arrector pili** (a-REK-tor PI-lē) **muscle** is attached to each hair follicle. When fur-covered animals are cold, this muscle contracts to raise the hair and trap a layer of warm air next to the skin. In humans, the muscle has no known thermoregulatory use because humans do not have enough hair to gain an insulation benefit. We do have arrector pili muscles, though, and their contracting when we are cold is what produces "gooseflesh."

- **Nails,** which protect the dorsal surface and tips of the fingers and toes, consist of tightly packed keratinized cells (**Figure 11.7**). The visible part of the nail, called the **nail body,** protects the underlying **nail bed** of the integument. Blood vessels underneath the nail body give the nail its pinkish color. The **free edge** of the nail body extends past the end of the digit. The **nail root** is at the base of the nail and is where new growth occurs. The **lunula** (LOO-nū-la; *luna,* moon) is a whitish portion of the proximal nail body where blood vessels do not show through the layer of keratinized cells. The epidermis around the nail is the **eponychium** (ep-ō-NIK-ē-um; *epi,* over + *onyx,* nail), what is commonly called the cuticle. At the cuticle the epidermis seals the nail with the nail groove and the raised nail fold. Under the free edge of the nail is the **hyponychium** (hī-pō-NIK-ē-um), a thicker region of the epidermis.

Clinical Application Acne

Many teenagers have dealt with skin blemishes called *acne.* During puberty, hormone levels increase, and sebaceous glands are activated to produce more sebum. If a gland's duct becomes blocked, sebum accumulates and causes inflammation, resulting in a pimple. A pimple with a white head indicates that a duct is blocked and full of sebum. A black head forms when an open sebaceous duct contains solid material infected with bacteria. ■

Figure 11.6 Structure of a Hair

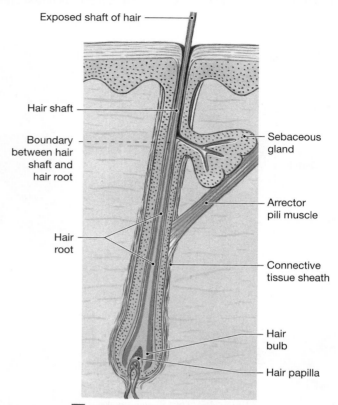

a Diagrammatic view of hair follicle

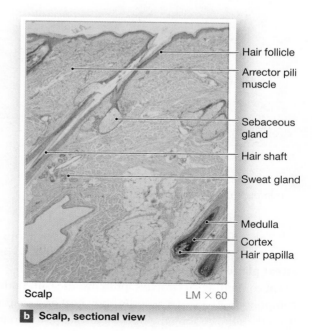

Scalp   LM × 60

b Scalp, sectional view

Figure 11.7 Structure of a Nail The prominent features of a typical fingernail.

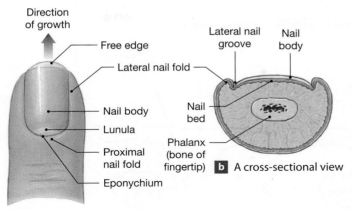

Direction of growth

Free edge

Lateral nail fold

Nail body

Lunula

Proximal nail fold

Eponychium

a A superficial view

Lateral nail groove

Nail body

Nail bed

Phalanx (bone of fingertip)

b A cross-sectional view

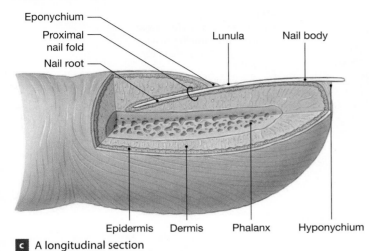

Eponychium

Proximal nail fold

Nail root

Lunula

Nail body

Epidermis   Dermis   Phalanx   Hyponychium

c A longitudinal section

QuickCheck Questions

2.1 List the accessory structures of the skin.

2.2 What are the two major types of glands in the skin?

Clinical Application Burns

Burns are classified by the damage they cause to the layers of the integument. First-degree burns injure cells of the epidermis, no deeper than the stratum germinativum. Sunburns and other topical burns are first-degree burns. Second-degree burns destroy the entire epidermis and portions of the dermis but do not injure hair follicles and glands in the dermis. This destruction of portions of the dermis causes blistering, and the wound is extremely painful. Third-degree burns penetrate completely through the integument, severely damaging epidermis, dermis, and subcutaneous structures. This type of wound cannot heal because the restorative layers of the epidermis are lost. To prevent infection in cases of third-degree burns and to reestablish the barrier formed by the skin, a skin graft is used to cover the wound. Nerves are usually damaged by third-degree burns, with the result that these more serious burns may not be as painful as first- and second-degree burns. ■

In the Lab 2

Materials

☐ Compound microscope

☐ Prepared slide of the scalp (cross section)

Procedures

1. On the scalp slide, locate a hair follicle.
  ■ What is the shape of the hair follicle? In which layer of the skin is it found?
  ■ Identify the hair shaft, cortex, and medulla.
  ■ Identify a sebaceous gland. Where does it empty its secretions?

2. Scan the dermis of the slide for a sudoriferous gland.
  ■ Look for small oval sections of a duct and follow them from the gland to the surface of the skin.

For additional practice, complete the ***Sketch to Learn*** activity (on p. 124). ■

Sketch to Learn

To reinforce the study of the integument, sketch the major components of the skin. Include a hair follicle and sebaceous and sweat glands in your drawing.

Sample Sketch

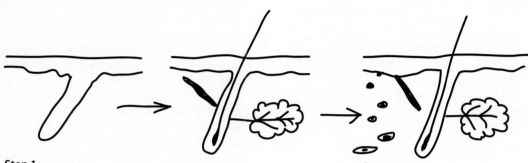

Step 1
- Draw a horizontal line for the surface of the skin.
- Add a wavy line with a long, thin pocket for a hair follicle.

Step 2
- Add a hair to the follicle.
- Attach an arrector pili muscle and a sebaceous gland to the follicle.

Step 3
- Draw a few ovals for sweat glands and smaller ovals leading to the surface for the duct.
- Label your sketch.

Your Sketch

Name _____

Date _____

Section _____

Integumentary System

A. Descriptions

Describe each skin structure.

1. sebaceous follicle

2. apocrine sweat gland

3. keratin

4. arrector pili

5. stratum corneum

6. stratum germinativum

7. reticular layer

8. subcutaneous layer

9. stratum lucidum

10. eccrine sweat gland

B. Short-Answer Questions

1. Describe the layers of epidermis in an area where the skin is thick.

2. How does the skin tan when exposed to sunlight?

3. List the types of sweat glands associated with the skin.

4. What is the function of arrector pili muscles in animals other than humans?

C. Labeling

Label the structures of the skin in **Figure 11.8**.

Figure 11.8 **Components of the Integumentary System**

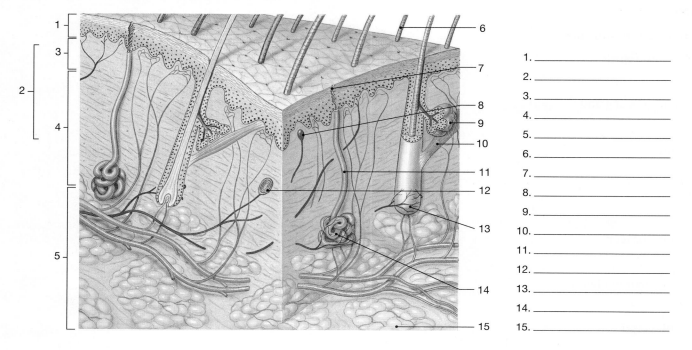

1. _____
2. _____
3. _____
4. _____
5. _____
6. _____
7. _____
8. _____
9. _____
10. _____
11. _____
12. _____
13. _____
14. _____
15. _____

D. Analysis and Application

1. What important function does the keratinized epidermis serve?

2. What is the main cause of acne, and in which part of the skin does it occur?

3. How are cells replaced in the epidermis?

4. What is the difference between sebaceous glands and sebaceous follicles?

E. Clinical Challenge

A nurse notices a spot on a patient's skin. How can one distinguish between a mass of skin cancer and a mole or freckle on the skin?

Body Membranes

Learning Outcomes

On completion of this exercise, you should be able to:

1. List and provide examples of the four types of body membranes.

2. Discuss the components of each type of body membrane.

3. Describe the histological organization of each type of body membrane.

The term *membrane* refers to a variety of anatomical structures, but in this exercise, **body membrane** refers to any sheet of tissue that wraps around an organ. Some body membranes cover structures and isolate them from the surrounding anatomy; other body membranes act as a barrier to prevent infections from spreading from one organ to another. Most body membranes produce a liquid that keeps the cells on the exposed, or free, surface of the membrane moist. Absorption may occur across these moist membranes.

Body membranes are composed of epithelium and connective tissue. Epithelium occurs on the exposed surface of the membrane and is supported by underlying connective tissue. There are four major types of membranes, as shown in **Figure 12.1**, classified by location. **Mucous membranes** occur where an opening is exposed to the external environment; **serous membranes** wrap around organs in the ventral body cavities; the **cutaneous membrane** is the skin; and **synovial** (sin-Ō-vē-ul) **membranes,** which do not contain true epithelium, line the cavities of movable joints.

Need More Practice and Review?

Build your knowledge—and confidence!—in the Study Area of MasteringA&P® at www.masteringaandp.com with Pre-lab Quizzes, Post-lab Quizzes, Practice Anatomy Lab™ (PAL™) 3.0 virtual anatomy practice tool, PhysioEx™ 9.0 laboratory simulations, and A&P Flix™ with Quizzes.

PAL | practice anatomy lab™ For this lab exercise, follow these navigation paths:
- PAL>Histology>Digestive System
- PAL>Histology>Integumentary System

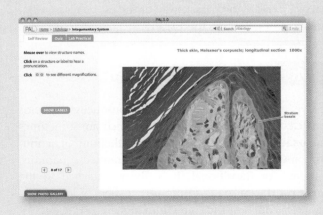

Figure 12.1 **Membranes**

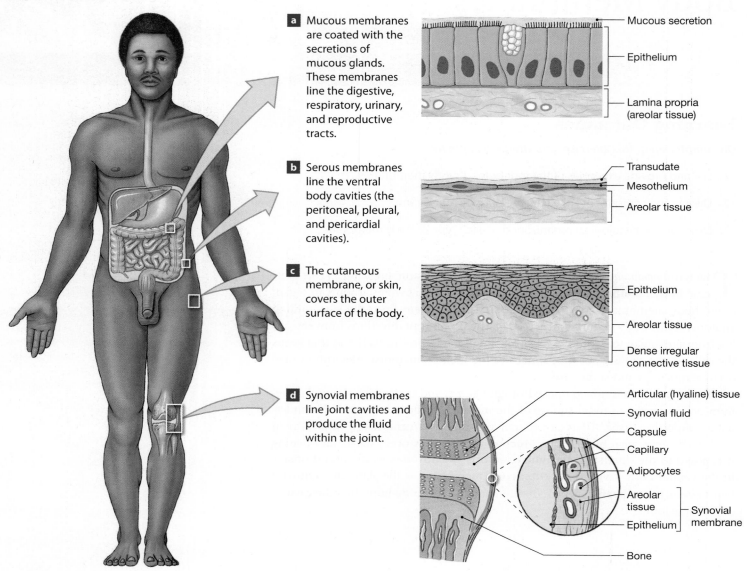

a Mucous membranes are coated with the secretions of mucous glands. These membranes line the digestive, respiratory, urinary, and reproductive tracts.

Mucous secretion
Epithelium
Lamina propria (areolar tissue)

b Serous membranes line the ventral body cavities (the peritoneal, pleural, and pericardial cavities).

Transudate
Mesothelium
Areolar tissue

c The cutaneous membrane, or skin, covers the outer surface of the body.

Epithelium
Areolar tissue
Dense irregular connective tissue

d Synovial membranes line joint cavities and produce the fluid within the joint.

Articular (hyaline) tissue
Synovial fluid
Capsule
Capillary
Adipocytes
Areolar tissue
Epithelium
} Synovial membrane
Bone

Make a Prediction

The esophagus is the muscular tube that connects the throat to the stomach. Consider the passageway of the esophagus—what type of body membrane lines this organ?

Lab Activity 1 Mucous Membranes

The digestive, respiratory, urinary, and reproductive systems are all protected by mucous membranes, which are sometimes called **mucosae** (singular *mucosa*). The epithelium of a mucous membrane may be simple columnar, stratified squamous, pseudostratified, or transitional. It is always attached to the **lamina propria** (LA-mi-na PRO-prē-uh), a sheet of loose connective tissue that anchors the epithelium in place (**Figure 12.2**).

Mucus, the thick liquid that protects the epithelium of a mucous membrane, is secreted either by goblet cells in the epithelium or by glands in the underlying **submucosa.** Mucus is viscous and contains a glycoprotein called **mucin,** salts, water, epithelial cells, and white blood cells. Mucin gives mucus its slippery and sticky attributes.

The mucous membrane of the digestive system has regional specializations. Where digestive contents are liquid, such as in the stomach and intestines, the epithelium of the mucous membrane is simple columnar with goblet cells. This mucous membrane functions in absorption and secretion. The mouth, pharynx, and rectum process either solid food or waste materials, both of which are abrasive to the epithelial lining. In these areas, the protective mucous membrane has a stratified squamous epithelium.

Figure 12.2 Mucous Membrane of the Esophagus

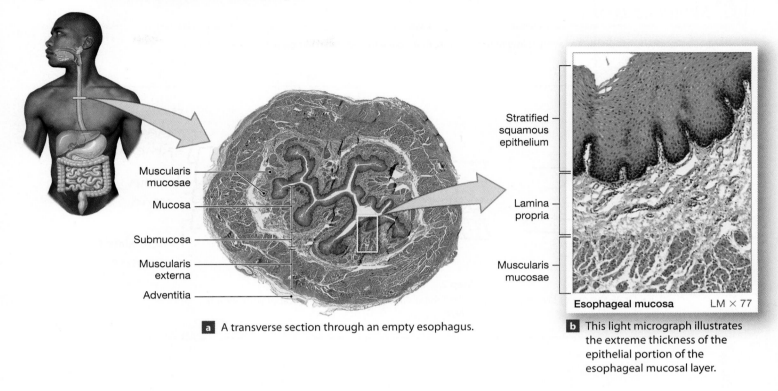

Muscularis mucosae
Mucosa
Submucosa
Muscularis externa
Adventitia

Stratified squamous epithelium
Lamina propria
Muscularis mucosae

Esophageal mucosa LM × 77

a A transverse section through an empty esophagus.

b This light micrograph illustrates the extreme thickness of the epithelial portion of the esophageal mucosal layer.

Most mucous membranes are constantly replacing epithelial cells. As materials move through a lumen, the exposed epithelial cells are scraped off the mucous membrane. Simple columnar cells in the small intestine, for instance, live for approximately 48 hours before they are replaced. The old cells are shed into the intestinal lumen and added to the feces.

The nasal cavity, trachea (windpipe), and large bronchi of the lungs are lined with a mucous membrane in which the epithelium is populated with goblet cells that secrete mucus to trap particulate matter in inhaled air. This epithelium is pseudostratified ciliated columnar and transports the dust-trapping mucus upward for removal from the respiratory system.

Portions of the male and female reproductive systems are lined with a ciliated mucous membrane to help in the transport of sperm or eggs.

The mucous membrane of the urinary bladder has a transitional epithelium that enables the bladder wall to stretch and recoil as the bladder fills and empties. Urine keeps the membrane moist and prevents dehydration.

QuickCheck Questions

1.1 Where do mucous membranes occur in the body?

1.2 What type of epithelium occurs in the mucous membrane of the nasal cavity?

In the Lab 1

Materials

☐ Compound microscope
☐ Prepared esophagus slide (transverse section)

Procedures

1. Place the slide on the microscope stage and focus at low magnification.
2. Move the slide to the center of the tissue section and locate the lumen, the open space of the esophagus.

- Notice the stratified squamous epithelium lining the folds along the wall. This is the epithelial part of the mucous membrane.
- Deep to the epithelium, note the lamina propria, which is mostly areolar connective tissue with numerous blood and lymphatic vessels.
- Deep to the lamina propria is a thick submucosal layer with esophageal glands that secrete mucus onto the surface of the esophagus. Scan the slide for a duct of an esophageal gland.

For additional practice, complete the *Sketch to Learn* activity (on p. 130). ■

Sketch to Learn

Let's sketch the esophagus as seen at low magnification. Refer back to your microscope with the esophagus slide at low power.

Sample Sketch

stratified
squamous
epithelium

lamina
propria

muscularis
mucosae

Step 1
• Draw the major layers.

Step 2
• Add details in each layer.

Step 3
• Label your sketch.

Your Sketch

Lab Activity 2 Serous Membranes

Serous membranes are double-layered membranes that cover internal organs and line the ventral body cavities to reduce friction between organs. The **visceral layer** of the membrane is in direct contact with the particular organ, and the **parietal layer** lines the wall of the cavity. Between the two layers of the membrane is a minute space filled with **serous fluid,** a slippery lubricant. Different from the mucus secreted by mucous membranes, serous fluid is a thin, watery secretion similar to blood plasma. A thin layer of specialized simple epithelium called **mesothelium** covers the exposed surface of the pleura where it faces the pleural cavity. Interstitial fluid from the underlying connective tissue transudes, or passes through, the mesothelium to form the serous fluid.

The three serous membranes of the body are the **pericardium** of the heart, the **pleurae** of the lungs, and the **peritoneum** of the abdominal organs (all discussed in Exercise 2). The pericardial and pleural serous membranes are detailed in **Figure 12.3**.

QuickCheck Questions

2.1 Where do serous membranes occur?

2.2 Describe the structure of a serous membrane.

In the Lab 2

Materials

☐ Compound microscope
☐ Prepared serous membrane slide (pleura)
☐ Laboratory models

Procedures

1. Place the slide on the microscope stage and focus at low power.

Figure 12.3 Serous Membranes of the Thoracic Cavity Each lung is enclosed in a pleural membrane; the heart is surrounded by the pericardium.

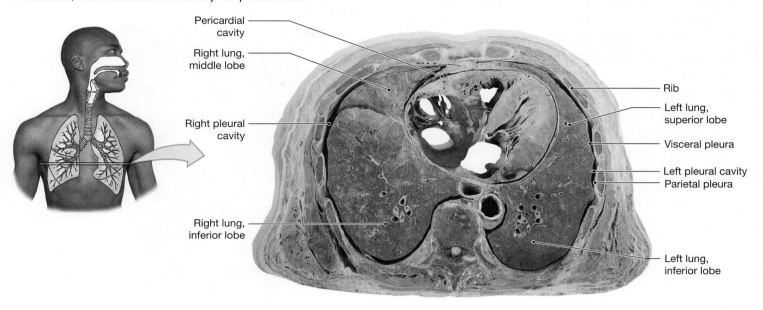

Pericardial cavity

Right lung, middle lobe

Right pleural cavity

Right lung, inferior lobe

Rib

Left lung, superior lobe

Visceral pleura

Left pleural cavity

Parietal pleura

Left lung, inferior lobe

2. Locate the surface of the tissue.

- On the surface is the epithelium called mesothelium. Describe the appearance of this tissue.

- Locate the underlying connective tissue components of the pleura.

3. *Draw It!* Draw a portion of the slide in the space provided here. Label the visceral layer, the serous cavity, and the mesothelium.

Serous membrane

4. Identify the three serous membranes on laboratory models. ◼

Lab Activity 3   Cutaneous Membrane

The cutaneous membrane is the skin, which consists of the epidermis and dermis that cover the exterior surface of the body. The epidermis consists of stratified squamous epithelium, and the dermis is made up of a variety of connective tissues. Protection against abrasion and against the entrance of microbes is a major function of the cutaneous membrane. Unlike all mucous, serous, and synovial membranes, which are kept moist, the cutaneous membrane is dry. A process called **keratinization** waterproofs the skin's surface. The epithelial cells of the cutaneous membrane synthesize the hard protein keratin, which kills the cells, leaving a dry protective layer of tough, scaly dead cells on the skin surface. The cells are eventually shed and are constantly replaced.

QuickCheck Questions

3.1 What are the two layers that make up the cutaneous membrane?

3.2 What type of epithelium is found in the cutaneous membrane?

In the Lab 3

Materials

☐ Skin model

☐ Compound microscope

☐ Prepared slide of the skin (transverse section)

Procedures

1. Study a model of the skin and distinguish among epidermis, dermis, and hypodermis. Which of these layers make up the cutaneous membrane?

2. Place the slide on the microscope stage and focus at low power.

3. Locate the epithelium of the epidermis.
   - Of what type of tissue is it composed?
   - Identify the connective tissue of the dermis.

4. *Draw It!* Draw a portion of the slide in the space provided here. Label the epidermis, stratified squamous epithelium, and dermis. ■

Transverse section of skin

Lab Activity 4   Synovial Membranes

Freely movable joints, such as the knee and elbow, are lined with a synovial membrane (Figure 12.4). The mobility of the joint is due primarily to the presence of a small joint cavity between the bones. An **articular cartilage** covers the surfaces of the bones in the joint cavity. The walls of the cavity are encapsulated with synovial membranes. This type of membrane is unique because the connective tissue is not covered with epithelium. In a synovial membrane, loose connective tissue produces a liquid that seeps from the tissue and fills the synovial cavity. This **synovial fluid**—a clear, lubricating solution containing mucin, salts, albumins, and nutrients—passes between a patchwork of lining cells similar to epithelium. Unlike epithelia, there is no basal lamina anchoring the lining cells to the underlying connective tissue. Fluid exchange between blood vessels and the loose connective tissue maintains and replenishes the synovial fluid. Movement helps synovial fluid to circulate in the joint; if a joint is immobilized for too long, both the cartilage and the synovial membrane may degenerate.

Figure 12.4 **The Structure of a Synovial Joint** A sectional view of the knee joint. The synovial membrane secretes fluid into the joint cavity.

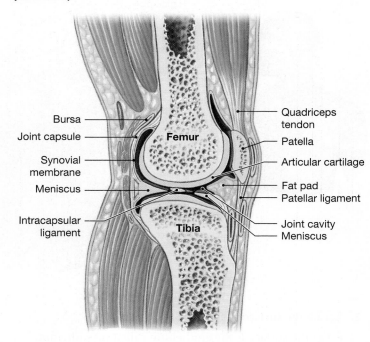

Joints such as the shoulder have additional synovial membranes called **bursae** (singular *bursa*), which cushion structures such as tendons and ligaments. Bursae also occur over bones where the skin is thin, such as the elbow, to reduce abrasion to the skin.

QuickCheck Questions

4.1 Where in the body do synovial membranes occur?

4.2 What is a bursa?

In the Lab   4

Materials

☐ Model of knee in sagittal section
☐ Longitudinally sectioned cow joint

Procedures

1. On the knee model identify the articular cartilage, joint cavity, synovial membrane, and bursae.

2. Examine the gross anatomy of the cow joint or the knee model.

3. Locate the articular cartilage at the ends of the articulating bones, the synovial membrane, and bursae. ■

Name _____

Date _____

Section _____

Body Membranes

A. Definitions

Define each of the following terms.

1. synovial membrane

2. parietal layer

3. serous fluid

4. lamina propria

5. goblet cell

6. visceral layer

7. bursa

8. articular cartilage

9. keratinization

10. mucous membrane

B. Short-Answer Questions

1. Where in the body do serous membranes occur?

2. What type of tissue occurs on the surface of all body membranes?

3. What layers of the skin constitute the cutaneous membrane?

C. Analysis and Application

1. How are serous and synovial fluids produced?

2. What is the function of goblet cells, and in which type of membrane do they occur?

3. A layer called the lamina propria occurs in which type of membrane?

D. Clinical Challenge

A patient is admitted to an emergency room with first- and second-degree burns on the anterior trunk and thighs. Consider the functions of the cutaneous membrane and describe the general care this patient requires.

Organization of the Skeletal System

Learning Outcomes

On completion of this exercise, you should be able to:

1. List the components of the axial skeleton and those of the appendicular skeleton.

2. Describe the gross anatomy of a long bone.

3. Describe the histological organization of compact bone and of spongy bone.

4. List the five shapes of bones and give an example of each type.

5. Describe the bone markings visible on the skeleton.

Lab Activities

1 Bone Structure   136

2 Histological Organization of Bone   137

3 The Skeleton   139

4 Bone Classification and Bone Markings   139

Clinical Application

Osteoporosis   137

The skeletal system serves many functions. Bones support the body's soft tissues and protect vital internal organs. Calcium, lipids, and other materials are stored in the bones, and blood cells are manufactured in the bones' red marrow. Bones serve as levers that allow the muscular system to produce movement or maintain posture. In this exercise, you will study the gross structure of bone and the individual bones of the skeletal system.

Two types of bone tissue are found in the skeleton: compact and spongy. **Compact bone,** which is also called **dense bone,** seals the outer surface of bones and is found wherever stress arrives from one direction on the bone. **Spongy bone,** or **cancellous tissue,** is found inside the compact-bone envelope.

Need More Practice and Review?

Build your knowledge—and confidence!—in the Study Area of MasteringA&P® at www.masteringaandp.com with Pre-lab Quizzes, Post-lab Quizzes, Practice Anatomy Lab™ (PAL™) 3.0 virtual anatomy practice tool, PhysioEx™ 9.0 laboratory simulations, and A&P Flix™ with Quizzes.

PAL | practice anatomy lab   For this lab exercise, follow these navigation paths:
- PAL>Anatomical Models>Axial Skeleton
- PAL>Anatomical Models>Appendicular Skeleton
- PAL>Histology>Connective Tissue

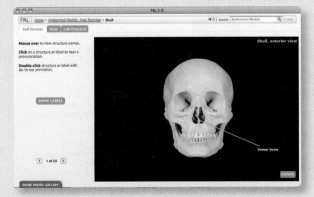

MasteringA&P®

Lab Activity 1   Bone Structure

Bones are encapsulated in a tough, fibrous membrane called the **periosteum** (per-ē-OS-tē-um). This membrane appears shiny and glossy and is sometimes visible on a chicken bone or on the bone in a steak. Histologically, the periosteum is composed of two layers: an outer fibrous layer, where muscle tendons and bone ligaments attach, and an inner cellular layer which produces cells called **osteoblasts** (OS-tē-ō-blasts) for bone growth and repair. **Osteocytes** are mature bone cells that maintain the mineral and protein components of bone matrix.

Long bones, such as the femur of the thigh, have a shaft, called the **diaphysis** (dī-AF-i-sis), with an **epiphysis** (ē-PIF-i-sis) on each end (**Figure 13.1**). The proximal epiphysis is on the superior end of the diaphysis, and the distal epiphysis is on the inferior end. Wherever an epiphysis articulates with another bone, a layer of hyaline cartilage, the **articular cartilage,** covers the epiphysis.

The wall of the diaphysis is made primarily of compact bone. The interior of the diaphysis is hollow, forming a space called the **medullary cavity** (or **marrow cavity**). This cavity is lined with spongy bone and is a storage site for **marrow,** a loose connective tissue. The marrow in long bones contains a high concentration of lipids and is called yellow marrow. A membrane called the **endosteum** (en-DOS-tē-um) lines the medullary cavity. **Osteoclasts** in the endosteum secrete carbonic acid, which dissolves bone matrix to tear down bone either so that it can be replaced with new, stronger bone in a

Figure 13.1 Bone Structure

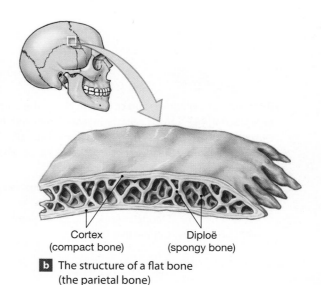

Spongy bone

Compact bone

Medullary cavity

Epiphysis

Metaphysis

Diaphysis (shaft)

Metaphysis

Epiphysis

a The structure of a representative long bone (the femur) in longitudinal section

Cortex (compact bone)

Diploë (spongy bone)

b The structure of a flat bone (the parietal bone)

process called **remodeling** or so that minerals stored in the bone can be released into the blood.

Between the diaphysis and either epiphysis is the **metaphysis** (me-TAF-i-sis). In a juvenile's bone, the metaphysis is called the **epiphyseal plate** and consists of a plate of hyaline cartilage that allows the bone to grow longer. By early adulthood, the rate of mitosis in the cartilage plate slows, and ossification fuses the epiphysis to the diaphysis. Bone growth stops when all the cartilage in the metaphysis has been replaced by bone. This bony remnant of the growth plate is now called the **epiphyseal line.**

Flat bones, such as the frontal and parietal bones of the skull, are thin bones with no marrow cavity. Flat bones have a layer of spongy bone sandwiched between layers of compact bone (Figure 13.1b). The compact bone layers are called the *external* and *internal tables* and are thick in order to provide strength for the bone. The spongy bone between the tables is called the **diploë** (DIP-lō-ē) and is filled with **red marrow,** a type of loose connective tissue made up of stem cells that produces red blood cells, platelets, and most of the white blood cells in the body.

QuickCheck Questions

1.1 What is the location of the two membranes found in long bones?

1.2 Where is spongy bone found?

In the Lab   1

Materials

- ☐ Preserved long bone or fresh long bone from butcher shop
- ☐ Blunt probe
- ☐ Disposable examination gloves
- ☐ Safety glasses

Procedures

1. Put on the safety glasses and examination gloves before you handle the bone.

2. Examine the long bone and locate the periosteum. Does it appear shiny? Are any tendons or ligaments attached to it? _____

3. If the bone has been sectioned, observe the internal bone tissue of the diaphysis. Is the bone tissue similar in all regions of the sectioned bone? _____

4. Locate an epiphysis and its articular cartilage. What is the function of the cartilage? _____

5. Locate the metaphysis and determine if the bone has an epiphyseal plate or an epiphyseal line. ■

Lab Activity 2   Histological Organization of Bone

Compact bone has supportive columns called either **osteons** or *Haversian systems* (**Figure 13.2**). Each osteon consists of many rings of calcified matrix called **concentric lamellae** (lah-MEL-lē; *lamella,* a thin plate). Between the lamellae, in small spaces in the matrix called **lacunae,** are mature bone cells called osteocytes. Bone requires a substantial supply of nutrients and oxygen. Nerves, blood vessels, and lymphatic vessels all pierce the periosteum and enter the bone in a **perforating canal** oriented perpendicular to the osteons. This canal interconnects with **central canals** positioned in the center of osteons. Radiating outward from a central canal

Clinical Application   Osteoporosis

As we age, our bones change. They become weaker and thinner, and they produce less collagen and are therefore less flexible. Calcium levels in the bone matrix decline, resulting in brittle bones. Osteopenia is the natural, age-related loss of bone mass that begins as early as 30 to 40 years of age in some individuals. The loss is a result of a decrease in osteoblast activity while osteoblasts continue to remain active. Bone degeneration beyond normal loss is called **osteoporosis** (os-tē-ō-po-RŌ-sis) and affects the epiphysis, vertebrae, and jaws, and leads to weak limbs, decrease in height, and loss of teeth. Bone fractures are common as spongy bone becomes more porous and unable to withstand stress. Osteoporosis is more common in women, with 29% of women 45 years or older having osteoporosis, but only 18% of males 45 years or older having this condition.

Osteoporosis is associated with an age-related decline in circulating sex hormones in the blood. Sex hormones stimulate osteoblasts to deposit calcium into new bone matrix. In menopausal women, decreasing levels of estrogen slow osteoblast activity, and one result is bone loss. As men age, hormone levels decline more gradually than in women, and as a result most men are able to maintain a healthy bone mass.

Exercising more and consuming adequate amounts of calcium can reduce the rate of bone degeneration and occurrence of osteoporosis in both men and women. However, increasing calcium intake is not enough to prevent osteoporosis. New bone matrix must be produced in order to maintain bone density. Hormone replacement therapy is sometimes prescribed to promote new bone growth in postmenopausal women. Unfortunately, many studies link hormone replacement therapy with blood clots in the lungs, uterine cancer, and other clinical complications. ■

Figure 13.2 **Bone Histology**

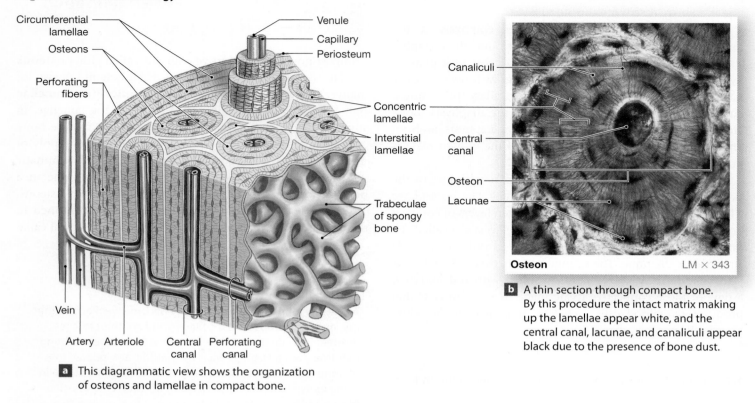

a This diagrammatic view shows the organization of osteons and lamellae in compact bone.

Osteon LM × 343

b A thin section through compact bone. By this procedure the intact matrix making up the lamellae appear white, and the central canal, lacunae, and canaliculi appear black due to the presence of bone dust.

are small diffusion channels called **canaliculi** (kan-a-LIK-ū-lē) that facilitate nutrient, gas, and waste exchange with the blood.

To maintain its strength and weight-bearing ability, bone tissue is continuously being remodeled in a process that leaves distinct structural features in compact bone. Old osteons are partially removed, and the concentric rings of lamellae are fragmented, resulting in **interstitial lamellae** (lah-MEL-lē) between intact osteons. Typically, the distal end of a bone is extensively remodeled throughout life, whereas areas of the diaphysis may never be remodeled. Other lamellae occur underneath the periosteum and wrap around the entire bone. These **circumferential lamellae** are added as a bone grows in diameter.

Unlike compact bone, spongy bone is not organized into osteons; instead, it forms a lattice, or meshwork, of bony struts called **trabeculae** (tra-BEK-ū-lē). Each trabecula is composed of layers of lamellae that are intersected with canaliculi. Filling the spaces between the trabeculae is red marrow, the tissue that produces most blood cells. Spongy bone is always sealed with a thin outer layer of compact bone.

QuickCheck Questions

2.1 What are the three types of lamellae found in bone and their characteristics?

2.2 How is blood supplied to an osteon?

In the Lab **2**

Materials

☐ Bone model

☐ Compound microscope

☐ Prepared slide of bone tissue (transverse section)

Procedures

1. Review the histology of bone in Figure 13.2.
2. Examine a bone model and locate each structure shown in Figure 13.2.
3. Obtain a prepared microscope slide of bone tissue. Most bone slides are a transverse section through bone that is ground very thin. This preparation process removes the bone cells but leaves the bone matrix intact for detailed studies.
4. At low magnification, observe the overall organization of the bone tissue. How many osteons can you locate?
   _____
5. Select an osteon and observe it at a higher magnification. Identify the central canal, canaliculi, and lacunae. What is the function of the canaliculi? _____
6. Locate an area of interstitial lamellae. How do these lamellae differ from the concentric lamellae? _____

For additional practice, complete the *Sketch to Learn* activity (on p. 139). ∎

 *Sketch to Learn*

Let's draw an osteon as observed at medium magnification with the microscope.

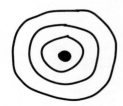 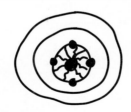

Sample Sketch

Your Sketch

Step 1
- Draw a large oval with several smaller ovals inside.
- Add a dark circle in the middle for the central canal.

Step 2
- Add wavy lines to each concentric lamella. The innermost one is shown as an example.
- Add small dark ovals for osteocytes.
- Label your sketch.

Lab Activity 3 The Skeleton

The adult skeletal system, shown in **Figure 13.3**, consists of 206 bones. Each bone is an organ and includes bone tissue, cartilage, and other connective tissues. The skeleton is organized into the axial and appendicular divisions. The **axial division,** which comprises 80 bones, includes the **skull, vertebral column, sternum, ribs,** and **hyoid bone.** The **appendicular division** (126 bones) consists of the **pectoral girdle, upper limbs, pelvic girdle,** and **lower limbs.** Each girdle attaches its respective limbs to the axial skeleton and allows limb mobility at the points of attachment.

Each side of the pectoral girdle includes a **scapula** (shoulder blade) and a **clavicle** (collar bone). Each upper limb consists of arm, forearm, wrist, and hand. The **humerus** is the arm bone, and the **ulna** and **radius** together form the forearm. The eight wrist bones, called **carpal bones,** articulate with the elongated **metacarpal bones** of the palm. The individual bones of the fingers are the **phalanges.**

The pelvic girdle is fashioned from two coxal bones, each of which is an aggregate of three bones: the superior **ilium** in the hip area, the **ischium** inferior to the ilium, and the **pubis** in the anterior pelvis. Each lower limb comprises thigh, knee-cap, leg, ankle, and foot. The **femur** is the thighbone and is the largest bone in the body. The two bones of the leg are a medial **tibia,** which bears most of the body weight, and a thin, lateral **fibula.** The **patella** occurs at the articulation between femur and tibia. The seven ankle bones are collectively called the **tarsal bones. Metatarsal bones** form the arch of the foot, and **phalanges** form the toes.

QuickCheck Questions

3.1 What are the two major divisions of the skeleton?

3.2 A rib belongs to which division of the skeletal system?

In the Lab 3

Material

☐ Articulated skeleton

Procedures

1. Using Figure 13.3 as a guide, locate the bones of the axial division of the skeleton. List the major components of the axial division.
2. Using Figure 13.3 as a guide, locate the major components of the appendicular division of the skeleton.
3. What bones are found in the shoulder and upper limb?
4. What three bones fuse to form the coxal bone?
5. List the bones of the lower limb. ■

Lab Activity 4 Bone Classification and Bone Markings

Bones may be grouped and classified according to shape (**Figure 13.4**). Already discussed in this exercise are **long bones,** which are greater in length than in width, and **flat bones,** which are thin and platelike. Bones of the arm, forearm, thigh, and leg are long bones. Bones of the wrist and ankle are

Figure 13.3 The Skeleton

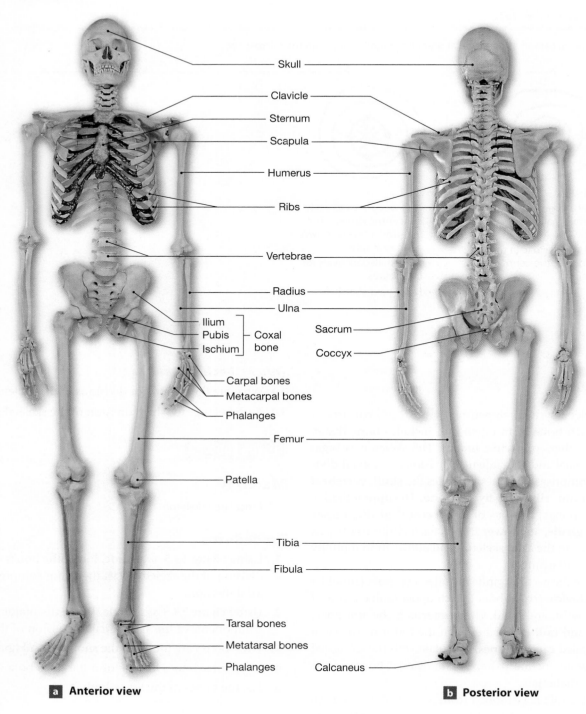

Skull
Clavicle
Sternum
Scapula
Humerus
Ribs
Vertebrae
Radius
Ulna
Ilium
Pubis — Coxal
Ischium — bone
Carpal bones
Metacarpal bones
Phalanges
Femur
Patella
Tibia
Fibula
Tarsal bones
Metatarsal bones
Phalanges

Sacrum
Coccyx
Calcaneus

a **Anterior view**

b **Posterior view**

short bones, almost as wide as they are long. The vertebrae of the spine are **irregular bones** that are not in any of the just-named categories. **Sesamoid bones** form inside tendons. The largest sesamoid bone is the patella, and it develops inside tendons anterior to the knee. **Sutural bones,** or **Wormian bones,** occur where the interlocking joints of the skull, called **sutures,** branch and isolate a small piece of bone. The number of sutural bones varies from one person to another and is not included when counting the number of bones in the skeletal system.

Each bone has certain anatomical features on its surface, called either **bone markings** or surface markings. A particular bone marking may be unique to a single bone or may occur throughout the skeleton. **Table 13.1** (on p. 142) illustrates examples of bone markings and organizes the markings

Figure 13.4 **Shapes of Bones**

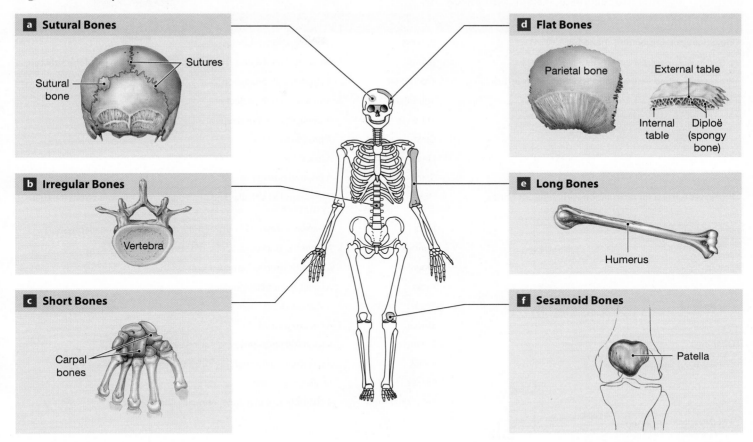

into five groups. The first group includes general anatomical structures, and the second group lists bony structures for tendon and ligament attachment. The third group contains structures that occur at sites of articulation with other bones. The last two groups include depressions and openings.

QuickCheck Questions

4.1 Give examples of two types of short bones.

4.2 Give an example of an irregular bone.

4.3 What is a foramen?

4.4 What is the neck of a bone?

In the Lab 4

Material

☐ Articulated skeleton

Procedures

1. Examine the articulated skeleton and determine how many of each type of bone is present.

- Long bones:
- Short bones:
- Flat bones:
- Irregular bones:
- Sesamoid bones:
- Sutural bones:

2. Using Table 13.1 for reference, locate on the skeleton:

- A *foramen* in the skull. Describe this structure. _____

- A *fossa* on the distal end of the humerus. How is the fossa different from a foramen? _____

- A *head* on the femur. Which other bones have a head? _____

- A *condyle* on two different bones. Describe this structure. _____

- A *tuberosity* on the proximal end of the humerus. What is the texture of this structure? _____

3. Locate one instance of each of the other marking types on the skeleton: process, ramus, trochanter, tubercle, crest, line, spine, neck, trochlea, facet, sulcus, canal, fissure, sinus. ■

Table 13.1   An Introduction to Skeletal Terminology

General Description	Anatomical Term	Definition
Elevations and projections (general)	Process	Any projection or bump
	Ramus	An extension of a bone making an angle with the rest of the structure
Processes formed where tendons or ligaments attach	Trochanter	A large, rough projection
	Tuberosity	A smaller, rough projection
	Tubercle	A small, rounded projection
	Crest	A prominent ridge
	Line	A low ridge
	Spine	A pointed process
Processes formed for articulation with adjacent bones	Head	The expanded articular end of an epiphysis, separated from the shaft by the neck
	Neck	A narrow connection between the epiphysis and the diaphysis
	Condyle	A smooth, rounded articular process
	Trochlea	A smooth, grooved articular process shaped like a pulley
	Facet	A small, flat articular surface
Depressions	Fossa	A shallow depression
	Sulcus	A narrow groove
Openings	Foramen	A rounded passageway for blood vessels or nerves
	Canal	A passageway through the substance of a bone
	Fissure	An elongated cleft
	Sinus or antrum	A chamber within a bone, normally filled with air

Trochanter
Head
Neck
Fissure
Ramus
Facet
Tubercle
Condyle

Femur

Sinus
Foramen
Process

Skull

Tubercle
Head
Sulcus
Neck
Tuberosity
Fossa
Trochlea
Condyle

Humerus

Crest
Fossa
Spine
Line
Foramen
Ramus

Pelvis

Name _____

Date _____

Section _____

Organization
of the Skeletal System

A. Definitions

Define each structure.

1. metaphysis

2. trabecula

3. articular cartilage

4. diaphysis

5. epiphyseal plate

6. periosteum

7. epiphysis

8. endosteum

9. medullary cavity

10. epiphyseal line

B. Matching

Match each bone with the correct division and part of the skeleton. Each question may have more than one answer, and each choice can be used more than once.

_____ **1.** scapula	**A.** axial division
_____ **2.** coxal bone	**B.** appendicular division
_____ **3.** patella	**C.** pectoral girdle
_____ **4.** hyoid	**D.** upper limb
_____ **5.** radius	**E.** pelvic girdle
_____ **6.** metacarpal	**F.** lower limb
_____ **7.** vertebra	
_____ **8.** clavicle	
_____ **9.** rib	
_____ **10.** femur	
_____ **11.** sternum	
_____ **12.** carpal	

C. Short-Answer Questions

1. List the components of the axial skeleton.

2. List the components of the appendicular skeleton.

3. Describe the five types of surface markings on bones.

4. List the different shapes of bones.

D. Labeling

1. Label Figure 13.5.

Figure 13.5 **Histology of Bone**

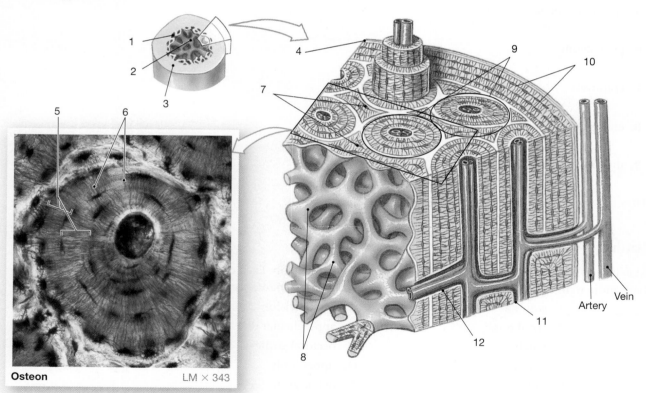

Osteon   LM × 343

1. _____	5. _____	9. _____
2. _____	6. _____	10. _____
3. _____	7. _____	11. _____
4. _____	8. _____	12. _____

2. Label **Figure 13.6**.

Figure 13.6 **Anterior View of the Skeleton**

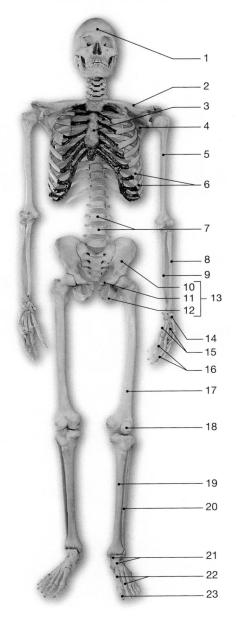

1. _____
2. _____
3. _____
4. _____
5. _____
6. _____
7. _____
8. _____
9. _____
10. _____
11. _____
12. _____
13. _____
14. _____
15. _____
16. _____
17. _____
18. _____
19. _____
20. _____
21. _____
22. _____
23. _____

E. Analysis and Application

1. Where does spongy bone occur in the skeleton?

2. How are the upper limbs attached to the axial skeleton?

3. Where does growth in length occur in a long bone?

F. Clinical Challenge

The result of an elderly woman's bone density test indicates that her bones are losing mass. What preventative measure can she take to slow her bone loss?

Axial Skeleton

Learning Outcomes

On completion of this exercise, you should be able to:

1. Identify the components of the axial skeleton.

2. Identify the cranial and facial bones of the skull.

3. Identify the surface features of the cranial and facial bones.

4. Describe the skull of a fetus.

5. Describe the five regions of the vertebral column and distinguish among the vertebrae of each region.

6. Identify the features of a typical vertebra.

7. Discuss the articulation of the ribs with the thoracic vertebrae.

8. Identify the components of the sternum.

The axial skeleton provides both a central framework for attachment of the appendicular skeleton and protection for the body's internal organs. The 80 bones of the axial skeleton include the skull, a thoracic cage made up of ribs and the sternum, and a flexible vertebral column with 24 vertebrae, 1 sacrum, and 1 coccyx (**Figure 14.1**). The 22 bones of the skull are organized into **facial bones** and 8 **cranial bones** that form the **cranium** (**Figure 14.2**). The six bones of the middle ear (three per ear) and the hyoid bone are referred to as the *associated bones* of skull.

Need More Practice and Review?

Build your knowledge—and confidence!—in the Study Area of MasteringA&P® at www.masteringaandp.com with Pre-lab Quizzes, Post-lab Quizzes, Practice Anatomy Lab™ (PAL™) 3.0 virtual anatomy practice tool, PhysioEx™ 9.0 laboratory simulations, and A&P Flix™ with Quizzes.

PAL | practice anatomy lab   For this lab exercise, follow these navigation paths:
  • PAL>Human Cadaver>Axial Skeleton
  • PAL>Anatomical Models>Axial Skeleton

Figure 14.1 The Axial Skeleton An anterior view of the entire skeleton, with the axial components highlighted. The numbers in the boxes indicate the number of bones in the adult skeleton.

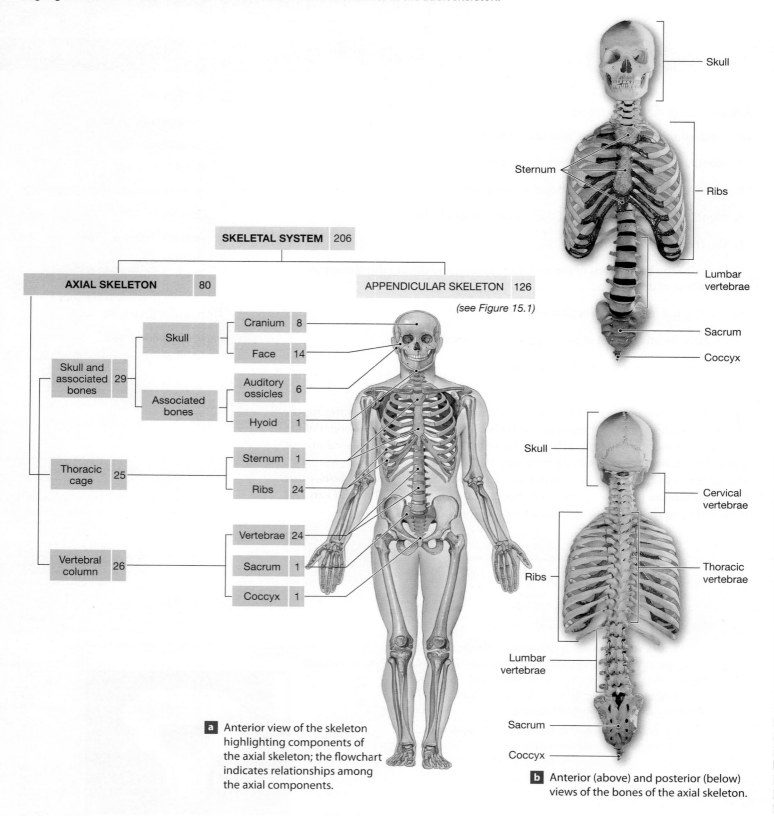

SKELETAL SYSTEM	206	

AXIAL SKELETON 80

APPENDICULAR SKELETON 126

(see Figure 15.1)

Skull and associated bones 29

- Skull
  - Cranium 8
  - Face 14
- Associated bones
  - Auditory ossicles 6
  - Hyoid 1

Thoracic cage 25
- Sternum 1
- Ribs 24

Vertebral column 26
- Vertebrae 24
- Sacrum 1
- Coccyx 1

a Anterior view of the skeleton highlighting components of the axial skeleton; the flowchart indicates relationships among the axial components.

Skull

Sternum

Ribs

Lumbar vertebrae

Sacrum

Coccyx

Skull

Cervical vertebrae

Ribs

Thoracic vertebrae

Lumbar vertebrae

Sacrum

Coccyx

b Anterior (above) and posterior (below) views of the bones of the axial skeleton.

Figure 14.2 Cranial and Facial Subdivisions of the Skull The seven associated bones are not illustrated.

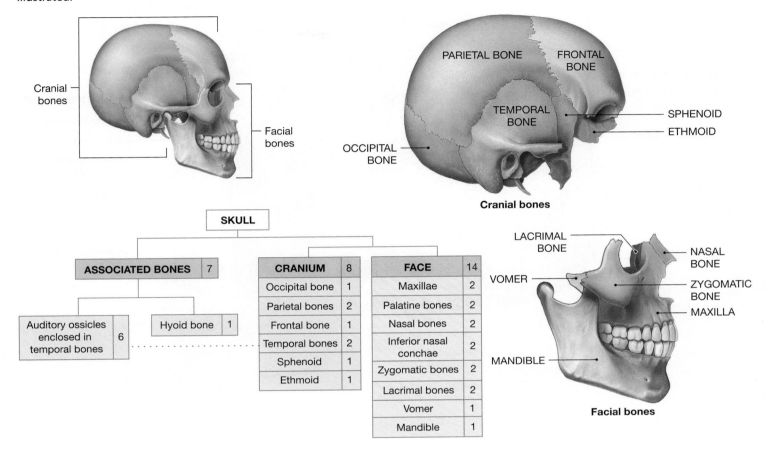

Cranial bones

Facial bones

PARIETAL BONE
FRONTAL BONE
TEMPORAL BONE
SPHENOID
ETHMOID
OCCIPITAL BONE

Cranial bones

LACRIMAL BONE
NASAL BONE
VOMER
ZYGOMATIC BONE
MAXILLA
MANDIBLE

Facial bones

SKULL		

ASSOCIATED BONES	7

Auditory ossicles enclosed in temporal bones	6

Hyoid bone	1

CRANIUM	8
Occipital bone	1
Parietal bones	2
Frontal bone	1
Temporal bones	2
Sphenoid	1
Ethmoid	1

FACE	14
Maxillae	2
Palatine bones	2
Nasal bones	2
Inferior nasal conchae	2
Zygomatic bones	2
Lacrimal bones	2
Vomer	1
Mandible	1

Lab Activity 1   Cranial Bones

The skull serves a wide variety of critical functions; it cradles the delicate brain and houses major sensory organs for vision, hearing, balance, taste, and smell. The skull is perforated with many holes called **foramina** where nerves and blood vessels pass to and from the brain and other structures of the head. Facial and cranial bones make sockets, called **orbits,** for the eyes. The joints of the skull are designed for strength instead of movement and only two joints can move: the jaw and the joint between the skull and the vertebral column.

The cranium, shown in **Figure 14.3**, has eight bones: one frontal bone, two parietal bones, two temporal bones, one occipital bone, one sphenoid, and one ethmoid (Figure 14.3a). The **frontal bone** of the cranium extends from the forehead posterior to the **coronal suture** and articulates (joins) with the two **parietal bones** on the lateral sides of the skull. The parietal bones are joined at their superior crest by the **sagittal suture** (Figure 14.3b). The two **temporal bones** are inferior to the parietal bones and are easy to identify by the canals where sound enters the ears. The temporal bone articulates with the parietal bone at the **squamous suture.** The posterior wall of the cranium is the **occipital bone,** which meets the parietals at the **lambdoid** (LAM-doyd) **suture,** also called the *occipitoparietal suture* (Figure 14.3c).

The **sphenoid** is the bat-shaped bone visible on the cranial floor, anterior to the temporal bone (Figure 14.3d). The sphenoid forms parts of the floor and lateral walls of the cranium and the posterolateral wall of the orbit. At the

Study Tip An Organized Approach to the Skull

The skull is perhaps the most challenging part of the skeleton to learn. The anatomy is small and very detailed and each bone has several surfaces. When faced with such a volume of material, take some time to survey the topic and formulate a study plan. With the skull, the study plan is: "Start big then go for details." First, start with the big picture—identify each cranial bone —then progress on to the detailed study of the individual cranial bones. ∎

Figure 14.3 Views of the Skull

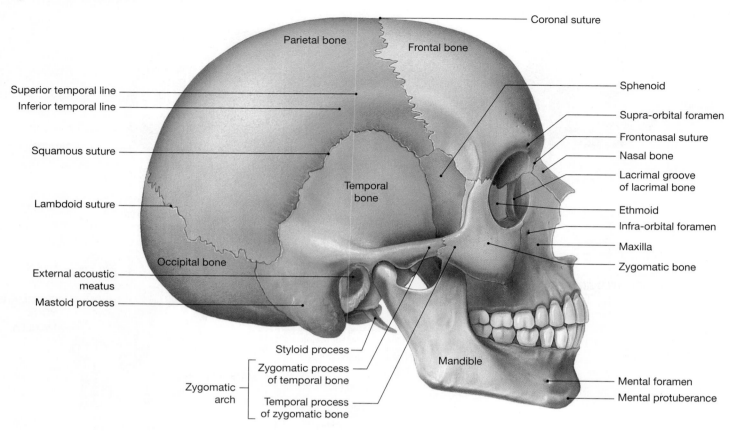

Coronal suture

Parietal bone

Frontal bone

Superior temporal line

Inferior temporal line

Squamous suture

Lambdoid suture

Occipital bone

External acoustic meatus

Mastoid process

Temporal bone

Sphenoid

Supra-orbital foramen

Frontonasal suture

Nasal bone

Lacrimal groove of lacrimal bone

Ethmoid

Infra-orbital foramen

Maxilla

Zygomatic bone

Styloid process

Zygomatic arch

Zygomatic process of temporal bone

Temporal process of zygomatic bone

Mandible

Mental foramen

Mental protuberance

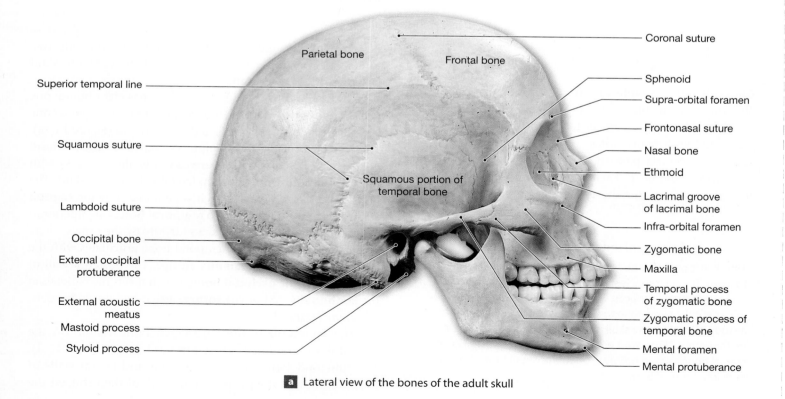

Coronal suture

Parietal bone

Frontal bone

Superior temporal line

Squamous suture

Lambdoid suture

Occipital bone

External occipital protuberance

External acoustic meatus

Mastoid process

Styloid process

Squamous portion of temporal bone

Sphenoid

Supra-orbital foramen

Frontonasal suture

Nasal bone

Ethmoid

Lacrimal groove of lacrimal bone

Infra-orbital foramen

Zygomatic bone

Maxilla

Temporal process of zygomatic bone

Zygomatic process of temporal bone

Mental foramen

Mental protuberance

a Lateral view of the bones of the adult skull

Figure 14.3   Views of the Skull *(continued)*

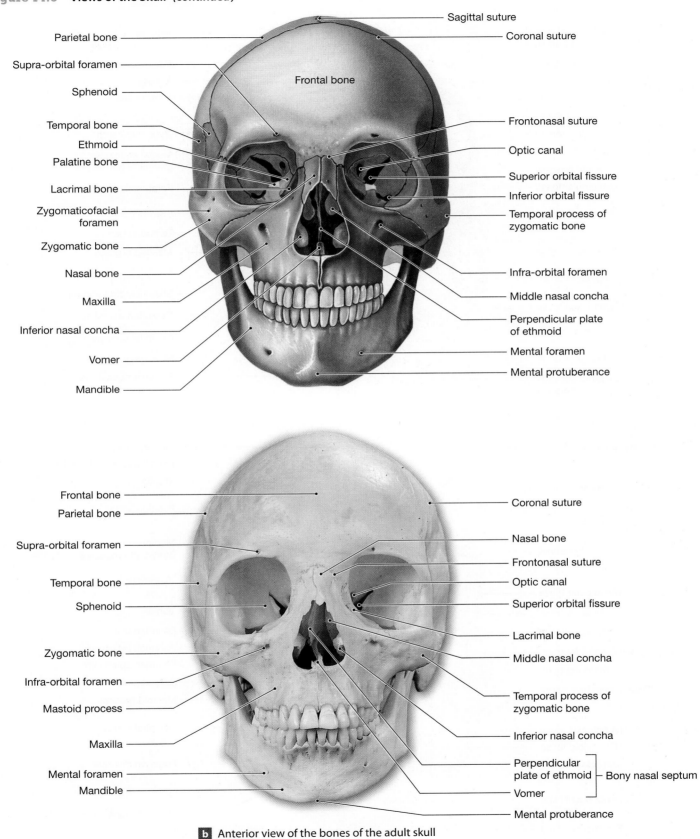

Sagittal suture

Coronal suture

Parietal bone

Supra-orbital foramen

Sphenoid

Frontal bone

Frontonasal suture

Temporal bone

Ethmoid

Optic canal

Palatine bone

Superior orbital fissure

Lacrimal bone

Inferior orbital fissure

Zygomaticofacial foramen

Temporal process of zygomatic bone

Zygomatic bone

Nasal bone

Infra-orbital foramen

Maxilla

Middle nasal concha

Inferior nasal concha

Perpendicular plate of ethmoid

Vomer

Mental foramen

Mandible

Mental protuberance

Frontal bone

Coronal suture

Parietal bone

Supra-orbital foramen

Nasal bone

Frontonasal suture

Temporal bone

Optic canal

Sphenoid

Superior orbital fissure

Zygomatic bone

Lacrimal bone

Infra-orbital foramen

Middle nasal concha

Mastoid process

Temporal process of zygomatic bone

Maxilla

Inferior nasal concha

Mental foramen

Perpendicular plate of ethmoid — Bony nasal septum

Mandible

Vomer

Mental protuberance

b Anterior view of the bones of the adult skull

Figure 14.3   Views of the Skull *(continued)*

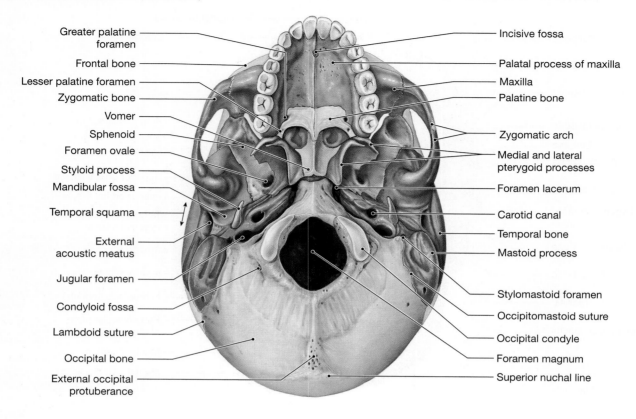

Greater palatine foramen
Frontal bone
Lesser palatine foramen
Zygomatic bone
Vomer
Sphenoid
Foramen ovale
Styloid process
Mandibular fossa
Temporal squama
External acoustic meatus
Jugular foramen
Condyloid fossa
Lambdoid suture
Occipital bone
External occipital protuberance

Incisive fossa
Palatal process of maxilla
Maxilla
Palatine bone
Zygomatic arch
Medial and lateral pterygoid processes
Foramen lacerum
Carotid canal
Temporal bone
Mastoid process
Stylomastoid foramen
Occipitomastoid suture
Occipital condyle
Foramen magnum
Superior nuchal line

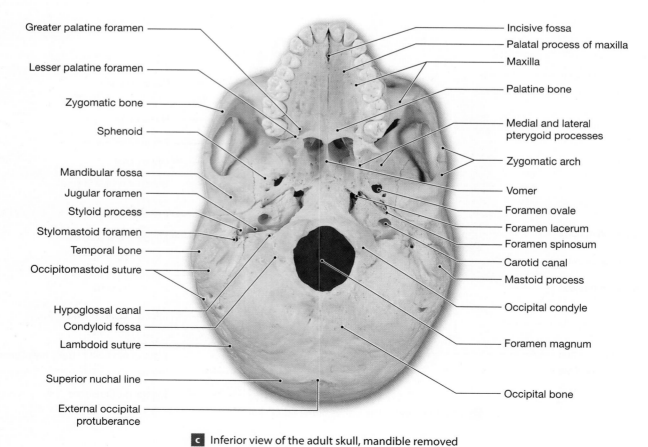

Greater palatine foramen
Lesser palatine foramen
Zygomatic bone
Sphenoid
Mandibular fossa
Jugular foramen
Styloid process
Stylomastoid foramen
Temporal bone
Occipitomastoid suture
Hypoglossal canal
Condyloid fossa
Lambdoid suture
Superior nuchal line
External occipital protuberance

Incisive fossa
Palatal process of maxilla
Maxilla
Palatine bone
Medial and lateral pterygoid processes
Zygomatic arch
Vomer
Foramen ovale
Foramen lacerum
Foramen spinosum
Carotid canal
Mastoid process
Occipital condyle
Foramen magnum
Occipital bone

c Inferior view of the adult skull, mandible removed

Figure 14.3 **Views of the Skull** *(continued)*

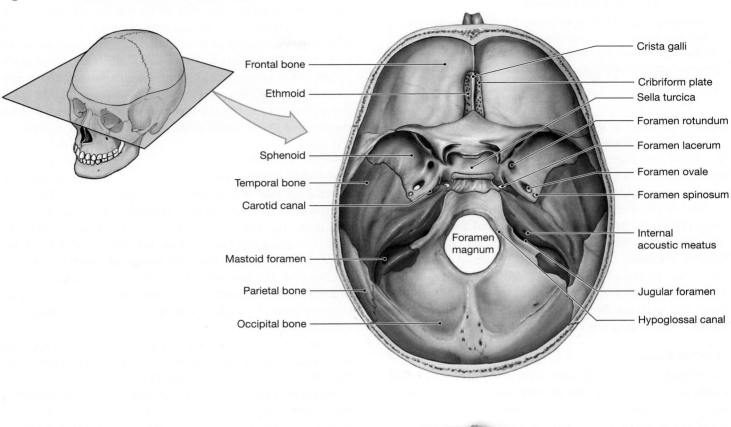

Frontal bone

Ethmoid

Sphenoid

Temporal bone

Carotid canal

Mastoid foramen

Parietal bone

Occipital bone

Crista galli

Cribriform plate

Sella turcica

Foramen rotundum

Foramen lacerum

Foramen ovale

Foramen spinosum

Foramen magnum

Internal acoustic meatus

Jugular foramen

Hypoglossal canal

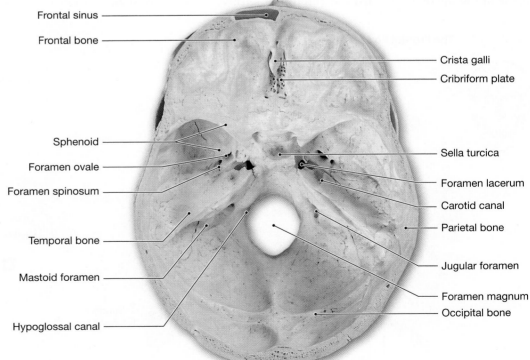

Frontal sinus

Frontal bone

Sphenoid

Foramen ovale

Foramen spinosum

Temporal bone

Mastoid foramen

Hypoglossal canal

Crista galli

Cribriform plate

Sella turcica

Foramen lacerum

Carotid canal

Parietal bone

Jugular foramen

Foramen magnum

Occipital bone

d Horizontal section through the skull showing the floor of the cranial cavity

Study Tip Locating the Ethmoid

Pinching the eye orbit is an easy way to locate the ethmoid. Insert your thumb halfway into one eye orbit of the study skull and your forefinger halfway into the other eye orbit. Gently pinch the bone deep between the orbits, which is the ethmoid. ■

superior margin the squamous and coronal sutures are connected by the **sphenoparietal suture.** The single **ethmoid** is a small, rectangular bone posterior to the bridge of the nose and anterior to the sphenoid (Figure 14.3d). The ethmoid contributes to the posteromedial wall of both orbits.

The floor of the cranium has three depressions called *fossae* (Figure 14.3d). The **anterior cranial fossa** is mainly the depression that forms the base of the frontal bone. Small portions of the ethmoid and sphenoid also contribute to the floor of this area. The **middle cranial fossa** is a depressed area extending over the sphenoid and the temporal and occipital bones. The **posterior cranial fossa** is found in the occipital bone.

Frontal Bone

The frontal bone forms the roof, walls, and floor of the anterior cranium (**Figure 14.4**). The **frontal squama** is the flattened expanse commonly called the forehead. In the midsagittal plane of the squama is the **frontal (metopic) suture,** where

the two frontal bones fuse in early childhood (typically by the time a child is eight years old). As natural remodeling of bone occurs, this suture typically disappears by age 30. The frontal bone forms the upper portion of the eye orbit. Superior to the orbit is the **supra-orbital foramen,** which on some skulls occurs not as a complete hole but as a small notch, the **supra-orbital notch.** In the anterior and medial regions of the orbit, the frontal bone forms the **lacrimal fossa,** an indentation for the lacrimal gland, which moistens and lubricates the eye.

Occipital Bone

The occipital bone, shown in **Figure 14.5**, forms the posterior floor and wall of the skull (Figure 14.5a). The most conspicuous structure of the occipital bone is the **foramen magnum,** the large hole where the spinal cord enters the skull and joins the brain. Along the lateral margins of the foramen magnum are flattened **occipital condyles** that articulate with the first vertebra of the spine. Passing superior to each occipital condyle is the **hypoglossal canal,** a passageway for the hypoglossal nerve, which controls muscles of the tongue and throat.

The occipital bone has many external surface marks that show where muscles and ligaments attach. The **external occipital crest** is a ridge that extends posteriorly from the foramen magnum to a small bump, the **external occipital protuberance.** Wrapping around the occipital bone lateral from the crest and protuberance are the **superior** and **inferior nuchal**

Figure 14.4 The Frontal Bone

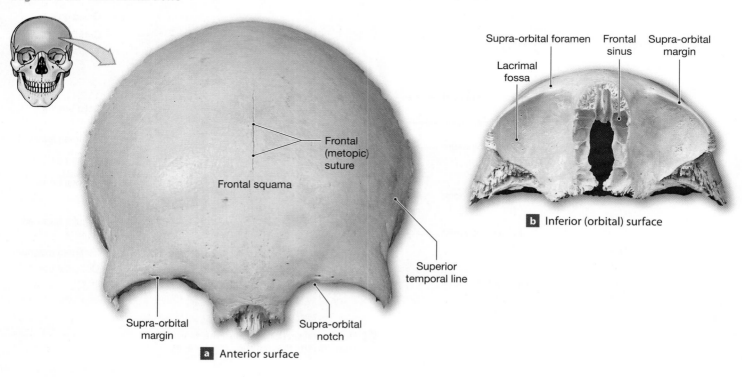

Supra-orbital foramen Frontal sinus Supra-orbital margin

Lacrimal fossa

Frontal (metopic) suture

Frontal squama

Superior temporal line

Supra-orbital margin

Supra-orbital notch

a Anterior surface

b Inferior (orbital) surface

Figure 14.5 The Occipital and Parietal Bones

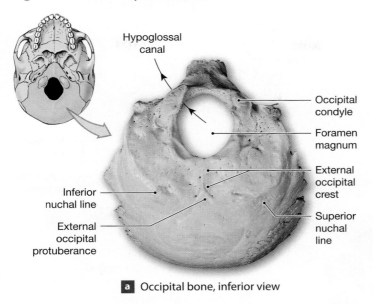

a Occipital bone, inferior view

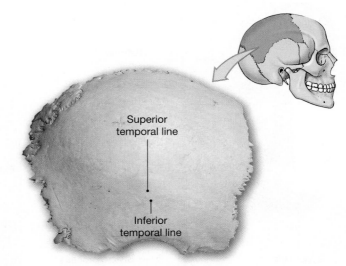

b Right parietal bone, lateral view

(NOO-kul) **lines,** surface marks indicating where muscles of the neck attach to the skull.

Parietal Bones

The two parietal bones form the posterior crest of the skull and are joined by the sagittal suture (Figure 14.5b). The bones are smooth and have few surface features. The low ridges of the **superior** and **inferior temporal lines** (Figure 14.5b) are superior to the squamous suture, where a muscle for chewing attaches. No major foramina pass through the parietal bones.

Temporal Bones

The two temporal bones constitute the inferior lateral walls of the skull and part of the floor of the middle cranial fossa (**Figure 14.6**). One of the most distinct features of a temporal bone is its articulation with the zygomatic bone of the face by the **zygomatic arch** (Figure 14.3a). The arch is a span of processes from two bones: the **zygomatic process** of the temporal bone and the temporal process of the zygomatic bone. Posterior to the zygomatic process is the region of the temporal bone called the **articular tubercle.** Immediately posterior to the articular tubercle is the **mandibular fossa,** a shallow depression where the mandible bone articulates with the temporal bone.

The broad, flattened superior surface of each temporal bone is the **squamous part.** The hole inferior to the squamous part is the **external acoustic meatus,** which conducts sound waves toward the eardrum. Directly posterior to the external acoustic meatus is the conical **mastoid process,**

where a muscle tendon that moves the head attaches. Within the mastoid process are many small, interconnected sinuses called **mastoid air cells.** The long, needlelike **styloid** (STĪ-loyd; *stylos*, pillar) **process** is located anteromedial to the mastoid process. Between the styloid and the mastoid processes is a small foramen, the **stylomastoid foramen,** where the facial nerve exits the cranium.

On the cranial floor, the large bony ridge of the temporal bone is the **petrous** (pet-rus; *petra*, a rock) **part,** which houses the organs for hearing and equilibrium and the tiny bones of the ear. The **internal acoustic meatus** is on the posteromedial surface of the petrous part. The union between the temporal and occipital bones creates an elongated **jugular foramen** that serves as a passageway for cranial nerves and the jugular vein, which drains blood from the brain. On the anterior side of the petrous part is the **carotid canal,** where the internal carotid artery enters the skull to deliver oxygenated blood to the brain.

The Sphenoid

The sphenoid is the base of the cranium and each cranial bone articulates with it. The sphenoid is visible from all views of the skull but the easiest aspect of this bone to work with is its superior surface exposed on the floor of the skull. On the anterior side, the sphenoid contributes to the lateral wall of the eye orbit; on the lateral side, it spans the floor of the cranium and braces the walls. The superior surface of the sphenoid is made up of two **lesser wings** and two **greater wings** on either side of the medial line, which give

Figure 14.6 The Temporal Bones The right temporal bone.

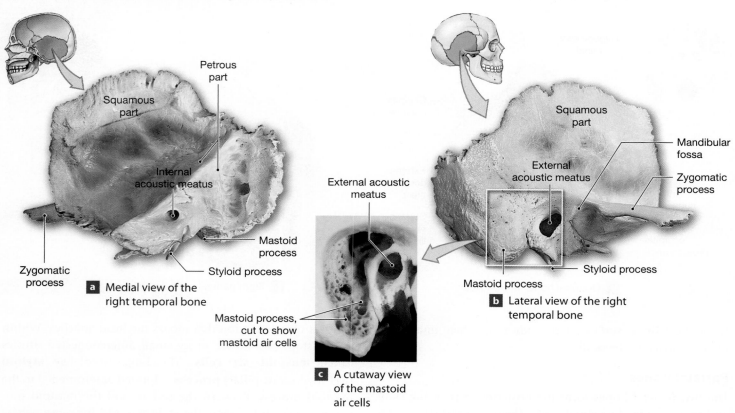

a Medial view of the right temporal bone

b Lateral view of the right temporal bone

c A cutaway view of the mastoid air cells

the bone the appearance of a bat (**Figure 14.7**). Each greater wing has an **orbital surface** contributing to the wall of the eye orbit. In the center of the sphenoid is the U-shaped **sella turcica** (TUR-si-kuh), commonly called the Turk's saddle. The depression in the sella turcica is the **hypophyseal** (hī-pō-FIZ-ē-ul) **fossa**, which contains the pituitary gland of the brain. The anterior part of the sella turcica is the **tuberculum sellae;** the posterior wall is the **posterior clinoid** (KLĪ-noyd; *kline*, a bed) **process.** The two **anterior clinoid processes** are the hornlike projections on either side of the tuberculum sellae. Extending vertically from the inferior surface of the sphenoid are the **pterygoid** (TER-i-goyd; *pterygion*, wing) **processes.** Each process divides into a **lateral plate** and a **medial plate,** where muscles of the mouth attach. At the base of each pterygoid process is a small **pterygoid canal** that serves as a passageway for nerves to the soft palate of the mouth.

Four pairs of foramina are aligned on either side of the sella turcica and serve as passageways for blood vessels and nerves. The oval **foramen ovale** (ō-VAH-lē; *oval*) and, posterior to it, the small **foramen spinosum** are passageways for parts of the trigeminal nerve of the head. The **foramen rotundum,** anterior to the foramen ovale, is the passageway for a major nerve of the face. Directly medial to the foramen ovale, where the sphenoid joins the temporal bone, is the **foramen**

lacerum (LA-se-rum; *lacerare*, to tear), where the auditory (eustachian) tube enters the skull. The sphenoid contribution of the foramen lacerum is visible in Figure 14.7 as the notch lateral to the posterior clinoid process. Frequently, the carotid canal merges with the nearby foramen lacerum to form a single passageway.

Superior to the foramen rotundum is a cleft in the sphenoid, the **superior orbital fissure,** where nerves to the ocular muscles pass. The **inferior orbital fissure** is the crevice at the inferior margin of the sphenoid. At the base of the anterior clinoid process is the **optic canal,** where the optic nerve enters the skull to carry visual signals to the brain. Medial to the optic canals is an **optic groove** that lies transverse on the tuberculum sellae.

Study Tip Using Foramina as Landmarks

Notice the positions of the foramina as they line up along the sphenoid. This pattern is very similar on all human skulls. Use the foramen ovale as a landmark, because it is easy to identify by its oval shape. Anterior to the foramen ovale is the foramen rotundum; posterior is the foramen spinosum. Medial to the foramen ovale is the foramen lacerum with the nearby carotid canal. ■

Figure 14.7 The Sphenoid

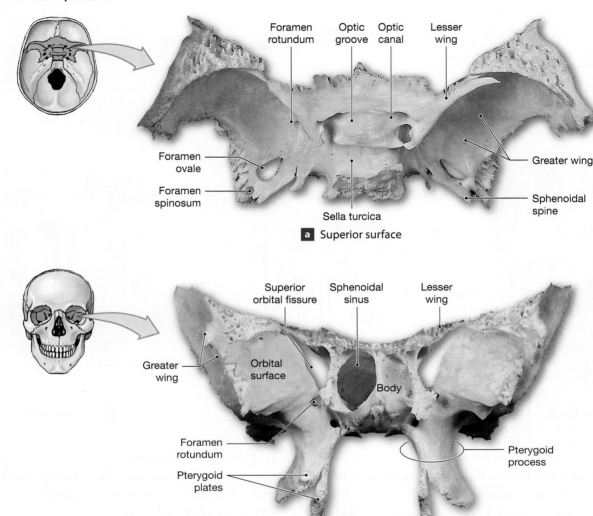

a Superior surface

b Anterior surface

The Ethmoid

The ethmoid (**Figure 14.8**) is a rectangular bone that is anterior to the sphenoid. It forms the medial orbital walls, the roof of the nose and part of the nasal septum, and the anteromedial cranial floor. On the superior surface is a vertical crest of bone called the **crista galli** (*crista*, crest + gal-lē, *gallus*, chicken; cock's comb), where membranes that protect and support the brain attach. At the base of the crista galli is a screenlike **cribriform** (*cribrum*, sieve) **plate** punctured by many small **olfactory foramina** that are passageways for branches of the olfactory nerve. The inferior ethmoid has a thin sheet of vertical bone, the **perpendicular plate,** which contributes to the septum of the nasal cavity. On each side of the perpendicular plate are the **lateral masses** that contain the **ethmoidal labyrinth,** which are full of connected **ethmoidal air cells,** also called the *ethmoidal sinuses*, which open into the nasal cavity. Extending inferiorly into the nasal cavity from the lateral masses are the **superior** and **middle nasal conchae.**

QuickCheck Questions

1.1 Where is the sella turcica located?

1.2 Describe the location of the ethmoid in the orbit of the eye.

1.3 Where are the squamous and petrous parts of the temporal bone located?

In the Lab 1

Materials

☐ Skull sectioned horizontally

☐ Disarticulated ethmoid

Figure 14.8 **The Ethmoid**

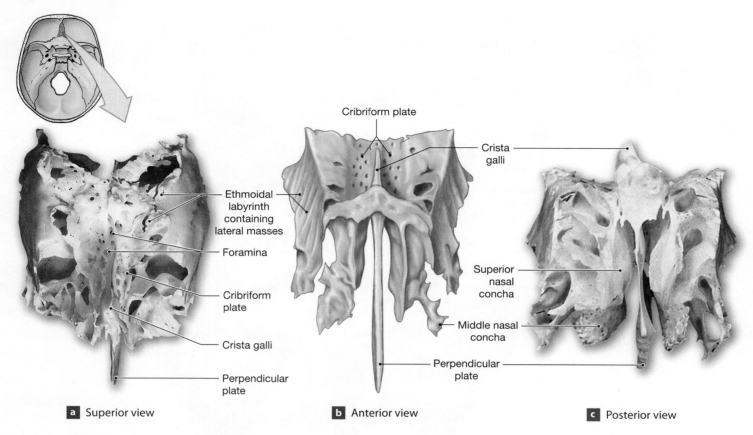

Cribriform plate

Crista galli

Ethmoidal labyrinth containing lateral masses

Foramina

Cribriform plate

Crista galli

Perpendicular plate

Superior nasal concha

Middle nasal concha

Perpendicular plate

a Superior view

b Anterior view

c Posterior view

Procedures

1. Review the cranial structures in Figures 14.3 through 14.8.
2. Locate the frontal bone on the skull.
   - Identify the frontal squama, supraorbital foramen, and lacrimal fossa.
   - Is the metopic suture visible on the skull?
3. Locate the parietal bones on the skull. Examine the lateral surface of a parietal bone and locate the superior and inferior temporal lines.
4. Identify the occipital bone on the skull.
   - Locate the foramen magnum, the occipital condyles, and the hypoglossal canal.
   - Locate the external occipital crest, external occipital protuberance, and superior and inferior nuchal lines.
5. Examine the temporal bones on the skull.
   - Locate the squamous and petrous parts, mastoid processes, and zygomatic processes. Can you feel the mastoid process on your own skull?
   - Find the mandibular fossa.
   - Identify the major passageways of the temporal bone: external and internal auditory meatuses, jugular foramen, and carotid canal.
   - Identify the styloid process and the stylomastoid foramen.
6. Examine the sphenoid and determine its borders with other bones.
   - Identify the lesser wings, greater wings, and sella turcica.
   - Observe the structure of the sella turcica, which includes the anterior, middle, and posterior clinoid processes.
   - Identify each foramen of the sphenoid: ovale, spinosum, rotundum, and lacerum.
   - Locate the optic canal and the superior and inferior orbital fissure.
   - On the inferior sphenoid, identify the pterygoid processes and the pterygoid plates.

7. Identify the ethmoid on the skull. Closely examine its location within the orbit.

- Observe on the floor of the skull the crista galli, cribriform plate, and olfactory foramina.

- Examine the perpendicular plate in the nasal cavity. Examine a disarticulated ethmoid and identify the lateral masses and the superior and middle nasal conchae. ■

Lab Activity 2 Facial Bones

The face is constructed of 14 bones: two nasal, two maxillae, two lacrimal, two zygomatic, two palatine, two inferior nasal conchae, the vomer, and the mandible (see Figure 14.3b). The small **nasal** bones form the bridge of the nose. Lateral to the nasals are the **maxillae**, or maxillary bones; these bones form the floor of the eye orbits and extend inferiorly to form the upper jaw. Below the eye orbits are the **zygomatic** bones, commonly called the cheekbones. At the bridge of the nose, lateral to each maxillary bone, are the small **lacrimal** bones of the medial eye orbitals. Through each lacrimal bone passes a small canal that allows tears to drain into the nasal cavity. The **inferior nasal conchae** (KONG-kē) are the lower shelves of bone in the nasal cavity. The other conchae in the nasal cavity are part of the ethmoid. The bone of the lower jaw is the **mandible.**

On the inferior surface of the skull, the **palatine** bones form the posterior roof of the mouth next to the last molar tooth (see Figure 14.3c). A thin bone called the **vomer** divides the nasal cavity.

The Maxillae

The paired maxillary bones, or *maxillae,* are the foundation of the face (Figure 14.9). Inferior to the orbit is the **infra-orbital fora-men.** The **alveolar** (al-VĒ-ō-lar) **process** consists of the U-shaped ridge where the upper teeth are embedded in the maxilla. From the inferior aspect, the **palatal process** of the maxilla is visible. This bony shelf forms the anterior hard palate of the mouth. At the anterior margin of the palatal process is the **incisive fossa.**

Zygomatic Bones

The zygomatic bones contribute to the inferior and lateral walls of the orbits (Figure 14.9). These bones also contribute to the floor and lateral walls of the orbit. Lateral and slightly inferior to the orbit is the small **zygomaticofacial foramen.** The posterior margin of the zygomatic bone narrows inferiorly to the **temporal process,** which joins the temporal bone's zygomatic process to complete the zygomatic arch.

Nasal Bones

The nasal bones form the bridge of the nose, and the maxilla separates them from the bones of the eye orbit (Figure 14.9). The superior margin of nasal bone articulates with the frontal

Figure 14.9 The Smaller Bones of the Face

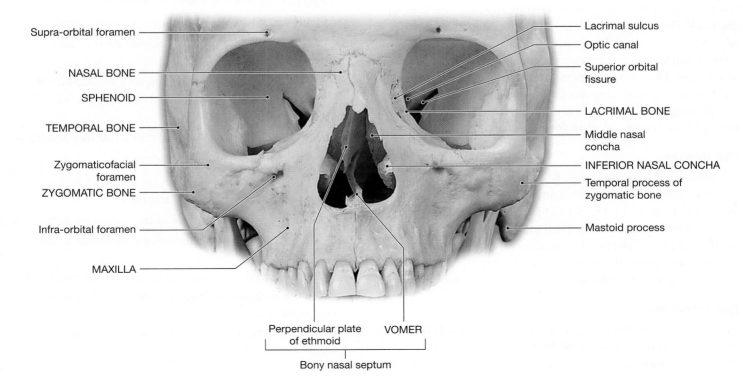

Supra-orbital foramen
NASAL BONE
SPHENOID
TEMPORAL BONE
Zygomaticofacial foramen
ZYGOMATIC BONE
Infra-orbital foramen
MAXILLA

Lacrimal sulcus
Optic canal
Superior orbital fissure
LACRIMAL BONE
Middle nasal concha
INFERIOR NASAL CONCHA
Temporal process of zygomatic bone
Mastoid process

Perpendicular plate of ethmoid
VOMER
Bony nasal septum

bone at the **frontonasal suture;** the posterior surface joins the ethmoid deep in the skull.

Lacrimal Bones

The lacrimal bones are the anterior portions of the medial orbital wall (Figure 14.9). Each lacrimal bone is named after the lacrimal glands that produce tears to lubricate and protect the eyeball. Tears flow medially across the eye and drain into the **lacrimal fossa,** which directs the lacrimal secretions toward the nasal cavity.

Inferior Nasal Conchae

The inferior nasal conchae are shelves that extend medially from the lower lateral portion of the nasal wall (Figure 14.9). They cause inspired air to swirl in the nasal cavity so that the moist mucous membrane lining can warm, cleanse, and moisten the air. Similar shelves of bone occur on the lateral walls of the ethmoid bone.

Palatine Bones

The palatine bones are posterior to the palatine processes of the maxilla. The palatine bones and maxillary bones fashion the roof of the mouth and separate the oral cavity from the nasal cavity (**Figure 14.10**). This separation of cavities allows us to chew and breathe at the same time. Only the inferior surfaces of the palatine bones are completely visible (see Figure 14.3c). The superior surface forms the floor of the nasal cavity and supports the base of the vomer. On the lateral margins of the bone are the **greater palatine foramen** and the **lesser palatine foramen** (Figure 14.3c).

The Vomer

The vomer is the inferior part of the **nasal septum,** the bony wall that partitions the nasal chamber into right and left nasal cavities (Figures 14.9 and 14.10). The vomer is also visible from the inferior aspect of the skull looking into the nasal cavities (Figure 14.3c). The nasal septum consists of two bones: the perpendicular plate of the ethmoid at the superior portion of the septum, and the vomer in the inferior part of the septum.

The Mandible

The mandible of the inferior jaw has a horizontal **body** that extends to a posterior **angle** where the bone turns to a raised projection, the **ramus** (**Figure 14.11**). The superior border of the

ramus terminates at the **mandibular notch** that has two processes extend upward: the anterior **coronoid** (kuh-RŌ-noyd) **process** and a posterior **condylar process,** also called the **mandibular condyle.** The smooth **articular surface** of the condylar **head** articulates in the mandibular fossa on the temporal bone at the **temporomandibular joint (TMJ).** The **alveolar process** of the mandible is the crest of bone where the lower teeth articulate with the mandible bone. Lateral to the chin, or **mental protuberance** (*mental,* chin), is the **mental foramen.** The medial mandibular surface features the **submandibular fossa,** a depression where the submandibular salivary gland rests against the bone. At the posterior end of the fossa is the **mandibular foramen,** a passageway for the sensory nerve from the lower teeth and gums.

QuickCheck Questions

2.1 Which facial bones contribute to the orbit of the eye?

2.2 Which facial bones form the roof of the mouth?

2.3 How does the mandible bone articulate with the cranium?

In the Lab 2

Material

☐ Skull

Procedures

1. Review the skeletal features of the face in Figures 14.3, 14.9, and 14.10.

2. Locate the maxillae on a skull.
   - Identify the infraorbital foramen below the orbit.
   - Locate the alveolar process, palatine process, and incisive fossa.
   - Feel your hard palate by placing your tongue on the roof of your mouth just behind your upper teeth.

3. Examine the palatine bones.
   - With which part of the maxillary bones do they articulate?
   - Identify the greater palatine foramen.

4. Identify the zygomatic bones.
   - Locate the zygomaticofacial foramen.
   - Locate the temporal process of the zygomatic arch.
   - Which part of the temporal bone contributes to the zygomatic arch?

5. Examine the lacrimal bone and identify the lacrimal fossa.

6. Locate the nasal bones. Which bone occurs between a nasal bone and a lacrimal bone?

7. Locate the vomer both in the inferior view of the skull and in the nasal cavity.

Figure 14.10 Sectional Anatomy of the Skull Medial view of a sagittal section through the skull.

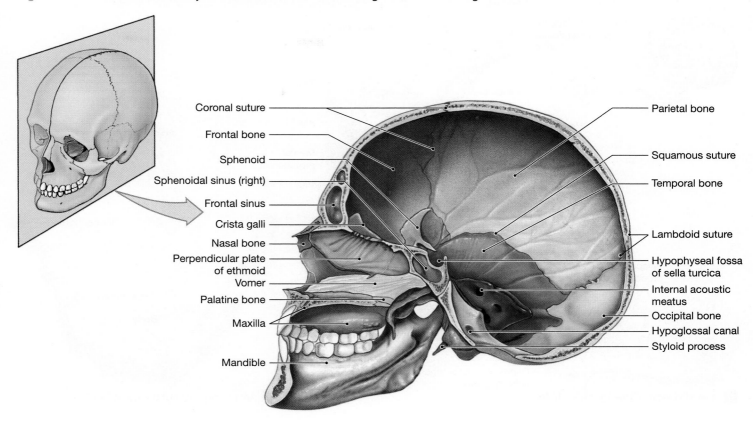

Coronal suture
Frontal bone
Sphenoid
Sphenoidal sinus (right)
Frontal sinus
Crista galli
Nasal bone
Perpendicular plate of ethmoid
Vomer
Palatine bone
Maxilla
Mandible

Parietal bone
Squamous suture
Temporal bone
Lambdoid suture
Hypophyseal fossa of sella turcica
Internal acoustic meatus
Occipital bone
Hypoglossal canal
Styloid process

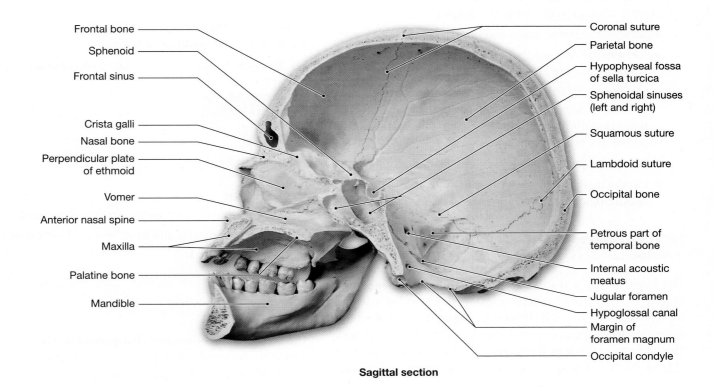

Frontal bone
Sphenoid
Frontal sinus
Crista galli
Nasal bone
Perpendicular plate of ethmoid
Vomer
Anterior nasal spine
Maxilla
Palatine bone
Mandible

Coronal suture
Parietal bone
Hypophyseal fossa of sella turcica
Sphenoidal sinuses (left and right)
Squamous suture
Lambdoid suture
Occipital bone
Petrous part of temporal bone
Internal acoustic meatus
Jugular foramen
Hypoglossal canal
Margin of foramen magnum
Occipital condyle

Sagittal section

Figure 14.11 **The Mandible and Hyoid Bone**

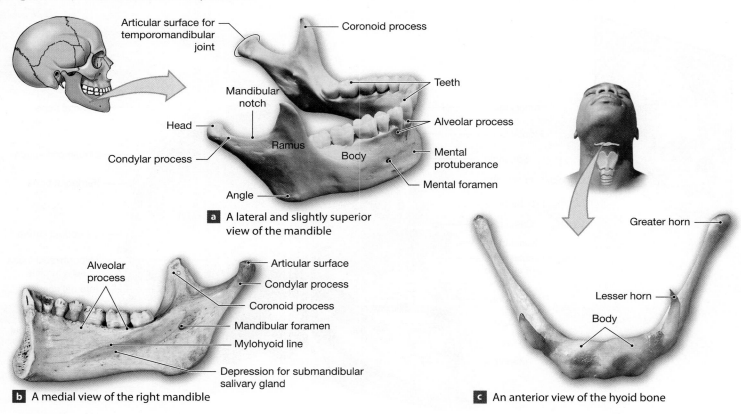

a A lateral and slightly superior view of the mandible

b A medial view of the right mandible

c An anterior view of the hyoid bone

8. Identify the inferior nasal conchae in the nasal cavity.

9. Examine the mandible. Disarticulate this bone from the skull if allowed to do so by your instructor.

   - Identify the body, angle, and ramus.

   - Identify the coronoid process, mandibular notch, and condylar process.

   - Note how the articular surface of the mandible articulates with the temporal bone at the temporomandibular joint. Open and close your mouth to feel this articulation.

   - Locate the alveolar process and the mental protuberance.

   - On the medial surface of the mandible, locate the mandibular groove and the mandibular foramen. ■

Lab Activity 3 | Hyoid Bone

The hyoid bone, a U-shaped bone inferior to the mandible (Figure 14.11c), is unique because it does not articulate with any other bones. The hyoid is difficult to palpate because the bone is surrounded by ligaments and muscles of the throat and neck. Two hornlike processes for muscle attachment occur on each side of the hyoid bone, an anterior **lesser horn** and a larger

posterior **greater horn.** These bony projections are also called the **lesser** and **greater cornua** (KOR-nū-uh; *cornu-*, horn).

QuickCheck Questions

3.1 Where is the hyoid bone located?

3.2 Does the hyoid bone articulate with other bones?

In the Lab 3

Material

☐ Articulated skeleton

Procedures

1. Examine the hyoid bone on an articulated skeleton.

2. Identify the greater and lesser horns of the hyoid bone. ■

Lab Activity 4 | Paranasal Sinuses of the Skull

The skull contains cavities called **paranasal sinuses** that connect with the nasal cavity (**Figure 14.12**). The sinuses lighten the skull and, like the nasal cavity, are lined with a mucous

Figure 14.12 Paranasal Sinuses

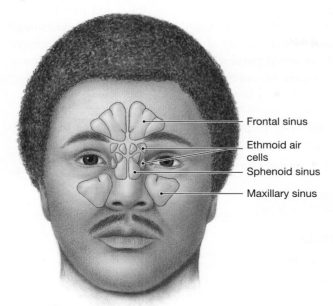

a This anterior view shows the general location of the four paranasal sinuses.

- Frontal sinus
- Ethmoid air cells
- Sphenoid sinus
- Maxillary sinus

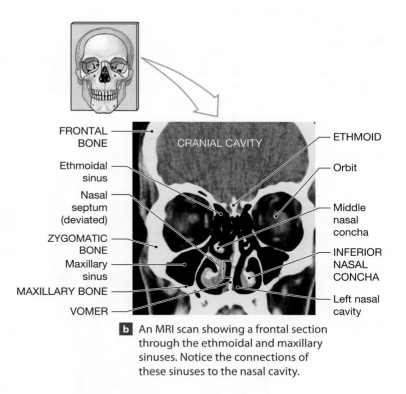

b An MRI scan showing a frontal section through the ethmoidal and maxillary sinuses. Notice the connections of these sinuses to the nasal cavity.

- FRONTAL BONE
- Ethmoidal sinus
- Nasal septum (deviated)
- ZYGOMATIC BONE
- Maxillary sinus
- MAXILLARY BONE
- VOMER
- CRANIAL CAVITY
- ETHMOID
- Orbit
- Middle nasal concha
- INFERIOR NASAL CONCHA
- Left nasal cavity

Clinical Application Sinus Congestion

In some individuals, allergies or changes in the weather can make the sinus membranes swell and secrete more mucus. The resulting congestion blocks connections with the nasal cavity, and the increased sinus pressure is felt as a headache. The sinuses also serve as resonating chambers for the voice, much like the body of a guitar amplifies its music, and when the sinuses and nasal cavity are congested, the voice sounds muffled. ■

membrane that cleans, warms, and moistens inhaled air. The **frontal sinus** extends laterally over the orbit of the eyes. The **sphenoidal sinus** is located in the sphenoid directly inferior to the sella turcica. The **ethmoid labyrinth** houses **ethmoidal air cells** that collectively constitute the **ethmoidal sinus.** Each maxilla contains a large **maxillary sinus** positioned lateral to the nasal cavity.

QuickCheck Questions

4.1 What are the names of the various paranasal sinuses?

4.2 What are the functions of the paranasal sinuses?

In the Lab 4

Material

☐ Skull (midsagittal section)

Procedures

1. Compare the frontal sinus on several sectioned skulls. Is the sinus the same size on each skull?

2. Locate the sphenoidal sinus on a sectioned skull. Under which sphenoidal structure is this sinus located?

3. Examine the maxillary sinus on a sectioned skull. How does the size of this sinus compare with the sizes of the other three sinuses?

4. Identify the ethmoidal air cells. What sinus do these cells collectively form? ■

Lab Activity 5 Fetal Skull

As the fetal skull develops, the cranium must remain flexible to accommodate the growth of the brain. This flexibility is possible because the cranial bones remain incompletely fused until after birth (Figure 14.13). Between wide developing sutures are expanses of fibrous connective tissue called **fontanels** (fon-tuh-NELZ). It is these so-called *soft spots* that allow the skull to expand as brain size increases and enable the skull to flex in order to squeeze through the birth canal during delivery. By the age of four to five years, the brain is nearly adult size, the fibrous connective tissue of the fontanels ossifies, and the cranial sutures interlock to securely support the articulating bones.

Four major fontanels are present at birth: the large **anterior fontanel** is between the frontal and parietal bones where

Figure 14.13 The Skull of an Infant The skull of an infant contains more individual bones than that of an adult. Many of the bones eventually fuse; thus, the adult skull has fewer bones. The flat bones of the skull are separated by areas of fibrous connective tissue, allowing for cranial expansion and the distortion of the skull during birth. The large fibrous areas are called fontanels. By about age four or five, these areas will disappear.

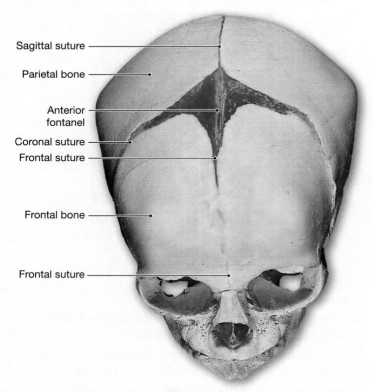

Sagittal suture
Parietal bone
Anterior fontanel
Coronal suture
Frontal suture
Frontal bone
Frontal suture

a Anterior/superior view

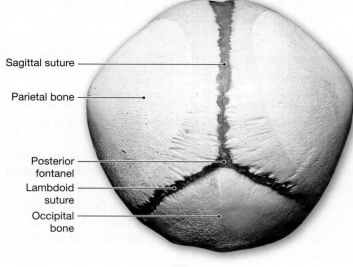

Sagittal suture
Parietal bone
Posterior fontanel
Lambdoid suture
Occipital bone

b Posterior view

the frontal, coronal, and sagittal sutures intersect; the **posterior fontanel** is at the juncture of the occipital and parietal bones where the lambdoid and sagittal sutures join; the **sphenoidal fontanel** is at the union of the coronal and squamous sutures; and the **mastoid fontanel** is posterolateral where the squamous and lambdoid sutures meet.

QuickCheck Questions

5.1 What does the presence of fontanels allow the fetal skull to do?

5.2 How long are fontanels present in the skull?

In the Lab 5

Materials

☐ Fetal skull

☐ Adult skull

Procedures

1. Identify each fontanel on a fetal skull, using Figure 14.13 as a guide.

2. Compare the fetal and adult skulls. Which has more bones? ∎

Lab Activity 6 Vertebral Column

The **vertebral column,** or **spine,** is a flexible chain of 26 bones; 24 vertebrae (singular: *vertebra*), the sacrum, and the coccyx. The column articulates at the superior end with the skull, the inferior portion with the pelvic girdle, and the ribs laterally. The bones of the vertebral column are grouped into five regions based on location and anatomical features (**Figure 14.14**). Starting at the superior end of the spine, the first seven vertebrae are the **cervical** vertebrae of the neck. Twelve **thoracic** vertebrae articulate with the ribs. The lower back has five **lumbar** vertebrae, and a single **sacrum** joining the hips is comprised of five fused **sacral vertebrae.** The **coccyx** (KOK-siks), commonly called the *tailbone*, is the inferior portion of the spine and consists of (usually) four fused **coccygeal vertebrae.**

The vertebral column is curved to balance the body weight while standing. Toward the end of gestation, the fetal spine develops **accommodation curves** in the thoracic and sacral regions, curves that provide space for internal organs in these regions. Because accommodation curves occur first they are also called **primary curves.** At birth, the accommodation curves are still forming, and the vertebral column is relatively straight. During early childhood, as the individual learns to hold the head up, crawl, and then walk, **compensation (secondary) curves** form in the cervical and lumbar regions

Figure 14.14 The Vertebral Column The major divisions of the vertebral column, showing the four spinal curvatures.

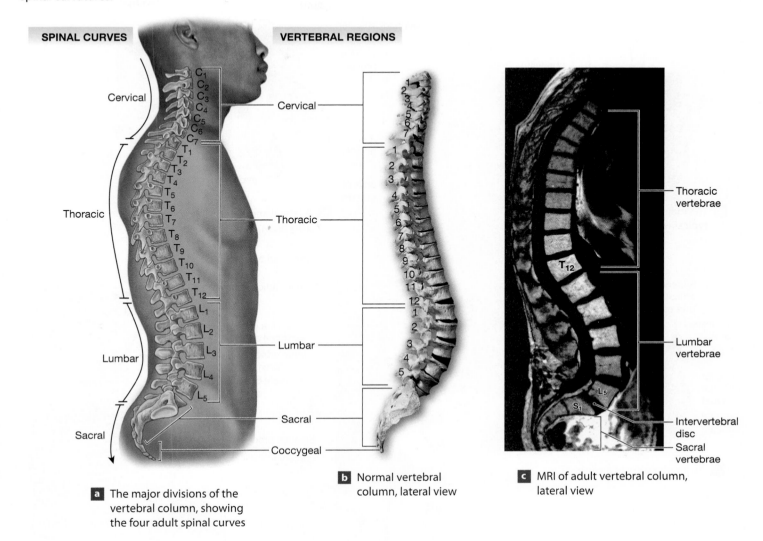

SPINAL CURVES

Cervical

Thoracic

Lumbar

Sacral

VERTEBRAL REGIONS

Cervical

Thoracic

Lumbar

Sacral

Coccygeal

a The major divisions of the vertebral column, showing the four adult spinal curves

b Normal vertebral column, lateral view

c MRI of adult vertebral column, lateral view

Thoracic vertebrae

Lumbar vertebrae

Intervertebral disc

Sacral vertebrae

to move the body weight closer to the body's axis for better balance. Once the child is approximately 10 years old, the spinal curves are established and the fully developed column has alternating secondary and primary curves.

In the cervical, thoracic, and lumbar regions are **intervertebral discs,** cushions of fibrocartilage between the articulating vertebrae. Each disc consists of an outer layer of strong fibrocartilage, the **annulus fibrosus,** surrounding a deeper mass, the **nucleus pulposus.** Water and elastic fibers in the gelatinous mass of the nucleus pulposus absorb stresses that arise between vertebrae whenever a person is either standing or moving.

Vertebral Anatomy

The anatomical features of a typical vertebra include a large, anterior, disc-shaped **vertebral body** (the *centrum*) and a posterior elongated **spinous process** (**Figure 14.15**). Lateral on

each side of the spinous process is a **transverse process.** The **lamina** (LA-mi-na) is a flat plate of bone between the transverse and spinous processes that forms the curved **vertebral arch.** The **pedicle** (PE-di-kul) is a strut of bone extending posteriorly from the vertebral body to a transverse process. The pedicle and lamina on each side form the wall of the large posterior **vertebral foramen,** which contributes to the spinal cavity where the spinal cord is housed. Inferior to the pedicle is an inverted U-shaped region called the **inferior vertebral notch.** Two articulating vertebrae contribute to fashion an **intervertebral foramen,** with the **inferior vertebral notch** of the superior vertebra joining the pedicle of the inferior vertebra. Spinal nerves pass through the intervertebral foramen to access the spinal cord.

The vertebral column moves much like a gooseneck lamp: Each joint moves only slightly, but the combination

Figure 14.15 Vertebral Anatomy The anatomy of a typical vertebra and the arrangement of articulations between vertebrae.

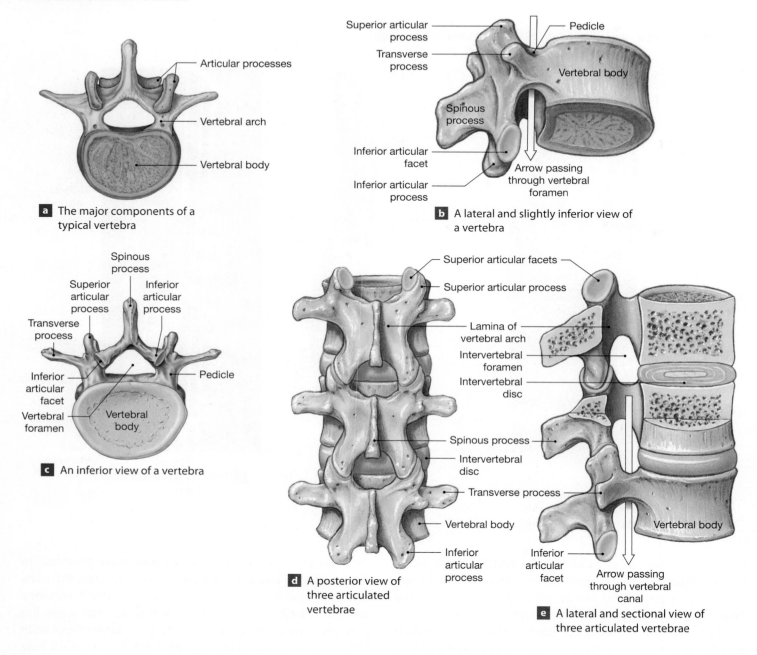

a The major components of a typical vertebra

b A lateral and slightly inferior view of a vertebra

c An inferior view of a vertebra

d A posterior view of three articulated vertebrae

e A lateral and sectional view of three articulated vertebrae

of all the individual movements permits the column a wide range of motion. Joints between adjacent vertebrae occur at smooth articular surfaces called *facets* that project from *articular processes*. The **superior articular process** is on the superior surface of the pedicle of each vertebra and has a **superior articular facet** at the posterior tip. The **inferior articular process** is a downward projection of the inferior lamina wall and has an **inferior articular facet** on the anterior tip. At a vertebral joint, the inferior articular facet of the superior vertebra glides across the superior articular facet of the inferior vertebra. The greatest movement of these joints is in the cervical region for head movement.

Cervical Vertebrae

The seven cervical vertebrae in the neck are recognizable by the presence of a **transverse foramen** on each transverse process (**Figure 14.16**). The vertebral artery travels up the neck through these foramina to enter the skull. The first two cervical vertebrae

Figure 14.16 The Cervical Vertebrae

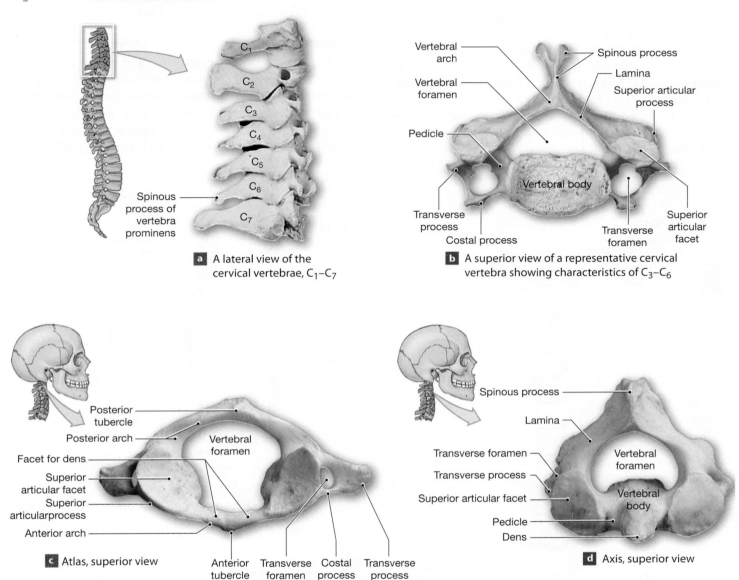

a A lateral view of the cervical vertebrae, C_1–C_7

b A superior view of a representative cervical vertebra showing characteristics of C_3–C_6

c Atlas, superior view

d Axis, superior view

are modified for special articulations with the skull. The tip of the spinous process is **bifid** (branched) in vertebrae C_2 through C_6. The last cervical vertebra, C_7, is called the **vertebra prominens** because of the broad tubercle at the end of the spinous process. The tubercle can be palpated at the base of the neck.

The first cervical vertebra, C_1, is called the **atlas** (Figure 14.16c), named after the Greek mythological character who carried the world on his shoulders. The atlas is the only vertebra that articulates with the skull. The superior articular facets of the atlas are greatly enlarged, and the occipital condyles of the occipital bone fit into the facets like spoons nested together. When you nod your head, the atlas remains stationary while the occipital condyles glide in the facets.

The atlas is unusual in that it lacks a vertebral body and a spinous process and has a very large vertebral foramen formed by the **anterior** and **posterior arches.** A small, rough **posterior tubercle** occurs where the spinous process normally resides. A long spinous process would interfere with occipitoatlas articulation.

The **axis** is the second cervical vertebra, C_2. It is specialized to articulate with the atlas. A peglike **dens** (DENZ; *dens,* tooth), or *odontoid process,* arises superiorly from the body of the axis (Figure 14.16d). It fits against the anterior wall of the vertebral foramen and provides the atlas with a pivot point for when the head is turned laterally and medially. A **transverse ligament** secures the atlas around the dens.

Thoracic Vertebrae

The 12 thoracic vertebrae, which articulate with the 12 pairs of ribs, are larger than the cervical vertebrae and increase in size as they approach the lumbar region. Most ribs attach to their thoracic vertebra at two sites on the vertebra: on a **transverse costal facet** at the tip of the transverse process and on a **costal facet** located on the posterior of the vertebral body (**Figure 14.17**). Two costal facets usually are present on the same vertebral body, a **superior costal facet** and an **inferior costal facet.** The costal facets are unique to the thoracic vertebrae, and there is variation in where these facets occur on the various thoracic vertebrae.

Lumbar Vertebrae

The five lumbar vertebrae are large and heavy in order to support the weight of the head, neck, and trunk. Compared with thoracic vertebrae, lumbar vertebrae have a wider body, a blunt and horizontal spinous process, and shorter transverse processes (**Figure 14.18**). The lumbar vertebral foramen is smaller than that in thoracic vertebrae. To prevent the back

Figure 14.17 **The Thoracic Vertebrae**

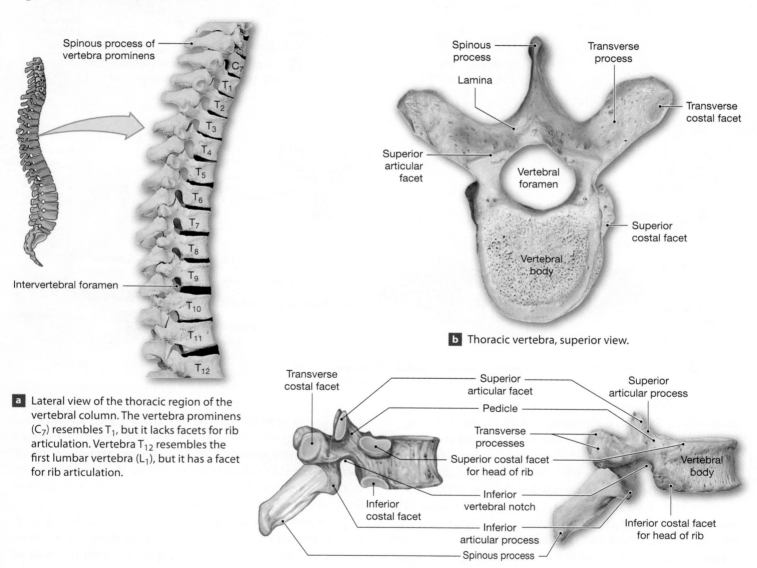

a Lateral view of the thoracic region of the vertebral column. The vertebra prominens (C_7) resembles T_1, but it lacks facets for rib articulation. Vertebra T_{12} resembles the first lumbar vertebra (L_1), but it has a facet for rib articulation.

b Thoracic vertebra, superior view.

c A representative thoracic vertebra, lateral view.

Figure 14.18 **The Lumbar Vertebrae**

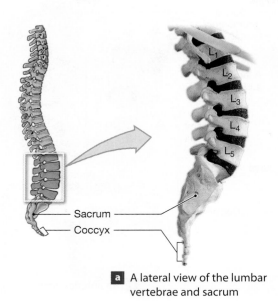

a A lateral view of the lumbar vertebrae and sacrum

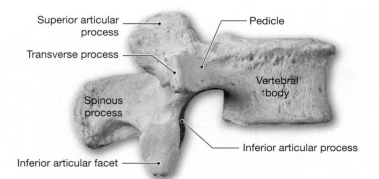

b A lateral view of a typical lumbar vertebra

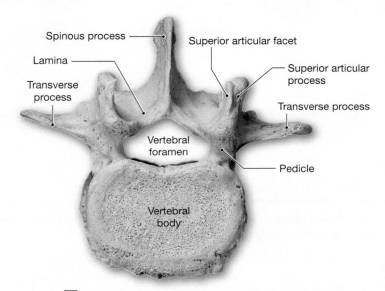

c A superior view of the same vertebra shown in part b

from twisting when objects are being lifted or carried, the lumbar superior articular process is turned medially and the lumbar inferior articular processes are oriented laterally to interlock the lumbar vertebrae. No facets or transverse foramina occur on the lumbar vertebrae.

Sacral and Coccygeal Vertebrae

As noted earlier, the sacrum is a single bony element composed of five fused sacral vertebrae (**Figure 14.19**). It articulates with the ilium of the pelvic girdle to form the posterior wall of the pelvis. Fusion of the sacral bones before birth consolidates the vertebral canal into the **sacral canal.** On the fifth sacral vertebra, the sacral canal opens as the **sacral hiatus** (hi-Ā-tus). Along the lateral margin of the fused vertebral bodies are **sacral foramina.** The spinous processes fuse to form an elevation called the **median sacral crest.** A **lateral sacral crest** extends from the lateral margin of the sacrum. The sacrum articulates with each pelvic bone at the large **auricular surface** on the

lateral border. Dorsal to this surface is the **sacral tuberosity,** where ligaments attach to support the **sacroiliac joint.**

The coccyx (Figure 14.19) articulates with the fifth fused sacral vertebra at the **coccygeal cornu.** There may be anywhere from three to five coccygeal bones, but most people have four.

QuickCheck Questions

6.1 What are the five major regions of the vertebral column and the number of vertebrae in each region?

6.2 What are three features found on all vertebrae?

In the Lab 6

Materials

☐ Articulated skeleton

☐ Articulated vertebral column

☐ Disarticulated vertebral column

Figure 14.19 **The Sacrum and Coccyx**

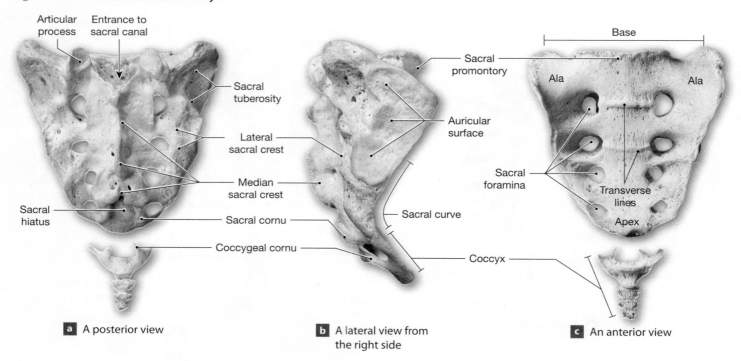

Articular process

Entrance to sacral canal

Sacral tuberosity

Lateral sacral crest

Median sacral crest

Sacral hiatus

Sacral cornu

Coccygeal cornu

a A posterior view

Sacral promontory

Auricular surface

Sacral curve

Coccyx

b A lateral view from the right side

Base

Ala

Ala

Sacral foramina

Transverse lines

Apex

c An anterior view

Sketch to Learn

Let's draw a lateral view of a thoracic vertebra and label the facets where articulations occur. Refer to a thoracic vertebra or Figure 14.17c while sketching.

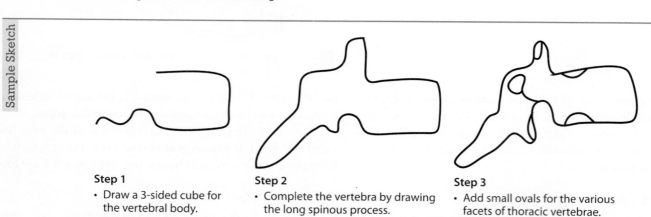

Sample Sketch

Step 1
- Draw a 3-sided cube for the vertebral body.
- Add a notch for the intervertebral foramen.

Step 2
- Complete the vertebra by drawing the long spinous process.

Step 3
- Add small ovals for the various facets of thoracic vertebrae.
- Label your sketch.

Your Sketch

Procedures

1. Review the vertebral anatomy presented in Figures 14.14 to 14.19.
2. Identify the four regions of the vertebral column on an articulated skeleton.
3. Describe the type of curves found in each region.
4. Describe the anatomy of a typical vertebra. Locate each feature on a disarticulated vertebra.
    - Distinguish the anatomical differences among cervical, thoracic, and lumbar vertebrae.
    - Identify the unique features of the atlas and the axis. How do these two vertebrae articulate with the skull and with each other?
    - Discuss how a lumbar vertebra differs from a thoracic vertebra.
5. Describe the anatomy of the sacrum and the coccyx. ■

Lab Activity 7 Thoracic Cage

The 12 pairs of ribs articulate with the thoracic vertebrae posteriorly and the sternum anteriorly to enclose the thoracic organs in a protective cage. In breathing, muscles move the ribs to increase or decrease the size of the thoracic cavity and cause air to move into or out of the lungs.

Sternum

The **sternum** is the flat bone located anterior to the thoracic region of the vertebral column. It is composed of three bony elements: a superior **manubrium** (ma-NOO-brē-um), a middle **sternal body,** and an inferior **xiphoid** (ZĪ-foyd) **process** (**Figure 14.20**). The manubrium is triangular and articulates with the first pair of ribs and the clavicle. Muscles that move the head and neck attach to the manubrium. The sternal body is elongated and receives the costal cartilage of ribs 2 through 7. The xiphoid process is shaped like an arrowhead and projects inferiorly off the sternal body. This process is cartilaginous until late adulthood, when it completely ossifies.

Ribs

Ribs, also called **costae,** are classified according to how they articulate with the sternum (Figure 14.20). The first seven pairs are called either **true ribs** or **vertebrosternal ribs** because their cartilage, the **costal cartilage,** attaches directly to the sternum. Rib pairs 8 through 12 are called **false ribs** or **vertebrochondral ribs** because their costal cartilage does not connect directly with the sternum but instead fuse with the

costal cartilage of rib 7. Rib pairs 11 and 12 are called **floating ribs** or **vertebral ribs** because they do not articulate with the sternum.

Each rib has a **head,** or **capitulum** (ka-PIT-ū-lum), and on the head are two **articular facets** for articulating with the costal facets of the rib's thoracic vertebra. The **tubercle** of the rib articulates with the transverse costal facet of the rib's vertebra. Between the head and tubercle is a slender **neck.**

Differences in the way ribs articulate with the thoracic vertebrae are reflected in variations in the vertebral costal facets. Vertebrae T_1 through T_8 all have paired costal facets, one superior and one inferior as noted in our previous thoracic discussion. The first rib articulates with a transverse costal facet of T_1. The second rib articulates with the inferior costal facet of T_1 and the superior costal facet of T_2. Ribs 3 through 9 continue this pattern of articulating with two adjacent costal facets. Vertebrae T_9 through T_{12} have a single costal facet on the vertebral body, and the ribs articulate entirely on the one costal facet. After each rib articulates on the single costal facet, the rib bends laterally and articulates on the transverse costal facet. Rib pairs 11 and 12 do not articulate on costal facets.

QuickCheck Questions

7.1 Which part of the sternum articulates with the clavicle?
7.2 Which ribs are true ribs, which are false ribs, and which are floating ribs?

In the Lab 7

Materials

- ☐ Articulated skeleton
- ☐ Articulated vertebral column with ribs
- ☐ Disarticulated vertebral column and ribs

Procedures

1. Review the anatomy in Figure 14.20.
2. Identify the manubrium, body, and xiphoid process of the sternum.
3. Discuss the anatomy of a typical rib.
    - How many pairs of ribs do human males have? How many pairs do human females have?
    - Describe the anatomical features involved in the articulation of a rib on a thoracic vertebra.
    - Identify the differences of articular facets along the thoracic region and relate this to how each rib articulates with the vertebrae. ■

Figure 14.20 The Thoracic Cage The thoracic cage is the articulated ribs and sternum.

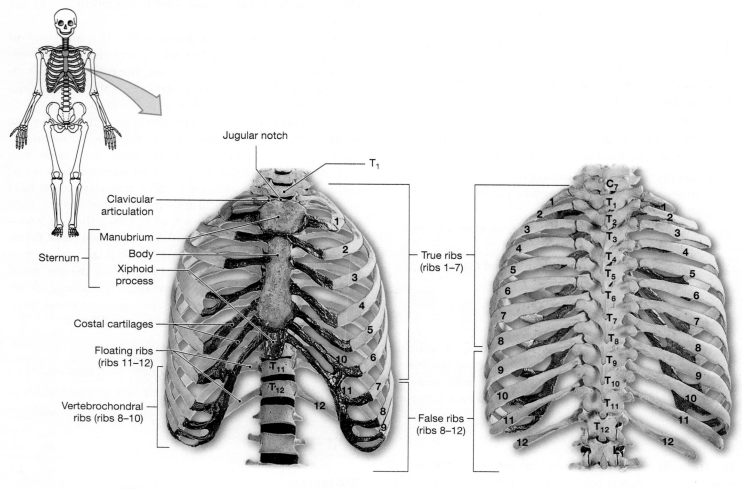

a Anterior view of the rib cage and sternum

b Posterior view of the rib cage

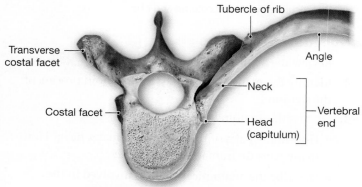

c A superior view of the articulation between a thoracic vertebra and the vertebral end of a left rib

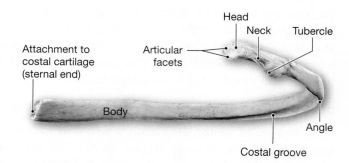

d A posterior and medial view showing major anatomical landmarks on an isolated left rib (rib 10)

Name _____

Date _____

Section _____

Axial Skeleton

A. Identification

Identify the bone that each structure occurs on.

1. sella turcica

2. crista galli

3. external acoustic meatus

4. foramen magnum

5. zygomatic process

6. condylar process

7. mandibular fossa

8. styloid process

9. coronoid process

10. jugular foramen

11. superior nuchal line

12. superior temporal line

B. Description

Describe each structure of the vertebral column and thoracic cage.

1. spinous process

2. transverse foramen

3. manubrium

4. capitulum

5. pedicle

6. xiphoid process

7. dens

8. vertebral foramen

9. vertebrochondral rib

10. costal facet

C. Labeling

1. Label **Figure 14.21**, an anterior view of the skull.

Figure 14.21 Anterior View of the Skull

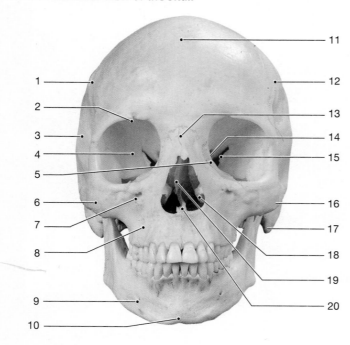

1. _____
2. _____
3. _____
4. _____
5. _____
6. _____
7. _____
8. _____
9. _____
10. _____
11. _____
12. _____
13. _____
14. _____
15. _____
16. _____
17. _____
18. _____
19. _____
20. _____

2. Label **Figure 14.22**, a lateral view of the skull.

Figure 14.22 **Lateral View of the Skull**

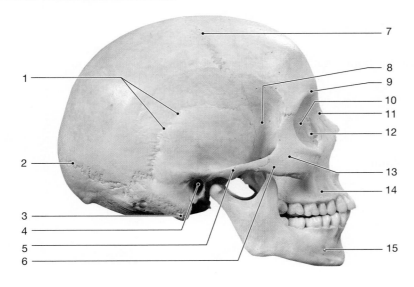

1. _____
2. _____
3. _____
4. _____
5. _____
6. _____
7. _____
8. _____
9. _____
10. _____
11. _____
12. _____
13. _____
14. _____
15. _____

3. Label **Figure 14.23**, a typical cervical vertebra.

Figure 14.23 **Typical Cervical Vertebra, Superior View**

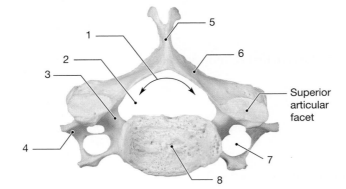

Superior articular facet

1. _____
2. _____
3. _____
4. _____
5. _____
6. _____
7. _____
8. _____

D. Short-Answer Questions

1. List the three primary components of the axial skeleton.

2. How many bones are found in the cranium and the face?

3. Describe the three cranial fossae and the bones that form the floor of each.

4. List the six primary sutures of the skull and the bones that articulate at each suture.

5. Describe the five regions of the vertebral column.

E. Analysis and Application

1. Name two passageways in the floor of the skull for major blood vessels that serve the brain.

2. Describe the skeletal features at the point where the vertebral column articulates with the skull.

3. Compare the articulation on the thoracic vertebrae of rib pairs 7 and 10.

F. Clinical Challenge

A patient is scheduled for surgery to correct a deviated nasal septum. Which bones and other facial features may be involved in this procedure?

Appendicular Skeleton

Learning Outcomes

On completion of this exercise, you should be able to:

1. Identify the bones and surface features of the pectoral girdle and upper limbs.

2. Articulate the clavicle with the scapula.

3. Articulate the scapula, humerus, radius, and ulna.

4. Identify the bones and surface features of the pelvic girdle and lower limbs.

5. Articulate the coxal bones with the sacrum to form the pelvis.

6. Articulate the coxa, femur, tibia, and fibula.

The appendicular skeleton provides the bony structure of the limbs, permitting us to move and to interact with our surroundings. It is attached to the vertebral column and sternum of the axial skeleton. The appendicular skeleton consists of a pectoral girdle and the attached upper limbs and a pelvic girdle and the attached lower limbs (**Figure 15.1**). The pectoral girdle is loosely attached to the axial skeleton, and as a result the shoulder joints have a great range of movement. The pelvic girdle is securely attached to the sacrum of the spine to support the weight of the body.

Need More Practice and Review?

Build your knowledge—and confidence!—in the Study Area of MasteringA&P® at www.masteringaandp.com with Pre-lab Quizzes, Post-lab Quizzes, Practice Anatomy Lab™ (PAL™) 3.0 virtual anatomy practice tool, PhysioEx™ 9.0 laboratory simulations, and A&P Flix™ with Quizzes.

PAL ᴾʳᵃᶜᵗⁱᶜᵉ ᵃⁿᵃᵗᵒᵐʸ ˡᵃᵇ   For this lab exercise, follow these navigation paths:
   • PAL>Human Cadaver>Appendicular Skeleton
   • PAL>Anatomical Models>Appendicular Skeleton

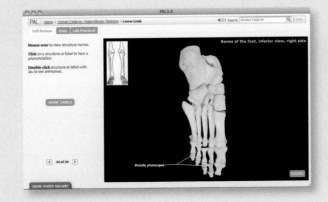

MasteringA&P®

Figure 15.1 The Appendicular Skeleton An anterior view of the skeleton detailing the appendicular components. The numbers in the boxes indicate the number of bones in each type or within each category.

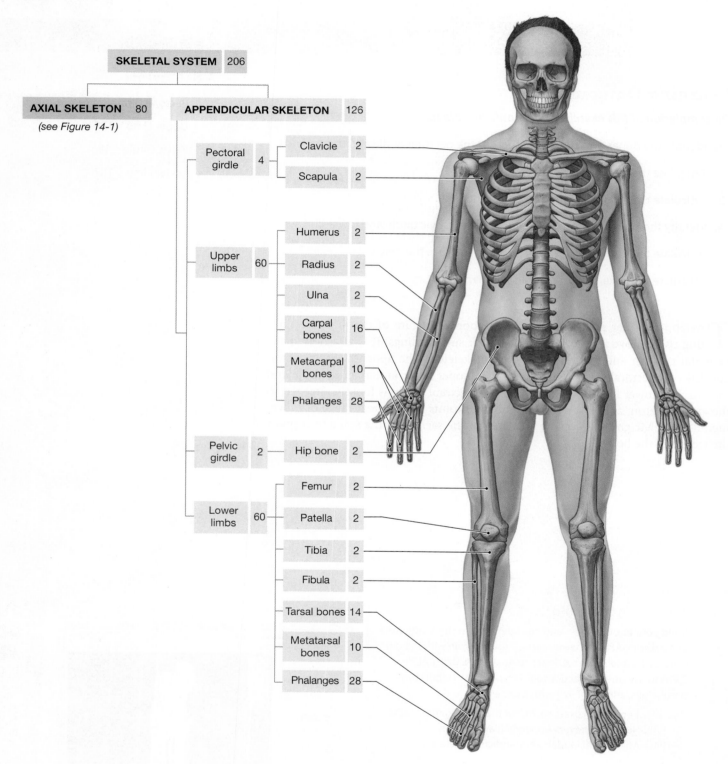

As you study the appendicular skeleton, keep in mind that each bone is one member of a left/right pair. Carefully observe the orientation of major surface features on the bones and use these features as landmarks for determining whether a given bone is from the left side of the body or from the right side.

Lab Activity 1 Pectoral Girdle

The **pectoral girdle** consist of four bones: two *clavicles*, commonly called collarbones, and two *scapulae*, the shoulder blades. These four bones are arranged in an incomplete ring that constitutes the bony architecture of the superior trunk. Each scapula rests against the posterior surface of the rib cage and against a clavicle, and provides an anchor for tendons of arm and shoulder muscles. The clavicles are like struts, providing support by connecting the scapulae to the sternum.

Clavicle

The S-shaped **clavicle** (KLAV-i-kul) is the only bony connection between the pectoral girdle and the axial skeleton. The **sternal end** articulates medially with the sternum, and laterally the flat **acromial** (a-KRO-mē-al) **end** joins the scapula (**Figure 15.2**). Inferiorly, toward the acromial end, where the clavicle bends, is the **conoid tubercle,** an attachment site for the coracoclavicular ligament. Near the inferior sternal end is the rough **costal tuberosity.**

The sternal end of the clavicle articulates lateral to the jugular notch on the manubrium of the sternum. The point where these two bones articulate is called the **sternoclavicular joint.** From this joint, the clavicle curves posterior and articulates with the scapula at the **acromioclavicular joint.**

Scapula

The **scapula** (SKAP-ū-la) is composed of a triangular **body** defined by long edges called the **superior, medial,** and **lateral borders** (**Figure 15.3**). The corners where the borders meet are the **superior, lateral,** and **inferior angles.** An indentation in the superior border is the **suprascapular notch.** The **subscapular fossa** is the smooth, triangular surface where the anterior surface of the scapula faces the ribs.

A prominent ridge, the **spine,** extends across the scapula body on the posterior surface and divides the convex surface into the **supraspinous fossa** superior to the spine and the **infraspinous fossa** inferiorly. At the lateral tip of the spine is the **acromion** (a-KRŌ-mē-on), which is superior to the **glenoid cavity** (also called the *glenoid fossa*) where the humerus articulates. Superior and inferior to the glenoid cavity are the **supraglenoid** and **infraglenoid tubercles** where the biceps brachii and triceps brachii muscles of the arm attach. Superior to the glenoid cavity is the beak-shaped **coracoid** (KOR-uh-koyd) **process.** The **scapular neck** is the ring of bone around the base of the coracoid process and the glenoid cavity.

QuickCheck Questions

1.1 Which bones form the pectoral girdle?

1.2 Where does the clavicle articulate with the axial skeleton?

Figure 15.2 The Clavicle

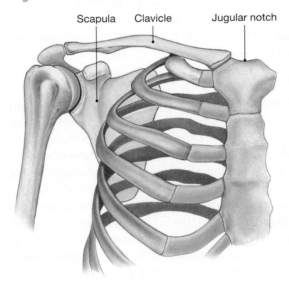

a The position of the clavicle within the pectoral girdle, anterior view.

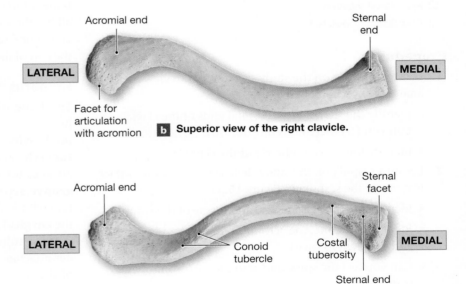

b Superior view of the right clavicle.

c Inferior view of the right clavicle. Stabilizing ligaments attach to the conoid tubercle and the costal tuberosity.

Figure 15.3 The Scapula

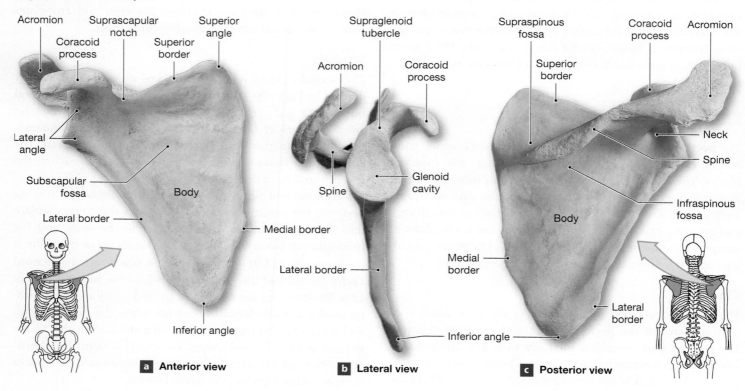

a Anterior view

b Lateral view

c Posterior view

Materials

☐ Articulated skeleton

☐ Disarticulated skeleton

Procedures

1. Locate a clavicle on the study skeleton and review the anatomy shown in Figure 15.2.

   ▪ Identify the sternal and acromial ends of the clavicle. Can you feel these ends on your own clavicles?

   ▪ Identify the conoid tubercle of the clavicle.

2. Locate a scapula on the study skeleton. Review the surface features of the scapula in Figure 15.3.

   ▪ Identify the borders, angles, and fossae of the scapula.

   ▪ Identify the spine, the acromion, the coracoid process, and the glenoid cavity.

   ▪ Can you feel the spine and acromion on your own scapula?

3. Place a clavicle from the disarticulated skeleton on your shoulder and determine how it would articulate with your scapula. ■

Each **upper limb,** also called an *upper extremity,* includes the bones of the arm, forearm, and hand—a total of 30 bones, with all but three of them in the wrist and hand. Note that the correct anatomical usage of the term *arm* is in reference to the **brachium,** the part of the upper limb between the shoulder and elbow.

Humerus

The bone of the **arm** (brachium) is the **humerus,** shown in **Figure 15.4**. The proximal **head** articulates with the glenoid cavity of the scapula. Lateral to the head is the **greater tubercle,** and medial to the head is the **lesser tubercle;** both are sites for muscle tendon attachment. The **intertubercular groove** separates the tubercles. Between the head and the tubercles is the **anatomical neck;** inferior to the tubercles is the **surgical neck.** Inferior to the greater tubercle is the rough **deltoid tuberosity,** where the deltoid muscle of the shoulder attaches. Along the diaphysis, at the inferior termination of the deltoid tuberosity, is the **radial groove,** a depression that serves as the passageway for the radial nerve.

The distal end of the humerus has a specialized **condyle** to accommodate two joints: the hingelike elbow joint and a pivot joint of the forearm, the latter used when doing

Figure 15.4 The Humerus

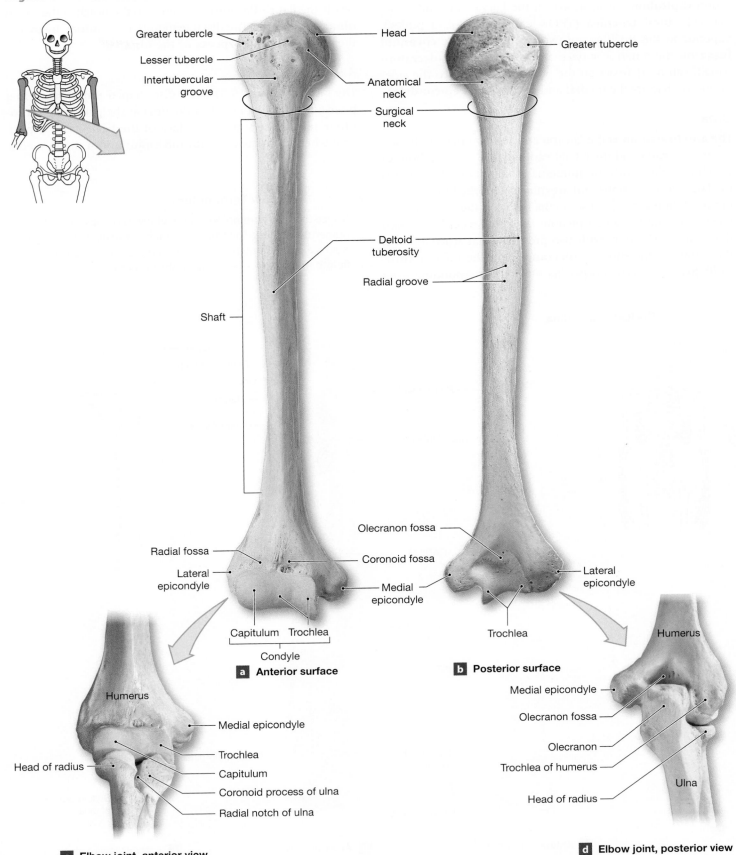

Greater tubercle

Lesser tubercle

Intertubercular groove

Head

Anatomical neck

Greater tubercle

Surgical neck

Shaft

Deltoid tuberosity

Radial groove

Radial fossa

Lateral epicondyle

Coronoid fossa

Olecranon fossa

Medial epicondyle

Lateral epicondyle

Capitulum   Trochlea

Condyle

Trochlea

a Anterior surface

b Posterior surface

Humerus

Medial epicondyle

Trochlea

Capitulum

Coronoid process of ulna

Radial notch of ulna

Head of radius

c Elbow joint, anterior view

Humerus

Medial epicondyle

Olecranon fossa

Olecranon

Trochlea of humerus

Head of radius

Ulna

d Elbow joint, posterior view

such movements as turning a doorknob. The condyle has a round **capitulum** (*capit,* head) on the lateral side and a medial cylindrical **trochlea** (TROK-lē-uh) (*trochlea,* a pulley). Superior to the trochlea are two depressions, the **coronoid fossa** on the anterior surface and the triangular **olecranon** (ō-LEK-ruh-non) **fossa** on the posterior surface. To the sides of the condyle are the **medial** and larger **lateral epicondyles.**

Ulna

The **antebrachium** is the *forearm* and has two parallel bones, the medial **ulna** and the lateral **radius** (**Figure 15.5**), both of which articulate with the humerus at the elbow. The ulna is the larger forearm bone and articulates with the humerus and radius. A fibrocartilage disc occurs between the ulna and the wrist. The ulna has a conspicuous U-shaped **trochlear notch** that is like a C clamp, with two processes that articulate with the humerus: the superior **olecranon** and the inferior **coronoid process.** Each process fits into its corresponding fossa

on the humerus. On the lateral surface of the coronoid process is the flat **radial notch.** Inferior to the notch is the rough **ulnar tuberosity.** The distal extremity is the **ulnar head** and the pointed **styloid process of the ulna.**

Radius

The radius (Figure 15.5) has a disc-shaped **radial head** that pivots in the radial notch of the ulna at the **proximal radio-ulnar joint.** The superior surface of the head has a depression where it articulates with the capitulum of the humerus.

Study Tip   Elbow Terminology

Notice that the terminology of the elbow is consistent in the humerus and ulna. The trochlear notch of the ulna fits into the trochlea of the humerus. The coronoid process and olecranon fit into their respective fossae on the humerus. ∎

Figure 15.5 **The Radius and Ulna**

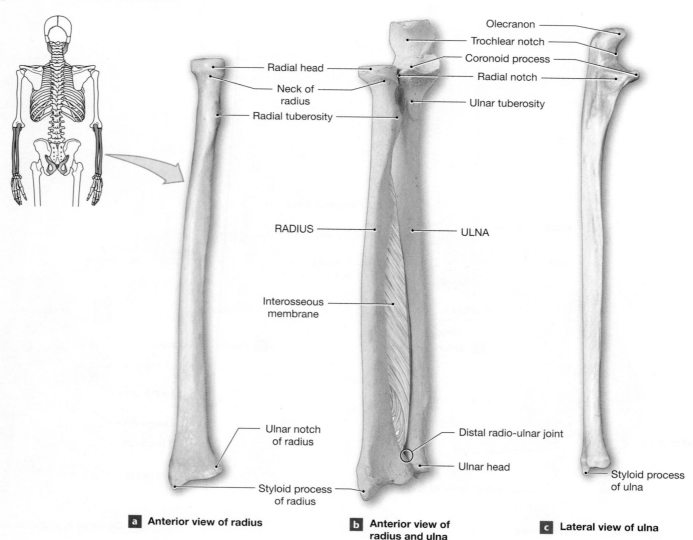

a **Anterior view of radius**

b **Anterior view of radius and ulna**

c **Lateral view of ulna**

Supporting the head is the **neck,** and inferior to the neck is the **radial tuberosity.** On the distal portion, the **styloid process of the radius** is larger and not as pointed as the styloid process of the ulna. The **ulnar notch** on the medial surface articulates with the ulna at the **distal radioulnar joint.** The **interosseous membrane** extends between the ulna and radius to support the bones.

Bones of the Hand

The **carpus** is the *wrist* and consists of eight **carpal** (KAR-pul) **bones** arranged in two rows of four, the **proximal** and **distal carpal bones.** An easy method of identifying the carpal bones is to use the anterior wrist and start with the carpal bone next to the styloid process of the radius (**Figure 15.6**). From this reference point moving medially, the proximal carpal bones are the **scaphoid bone, lunate bone, triquetrum,** and small **pisiform** (PIS-i-form) **bone.** Returning on the lateral side, the four distal carpal bones are the **trapezium, trapezoid bone, capitate bone,** and **hamate bone.** The hamate bone has a process called the **hook of the hamate.**

The five long bones of the palm are **metacarpal bones.** Each metacarpal bone is numbered with a roman numeral, with the lateral metacarpal bone that articulates with the thumb being digit I.

The 14 bones of the fingers are called **phalanges.** Digits II, III, IV, and V each have a **proximal, middle,** and **distal phalanx.** The thumb, or **pollex,** has only proximal and distal phalanges.

QuickCheck Questions

2.1 List the three bones that constitute the arm and forearm.

2.2 What are the three major groups of bones in the hand?

In the Lab 2

Materials

☐ Articulated skeleton

☐ Disarticulated skeleton

Procedures

1. Review the anatomy of the humerus in Figure 15.4.

2. Locate a humerus of your study skeleton and review its surface features.

   ▪ Identify the head, surgical and anatomical necks, tubercles, and intertubercular groove.

   ▪ Identify the deltoid tuberosity and radial groove.

Figure 15.6 Bones of the Hand

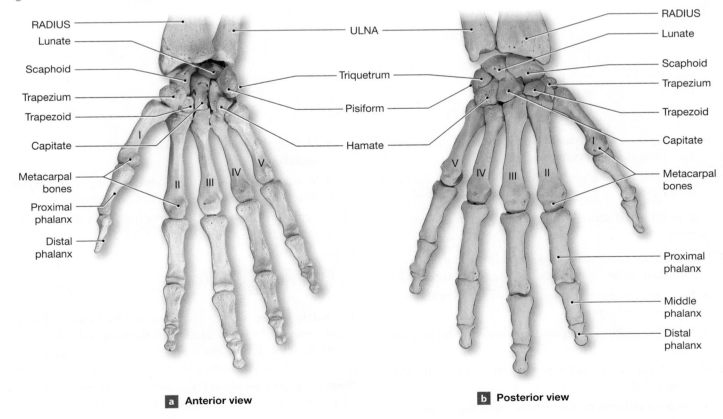

a Anterior view

b Posterior view

Sketch to Learn

Make a quick sketch of the distal end of the humerus to help you remember all those bony features.

Sample Sketch

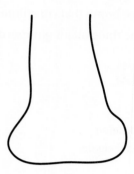 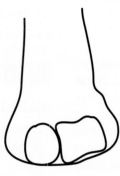 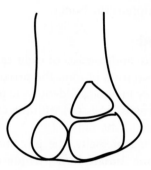

Step 1
- Draw a bell shape (the sides represent the epicondyles).

Step 2
- Add a circle for the capitulum.
- Sketch in a rectangle with top and bottom lines curving inward. This represents the trochlea.

Step 3
- Add a triangle fossa above the trochlea.
- Label your sketch.

Your Sketch

- Identify the epicondyles and the condyle. Can you feel the epicondyles on your own humerus?
- Identify the capitulum, trochlea, and fossae of the distal humerus.

For additional practice, complete the *Sketch to Learn* activity.

3. Locate an ulna and radius of your study skeleton and review their surface features using Figure 15.5 as reference.
   - Study the processes that fit into the corresponding fossae of the humerus.
   - Identify the articulating anatomy between the ulna and radius.

4. Review Figure 15.6 and locate the bones of the hand on the study skeleton.
   - Distinguish between the carpals, metacarpals, and phalanges.

- Identify the four proximal carpals and the four distal carpals.
- Identify the metacarpals and the phalanges. Do all the fingers have the same number of phalanges?

5. Articulate the bones of the upper limb with those of the pectoral girdle. ■

Lab Activity 3 Pelvic Girdle

The **pelvic girdle** is made up of the two hipbones, called the **coxal bones,** which articulate with the vertebral column and attach the lower limbs. The coxal bone, also called the *os coxae* (plural *ossa coxae*) is formed by the fusion of three bones: the **ilium** (IL-ē-um), **ischium** (IS-kē-um), and **pubis** (PŪ-bis). By 20 to 25 years of age, these three bones ossify and fuse into a single os coxae, but the three bones are still referred to and used to name related structures. The **pelvis** is the bony ring of the two coxal bones of the appendicular skeleton and the sacrum and coccyx of the axial skeleton.

Coxal Bone

A conspicuous feature of the coxal bone is the deep socket, the **acetabulum** (a-se-TAB-ū -lum), where the head of the femur articulates (**Figure 15.7**). The smooth inner wall of the acetabulum is the C-shaped **lunate surface.** The center of the acetabulum is the **acetabular fossa.** The anterior and inferior rims of the acetabulum are not continuous; instead, there is an open gap between them, the **acetabular notch.**

The superior ridge of the ilium is the **iliac crest.** It is shaped like a shovel blade, with the **anterior** and **posterior superior iliac spines** at each end. The large indentation below the posterior superior iliac spine is the **greater sciatic** (sī-A-tik) **notch.** A conspicuous feature on the posterior iliac crest is the rough **auricular surface,** where the **sacroiliac joint** attaches the pelvic girdle to the sacrum of the axial skeleton. On the flat expanse of the ilium are ridges, the **anterior, posterior,** and **inferior gluteal lines,** which are attachment sites for muscles that move the femur.

The ischium is the bone we sit on. The greater sciatic notch terminates at a bony point, the **ischial spine.** Inferior to this spine is the **lesser sciatic notch.** The **ischial tuberosity** is in

Figure 15.7 The Pelvic Girdle

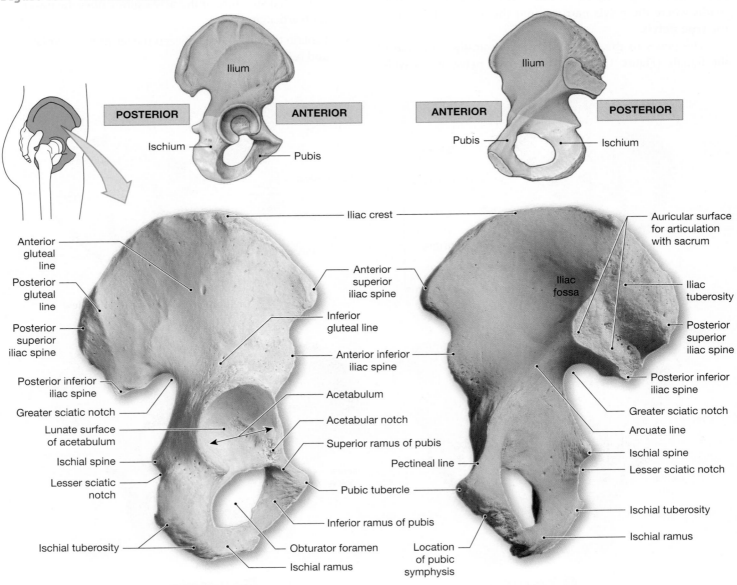

a Right hip bone, lateral view

b Right hip bone, medial view

the most inferior portion of the ischium and is a site for muscle attachment. The **ischial ramus** extends from the tuberosity and fuses with the pubis bone.

The pubis forms the anterior portion of the coxal bone. The most anterior region of the pubis is the pointed **pubic tubercle.** The **superior ramus** of the pubis is above the tubercle and extends to the ilium. On the medial surface, the superior ramus narrows to a rim called the **pectineal line** of the pubis. The **inferior ramus** joins the ischial ramus, creating the **obturator** (OB-tū-rā-tor) **foramen.**

The pelvis has three articulations; two **sarcoiliac joints** between the coxal bones and sacrum, and the **pubic symphysis,** a strong joint of fibrocartilage holding the pubis bones together (**Figure 15.8**). On the medial surface of each os coxae, the **iliac fossa** forms the wall of the upper pelvis, called the **false pelvis.** The **arcuate line** on this same surface marks where the pelvis narrows into the lower pelvis, called the **true pelvis.**

The pelvis of the male differs anatomically from that of the female (Figure 15.8c, d). The female pelvis has a wider

pelvic outlet, which is the space between the two ischii. The circle formed by the top of the pelvis, called the **brim,** defines the **pelvic inlet.** This opening is wider and rounder in females. Additionally, the **pubic angle** at the pubis symphysis is wider in the female and more U-shaped. This angle is V-shaped in the male. The wider female pelvis provides a larger passageway for childbirth.

Make a Prediction

How much does the pelvic girdle move in comparison to the pectoral girdle? Support your prediction with anatomical observations of both girdles.

QuickCheck Questions

3.1 Which bones make up the pelvic girdle?

3.2 With what structure of the pelvic girdle does the lower limb articulate?

3.3 Explain the difference between the terms *pelvic girdle* and *pelvis.*

Figure 15.8 The Pelvis

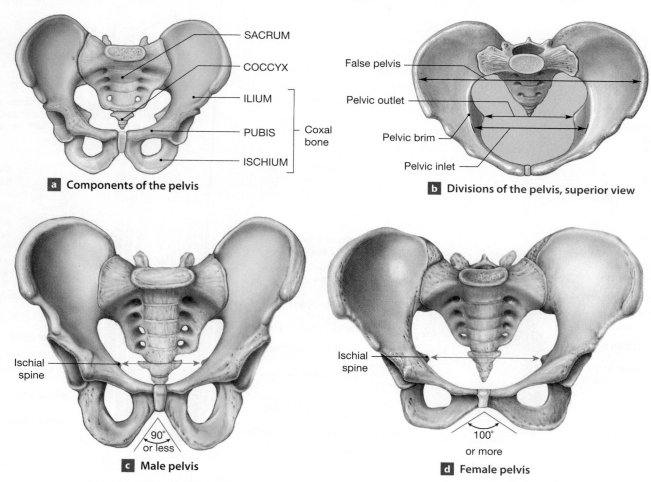

a   **Components of the pelvis**

- SACRUM
- COCCYX
- ILIUM
- PUBIS } Coxal bone
- ISCHIUM

b   **Divisions of the pelvis, superior view**

- False pelvis
- Pelvic outlet
- Pelvic brim
- Pelvic inlet

c   **Male pelvis**

Ischial spine

90° or less

d   **Female pelvis**

Ischial spine

100° or more

In the Lab 3

Materials

☐ Articulated skeleton
☐ Disarticulated skeleton

Procedures

1. Locate a coxal bone on your study skeleton and review the anatomy in Figures 15.7 and 15.8.
2. Identify the ilium, ischium, and pubis bones.
   - Are sutures visible where these bones fused?
   - Locate the acetabulum and obturator foramen.
   - Identify other features of the ilium.
3. Trace along the iliac crest and down the posterior surface.
   - Identify the greater and lesser sciatic notches and the ischial spine.
   - What is the large rough area on the inferior ischium called?
   - Identify other features of the ischium and pubis.
4. Locate the auricular surface of the sacroiliac joint and the pubic symphysis.
5. Articulate the two coxal bones and the sacrum to form the pelvis.
6. Examine the pelvis on several articulated skeletons in the laboratory. How can you distinguish a male pelvis from a female pelvis? ■

Lab Activity 4 Lower Limb

Each **lower limb,** also called the *lower extremity,* includes the bones of the thigh, knee, leg, and foot—a total of 30 bones. Recall that the term *leg* refers not to the entire lower limb but only to the region between the knee and ankle. Superior to the leg is the *thigh*.

The Femur

The **femur** is the largest bone of the skeleton (**Figure 15.9**). It supports the body's weight and bears the stress from the leg. The smooth, round **head** fits into the acetabulum of the coxal bone and permits the femur a wide range of movement. The depression on the head is the **fovea capitis,** where the *ligamentum capitis femoris* stabilizes the hip joint during movement. A narrow **neck** joins the head to the proximal shaft. Lateral to the head is a large stump, the **greater trochanter** (trō-KAN-ter); on the inferiomedial surface is the **lesser trochanter.** These large processes are attachment sites for tendons of powerful hip and thigh muscles. On the anterior surface of the femur, between the trochanters, is the **intertrochanteric line,** where the *iliofemoral ligament* inserts to encase the hip joint. Posteriorly, the **intertrochanteric crest** lies between the trochanters.

On the lateral side of the intertrochanteric crest, the **gluteal tuberosity** continues inferiorly and joins with the medial **pectineal line** of the femur as the **linea aspera,** a rough line for thigh muscle attachment. Toward the distal end of the femur, the linea aspera divides into the **medial** and **lateral supracondylar ridges** encompassing a flat triangle called the **popliteal surface.** The medial supracondylar ridge terminates at the **adductor tubercle.**

The largest condyles of the skeleton are the **lateral** and **medial condyles** of the femur, which articulate with the tibial head. The condyles are separated posteriorly by the **intercondylar fossa.** A smooth **patellar surface** spans the condyles and serves as a gliding platform for the patella. To the sides of the condyles are **lateral** and **medial epicondyles.**

The Patella

The **patella** is the kneecap and protects the knee joint during movement. It is a sesamoid bone encased in the distal tendons of the anterior thigh muscles. The superior border of the patella is the flat **base;** the **apex** is at the inferior tip (**Figure 15.10**). Along the base is the attachment site of the quadriceps muscle tendons that straighten (*extend*) the leg. The patellar ligament joins around the apex of the bone. Tendons attach to the rough anterior surface, and the smooth posterior facets glide over the condyles of the femur. The **medial facet** is narrower than the **lateral facet.**

The Tibia

The **tibia** (TI-bē-uh) is the large medial bone of the leg (**Figure 15.11**). The proximal portion of the tibia flares to develop the **lateral** and **medial condyles,** which articulate with the corresponding femoral condyles. Separating the tibial condyles is a ridge of bone, the **intercondylar eminence.** This eminence has two projections, the **medial** and **lateral tubercles,** that fit into the intercondylar fossa of the femur. On the anterior surface of the tibia, inferior to the condyles, is the large **tibial tuberosity,** where the patellar

> **Study Tip Patella Pointers**
>
> It is easy to distinguish a right patella from a left one. Lay the bone on its facets, and point the apex away from you. Notice that the bone leans to one side. Because the lateral facet is larger, the bone will tilt and lean on that facet. Therefore, if the patella leans to the left, it is a left patella. ■

Figure 15.9 The Femur

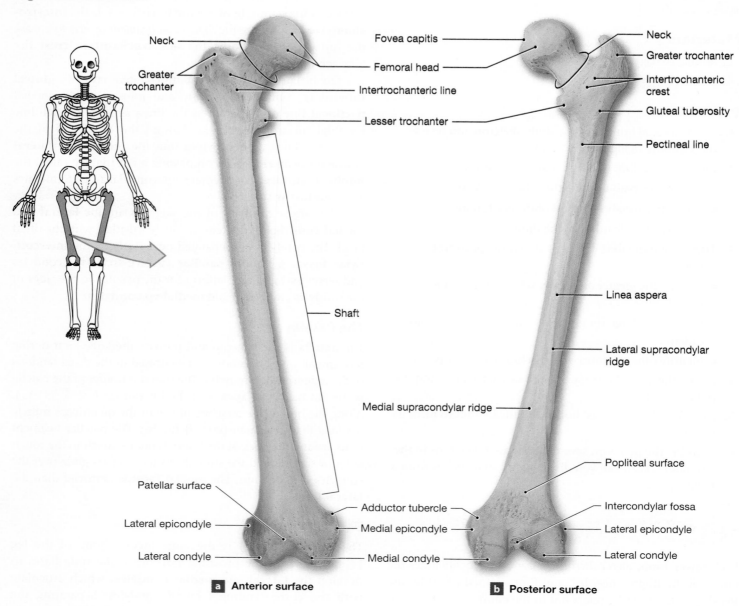

Neck

Greater trochanter

Fovea capitis

Femoral head

Intertrochanteric line

Lesser trochanter

Neck

Greater trochanter

Intertrochanteric crest

Gluteal tuberosity

Pectineal line

Shaft

Linea aspera

Lateral supracondylar ridge

Medial supracondylar ridge

Popliteal surface

Patellar surface

Lateral epicondyle

Lateral condyle

Adductor tubercle

Medial epicondyle

Medial condyle

Intercondylar fossa

Lateral epicondyle

Lateral condyle

a Anterior surface

b Posterior surface

Figure 15.10 The Patella The right patella.

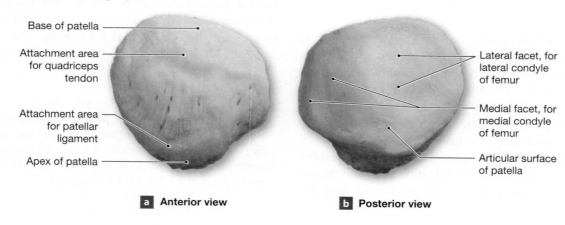

Base of patella

Attachment area for quadriceps tendon

Attachment area for patellar ligament

Apex of patella

Lateral facet, for lateral condyle of femur

Medial facet, for medial condyle of femur

Articular surface of patella

a Anterior view

b Posterior view

Figure 15.11 The Tibia and Fibula Bones of the right leg.

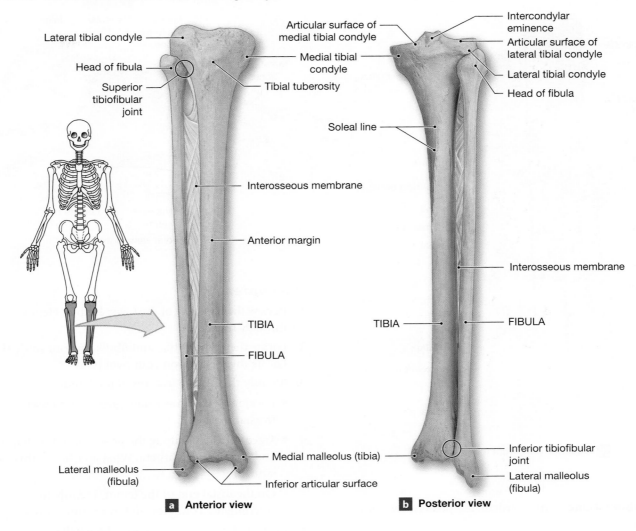

Lateral tibial condyle

Head of fibula

Superior
tibiofibular
joint

Medial tibial
condyle

Tibial tuberosity

Interosseous membrane

Anterior margin

TIBIA

FIBULA

Lateral malleolus
(fibula)

Medial malleolus (tibia)

Inferior articular surface

a Anterior view

Articular surface of
medial tibial condyle

Intercondylar
eminence

Articular surface of
lateral tibial condyle

Lateral tibial condyle

Head of fibula

Soleal line

Interosseous membrane

TIBIA

FIBULA

Inferior tibiofibular
joint

Lateral malleolus
(fibula)

b Posterior view

ligament attaches. Along most of the length of the ante-rior shaft is the **anterior margin,** a ridge commonly called the *shin.* The distal tibia is constructed to articulate with the ankle. A large wedge, the **medial malleolus** (ma-LĒ-rō-lus) **of the tibia,** stabilizes the ankle joint. The inferior **articular surface** is smooth so that it can slide over the talus of the ankle. Posteriorly, the proximal tibial shaft has a rough line, the **popliteal line,** where leg muscles attach.

The Fibula

The **fibula** (FIB-ū-la) is the slender bone lateral to the tibia (Fig-ure 15.11). The proximal and distal regions of the fibula appear very similar at first, but closer examination reveals the proximal head to be more rounded (less pointed) than the distal **lateral malleolus of the fibula.** The head of the fibula articulates below the lateral condyle of the tibia at the **superior tibiofibular joint.** The distal articulation creates the **inferior tibiofibular joint.**

Study Tip Hands and Feet

Because their names are so similar, it is easy to confuse the carpal and metacarpal bones of the wrist and hand with the tarsal and metatarsal bones of the ankle and foot. Just remember, when you listen to music you *clap* your *carpal bones* and *tap* your *tarsal bones*! ∎

Bones of the Foot

The ankle is formed by seven **tarsal bones** (Figure 15.12). One of them, the **talus** (TĀ-lus), sits on top of the heel bone, the **calcaneus** (kal-KĀ-nē-us), and articulates with the tibia and the lateral malleolus of the fibula. Anterior to the talus is the tarsal bone called the **navicular bone,** which articu-lates with the **medial, intermediate,** and **lateral cuneiform** (kū-NĒ-i-form) **bones.** Lateral to the lateral cuneiform bone

Figure 15.12 Bones of the Foot

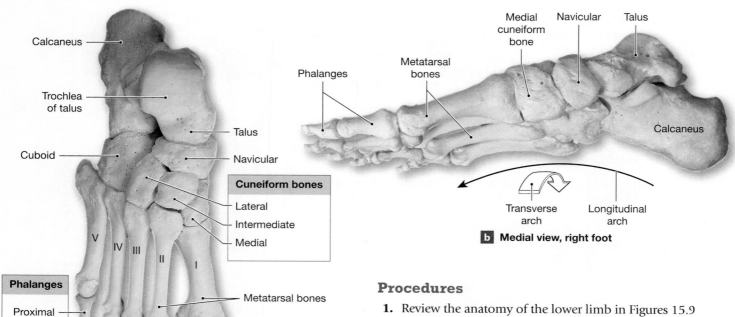

a **Superior view, right foot**

b **Medial view, right foot**

is the **cuboid bone,** which articulates posteriorly with the calcaneus.

The arch of the foot is formed by five **metatarsal bones.** Each metatarsal bone is named with a roman numeral, with the medial metatarsal that articulates with the big toe being designated as toe I.

The 14 bones of the toes are called **phalanges.** Like the fingers of the hand, toes II through V have a **proximal, middle, and distal phalanx.** The big toe, or **hallux,** has only a proximal and a distal phalanx.

QuickCheck Questions

4.1 What are the bones of the thigh, knee, and leg?

4.2 What are the three major groups of bones in the foot?

In the Lab   4

Materials

☐ Articulated skeleton
☐ Disarticulated skeleton

Procedures

1. Review the anatomy of the lower limb in Figures 15.9 through 15.12.
2. Locate the femur, tibia, and fibula on your study skeleton. Locate these bones on your own body.
3. Identify the surface features of the femur.
   - Locate the head, neck, and greater and lesser trochanters.
   - Trace your hand along the posterior of the diaphysis and feel the linea aspera. What attaches to this rough structure?
   - On the distal end of the femur, identify the epicondyles, condyles, and intercondylar fossa.
4. Identify the surface features of the patella.
   - Examine the two facets of the patella.
   - How can the facets be used to determine whether the patella is from a right leg or a left one?
5. Review the anatomy of the tibia and fibula.
   - What is the ridge on the tibial head called?
   - On the tibia, locate the condyles and the tibial tuberosity.
   - On the distal tibia, locate the medial malleolus.
   - Locate the lateral malleolus of the fibula. How does its shape differ from the fibular head?
6. Identify the bones of the foot.
   - Identify the tarsals, metatarsals, and phalanges.
   - Which tarsal bone directly receives body weight?
   - Which bones form the arch of the foot?
   - Do the toes all have the same number of phalanges?
7. Articulate the bones of the lower limb with the pelvis. ◼

Name _____

Date _____

Section _____

Review & Practice Sheet

EXERCISE

15

Appendicular Skeleton

A. Matching

Match each surface feature in the left column with its correct bone from the right column.
Each choice from the right column may be used more than once.

_____	**1.** acromion	**A.**	clavicle
_____	**2.** intercondylar fossa	**B.**	patella
_____	**3.** trochlea	**C.**	fibula
_____	**4.** glenoid cavity	**D.**	humerus
_____	**5.** ulnar notch	**E.**	femur
_____	**6.** deltoid tuberosity	**F.**	scapula
_____	**7.** greater trochanter	**G.**	tibia
_____	**8.** sternal end	**H.**	radius
_____	**9.** lateral malleolus		
_____	**10.** linea aspera		
_____	**11.** capitulum		
_____	**12.** medial malleolus		
_____	**13.** intercondylar eminence		
_____	**14.** base		

B. Short-Answer Questions

1. List the bones of the pectoral girdle and the upper limb.

2. List the bones of the pelvic girdle and the lower limb.

3. Compare the pelvis of males and females.

4. Which bony process acts like a doorstop to prevent excessive movement of the elbow?

5. On what two structures does the radial head pivot during movements such as turning a doorknob?

6. How are the carpal bones arranged in the wrist?

7. Where is the deltoid tuberosity located?

8. Which appendicular bones have a styloid process?

9. Which bone of the ankle articulates with the tibia?

10. What are the major features of the proximal portion of the femur?

11. Do the toes all have the same number of phalanges?

12. Which bones form the arch of the foot?

13. What is the ridge on the tibial head called?

14. Where is the glenoid cavity located?

C. Labeling

1. Label the surface features of the scapula in Figure 15.13.

Figure 15.13 **Posterior Surface of Right Scapula**

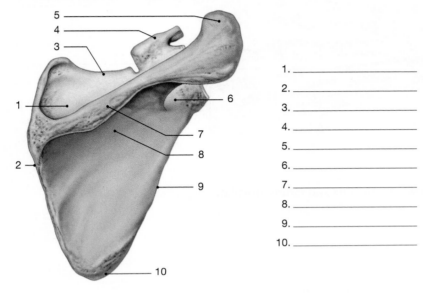

1. _____
2. _____
3. _____
4. _____
5. _____
6. _____
7. _____
8. _____
9. _____
10. _____

2. Label the surface features of the coxal bone in Figure 15.14.

Figure 15.14 **Right Coxal Bone**

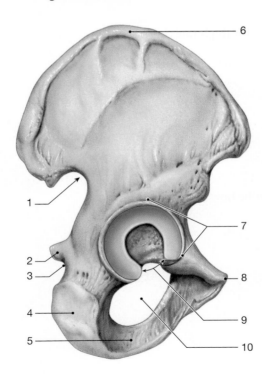

1. _____
2. _____
3. _____
4. _____
5. _____
6. _____
7. _____
8. _____
9. _____
10. _____

3. Label the surface features of the foot in **Figure 15.15**.

Figure 15.15 Right Foot

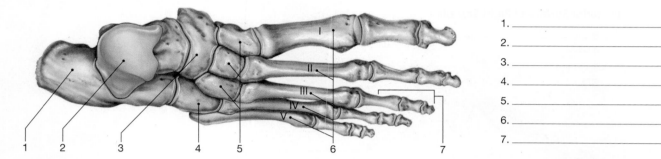

1. _____
2. _____
3. _____
4. _____
5. _____
6. _____
7. _____

D. Analysis and Application

1. Describe the condyle of the humerus where the ulna and radius articulate.

2. Compare the bones of the hand with the bones of the foot.

3. Name a tuberosity for shoulder muscle attachment and a tuberosity for thigh muscle attachment.

E. Clinical Challenge

The clavicle can break when catching a fall with outstretched hands. Describe how the impact on the hands could cause damage to the clavicle.

Articulations

Learning Outcomes

On completion of this exercise, you should be able to:

1. List the three types of functional joints and give an example of each.

2. List the four types of structural joints and give an example of each.

3. Describe the three types of diarthroses and the movement each produces.

4. Describe the anatomy of a typical synovial joint.

5. Describe and demonstrate the various movements of synovial joints.

Arthrology is the study of the structure and function of **joints;** a joint is defined as any location where two or more bones articulate. (In anatomic terminology, a synonym for *joint* is **articulation.**) If you were asked to identify joints in your body, you would most likely name those that allow a large range of movement, such as your knee or hip joint. In large-range joints like these, a cavity between the two bones of the joint permits free movement. In some joints, however, the bones are held closely together, a condition that allows no movement; an example of this type of nonmoving joint is found in the bones of the cranium.

Some individuals have more movement in a particular joint than most other people and are called "double jointed." Of course, they do not have two joints; the additional movement is a result of either the anatomy of the articulating bones or the position of tendons and ligaments around the joint.

Need More Practice and Review?

Build your knowledge—and confidence!—in the Study Area of MasteringA&P® at www.masteringaandp.com with Pre-lab Quizzes, Post-lab Quizzes, Practice Anatomy Lab™ (PAL™) 3.0 virtual anatomy practice tool, PhysioEx™ 9.0 laboratory simulations, and A&P Flix™ with Quizzes.

PAL | practice anatomy lab   For this lab exercise, follow these navigation paths:
- PAL>Human Cadaver>Joints
- PAL>Anatomical Models>Joints

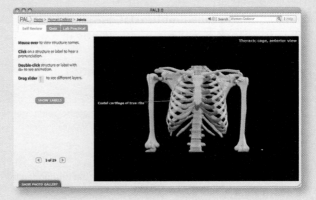

Lab Activity 1   Joint Classification

Two classification schemes are commonly used for articulations. The functional scheme groups joints by the amount of movement permitted, and the structural scheme groups joints by the type of connective tissue between the articulating bones. The three kinds of functional joints permit no, some, or free movement of the articulating bones. Four types of structural joints occur: bony fusion, fibrous, cartilaginous, and synovial.

Functional Classification

The functional classification scheme divides joints into three groups: immovable joints, the *synarthroses;* semimovable joints, the *amphiarthroses;* and freely movable joints, the *diarthroses.* **Table 16.1** summarizes these three groups.

1. **Synarthroses** (sin-ar-THRŌ-sēz; *syn-*, together + *arthros*, joint) are immovable joints in which the bones are either closely fitted together or surrounded by a strong ligament. Sutures of the skull are synarthroses.

2. **Amphiarthroses** (am-fē-ar-THRŌ-sēz) are joints held together by strong connective tissue; they are capable of only minimal movement. Examples of an amphiarthrosis include the joint between the tibia and fibula and the joints between vertebral bodies.

3. **Diarthroses** (dī-ar-THRŌ-sēz) are joints in which the bones are separated by a small membrane-lined cavity. The cavity allows a wide range of motion which makes diarthroses freely movable **synovial** (sin-NŌ-vē-ul) joints. Movements are classified according to the number of planes through which the bones move.

   - **Monaxial** (mon-AX-ē-ul) joints, like the elbow, move in one plane.

   - **Biaxial** (bī-AX-ē-ul) joints allow movement in two planes; move your wrist up and down and side to side to demonstrate biaxial movement.

   - **Triaxial** (trī-AX-ē-ul) joints occur in the ball-and-socket joints of the shoulder and hip and permit movement in three planes.

   - **Nonaxial** joints, also called **multiaxial,** are glide joints where the articulating bones can move slightly in a

variety of directions. The anatomy of a diarthrotic joint is examined in more detail later in this exercise.

Structural Classification

The structural classification scheme for joints is important when discussing joint anatomy rather than movement. As **Table 16.2** summarizes, four types of structural joints occur in the skeleton.

1. **Bony fusion** occurs where bones have fused together, and this type of joint permits no movement. These joints are also called **synostoses** (sin-os-TŌ-sēz; *-osteo*, bone) and occur in the frontal bone, coxal bones, and mandible bone. A good example of a synostosis is the frontal bone. Humans are born with two frontal bones that, by the age of eight, fuse into a single frontal bone. The old articulation site is then occupied by bony tissue to form a bony fusion joint. The joint between the diaphysis and either epiphysis of a mature long bone is also a synostosis.

2. **Fibrous joints** are synarthroses that have fibrous connective tissue between the articulating bones, and as a result little to no movement occurs in these strong joints. There are three main types of fibrous joints: suture, the gomphosis, and syndesmosis.

   - **Sutures** (*sutura*, a sewing together) occur in the skull wherever the bones interlock. This strong synarthrosis has no movement.

   - A **gomphosis** (gom-FŌ-sis; *gompho*, a peg or nail) is characterized by the insertion of a conical process into a socket in the alveolar bone of the jaw. The gomphosis is the joint between the tooth and the socket of alveolar bone of the jaw. The fibrous periodontal ligament lined the joint and holds the tooth firmly in place and permits no movement.

   - **Syndesmoses** (sin-dez-MŌ-sēz; *syn-*, together + *desmo-*, band) occur between the parallel bones of the forearm and leg. A ligament of fibrous connective tissue forms a strong band that wraps around the bones. The syndesmosis thus formed prevents excessive movement in the joint.

Table 16.1	A Functional Classification of Articulation	
Functional Category	**Description**	**Examples**
Synarthrosis	Strong joint with no movement. Bones are held together with fibrous connective tissue or cartilage.	Sutures of the skull, fusion of frontal bones
Amphiarthrosis	Strong joint with limited movement.	Between the tibia and fibula, between right and left halves of pelvis; between adjacent vertebral bodies
Diarthrosis	Complex joint with free movement that is bounded by joint capsule containing synovial fluid.	Elbow, ankle, wrist, shoulder, hip

Table 16.2	**A Structural Classification of Articulations**		
Structure	**Type**	**Functional Category**	**Example**
Bony fusion	Synostosis (illustrated)	Synarthrosis	Metopic suture (fusion) — Frontal bone
Fibrous joint	Suture (illustrated) Gomphosis Syndesmosis	Synarthrosis Synarthrosis Amphiarthrosis	Lambdoid suture — Skull
Cartilaginous joint	Synchondrosis Symphysis (illustrated)	Synarthrosis Amphiarthrosis	Pubic symphysis — Pelvis
Synovial joint	Monaxial Biaxial Triaxial (illustrated)	Diarthroses	Synovial joint

3. **Cartilaginous joints,** as their name implies, have cartilage between the bones. The type of cartilage—hyaline or fibrocartilage—determines the type of cartilaginous joint.

   - **Symphyses** are amphiarthroses characterized by the presence of fibrocartilage between the articulating bones. The intervertebral disks, for instance, construct a symphysis between any two articulating vertebrae. Another symphysis in the body occurs where the coxal bones unite at the pubis. This strong joint, called the pubic symphysis, limits flexion of the pelvis. During childbirth, a hormone softens the fibrocartilage to widen the pelvic bowl.

   - **Synchondroses** (sin-kon-DRŌ-sēz; *syn-*, together + *condros,* cartilage) are synarthroses that have cartilage between the bones making up the joints. Two examples of this type of synarthrosis are the epiphyseal plate in a child's long bones and the cartilage between the ribs and sternum.

4. **Synovial joints** have a joint cavity lined by a **synovial membrane.** All the free-moving joints—in other words, the diarthroses—are synovial joints. The four types are the monaxial, biaxial, triaxial, and multiaxial joints, described earlier.

Clinical Application Arthritis

Arthritis, a disease that destroys synovial joints by damaging the articular cartilage, comes in two forms. **Rheumatoid arthritis** is an autoimmune disease that occurs when the body's immune system attacks the cartilage and synovial membrane of the joint. As the disease progresses, the joint cavity is eliminated and the articulating bones fuse, which results in painful disfiguration of the joint and loss of joint function. **Osteoarthritis** is a degenerative joint disease that often occurs due to age and wearing of the joint tissues. The articular cartilage is damaged, and bone spurs may project into the joint cavity. Osteoarthritis tends to occur in the knee and hip joints, whereas rheumatoid arthritis is more common in the smaller joints of the hand. ■

QuickCheck Questions

1.1 What is the difference between the functional classification scheme for joints and the structural classification scheme?

1.2 What are the three types of functional joints and how much movement does each allow?

1.3 What are the four types of structural joints and the type of connective tissue found in each?

In the Lab   1

Material

☐ Articulated skeleton

Procedures

1. Locate on an articulated skeleton or on your body a joint from each functional group and one from each structural group.

2. Identify on your body and give an example of each of the following joints.

   ▪ Synarthrosis _____

   ▪ Amphiarthrosis _____

   ▪ Diarthrosis _____

   ▪ Syndesmosis _____

   ▪ Synchondrosis _____

   ▪ Synostosis _____

   ▪ Symphysis _____

   ▪ Suture _____

3. Identify two monaxial joints, two biaxial joints, and two triaxial joints on your body.

   ▪ Monaxial joints _____

   ▪ Biaxial joints _____

   ▪ Triaxial joints _____   ▪

Lab Activity 2   Structure of Synovial Joints

The wide range of motion of synovial joints is attributed to the small **joint cavity** between articulating bones (**Figure 16.1**). When you consider how a door can swing open even though there is only a small space between the metal pieces of the hinges, you can appreciate how a joint cavity permits free movement of a joint. The epiphyses are capped with **articular cartilage,** a slippery gelatinous surface of hyaline cartilage that protects the epiphyses and prevents the bones from making contact across the joint cavity. A membrane called the **synovial membrane** lines the cavity and produces **synovial fluid.** Injury to a joint may cause inflammation of the membrane and lead to excessive fluid production.

A **bursa** (BUR-sa; *bursa,* a pouch) is similar to a synovial membrane except that, instead of lining a joint cavity, the bursa provides padding between bones and other structures. The periosteum of each bone is continuous with the strong **articular capsule** that encases the joint.

As mentioned previously, all diarthrotic joints are capable of free movement. This large range of motion is due to the anatomical organization of the joint: Between the bones of every diarthrotic joint is a cavity lined with a synovial membrane.

Figure 16.1 Structure of a Synovial Joint

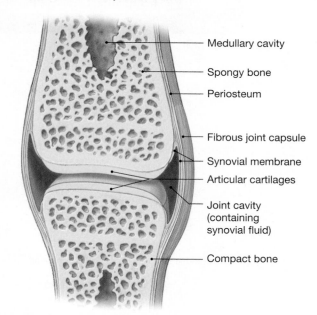

a Synovial joint, sagittal section

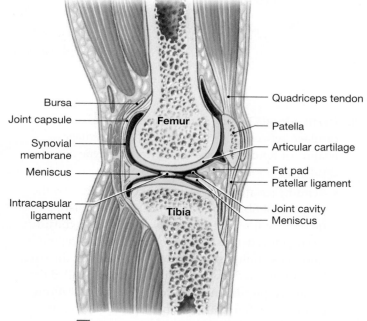

b Knee joint, sagittal section

QuickCheck Questions

2.1 Where is the synovial membrane located in a joint?

2.2 Where is cartilage found in a synovial joint?

In the Lab   2

Material

☐ Fresh beef joint

Procedures

1. Review the anatomy of the synovial joint in Figure 16.1.
2. On the fresh beef joint, locate and describe the joint cavity.
3. Identify the articular cartilage and articular capsule. ■

Lab Activity 3 Types of Diarthroses

Six types of diarthroses (synovial joints) occur in the skeleton. Each type permits a certain amount of movement owing to the joining surfaces of the articulating bones. **Figure 16.2** details each type of joint and includes a mechanical representation of each joint to show planes of motion.

- **Gliding joints** (also called *plane joints*) are common where flat articular surfaces, such as in the wrist, slide by neighboring bones. The movement is typically nonaxial. In addition to the wrist, glide joints also occur between bones of the sternum and between the tarsals. When you place your open hand, palm facing down, on your desktop and press down hard, you can observe the gliding of your wrist bones.
- **Hinge joints** are monaxial, operating like a door hinge, and are located in the elbows, fingers, toes, and knees. Bending your legs at the knees and your arms at the elbows is possible because of your hinge joints.
- **Pivot joints** are monaxial joints that permit one bone to rotate around another. Shake your head "no" to operate the pivot joint between your first two cervical vertebrae. The first cervical vertebra (the atlas) pivots around the second cervical vertebra (the axis).
- **Ellipsoid joints** (also called *condyloid joints*) are characterized by a convex surface of one bone that articulates in a concave depression of another bone. This concave-to-convex spooning of articulating surfaces permits biaxial movement. The articulation between the bones of the forearm and wrist is an ellipsoidal joint.
- The **saddle joint** is a biaxial joint found only at the junction between the thumb metacarpus and the trapezium bone of the wrist. Place a finger on your lateral wrist and feel the saddle joint move as you touch your little finger with your thumb. This joint permits you to oppose your thumb to grasp and manipulate objects in your hand.
- **Ball-and-socket joints** occur where a spherical head of one bone fits into a cup-shaped fossa of another bone, as in the joint between the humerus and the scapula. This triaxial joint permits dynamic movement in many planes.

QuickCheck Questions

3.1 What are the six types of diarthroses?

3.2 What type of diarthrosis is a knuckle joint?

In the Lab 3

Material

☐ Articulated skeleton

Procedures

1. Locate each type of synovial joint on an articulated skeleton or on your body. On the skeleton, notice how the structure of the joining bones determines the amount of joint movement.
2. Give an example of each type of synovial joint.
   - Gliding _____
   - Hinge _____
   - Pivot _____
   - Ellipsoidal _____
   - Saddle _____
   - Ball-and-socket _____ ■

Lab Activity 4 Skeletal Movement at Diarthrotic Joints

The diversity of bone shapes and joint types permits the skeleton to move in a variety of ways. **Figure 16.3** illustrates angular movements, which occur either front to back in the anterior/posterior plane or side to side in the lateral plane. **Figure 16.4** illustrates rotational movements. For clarity, these figures include a small dot at the joint where a demonstrated movement is described. **Table 16.3** summarizes articulations of the axial skeleton and **Table 16.4** summarizes articulations of the appendicular divisions.

- **Flexion** is movement that decreases the angle between the articulating bones, and **extension** is movement that increases the angle between the bones (Figure 16.3a). Hang your arm down at your side in anatomical position. Now flex your arm by moving the elbow joint. Your hand should be up by your shoulder. Notice how close the antebrachium is to the brachium and how the angle between them has decreased. Is your flexed arm still in anatomical position? Now extend your arm to return it to anatomical position. How has the angle changed? **Hyperextension** moves the body beyond anatomical position.
- **Abduction** is movement away from the midline of the body (Figures 16.3b, c). **Adduction** is movement toward the midline. Notice how you move your arm at the shoulder for these two motions. Practice this movement first with your shoulder joint and then with your wrist joint.

Figure 16.2 Movements at Synovial Joints The types of movement permitted are illustrated on the left anatomically and on the right by a mechanical model.

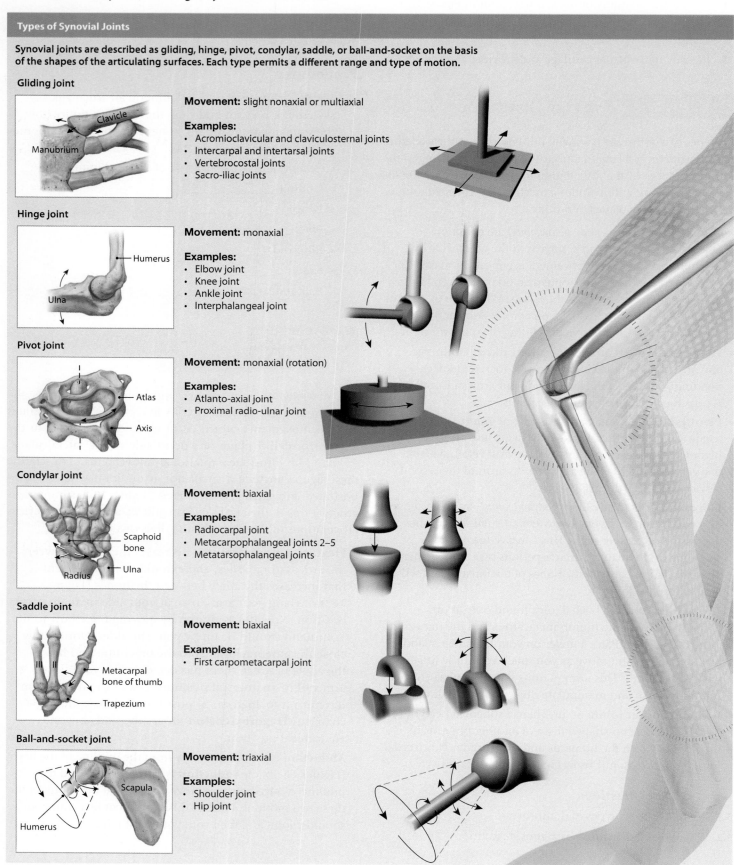

Types of Synovial Joints

Synovial joints are described as gliding, hinge, pivot, condylar, saddle, or ball-and-socket on the basis of the shapes of the articulating surfaces. Each type permits a different range and type of motion.

Gliding joint

Clavicle
Manubrium

Movement: slight nonaxial or multiaxial

Examples:
• Acromioclavicular and claviculosternal joints
• Intercarpal and intertarsal joints
• Vertebrocostal joints
• Sacro-iliac joints

Hinge joint

Humerus
Ulna

Movement: monaxial

Examples:
• Elbow joint
• Knee joint
• Ankle joint
• Interphalangeal joint

Pivot joint

Atlas
Axis

Movement: monaxial (rotation)

Examples:
• Atlanto-axial joint
• Proximal radio-ulnar joint

Condylar joint

Scaphoid bone
Radius
Ulna

Movement: biaxial

Examples:
• Radiocarpal joint
• Metacarpophalangeal joints 2–5
• Metatarsophalangeal joints

Saddle joint

III II
Metacarpal bone of thumb
Trapezium

Movement: biaxial

Examples:
• First carpometacarpal joint

Ball-and-socket joint

Scapula
Humerus

Movement: triaxial

Examples:
• Shoulder joint
• Hip joint

Figure 16.3 Angular Movements Examples of angular movements that change the angle between the two bones making up a joint. The red dots indicate the locations of the joints involved in the illustrated movement.

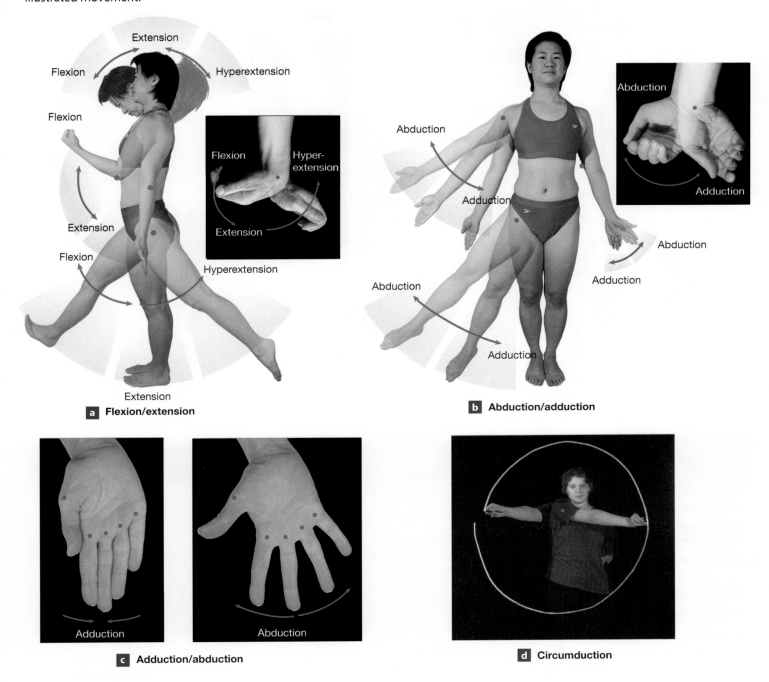

a **Flexion/extension**

b **Abduction/adduction**

c **Adduction/abduction**

d **Circumduction**

- **Circumduction** is circular movement at a ball-and-socket joint (Figure 16.3d). During this movement, motion of the proximal region of the upper limb is relatively stationary while the distal portion traces a wide circle in the air.

- **Rotation** is a turning movement of bones at a joint (Figure 16.4a). **Left rotation** or **right rotation** occur when the head is turned, as in shaking to indicate "no."

Lateral rotation and **medial rotation** of the limbs occur at ball-and-socket joints and at the radioulnar joint. These movements turn the rounded head of one bone in the socket of another bone.

- **Supination** (soo-pi-NĀ-shun) is movement that moves the palm into the anatomical position (Figure 16.4b). **Pronation** (pro-NĀ-shun) is movement that moves the

Figure 16.4 Rotational Movements Examples of motion in which a body part rotates.

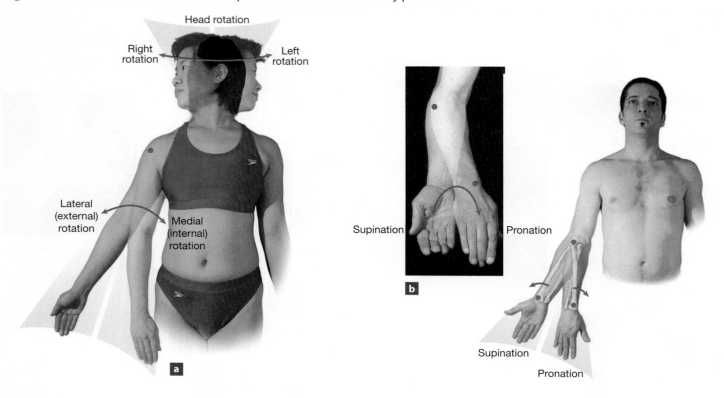

Table 16.3	Articulations of the Axial Skeleton		
Element	**Joint**	**Type of Articulation**	**Movements**
SKULL			
Cranial and facial bones of skull	Various	Synarthroses (suture or synostosis)	None
Maxillary bone/teeth and mandible/teeth	Alveolar	Synarthrosis (gomphosis)	None
Temporal bone/mandible	Temporomandibular	Combined gliding joint and hinge diarthrosis	Elevation, depression, and lateral gliding
VERTEBRAL COLUMN			
Occipital bone/atlas	Atlanto-occipital	Ellipsoidal diarthrosis	Flexion/extension
Atlas/axis	Atlanto-axial	Pivot diarthrosis	Rotation
Other vertebral elements	Intervertebral (*between vertebral bodies*)	Amphiarthrosis (symphysis)	Slight movement
	Intervertebral (*between articular processes*)	Gliding diarthrosis	Slight rotation and flexion/extension
L$_5$/sacrum	Between L$_5$ body and sacral body	Amphiarthrosis (symphysis)	Slight movement
	Between inferior articular processes of L$_5$ and articular processes of sacrum	Gliding diarthrosis	Slight flexion/extension
Sacrum/coxal bone	Sacroiliac	Gliding diarthrosis	Slight movement
Sacrum/coccyx	Sacrococcygeal	Gliding diarthrosis (*may become fused*)	Slight movement
Coccygeal bones		Synarthrosis (synostosis)	No movement

(continued)

Table 16.3	Articulations of the Axial Skeleton *(continued)*		
Element	**Joint**	**Type of Articulation**	**Movements**
THORACIC CAGE			
Bodies of T$_1$–T$_{12}$ and heads of ribs	Costovertebral	Gliding diarthrosis	Slight movement
Transverse processes of T$_1$–T$_{10}$	Costovertebral	Gliding diarthrosis	Slight movement
Ribs and costal cartilages		Synarthrosis (synchondrosis)	No movement
Sternum and first costal cartilage	Sternocostal (1st)	Synarthrosis (synchondrosis)	No movement
Sternum and costal cartilages 2–7	Sternocostal (2nd–7th)	Gliding diarthrosis*	Slight movement

*Commonly converts to synchondrosis in elderly individuals.

Table 16.4	Articulations of the Appendicular Skeleton		
Element	**Joint**	**Type of Articulation**	**Movements**
ARTICULATIONS OF THE PECTORAL GIRDLE AND UPPER LIMB			
Sternum/clavicle	Sternoclavicular	Gliding diarthrosis*	Protraction/retraction, elevation/depression, slight rotation
Scapula/clavicle	Acromioclavicular	Gliding diarthrosis	Slight movement
Scapula/humerus	Shoulder, or glenohumeral	Ball-and-socket diarthrosis	Flexion/extension, adduction/abduction, circumduction, rotation
Humerus/ulna and humerus/radius	Elbow (humeroulnar and humeroradial)	Hinge diarthrosis	Flexion/extension
Radius/ulna	Proximal radioulnar	Pivot diarthrosis	Rotation
	Distal radioulnar	Pivot diarthrosis	Pronation/supination
Radius/carpal bones	Radiocarpal	Ellipsoidal diarthrosis	Flexion/extension, adduction/abduction, circumduction
Carpal bone to carpal bone	Intercarpal	Gliding diarthrosis	Slight movement
Carpal bone to metacarpal bone (I)	Carpometacarpal of thumb	Saddle diarthrosis	Flexion/extension, adduction/abduction, circumduction, opposition
Carpal bone to metacarpal bone (II–V)	Carpometacarpal	Gliding diarthrosis	Slight flexion/extension, adduction/abduction
Metacarpal bone to phalanx	Metacarpophalangeal	Ellipsoidal diarthrosis	Flexion/extension, adduction/abduction, circumduction
Phalanx/phalanx	Interphalangeal	Hinge diarthrosis	Flexion/extension
ARTICULATIONS OF THE PELVIC GIRDLE AND LOWER LIMB			
Sacrum/ilium os coxae	Sacroiliac	Gliding diarthrosis	Slight movement
Coxal bone	Pubic symphysis	Amphiarthrosis (symphysis)	None†
Coxal bone/femur	Hip	Ball-and-socket diarthrosis	Flexion/extension, adduction/abduction, circumduction, rotation
Femur/tibia	Knee	Complex, functions as hinge	Flexion/extension, limited rotation
Tibia/fibula	Tibiofibular (proximal)	Gliding diarthrosis	Slight movement
	Tibiofibular (distal)	Gliding diarthrosis and amphiarthrotic syndesmosis	Slight movement
Tibia and fibula with talus	Ankle, or talocrural	Hinge diarthrosis	Flexion/extension (dorsiflexion/plantar flexion)
Tarsal bone to tarsal bone	Intertarsal	Gliding diarthrosis	Slight movement
Tarsal bone to metatarsal bone	Tarsometatarsal	Gliding diarthrosis	Slight movement
Metatarsal bone to phalanx	Metatarsophalangeal	Ellipsoidal diarthrosis	Flexion/extension, adduction/abduction
Phalanx/phalanx	Interphalangeal	Hinge diarthrosis	Flexion/extension

*A "double gliding joint," with two joint cavities separated by an articular cartilage.
†During pregnancy, hormones weaken the symphysis and permit movement important to childbirth.

palm to face posteriorly. During these two motions, the humerus serves as a foundation for the radius to pivot around the ulna.

The following specialized motions are illustrated in **Figure 16.5**.

- **Eversion** (ē-VER-zhun) is lateral movement of the ankle to move the foot so that the toes point away from the body's midline. Moving the sole medially so that the toes point toward the midline is **inversion;** the foot moves "in." Eversion and inversion are commonly mistaken for pronation and supination of the ankle.

- Two other terms describing ankle movement are dorsiflexion and plantar flexion. **Dorsiflexion** is the joint movement that permits you to walk on your heels, which means the soles of your feet are raised up off the floor and the angle between the ankle and the bones of the leg is decreased. **Plantar flexion** (*plantar*, sole) moves the foot so that you can walk on your tiptoes; here the angle between the ankle and the tibia/fibula is increased.

- **Opposition** is touching the thumb pad with the pad of the little finger.

- **Retraction,** which means to take back, moves structures posteriorly out of the anatomical position, as when the mandible is moved posteriorly to demonstrate an overbite.

Protraction moves a structure anteriorly, as when you jut your mandible forward.

- **Depression** lowers bones. This motion occurs, for instance, when you lower your mandible bone to take a bite of food. Closing your mouth is **elevation** of the mandible bone.

Study Tip Movements of the Upper Limb

To see the difference between medial and lateral rotation, start with your right upper limb in the anatomical position and then flex your right elbow until the forearm is parallel to the floor. Keeping your forearm parallel to the floor, move your right hand until it hits your torso; the movement of your humerus at the shoulder when you do this is medial rotation. Still keeping your right forearm parallel to the floor, now swing your right hand away from your torso. In this motion, the humerus is rotating laterally.

To see the difference between supination and pronation, return your right upper limb to the anatomical position and again flex the elbow to bring your right forearm parallel to the floor. Now twist your hand as if turning a doorknob and observe the movement of the forearm and hand. Twisting your hand until the palm faces the floor is pronation; twisting your hand back until the palm faces the ceiling is supination. ■

Figure 16.5 Special Movements Special movements occur at specific joints.

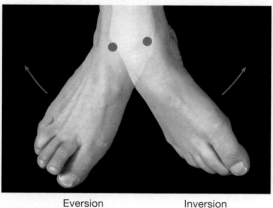

Eversion   Inversion

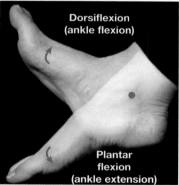

Dorsiflexion (ankle flexion)
Plantar flexion (ankle extension)

Opposition

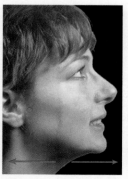

Retraction   Protraction

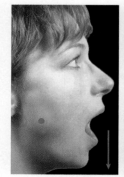

Depression

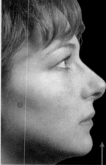

Elevation

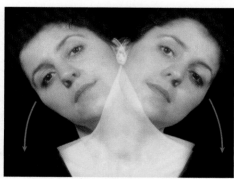

Lateral flexion

Sketch to Learn

Let's sketch the knee joint from a superior view and detail the seven ligaments that surround this joint.

Sample Sketch

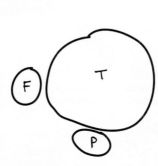 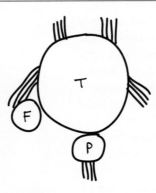 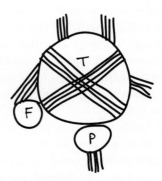

Step 1
- Draw a large oval for the tibial head and two smaller ovals for the fibula and patella.
- Place a letter in each oval to identify the bones.

Step 2
- Use a series of lines to represent each popliteal ligament.
- Add the two collateral ligaments.
- Add the patellar ligament.

Step 3
- Include the anterior and posterior cruciate ligaments.
- Label your sketch.

Your Sketch

- **Lateral flexion** is the bending of the vertebral column from side to side. Most of the movement occurs in the cervical and lumbar regions.

QuickCheck Questions

4.1 How does flexion differ from extension?

4.2 How does hyperextension differ from extension?

4.3 What are pronation and supination?

4.4 How does dorsiflexion differ from plantar flexion?

In the Lab   4

Material

☐ Articulated skeleton

Procedures

1. Use an articulated skeleton or your body to demonstrate each of the movements in Figures 16.3, 16.4, and 16.5.

2. Label the movements illustrated in Figures 16.3 and 16.4.

3. Give an example of each of the following movements.
   - Abduction _____
   - Extension _____
   - Hyperextension _____
   - Pronation _____
   - Supination _____
   - Depression _____
   - Retraction _____
   - Lateral rotation _____

For additional practice, complete the ***Sketch to Learn*** activity ▪

Lab Activity 5   Selected Synovial Joints: Elbow and Knee Joints

The elbow is a hinge joint involving humeroradial and humeroulnar articulations. (Within the elbow complex is also the radioulnar joint, which allows the radius to pivot during supination and pronation.) The morphology of the articulating bones and a strong articular capsule and ligaments result in a strong and highly movable elbow. **Radial** and **ulnar collateral ligaments** reinforce the lateral aspects of the joint, and the **annular ligament** holds the radial head in position to pivot (**Figure 16.6**).

The knee is a hinge joint that permits flexion and extension of the leg. Most support for the knee is provided by seven bands of ligaments that encase the joint (**Figure 16.7**). Cushions of fibrocartilage, the **lateral meniscus** (men-IS-kus; *meniskos*, a crescent) and the **medial meniscus,** pad the area between the condyles of the femur and tibia. Areas where tendons move against the bones in the knee are protected with bursae.

The seven ligaments of the knee occur in three pairs and a single patellar ligament. **Tibial** and **fibular collateral ligaments** provide medial and lateral support when a person is standing. Two **popliteal ligaments** extend from the head of the femur to the fibula and tibia to support the posterior of the knee. The **anterior** and **posterior cruciate ligaments** are inside the articular capsule. The cruciate (*cruciate,* crosslike) ligaments originate on the tibial head and cross each other as they pass through the intercondylar fossa of the femur. The **patellar ligament** attaches the inferior aspect of the patella to the tibial tuberosity, adding anterior support to the knee. The large quadriceps tendon is attached to the superior margin of the patella. Cords of ligaments called the **patellar retinaculae** contribute to anterior support of the knee.

QuickCheck Questions

5.1 What structure reinforces the radial head?

5.2 How many ligaments are in the knee?

5.3 What structures cushion the knee?

In the Lab   5

Materials

☐ Articulated skeleton
☐ Elbow model
☐ Knee model

Figure 16.6   **The Right Elbow**   The right elbow joint.

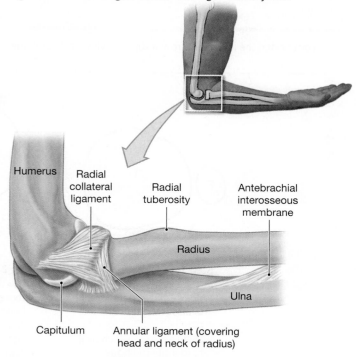

Humerus   Radial collateral ligament   Radial tuberosity   Antebrachial interosseous membrane

Radius

Ulna

Capitulum   Annular ligament (covering head and neck of radius)

a   Lateral view

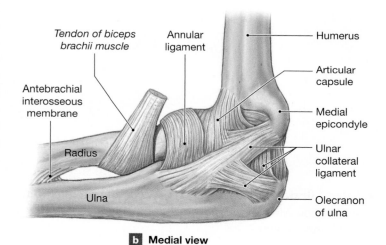

Tendon of biceps brachii muscle   Annular ligament   Humerus

Antebrachial interosseous membrane

Articular capsule

Radius

Medial epicondyle

Ulnar collateral ligament

Ulna

Olecranon of ulna

b   Medial view

Procedures

1. Examine the elbow of the articulated skeleton and review the skeletal anatomy of the joint.

2. Locate the annular and collateral ligaments on the elbow model.

3. Review the skeletal components of the knee joint on the articulated skeleton.

4. On the knee model, examine the relationship of the ligaments and determine how each supports the knee. Note how the menisci cushion between the bones. ■

Figure 16.7 The Knee Joint The right knee joint.

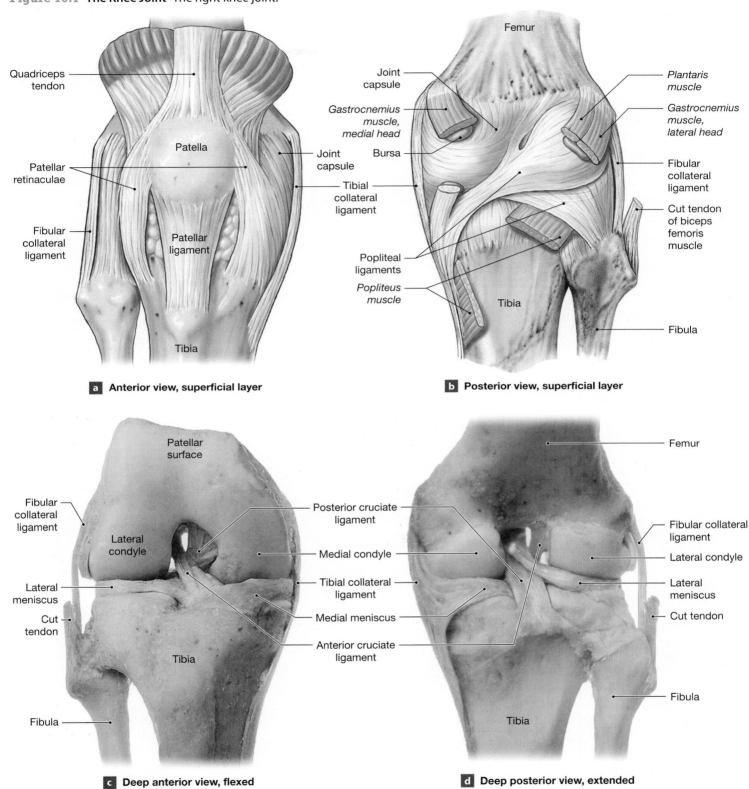

a Anterior view, superficial layer

b Posterior view, superficial layer

c Deep anterior view, flexed

d Deep posterior view, extended

Name _____

Date _____

Section _____

Articulations

A. Matching

Match each joint in the left column with its correct description from the right.

_____	**1.** pivot	**A.** forearm-to-wrist joint
_____	**2.** symphysis	**B.** joint between parietal bones
_____	**3.** ball and socket	**C.** rib-to-sternum joint
_____	**4.** gomphosis	**D.** joint between vertebral bodies
_____	**5.** hinge	**E.** femur-to-coxal bone joint
_____	**6.** suture	**F.** phalangeal joint
_____	**7.** synostosis	**G.** distal tibia-to-fibula joint
_____	**8.** syndesmosis	**H.** atlas-to-axis joint
_____	**9.** condyloid	**I.** fused frontal bones
_____	**10.** synchondrosis	**J.** joint holding tooth in a socket

B. Matching

Match each movement in the left column with its correct description from the right column.

_____	**1.** retraction	**A.** movement away from midline
_____	**2.** dorsiflexion	**B.** movement to turn foot outward
_____	**3.** eversion	**C.** palm moved to face posteriorly
_____	**4.** inversion	**D.** palm moved to face anteriorly
_____	**5.** pronation	**E.** movement to posterior plane
_____	**6.** plantar flexion	**F.** movement to stand on tiptoes
_____	**7.** protraction	**G.** movement in anterior plane
_____	**8.** supination	**H.** movement to turn foot inward
_____	**9.** adduction	**I.** movement to stand on heels
_____	**10.** abduction	**J.** movement toward midline

C. Labeling

1. Label the anatomy of a synovial joint in **Figure 16.8**.

2. Label the six numbered joint movements in **Figure 16.9**.

Figure 16.8 Synovial Joint

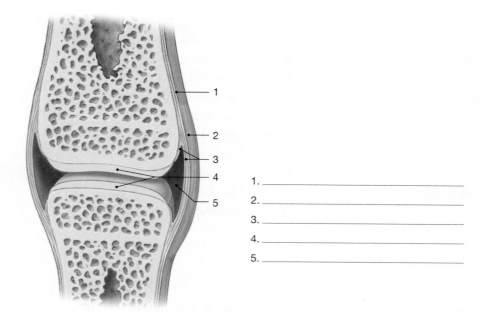

1. _____
2. _____
3. _____
4. _____
5. _____

Figure 16.9 Angular Movements

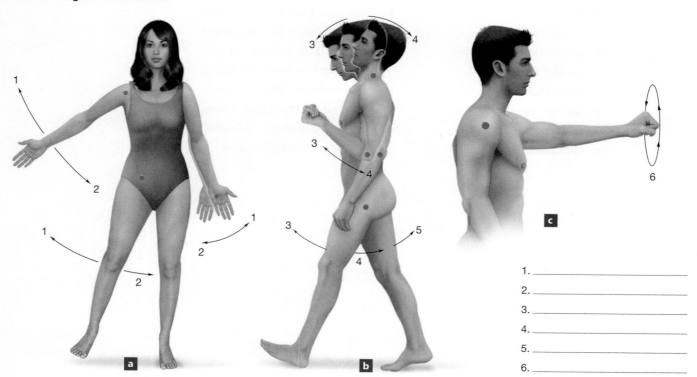

1. _____
2. _____
3. _____
4. _____
5. _____
6. _____

3. Label the four numbered joint movements in **Figure 16.10**.

Figure 16.10 Rotational Movements

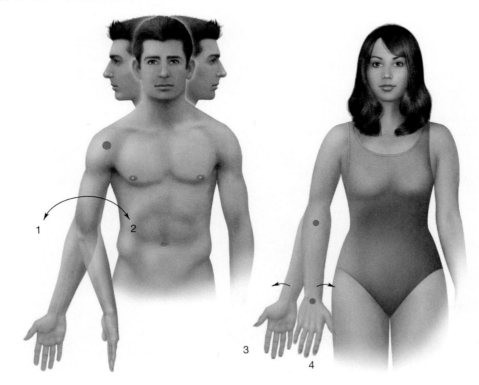

1. _____

2. _____

3. _____

4. _____

D. Description

Describe the joints and movements involved in:

   1. walking

   2. throwing a ball

   3. turning a doorknob

   4. crossing your legs while sitting

5. shaking your head "no"

6. chewing food

E. Short-Answer Questions

1. Describe the three types of functional joints.

2. What factors limit the range of movement of a joint?

3. Describe the four types of structural joints.

4. List the seven ligaments of the knee and how each supports the joint.

F. Analysis and Application

1. Describe how the articulating bones of the elbow prevent hyperextension of this joint.

2. Which joint is unique to the hand, and how does this joint move the hand?

3. Which structural feature enables diarthrotic joints to have free movement?

G. Clinical Challenge

1. Why do bones fuse in joints damaged by rheumatoid arthritis?

2. Why is the lateral meniscus often associated with a knee injury?

Organization
of Skeletal Muscles

Learning Outcomes

On completion of this exercise, you should be able to:

1. Describe the basic functions of skeletal muscles.

2. Describe the organization of a skeletal muscle.

3. Describe the microanatomy of a muscle fiber.

4. Discuss and provide examples of a lever system.

5. Understand the rules that determine the names of some muscles.

E very time you move some part of your body, either consciously or unconsciously, you use muscles. There are three kinds of muscle tissue: skeletal, smooth, and cardiac (refer to Exercise 9 for a detailed description). Skeletal muscles are primarily responsible for **locomotion,** or movement of the body. Locomotions such as rolling your eyes, writing your name, and speaking are the result of highly coordinated muscle contractions. Other functions of skeletal muscle include maintenance of posture and body temperature and support of soft tissues, as with the muscles of the abdomen.

In addition to the ability to contract, muscle tissue has several other unique characteristics. Like nerve tissue, muscle tissue is **excitable** and, in response to a stimulus, produces electrical impulses called **action potentials.** Muscle tissue can stretch and is therefore **extensible.** When the ends of a stretched muscle are released, it recoils to its original size, like a rubber band. This property is called **elasticity.**

Need More Practice and Review?

Build your knowledge—and confidence!—in the Study Area of MasteringA&P® at www.masteringaandp.com with Pre-lab Quizzes, Post-lab Quizzes, Practice Anatomy Lab™ (PAL™) 3.0 virtual anatomy practice tool, PhysioEx™ 9.0 laboratory simulations, and A&P Flix™ with Quizzes.

PAL   For this lab exercise, follow these navigation paths:
- PAL>Anatomical Models>Muscular System
- PAL>Histology>Muscular System

A&PFlix   For this lab exercise, go to these topics:
- Events at the Neuromuscular Junction
- Excitation-Contraction Coupling

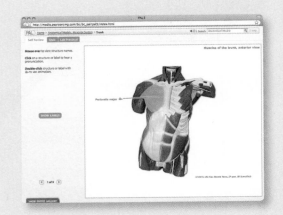

MasteringA&P®

Lab Activity 1   Skeletal Muscle Organization

Connective Tissue Coverings

Connective tissues support and organize skeletal muscles and attach them to bones. Three layers of connective tissue partition a muscle. Superficially, a collagenous connective tissue layer called the **epimysium** (ep-i-MĪZ-ē-um; *epi*, on + *mys*, muscle) covers the muscle and separates it from neighboring structures (**Figure 17.1**). The epimysium folds into the muscle as the **perimysium** (per-i-MĪZ-ē-um; *peri-*, around), and divides the muscle fibers into groups called **fascicles** (FAS-i-kl). Connective tissue fibers of the perimysium extend deep into the fascicles, as the **endomysium** (en-dō-MĪZ-ē-um; *endo-*, inside), and surround each muscle fiber (cell). The parallel, threadlike fascicles can be easily seen when a muscle is teased apart with a probe.

Figure 17.1 The Organization of Skeletal Muscles A skeletal muscle consists of fascicles (bundles of muscle fibers) enclosed by the epimysium. The bundles are separated by the connective tissue fibers of the perimysium, and within each bundle the muscle fibers are surrounded by the endomysium. Each muscle fiber has many superficial nuclei, as well as mitochondria and other organelles (see Figure 17.2).

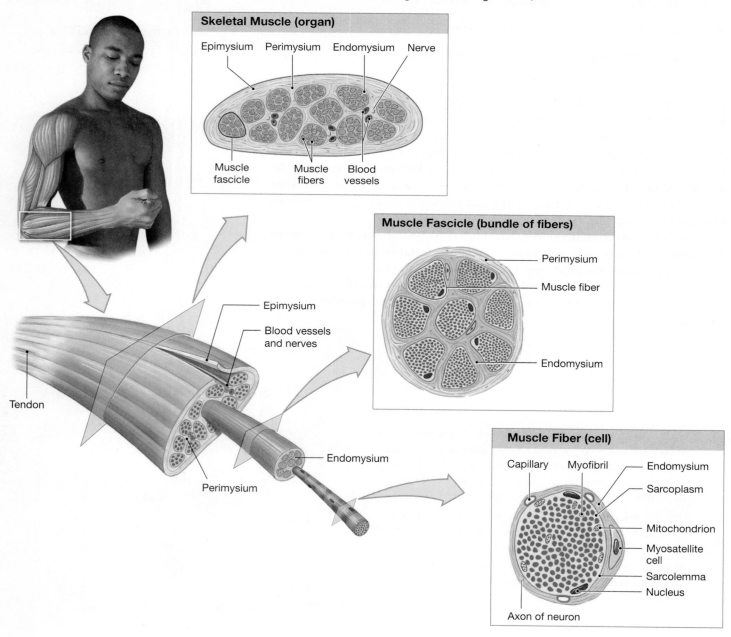

The connective tissues of the muscle interweave and combine as the **tendon** at each end of the muscle. The fibers of the tendon and the bone's periosteum interlace to firmly attach the tendon to the bone. When the muscle fibers contract and generate tension, they transmit this force through the connective tissue layers to the tendon, which pulls on the associated bone and produces movement. As the muscle contracts, its two ends move closer together and the central part of the muscle, called the **belly,** increases in diameter.

Structure of a Skeletal Muscle Fiber

Each muscle fiber is a composite of many cells that fused into a single cell during embryonic development. The cell membrane of a muscle fiber is called the **sarcolemma** (sar-cō-LEM-uh; *sarkos,* flesh + *lemma,* husk) and the cytoplasm is called **sarcoplasm** (**Figure 17.2**). Many **transverse tubules,** also called *T tubules,* connect the sarcolemma to the interior of the muscle fiber. The function of these tubules is to pass contraction stimuli to deeper regions of the muscle fiber.

Inside the muscle fiber are proteins arranged in thousands of rods, called **myofibrils,** that extend the length of the fiber. Each myofibril is surrounded by the **sarcoplasmic reticulum,** a modified endoplasmic reticulum where calcium ions are stored. Branches of the sarcoplasmic reticulum fuse to form large calcium ion storage chambers called **terminal cisternae** (sis-TUR-nē), which lie adjacent to the transverse tubules. A **triad** is a "sandwich" consisting of a transverse tubule plus the terminal cisterna on either side of the tubule. In order for a muscle to contract, calcium ions must be released from the cisternae; the transverse tubules stimulate this ion release. When a muscle relaxes, protein carriers in the sarcoplasmic reticulum transport calcium ions back into the cisternae.

Each myofibril consists of several kinds of proteins arranged in about 3000 **thin filaments** and 1500 **thick filaments** (**Figure 17.3**). During contraction, thick and thin protein molecules interact to produce tension and shorten the muscle. The thin filaments are mostly composed of the protein **actin,** and the thick filaments are made of the protein **myosin.**

Figure 17.2 The Structure of a Skeletal Muscle Fiber The internal organization of a muscle fiber.

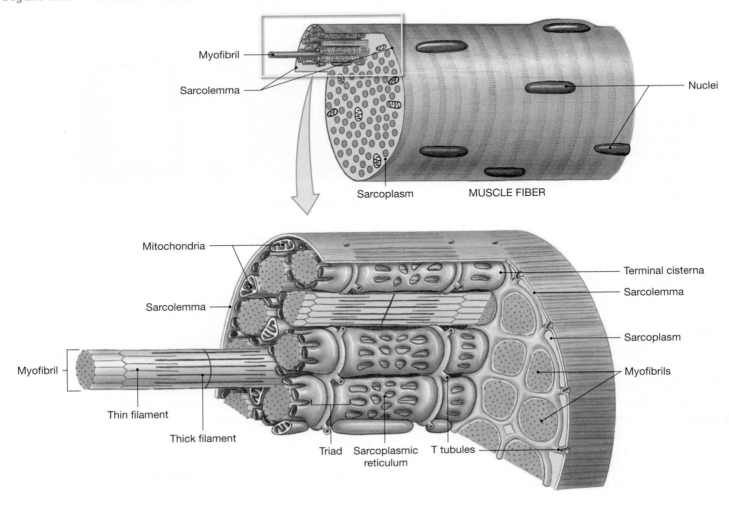

Figure 17.3 Sarcomere Structure

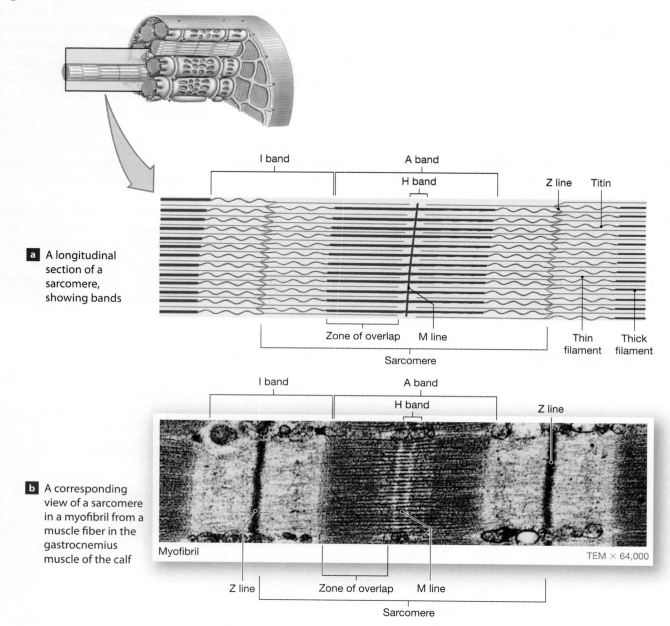

a A longitudinal section of a sarcomere, showing bands

b A corresponding view of a sarcomere in a myofibril from a muscle fiber in the gastrocnemius muscle of the calf

The filaments are arranged in repeating patterns called **sarcomeres** (SAR-kō-mĕrz; *sarkos*, flesh + *meros*, part) along a myofibril. The thin filaments connect to one another at the **Z lines** on each end of the sarcomere. Each Z line is made of a protein called **actinin.** Areas near the Z line that contain only thin filaments are **I bands.** Between I bands in a sarcomere is the **A band,** an area containing both thin and thick filaments. The edges of the A band are the **zone of overlap** where the thick and thin filaments bind during muscle contraction. The middle region of the A band is the **H zone** and contains only thick filaments. A dense **M line** in the center of the A band attaches the thick filaments. Because the thick and thin filaments do not overlap one another completely, some areas of the sarcomere appear lighter than others. This organization results in the striated (striped) appearance of skeletal muscle tissue visible in Figure 17.3b.

A thin filament consists of two intertwined strands of actin (**Figure 17.4**). Four protein components make up the actin strands: G actin, F actin, nebulin, and active sites. The **G actins** are individual spherical molecules, like pearls on a necklace. Approximately 300 to 400 G actins twist together into an **F-actin strand.** The G actins are held in position along the strand by

Figure 17.4 Thick and Thin Filaments

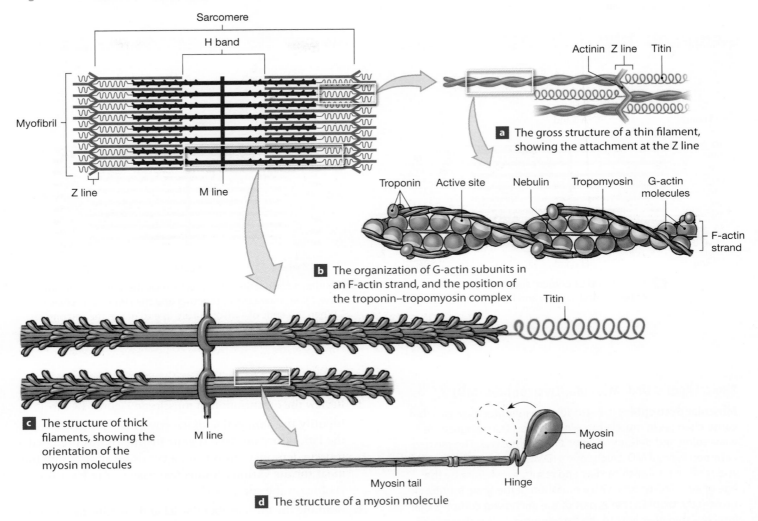

a The gross structure of a thin filament, showing the attachment at the Z line

b The organization of G-actin subunits in an F-actin strand, and the position of the troponin–tropomyosin complex

c The structure of thick filaments, showing the orientation of the myosin molecules

d The structure of a myosin molecule

the protein **nebulin,** much like the string of a necklace holds the beads in place. On each G-actin molecule is an **active site** where myosin molecules in the thick filaments bind during contraction. Associated with the actin strands are two other proteins, **tropomyosin** (trō-pō-MĪ-ō-sin; *trope,* turning) and **troponin** (TRŌ-pō-nin). Tropomyosin follows the twisted actin strands and blocks active sites to regulate muscle contraction. Troponin holds tropomyosin in position and has binding sites for calcium ions. When calcium ions are released into the sarcoplasm, they bind to and cause troponin to change shape. This change in shape moves tropomyosin away from the binding sites, exposing the sites to myosin heads so that the interactions necessary for contraction can take place.

A thick filament is made of approximately 300 subunits of myosin. Each subunit consists of two strands: two intertwined **tail** regions and two globular **heads.** Bundles of myosin subunits, with the tails parallel to one another and the heads projecting outward, constitute a thick filament. A protein called **titin** attaches

the thick filament to the Z line on the end of the sarcomere. Each myosin head contains a **binding site** for actin and a region that functions as an **ATPase enzyme.** This portion of the head splits an ATP molecule and absorbs the released energy to bind to a thin filament and then pivot and slide the thin filament inward.

When a muscle fiber contracts, the thin filaments are pulled deep into the sarcomere (**Figure 17.5**). As the thin filaments slide inward, the I band and H zone become smaller. Each myofibril consists of approximately 10,000 sarcomeres that are joined end to end. During contraction, the sarcomeres compress and the myofibril shortens and pulls on the sarcolemma, causing the muscle fiber to shorten as well.

QuickCheck Questions

1.1 Describe the connective tissue organization of a muscle.

1.2 What is the relationship between myofibrils and sarcomeres?

1.3 Where are calcium ions stored in a muscle fiber?

Figure 17.5 Changes in the Appearance of a Sarcomere During the Contraction of a Skeletal Muscle Fiber

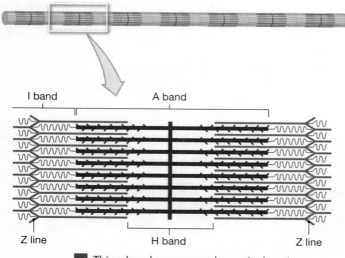

I band   A band

Z line   H band   Z line

a This relaxed sarcomere shows the location of the A band, Z lines, and I band.

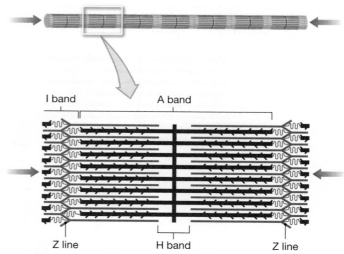

I band   A band

Z line   H band   Z line

b During a contraction, the A band stays the same width, but the Z lines move closer together and the I band gets smaller. When the ends of a myofibril are free to move, the sarcomeres shorten simultaneously and the ends of the myofibril are pulled toward its center.

Clinical Application Muscular Dystrophies (MD)

Muscular dystrophies (DIS-trō-fēz) are inherited diseases that cause changes in muscle proteins that result in progressive weakening and deterioration of skeletal muscles. The most common type of MD, Duchenne's (DMD), starts in childhood and is often fatal with cardiac and respiratory failure by the age of 20. DMD occurs with a mutation of the gene for muscle protein **dystrophin** that is part of the anchoring complex that attaches thin filaments to the sarcolemma. In the absence of normal dystrophin molecules, calcium channels in the sarcoplasmic reticulum remain open. Elevated calcium levels gradually deteriorate muscle proteins important in contraction. The dystrophin gene is located on the X chromosome. Mothers can carry the mutated MD form and have a 50 percent chance of passing it on to each of their male offspring. Steroids are used to slow degeneration and inflammation of muscles; however, treatment for the disease is still in the research phase. ■

In the Lab   1

Materials

☐ Muscle model
☐ Muscle fiber model
☐ Round steak or similar cut of meat
☐ Preserved muscle tissue
☐ Dissecting microscope

Procedures

1. Review the organization of muscles in Figures 17.1 to 17.5.

2. Identify the connective tissue coverings of muscles on the laboratory models. If your instructor has prepared a muscle demonstration from a cut of meat, examine the meat for the various connective tissues. Are fascicles visible on the specimen?

3. Examine the muscle fiber model and identify each feature. Describe the location of the sarcoplasmic reticulum, myofibrils, sarcomeres, and filaments.

4. Examine a specimen of preserved muscle tissue by placing the tissue in saline solution and then teasing the muscle apart using tweezers and a probe. Notice how the fascicles appear as strands of muscle tissue. Examine the fascicles under a dissecting microscope. How are they arranged in the muscle?

For additional practice, complete the *Sketch to Learn* Activity (on p. 219). ■

Lab Activity 2   The Neuromuscular Junction

Each skeletal muscle fiber is controlled by a nerve cell called a **motor neuron.** To excite the muscle fiber, the motor neuron releases a chemical message called **acetylcholine** (as-ē-til-KŌ-lēn), abbreviated ACh. The motor neuron and the

Sketch to Learn

Let's draw a relaxed sarcomere and label each region.

Sample Sketch

Step 1
- Draw 2 stacks of 3 horizontal lines. Be sure to keep the lines the same length.

Step 2
- Connect the outer edges of lines together for the 2 stacks.
- Add thick lines between the thin ones.

Step 3
- Add small bristle-like heads on thick filaments.
- Label your drawing.

Your Sketch

muscle fiber meet at a **neuromuscular junction** (**Figure 17.6**), also called a **myoneural junction.** The end of the neuron, called the *axon,* expands to form a bulbous **synaptic terminal,** also called a **synaptic knob.** In the synaptic terminal are **synaptic vesicles** that contain ACh. A small gap, the **synaptic cleft,** separates the synaptic terminal from a folded area of the sarcolemma called the **motor end plate.** At the motor end plate, the sarcolemma releases into the synaptic cleft the enzyme **acetylcholinesterase (AChE),** which prevents over-stimulation of the muscle fiber by deactivating ACh.

An Overview of Muscle Contraction

When a nerve impulse, called an **action potential,** travels down a neuron and reaches the synaptic terminal, the synaptic vesicles release ACh into the synaptic cleft. The ACh diffuses across the cleft and binds to ACh receptors embedded in the sarcolemma at the motor end plate of the muscle fiber. This binding of the chemical stimulus causes the sarcolemma to generate an action potential. The potential spreads across the sarcolemma and down transverse tubules, causing calcium ions to be released from the sarcoplasmic reticulum into the sarcoplasm of the muscle fiber. The calcium ions bind to troponin, which in turn moves tropomyosin and exposes the active sites on the G actins. The myosin heads attach to the active sites and ratchet the thin filaments inward, much like a tug-of-war team pulling on a rope. As thin filaments slide into the H zone, the sarcomere shortens. The additive effect of the shortening of many sarcomeres along the myofibril results in a decrease in the length of the myofibril and contraction of the muscle.

In a relaxed muscle fiber, the calcium ion concentration in the sarcoplasm is minimal. When the muscle fiber is stimulated, calcium channels in the sarcoplasmic reticulum open and calcium ions rapidly flow down the concentration gradient into the sarcoplasm. Because each myofibril is surrounded by sarcoplasmic reticulum, calcium ions are quickly and efficiently released among the thick and thin filaments of the myofibril. For the muscle to relax, calcium-ion pumps in the sarcoplasmic reticulum actively transport calcium ions out of the sarcoplasm and into the cisternae.

QuickCheck Questions

2.1 What molecules are in the synaptic vesicles?

2.2 Where is the motor end plate?

Figure 17.6 Skeletal Muscle Innervation

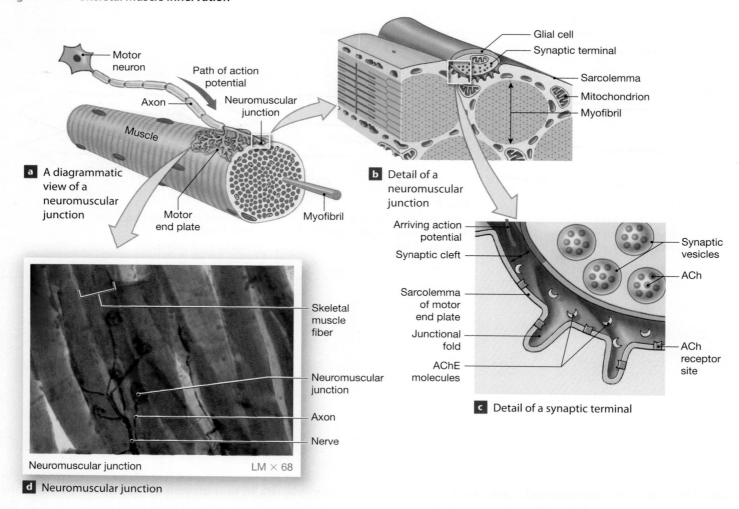

a A diagrammatic view of a neuromuscular junction

Motor neuron

Axon

Muscle

Path of action potential

Neuromuscular junction

Motor end plate

Myofibril

b Detail of a neuromuscular junction

Glial cell

Synaptic terminal

Sarcolemma

Mitochondrion

Myofibril

c Detail of a synaptic terminal

Arriving action potential

Synaptic cleft

Sarcolemma of motor end plate

Junctional fold

AChE molecules

Synaptic vesicles

ACh

ACh receptor site

d Neuromuscular junction

Skeletal muscle fiber

Neuromuscular junction

Axon

Nerve

Neuromuscular junction   LM × 68

In the Lab 2

Materials

☐ Compound microscope
☐ Neuromuscular junction slide

Procedures

1. Review the structure of the neuromuscular junction in Figure 17.6.

2. Examine the slide of the neuromuscular junction at low and medium powers. Identify the long, dark, threadlike structures and the oval disks. Describe the appearance of the muscle fibers.

3. **Draw It!** In the space provided, sketch several muscle fibers and their neuromuscular junctions. ■

Muscle fibers

Lab Activity 3   Naming Muscles

Numerous methods are used to name skeletal muscles (**Table 17.1**). One method names muscles according to either the bones they attach to or the region of the body in which they are found. For example, the temporalis muscle is found on the temporal bone, and the rectus abdominis muscle forms the anterior muscular wall of the abdomen. Another easily identifiable muscle is the sternocleido-mastoid, which is attached to the sternum (*sterno-*), the

Table 17.1	Muscle Terminology		
Terms Indicating Specific Regions of the Body	**Terms Indicating Position, Direction, or Fascicle Organization**	**Terms Indicating Structural Characteristics of the Muscle**	**Terms Indicating Actions**
Abdominis (abdomen)	Anterior (front)	**NATURE OF ORIGIN**	**GENERAL**
Anconeus (elbow)	Externus (superficial)	Biceps (two heads)	Abductor
Auricularis (auricle of ear)	Extrinsic (outside)	Triceps (three heads)	Adductor
Brachialis (brachium)	Inferioris (inferior)	Quadriceps (four heads)	Depressor
Capitis (head)	Internus (deep internal)		Extensor
Carpi (wrist)	Intrinsic (inside)	**SHAPE**	Flexor
Cervicis (neck)	Lateralis (lateral)	Deltoid (triangle)	Levator
Cleido-/-clavius (clavicle)	Medialis/medius (medial middle)	Orbicularis (circle)	Pronator
Coccygeus (coccyx)	Oblique	Pectinate (comblike)	Rotator
Costalis (ribs)	Posterior (back)	Piriformis (pear-shaped)	Supinator
Cutaneous (skin)	Profundus (deep)	Platy- (flat)	Tensor
Femoris (femur)	Rectus (straight parallel)	Pyramidal (pyramid)	
Genio- (chin)	Superficialis (superficial)	Rhomboid	**SPECIFIC**
Glosso-/-glossal (tongue)	Superions (superior)	Serratus (serrated)	Buccinator (trumpeter)
Hallucis (great toe)	Transversus (transverse)	Splenius (bandage)	Risorius (laughter)
Ilio- (ilium)		Teres (long and round)	Sartorius (like a tailor)
Inguinal (groin)		Trapezius (trapezoid)	
Lumborum (lumbar region)			
Nasalis (nose)		**OTHER STRIKING FEATURES**	
Nuchal (back of neck)		Alba (white)	
Oculo- (eye)		Brevis (short)	
Oris (mouth)		Gracilis (slender)	
Palpebrae (eyelid)		Lata (wide)	
Pollicis (thumb)		Latissimus (widest)	
Popliteus (posterior to knee)		Longissimus (longest)	
Psoas (loin)		Longus (long)	
Radialis (radius)		Magnus (large)	
Scapularis (scapula)		Major (larger)	
Temporalis (temples)		Maximus (largest)	
Thoracis (thoracic region)		Minimus (smallest)	
Tibialis (tibia)		Minor (smaller)	
Ulnaris (ulna)		Tendinosus (tendinous)	
Uro- (urinary)		Vastus (great)	

clavicle (*cleido-*), and the mastoid process of the temporal bone. The size of a muscle or the direction of fibers in a muscle is often reflected in its name. Many muscles have multiple origins. Look for the prefixes *bi-* for two, *tri-* for three, and *quad-* for four origins. Anatomists also conceive names based on muscle shape. The deltoid has a broad origin and inserts on a very narrow region of the humerus. This gives this muscle a triangular, or *deltoid*, shape; hence the name.

Many muscles are named based on how they move the body. The name *flexor carpi ulnaris* appears complex, but it is really quite easy to understand if you examine it step by step. *Flexor* means the muscle flexes something. *Carpi* refers to carpals, the bones of the wrist, and *ulnaris* suggests that the muscle flexes the carpi on the medial side of the wrist, where the ulna is located. Therefore, the flexor carpi ulnaris is a muscle that flexes and adducts the wrist.

Make a Prediction

What does the muscle name *biceps femoris* mean?

QuickCheck Questions

3.1 Give two examples of how muscles are named.

3.2 What does the name *sternocleidomastoid* mean?

In the Lab   3

Material

☐ Torso model

Procedures

1. Review the muscle terminology in Table 17.1.
2. Using the names of the following muscles, locate each muscle on the torso model.
   - Rectus abdominis
   - Gluteus maximus
   - Deltoid ■

Name _____

Date _____

Section _____

Organization
of Skeletal Muscles

A. Definitions

Write a definition for each term.

1. sarcomere

2. epimysium

3. perimysium

4. endomysium

5. myofibril

6. striations

7. sarcolemma

8. transverse tubule

9. sarcoplasmic reticulum

10. actin

11. myosin

12. fascicle

B. Matching

Match each term in the left column with its correct description from the right column.

_____	1. glossal	A.	mouth
_____	2. cleido	B.	clavicle
_____	3. scapularis	C.	moves away
_____	4. abductor	D.	great
_____	5. oris	E.	tongue
_____	6. brevis	F.	moves toward
_____	7. adductor	G.	eye
_____	8. oculi	H.	scapula
_____	9. vastus	I.	tenses
_____	10. rectus	J.	head
_____	11. tensor	K.	short
_____	12. capitis	L.	straight

C. Drawing

1. ***Draw It!*** Draw the organization of a skeletal muscle and show the various types of connective tissue, fascicles, and muscle fiber.

2. ***Draw It!*** Draw a sarcomere and label each band and zone.

D. Short-Answer Questions

1. Describe the structure of a fascicle, including the connective tissue covering around and within the fascicle.

2. How does a motor neuron stimulate a muscle fiber to contract?

3. Describe the structure of a sarcomere.

E. Labeling

Label the structure of the muscle fiber in **Figure 17.7**.

Figure 17.7 **Structure of a Muscle Fiber**

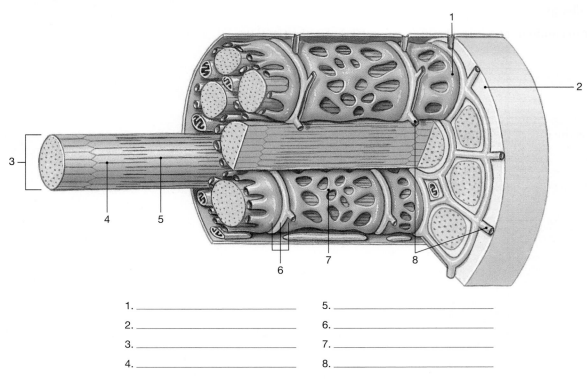

1. _____   5. _____

2. _____   6. _____

3. _____   7. _____

4. _____   8. _____

F. Analysis and Application

1. What gives skeletal muscle fibers their striations?

2. Describe the role of each thin-filament protein and each thick-filament protein in muscle contraction.

3. Many insecticides contain a compound that is an acetylcholinesterase inhibitor. How would exposure to this poison affect skeletal muscles in a human?

G. Clinical Challenge

1. How can children inherit Duchenne's muscular dystrophy (DMD)?

2. What is the role of the protein dystrophin in normal muscle function and in DMD?

Muscles of the Head and Neck

Learning Outcomes

On completion of this exercise, you should be able to:

1. Identify the origin, insertion, and action of the muscles used for facial expression and mastication.

2. Identify the origin, insertion, and action of the muscles that move the eye.

3. Identify the origin, insertion, and action of the muscles that move the tongue, head, and anterior neck.

Muscles are organized into the axial and appendicular divisions to reflect their attachment to either the axial or appendicular skeleton. Axial muscles include the muscles of the head and neck (which are covered in this exercise), and muscles of the vertebral column, abdomen, and pelvis (which are presented in Exercise 19). Appendicular muscles are muscles of the pectoral girdle and upper limb and the pelvic girdle and lower limb. (The appendicular musculature is covered in Exercises 20 and 21.)

The movement and attachments of a muscle are often reflected in its name. Each muscle causes a movement, called the **action,** that depends on many factors, especially the shape of the attached bones. For a muscle to produce a smooth, coordinated action, one end of it must serve as an attachment site while the other end moves the intended bone. The relatively stationary part of the muscle is called the **origin.** The opposite end of the muscle, the part that moves the bone, is called the **insertion.** As the muscle contracts, the insertion moves toward the origin to generate a pulling force and cause the muscle's action. Muscles can generate only a pulling force; they can never push. Usually, when one muscle, called

Need More Practice and Review?

Build your knowledge—and confidence!—in the Study Area of MasteringA&P® at www.masteringaandp.com with Pre-lab Quizzes, Post-lab Quizzes, Practice Anatomy Lab™ (PAL™) 3.0 virtual anatomy practice tool, PhysioEx™ 9.0 laboratory simulations, and A&P Flix™ with Quizzes.

PAL   For this lab exercise, follow these navigation paths:
- PAL>Human Cadaver>Muscular System>Head and Neck
- PAL>Anatomical Models>Muscular System>Head and Neck

A&PFlix   For this lab exercise, go to these topics:
- Origins, Insertions, Actions, and Innervations
- Group Muscle Actions and Joints

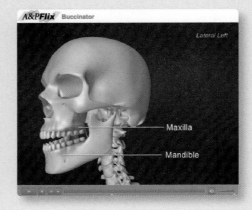

an **agonist,** pulls in one direction, an **antagonistic** muscle pulls in the opposite direction to produce resistance and promote smooth movement. **Synergists** are muscles that work together and are often classified together in a **muscle group,** such as the oblique group of the abdomen.

The muscles of the head and neck produce a wide range of motions for making facial expressions, processing food, producing speech, and positioning the head. The names of these muscles usually indicate either the bone to which a muscle is attached or the structure a muscle surrounds. In this exercise you will identify the major muscles used for facial expression and mastication, the muscles that move the eyes, and those that position the head and neck. As you study each group, attempt to find the general location of each muscle on your body. Contract the muscle and observe its action as your body moves.

Study Tip Muscle Modeling

Your fingers and hands can be used to simulate the origin, insertion, and action of muscles. For example, place the base of your right index finger on your right zygomatic bone and the tip of the index finger at the right corner of your mouth. The finger now represents the zygomatic major muscle, which elevates the edge of the mouth. The base of the finger represents the muscle's origin at the zygomatic bone, and the tip represents the insertion. When you flex your finger and elevate your mouth, you are mimicking the major action of this muscle. Smile! ■

Lab Activity 1   Muscles of Facial Expression

The muscles of facial expression are those associated with the mouth, eyes, nose, ears, scalp, and neck. These muscles are unique in that one or both attachments are to the dermis of the skin rather than to a bone. Refer to **Figure 18.1** and **Table 18.1** for details on the origin, insertion, and actions of these muscles.

Make a Prediction

The face has two sphincter muscles. Where are they and what action does each perform?

Figure 18.1 Muscles of Facial Expression Anterior view.

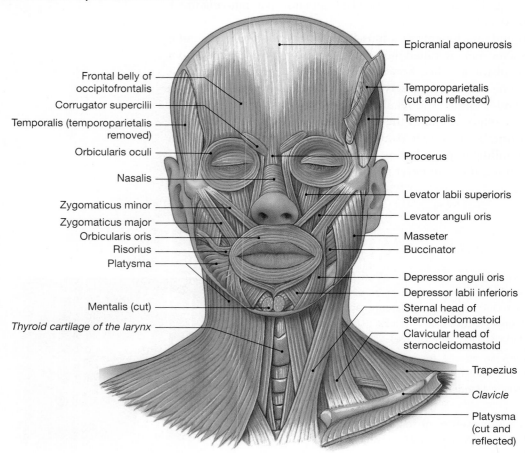

Epicranial aponeurosis

Frontal belly of occipitofrontalis

Corrugator supercilii

Temporalis (temporoparietalis removed)

Orbicularis oculi

Nasalis

Zygomaticus minor

Zygomaticus major

Orbicularis oris

Risorius

Platysma

Mentalis (cut)

Thyroid cartilage of the larynx

Temporoparietalis (cut and reflected)

Temporalis

Procerus

Levator labii superioris

Levator anguli oris

Masseter

Buccinator

Depressor anguli oris

Depressor labii inferioris

Sternal head of sternocleidomastoid

Clavicular head of sternocleidomastoid

Trapezius

Clavicle

Platysma (cut and reflected)

Anterior view

Scalp

The **occipitofrontalis muscle** is the major muscle of the scalp, which is called the *epicranium*. It consists of two muscle bellies, the **frontal belly** and the **occipital belly,** which are separated by a flat sheet of connective tissue attached to the scalp called the **epicranial aponeurosis** (ep-i-KRĀ-nē-ul ap-ō-nū-RŌ-sis; *epi-*, on + *kranion*, skull). The frontal belly of the occipitofrontalis muscle is the broad anterior muscle on the forehead that covers the frontal bone. It originates at the epicranial aponeurosis and inserts on the superior margin of the eye orbit, near the eyebrow and on the bridge of the nose. The actions of the frontal belly include wrinkling the forehead, raising the eyebrows, and pulling the scalp forward. The occipital belly of the occipitofrontalis muscle covers the posterior of the skull. It originates on the superior nuchal line and inserts on the epicranial aponeurosis. This muscle tenses and retracts the scalp, an action difficult for most people to isolate and perform.

Ear

The **temporoparietalis** muscle occurs on the lateral sides of the epicranium. The muscle is cut and reflected (pulled up) in Figure 18.1 to illustrate deeper muscles of the epicranium. The action of the temporoparietalis is to tense the scalp and move the auricle (flap) of the ear. The origin and insertion for this muscle are on the epicranial aponeurosis.

Eye

Muscles of the face that surround the eyes wrinkle the brow and move the eyelids. Muscles that move the eyeball are covered in an upcoming activity in this exercise.

The sphincter muscle of the eye is the **orbicularis oculi** (or-bik-ū-LA-ris OK-ū-lī). It arises from the medial wall of the eye orbit, and its fibers form a band of muscle that passes around the circumference of the eye, which serves as the insertion. The muscle acts to close the eye, as during an exaggerated blink. The **corrugator supercilii** muscle is a small muscle that originates on the orbital rim of the frontal bone and inserts on the eyebrow. It acts to pull the skin inferiorly and wrinkles the forehead into a frown. The **levator palpebrae superioris** muscle inserts on and elevates the upper eyelid. (This muscle is not visible in Figure 18.1. See Figure 18.2.)

Nose

The human nose has limited movement, and the related muscles serve mainly to change the shape of the nostrils. The **procerus muscle** has a vertical orientation over the nasal bones; the **nasalis muscle** horizontally spans the inferior nasal bridge.

Mouth

The **buccinator** (BUK-si-nā-tor) **muscle** is the horizontal muscle spanning between the jaws. It compresses the cheeks when you are eating or sucking on a straw. The **orbicularis oris muscle** is a sphincter muscle that inserts on the skin surrounding the mouth. This muscle shapes the lips for a variety of functions, including speech, food manipulation, and facial expressions, and purses the lips together for a kiss. The **levator labii superioris muscle** is lateral to the nose and inserts on the superolateral edge of the orbicularis oris muscle. As its name implies, the levator labii superioris muscle elevates the upper lip. Muscles that act on the lower lip are inferior to the mouth. The **depressor anguli oris muscle** inserts on the skin at the angle of the mouth to depress the corners of the mouth. The **depressor labii inferioris muscle** is medial to the anguli muscle and inserts along the edge of the lower lip to depress the lower lip. On the medial chin is the **mentalis muscle,** which elevates and protrudes the lower lip.

The **risorius muscle** is a narrow muscle that inserts on the angle of the mouth. When it contracts, the risorius muscle pulls and produces a grimace-like tensing of the mouth. Although the term *risorius* refers to a smile, the muscle is probably more associated with the expression of pain than pleasure. In the disease tetanus, the risorius is involved in the painful contractions that pull the corners of the mouth back into "lockjaw."

The **zygomaticus major** and **zygomaticus minor muscles** originate on the zygomatic bone and insert on the skin and corners of the mouth. These muscles retract and elevate the corners of the mouth when you smile.

Neck

The **platysma** (pla-TIZ-muh; *platy*, flat) is a thin, broad muscle covering the sides of the neck. It originates on the fascia covering the pectoralis and deltoid muscles and extends upward to insert on the inferior edge of the mandible. Some of the fibers of the platysma also extend into the fascia and muscles of the lower face. The platysma depresses the mandible and the soft structures of the lower face, resulting in an expression of horror and disgust.

QuickCheck Questions

1.1 What are the two facial muscles that are circular?

1.2 What are the muscles associated with the epicranial aponeurosis?

In the Lab 1

Materials

☐ Head model

☐ Muscle chart

Table 18.1   ORIGINS AND INSERTIONS Muscles of Facial Expression (See Figure 18.1)

Region/Muscle	Origin	Insertion	Action	Innervation
MOUTH				
Buccinator	Alveolar processes of maxilla and mandible	Blends into fibers of orbicularis oris	Compresses cheeks	Facial nerve (N VII)*
Depressor labii inferioris	Mandible between the anterior midline and the mental foramen	Skin of lower lip	Depresses lower lip	As above
Levator labii superioris	Inferior margin of orbit, superior to the infraorbital foramen	Orbicularis oris	Elevates upper lip	As above
Levator anguli oris	Maxilla below the infraorbital foramen	Corner of mouth	Elevates mouth corner	As above
Mentalis	Incisive fossa of mandible	Skin of chin	Elevates and protrudes lower lip	As above
Orbicularis oris	Maxilla and mandible	Lips	Compresses, purses lips	As above
Risorius	Fascia surrounding parotid salivary gland	Angle of mouth	Draws corner of mouth to the side	As above
Depressor anguli oris	Anterolateral surface of mandibular body	Skin at angle of mouth	Depresses corner of mouth	As above
Zygomaticus major	Zygomatic bone near zygomaticomaxillary suture	Angle of mouth	Retracts and elevates corner of mouth	As above
Zygomaticus minor	Zygomatic bone posterior to zygomaticotemporal suture	Upper lip	Retracts and elevates upper lip	As above
EYE				
Corrugator supercilii	Orbital rim of frontal bone near nasal suture	Eyebrow	Pulls skin inferiorly and anteriorly; wrinkles brow	As above
Levator palpebrae superioris (*Figure 18.2*)	Tendinous band around optic foramen	Upper eyelid	Elevates upper eyelid	Oculomotor nerve (N III)**
Orbicularis oculi	Medial margin of orbit	Skin around eyelids	Closes eye	Facial nerve (N VII)
NOSE				
Procerus	Nasal bones and lateral nasal cartilages	Aponeurosis at bridge of nose and skin of forehead	Moves nose, changes position and shape of nostrils	As above
Nasalis	Maxilla and alar cartilage of nose	Bridge of nose	Compresses bridge, depresses tip of nose; elevates corners of nostrils	As above
EAR				
Temporoparietalis	Fascia around external ear	Epicranial aponeurosis	Tenses scalp, moves auricle of ear	As above
SCALP (EPICRANIUM)				
Occipitofrontalis frontal belly	Epicranial aponeurosis	Skin of eyebrow and bridge of nose	Raises eyebrows, wrinkles forehead	As above
Occipital belly	Epicranial aponeurosis	Epicranial aponeurosis	Tenses and retracts scalp	As above
NECK				
Platysma	Superior thorax between cartilage of 2nd rib and acromion of scapula	Mandible and skin of cheek	Tenses skin of neck; depresses mandible	As above

*An uppercase N and Roman numerals refer to a cranial nerve.
**This muscle originates in association with the extrinsic eye muscles, so its innervation is unusual.

Procedures

1. Review the muscles of the head in Figure 18.1.
2. Examine the head model and/or the muscle chart, and locate each muscle described in the preceding paragraphs.
3. Find the general location of the muscles of facial expression on your face. Practice the action of each muscle and observe how your facial expression changes. ■

Lab Activity 2 Muscles of the Eye

The **extrinsic muscles** of the eye, also called **extraocular eye muscles** or **oculomotor muscles,** are the muscles that move the eyeballs. (In general, any muscle located outside the structure it controls is called an **extrinsic muscle,** and any muscle inside the structure it controls is referred to as an **intrinsic muscle.**) The extraocular muscles insert on the *sclera*, which is the white, fibrous covering of the eye. **Intrinsic eye muscles** are involved in focusing the eye for vision. (These muscles are discussed in Exercise 29.)

Six extraocular eye muscles control eye movements (**Figure 18.2** and **Table 18.2**). The **superior rectus, inferior rectus, medial rectus,** and **lateral rectus** muscles are straight muscles that move the eyeball in the superior, inferior, medial, and lateral directions, respectively. They originate around the optic foramen in the eye orbit and insert on the sclera. The **superior** and **inferior oblique** muscles attach diagonally on the eyeball. The superior oblique muscle has a tendon passing through a trochlea (pulley) located on the upper orbit. This muscle rolls the eye downward, and the inferior oblique muscle rolls the eye upward.

QuickCheck Questions

2.1 What are the four rectus muscles of the eye?

2.2 Which eye muscle passes through a pulleylike structure?

Figure 18.2 Extrinsic Eye Muscles

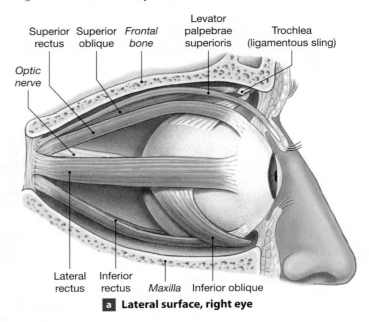

a Lateral surface, right eye

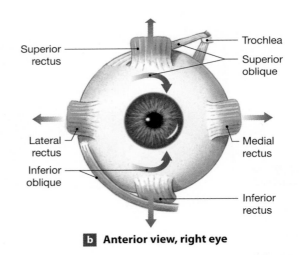

b Anterior view, right eye

Table 18.2	*ORIGINS AND INSERTIONS* Extrinsic Eye Muscles (See Figure 18.2)			
Muscle	**Origin**	**Insertion**	**Action**	**Innervation**
Inferior rectus	Sphenoid around optic canal	Inferior, medial surface of eyeball	Eye looks down	Oculomotor nerve (N III)
Medial rectus	As above	Medial surface of eyeball	Eye looks medially	As above
Superior rectus	As above	Superior surface of eyeball	Eye looks up	As above
Lateral rectus	As above	Lateral surface of eyeball	Eye looks laterally	Abducens nerve (N VI)
Inferior oblique	Maxilla at anterior portion of orbit	Inferior, lateral surface of eyeball	Eye rolls, looks up and laterally	Oculomotor nerve (N III)
Superior oblique	Sphenoid around optic canal	Superior, lateral surface of eyeball	Eye rolls, looks down and laterally	Trochlear nerve (N IV)

Sketch to Learn

The muscles of the eye are positioned on the compass points. Let's draw an eyeball with the extraocular muscles included.

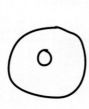 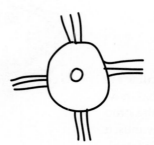 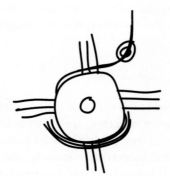

Step 1
- Draw an eyeball.

Step 2
- Add some lines "at the compass points" to represent the rectus muscles.

Step 3
- Draw the superior oblique and inferior oblique muscles.
- Label your drawing.

Your Sketch

In the Lab   2

Materials

- ☐ Eye model
- ☐ Eye muscle chart

Procedures

1. Review the muscles of the eye in Figure 18.2.
2. Examine the eye model and/or the eye muscle chart, and locate each extrinsic eye muscle.
3. Practice the action of each eye muscle by moving your eyeballs.

For additional practice, complete the *Sketch to Learn* activity. ■

Lab Activity   3   Muscles of Mastication

The muscles involved in mastication depress and elevate the mandible to open and close the jaws and grind the teeth against the food (**Figure 18.3** and **Table 18.3**). The **masseter** (MAS-se-tur) muscle is a short, thick muscle originating on the zygomatic arch and inserting on the angle and the ramus of the mandible. The **temporalis** (tem-pō-RA-lis) muscle covers almost the entire temporal fossa. This muscle has its origin on the temporal lines of the cranium and inserts on the coronoid process of the mandible.

Deep to the masseter and other cheek muscles are the **lateral** and **medial pterygoid** (TER-i-goyd; *pterygoin,* wing) muscles, which assist in mastication by elevating and depressing the mandible and moving the mandible from side to side, an action called *lateral excursion.*

Study Tip   The Mighty Masseter

Put your fingertips at the angle of your jaw and clench your teeth. You should feel the masseter bunch up as it forces the teeth together. ■

Figure 18.3 Muscles of Mastication

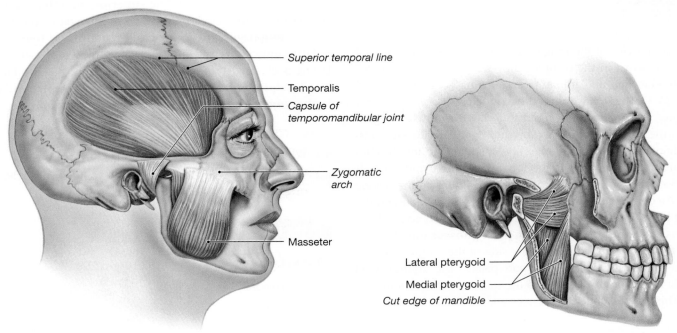

Superior temporal line

Temporalis

Capsule of temporomandibular joint

Zygomatic arch

Masseter

Lateral pterygoid

Medial pterygoid

Cut edge of mandible

a Lateral view. The temporalis muscle passes medial to the zygomatic arch to insert on the coronoid process of the manible. The masseter inserts on the angle and lateral surface of the mandible.

b Lateral view, pterygoid muscles exposed. The location and orientation of the pterygoid muscles can be seen after the overlying muscles, along with a portion of the mandible, are removed.

Table 18.3	*ORIGINS AND INSERTIONS* Muscles of Mastication (See Figure 18.3)			
Muscle	**Origin**	**Insertion**	**Action**	**Innervation**
Masseter	Zygomatic arch	Lateral surface of mandibular ramus	Elevates mandible and closes the jaws	Trigeminal nerve (N V), mandibular branch
Temporalis	Along temporal lines of skull	Coronoid process of mandible	Elevates mandible	As above
Pterygoids (medial and lateral)	Lateral pterygoid plate	Medial surface of mandibular ramus	*Medial:* Elevates the mandible and closes the jaws, or performs lateral excursion	As above
			Lateral: Opens jaws, protrudes mandible, or performs lateral excursion	As above

QuickCheck Questions

3.1 To which bones do the muscles for mastication attach?

3.2 Which muscle protracts the mandible?

In the Lab   3

Materials

☐ Head model

☐ Muscle chart

Procedures

1. Review the mastication muscles in Table 18.3 and Figures 18.1 and 18.3.

2. Examine the head model and/or the muscle chart, and locate each mastication muscle described in this activity.

3. Find the general location of the muscles of mastication on your face. Practice the action of each muscle and observe how your mandible moves. ■

Lab Activity 4 Muscles of the Tongue and Pharynx

Extrinsic muscles of the tongue constitute the floor of the oral cavity and assist in the complex movements of the tongue for speech, chewing, and initiating swallowing (**Figure 18.4** and **Table 18.4**). The root word for these muscles is *glossus*, Greek for "tongue." Each prefix in the name indicates the muscle's origin.

In the anterior floor of the mouth, the **genioglossus** muscle originates on the medial mandibular surface around the chin and inserts on the body of the tongue and the hyoid bone. It depresses and protracts the tongue, as in initiating the licking of an ice cream cone. The **hyoglossus** muscle originates on the hyoid bone, inserts on the side of the tongue, and acts to both depress and retract the tongue. The **palatoglossus** muscle arises from the soft palate, inserts on the side of the tongue, elevates the tongue, and depresses the soft palate. The **styloglossus** muscle has its origin superior to the tongue on the styloid process. This muscle retracts the tongue and elevates its sides.

Muscles of the pharynx are involved in swallowing (Figure 18.4b and **Table 18.5**). The **superior, middle,** and **inferior constrictor** muscles constrict the pharynx to push food into the esophagus. The **levator veli palatini** and **tensor veli palatini** muscles elevate the soft palate during swallowing. The larynx is elevated by the **palatopharyngeus** (pal-āt-ō-far-IN-jē-us), **salpingopharyngeus** (sal-pin-gō-far-IN-jē-us), and **stylopharyngeus** muscles, and by some of the neck muscles.

QuickCheck Questions

4.1 What does the word *glossus* mean?

4.2 Where do the styloglossus and the hyoglossus muscles originate?

In the Lab 4

Materials

☐ Head model
☐ Muscle chart

Figure 18.4 Muscles of the Tongue and Pharynx

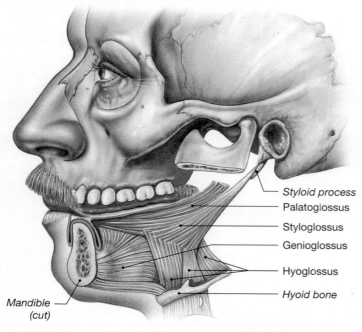

- Styloid process
- Palatoglossus
- Styloglossus
- Genioglossus
- Hyoglossus
- Hyoid bone

Mandible (cut)

a The left mandibular ramus has been removed to show the extrinsic muscles on the left side of the tongue.

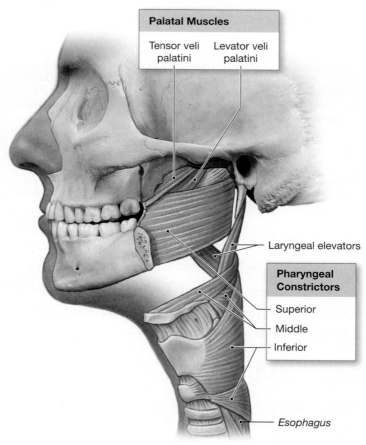

Palatal Muscles

Tensor veli palatini   Levator veli palatini

Laryngeal elevators

Pharyngeal Constrictors

- Superior
- Middle
- Inferior

Esophagus

b Pharyngeal muscles for swallowing are shown in lateral view.

Table 18.4	*ORIGINS AND INSERTIONS* Muscles of the Tongue (See Figure 18.4a)			
Muscle	**Origin**	**Insertion**	**Action**	**Innervation**
Genioglossus	Medial surface of mandible around chin	Body of tongue, hyoid bone	Depresses and protracts tongue	Hypoglossal nerve (N XII)
Hyoglossus	Body and greater horn of hyoid bone	Side of tongue	Depresses and retracts tongue	As above
Palatoglossus	Anterior surface of soft palate	As above	Elevates tongue, depresses soft palate	Internal branch of accessory nerve (N XI)
Styloglossus	Styloid process of temporal bone	Along the side to tip and base of tongue	Retracts tongue, elevates side	Hypoglossal nerve (N XII)

Table 18.5	*ORIGINS AND INSERTIONS* Muscles of the Pharynx (See Figure 18.4b)			
Muscle	**Origin**	**Insertion**	**Action**	**Innervation**
PHARYNGEAL CONSTRICTORS				
Superior constrictor	Pterygoid process of sphenoid, medial surfaces of mandible	Median raphe attached to occipital bone	Constricts pharynx to propel bolus into esophagus	Branches of pharyngeal plexus (N X)
Middle constrictor	Horns of hyoid bone	Median raphe	As above	As above
Inferior constrictor	Cricoid and thyroid cartilages of larynx	As above	As above	As above
LARYNGEAL ELEVATORS*				
	Ranges from soft palate, to cartilage around inferior portion of auditory tube, to styloid process of temporal bone	Thyroid cartilage	Elevate larynx	Branches of pharyngeal plexus (N IX and N X)
PALATAL MUSCLES				
Levator veli palatini	Petrous part of temporal bone; tissues around the auditory tube	Soft palate	Elevates soft palate	Branches of pharyngeal plexus (N X)
Tensor veli palatini	Sphenoidal spine; tissues around the auditory tube	As above	As above	Trigeminal nerve (N V)

*Refers to the palatopharyngeus, salpingopharyngeus, and stylopharyngeus, assisted by the thyrohyoid, geniohyoid, stylohyoid, and hyoglossus muscles, discussed in Tables 18.4 and 18.6.

Procedures

1. Review the extrinsic muscles of the tongue in Figure 18.4 and Table 18.4.
2. Examine the head model and/or the muscle chart and identify each muscle of the tongue.
3. Practice the action of each tongue muscle. The ability to curl your tongue with the styloglossus is genetically controlled by a single gene. Individuals with the dominant gene are "rollers" and can curl the tongue. Those with the recessive form of the gene are "nonrollers." Is it possible for nonrollers to learn how to roll the tongue? Are you, your parents, or your children rollers or nonrollers?
4. Review the muscles of the pharynx in Figure 18.4b and Table 18.5.
5. Locate each pharyngeal muscle on the head model and/or muscle chart.
6. Place your finger on your larynx (Adam's apple) and swallow. Which muscles caused the larynx to move? ■

Lab Activity 5 Muscles of the Anterior Neck

Muscles of the anterior neck, which stabilize and move the neck, act on the mandible and the hyoid bone (**Figure 18.5** and **Table 18.6**). The principal muscle of the anterior neck is the **sternocleidomastoid** (ster-nō-klī-dō-MAS-toyd) muscle. This long, slender muscle occurs on both sides of the neck and is named after its points of attachment on the sternum, clavicle, and mastoid process of the temporal bone. When

Figure 18.5 Muscles of the Anterior Neck Muscles of the anterior neck adjust the position of the larynx, mandible, and floor of the mouth and establish a foundation for tongue and pharyngeal muscles.

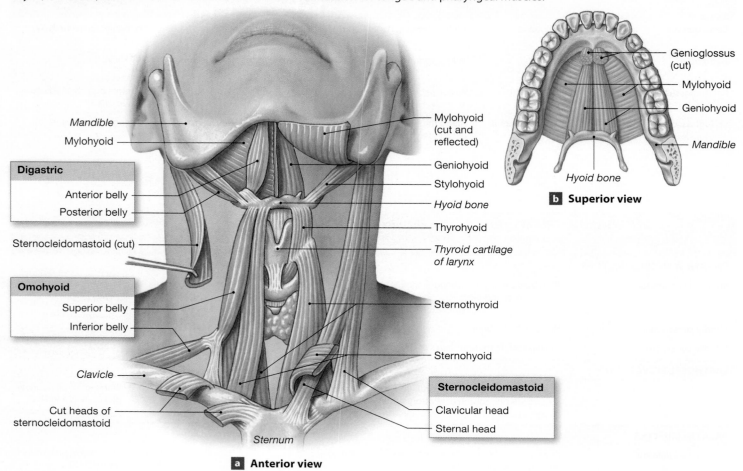

a Anterior view

b Superior view

the sternocleidomastoid muscles on the two sides of the neck contract, they act together to flex the neck; when only one sternocleidomastoid muscle contracts, it bends the head toward the shoulder and turns the face to the opposite side.

The *suprahyoid muscles* are a group of neck muscles that originate superior to, and act on, the hyoid bone. The suprahyoid muscle known as the **digastric muscle** has two parts: The **anterior belly** originates on the inferior surface of the mandible near the chin, and the **posterior belly** arises on the mastoid process of the temporal bone. The bellies insert on the hyoid bone and form a muscular swing that elevates the hyoid bone or depresses the mandible. The **mylohyoid** muscle is a wide muscle posterior to the anterior belly of the digastric muscle. The mylohyoid muscle elevates the hyoid bone or depresses the mandible. Deep and medial to the mylohyoid muscle is the **geniohyoid** muscle, which depresses the mandible, elevates the larynx, and can also retract the hyoid

bone. The **stylohyoid** muscle originates on the styloid process of the temporal bone, inserts on the hyoid bone, and elevates the hyoid bone and the larynx.

The *infrahyoid muscles* are a group of neck muscles that arise inferior to the hyoid bone, and their actions depress that bone and the larynx. The infrahyoid called the **omohyoid** (ō-mō-HĪ-ōyd) muscle has two bellies that meet at a central tendon attached to the clavicle and the first rib. Medial to the omohyoid muscle is the straplike **sternohyoid** muscle, which originates on the sternal end of the clavicle. Deep to the sternohyoid is the **sternothyroid** muscle, which arises on the manubrium of the sternum and inserts on the thyroid cartilage of the larynx. The omohyoid, sternohyoid, and sternothyroid muscles depress the hyoid bone and larynx. The **thyrohyoid** muscle originates on the thyroid cartilage of the larynx and inserts on the hyoid bone. It depresses this bone and elevates the larynx.

Table 18.6	ORIGINS AND INSERTIONS Anterior Muscles of the Neck (See Figure 18.5)			
Muscle	Origin	Insertion	Action	Innervation
Digastric	Two bellies *anterior* from inferior surface of mandible at chin; *posterior* from mastoid region of temporal bone	Hyoid bone	Depresses mandible or elevates larynx	*Anterior belly:* Trigeminal nerve (N V), mandibular branch *Posterior belly:* Facial nerve (N VII)
Geniohyoid	Medial surface of mandible at chin	Hyoid bone	As above and pulls hyoid bone anteriorly	Cervical nerve C_1 via hypoglossal nerve (N XII)
Mylohyoid	Mylohyoid line of mandible	Median connective tissue band (raphe) that runs to hyoid bone	Elevates floor of mouth and hyoid bone or depresses mandible	Trigeminal nerve (N V), mandibular branch
Omohyoid (superior and inferior bellies united at central tendon anchored to clavicle and first rib)	Superior border of scapula near scapular notch	Hyoid bone	Depresses hyoid bone and larynx	Cervical spinal nerves C_2–C_3
Sternohyoid	Clavicle and manubrium	Hyoid bone	As above	Cervical spinal nerves C_1–C_3
Sternothyroid	Dorsal surface of manubrium and first costal cartilage	Thyroid cartilage of larynx	As above	As above
Stylohyoid	Styloid process of temporal bone	Hyoid bone	Elevates larynx	Facial nerve (N VII)
Thyrohyoid	Thyroid cartilage of larynx	Hyoid bone	Elevates thyroid, depresses hyoid bone	Cervical spinal nerves C_1–C_2 via hypoglossal nerve (N XII)
Sternocleidomastoid	Two bellies: *clavicular head* attaches to sternal end of clavicle; *sternal head* attaches to manubrium	Mastoid region of skull and lateral portion of superior nuchal line	Together, they flex the neck; alone, one side bends head toward shoulder and turns face to opposite side	Accessory nerve (N XI) and cervical spinal nerves (C_2–C_3) of cervical plexus

QuickCheck Questions

5.1 Where does the sternocleidomastoid muscle attach?

5.2 What is the suffix in the names of muscles that insert on the hyoid bone?

5.3 Where is the digastric muscle located?

In the Lab 5

Materials

☐ Head–torso model
☐ Muscle chart

Procedures

1. Review the anterior neck muscles in Figure 18.5 and Table 18.6.
2. Locate each muscle on the head–torso model and/or the muscle chart.
3. Produce the actions of your suprahyoid and infrahyoid muscles and observe how your larynx moves.
4. Locate the sternocleidomastoid on the head–torso model and/or on the muscle chart.
5. Contract your sternocleidomastoid on one side and observe your head movement. Next, contract both sides and note how your head flexes.
6. Rotate your head until your chin almost touches your right shoulder and locate your left sternocleidomastoid just above the manubrium of the sternum. ■

Name _____

Date _____

Section _____

Muscles of the Head and Neck

A. Matching

Match each term in the left column with its correct description from the right column.

_____ 1. orbicularis oculi

_____ 2. buccinator

_____ 3. stylohyoid

_____ 4. masseter

_____ 5. frontal belly of occipitofrontalis

_____ 6. platysma

_____ 7. corrugator supercilii

_____ 8. zygomaticus major

_____ 9. occipital belly of occipitofrontalis

_____ 10. levator labii superioris

_____ 11. digastric

_____ 12. risorius

A. retracts scalp

B. elevates upper lip

C. thin muscle on sides of neck, depresses jaw

D. attached to styloid process and hyoid bone

E. tenses angle of mouth laterally

F. elevates corner of mouth

G. elevates jaw

H. two-bellied neck muscle

I. wrinkles forehead

J. tenses cheeks

K. protracts scalp

L. closes eye

B. Descriptions

Describe the location of each of the following muscles.

1. masseter

2. sternocleidomastoid

3. sternohyoid

4. orbicularis oris

5. zygomaticus minor

6. platysma

7. risorius

8. temporoparietalis

9. superior constrictor

10. digastric

C. Labeling

1. Label the eye muscles in **Figure 18.6**.

Figure 18.6 Muscles of the Right Eye Anterior view.

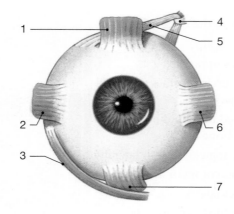

1. _____
2. _____
3. _____
4. _____
5. _____
6. _____
7. _____

2. Label the muscles of the head in **Figure 18.7**.

Figure 18.7 Muscles of the Head and Neck Lateral view.

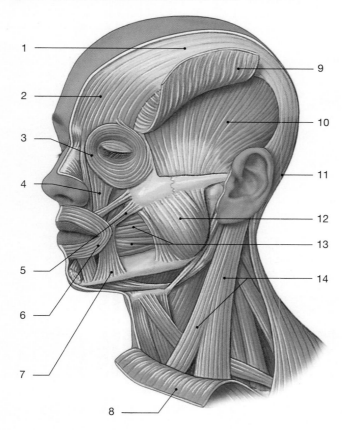

1. _____
2. _____
3. _____
4. _____
5. _____
6. _____
7. _____
8. _____
9. _____
10. _____
11. _____
12. _____
13. _____
14. _____

3. Label the muscles of the anterior neck in **Figure 18.8**.

Figure 18.8 **Muscles of the Anterior Neck** Anterior view.

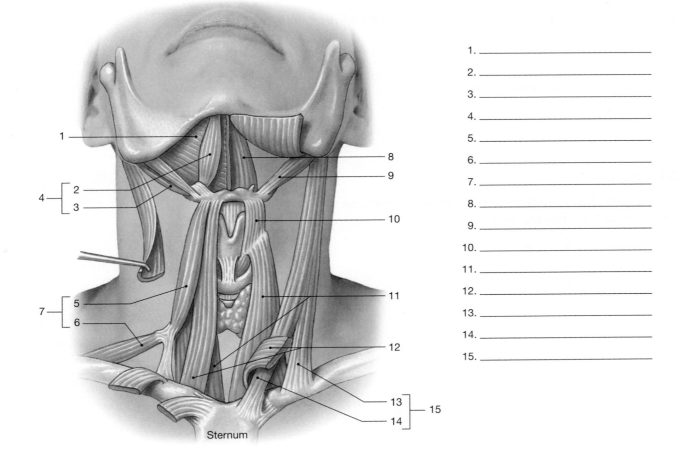

Sternum

1. _____
2. _____
3. _____
4. _____
5. _____
6. _____
7. _____
8. _____
9. _____
10. _____
11. _____
12. _____
13. _____
14. _____
15. _____

D. Analysis and Application

1. Describe the position and action of the muscles of mastication. Which muscles oppose the
   action of these muscles?

2. What anatomical descriptions are in the name *epicranial aponeurosis*?

3. Describe the movement produced by each extraocular eye muscle.

4. Explain how the muscles of the tongue and anterior neck are named.

5. Describe the actions of the digastric muscle.

E. Clinical Challenge

Mary suffers with temporomandibular joint (TMJ) syndrome. Describe the bones at this articulation and the muscles that act on them.

Muscles of the Vertebral Column, Abdomen, and Pelvis

Learning Outcomes

On completion of this exercise, you should be able to:

1. Locate the muscles of the vertebral column, abdomen, and pelvis on laboratory models and charts.

2. Identify on the models the origin, insertion, and action of the muscles of the vertebral, abdominal, and pelvic regions.

3. Demonstrate or describe the action of the major muscles of the vertebral column, abdomen, and pelvis.

The body torso has both axial and appendicular muscles. Axial muscles of the torso are the muscles that act on the vertebral column, abdomen, and pelvis. The muscles that flex, extend, and support the spine are on the posterior surface of the vertebral column. Oblique and rectus muscles occur in the neck and the abdomen. The primary functions of the abdominal muscles are to support the abdomen, viscera, and lower back, and to move the legs. Muscles of the pelvic region form the floor and walls of the pelvis and support local organs of the reproductive and digestive systems. The appendicular muscles of the torso are the large chest and back muscles that act on the shoulder and arm. (These muscles are covered in Exercise 20.)

Lab Activity 1   Muscles of the Vertebral Column

The muscles of the back are organized into three layers: *superficial, intermediate,* and *deep* (**Figure 19.1** and **Table 19.1**, p. 246). Except for two superficial muscles that act on the appendicular skeleton, the trapezius and latissimus dorsi muscles (discussed in Exercise 20), all the back muscles move the vertebral column. The superficial vertebral muscles move

Need More Practice and Review?

Build your knowledge—and confidence!—in the Study Area of MasteringA&P® at www.masteringaandp.com with Pre-lab Quizzes, Post-lab Quizzes, Practice Anatomy Lab™ (PAL™) 3.0 virtual anatomy practice tool, PhysioEx™ 9.0 laboratory simulations, and A&P Flix™ with Quizzes.

PAL   For this lab exercise, follow these navigation paths:
- PAL>Human Cadaver>Muscular System>Trunk
- PAL>Anatomical Models>Muscular System>Trunk

A&PFlix   For this lab exercise, go to these topics:
- Origins, Insertions, Actions, and Innervations
- Group Muscle Actions and Joints

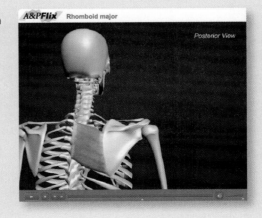

MasteringA&P®

Figure 19.1 Muscles of the Vertebral Column These muscles adjust the position of the vertebral column, head, neck, and ribs.

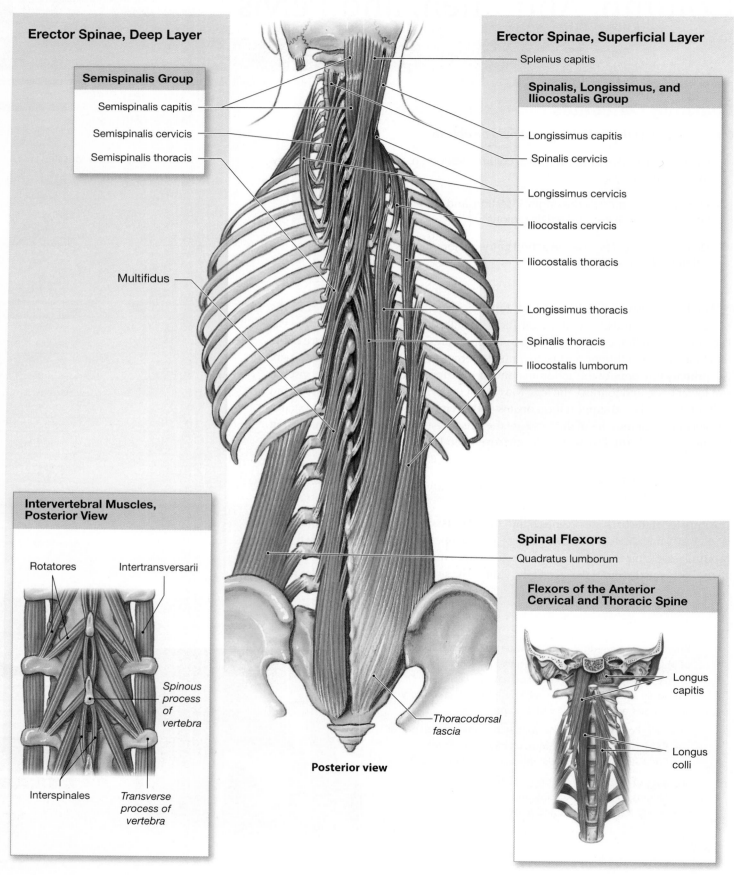

Erector Spinae, Deep Layer

Semispinalis Group

Semispinalis capitis

Semispinalis cervicis

Semispinalis thoracis

Multifidus

Erector Spinae, Superficial Layer

Splenius capitis

Spinalis, Longissimus, and Iliocostalis Group

Longissimus capitis

Spinalis cervicis

Longissimus cervicis

Iliocostalis cervicis

Iliocostalis thoracis

Longissimus thoracis

Spinalis thoracis

Iliocostalis lumborum

Intervertebral Muscles, Posterior View

Rotatores   Intertransversarii

Spinous process of vertebra

Interspinales   *Transverse process of vertebra*

Spinal Flexors

Quadratus lumborum

Flexors of the Anterior Cervical and Thoracic Spine

Longus capitis

Longus colli

Thoracodorsal fascia

Posterior view

the head and neck. The intermediate muscles are in long bands that stabilize and extend the vertebral column. The deep layer consists of small muscles that connect adjacent vertebrae to each other.

Most of the vertebral muscles are *extensor muscles* that extend the vertebral column and in doing so resist the downward pull of gravity. The extensors are located on the back, posterior to the spine. *Flexor muscles* are muscles that cause flexion; the vertebral flexors are positioned lateral to the vertebral column. There are few flexor muscles for the vertebral column because most of the body's mass is positioned anterior to the vertebral column, and consequently the force of gravity naturally pulls the column to flex.

Many vertebral muscles are named after their insertion to assist with grouping and identification. Muscles that insert on the skull include *capitis* in their name. Muscles that insert on the neck are called *cervicis,* those that insert on the thoracic vertebrae are *thoracis,* and those that insert on the lumbar are *lumborum.*

Superficial Layer

The superficial vertebral muscles are the **splenius capitis muscle** (Figure 19.1) and the **splenius cervicis muscle.** When the two left splenius muscles and the two right ones contract in concert, the neck is extended. When the splenius capitis and splenius cervicis muscles on only one side of the neck contract, the neck is rotated laterally and flexed.

Intermediate Layer

The **erector spinae group** of muscles forms the intermediate layer of the back musculature. This group is made up of three subgroups: *spinalis, longissimus,* and *iliocostalis.* The **spinalis cervicis** muscles extend the neck, and the **spinalis thoracis** muscles extend the vertebral column.

The **longissimus capitis** and **longissimus cervicis** muscles act on the neck (Figure 19.1). When either both longissimus capitis muscles or both longissimus cervicis muscles contract, the head is extended. When only one longissimus capitis or one longissimus cervicis contracts, the neck is flexed and rotated laterally. The **longissimus thoracis** muscles extend the vertebral column, and when only one of these muscles contracts, the column is flexed laterally.

The **iliocostalis cervicis, iliocostalis thoracis,** and **iliocostalis lumborum** muscles all extend the neck and vertebral column and stabilize the thoracic vertebrae.

Deep Layer

The back muscles that make up the deep layer are collectively called the *transversospinalis muscles;* they interconnect and support the vertebrae. The various types in this layer are the semispinalis group and the multifidus, rotatores, interspinales, and intertransversarii muscles.

The **semispinalis capitis** muscles extend the neck when both of them contract; if only one semispinalis capitis contracts, it extends and laterally flexes the neck and turns the head to the opposite side. The **semispinalis cervicis** muscles extend the vertebral column when both contract and rotate the column to the opposite side when only one of them contracts. The **semispinalis thoracis** muscles work in the same way.

The **multifidus muscles** are a deep band of muscles that span the length of the vertebral column. Each portion of the band originates either on the sacrum or on a transverse process of a vertebra and inserts on the spinous process of a vertebra that is three or four vertebrae superior to the origin.

Between transverse processes are the **rotatores cervicis, rotatores thoracis,** and **rotatores lumborum** muscles, each named after the vertebra of origin (Figure 19.1). The multifidus and rotatores muscles act with the semispinalis thoracis to extend and flex the vertebral column. Spanning adjacent spinous processes are **interspinales muscles,** which extend the vertebral column. Contiguous transverse processes have **intertransversarii muscles,** which laterally flex the column.

Spinal Flexors

The spinal flexor muscles are located along the lateral and anterior surfaces of the vertebrae. The **longus capitis** (Figure 19.1) muscles are visible as bands along the anterior margin of the vertebral column that insert on the occipital bone and flex the neck; when only one longus capitis contracts, it rotates the head to the side of contraction. The **longus colli** muscles insert on the cervical vertebrae, flex and rotate the neck, and limit extension. The **quadratus lumborum** muscles originate on the iliac crest and the iliolumbar ligament, and insert on the inferior border of the 12th pair of ribs and the transverse processes of the lumbar vertebrae. These muscles flex the vertebral column; when only one quadratus lumborum contracts, the column is flexed laterally toward the side of contraction.

QuickCheck Questions

1.1 Name the three muscles of the longissimus group and describe the action of each.

1.2 Which muscle inserts on the 12th pair of ribs?

In the Lab 1

Materials

☐ Torso model
☐ Muscle chart

Procedures

1. Review the muscles of the vertebral column in Figure 19.1 and Table 19.1.

2. Examine the back of the torso model and identify the superficial vertebral muscles. Note the insertion of each muscle.

3. Distinguish the various erector spinae muscles in the intermediate layer of vertebral muscles. Note the insertion of each muscle group.

Table 19.1 ORIGINS AND INSERTIONS Muscles of the Vertebral Column (See Figure 19.1)

Group and Muscle(s)	Origin	Insertion	Action	Innervation
SUPERFICIAL LAYER				
Splenius (splenius capitis, splenius cervicis)	Spinous processes and ligaments connecting inferior cervical and superior thoracic vertebrae	Mastoid process, occipital bone of skull, and superior cervical vertebrae	Together, the two sides extend neck; alone, each rotates and laterally flexes neck to that side	Cervical spinal nerves
INTERMEDIATE LAYER				
Erector spinae				
Spinalis group Spinalis cervicis	Inferior portion of ligamentum nuchae and spinous process of C_7	Spinous process of axis	Extends neck	As above
Spinalis thoracis	Spinous processes of inferior thoracic and superior lumbar vertebrae	Spinous processes of superior thoracic vertebrae	Extends vertebral column	Thoracic and lumbar spinal nerves
Longissimus group Longissimus capitis	Transverse processes of inferior cervical and superior thoracic vertebrae	Mastoid process of temporal bone	Together, the two sides extend head; alone, each rotates and laterally flexes neck to that side	Cervical and thoracic spinal nerves
Longissimus cervicis	Transverse processes of superior thoracic vertebrae	Transverse processes of middle and superior cervical vertebrae	As above	As above
Longissimus thoracis	Broad aponeurosis and transverse processes of inferior thoracic and superior lumbar vertebrae; joins iliocostalis	Transverse processes of superior vertebrae and inferior surfaces of ribs	Extends vertebral column; alone, each produces lateral flexion to that side	Thoracic and lumbar spinal nerves
Iliocostalis group Iliocostalis cervicis	Superior borders of vertebrosternal ribs near the angles	Transverse processes of middle and inferior cervical vertebrae	Extends or laterally flexes neck, elevates ribs	Cervical and superior thoracic spinal nerves
Iliocostalis thoracis	Superior borders of inferior seven ribs medial to the angles	Upper ribs and transverse process of last cervical vertebra	Stabilizes thoracic vertebrae in extension	Thoracic spinal nerves
Iliocostalis lumborum	Iliac crest, sacral crests, and spinous processes	Inferior surfaces of inferior seven ribs near their angles	Extends vertebral column, depresses ribs	Inferior thoracic and lumbar spinal nerves
DEEP LAYER (TRANSVERSOSPINALIS)				
Semispinalis group Semispinalis capitis	Articular processes of inferior cervical and transverse processes of superior thoracic vertebrae	Occipital bone, between nuchal lines	Together, the two sides extend head; alone, each extends and laterally flexes neck	Cervical spinal nerves
Semispinalis cervicis	Transverse processes of T_1–T_5 or T_6	Spinous processes of C_2–C_5	Extends vertebral column and rotates toward opposite side	As above
Semispinalis thoracis	Transverse processes of T_6–T_{10}	Spinous processes of C_5–T_4	As above	Thoracic spinal nerves
Multifidus	Sacrum and transverse processes of each vertebra	Spinous processes of the third or fourth more superior vertebra	As above	Cervical, thoracic, and lumbar spinal nerves
Rotatores	Transverse processes of each vertebra	Spinous processes of adjacent, more superior vertebra	As above	As above
Interspinales	Spinous processes of each vertebra	Spinous processes of more superior vertebra	Extends vertebral column	As above
Intertransversarii	Transverse processes of each vertebra	Transverse processes of more superior vertebra	Laterally flexes the vertebral column	As above
SPINAL FLEXORS				
Longus capitis	Transverse processes of cervical vertebrae	Base of the occipital bone	Together, the two sides flex the neck; alone, each rotates head to that side	Cervical spinal nerves
Longus colli	Anterior surfaces of cervical and superior thoracic vertebrae	Transverse processes of superior cervical vertebrae	Flexes or rotates neck; limits hyperextension	As above
Quadratus lumborum	Iliac crest and iliolumbar ligament	Last rib and transverse processes of lumbar vertebrae	Together, they depress ribs; alone, each side laterally flexes vertebral column	Thoracic and lumbar spinal nerves

4. Identify the transversospinalis muscles associated with the individual vertebrae on the torso model and/or muscle chart.

5. Locate the flexor muscles of the vertebral column on the torso model and/or muscle chart.

6. Extend and flex your vertebral column and consider the muscles producing each action. ■

Lab Activity 2   Oblique and Rectus Muscles

Muscles between the vertebral column and the anterior midline are grouped into either the *oblique* (slanted) or *rectus* muscle groups. As the names imply, the oblique muscles are slanted relative to the body's vertical central axis and the rectus muscles are oriented either parallel or perpendicular to this axis. Both muscle groups are found in the cervical, thoracic, and abdominal regions (**Figure 19.2** and **Table 19.2**). All these muscles support the vertebral column, provide resistance against the erector spinae muscles, move the ribs during respiration, and constitute the abdominal wall. Another major action of these muscles is to increase intra-abdominal pressure during urination, defecation, and childbirth.

Make a Prediction

In the abdomen, what muscle is superficial to the internal oblique muscle?

Oblique Muscles

The oblique muscles of the neck, collectively called the *scalene group,* are the **anterior, middle,** and **posterior scalene muscles** (see Figure 19.1). Each originates on the transverse process of a cervical vertebra and inserts on a first or second rib. When the ribs are held in position, the scalene muscles flex the neck. When the neck is stationary, they elevate the ribs during inspiration.

Oblique muscles of the thoracic region include the intercostal and transversus thoracis muscles (Figure 19.2). The intercostal muscles are located between the ribs and, along with the diaphragm, change the size of the chest for breathing. The superficial **external intercostal muscles** and the deep **internal intercostal muscles** span the gaps between the ribs. These muscles are difficult to palpate because they are deep to other chest muscles. The **transversus thoracis muscle** lines the posterior surfaces of the sternum and the cartilages of the ribs. The muscle is covered by the serous membrane of the lungs (pleura). It depresses the ribs.

The serratus posterior muscles insert on the ribs and assist the intercostal muscles in moving the rib cage. The **superior serratus posterior muscle** elevates the ribs, and the **inferior serratus posterior muscle** (Figure 19.2b) pulls the rib inferiorly and opposes the diaphragm.

The abdomen has layers of oblique and rectus muscles organized in crossing layers, much like the laminar structure of a sheet of plywood (**Figure 19.3**; also Figure 19.2). On the lateral abdominal wall is the thin, membranous **external oblique muscle.** This muscle originates on the external and inferior borders of ribs 5 through 12 and inserts on the external oblique aponeurosis that extends to the iliac crest and to a midsagittal fibrous line called the **linea alba.** The **internal oblique muscle** lies deep and at a right angle to the external oblique muscle. The internal oblique muscle arises from the thoracolumbar fascia and iliac crest and inserts on the inferior surfaces of the lower ribs and costal cartilages, the linea alba, and the pubis. Both the external and internal oblique muscles compress and flex the abdomen, depress the ribs, and rotate the vertebral column.

The **transversus abdominis muscle,** located deep to the internal oblique muscle, originates on the lower ribs, the iliac crest, and the thoracolumbar fascia and inserts on the linea alba and the pubis. It contracts with the other abdominal muscles to compress the abdomen.

Rectus Muscles

Rectus muscles are found in the cervical, thoracic, and abdominal regions of the body. Those of the cervical region are the suprahyoid and infrahyoid muscles (refer to Exercise 18).

The **diaphragm** is a sheet of muscle that forms the thoracic floor and separates the thoracic cavity from the abdominopelvic cavity (Figure 19.2). The diaphragm originates at many points along its edges, and the muscle fibers meet at a central tendon. Contracting the diaphragm to expand the thoracic cavity is the muscular process by which air is inhaled into the lungs.

The **rectus abdominis muscle** is the vertical muscle along the midline of the abdomen between the pubic symphysis and the xiphoid process of the sternum. This muscle is divided by the linea alba. A well-developed rectus abdominis muscle has a washboard appearance because transverse bands of collagen called **tendinous inscriptions** separate the muscle into many segments. Bodybuilders often call the rectus abdominis the "six pack" because of the bulging segments of the muscle. During exercise, the rectus abdominis flexes and compresses the vertebral column and depresses the ribs for forced exhalation that occurs during increased activity.

Study Tip Fiber Orientation

Find the external oblique and internal oblique muscles on a muscle model, and notice the difference in the way the muscle fibers are oriented. The fibers of the external oblique muscle flare laterally as they are traced from bottom to top, whereas those of the internal oblique muscle are directed medially. Just remember: From bottom up the externals flare out and internals go in. This tip is also useful in examining the external and internal intercostal muscles between the ribs. By the way, the intercostal muscles of beef and pork are the barbecue "ribs" that you might enjoy. ■

Figure 19.2 Oblique and Rectus Muscles and the Diaphragm Oblique muscles compress underlying structures between the vertebral column and the ventral midline; rectus muscles are flexors of the vertebral column.

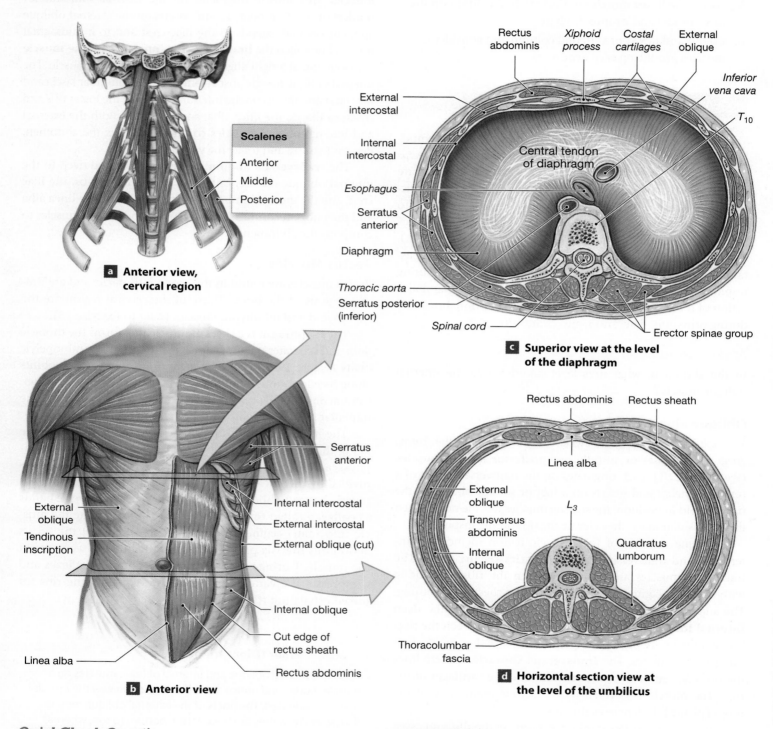

a Anterior view, cervical region

Scalenes
Anterior
Middle
Posterior

Rectus abdominis   *Xiphoid process*   *Costal cartilages*   External oblique
Inferior vena cava
External intercostal
Internal intercostal
Esophagus
Central tendon of diaphragm
T_{10}
Serratus anterior
Diaphragm
Thoracic aorta
Serratus posterior (inferior)
Spinal cord
Erector spinae group

c Superior view at the level of the diaphragm

Serratus anterior
External oblique
Tendinous inscription
Internal intercostal
External intercostal
External oblique (cut)
Internal oblique
Cut edge of rectus sheath
Linea alba
Rectus abdominis

b Anterior view

Rectus abdominis   Rectus sheath
Linea alba
External oblique
L_3
Transversus abdominis
Internal oblique
Quadratus lumborum
Thoracolumbar fascia

d Horizontal section view at the level of the umbilicus

QuickCheck Questions

2.1 What is the basic difference between muscles classified as oblique and those classified as rectus?

2.2 Describe all the muscles of the abdomen wall.

2.3 Why is the rectus abdominis muscle nicknamed the "six pack"?

In the Lab   2

Materials

☐ Head and neck model
☐ Torso model
☐ Muscle chart

Table 19.2 *ORIGINS AND INSERTIONS* Oblique and Rectus Muscles (See Figure 19.2)

Group and Muscle(s)	Origin	Insertion	Action	Innervation*
OBLIQUE GROUP				
Cervical region				
Scalenes (anterior, middle, and posterior)	Transverse and costal processes of cervical vertebrae	Superior surfaces of first two ribs	Elevate ribs or flex neck	Cervical spinal nerves
Thoracic region				
External intercostals	Inferior border of each rib	Superior border of more inferior rib	Elevate ribs	Intercostal nerves (branches of thoracic spinal nerves)
Internal intercostals	Superior border of each rib	Inferior border of the preceding rib	Depress ribs	As above
Transversus thoracis	Posterior surface of sternum	Cartilages of ribs	As above	As above
Serratus posterior superior (Figure 19.2b)	Spinous processes of C_7–T_3 and ligamentum nuchae	Superior borders of ribs 2–5 near angles	Elevates ribs, enlarges thoracic cavity	Thoracic nerves (T_1–T_4)
Serratus posterior inferior	Aponeurosis from spinous processes of T_{10}–L_3	Inferior borders of ribs 8–12	Pulls ribs inferiorly; also pulls outward, opposing diaphragm	Thoracic nerves (T_9–T_{12})
Abdominal region				
External oblique	External and inferior borders of ribs 5–12	Linea alba and iliac crest	Compresses abdomen, depresses ribs, flexes or bends spine	Intercostal, iliohypogastric, and ilioinguinal nerves
Internal oblique	Thoracolumbar fascia and iliac crest	Inferior ribs, xiphoid process, and linea alba	As above	As above
Transversus abdominis	Cartilages of ribs 6–12, iliac crest, and thoracolumbar fascia	Linea alba and pubis	Compresses abdomen	As above
RECTUS GROUP				
Cervical region	See muscles in Table 19.1			
Thoracic region				
Diaphragm	Xiphoid process, cartilages of ribs 4–10, and anterior surfaces of lumbar vertebrae	Central tendinous sheet	Contraction expands thoracic cavity, compresses abdominopelvic cavity	Phrenic nerves (C_3–C_5)
Abdominal region				
Rectus abdominis	Superior surface of pubis around symphysis	Inferior surfaces of costal cartilages (ribs 5–7) and xiphoid process	Depresses ribs, flexes vertebral column, compresses abdomen	Intercostal nerves (T_7–T_{12})

*Where appropriate, spinal nerves involved are given in parentheses.

Procedures

1. Review the oblique and rectus muscles in Figures 19.1 to Figure 19.3 and in Table 19.2.
2. Examine the head and neck model and distinguish each muscle of the scalene group.
3. Locate the intercostal muscles on the torso model and note differences in the orientation of the fibers of each muscle.
4. Identify each abdominal muscle on the torso model and/or on the muscle chart.
5. Locate the general position of each oblique muscle and the rectus muscles on the torso model.

For additional practice, complete the *Sketch to Learn* activity (on p. 250). ■

Lab Activity 3 Muscles of the Pelvic Region

The pelvic floor and wall form a bowl that supports the organs of the reproductive and digestive systems. The floor mainly consists of the **coccygeus muscle** and the **levator ani muscle,** peritoneal muscles, and muscles associated with the reproductive organs (**Figure 19.4** and **Table 19.3**). The coccygeus muscle originates on the ischial spine, passes posteriorly, and inserts on the lateral and inferior borders of the sacrum. The levator ani muscle, anterior to the coccygeus muscle, originates on the inside edge of the pubis and the ischial spine and inserts on the coccyx.

Figure 19.3 **Dissectional View of Muscles of the Trunk**

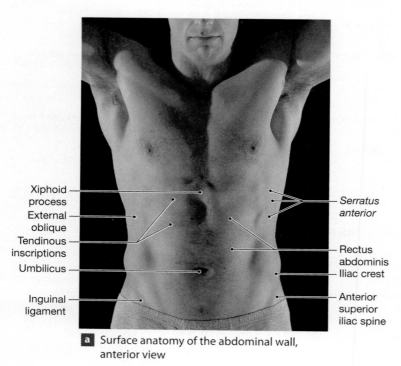

Xiphoid process
External oblique
Tendinous inscriptions
Umbilicus
Inguinal ligament

Serratus anterior
Rectus abdominis
Iliac crest
Anterior superior iliac spine

a Surface anatomy of the abdominal wall, anterior view

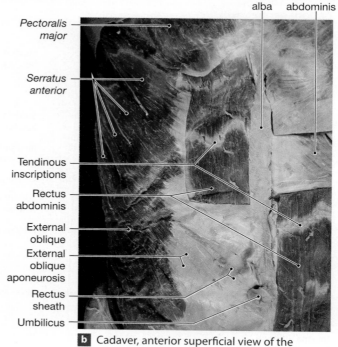

Linea alba   Transversus abdominis

Pectoralis major
Serratus anterior
Tendinous inscriptions
Rectus abdominis
External oblique
External oblique aponeurosis
Rectus sheath
Umbilicus

b Cadaver, anterior superficial view of the abdominal wall

 Sketch to Learn!

Let's do a simple sketch of the abdominal muscles and show the overlapping muscles.

Sample Sketch

Step 1
- Draw three vertical lines for the rectus abdominis muscle.
- Add horizontal lines on each side for the transverse abdominis.

Step 2
- Add the internal oblique muscles slanted toward the rectus abdominis.

Step 3
- Sketch the external obliques as lines slanted away from the midline.
- Label your drawing.

Your Sketch

Figure 19.4 Muscles of the Pelvic Floor The muscles of the pelvic floor form the urogenital triangle and the anal triangle to support organs of the pelvic cavity, flex the sacrum and coccyx, and control material movement through the urethra and anus.

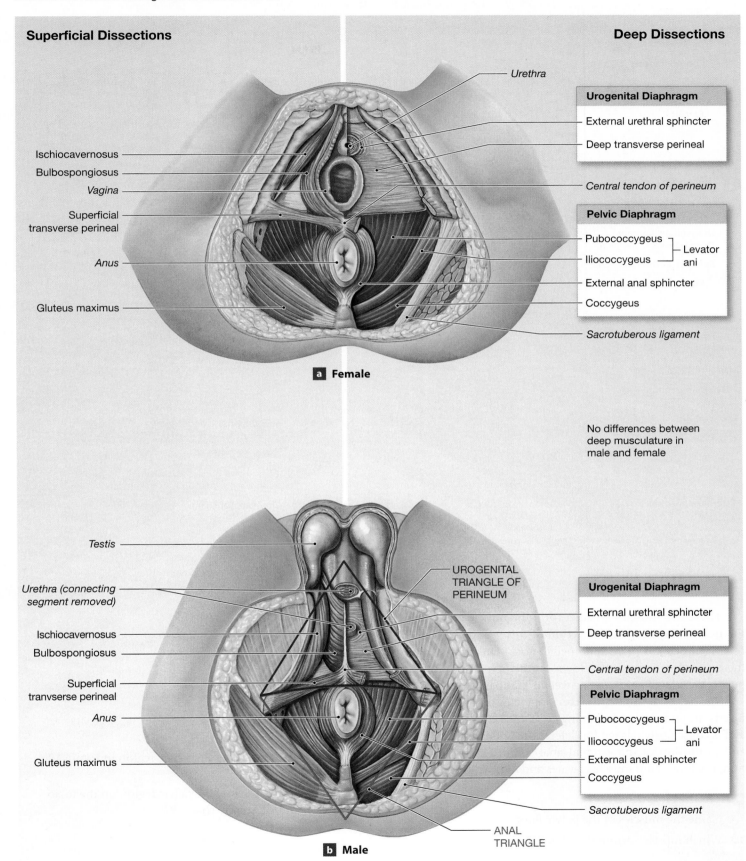

Superficial Dissections

Deep Dissections

Urethra

Urogenital Diaphragm

External urethral sphincter

Deep transverse perineal

Ischiocavernosus

Bulbospongiosus

Vagina

Superficial transverse perineal

Central tendon of perineum

Pelvic Diaphragm

Pubococcygeus ⎤
                    ⎬ Levator ani
Iliococcygeus ⎦

Anus

External anal sphincter

Coccygeus

Gluteus maximus

Sacrotuberous ligament

a Female

No differences between deep musculature in male and female

Testis

Urethra (connecting segment removed)

UROGENITAL TRIANGLE OF PERINEUM

Urogenital Diaphragm

External urethral sphincter

Deep transverse perineal

Ischiocavernosus

Bulbospongiosus

Superficial transverse perineal

Central tendon of perineum

Pelvic Diaphragm

Pubococcygeus ⎤
                    ⎬ Levator ani
Iliococcygeus ⎦

Anus

External anal sphincter

Coccygeus

Gluteus maximus

Sacrotuberous ligament

ANAL TRIANGLE

b Male

Table 19.3 *ORIGINS AND INSERTIONS* Muscles of the Pelvic Floor (See Figure 19.4)

Group and Muscle(s)	Origin	Insertion	Action	Innervation*
UROGENITAL TRIANGLE				
Bulbospongiosus:				
Superficial muscles Males	Collagen sheath at base of penis; fibers cross over urethra	Median raphe and central tendon of perineum	Compresses base and stiffens penis; ejects urine or semen	Pudendal nerve, perineal branch (S_2–S_4)
Females	Collagen sheath at base of clitoris; fibers run on either side of urethral and vaginal opening	Central tendon of perineum	Compresses and stiffens clitoris; narrows vaginal opening	As above
Ischiocavernosus	Ischial ramus and tuberosity	Pubic symphysis anterior to base of penis or clitoris	Compresses and stiffens penis or clitoris	As above
Superficial transverse perineal	Ischial ramus	Central tendon of perineum	Stabilizes central tendon of perineum	As above
Urogenital diaphragm Deep transverse perineal	Ischial ramus	Median raphe of urogenital diaphragm	As above	As above
External urethral sphincter:				
Deep muscles Males	Ischial and pubic rami	To median raphe at base of penis; inner fibers encircle urethra	Closes urethra; compresses prostate and bulbourethral glands	As above
Females	Ischial and pubic rami	To median raphe; inner fibers encircle urethra	Closes urethra; compresses vagina and greater vestibular glands	As above
ANAL TRIANGLE				
Pelvic diaphragm:				
Coccygeus	Ischial spine	Lateral, inferior borders of sacrum and coccyx	Flexes coccygeal joints; tenses and supports pelvic floor	Inferior sacral nerves (S_4–S_5)
Levator ani Iliococcygeus	Ischial spine, pubis	Coccyx and median raphe	Tenses floor of pelvis; flexes coccygeal joints; elevates and retracts anus	Pudendal nerve (S_2–S_4)
Pubococcygeus	Inner margins of pubis	As above	As above	As above
External anal sphincter	Via tendon from coccyx	Encircles anal opening	Closes anal opening	Pudendal nerve, hemorrhoidal branch (S_2–S_4)

*Where appropriate, spinal nerves involved are given in parentheses.

These two muscles together form the muscle group called the **pelvic diaphragm.** The action of this group is to flex the coccyx muscle and tense the pelvic floor. During pregnancy, the expanding uterus bears down on the pelvic floor, and the pelvic diaphragm supports the weight of the fetus.

The **external anal sphincter** originates on the coccyx and inserts around the anal opening. This muscle closes the anus and is consciously relaxed for defecation. Following depression and protrusion of the external anal sphincter during defecation, the levator ani muscle elevates and retracts the anus.

QuickCheck Questions

3.1 Name the muscles of the pelvic floor.

3.2 Which muscle surrounds the anus?

In the Lab   3

Materials

☐ Torso model
☐ Muscle chart

Procedures

1. Review the muscles of the pelvic region in Figure 19.4 and Table 19.3.

2. Locate each muscle of the pelvic region on the torso model and/or muscle chart. ■

Name _____

Date _____

Section _____

Muscles of the Vertebral Column, Abdomen, and Pelvis

A. Matching

Match each term in the left column with its correct description from the right column.

_____ 1. rectus abdominis

_____ 2. quadratus lumborum

_____ 3. levator ani

_____ 4. transverse abdominis

_____ 5. external oblique

_____ 6. external intercostal

_____ 7. linea alba

_____ 8. longissimus cervicis

_____ 9. diaphragm

_____ 10. internal intercostal

_____ 11. external anal sphincter

_____ 12. internal oblique

A. neck muscles that extend head

B. superficial lateral muscle of abdomen

C. middle lateral muscle layer of abdomen

D. major muscle of inhalation

E. abdominal muscle with horizontal fibers

F. muscle at trunk midline that compresses abdomen

G. circular muscle in pelvic floor

H. found between ribs; elevates rib cage

I. fibrous line located along midline of trunk

J. elevates anal sphincter

K. posterior vertebral muscle that flexes spine

L. found between ribs; depresses rib cage

B. Descriptions

Describe the location of each of the following muscles.

1. splenius cervicis

2. longissimus thoracis

3. multifidus

4. anterior scalene

5. internal intercostal

6. transverse abdominis

7. coccygeus

8. diaphragm

C. Labeling

Label the muscles that act on the vertebral column in **Figure 19.5**.

Figure 19.5 Muscles of the Vertebral Column Posterior view.

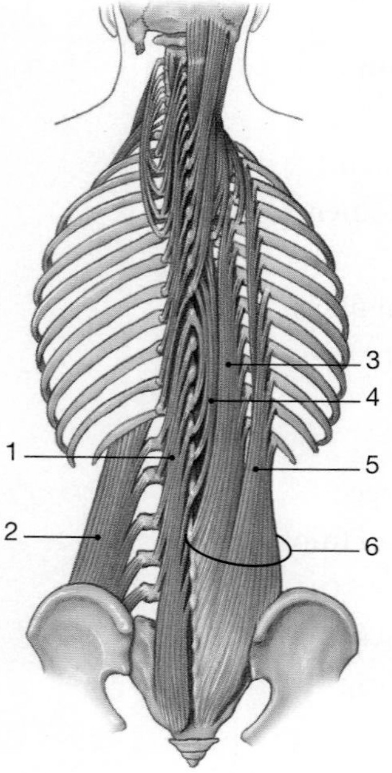

1. _____
2. _____
3. _____
4. _____
5. _____
6. _____

D. Analysis and Application

1. The anterior abdominal wall lacks bone. This being true, on what structure do the abdominal muscles insert?

2. Describe the longissimus muscle group of the vertebral column.

E. Clinical Challenge

1. A patient is admitted to the surgery ward for an appendectomy. Describe the layers of muscles the surgeon must cut in order to reach the appendix.

2. Frank uses improper body position to lift a heavy box and strains the muscles in his lumbar region. Which muscles are most likely to be involved in this injury?

Muscles of the Pectoral Girdle and Upper Limb

Learning Outcomes

On completion of this exercise, you should be able to:

1. Locate the muscles of the pectoral and upper limb on lab models and charts.

2. Identify on the models the origin, insertion, and action of the muscles of the shoulder and upper limb.

3. Demonstrate or describe the action of the major muscles of the scapula and upper limb.

The appendicular musculature supports and moves the pectoral girdle and upper limb and the pelvic girdle and lower limb. Many of the muscles of the pectoral and pelvic girdles are on the body trunk but move appendicular bones (**Figure 20.1**). For example, the largest muscle that moves the arm, the latissimus dorsi, is located on the lumbus.

Muscles of the pectoral girdle and upper limb are covered in this exercise.

Lab Activity 1   Muscles That Move the Pectoral Girdle

The muscles of the pectoral girdle support and position the scapula and clavicle and help maintain the articulation between the humerus and scapula (**Figure 20.2**). The shoulder joint is the most movable and least stable joint of the body, and many of the surrounding muscles help keep the humerus articulated in the scapula. Origin, insertion, action, and innervation for these muscles are detailed in **Table 20.1**.

Need More Practice and Review?

Build your knowledge—and confidence!—in the Study Area of MasteringA&P® at www.masteringaandp.com with Pre-lab Quizzes, Post-lab Quizzes, Practice Anatomy Lab™ (PAL™) 3.0 virtual anatomy practice tool, PhysioEx™ 9.0 laboratory simulations, and A&P Flix™ with Quizzes.

PAL   For this lab exercise, follow these navigation paths:
- PAL>Human Cadaver>Muscular System>Upper Limb
- PAL>Anatomical Models>Muscular System>Upper Limb

A&PFlix   For this lab exercise, go to these topics:
- Origins, Insertions, Actions, and Innervations
- Group Muscle Actions and Joints

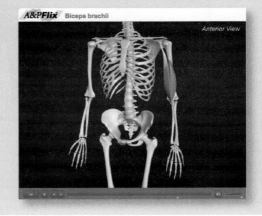

Figure 20.1 Superficial and Deep Muscles of the Neck, Shoulder, and Back A posterior view of many of the major muscles of the neck, trunk, and proximal portions of the upper limbs.

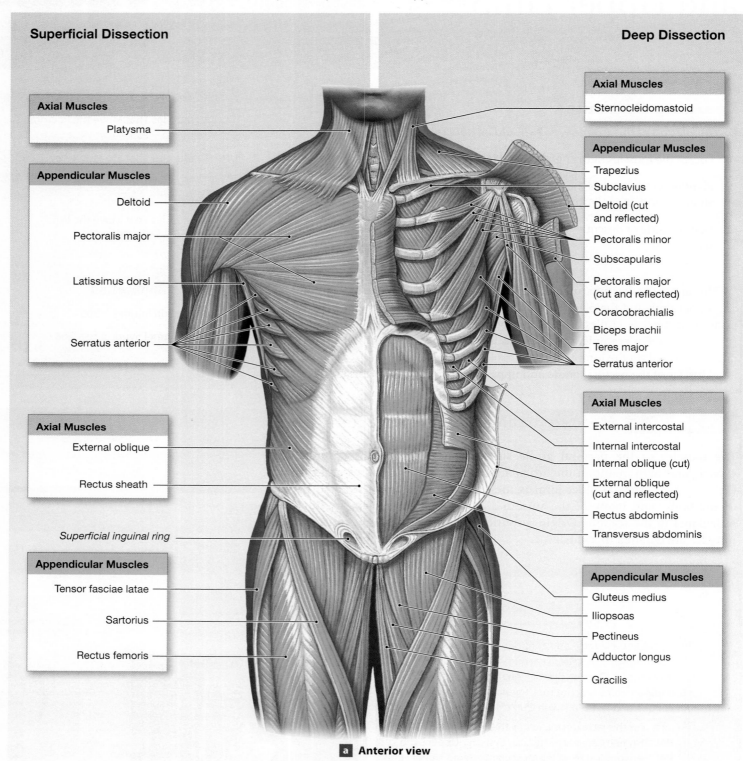

Superficial Dissection

Deep Dissection

Axial Muscles

Platysma

Appendicular Muscles

Deltoid

Pectoralis major

Latissimus dorsi

Serratus anterior

Axial Muscles

External oblique

Rectus sheath

Superficial inguinal ring

Appendicular Muscles

Tensor fasciae latae

Sartorius

Rectus femoris

Axial Muscles

Sternocleidomastoid

Appendicular Muscles

Trapezius

Subclavius

Deltoid (cut and reflected)

Pectoralis minor

Subscapularis

Pectoralis major (cut and reflected)

Coracobrachialis

Biceps brachii

Teres major

Serratus anterior

Axial Muscles

External intercostal

Internal intercostal

Internal oblique (cut)

External oblique (cut and reflected)

Rectus abdominis

Transversus abdominis

Appendicular Muscles

Gluteus medius

Iliopsoas

Pectineus

Adductor longus

Gracilis

a Anterior view

Figure 20.1 Superficial and Deep Muscles of the Neck, Shoulder, and Back *(continued)*

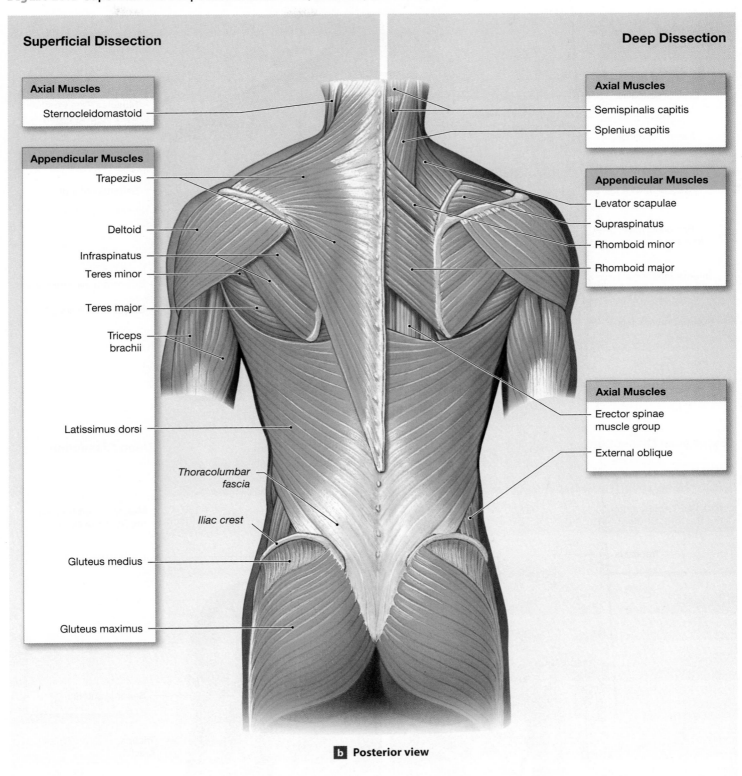

Superficial Dissection

Deep Dissection

Axial Muscles

Sternocleidomastoid

Appendicular Muscles

Trapezius

Deltoid

Infraspinatus

Teres minor

Teres major

Triceps brachii

Latissimus dorsi

Thoracolumbar fascia

Iliac crest

Gluteus medius

Gluteus maximus

Axial Muscles

Semispinalis capitis

Splenius capitis

Appendicular Muscles

Levator scapulae

Supraspinatus

Rhomboid minor

Rhomboid major

Axial Muscles

Erector spinae muscle group

External oblique

b **Posterior view**

Figure 20.2 Superficial and Deep Muscles of the Trunk and Proximal Portion of the Limbs Anterior view of the axial muscles of the trunk and the appendicular muscles associated with the pectoral and pelvic girdles and the proximal portion of the upper and lower limbs.

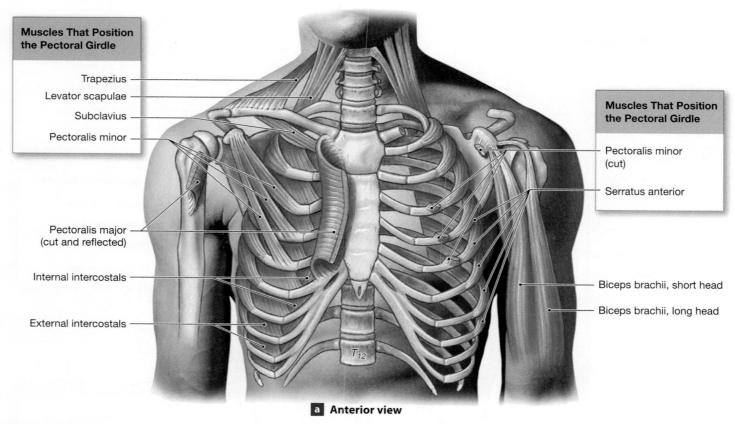

Muscles That Position the Pectoral Girdle

Trapezius

Levator scapulae

Subclavius

Pectoralis minor

Pectoralis major (cut and reflected)

Internal intercostals

External intercostals

T_{12}

Muscles That Position the Pectoral Girdle

Pectoralis minor (cut)

Serratus anterior

Biceps brachii, short head

Biceps brachii, long head

a Anterior view

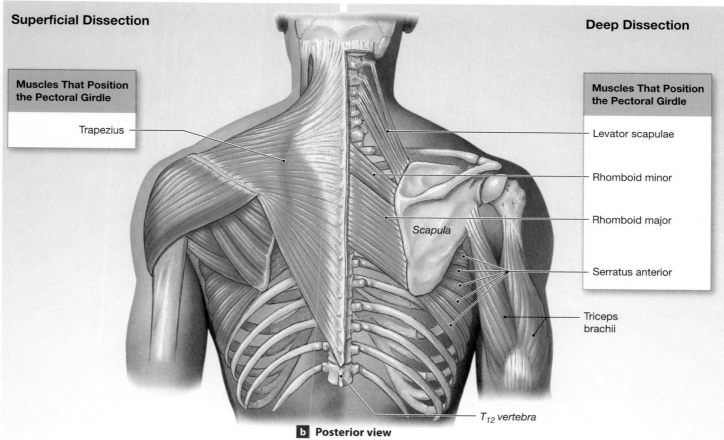

Superficial Dissection

Deep Dissection

Muscles That Position the Pectoral Girdle

Trapezius

Muscles That Position the Pectoral Girdle

Levator scapulae

Rhomboid minor

Rhomboid major

Scapula

Serratus anterior

Triceps brachii

T_{12} *vertebra*

b Posterior view

Table 20.1	*ORIGINS AND INSERTIONS* Muscles That Position the Pectoral Girdle (See Figures 20.1 and 20.2)			
Muscle	**Origin**	**Insertion**	**Action**	**Innervation***
Levator scapulae	Transverse processes of first four cervical vertebrae	Vertebral border of scapula near superior angle	Elevates scapula	Cervical nerves C_3–C_4 and dorsal scapular nerve (C_5)
Pectoralis minor	Anterior and superior surfaces of ribs 3–5	Coracoid process of scapula	Depresses and protracts shoulder; rotates scapula so glenoid cavity moves inferiorly (downward rotation); elevates ribs if scapula is stationary	Medial pectoral nerve (C_8, T_1)
Rhomboid major	Spinous processes of superior thoracic vertebrae	Vertebral border of scapula from spine to inferior angle	Adducts scapula and performs downward rotation	Dorsal scapular nerve (C_5)
Rhomboid minor	Spinous processes of vertebrae C_7–T_1	Vertebral border of scapula near spine	As above	As above
Serratus anterior	Anterior and superior margins of ribs 1–8 or 1–9	Anterior surface of vertebral border of scapula	Protracts shoulder, rotates scapula so glenoid cavity moves superiorly (upward rotation)	Long thoracic nerve (C_5–C_7)
Subclavius	First rib	Clavicle (inferior border)	Depresses and protracts shoulder	Nerve to subclavius (C_5–C_6)
Trapezius	Occipital bone, ligamentum nuchae, and spinous processes of thoracic vertebrae	Clavicle and scapula (acromion and scapular spine)	Depends on active region and state of other muscles; may (1) elevate, retract, depress, or rotate scapula upward, (2) elevate clavicle, or (3) extend neck	Accessory nerve (N XI) and cervical spinal nerves (C_3–C_4)

*Where appropriate, spinal nerves involved are given in parentheses.

On the anterior of the trunk, the **subclavius** (sub-KLĀ-vē-us) muscle is inferior to the clavicle (Figure 20.2a). It arises from the first rib, inserts on the underside of the clavicle, and depresses and protracts the clavicle. The **serratus** (ser-Ā-tus; *serratus*, a saw) **anterior** muscle appears as fan-shaped wedges on the side of the chest. This arrangement gives the muscle a sawtooth appearance similar to that of a bread knife with its *serrated* cutting edge. The muscle protracts the shoulder and rotates the scapula upward.

The **pectoralis** (pek-tō-RĀ-lis; *pectus*; chest) **minor** muscle is a deep muscle of the anterior trunk. Its origin is the anterior surfaces and superior margins of ribs 3 through 5, and it inserts on the coracoid process of the scapula. The function of this muscle is to pull the top of the scapula forward and depress the shoulders. It also elevates the ribs during forced inspiration, as during strenuous exercise.

The large, diamond-shaped muscle of the upper back is the **trapezius** (tra-PĒ-zē-us) muscle. It spans the gap between the scapulae and extends from the lower thoracic vertebrae to the back of the head (Figure 20.2b). The superior portion of the trapezius originates at three places: on the occipital bone; on the **ligamentum nuchae** (li-guh-MEN-tum NŪ-kē; *nucha*, nape), which is a ligament extending from the cervical vertebrae to the occipital bone; and on the spinous processes of thoracic vertebrae. It inserts on the clavicle and on the acromion and scapular spine of the scapula. Because the trapezius has origins superior and inferior to its insertion, it may elevate, depress, retract, or rotate the scapula and/or the

clavicle upward. The trapezius also can extend the neck. Deep to the trapezius are the **rhomboid** (rom-boyd) **major** and **rhomboid minor** muscles, which extend between the upper thoracic vertebrae and the scapula. The rhomboid muscles adduct the scapula and rotate it downward. The **levator scapulae** (lē-VĀ-tor SKAP-ū-lē; *levator*, lifter) muscle originates on cervical vertebrae 1 through 4 and inserts on the superior border of the scapula. As its name implies, it elevates the scapula.

QuickCheck Questions

1.1 Describe the actions of the trapezius muscle.

1.2 Describe the action of the rhomboid muscles.

In the Lab   1

Materials

☐ Torso model

☐ Articulated skeleton

☐ Muscle chart

Procedures

1. Review the anterior and posterior muscles of the chest in Table 20.1 and Figures 20.1 and 20.2.

2. Identify each muscle on the torso model and the muscle chart.

3. Examine an articulated skeleton and note the origin, insertion, and action of the major muscles that act on the shoulder.

4. Locate the position of these muscles on your body and practice each muscle's action. ■

Lab Activity 2   Muscles That Move the Arm

Muscles that move the arm originate either on the scapula or on the vertebral column, span the ball-and-socket joint of the shoulder, and insert on the humerus to abduct, adduct, flex, or extend the arm (**Figure 20.3**). Refer to **Table 20.2** for details on origin, insertion, action, and innervation for these muscles.

The **coracobrachialis** (kor-uh-kō-brā-kē-AL-is) muscle is a small muscle that originates on the coracoid process of the scapula and adducts and flexes the shoulder (Figure 20.3a). The largest muscle of the chest is the **pectoralis major** muscle, which covers most of the upper rib cage on the two sides of the chest, and is one of the main muscles that move the arm. It originates on the clavicle, on the body of the sternum, and on costal cartilages for ribs 2 through 6, and inserts on the humerus at the greater tubercule and lateral surface of the intertubercular groove. This muscle flexes, adducts, and medially rotates the arm. In females, the breasts cover the inferior part of the pectoralis major muscle. Lateral to the pectoralis major muscle is the **deltoid** (DEL-toyd; *delta*, triangular) muscle, the triangular muscle of the shoulder. It originates on the anterior edge of the clavicle, on the inferior margins of the scapular spine, and on the acromion process of the scapula. The deltoid inserts on the deltoid tuberosity and is the major abductor of the humerus.

The **subscapularis** (sub-skap-ū-LAR-is) muscle is deep to the scapula next to the posterior surface of the rib cage (Figure 20.2). It originates on the subscapular fossa, inserts on the lesser tubercle of the humerus, and medially rotates the shoulder.

The **latissimus dorsi** (la-TIS-i-mus DOR-sē; *lati*, broad) muscle is the large muscle wrapping around the lower back (Figure 20.3b). This muscle has a broad origin from the sacral and lumbar vertebrae up to the sixth thoracic vertebra and sweeps up and inserts on the humerus. The latissimus dorsi muscle extends, adducts, and medially rotates the arm.

Two muscles occur superior and inferior to the spine of the posterior scapular surface. The **supraspinatus** (sū-pra-spī-NĀ-tus; *supra*, above + *spin*, spine) muscle originates on the supraspinous fossa, the depression located superior to the scapular spine (Figure 20.3b). It abducts the shoulder. The **infraspinatus** (inf-ra-spī-NĀ-tus; *infra*, below) muscle arises from the infraspinous fossa of the scapula and inserts on the greater tubercle of the humerus to laterally rotate the humerus at the shoulder.

Clinical Application Rotator Cuff Injuries

Four shoulder muscles—the supraspinatus, infraspinatus, teres minor, and subscapularis muscles—all act to position the head of the humerus firmly in the glenoid fossa to prevent dislocation of the shoulder. Collectively these muscles are called the **rotator cuff.** Remember the acronym **SITS** (supraspinatus, infraspinatus, teres minor, subscapularis) for the rotator cuff muscles. Although part of the rotator cuff, the supraspinatus is not itself a rotator; rather, it is an abductor. You may be familiar with rotator cuff injuries if you are a baseball fan. The windup and throw of a pitcher involve circumduction of the humerus. This motion places tremendous stress on the shoulder joint and on the rotator cuff—stress that can cause premature degeneration of the joint. To protect the shoulder joint and muscles, bursal sacs are interspersed between the tendons of the rotator cuff muscles and the neighboring bony structures. Repeated friction on the bursae may result in an inflammation called *bursitis*. ■

The **teres** (TER-ēs; *teres*, round) **major** muscle is a thick muscle that arises on the inferior angle of the posterior surface of the scapula. The muscle converges up and laterally into a flat tendon that ends on the anterior side of the humerus. On the lateral border of the scapula is the small and flat **teres minor** muscle. The teres major muscle extends, adducts, and medially rotates the humerus; the teres minor muscle laterally rotates the humerus at the shoulder.

QuickCheck Questions

2.1 Which muscles adduct the arm?

2.2 Which muscle flexes the arm?

In the Lab 2

Materials

- ☐ Torso model
- ☐ Upper limb model
- ☐ Articulated skeleton
- ☐ Muscle chart

Procedures

1. Review the muscles that move the arm in Table 20.2 and Figures 20.1 through 20.3.

2. Locate each muscle that moves the arm on the torso model, upper limb model, and muscle chart.

3. Examine the articulated skeleton and note the origin, insertion, and action of the major muscles that act on the arm.

4. Locate the general position of each arm muscle on your body. Contract each muscle and observe how your arm moves. ■

Figure 20.3 Muscles That Move the Arm Muscles that move the arm are located on the trunk and insert on the proximal portions of the humerus.

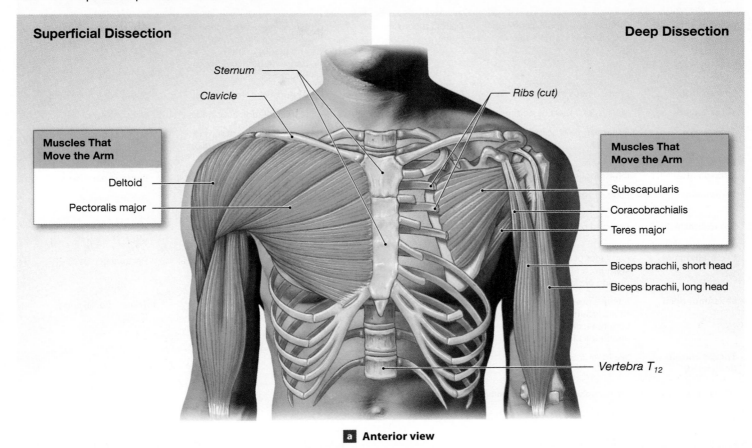

a Anterior view

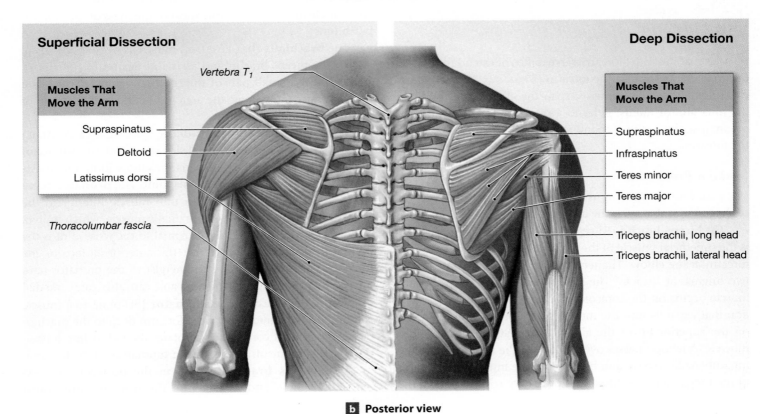

b Posterior view

Table 20.2	ORIGINS AND INSERTIONS	Muscles That Move the Arm (See Figures 20.1, 20.2, and 20.3)		
Muscle	**Origin**	**Insertion**	**Action**	**Innervation***
Deltoid	Clavicle and scapula (acromion and adjacent scapular spine)	Deltoid tuberosity of humerus	*Whole muscle:* abduction at shoulder; *anterior part:* flexion and medial rotation, *posterior part:* extension and lateral rotation	Axillary nerve (C₅–C₆)
Supraspinatus	Supraspinous fossa of scapula	Greater tubercle of humerus	Abduction at the shoulder	Suprascapular nerve (C₅)
Subscapularis	Subscapular fossa of scapula	Lesser tubercle of humerus	Medial rotation at shoulder	Subscapular nerves (C₅–C₆)
Teres major	Inferior angle of scapula	Passes medially to reach the medial lip of intertubercular groove of humerus	Extension, adduction, and medial rotation at shoulder	Lower subscapular nerve (C₅–C₆)
Infraspinatus	Infraspinous fossa of scapula	Greater tubercle of humerus	Lateral rotation at shoulder	Suprascapular nerve (C₅–C₆)
Teres minor	Lateral border of scapula	Passes laterally to reach the greater tubercle of humerus	Lateral rotation at shoulder	Axillary nerve (C₅)
Coracobrachialis	Coracoid process	Medial margin of shaft of humerus	Adduction and flexion at shoulder	Musculocutaneous nerve (C₅–C₇)
Pectoralis major	Cartilages of ribs 2–6, body of sternum, and inferior, medial portion of clavicle	Crest of greater tubercle and lateral lip of intertubercular groove of humerus	Flexion, adduction, and medial rotation at shoulder	Pectoral nerves (C₅–T₁)
Latissimus dorsi	Spinous processes of inferior thoracic and all lumbar vertebrae, ribs 8–12, and thoracolumbar fascia	Floor of intertubercular groove of the humerus	Extension, adduction, and medial rotation at shoulder	Thoracodorsal nerve (C₆–C₈)
Triceps brachii (long head)	*See Table 20.3*			

*Where appropriate, spinal nerves involved are given in parentheses.

Lab Activity 3 Muscles That Move the Forearm

Muscles that move the forearm serve to flex or extend the elbow, or pronate and supinate the forearm (**Figure 20.4**). These muscles originate on the humerus, span the elbow, and insert on the ulna and/or radius. Refer to **Table 20.3** for details on the origin, insertion, action, and innervation for muscles that move the forearm.

Make a Prediction

Use your knowledge of muscle actions and predict the action of the muscles that span the anterior of the elbow.

The **biceps brachii** (BĪ-ceps BRĀ-kē-ī) muscle (Figure 20.4) is the superficial muscle of the anterior brachium that flexes the forearm at the elbow. The term *biceps* refers to the presence of two origins, or "heads." The **short head** of the biceps brachii muscle begins on the coracoid process of the scapula as a tendon that expands into the muscle belly. The **long head** arises on the superior lip of the glenoid fossa at the supraglenoid tubercle. A tendon passes over the top of the humerus into the intertubercular groove and blends into the muscle. The tendon of the long head is enclosed in a protective covering called the

intertubercular synovial sheath. The two heads of the biceps brachii muscle fuse and constitute most of the mass of the anterior brachium.

The **brachialis** (brā-kē-AL-is) muscle is deep to the distal end of the biceps brachii and assists in flexion of the elbow. You can feel a small part of the brachialis muscle when you flex your arm and palpate the area just lateral to the tendon of the biceps brachii muscle.

The superficial **brachioradialis** (brā-kē-ō-rā-dē-AL-is) muscle is easily felt on the lateral side of the anterior surface of the forearm (Figure 20.4). It spans the elbow joint and assists the biceps brachii in flexion of this joint. The **pronator teres** (PRŌ-nā-tōr TE-rēs) muscle is a thin muscle inferior to the elbow and medial to the brachioradialis muscle, which it dives under to insert on the radius. Proximal to the wrist joint is the **pronator quadratus** muscle on the anterior surface of the forearm. This muscle acts as a synergist to the pronator teres muscle in pronating the forearm and can also cause medial rotation of the forearm. The **supinator** (SŪ-pi-nā-tor) muscle is found on the lateral side of the forearm deep to the brachioradialis muscle. It contracts and rotates the radius into a position parallel to the ulna, resulting in supination of the forearm.

The **triceps brachii** muscle on the posterior arm extends the elbow and is therefore the principal antagonist

Figure 20.4 Muscles That Move the Forearm and Hand—Anterior View Muscles of the right upper limb.

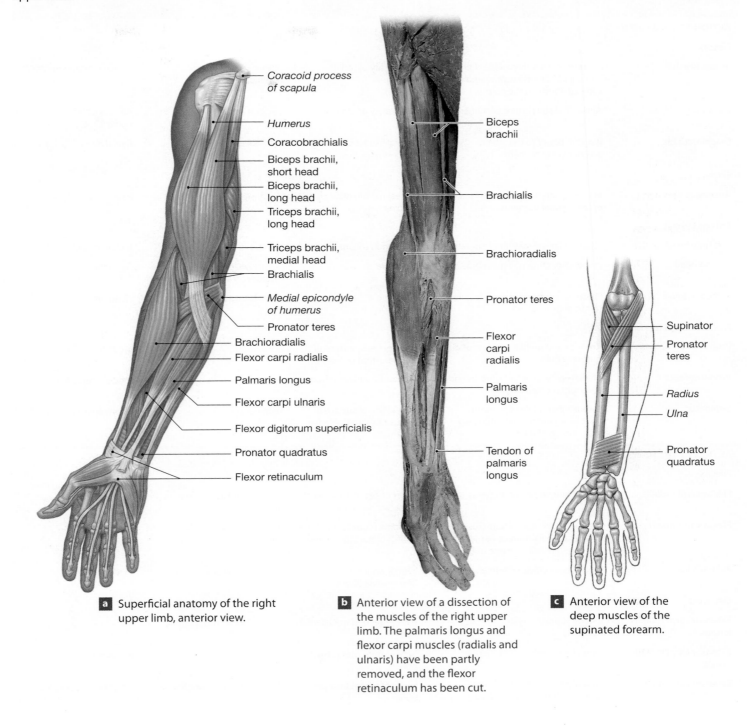

Coracoid process of scapula

Humerus

Coracobrachialis

Biceps brachii, short head

Biceps brachii, long head

Triceps brachii, long head

Triceps brachii, medial head

Brachialis

Medial epicondyle of humerus

Pronator teres

Brachioradialis

Flexor carpi radialis

Palmaris longus

Flexor carpi ulnaris

Flexor digitorum superficialis

Pronator quadratus

Flexor retinaculum

Biceps brachii

Brachialis

Brachioradialis

Pronator teres

Flexor carpi radialis

Palmaris longus

Tendon of palmaris longus

Supinator

Pronator teres

Radius

Ulna

Pronator quadratus

a Superficial anatomy of the right upper limb, anterior view.

b Anterior view of a dissection of the muscles of the right upper limb. The palmaris longus and flexor carpi muscles (radialis and ulnaris) have been partly removed, and the flexor retinaculum has been cut.

c Anterior view of the deep muscles of the supinated forearm.

to the biceps brachii and brachialis muscles (**Figure 20.5**). The muscle arises from three heads, called the **long, lateral,** and **medial heads,** which merge into a common tendon that begins near the middle of the muscle and inserts on the olecranon process of the ulna. At the posterior lateral humerus is the small **anconeus** (ang-KŌ-nē-ūs; *ankon,*

elbow) muscle, which assists the triceps brachii muscle in extending the elbow.

QuickCheck Questions

3.1 Which muscles are antagonistic to the triceps brachii?

3.2 Which muscles are involved when you turn a doorknob?

Table 20.3   *ORIGINS AND INSERTIONS* **Muscles That Move the Forearm and Hand (See Figures 20.4 to 20.6)**

Muscle	Origin	Insertion	Action	Innervation
ACTION AT THE ELBOW				
Flexors				
Biceps brachii	*Short head* from the coracoid process; *long head* from the supraglenoid tubercle (both on the scapula)	Tuberosity of radius	Flexion at elbow and shoulder; supination	Musculocutaneous nerve $(C_5–C_6)$
Brachialis	Anterior, distal surface of humerus	Tuberosity of ulna	Flexion at elbow	As above and radial nerve $(C_7–C_8)$
Brachioradialis	Ridge superior to the lateral epicondyle of humerus	Lateral aspect of styloid process of radius	As above	Radial nerve $(C_5–C_6)$
Extensors				
Anconeus	Posterior, inferior surface of lateral epicondyle of humerus	Lateral margin of olecranon on ulna	Extension at elbow	Radial nerve $(C_7–C_8)$
Triceps brachii				
lateral head	Superior lateral margin of humerus	Olecranon of ulna	As above	Radial nerve $(C_6–C_8)$
long head	Infraglenoid tubercle of scapula	As above	As above, plus extension and adduction at the shoulder	As above
medial head	Posterior surface of humerus inferior to radial groove	As above	Extension at elbow	As above
Pronators/Supinators				
Pronator quadratus	Anterior and medial surfaces of distal portion of ulna	Anterolateral surface of distal portion of radius	Pronation	Median nerve $(C_8–T_1)$
Pronator teres	Medial epicondyle of humerus and coronoid process of ulna	Midlateral surface of radius	As above	Median nerve $(C_6–C_7)$
Supinator	Lateral epicondyle of humerus, annular ligament, and ridge near radial notch of ulna	Anterolateral surface of radius distal to the radial tuberosity	Supination	Deep radial nerve $(C_6–C_8)$
ACTION AT THE HAND				
Flexors				
Flexor carpi radialis	Medial epicondyle of humerus	Bases of second and third metacarpal bones	Flexion and abduction at wrist	Median nerve $(C_6–C_7)$
Flexor carpi ulnaris	Medial epicondyle of humerus; adjacent medial surface of olecranon and anteromedial portion of ulna	Pisiform, hamate, and base of fifth metacarpal bone	Flexion and adduction at wrist	Ulnar nerve $(C_8–T_1)$
Palmaris longus	Medial epicondyle of humerus	Palmar aponeurosis and flexor retinaculum	Flexion at wrist	Median nerve $(C_6–C_7)$
Extensors				
Extensor carpi radialis longus	Lateral supracondylar ridge of humerus	Base of second metacarpal bone	Extension and abduction at wrist	Radial nerve $(C_6–C_7)$
Extensor carpi radialis brevis	Lateral epicondyle of humerus	Base of third metacarpal bone	As above	As above
Extensor carpi ulnaris	Lateral epicondyle of humerus; adjacent dorsal surface of ulna	Base of fifth metacarpal bone	Extension and adduction at wrist	Deep radial nerve $(C_6–C_8)$

In the Lab   3

Materials

- ☐ Torso model
- ☐ Upper limb model
- ☐ Articulated skeleton
- ☐ Muscle chart

Procedures

1. Review the muscles of the forearm in Figures 20.4 and 20.5 and in Table 20.3.
2. Identify each muscle on the torso and upper limb models and on the muscle chart.

Figure 20.5 Muscles That Move the Forearm, Wrist, and Hand—Posterior View Muscles of the right limb, posterior view.

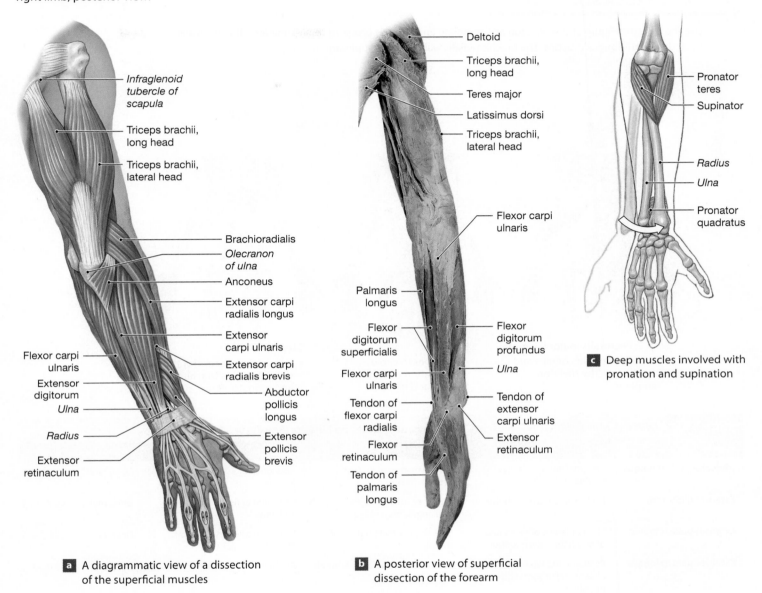

Infraglenoid tubercle of scapula
Triceps brachii, long head
Triceps brachii, lateral head
Brachioradialis
Olecranon of ulna
Anconeus
Extensor carpi radialis longus
Extensor carpi ulnaris
Extensor carpi radialis brevis
Abductor pollicis longus
Extensor pollicis brevis
Flexor carpi ulnaris
Extensor digitorum
Ulna
Radius
Extensor retinaculum

a A diagrammatic view of a dissection of the superficial muscles

Deltoid
Triceps brachii, long head
Teres major
Latissimus dorsi
Triceps brachii, lateral head
Flexor carpi ulnaris
Palmaris longus
Flexor digitorum superficialis
Flexor digitorum profundus
Flexor carpi ulnaris
Ulna
Tendon of flexor carpi radialis
Tendon of extensor carpi ulnaris
Flexor retinaculum
Extensor retinaculum
Tendon of palmaris longus

b A posterior view of superficial dissection of the forearm

Pronator teres
Supinator
Radius
Ulna
Pronator quadratus

c Deep muscles involved with pronation and supination

3. Examine the articulated skeleton and note the origin, insertion, and action of the muscles that act on the forearm.

4. On your body, locate the general position of each muscle involved with movement of the forearm. Flex and extend your elbow joint and watch the action of the muscles on your arm. ■

Lab Activity 4 Muscles That Move the Wrist and Hand

The muscles of the wrist and hand can be organized into two groups based on location: extrinsic muscles in the forearm and intrinsic muscles in the hand. The extrinsic muscles flex and extend the wrist and fingers, and the intrinsic muscles control fine finger and thumb movements. Refer to Table 20.3 as well as **Table 20.4** and **Table 20.5** for descriptions of the origins, insertions, actions, and innervation for muscles of the wrist and hand.

The flexor muscles that move the wrist and hand are on the anterior forearm and the extensor muscles are on the posterior forearm. The brachioradialis muscle is between the flexor and extensor muscles of the forearm and is a good anatomical landmark. At the wrist, the long tendons of the flexor muscles are supported and stabilized by a wide sheath called the **flexor retinaculum** (ret-i-NAK-ū-lum; *retinaculum*, a halter or band). Many of the extensor muscles on the posterior forearm originate from a common tendon on the lateral epicondyle of the humerus.

Sketch to Learn

Use the provided figure of the skeleton as a template and add the pronator teres, supinator, brachioradialis, and palmaris longus muscles. The brachioradialis will cover the supinator muscle.

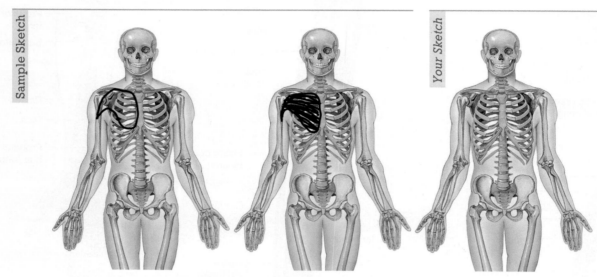

Sample Sketch

Your Sketch

Step 1: Pectoralis major example
- Trace the muscle's origins and insertion on the skeleton template.

Step 2
- Add lines to show orientation of muscle fibers and fascicles.

Step 3
- Repeat the process for the deltoid, biceps brachii, brachialis, and pronator teres.

Table 20.4	*ORIGINS AND INSERTIONS* Muscles That Move the Hand and Fingers (See Figure 20.6)			
Muscle	**Origin**	**Insertion**	**Action**	**Innervation***
Abductor pollicis longus	Proximal dorsal surfaces or ulna and radius	Lateral margin of first metacarpal bone	Abduction at joints of thumb and wrist	Deep radial nerve (C_6–C_7)
Extensor digitorum	Lateral epicondyle of humerus	Posterior surfaces of the phalanges, fingers 2–5	Extension at finger joints and wrist	Deep radial nerve (C_6–C_8)
Extensor pollicis brevis	Shaft of radius distal to origin of adductor pollicis longus	Base of proximal phalanx of thumb	Extension at joints of thumb; abduction at wrist	Deep radial nerve (C_6–C_7)
Extensor pollicis longus	Posterior and lateral surfaces of ulna and interosseous membrane	Base of distal phalanx of thumb	As above	Deep radial nerve (C_6–C_8)
Extensor indicis	Posterior surface of ulna and interosseous membrane	Posterior surface of phalanges of index finger (2), with tendon of extensor digitorum	Extension and adduction at joints of index finger	As above
Extensor digiti minimi	Via extensor tendon to lateral epicondyle of humerus and from intermuscular septa	Posterior surface of proximal phalanx of little finger (5)	Extension at joints of little finger	As above
Flexor digitorum superficialis	Medial epicondyle of humerus; adjacent anterior surfaces of ulna and radius	Midlateral surfaces of middle phalanges of fingers 2–5	Flexion at proximal interphalangeal, metacarpophalangeal, and wrist joints	Median nerve (C_7–T_1)
Flexor digitorum profundus	Medial and posterior surfaces of ulna, medial surface of coronoid process, and interosseus membrane	Bases of distal phalanges of fingers 2–5	Flexion at distal interphalangeal joints and, to a lesser degree, proximal interphalangeal joints and wrist	Palmar interosseous nerve, from median nerve, and ulnar nerve (C_8–T_1)
Flexor pollicis longus	Anterior shaft of radius, interosseous membrane	Base of distal phalanx of thumb	Flexion at joints of thumb	Median nerve (C_8–T_1)

*Where appropriate, spinal nerves involved are given in parentheses.

Table 20.5   *ORIGINS AND INSERTIONS* Intrinsic Muscles of the Hand (See Figures 20.6 and 20.7)

Muscle	Origin	Insertion	Action	Innervation*
Adductor pollicis	Metacarpal and carpal bones	Proximal phalanx of thumb	Adduction of thumb	Ulnar nerve; deep branch (C_8–T_1)
Opponens pollicis	Trapezium and flexor retinaculum	First metacarpal bone	Opposition of thumb	Median nerve (C_6–C_7)
Palmaris brevis	Palmar aponeurosis	Skin of medial border of hand	Moves skin on medial border toward midline of palm	Ulnar nerve, superficial branch (C_8)
Abductor digiti minimi	Pisiform	Proximal phalanx of little finger	Abduction of little finger and flexion at its metacarpophalangeal joint	Ulnar nerve, deep branch (C_8–T_1)
Abductor pollicis brevis	Transverse carpal ligament, scaphoid and trapezium	Radial side of base of proximal phalanx of thumb	Abduction of thumb	Median nerve (C_6–C_7)
Flexor pollicis brevis	Flexor retinaculum, trapezium, capitate, and ulnar side of first metacarpal bone	Radial and ulnar sides of proximal phalanx of thumb	Flexion and adduction of thumb	Branches of median and ulnar nerves
Flexor digiti minimi brevis	Hamate	Proximal phalanx of little finger	Flexion at joints of little finger	Ulnar nerve deep branch (C_8–T_1)
Opponens digiti minimi	As above	Fifth metacarpal bone	Opposition of fifth metacarpal bone	As above
Lumbrical (4)	Tendons of flexor digitorum profundus	Tendons of extensor digitorum to digits 2–5	Flexion at metacarpophalangeal joints 2–5; extension at proximal and distal interphalangeal joints, digits 2–5	No. 1 and no. 2 by median nerve; no. 3 and no. 4 by ulnar nerve; deep branch
Dorsal interosseus (4)	Each originates from opposing faces of two metacarpal bones (I and II, II and III, and IV, IV, and V)	Bases of proximal phalanges of fingers 2–4	Adduction at metacarpophalangeal joints of fingers 2 and 4; flexion at metacarpophalangeal joints; extension at interphalangeal joints	Ulner nerve, deep branch (C_8–T_1)
Palmar interosseus** (3–4)	Sides of metacarpal bones II, IV, and V	Bases of proximal phalanges of fingers 2, 4, and 5	Adduction at metacarpophalangeal joints of fingers 2, 4, and 5; flexion at metacarpophalangeal joints; extension at interphalangeal joints	As above

*Where appropriate, spinal nerves involved are given in parentheses.

**The deep, medial portion of the flexor pollicis brevis originating on the first metacarpal bone is sometimes called the *first palmar interosseus muscle;* it inserts on the ulnar side of the phalanx and is innervated by the ulnar nerve.

Study Tip **Use a Reference Muscle to Remember the Forearm**

Here's a quick method to learn the muscles of the forearm. Follow the tendon of the palmaris longus muscle from the middle of flexor retinaculum to the belly of the muscle. Next, identify the flexor carpi radialis longus and the flexor carpi ulnaris muscles on each side of the palmaris longus by remembering the radius bone is medial and the ulna is lateral. On the posterior forearm, trace the tendon on the middle finger toward the belly of the extensor digitorum muscle. Now identify the other extensor muscles on each side. ∎

Tendons of these muscles are secured across the posterior aspect of the wrist by the **extensor retinaculum** (Figure 20.6).

Medial to the brachioradialis is the **flexor carpi radialis** muscle, the flexor muscle closest to the radius. The fibers of this muscle blend into a long tendon that inserts on the second and third metacarpals. The **palmaris longus** muscle is medial to the flexor carpi radialis muscle and is easy to locate by its tendon that inserts on the flexor retinaculum. Medial to the palmaris longus muscle is the **flexor carpi ulnaris** muscle. This muscle rests on the ulnar side of the forearm and inserts on the pisiform and hamate bones of the carpus and on the base of metacarpal IV. The **flexor digitorum superficialis** muscle is located deep to the superficial flexors of the hand. It has four tendons that insert on the midlateral surface of the middle phalanges of fingers 2 through 5. Deeper flexors are also shown in Figure 20.6.

Posterior to the brachioradialis muscle, the long **extensor carpi radialis longus** muscle is the only extensor that does not originate on a tendon attached to the humerus lateral epicondyle. Instead, it arises from the humerus just proximal to the lateral epicondyle, although a few fibers do extend

Figure 20.6 Muscles That Move the Wrist, Hand, and Fingers Middle and deep muscle layers of the right forearm.

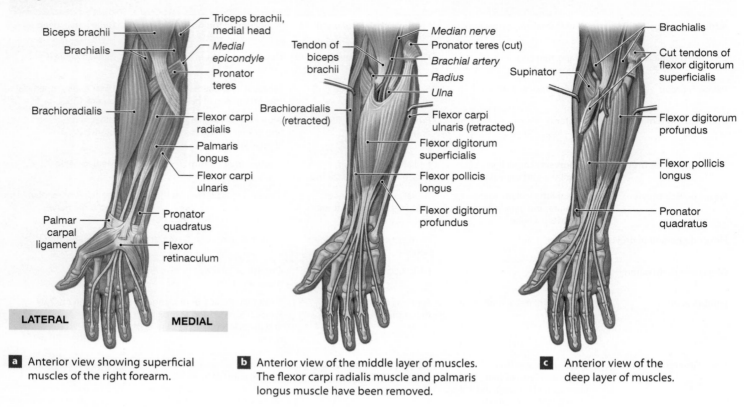

a Anterior view showing superficial muscles of the right forearm.

b Anterior view of the middle layer of muscles. The flexor carpi radialis muscle and palmaris longus muscle have been removed.

c Anterior view of the deep layer of muscles.

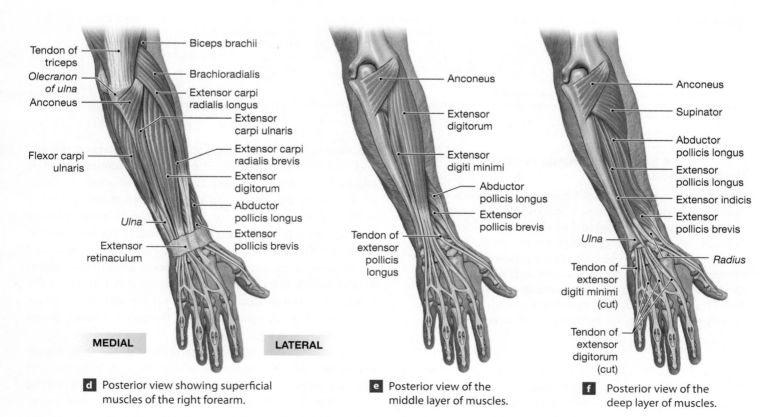

d Posterior view showing superficial muscles of the right forearm.

e Posterior view of the middle layer of muscles.

f Posterior view of the deep layer of muscles.

Clinical Application Carpal Tunnel Syndrome

The tendons of the flexor digitorum superficialis muscle pass through a narrow valley, the *carpal tunnel*, bounded by carpal bones. A protective synovial sheath lubricates the tendons in the tunnel, but repeated flexing of the hand and fingers, such as with prolonged typing or piano playing, causes the sheath to swell and compress the median nerve. Pain and numbness occur in the palm during flexion, a condition called *carpal tunnel syndrome*. ◼

from the common tendon. Inferior to the longus muscle is the **extensor carpi radialis brevis** muscle. The carpi muscles extend and abduct the wrist. The **extensor digitorum** muscle is medial to the extensor carpi radialis muscles and is easy to identify by the three or four tendons that insert on the posterior surface of the phalanges of fingers 2 through 5. Lateral to the digitorum muscle is the **extensor carpi ulnaris** muscle. Deeper extensor muscles are shown in Figure 20.6.

Muscles of the Hand

The masses of tissue at the base of the thumb and along the medial margin of the hand are called **eminences.** See Tables 20.4 and 20.5 for details on the origins, insertions, and actions of these muscles. The **thenar** (THĒ-nar; *thenar*, palm) eminence of the thumb consists of several muscles (**Figure 20.7**). The most medial of the thenar muscles is the **flexor pollicis brevis** (POL-i-sis; *pollex*, thumb, BREV-is; *brevis*, short) muscle, which flexes and adducts the thumb. Lateral to this flexor is the **abductor pollicis brevis** muscle, which abducts the thumb. The most lateral thenar muscle is the **opponens pollicis** muscle, which opposes the thumb toward the little finger.

The **adductor pollicis** muscle is often not considered part of the thenar eminence, as it is found just medial to the flexor pollicis brevis muscle and deep in the web of tissue between the thumb and palm. This muscle adducts the thumb and opposes the action of the abductor pollicis brevis muscle.

The **hypothenar eminence** is fleshy mass on the medial side of the palm at the base of the little finger and consists of three muscles (Figure 20.7). The most lateral is the **opponens digiti minimi** muscle, which opposes the little finger toward the thumb. Medial to this muscle is the **flexor digiti minimi brevis** muscle, which flexes the little finger. The most medial muscle of the hypothenar eminence is the **abductor digiti minimi** muscle, which abducts the little finger.

You should note that no muscles originate on the fingers. Instead, the phalanges of the fingers serve as insertion points for muscles whose origins are more proximal.

QuickCheck Questions

4.1 What is the general action of the muscles on the posterior forearm?

4.2 What are the muscles of the thenar eminence?

In the Lab 4

Materials

- ☐ Torso model
- ☐ Upper limb model
- ☐ Articulated skeleton
- ☐ Muscle chart

Procedures

1. Review the muscles of the wrist and hand in Figures 20.4 through 20.7 and in Tables 20.3 through 20.5.

2. Examine the articulated skeleton and note the origin, insertion, and action of the muscles that act on the wrist and hand.

3. On your body, locate the tendons of the extensor digitorum on the posterior of the hand. Also identify on your forearm the general position of each muscle involved with movement of the wrist and hand. Contract each muscle and observe the action of your wrist. ◼

Figure 20.7 Muscles of the Wrist and Hand Anatomy of the right wrist and hand.

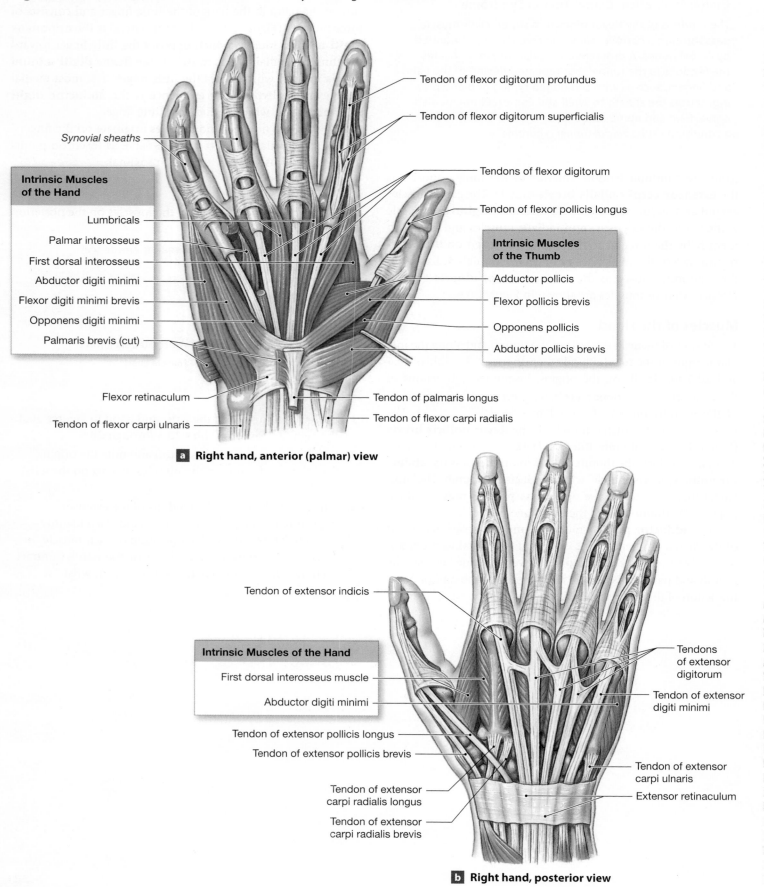

Tendon of flexor digitorum profundus

Tendon of flexor digitorum superficialis

Synovial sheaths

Tendons of flexor digitorum

Intrinsic Muscles of the Hand

Tendon of flexor pollicis longus

Lumbricals

Palmar interosseus

First dorsal interosseus

Abductor digiti minimi

Flexor digiti minimi brevis

Opponens digiti minimi

Palmaris brevis (cut)

Intrinsic Muscles of the Thumb

Adductor pollicis

Flexor pollicis brevis

Opponens pollicis

Abductor pollicis brevis

Flexor retinaculum

Tendon of palmaris longus

Tendon of flexor carpi ulnaris

Tendon of flexor carpi radialis

a Right hand, anterior (palmar) view

Tendon of extensor indicis

Intrinsic Muscles of the Hand

First dorsal interosseus muscle

Abductor digiti minimi

Tendons of extensor digitorum

Tendon of extensor digiti minimi

Tendon of extensor pollicis longus

Tendon of extensor pollicis brevis

Tendon of extensor carpi radialis longus

Tendon of extensor carpi ulnaris

Extensor retinaculum

Tendon of extensor carpi radialis brevis

b Right hand, posterior view

Name _____

Date _____

Section _____

Muscles of the Pectoral Girdle and Upper Limb

A. Matching

Match each term in the left column with its correct description from the right column.

_____	**1.** opponens digiti minimi	**A.** tenses palmar fascia and flexes wrist
_____	**2.** palmaris longus	**B.** major pronator of arm
_____	**3.** pronator teres	**C.** flexes and adducts wrist
_____	**4.** flexor carpi ulnaris	**D.** opposes thumb
_____	**5.** extensor carpi ulnaris	**E.** major supinator of forearm
_____	**6.** extensor digitorum	**F.** extends and adducts wrist
_____	**7.** extensor carpi radialis	**G.** band of connective tissue on flexor tendons
_____	**8.** supinator	**H.** brings little finger toward thumb
_____	**9.** flexor retinaculum	**I.** extends fingers
_____	**10.** opponens pollicis	**J.** extends and abducts wrist

B. Descriptions

Describe the location of each of the following muscles.

1. triceps brachii

2. infraspinatus

3. teres minor

4. biceps brachii

5. supraspinatus

6. brachialis

7. coracobrachialis

8. teres major

9. deltoid

10. subscapularis

C. Labeling

1. Label the muscles that move the pectoral girdle in **Figure 20.8**.

Figure 20.8 Muscles That Move the Pectoral Girdle Anterior view.

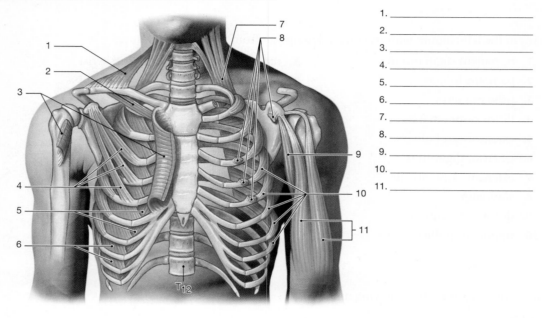

1. _____
2. _____
3. _____
4. _____
5. _____
6. _____
7. _____
8. _____
9. _____
10. _____
11. _____

2. Label the muscles that move the arm in **Figure 20.9**.

Figure 20.9 Muscles That Move the Arm Posterior view.

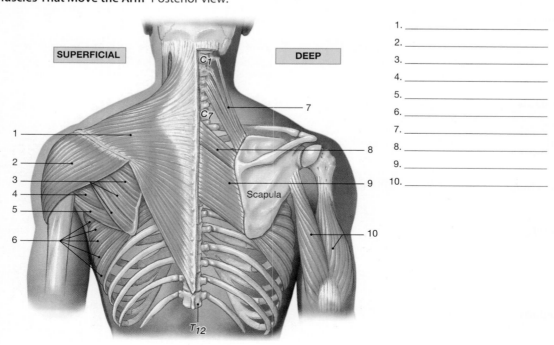

1. _____
2. _____
3. _____
4. _____
5. _____
6. _____
7. _____
8. _____
9. _____
10. _____

3. Label the muscles that move the forearm and hand in **Figure 20.10**.

Figure 20.10 **Muscles That Move the Forearm and Hand** Anterior view.

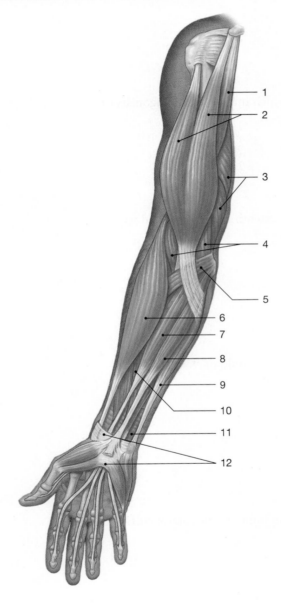

1. _____

2. _____

3. _____

4. _____

5. _____

6. _____

7. _____

8. _____

9. _____

10. _____

11. _____

12. _____

D. Short-Answer Questions

1. Describe the muscles involved in turning the hand, as when twisting a doorknob back and forth.

2. Name the muscles responsible for flexing the arm. Which muscles are antagonists to these flexors?

3. Name a muscle for each movement of the wrist: flex, extend, abduct, and adduct.

E. Analysis and Application

A brace placed on your wrist to treat carpal tunnel syndrome would prevent what wrist action? What would you accomplish by limiting this action?

F. Clinical Challenge

How would a dislocated shoulder also potentially result in injury to the rotator cuff?

Muscles of the Pelvic Girdle and Lower Limb

Learning Outcomes

On completion of this exercise, you should be able to:

1. Locate the muscles of the pelvic girdle and lower limb on lab models and charts.

2. Identify on the models the origin, insertion, and action of the muscles of the pelvic girdle and lower limb.

3. Demonstrate or describe the action of the major muscles of the pelvic girdle and lower limb.

The muscles of the pelvis help support the mass of the body and stabilize the pelvic girdle. Leg muscles move the thigh, knee, and foot. Flexors of the knee are on the posterior thigh, and knee extensors are anterior. Muscles that abduct the thigh are on the lateral side of the thigh, and the adductors are on the medial thigh.

Lab Activity 1   Muscles That Move the Thigh

Unlike the articulations between the axial skeleton and the pectoral girdle, which give this region great mobility, the articulations between the axial skeleton and the pelvic girdle limit movement of the hips. (Axial muscles that move the pelvic girdle are discussed in Exercise 19.) Muscles that move the thigh insert on the femur and cause movement at the ball-and-socket joint. These muscles are organized into four groups: gluteal, lateral rotator, adductor, and iliopsoas. Refer to **Figure 21.1** and **Table 21.1** (on p. 278) for details on these muscles.

Need More Practice and Review?

Build your knowledge—and confidence!—in the Study Area of MasteringA&P® at www.masteringaandp.com with Pre-lab Quizzes, Post-lab Quizzes, Practice Anatomy Lab™ (PAL™) 3.0 virtual anatomy practice tool, PhysioEx™ 9.0 laboratory simulations, and A&P Flix™ with Quizzes.

PAL ⟨practice anatomy lab⟩   For this lab exercise, follow these navigation paths:
- PAL>Human Cadaver>Muscular System>Lower Limb
- PAL>Anatomical Models>Muscular System>Lower Limb

A&PFlix   For this lab exercise, go to these topics:
- Origins, Insertions, Actions, and Innervations
- Group Muscle Actions and Joints

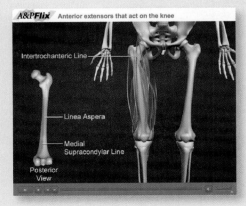

A&PFlix   Anterior extensors that act on the knee

Intertrochanteric Line

Linea Aspera

Medial Supracondylar Line

Posterior View

MasteringA&P®

Figure 21.1 Anterior Muscles That Move the Thigh The gluteal and lateral rotator muscle groups of the right hip.

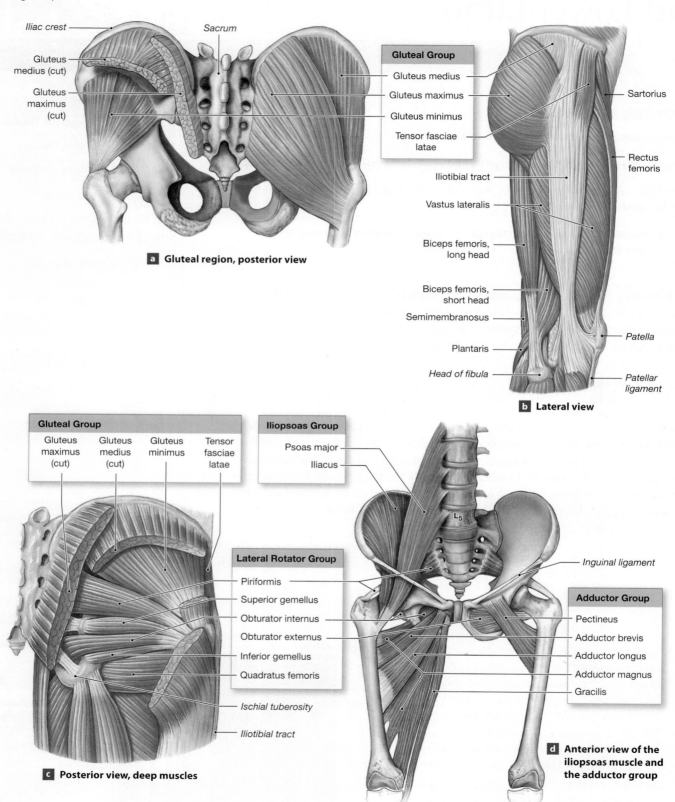

Iliac crest
Sacrum

Gluteus medius (cut)

Gluteus maximus (cut)

Gluteal Group
Gluteus medius
Gluteus maximus
Gluteus minimus
Tensor fasciae latae

Sartorius

Rectus femoris

Iliotibial tract

Vastus lateralis

Biceps femoris, long head

Biceps femoris, short head

Semimembranosus

Plantaris

Head of fibula

Patella

Patellar ligament

a **Gluteal region, posterior view**

b **Lateral view**

Gluteal Group

| Gluteus maximus (cut) | Gluteus medius (cut) | Gluteus minimus | Tensor fasciae latae |

Iliopsoas Group
Psoas major
Iliacus

L₅

Lateral Rotator Group
Piriformis
Superior gemellus
Obturator internus
Obturator externus
Inferior gemellus
Quadratus femoris

Ischial tuberosity

Iliotibial tract

Inguinal ligament

Adductor Group
Pectineus
Adductor brevis
Adductor longus
Adductor magnus
Gracilis

c **Posterior view, deep muscles**

d **Anterior view of the iliopsoas muscle and the adductor group**

Gluteal Group

The posterior muscles originating on the ilium of the pelvis are the three gluteal muscles that constitute the buttocks. The most superficial and prominent is the **gluteus maximus** muscle (Figure 21.1). It is a large, fleshy muscle and is easily located as the major muscle of the buttocks. Its muscle fibers pass inferiorolaterally and insert on a thick band of tendon called the **iliotibial** (il-ē-ō-TIB-ē-ul) **tract** that attaches to the lateral condyle of the tibia.

The **gluteus medius** muscle originates on the iliac crest and on the lateral surface of the ilium, and gathers laterally into a thick tendon that inserts posteriorly on the greater trochanter of the femur. The **gluteus minimus** muscle begins on the lateral surface of the ilium, tucked under the origin of the gluteus medius muscle. The fibers of the gluteus minimus muscle also pass laterally to insert on the anterior surface of the greater trochanter. Both the gluteus medius muscle and the gluteus minimus muscle abduct and medially rotate the thigh.

The **tensor fasciae latae** (TEN-sor FAH-shē-āy LAH-tāy) muscle is a small muscle on the proximal part of the lateral thigh. It originates on the iliac crest and on the outer surface of the anterior superior iliac spine. It is a gluteal muscle because it shares its insertion on the iliotibial tract with the gluteus maximus. As the name implies, the tensor fasciae latae muscle tenses the fascia of the thigh and helps stabilize the pelvis on the femur. The muscle also abducts and medially rotates the thigh.

Lateral Rotator Group

The lateral rotator group consists of the obturator internus and externus muscles and the piriformis, gamellus, and quadratus femoris muscles (Figure 21.1b–d). All of these muscles rotate the thigh laterally, and the piriformis muscle also abducts the thigh. Both the **obturator internus** muscle and the **obturator externus** muscle originate along the medial and lateral edges of the obturator foramen of the os coxae and insert on the trochanteric fossa, a shallow depression on the medial side of the greater trochanter of the femur.

The **piriformis** (pir-i-FOR-mis) muscle arises from the anterior and lateral surfaces of the sacrum and inserts on the greater trochanter of the femur. Inferior to the piriformis is the **quadratus femoris** muscle. Its origin is on the lateral surface of the ischial tuberosity and inserts on the femur between the greater and lesser trochanters.

The **superior gemellus** muscle and **inferior gemellus** muscle are deep to the gluteal muscles. These small rotators originate on the ischial spine and ischial tuberosity and insert on the greater trochanter with the tendon of the obturator internus. Both muscles rotate the thigh laterally.

Iliopsoas Group

The iliopsoas (il-ē-ō-SŌ-us) group consists of two muscles, the psoas major and the iliacus (Figure 21.1d). The **psoas** (SŌ-us) **major** muscle originates on the body and transverse processes of vertebrae T_{12} through L_5. The muscle sweeps inferiorly, passing between the femur and the ischial ramus, and inserts on the lesser trochanter of the femur. The **iliacus** (il-Ē-ah-kus) muscle originates on the iliac fossa on the medial portion of the ilium and joins the tendon of the psoas major muscle. The psoas major and iliacus muscles work together to flex the thigh, bringing its anterior surface toward the abdomen.

Adductor Group

Muscles that adduct the thigh are organized into the adductor group and the pectineus and gracilis muscles. The **pectineus** (pek-TIN-ē-us) muscle is another superficial adductor muscle of the medial thigh (Figure 21.1d). It is located next to the iliacus muscle. It originates along the superior ramus of the pubic bone and inserts on the pectineal line of the femur.

The **gracilis** (GRAS-i-lis) muscle is the most superficial of the thigh adductors and is located at the midline of the medial thigh (**Figure 21.2**). It arises from the superior ramus of the pubic bone, near the symphysis, extends inferiorly along the medial surface of the thigh, and inserts just medial to the insertion of the sartorius near the tibial tuberosity. Because it passes over both the hip and knee joints, it acts to adduct and medially rotate the thigh and flex the knee.

Three additional adductor muscles originate on the inferior pubis and insert on the posterior femur and are powerful adductors of the thigh (Figure 21.2). They also allow the thigh to flex and rotate medially. The **adductor magnus** muscle is the largest of the adductor muscles. It arises on the inferior ramus of the pubis and the ischial tuberosity and inserts along the length of the linea aspera of the femur. It is easily observed on a leg model if the superficial muscles are removed. Superficial to the adductor magnus is the **adductor longus** muscle. Not visible on the surface is the **adductor brevis** muscle, which is positioned superior and posterior to the adductor longus muscle (see Figure 21.1d).

> **Study Tip Learning by Anatomical Association**
>
> An easy method for remembering the superficial muscles of the medial thigh is to reference them to other regional muscles. Locate the gracilis muscle in the midline of the thigh. Anterior to the gracilis is the adductor longus, which is next to the long sartorius muscle (described in the next activity). Posterior to the gracilis is the adductor magnus, which is by the gluteus maximus. ∎

Table 21.1   *ORIGINS AND INSERTIONS* Muscles That Move the Thigh (See Figures 21.1 and 21.2)

Group and Muscle(s)	Origin	Insertion	Action	Innervation*
GLUTEAL GROUP				
Gluteus maximus	Iliac crest, posterior gluteal line, and lateral surface of ilium; sacrum, coccyx, and thoracolumbar fascia	Iliotibial tract and gluteal tuberosity of femur	Extension and lateral rotation at hip	Inferior gluteal nerve (L_5–S_2)
Gluteus medius	Anterior iliac crest of ilium, lateral surface between posterior and anterior gluteal lines	Greater trochanter of femur	Abduction and medial rotation at hip	Superior gluteal nerve (L_4–S_1)
Gluteus minimus	Lateral surface of ilium between inferior and anterior gluteal lines	As above	As above	As above
Tensor fasciae latae	Iliac crest and lateral surface of anterior superior iliac spine	Iliotibial tract	Flexion and medial rotation at hip; tenses fascia lata, which laterally supports the knee	As above
LATERAL ROTATOR GROUP				
Obturators (externus and internus)	Lateral and medial margins of obturator foramen	Trochanteric fossa of femur (externus); medial surface of greater trochanter (internus)	Lateral rotation at hip	Obturator nerve (externus: L_3–L_4) and special nerve from sacral plexus (internus: L_5–S_2)
Piriformis	Anterolateral surface of sacrum	Greater trochanter of femur	Lateral rotation and abduction at hip	Branches of sacral nerves (S_1–S_2)
Gemelli (superior and inferior)	Ischial spine and tuberosity	Medial surface of greater trochanter with tendon of obturator internus	Lateral rotation at hip	Nerves to obturator internus and quadratus femoris
Quadratus femoris	Lateral border of ischial tuberosity	Intertrochanteric crest of femur	As above	Special nerve from sacral plexus (L_4–S_1)
ADDUCTOR GROUP				
Adductor brevis	Inferior ramus of pubis	Linea aspera of femur	Adduction, flexion, and medial rotation at hip	Obturator nerve (L_3–L_4)
Adductor longus	Inferior ramus of pubis anterior to adductor brevis	As above	As above	As above
Adductor magnus	Inferior ramus of pubis posterior to adductor brevis and ischial tuberosity	Linea aspera and adductor tubercle of femur	Adduction at hip; superior part produces flexion and medial rotation; inferior part produces extension and lateral rotation	Obturator and sciatic nerves
Pectineus	Superior ramus of pubis	Pectineal line inferior to lesser trochanter of femur	Flexion medial rotation and adduction at hip	Femoral nerve (L_2–L_4)
Gracilis	Inferior ramus of pubis	Medial surface of tibia inferior to medial condyle	Flexion at knee; adduction and medial rotation at hip	Obturator nerve (L_3–L_4)
ILIOPSOAS GROUP				
Iliacus	Iliac fossa of ilium	Femur distal to lesser trochanter; tendon fused with that of psoas major	Flexion at hip	Femoral nerve (L_2–L_3)
Psoas major	Anterior surfaces and transverse processes of vertebrae (T_{12}–L_5)	Lesser trochanter in company with iliacus	Flexion at hip or lumbar intervertebral joints	Branches of the lumbar plexus (L_2–L_3)

*Where appropriate, spinal nerves involved are given in parentheses.

Figure 21.2 Medial Muscles That Move the Leg Medial view of the muscles of the right thigh.

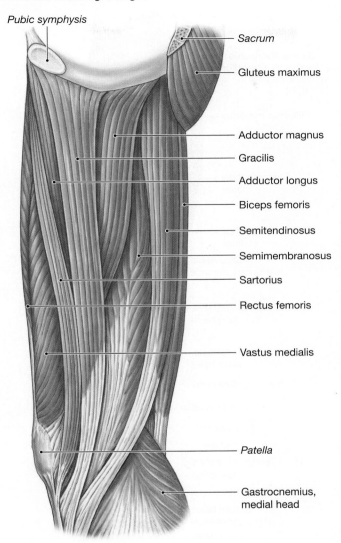

Pubic symphysis

Sacrum

Gluteus maximus

Adductor magnus

Gracilis

Adductor longus

Biceps femoris

Semitendinosus

Semimembranosus

Sartorius

Rectus femoris

Vastus medialis

Patella

Gastrocnemius, medial head

QuickCheck Questions

1.1 Where are the abductors of the thigh located?

1.2 What is the iliotibial tract?

1.3 Name two muscles that rotate the thigh.

In the Lab 1

Materials

☐ Torso model
☐ Lower limb model
☐ Articulated skeleton
☐ Muscle chart

Procedures

1. Review the pelvic and gluteal muscles in Figures 21.1 and 21.2 and in Table 21.1.

2. Identify each muscle on the torso and lower limb models and on the muscle chart.

3. On the lower limb model, observe how the gluteal muscles and the tensor fasciae latae muscle insert on the lateral portion of the femur.

4. Locate as many of your own thigh muscles as possible. Practice the actions of the muscles and observe how your lower limb moves.

5. Examine the articulated skeleton and note the origin, insertion, and action of the major muscles that act on the thigh. ■

Lab Activity 2 Muscles That Move the Leg

Muscles that flex and extend the leg at the knee joint are on the posterior and anterior sides of the femur. Refer to **Table 21.2** (on p. 282) for details on these muscles. Some of these muscles originate on the pelvis and cross both the hip and the knee joints and can therefore also move the thigh.

The major muscles of the posterior thigh are collectively called the **hamstrings.** They all have a common origin on the ischial tuberosity and flex the knee. The **biceps femoris** muscle is the lateral muscle of the posterior thigh (**Figure 21.3**). It has two heads and two origins, one on the ischial tuberosity and a second on the linea aspera of the femur. The two heads merge to form the belly of the muscle and insert on the lateral condyle of the tibia and the head of the fibula. Because this muscle spans both the hip and knee joints, it can extend the thigh and flex the knee. Medial to the biceps femoris muscle is the **semitendinosus** (sem-ē-ten-di-NŌ-sus) muscle. It is a long muscle that passes the posterior knee to insert on the proximomedial surface of the tibia near the insertion of the gracilis. The **semimembranosus** (sem-ē-mem-bra-NŌ-sus) muscle is medial to the semitendinosus muscle and inserts on the medial tibia. These muscles cross both the hip joint and the knee joint and extend the thigh and flex the knee. The hamstrings are therefore antagonists to the quadriceps muscles. When the thigh is flexed and drawn up toward the pelvis, the hamstrings extend the thigh.

The extensors of the leg are collectively called either the **quadriceps** muscles or the **quadriceps femoris.** They make up the bulk of the anterior mass of the thigh and are consequently easy to locate. The largest muscle in the group, the **rectus femoris** muscle (**Figure 21.4**), is located along the midline of the anterior surface of the thigh. Covering almost the entire medial surface of the femur is the **vastus medialis** muscle. The **vastus lateralis** muscle is located on the lateral side of the rectus femoris muscle, and the **vastus intermedius** muscle is directly deep to the rectus femoris muscle. The quadriceps muscles converge on a patellar tendon and insert on the

Figure 21.3 **Posterior Muscles That Move the Leg**

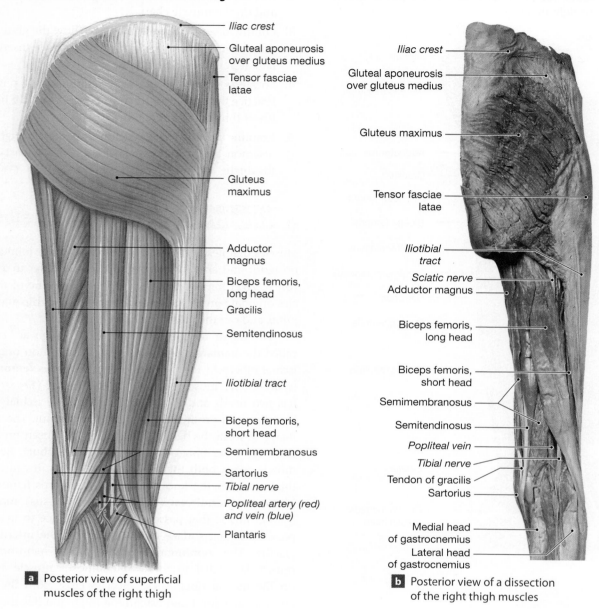

a Posterior view of superficial muscles of the right thigh

b Posterior view of a dissection of the right thigh muscles

tibial tuberosity. Because the rectus femoris muscle crosses two joints, the hip and knee, it allows the hip to flex and the leg to extend.

The **sartorius** (sar-TOR-ē-us; *sartor*, a tailor) muscle is a thin, ribbonlike muscle originating on the anterior superior iliac spine and passing inferiorly, cutting obliquely across the thigh (Figure 21.3b). It is the longest muscle in the body. It crosses the knee joint to insert on the medial surface of the tibia near the tibial tuberosity. This muscle is a flexor of the knee and thigh and a lateral rotator of the thigh. Figures 21.3b and 21.4b show the anterior and posterior thigh muscles in superficial

dissection for comparison with the muscles illustrated in Figures 21.3a and 21.4a.

A small muscle on the posterior of the knee assists in flexing the knee. The **popliteus** (pop-LI-tē-us) muscle crosses from its origin on the lateral condyle of the femur to insert on the posterior surface of the tibial shaft.

QuickCheck Questions

2.1 Name all the quadriceps muscles.

2.2 Name all the hamstrings.

Figure 21.4 Anterior Muscles That Move the Leg

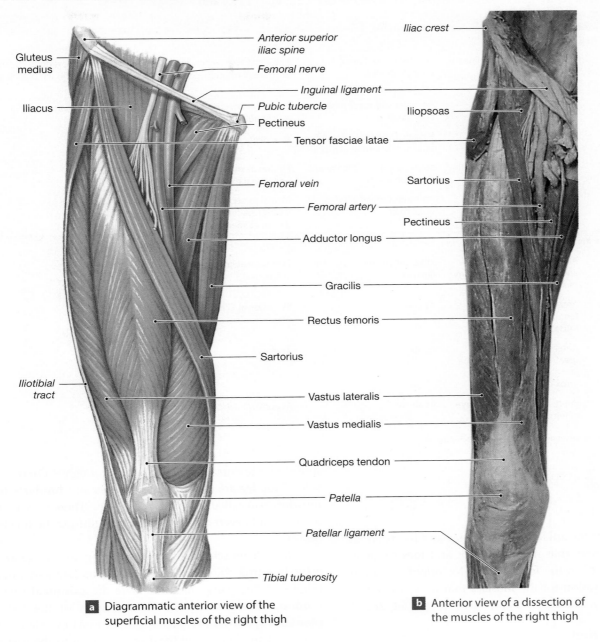

Gluteus medius

Iliacus

Anterior superior iliac spine

Femoral nerve

Inguinal ligament

Pubic tubercle

Pectineus

Tensor fasciae latae

Femoral vein

Femoral artery

Adductor longus

Gracilis

Rectus femoris

Sartorius

Iliotibial tract

Vastus lateralis

Vastus medialis

Quadriceps tendon

Patella

Patellar ligament

Tibial tuberosity

a Diagrammatic anterior view of the superficial muscles of the right thigh

Iliac crest

Iliopsoas

Sartorius

Pectineus

b Anterior view of a dissection of the muscles of the right thigh

In the Lab   2

Materials

☐ Torso model
☐ Lower limb model
☐ Articulated skeleton
☐ Muscle chart

Procedures

1. Review the muscles that move the leg in Figures 21.2 through 21.4. On the torso and lower limb models and the muscle chart, identify the muscles that move the leg, categorizing each muscle as being a flexor, extensor, adductor, or abductor.

2. Flex your knee and feel the tendons of the semimembranosus and semitendinosus muscles, located just above the posterior knee on the medial side. Similarly, on the lateral side of the knee, just above the fibular head, the tendon of the biceps femoris muscle can be palpated.

3. Examine the articulated skeleton and note the origin, insertion, and action of the major muscles that act on the thigh, knee, and leg. ■

Table 21.2   *ORIGINS AND INSERTIONS* Muscles That Move the Leg (See Figures 21.3 and 21.4)

Muscle	Origin	Insertion	Action	Innervation*
FLEXORS OF THE KNEE				
Biceps femoris	Ischial tuberosity and linea aspera of femur	Head of fibula, lateral condyle of tibia	Flexion at knee; extension and lateral rotation at hip	Sciatic nerve; tibial portion (S_1–S_3; to long head) and common fibular branch (L_5–S_2; to short head)
Semimembranosus	Ischial tuberosity	Posterior surface of medial condyle of tibia	Flexion at knee; extension and medial rotation at hip	Sciatic nerve (tibial portion; L_5–S_2)
Semitendinosus	As above	Proximal, medial surface of tibia near insertion of gracilis	As above	As above
Sartorius	Anterior superior iliac spine	Medial surface of tibia near tibial tuberosity	Flexion at knee; flexion and lateral rotation at hip	Femoral nerve (L_2–L_3)
Popliteus	Lateral condyle of femur	Posterior surface of proximal tibial shaft	Medial rotation of tibia (or lateral rotation of femur); flexion at knee	Tibial nerve (L_4–S_1)
EXTENSORS OF THE KNEE				
Rectus femoris	Anterior inferior iliac spine and superior acetabular rim of ilium	Tibial tuberosity via patellar ligament	Extension at knee; flexion at hip	Femoral nerve (L_2–L_4)
Vastus intermedius	Anterolateral surface of femur and linea aspera (distal half)	As above	Extension at knee	As above
Vastus lateralis	Anterior and inferior to greater trochanter of femur and along linea aspera (proximal half)	As above	As above	As above
Vastus medialis	Entire length of linea aspera of femur	As above	As above	As above

*Where appropriate, spinal nerves involved are given in parentheses.

Lab Activity 3   Muscles That Move the Ankle and Foot

Muscles that move the ankle arise on the leg and insert on the tarsal bones. Muscles that move the foot and toes originate either on the leg or in the foot. Details for origin, insertion, action, and innervation for the muscles that move the ankle, foot, and toes are in **Tables 21.3** and **21.4** (on pp. 284, 286).

Make a Prediction

Consider the action of flexing the toes and then predict the location of the contracting flexor muscle.

The **tibialis** (tib-ē-A-lis) **anterior** muscle is located on the anterior side of the leg (**Figure 21.5**). This muscle is easy to locate as the lateral muscle mass of the shin on the anterior edge of the tibia bone. Its tendon passes over the dorsal surface of the foot, and the muscle dorsiflexes and inverts the foot. Two extensor muscles arise on the anterior leg and insert on the various phalanges of the foot. The **extensor hallucis** (HAL-i-sis; *hallux*, great toe) **longus** muscle (not shown in any illustration here) is lateral and deep to the tibialis anterior muscle. Lateral to the extensor hallucis longus muscle is the **extensor digitorum longus** muscle with four tendons that

spread on the dorsal surface of toes 2 through 5. On the lateral side of the leg are the **fibularis longus** and **fibularis brevis** muscles, also called the *peroneus* muscles. These muscles insert on the foot to evert the foot by laterally turning the sole to face outward.

The calf muscles of the posterior leg are the **gastrocnemius** (gas-trok-NĒ-mē-us) and the **soleus** (SŌ-lē-us) muscles (Figure 21.5d). These muscles share the **calcaneal** (Achilles) **tendon,** which inserts on the calcaneus of the foot. The **plantaris** (plan-TĀR-is; *planta*, sole of foot) muscle is a short muscle of the lateral popliteal region, deep to the gastrocnemius muscle. The plantaris muscle has a long tendon that inserts on the posterior of the calcaneus. The gastrocnemius, soleus, and plantaris muscles plantarflex the ankle; the soleus is also a postural muscle for support while standing.

Deep to the soleus muscle is the **tibialis posterior** muscle (Figure 21.5e), which adducts and inverts the foot and plantar flexes the ankle. Its tendon passes medially to the calcaneus and inserts on the plantar surface of the navicular and cuneiform bones and metatarsals II, III, and IV. The **flexor hallucis longus** muscle begins lateral to the origin of the tibialis posterior muscle on the fibular shaft. Its tendon runs parallel to that of the tibialis posterior muscle, passes medial to the calcaneus, and inserts on the plantar surface of the distal phalanx of the

Figure 21.5 Muscles That Move the Ankle, Foot, and Toes Relationships among the muscles of the right leg and foot.

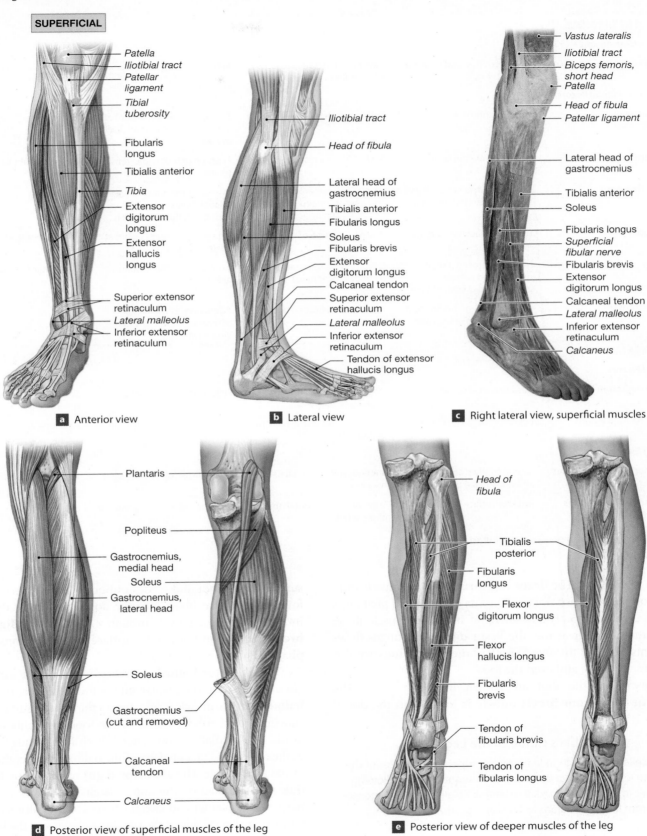

SUPERFICIAL

a Anterior view

Patella
Iliotibial tract
Patellar ligament
Tibial tuberosity
Fibularis longus
Tibialis anterior
Tibia
Extensor digitorum longus
Extensor hallucis longus
Superior extensor retinaculum
Lateral malleolus
Inferior extensor retinaculum

b Lateral view

Iliotibial tract
Head of fibula
Lateral head of gastrocnemius
Tibialis anterior
Fibularis longus
Soleus
Fibularis brevis
Extensor digitorum longus
Calcaneal tendon
Superior extensor retinaculum
Lateral malleolus
Inferior extensor retinaculum
Tendon of extensor hallucis longus

c Right lateral view, superficial muscles

Vastus lateralis
Iliotibial tract
Biceps femoris, short head
Patella
Head of fibula
Patellar ligament
Lateral head of gastrocnemius
Tibialis anterior
Soleus
Fibularis longus
Superficial fibular nerve
Fibularis brevis
Extensor digitorum longus
Calcaneal tendon
Lateral malleolus
Inferior extensor retinaculum
Calcaneus

d Posterior view of superficial muscles of the leg

Plantaris
Popliteus
Gastrocnemius, medial head
Soleus
Gastrocnemius, lateral head
Soleus
Gastrocnemius (cut and removed)
Calcaneal tendon
Calcaneus

e Posterior view of deeper muscles of the leg

Head of fibula
Tibialis posterior
Fibularis longus
Flexor digitorum longus
Flexor hallucis longus
Fibularis brevis
Tendon of fibularis brevis
Tendon of fibularis longus

Table 21.3 *ORIGINS AND INSERTIONS* Extrinsic Muscles That Move the Foot and Toes (See Figure 21.5)				
Muscle	**Origin**	**Insertion**	**Action**	**Innervation***
ACTION AT THE ANKLE				
Flexors (Dorsiflexors)				
Tibialis anterior	Lateral condyle and proximal shaft of tibia	Base of first metatarsal bone and medial cuneiform bone	Flexion (dorsiflexion) at ankle; inversion of foot	Deep fibular nerve (L_4–S_1)
Extensors (Plantarflexors)				
Gastrocnemius	Femoral condyles	Calcaneus via calcaneal tendon	Extension (plantar flexion) at ankle; inversion of foot; flexion at knee	Tibial nerve (S_1–S_2)
Fibularis brevis	Midlateral margin of fibula	Base of fifth metatarsal bone	Eversion of foot and extension (plantar flexion) at ankle	Superficial fibular nerve (L_4–S_1)
Fibularis longus	Lateral condyle of tibia, head and proximal shaft of fibula	Base of fifth metatarsal bone and medial cuneiform bone	Eversion of foot and extension (plantar flexion) at ankle; supports longitudinal arch	As above
Plantaris	Lateral supracondylar ridge	Posterior portion of calcaneus	Extension (plantar flexion) at ankle; flexion at knee	Tibial nerve (L_4–S_1)
Soleus	Head and proximal shaft of fibula and adjacent posteromedial shaft of tibia	Calcaneus via calcaneal tendon (with gastrocnemius)	Extension (plantar flexion) at ankle	Sciatic nerve, tibial branch (S_1–S_2)
Tibialis posterior	Interosseous membrane and adjacent shafts of tibia and fibula	Tarsal and metatarsal bones	Adduction and inversion of foot; extension (plantar flexion) at ankle	As above
ACTION AT THE TOES				
Digital Flexors				
Flexor digitorum longus	Posteromedial surface of tibia	Inferior surfaces of distal phalanges, toes 2–5	Flexion at joints of toes 2–5	Sciatic nerve, tibial branch (L_5–S_1)
Flexor hallucis longus	Posterior surface of fibula	Inferior surface, distal phalanx of great toe	Flexion at joints of great toe	As above
Digital Extensors				
Extensor digitorum longus	Lateral condyle of tibia, anterior surface of fibula	Superior surfaces of phalanges, toes 2–5	Extension at joints of toes 2–5	Deep fibular nerve (L_4–S_1)
Extensor hallucis longus	Anterior surface of fibula	Superior surface, distal phalanx of great toe	Extension at joints of great toe	As above

*Where appropriate, spinal nerves involved are given in parentheses.

hallux, or great toe. The **flexor digitorum longus** muscle originates on the posterior tibia and inserts on the distal phalanges of toes 2 through 5. The flexor hallucis longus muscle flexes the joints of the great toe; the flexor digitorum longus flexes the joints of toes 2 through 5. Both of these flexor muscles also dorsiflex the ankle and evert the foot.

Muscles of the foot are shown in **Figure 21.6**. The **extensor digitorum brevis** muscle is located on the dorsal

Study Tip The Tibia's Guide to the Leg

An excellent approach to learning the superficial muscles of the leg is to locate the tibia bone and then sequence the following muscles in order from medial to lateral: tibialis anterior, extensor digitorum longus, fibularis longus, and gastrocnemius. ■

surface of the foot and passes obliquely across the foot with four tendons that insert into the dorsal surface of the proximal phalanges of toes 1 through 4. The **flexor digitorum brevis** muscle on the plantar surface inserts tendons on the phalanges of toes 2 through 5.

The **abductor hallucis** muscle is found on the inner margin of the foot on the plantar side of the calcaneus. The **flexor hallucis brevis** muscle originates on the plantar surface of the cuneiform and cuboid bones of the foot and splits into two heads, one medial and one lateral. Each head sends a tendon to the base of the first phalanx of the hallux, to either the lateral or the medial side. The **abductor digiti minimi** muscle of the little toe is located on the outer margin of the foot and originates on the plantar and lateral surfaces of the calcaneus. It inserts on the lateral side of the proximal phalanx of the little toe.

Figure 21.6 Muscles of the Foot

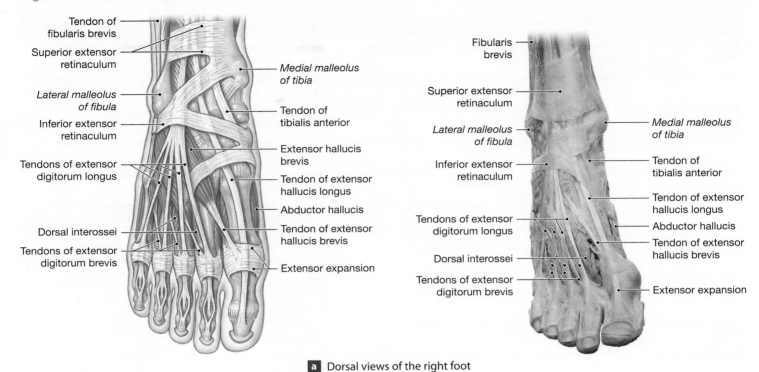

Tendon of fibularis brevis

Superior extensor retinaculum

Lateral malleolus of fibula

Inferior extensor retinaculum

Tendons of extensor digitorum longus

Dorsal interossei

Tendons of extensor digitorum brevis

Medial malleolus of tibia

Tendon of tibialis anterior

Extensor hallucis brevis

Tendon of extensor hallucis longus

Abductor hallucis

Tendon of extensor hallucis brevis

Extensor expansion

Fibularis brevis

Superior extensor retinaculum

Lateral malleolus of fibula

Inferior extensor retinaculum

Tendons of extensor digitorum longus

Dorsal interossei

Tendons of extensor digitorum brevis

Medial malleolus of tibia

Tendon of tibialis anterior

Tendon of extensor hallucis longus

Abductor hallucis

Tendon of extensor hallucis brevis

Extensor expansion

a Dorsal views of the right foot

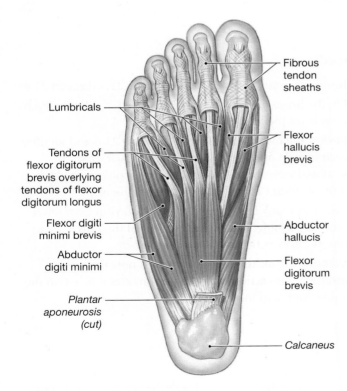

Lumbricals

Tendons of flexor digitorum brevis overlying tendons of flexor digitorum longus

Flexor digiti minimi brevis

Abductor digiti minimi

Plantar aponeurosis (cut)

Fibrous tendon sheaths

Flexor hallucis brevis

Abductor hallucis

Flexor digitorum brevis

Calcaneus

b Plantar (inferior) view, superficial layer of the right foot

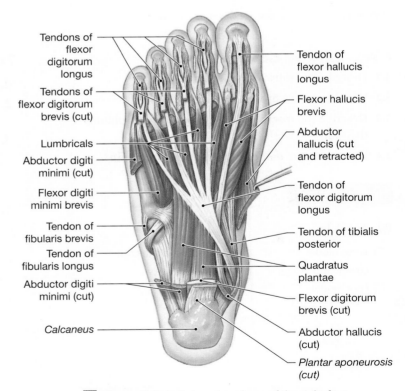

Tendons of flexor digitorum longus

Tendons of flexor digitorum brevis (cut)

Lumbricals

Abductor digiti minimi (cut)

Flexor digiti minimi brevis

Tendon of fibularis brevis

Tendon of fibularis longus

Abductor digiti minimi (cut)

Calcaneus

Tendon of flexor hallucis longus

Flexor hallucis brevis

Abductor hallucis (cut and retracted)

Tendon of flexor digitorum longus

Tendon of tibialis posterior

Quadratus plantae

Flexor digitorum brevis (cut)

Abductor hallucis (cut)

Plantar aponeurosis (cut)

c Plantar (inferior) view, deep layer of the right foot

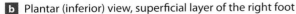

Table 21.4 *ORIGINS AND INSERTIONS* **Intrinsic Muscles of the Foot (See Figure 21.6)**

Muscle	Origin	Insertion	Action	Innervation*
Extensor digitorum brevis	Calcaneus (superior and lateral surfaces)	Dorsal surfaces of toes 1–4	Extension at metatarsophalangeal joints of toes 1–4	Deep fibular nerve (L_5–S_1)
Abductor hallucis	Calcaneus (tuberosity on inferior surface)	Medial side of proximal phalanx of great toe	Abduction at metatarsophalangeal joint of great toe	Medial plantar nerve (L_4–L_5)
Flexor digitorum brevis	As above	Sides of middle phalanges, toes 2–5	Flexion at proximal interphalangeal joints of toes 2–5	As above
Abductor digiti minimi	As above	Lateral side of proximal phalanx, toe 5	Abduction at metatarsophalangeal joint of toe 5	Lateral plantar nerve (L_4–L_5)
Quadratus plantae	Calcaneus (medial, inferior surfaces)	Tendon of flexor digitorum longus	Flexion at joints of toes 2–5	As above
Lumbrical (4)	Tendons of flexor digitorum longus	Insertions of extensor digitorum longus	Flexion at metatarsophalangeal joints; extension at proximal interphalangeal joints of toes 2–5	Medial plantar nerve (1), lateral plantar nerve (2–4)
Flexor hallucis brevis	Cuboid and lateral cuneiform bones	Proximal phalanx of great toe	Flexion at metatarsophalangeal joints of great toe	Medial plantar nerve (L_4–L_5)
Adductor hallucis	Bases of metatarsal bones II–IV and plantar ligaments	As above	Adduction at metatarsophalangeal joint of great toe	Lateral plantar nerve (S_1–S_2)
Flexor digiti minimi brevis	Base of metatarsal bone V	Lateral side of proximal phalanx of toe 5	Flexion at metatarsophalangeal joint of toe 5	As above
Dorsal interosseus (4)	Sides of metatarsal bones	Medial and lateral sides of toe 2; lateral sides of toes 3 and 4	Abduction at metatarsophalangeal joints of toes 3 and 4	As above
Plantar interosseus (3)	Bases and medial sides of metatarsal bones	Medial sides of toes 3–5	Adduction at metatarsophalangeal joints of toes 3–5	As above

*Where appropriate, spinal nerves involved are given in parentheses.

QuickCheck Questions

3.1 Describe the muscles of the calf.

3.2 Which muscles move the great toe?

3.3 Describe the insertions of the muscles that plantar flex the foot.

3.4 What does the name *flexor hallucis brevis* mean?

In the Lab 3

Materials

☐ Torso model

☐ Lower limb model

☐ Foot model

☐ Articulated skeleton

☐ Muscle chart

Procedures

1. Review the muscles of the leg in Figures 21.5 through 21.6.

2. On the lower limb model and muscle chart, identify each muscle on the leg.

3. Review the muscles of the foot in Figure 21.6 and identify each muscle on the foot model and muscle chart. If sectional views of the lower limb are available, study the muscles found in each muscle compartment.

4. Locate as many leg and foot muscles on your own lower limb as possible. Practice the actions of the muscles and observe how your leg and foot move.

5. Examine the articulated skeleton and note the origin, insertion, and action of the major muscles that act on the ankle, foot, and toes. ■

Clinical Application Compartment Syndrome

Muscles on the upper and lower limbs are surrounded by the deep fascia and isolated in saclike muscle **compartments.** These compartments separate the various muscles into anterior, posterior, lateral, and deep groups that have similar muscle actions. Within a muscle compartment are arteries, veins, nerves, and other structures.

Lower limb compartments and the muscles, blood vessels, and nerves in each compartment are illustrated in **Figure 21.7.**

Examine the figure and observe how superficial and deep muscles of the limb are in different compartments.

Treating a limb injury includes watching for blood trapped in a muscle compartment. Bleeding increases pressure in the compartment and causes compression of local nerves and blood vessels. If the compression persists beyond four to six hours, permanent damage to nerve and muscle tissue may occur, a condition called *compartment syndrome*. To prevent compartment syndrome, drains are inserted into wounds to remove blood and other liquids both from the muscle and from the compartment. ∎

Figure 21.7 Muscle Compartments of the Lower Limb

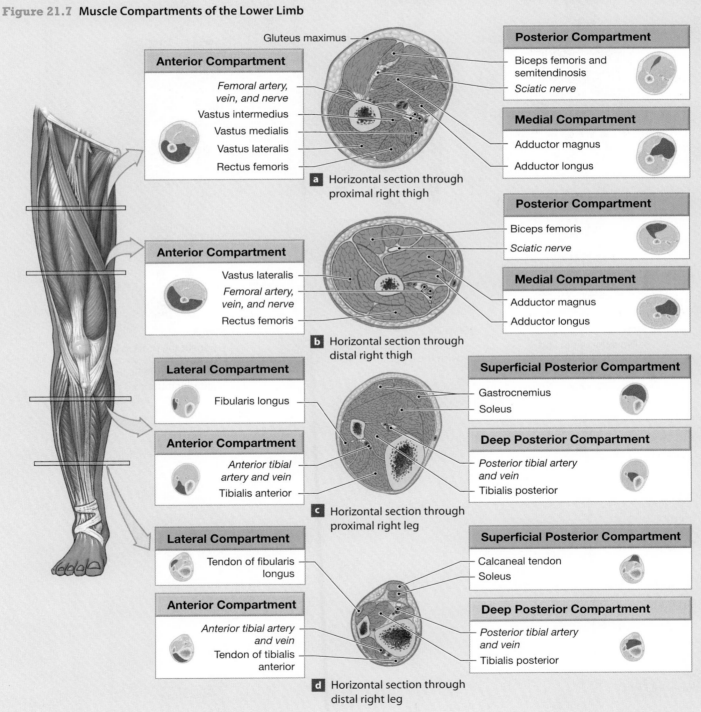

Gluteus maximus

Anterior Compartment

Femoral artery, vein, and nerve
Vastus intermedius
Vastus medialis
Vastus lateralis
Rectus femoris

Posterior Compartment

Biceps femoris and semitendinosis
Sciatic nerve

Medial Compartment

Adductor magnus
Adductor longus

a Horizontal section through proximal right thigh

Anterior Compartment

Vastus lateralis
Femoral artery, vein, and nerve
Rectus femoris

Posterior Compartment

Biceps femoris
Sciatic nerve

Medial Compartment

Adductor magnus
Adductor longus

b Horizontal section through distal right thigh

Lateral Compartment

Fibularis longus

Anterior Compartment

Anterior tibial artery and vein
Tibialis anterior

Superficial Posterior Compartment

Gastrocnemius
Soleus

Deep Posterior Compartment

Posterior tibial artery and vein
Tibialis posterior

c Horizontal section through proximal right leg

Lateral Compartment

Tendon of fibularis longus

Anterior Compartment

Anterior tibial artery and vein
Tendon of tibialis anterior

Superficial Posterior Compartment

Calcaneal tendon
Soleus

Deep Posterior Compartment

Posterior tibial artery and vein
Tibialis posterior

d Horizontal section through distal right leg

Sketch to Learn

Use the provided figure of the skeleton and add the vastus lateralis, vastus medialis, and tibialis anterior muscles. The rectus femoris muscle is shown as an example.

Sample Sketch

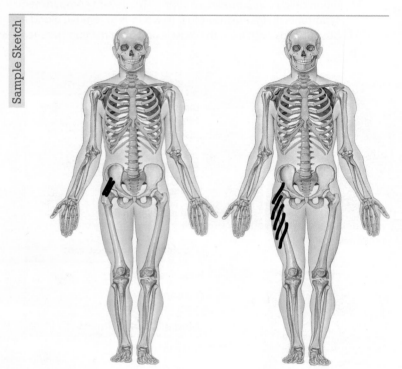

Your Sketch

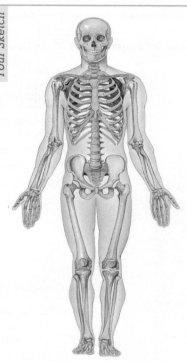

Step 1
• Draw the origin and insertion of the rectus femoris muscle.

Step 2
• Connect the two ends with the muscle's belly.

Step 3
• Repeat the process for the vastus lateralis, vastus medialis, and tibialis anterior muscles.

Name _____

Date _____

Section _____

Muscles of the Pelvic Girdle and Lower Limb

A. Descriptions

Describe the location of each of the following muscles.

1. sartorius

2. semitendinosus

3. psoas major

4. adductor magnus

5. gracilis

6. tensor fasciae latae

7. fibularis longus

8. vastus intermedius

B. Labeling

1. Label each muscle in **Figure 21.8**.

Figure 21.8 An Overview of the Major Anterior Skeletal Muscles

1. _____

2. _____

3. _____

4. _____

5. _____

6. _____

7. _____

8. _____

9. _____

10. _____

11. _____

12. _____

13. _____

14. _____

15. _____

16. _____

17. _____

18. _____

19. _____

20. _____

21. _____

22. _____

23. _____

24. _____

25. _____

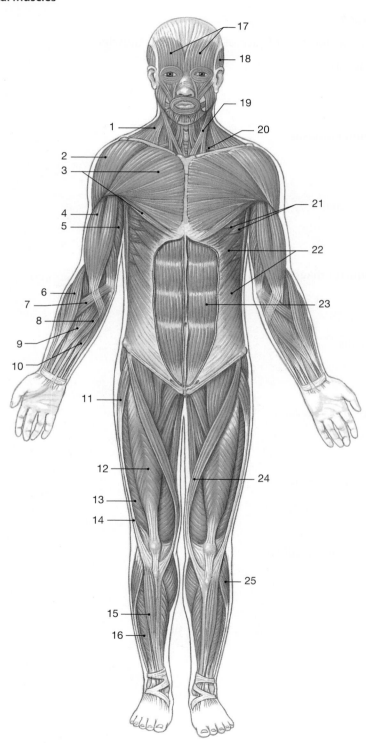

2. Label each muscle in **Figure 21.9**.

Figure 21.9 **An Overview of the Major Posterior Skeletal Muscles**

1. _____

2. _____

3. _____

4. _____

5. _____

6. _____

7. _____

8. _____

9. _____

10. _____

11. _____

12. _____

13. _____

14. _____

15. _____

16. _____

17. _____

18. _____

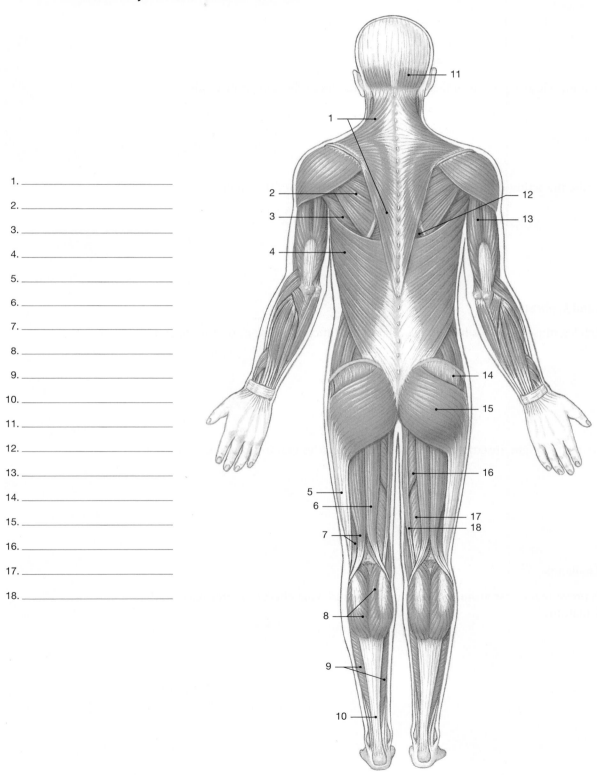

C. Short-Answer Questions

1. Describe how the hamstring muscle group moves the leg.

2. Which muscle group is the antagonist to the muscles of the hamstring group?

3. Describe the action of the abductor and adductor muscles of the thigh.

D. Analysis and Application

1. Which leg muscles serve a function similar to the function of the arm's rotator cuff muscles?

2. Describe the origin, insertion, and action of the muscles that invert and evert the foot.

E. Clinical Challenge

How can pressure increase around injured muscles, and what effect does this have on the regional anatomy?

Muscle Physiology

Learning Outcomes

On completion of this exercise, you should be able to:

1. Explain the differences among a twitch, wave summation, incomplete tetanus, and complete tetanus.

2. Describe how a muscle fatigues.

3. Explain the differences between isometric and isotonic contractions.

4. BIOPAC: Observe and record skeletal muscle tonus measured against a baseline activity level associated with the resting state.

5. BIOPAC: Observe and record how motor unit recruitment changes as the power of a skeletal muscle contraction increases.

6. BIOPAC: Record the force produced by clenched muscles, EMG, and integrated EMG when inducing fatigue.

M uscle and nerve tissues are excitable tissues that produce self-propagating electrical impulses called **action potentials.** These electrical impulses result from the movement of sodium and potassium ions through specific protein channels in the cell membrane. When a muscle fiber or a neuron is at rest, the net electrical charge inside the cell is different from the net charge outside the cell. This electrical difference is measured in millivolts (mV) and is called the **resting membrane potential.** Resting potential values differ from one type of cell to another. Typical values at the inner membrane surface are –70 mV for a neuron and –85 mV for skeletal muscles.

Lab Activities

Clinical Application

Need More Practice and Review?

Build your knowledge—and confidence!—in the Study Area of MasteringA&P® at www.masteringaandp.com with Pre-lab Quizzes, Post-lab Quizzes, Practice Anatomy Lab™ (PAL™) 3.0 virtual anatomy practice tool, PhysioEx™ 9.0 laboratory simulations, and A&P Flix™ with Quizzes.

PhysioEx™ 9.0 For this lab exercise, go to this topic:
• PhysioEx Exercise 2: Skeletal Muscle Physiology

A&PFlix For this lab exercise, go to these topics:
• Excitation-Contraction Coupling
• The Cross-Bridge Cycle

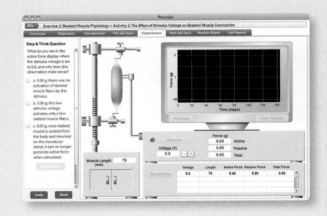

MasteringA&P®

When a neuron stimulates a muscle, the nerve action potential causes the neuron to release specific chemicals, collectively called **neurotransmitters,** that cause an action potential and thus contraction in the muscle fiber. Sodium channels open in the sarcolemma of the muscle fiber and sodium ions flood into the fiber, causing the sarcolemma to **depolarize,** a term used when the membrane becomes less negative. At the peak of depolarization, the sarcolemma is at +30 mV. At this millivoltage, the sodium channels close and potassium channels open. Potassium ions exit the fiber, and the fiber **repolarizes** to the resting potential. In summary, an electrical signal in the neuron causes release of a chemical signal, the neurotransmitter, which causes an electrical signal in the muscle fiber that results in contraction. In this exercise you will investigate a variety of muscle contractions.

Your laboratory may be equipped with a physiograph, an instrument that electrically stimulates muscles and records the characteristics of the contraction. Lab Activities 1 through 4 of this exercise provide the background physiology necessary to perform such investigations. Lab Activities 5 and 6 utilize the Biopac Student Lab physiograph to produce and interpret human electromyographs.

Lab Activity 1   Biochemical Nature of Muscle Contraction

In this activity preserved muscle tissue, prepared by a biological supply company, is used to demonstrate the biochemical nature of muscle contraction. The muscle tissue is glycerinated to denature the regulatory proteins of the muscle tissue so chemical interactions between thin and thick filaments can be observed. Although their roles are not fully understood, salt solutions of KCl and $MgCl_2$ are important in the utilization of ATP by muscle fibers. The head is an ATPase and hydrolyzes ATP to ADP plus released energy.

The experiment involves applying different combinations of ATP and salts to the glycerinated muscle fibers and observing muscle contraction.

Make a Prediction

What is the role of ATP in muscle contraction?

QuickCheck Question

1.1 What is a glycerinated muscle preparation?

In the Lab   1

Materials

- ☐ Muscle preparation (glycerinated muscle from biological supply company)
- ☐ Glycerol (supplied with muscle preparation)
- ☐ ATP solution (supplied with muscle preparation)
- ☐ ATP, KCl, and $MgCl_2$ solution (supplied with muscle preparation)
- ☐ Dissecting microscope
- ☐ Clean microscope slides
- ☐ Pipette or eye dropper
- ☐ Clean teasing needles
- ☐ Millimeter ruler

Procedures

1. Label three clean microscope slides A, B, and C.

2. Place a sample of the muscle under the dissecting microscope, and use a teasing needle to gently pry the fibers apart. Separate two or three fibers and transfer this group of fibers to slide A. Repeat this teasing process three more times, placing one group of fibers on slide B, and the last on slide C. Add a drop of glycerol to each slide to prevent dehydration of the fibers. Do not add coverslips to the slides.

3. **Slide A:** Examine the fibers with the dissecting microscope. Place a millimeter ruler under the slide, measure the length of the fibers, and record your measurement in **Table 22.1.** Add a drop of ATP solution to the fibers under the dissecting microscope and observe their response. After 30 to 45 seconds, measure the length of the fibers and record your measurement in Table 22.1.

4. **Slide B:** Repeat step 3 with slide B, this time adding the salt solution KCl and $MgCl_2$ instead of the ATP. Record your two measurements in Table 22.1.

5. **Slide C:** Repeat step 3 with slide C, this time using both the ATP solution and the salt solution. Record your two measurements in Table 22.1.

6. Dispose of the muscle preparations as indicated by your laboratory instructor. ■

Lab Activity 2   Types of Muscle Contraction

In the preceding section, muscle contraction was presented in terms of the events occurring inside a single muscle fiber. Muscle fibers do not act individually, however, because a single motor neuron controls multiple fibers. A motor neuron innervating the large muscles of the thigh, for example, stimulates

Table 22.1	Lengths of Resting and Contracting Muscle Fibers		
Microscope Slide	**Initial Length of Fibers**	**Substance Added to Slide**	**Contracted Length of Fibers**
Slide A	_____	_____	_____
Slide B	_____	_____	_____
Slide C	_____	_____	_____

more than 1000 fibers. Any group of fibers controlled by the same neuron is called a **motor unit** and can be considered a "muscle team" that contracts together when stimulated by the neuron. The muscle fibers are said to be "on" for contraction or "off" for relaxation, a concept called the **all-or-none principle.**

The type of contraction a muscle fiber undergoes is determined by the frequency at which the fiber is stimulated by its motor neuron. If a single action potential occurs in the neuron, only a small amount of ACh will be released and the muscle fiber will twitch. A **twitch** is a single stimulation-contraction-relaxation event in the fiber. **Figure 22.1** displays recordings of muscle contractions called **myograms.** The figure compares the

Figure 22.1 The Twitch and the Development of Tension

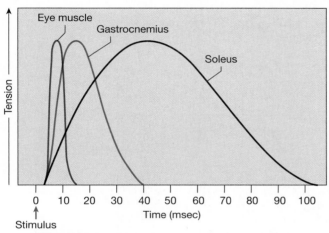

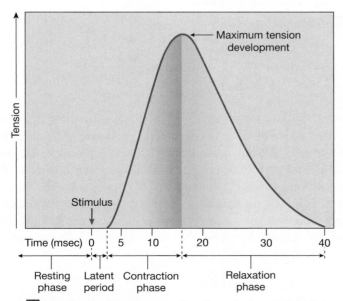

a This myogram shows the differences in tension over time for a twitch in different skeletal muscles.

b The details of tension over time for a single twitch in the gastrocnemius muscle. Notice the presence of a latent period, which corresponds to the time needed for the conduction of an action potential and the subsequent release of calcium ions by the sarcoplasmic reticulum.

twitching of different muscles. Each twitch has three sections: latent period, contraction phase, and relaxation phase.

The **latent period** is the time from the initial stimulation to the start of muscle contraction. During this brief period, the fiber is stimulated by ACh, releases calcium ions, exposes active sites, and attaches cross-bridges. No tension is produced during this period, and no pivoting occurs. The **contraction phase** involves shortening of the fiber and the production of muscle tension, or force. During this phase, myosin heads are pivoting and cycling through the attach-pivot-detach-return sequence. As more calcium ions enter the sarcoplasm, more cross-bridges are formed, and tension increases as the thick filaments pull the thin filaments toward the center of the sarcomere. The **relaxation phase** occurs as AChE inactivates ACh and calcium ions are returned to the sarcoplasmic reticulum. The thin filaments passively slide back to their resting positions, and muscle tension decreases.

The myograms in **Figure 22.2** show the effects of repeated stimulation. Figure 22.2a is a recording of **treppe** (TREP-eh), which is muscle contraction with complete relaxation between each stimulus. As the muscle is repeatedly stimulated, calcium ions accumulate in the cytosol, causing the first 30 to 50 contractions to increase in tension. The rest of Figure 22.2 illustrates the effect of increasing the frequency at which a skeletal muscle is stimulated. If the muscle is stimulated a second time before it has completely relaxed from a first stimulation, the two contractions are summed. This phenomenon, called **wave summation** (shown in Figure 22.2b), results in an increase in tension with each summation. Because the thin filaments have not returned to their resting length when the muscle is stimulated again, contractile force increases as more calcium ions are released into the sarcoplasm and more cross-bridges attach and pivot.

If the frequency of stimulation is increased further, the muscle produces peak tension with short cycles of relaxation. This type of contraction is called **incomplete tetanus** (Figure 22.2c). If the rate of stimulation is such that the relaxation phase is completely eliminated, the contraction type is **complete tetanus** (Figure 22.2d). During complete tetanus, peak tension is produced for a sustained period of time and results in a smooth, strong contraction. Most muscle work is accomplished by complete tetanus.

Because all muscle work requires complete tetanus, the type of contraction does not determine the overall tension a muscle produces. Muscle strength is varied through the number of motor units activated. The process called **recruitment** stimulates more motor units to carry or move a load placed on a muscle. As recruitment occurs, more muscle fibers are turned on and contract, and thus tension increases. Imagine holding a book in your outstretched hand. If another book is added to the load, additional motor units are recruited to increase the muscle tension to support the added weight.

Muscle fibers cannot contract indefinitely, and eventually they become fatigued. The force of contraction decreases as fibers

Figure 22.2 Effects of Repeated Stimulations

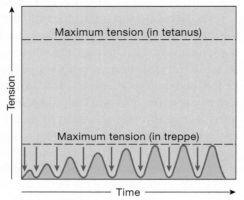

a Treppe. Treppe is an increase in peak tension with each successive stimulus delivered shortly after the completion of the relaxation phase of the preceding twitch.

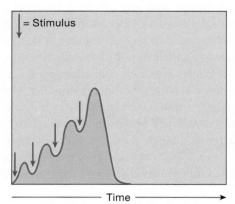

b Wave summation. Wave summation occurs when successive stimuli arrive before the relaxation phase has been completed.

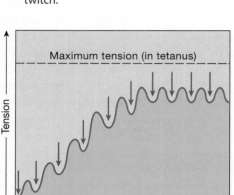

c Incomplete tetanus. Incomplete tetanus occurs if the stimulus frequency increases further. Tension production rises to a peak, and the periods of relaxation are very brief.

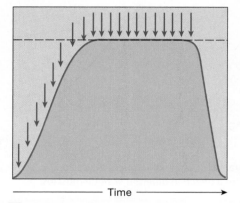

d Complete tetanus. During complete tetanus, the stimulus frequency is so high that the relaxation phase is eliminated; tension plateaus at maximal levels.

Clinical Application Tetanus

Surely you have had a tetanus shot. Why is the injection called a tetanus shot when tetanus is a type of muscle contraction? Often an injury introduces bacteria, *Clostridium tetani*, into the wound. The bacteria produce a toxin that binds to ACh receptors in skeletal muscles and stimulates the muscles to contract. The enzyme AChE cannot inactivate the bacterial toxin, and the muscle remains in a painful tetanic contraction for an extended period of time. Muscles for mastication are often affected by the toxin; hence the common name "lockjaw" for the symptom. As a preventative measure, a tetanus shot contains human tetanus immune globulins that prevent the *Clostridium* from surviving in the body and producing toxins. Actual treatment for the infection and toxin is usually ineffective in preventing tetanus. Is your tetanus booster shot current? ∎

lose the ability to maintain complete tetanic contractions. Fatigue is caused by a decrease in cellular energy and oxygen sources in the muscle and an accumulation of waste products. During intense muscle contraction, such as lifting a heavy object, the fibers become fatigued because of low ATP levels and a buildup of lactic acid, a by-product of anaerobic respiration. Joggers experience muscle fatigue as secondary energy reserves are depleted and damage accumulates in muscle fibers, especially in the sarcoplasmic reticulum and calcium-regulating mechanisms.

QuickCheck Questions

2.1 What are the stages of a muscle twitch?

2.2 What is the difference between incomplete tetanus and complete tetanus?

2.3 Why do skeletal muscles fatigue?

In the Lab 2

Materials

☐ Heavy object
☐ Stop watch

Procedures

1. Set the stop watch to zero, and record this time in Table 22.2 under Trial 1.

2. Extend your non-dominant arm (opposite side of writing hand) straight out in front of you, parallel to the floor, and load your arm by placing the heavy object in your hand. *Immediately* start the stop watch.

3. Hold the object with your arm straight for as long as possible. Once your arm starts to shake or your muscles ache, put the object down, and stop the stop watch. Record the end time for Trial 1 in Table 22.2.

4. Rest for one minute, and then repeat steps 1 through 3 as Trial 2.

5. Rest for another minute and then repeat steps 1 through 3 as Trial 3.

6. Calculate the total time in seconds required for each trial. Enter these values in the Duration column in Table 22.2.

7. Plot on the provided grid the total time until fatigue for each trial. Label the horizontal axis "Trial" and the vertical axis "Time in seconds."

8. Interpret your experimental data. Why is there a difference in the time to fatigue for each trial? ■

Table 22.2 Muscle Fatigue Demonstration

Trial	Start Time (seconds)	End Time (seconds)	Duration (seconds)
1	_____	_____	_____
2	_____	_____	_____
3	_____	_____	_____

Lab Activity 3   Isometric and Isotonic Contractions

Two major types of complete tetanic contractions occur: isometric and isotonic (**Figure 22.3**). **Isometric** (*iso-*, same + *metric*, length) **contractions** occur when the muscle length is relatively constant but muscle tension changes. Because length is constant, no body movement occurs. Muscles for maintaining posture use isometric contractions to support the body weight. **Isotonic** (*tonic*, tone) **contractions** involve constant tension while the length of the muscle changes. Consider picking up a book and then flexing your arm so that you move the book up to your shoulder. Once you are holding the book, your arm muscles are "loaded" with the weight of the book. As you flex your arm, muscle tension changes minimally while muscle length varies greatly.

Make a Prediction

What kind of muscle contraction occurs when you try to pick up something that is too heavy to lift?

QuickCheck Questions

3.1 Define and give an example of an isotonic contraction.

3.2 Define and give an example of an isometric contraction.

In the Lab 3

Materials

☐ Heavy object
☐ Ruler calibrated in millimeters

Procedures

1. Extend your arm straight out in front of you, parallel to the floor.

2. Palpate your extended biceps brachii muscle to feel the tension of the muscle. Also notice the length of the muscle. Record your observations in the Trial 1 column in Table 22.3.

3. Load your extended arm by placing the heavy object in your hand. Palpate your biceps brachii again, and notice the degree of muscle tension and the length of the muscle. Record your observations in the Trial 2 column in Table 22.3.

Figure 22.3 Isotonic and Isometric Contractions

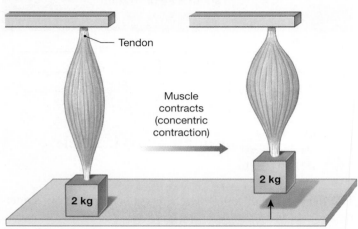

Tendon

Muscle contracts (concentric contraction)

2 kg

2 kg

a In this experiment, a muscle is attached to a weight less than its peak tension capabilities. On stimulation, it develops enough tension to lift the weight. Tension remains constant for the duration of the contraction, although the length of the muscle changes. This is an example of isotonic contraction.

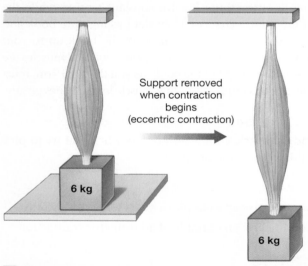

Support removed when contraction begins (eccentric contraction)

6 kg

6 kg

b In this eccentric contraction, the muscle elongates as it generates tension.

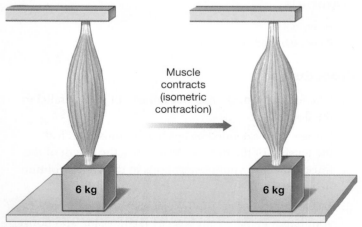

Muscle contracts (isometric contraction)

6 kg

6 kg

c The same muscle is attached to a weight that exceeds its peak tension capabilities. On stimulation, tension will rise to a peak, but the muscle as a whole cannot shorten. This is an isometric contraction.

Table 22.3	Isometric and Isotonic Contractions		
	Trial 1	**Trial 2**	**Trial 3**
Tension	_____	_____	_____
Length	_____	_____	_____
Type of contraction	_____	_____	_____

4. Gently squeeze the loaded biceps brachii and repeatedly flex and extend your arm six times, moving the heavy object 4 to 6 inches each time. When you are done with this motion, record your muscle tension and length observations in the Trial 3 column in Table 22.3.

5. Describe each type of muscle contraction to complete Table 22.3. ∎

Lab Activity 4 Muscles as Lever Systems

The property of contractility enables the body to move all its various parts, particularly where muscles attach to bones. The combination of muscles, bones, and joints forms a **lever system,** which is a way of increasing the amount of **work** a given force can do. If you have to lift, for example, a 100-pound object, you probably cannot exert enough force to do the lifting directly. However, if the object is sitting on one end of a playground seesaw, it can now be easily lifted by pulling down the seesaw. The seesaw provided *leverage*, and a seesaw, like a muscle, is a lever system.

A lever system consists of a rigid rod, the **lever** that moves around a pivot point called the **fulcrum (F).** The muscle effort required to move the lever is the applied force (AF). The **load (L),** also called the resistance, of the system is the weight of the lever or the weight of any object resting on the lever. In a muscle-bone lever system, the body weight is the load, the bone is the lever, the joint is the fulcrum, and the muscle supplies the applied force (**Figure 22.4**).

There are three principal classes of levers, determined by the relative positions of the resistance, the fulcrum, and the point where force is applied. Most lever systems in the human body are third class, where the force is applied between the fulcrum and the load. For example, the biceps moves the arm by applying a force between the load (the mass of the hand) and the elbow joint.

The position of the fulcrum along the lever arm can dramatically change the range of motion of a lever system and amount of work produced. For example, if

Figure 22.4 **The Three Classes of Levers**

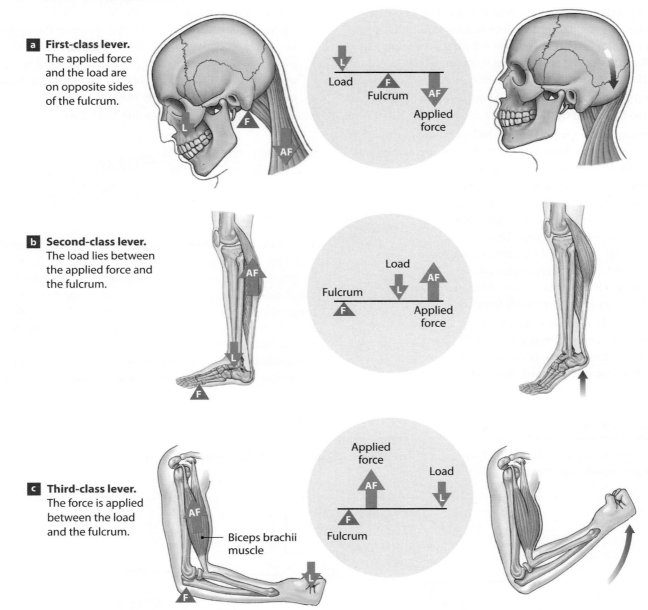

a **First-class lever.**
The applied force and the load are on opposite sides of the fulcrum.

b **Second-class lever.**
The load lies between the applied force and the fulcrum.

c **Third-class lever.**
The force is applied between the load and the fulcrum.

Biceps brachii muscle

the point where a muscle attaches to a bone is close to a joint, the muscle will be able to produce a great range of motion. However, this large range comes at a cost: a reduction loss in the work done by that muscle. Conversely, if another muscle of the same strength attaches to a bone farther from the joint, this second muscle can generate more work, but with a limited range of motion. This principle is called **leverage.**

QuickCheck Questions

4.1 What is a fulcrum?

4.2 How are muscles attached to bone to gain more leverage?

In the Lab 4

Materials

☐ Textbook
☐ Two pencils

Procedures

1. Review the organization of lever systems in Figure 22.4. You are going to construct a simple first-class lever system to demonstrate leverage.

2. Lay the book down on the top of a desk and place one pencil parallel to and right alongside the book spine; this pencil is the fulcrum of your lever system. Put the sharpened end of the second pencil under the spine, and rest the body of the second pencil on top of the fulcrum pencil; this second pencil is the lever.

3. Slide the fulcrum pencil away from the book until it is close to the eraser of the lever pencil.

4. Using care to prevent breaking the lever pencil, push down on it by applying a force at the eraser end. Notice how difficult it is to move the resistance (the book) with the fulcrum at the far end of the lever.

5. Slide the fulcrum pencil toward the spine of the book, to a point about midway along the lever pencil, and push down on the lever pencil as before. This time the book is easily raised. What you have demonstrated is the principle of leverage. When the fulcrum (the point of contact between the two pencils) is closer to the resistance (the book), it is easier for you to apply enough force to raise the book.

6. Next, place the pencils as they were in step 3, with the fulcrum far from the resistance, and then remove the book. Now press down on the eraser end of the lever pencil and note that, without the book, the tip moves easily.

7. Shift the fulcrum pencil toward the tip of the lever pencil. Now press down on the lever pencil. Again the tip moves easily, but this time the distance the tip moves is shorter. The force on the lever is used to move either a heavy weight over a short distance or a small weight over a greater distance. Strength is sacrificed for range of movement. Muscle systems work in a similar manner.

For additional practice, complete the *Sketch to Learn* activity. ■

Lab Activity 5 BIOPAC
Electromyography—Standard and Integrated EMG Activity

When a muscle or nerve is stimulated, it responds by producing action potentials. An action potential in a muscle fiber activates the physiological events of contraction. Because the electrical stimulation is also passively conducted to the body surface by surrounding tissues, sensors placed on the skin can detect the electrical activity produced by a muscle. The impulse produced by individual muscle fibers is minimal, but the combination of impulses from thousands of stimulated fibers produces a measurable electrical change in the overlying skin. The sensors,

 Sketch to Learn

To reinforce our understanding of levers, let's draw a first-class and a second-class lever. The first-class lever is shown in the sketch to get us started.

Sample Sketch

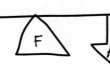

Step 1
• Draw a horizontal line for the lever.
• Add a triangle in the middle of the lever as the fulcrum.

Step 2
• Draw a load arrow pushing down on the lever.
• Draw an applied force arrow pulling down on the lever.

Your Sketch

⚠️ **Safety Alert:** Electrodes and Transducers

The Biopac Student Lab system is safe and easy to use, but be sure to follow the procedures as outlined in the laboratory activities. Under no circumstances should you deviate from the experimental procedures. Exercise extreme caution when using the electrodes and transducers with other equipment that also uses electrodes or transducers that may make electrical contact with you or your laboratory partner. Always assume that a current exists between any two electrodes or electrical contact points. ▲

which are electrodes, are connected to an amplifier that passes the signals to a recorder that then produces an **electromyogram (EMG),** a graph of the muscle's electrical activity. (You are probably familiar with electrical tracings of the heart's electrical activity done in an electrocardiogram, or ECG.)

In this laboratory activity, you will use the Biopac Student Lab system to detect, record, and analyze a series of muscle impulses. The system consists of three main components: sensors (both electrodes and transducers) that detect electrical impulses and other physiological phenomena; an acquisition unit, which collects and amplifies data from the sensors; and a computer with software to record and interpret the EMG data. After applying electrodes to the skin over your forearm muscles, you will clench your fist repeatedly with increasing force, and the Biopac system will produce an EMG of the muscle impulses. Each time you increase the force of a fist clench, the muscles recruit additional motor units to contract and produce more tension. This EMG laboratory activity is organized into four major sections. Section 1, Setup, describes where to plug in the electrode leads and how to apply the skin electrodes. Section 2, Calibration, adjusts the hardware so that it can collect accurate physiological data. Section 3, Data Recording, describes how to record the fist clench impulses once the hardware is calibrated. After you have saved the muscle data to a computer disk, Section 4, Data Analysis, instructs you how to use the software tools to interpret and evaluate the EMG.

QuickCheck Questions

5.1 What are electrodes?

5.2 What is the purpose of calibrating the BIOPAC hardware?

In the Lab 5

Materials

☐ BIOPAC acquisition unit (MP36/35/30)
☐ BIOPAC software: Biopac Student Lab (BSL) v3.7.5 or better
☐ BIOPAC electrode lead set (SS2L)
☐ BIOPAC disposable vinyl electrodes (EL503), 6 electrodes per subject
☐ BIOPAC headphones (OUT1)
☐ BIOPAC electrode gel (GEL1) and abrasive pad (ELPAD) or skin cleanser or alcohol prep
☐ Computer: PC Windows 7, Vista, or XP; Mac OS X 10.4-10.6

Procedures

Section 1: Setup

1. Turn on your computer but keep the BIOPAC MP36/35/30 unit off.

2. Plug the equipment in as shown in **Figure 22.5**: the electrode lead (SS2L) into CH 1 and the headphones (OUT1) into the back of the MP unit. Turn on the MP36/35/30 unit.

3. Attach six electrodes either to your own or your partner's forearms, three to each forearm as shown in **Figure 22.6**. The dominant forearm will be forearm 1, and the non-dominant forearm will be forearm 2. For optimal signal quality, you should place the electrodes on the skin at least five minutes before starting the calibration section.

4. Attach the electrode lead set (SS2L) to the electrodes on forearm 1. Make sure the electrode lead colors match those shown in Figure 22.6. Each pinch connector works like a small clothespin, but will latch onto the nipple of the electrode only from one side of the connector.

5. Start the Biopac Student Lab program on your computer and then choose lesson "L01-EMG-1." Follow the instructions on the screen to complete the setup process.

Section 2: Calibration

This series of steps establishes the hardware's internal parameters and is critical for optimum performance.

1. On the computer screen, click on Calibrate.

Figure 22.5 BIOPAC Cable Setup

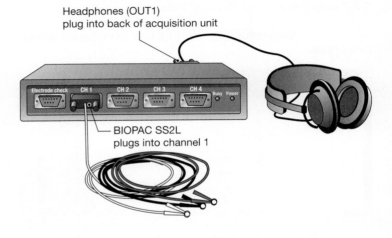

Headphones (OUT1) plug into back of acquisition unit

BIOPAC SS2L plugs into channel 1

Figure 22.6 Electrode Placement and Lead Attachment Carefully note the location of each electrode and the color of each attached lead. The fist is clenched for calibration procedure.

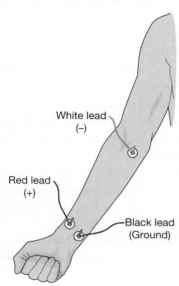

White lead
(–)

Red lead
(+)

Black lead
(Ground)

2. Clench your forearm 1 fist as hard as possible for two to three seconds and release. The calibration will last eight seconds and will stop automatically. (You do not need to keep your fist clenched for the whole eight seconds.)

3. Your computer screen should resemble **Figure 22.7**. Repeat Calibration steps 1 and 2 if your screen does not show a burst during the time the fist was clenched.

Section 3: Data Recording

You will record EMG activity data for two segments: segment 1 from forearm 1/dominant and segment 2 from forearm 2/ non-dominant. To work efficiently, read through the rest of this activity so that you will know what to do before recording. Screen prompts are similar in the recent version of the Biopac Student Lab (BSL) software. BSL 4 uses 'Continue' and 'Record' buttons to allow review/prep between segments; BSL 3.7.5– 3.7.7 uses 'Record' and 'Resume' buttons.

Figure 22.7 Calibration EMG Your calibration myogram should look similar.

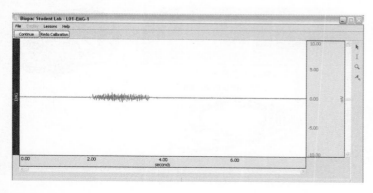

Figure 22.8 Selection of EMG Cluster for Analysis Data for the highlighted EMG are displayed in the small measurement boxes.

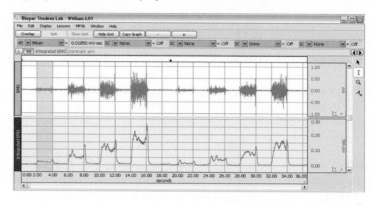

Segment 1: Forearm 1, Dominant

1. On your computer, click on Continue and when ready click on Record.

2. Clench your fist and hold for two seconds. Release the clench and wait two seconds. Repeat the clench-release-wait sequence while increasing the force in each sequence by equal increments so that the fourth clench uses maximum force.

3. On your computer, click on Suspend, and then review the recording on the screen. If your recording looks different than **Figure 22.8**, click on Redo and then click on Yes, and repeat.

4. Remove the electrode cable pinch connectors from the dominant arm.

Segment 2: Forearm 2, Non-dominant

5. Attach the electrode lead set (SS2L) pinch connectors to the electrodes on the non-dominant arm, again matching lead colors as shown in Figure 22.6.

6. On the computer, click on Continue/Record. A marker labeled "Non-dominant" will automatically be inserted when you do this.

7. Repeat four cycles of clench-release-wait, holding each clench for two seconds and waiting two seconds after release before beginning the next cycle. Increase the strength of your clench by the same amount for each cycle, with the fourth clench having the maximum force.

8. On your computer, click on Suspend, and then review the recording on the screen. If your recording looks different than Figure 22.8, click on Redo and then click on Yes, and repeat.

9. Click on Stop. Click YES to stop data recording or click NO to repeat the recording. If you want to listen to the EMG signal, go to step 10; to end go to step 12.

10. Listening to the EMG can be a valuable tool in detecting muscle abnormalities and is performed here for general interest. Put on the headphones, and click on Listen. The volume may be loud so position the headphones slightly off your ears as a precaution.

11. Experiment by changing the clench force during the clench-release-wait cycles as you watch the screen and listen. You will hear the EMG signal through the headphones as it is being displayed on the screen. The screen will display two channels: CH 1 EMG and CH 40 Integrated EMG. The data on the screen will not be saved. The signal will run until you press STOP.

12. Click on Done. To "Record from another subject," return to Section 1: Setup and attach electrodes and leads. Enter a new student filename and run the program again.

13. Remove the electrode cable pinch connectors. Peel the electrodes from both arms, and dispose of them. Use soap and water to wash the electrode gel residue from your skin. The electrodes may leave a slight ring on the skin for a few hours; this is quite normal.

Section 4: Data Analysis

1. Enter the Review Saved Data mode, and choose the correct file. Note Channel Number (CH) settings in the small menu boxes, shown in Figure 22.8 as the string of small squares across the top containing the numbers, left to right, 1, 1, 1, and 40.

Channel	Displays
CH 1	Raw EMG
CH 40	Integrated EMG

2. Set up your display window for optimal viewing of the first data segment (forearm 1, dominant). Figure 22.8 shows a sample display of this segment. The following tools help you adjust the data window.

Autoscale	Zoom Previous
Horizontal	Horizontal (Time) scroll bar
Autoscale	Vertical (Amplitude) scroll bar
Waveforms	Overlap button
Zoom tool	Split button

3. Set up the measurement boxes as follows:

Channel	Measurement
CH 40	Mean

The measurement boxes are above the marker region in the data window. Each measurement has three sections: channel number, measurement type, and

result. The first two sections are pull-down menus that are activated when you click on them. The following is a brief description of these measurements, where "selected area" is the area selected by the I-beam tool (including endpoints).

mean displays the average value in the selected area.

4. Using the I-beam cursor, select an area enclosing the first EMG cluster, as, for instance, the black area on the left in Figure 22.8. Record your data in **Table 22.4** in section A of the Electromyography—Standard and Integrated EMG Activity Review & Practice Sheet.

5. Using the I-beam cursor, measure the distance between the EMG bursts. Record your data in **Table 22.5** in Section A of the Electromyography—Standard and Integrated EMG Activity Review & Practice Sheet.

6. Repeat steps 4 and 5 on each successive EMG cluster.

7. Scroll to the second recording segment, which is for forearm 2 (non-dominant) and begins after the first marker, and repeat steps 4, 5, and 6 for the forearm 2 data.

8. Save or print the data file. You may save the data, save notes that are in the software journal, or print the data file. Exit the program.

9. Complete Electromyography—Standard and Integrated EMG Activity Review & Practice Sheet. ∎

Lab Activity 6 BIOPAC
Electromyography—Motor Unit Recruitment and Fatigue

In this activity, you will examine motor unit recruitment and skeletal muscle fatigue by combining electromyography with dynamometry (*dyno-*, power + *meter*, measure). **Dynamometry** is the measurement of power, and the graphic record derived from the use of a dynamometer is called a **dynagram.** In this activity, the contraction power of clenching muscles will be determined by a **hand dynamometer** equipped with an electronic transducer for recording. For more background information, you should review

! Safety Alert: Electrodes and Transducers

The Biopac Student Lab system is safe and easy to use, but be sure to follow the procedures as outlined in the laboratory activities. Under no circumstances should you deviate from the experimental procedures. Exercise extreme caution when using the electrodes and transducers with other equipment that also uses electrodes or transducers that may make electrical contact with you or your laboratory partner. Always assume that a current exists between any two electrodes or electrical contact points. ▲

the physiology of muscle stimulation, contraction, and recruitment in the first three laboratory activities of this exercise.

This recruitment and fatigue laboratory activity is organized into the same four major sections described earlier for the EMG laboratory activity: Setup, Calibration, Data Recording, and Data Analysis.

QuickCheck Questions

6.1 What does a dynamometer measure?

6.2 What is motor unit recruitment?

In the Lab | 6

Materials

- ☐ BIOPAC acquisition unit (MP36/35/30)
- ☐ Computer: PC Windows 7, Vista, or XP; Mac OS X 10.4-10.6
- ☐ BIOPAC electrode lead set (SS2L)
- ☐ BIOPAC disposable vinyl electrodes (EL503), 6 electrodes per subject
- ☐ BIOPAC headphones (OUT1)
- ☐ BIOPAC hand dynamometer (SS25L)
- ☐ BIOPAC electrode gel (GEL1) and abrasive pad (ELPAD) or skin cleanser or alcohol prep
- ☐ BIOPAC software: Biopac Student Lab (BSL) v3.7.5 or better

Procedures

Section 1: Setup

1. Turn on your computer.
2. Make sure the BIOPAC MP36/35/30 unit is off.
3. Plug the equipment in as shown in **Figure 22.9**: the electrode lead (SS2L) into CH 1, the dynamometer (SS25L) into CH 2, and the headphones (OUT1) into the back of the MP unit.
4. Turn on the BIOPAC MP36/35/30 unit.
5. Attach six electrodes either to your own forearms or your partner's forearms, three on each forearm as shown in **Figure 22.10**. The dominant forearm will be forearm 1, and the non-dominant forearm will be forearm 2. For optimal signal quality, you should place the electrodes on the skin at least five minutes before starting the Calibration section.
6. Attach the electrode lead set (SS2L) to the electrodes on forearm 1. Make sure the electrode lead colors match those shown in Figure 22.10. (Each pinch connector works like a small clothespin, but will latch onto the nipple of the electrode only from one side of the connector.)

Figure 22.9 Equipment Setup Carefully note where each piece is plugged into the main unit.

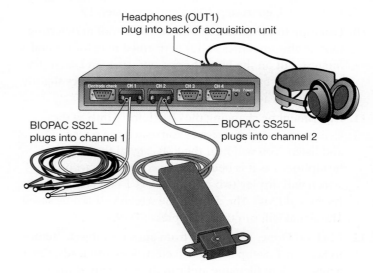

Figure 22.10 Holding the Hand Dynamometer Carefully note the location of each electrode and the color of each attached lead. The fist is clenched for calibration procedure. Also note the position of the dynagrip sensor relative to the hand.

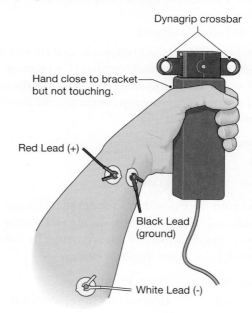

7. Start the Biopac Student Lab program on your computer. Choose lesson "L02-EMG-2," and click OK and type in a filename, using a unique identifier such as your or your partner's nickname or student ID number.
8. Click OK to end the Setup section.

Section 2: Calibration

This series of steps establishes the hardware's internal parameters and is critical for optimum performance.

1. Set the dynamometer down, and on the computer screen, click on Calibrate. To get an accurate calibration, there must be no force on the dynamometer transducer. Follow the instructions in the Calibrate dialog box, and click OK when ready.

2. Grasp the dynamometer with the hand of forearm 1 and click OK. If using SS25L, see Figure 22.10 for proper grasp—place hand as close to the crossbar as possible without touching the crossbar. If using SS25LA, place the short grip bar against the palm, toward the thumb, and wrap fingers to center the force.

   Important: The dynamometer should be in the same position for all measurements from each arm. Note the hand grasp position used here and try to replicate it exactly in all subsequent steps in this activity.

3. Wait two seconds. Then clench the hand dynamometer as hard as possible for about four seconds and release. Wait for the calibration to stop.

4. Your computer screen should resemble Figure 22.11. Repeat Calibration steps 1 through 3 if your screen does not show a burst on each channel while the dynamometer was clenched.

Section 3: Data Recording

You will record data for four segments: forearm 1 motor unit recruitment, forearm 1 fatigue, forearm 2 motor unit recruitment, and forearm 2 fatigue. To work efficiently, read through the rest of this activity so that you will know what to do before recording. Screen prompts are similar in the recent version of the Biopac Student Lab (BSL) software. BSL 4 uses 'Continue' and 'Record' buttons to allow review/prep between segments; BSL 3.7.5–3.7.7 uses 'Record' and 'Resume' buttons.

Segment 1: Forearm 1 Motor Unit Recruitment

1. You will complete a series of clenches, and should try to increase clench force by one grid line per clench.

Figure 22.11 Calibration Recording Your calibration myogram should look similar.

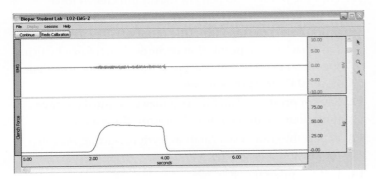

BSL 3.7.5–3.7.7 users will see an Assigned Increment Level in the journal (the BSL software calculates this level during your grip force calibration) and should increase clench force by the assigned increment for each cycle until maximum clench force is obtained. For example, if your Assigned Increment Level is 5 kg, start at a force of 5 kg and repeat cycles at forces of 10 kg and 15 kg.

Force Calibration	Assigned Increment Level
0–25 kg	5 kg
25–50 kg	10 kg
> 50 kg	20 kg

2. Click on Continue and when ready click on Record.

3. Clench your fist and hold for two seconds. Release the clench and wait two seconds. Repeat 3 times, with increasing force.

4. On your computer, click on Suspend, and then review the recording of the screen. If it looks different than Figure 22.12, click on REDO and repeat step 3.

Segment 2: Forearm 1 Fatigue

5. Click on Continue and when ready click on Record. A marker labeled "Dominant arm: Continued clench at maximum force" will automatically be inserted when you do this.

6. Clench the dynamometer with your maximum force. Note this force and try to maintain it. When the maximum clench force displayed on the screen has decreased by more than 50 percent, click on Suspend and review the data on the screen. Repeat recording if necessary.

7. Remove the electrode cable pinch connectors from the dominant forearm.

Segment 3: Forearm 2 Motor Unit Recruitment

8. Attach the electrode lead set (SS2L) pinch connectors to the electrodes on the non-dominant forearm, again matching lead colors as shown in Figure 22.10.

Figure 22.12 Motor Unit Recruitment Note the increase in tension as recruitment occurs.

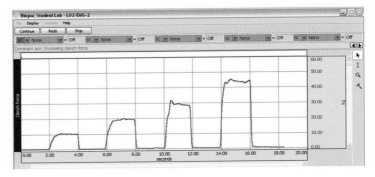

9. Click Continue/Record and repeat all the steps of Segment 1 (Forearm 1 Motor Unit Recruitment).

Segment 4: Forearm 2 Fatigue

Repeat all the steps of Segment 2 (Forearm 1 Fatigue).

10. Click on Stop. Click YES to stop data recording or click NO to repeat the recording. If you want to listen to the EMG signal, go to step 11; to end go to step 12.

11. Listening to the EMG can be a valuable tool in detecting muscle abnormalities and is performed here for general interest. Put on the headphones, and click on Listen. The volume may be loud so position the headphones slightly off your ears as a precaution. Experiment by changing the clench force during the clench-release-wait cycles as you watch the screen and listen. You will hear the EMG signal through the headphones as it is being displayed on the screen. The screen will display two channels: CH 1 EMG and CH 41 Clench Force. The data on the screen will not be saved. The signal will run until you press STOP and end the listening-to-EMG. To listen again, or to have another person listen, click Redo.

12. Click Done. A pop-up window will appear. Make your choice, and continue as directed. If choosing the "Record from another subject" option return to the Setup section and correctly attach electrodes and connect leads. Setup a new student filename and run the program again. Remove the electrode cable pinch connectors. Peel the electrodes from both forearms and dispose of them. Use soap and water to wash the electrode gel residue from your skin after the experiment is done. The electrodes may leave a slight ring on the skin for a few hours; this is quite normal.

Section 4: Data Analysis

1. Enter the Review Saved Data mode, and choose the correct file.

Channel	Displays
CH 1	EMG
CH 40	Integrated EMG
CH 41	Clench Force

2. Note your force increment in the Data Report (estimate from grid values or as noted in journal).

3. Set up your display window for optimal viewing of the first data segment (Dominant arm: increasing clench force). The following tools help you adjust the data window.

Autoscale horizontal	Horizontal (Time) scroll bar
Autoscale waveforms	Vertical (Amplitude) scroll bar
Zoom tool	Zoom previous

4. Set up the measurement boxes as follows:

Channel	Measurement
CH 41	Mean
CH 40	Mean
CH 41	Value
CH 40	Delta T

The measurement boxes are above the marker region in the data window. Each measurement has three sections: channel number, measurement type, and result. The first two sections are pull-down menus that are activated when you click on them. The following is a brief description of these measurements, where "selected area" is the area selected by the I-beam tool (including endpoints).

mean displays the average value in the selected area.

Value displays the amplitude for the channel at the point selected by the cursor. If a single point is selected, the value is for that point; if an area is selected, the value is the endpoint of the selected area.

Delta T displays the amount of time in the selected segment (difference in time between the endpoints of the selected area).

5. Using the I-beam cursor, select an area on the plateau phase of the first clench. Record your data in **Table 22.6** in section A of the Electromyography—Motor Unit Recruitment and Fatigue Review & Practice Sheet.

6. Repeat step 5 on the plateau phase of each successive clench. Record your data in **Table 22.7** in section A of the Electromyography—Motor Unit Recruitment and Fatigue Review & Practice Sheet.

7. Scroll to the second recording segment. This begins after the first marker and represents the continuous maximum clench.

8. Using the I-beam cursor, select a point of maximal clench force immediately following the start of segment 2. Record your data in section A of the Electromyography—Motor Unit Recruitment and Fatigue Review & Practice Sheet.

9. Calculate 50 percent of the maximum clench force from step 8.

10. Find the point of 50 percent maximum clench force by using the I-beam cursor, and leave the cursor at this point. Select the area from the point of 50 percent clench force back to the point of maximum clench force by using the I-beam cursor and dragging. Note the time to fatigue (CH 40 Delta T) measurement.

11. Save or print the data file. You may save the data, save notes that are in the journal, or print the data file.

12. Repeat the entire Data Analysis section, starting with step 1, for forearm 2 (segments 3 and 4).

13. Exit the program. ∎

Name _____

Date _____

Section _____

Muscle Physiology

A. Matching

Match each term in the left column with its correct description from the right column.

_____ **1.** isometric contraction
_____ **2.** complete tetanus
_____ **3.** repolarization
_____ **4.** all-or-none principle
_____ **5.** fatigue
_____ **6.** isotonic contraction
_____ **7.** depolarization
_____ **8.** treppe
_____ **9.** latent period
_____ **10.** wave summation
_____ **11.** incomplete tetanus
_____ **12.** twitch
_____ **13.** recruitment
_____ **14.** acetylcholine

A. fusion of twitches
B. tension changes more than length
C. increase in twitch tension
D. shift in transmembrane potential toward 0 mV
E. contraction with no relaxation cycles
F. response to single stimulus
G. length changes more than tension
H. muscle fibers either "on" or "off"
I. contraction with rapid relaxation cycles
J. neurotransmitter
K. time prior to tension
L. reduction in contraction and performance
M. return of membrane to resting potential
N. activating more motor units

B. Short-Answer Questions

Describe each of the following types of muscle contraction.

a. wave summation

b. treppe

c. complete tetanus

d. twitch

e. incomplete tetanus

C. Drawing

Draw It! Draw and label a twitch myogram and a myogram showing wave summation.

Twitch myogram   Ware summation

D. Analysis and Application

1. Describe the events taking place at the neuromuscular junction during muscle stimulation.

2. Distinguish between isometric and isotonic contractions.

3. Explain why muscles become fatigued.

4. Describe the lever system of the calf muscle attached at the heel of the foot. Where are the fulcrum and lever located? What is the load and what applies the force?

E. Clinical Challenge

While remodeling his house, Mike accidentally touches a live electrical wire and receives an electrical shock. How does this shock stimulate his muscles to contract?

Name _____

Date _____

Section _____

BIOPAC:
Electromyography—Standard
and Integrated EMG Activity

A. Data and Calculations

Subject Profile

Name _____   Height _____

Age _____   Weight _____

Gender _____

1. **EMG Measurements:** Complete **Tables 22.4** and **22.5** by using the data obtained during the EMG I experiment. Refer to the computer journal for recorded data.

Table 22.4	EMG Measurements	
	Forearm 1 (Dominant arm)	Forearm 2 (Non-dominant arm)
Cluster Number	[CH 40] Mean	[CH 40] Mean
1		
2		
3		
4		

Table 22.5	Tonus Measurements	
Between Cluster Numbers	Forearm 1 (Dominant arm) [CH 40] Mean	Forearm 2 (Non-dominant arm) [CH 40] Mean
1-2		
2-3		
3-4		

Note: "Clusters" are the EMG bursts associated with each clench.

2. Use the mean measurements from Tables 22.4 and 22.5 to compute the percentage increase in EMG activity recorded between the weakest clench and the strongest clench of forearm 1.

Calculation: Answer: _____%

B. Analysis and Application

1. Does there appear to be any difference in tonus between the two forearm clench muscles? Would you expect to see a difference? Does the subject's gender influence your expectations? Explain.

2. Compare the mean measurement for the right and left maximum clench EMG cluster. Are they the same or different? Which one suggests the greater clench strength? Explain.

3. What factors in addition to gender contribute to observed differences in clench strength?

4. Explain the source of signals detected by the EMG electrodes.

5. What does the term *motor unit recruitment* mean?

6. Define electromyography.

Name _____

Date _____

Section _____

BIOPAC:
Electromyography—Motor
Unit Recruitment and Fatigue

A. Data and Calculations

Subject Profile

Name _____ Height _____

Age _____ Weight _____

Gender _____

Dominant forearm (right or left) _____

1. Complete **Table 22.6** using data from segments 1 and 3. In the "Assigned Increment Level (kg)" column, note the force increment assigned for your recording under peak 1; the increment was pasted into the software journal and should be transferred to Table 22.6 from step 2 of the Data Analysis section of Laboratory Activity 5. For subsequent peaks, add the increment (that is, 5-10-15 kg or 10-20-30 kg). You may not need nine peaks to reach max.

Table 22.6	Data on Motor Unit Recruitment from Segments 1 and 3				
		Forearm 1 (Dominant)		**Forearm 2**	
Peak Number	**Assigned Increment Level (kg)**	**Force at Peak [CH 41] Mean (kg)**	**Int. EMG [CH 40] Mean (mV)**	**Force at peak [CH 41] Mean (kg)**	**Int. EMG [CH 40] Mean (mV)**
1	kg				
2	kg				
3	kg				
4	kg				
5	kg				
6	kg				
7	kg				
8	kg				
9	kg				

2. Complete **Table 22.7** using data from segment 2 and 4.

Table 22.7	Data on Fatigue from Segments 2 and 4				
Forearm 1 (Dominant)				**Forearm 2**	
Maximum Clench Force	**50% of Max Clench Force**	**Time to Fatigue**	**Maximum Clench Force**	**50% Max Clench Force**	**Time to Fatigue**
CH 41 value	calculate	CH 40 Delta T	CH 41 value	calculate	CH 40 Delta T

B. Analysis and Application

1. Is the strength of your right arm different from the strength of your left arm?

2. Is there a difference in the absolute values of force generated by males and females in your class? What might explain any differences?

3. When you are holding an object, does the number of motor units in use remain the same? Are the same motor units used for as long as you hold the object?

4. As you fatigue, the force exerted by your muscles decreases. What physiological processes explain this decline in strength?

Organization of the Nervous System

Learning Outcomes

On completion of this exercise, you should be able to:

1. Outline the organization of the nervous system.

2. List six types of glial cells and describe a basic function of each type.

3. Describe the cellular anatomy of a neuron.

4. Discuss how a neuron communicates with other cells.

5. Describe the organization and distribution of spinal nerves.

6. BIOPAC: Describe how learning a task influences reaction time.

7. BIOPAC: Compare reaction times for fixed-interval and pseudorandom presentations of a stimulus.

8. BIOPAC: Calculate group mean, variance, and standard deviation values for a data set.

The nervous system orchestrates body functions to maintain homeostasis. To accomplish this control, the nervous system must perform three vital tasks. First, it must detect changes in and around the body. For this task, sensory receptors monitor environmental conditions and encode information about environmental changes as electrical impulses. Second, it must process incoming sensory information and generate an appropriate motor response to adjust the activity of muscles and glands. Third, it must orchestrate and integrate all sensory and motor activities so that homeostasis is maintained.

The nervous system is divided into two main components (Figure 23.1): the **central nervous system (CNS),** which consists of the brain and spinal cord, and

Need More Practice and Review?

Build your knowledge—and confidence!—in the Study Area of MasteringA&P® at www.masteringaandp.com with Pre-lab Quizzes, Post-lab Quizzes, Practice Anatomy Lab™ (PAL™) 3.0 virtual anatomy practice tool, PhysioEx™ 9.0 laboratory simulations, and A&P Flix™ with Quizzes.

PAL | practice anatomy lab   For this lab exercise, follow these navigation paths:
  • PAL>Histology>Nervous Tissue

PhysioEx™ 9.0 For this lab exercise, go to this topic:
  • PhysioEx Exercise 3: Neurophysiology of Nerve Impulses

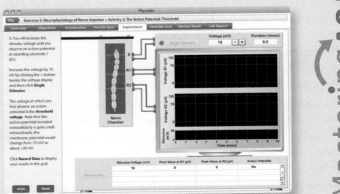

MasteringA&P®

Figure 23.1 **An Overview of the Nervous System** Note the relationship between the central and peripheral nervous systems and the function and components of the afferent and efferent divisions of the latter.

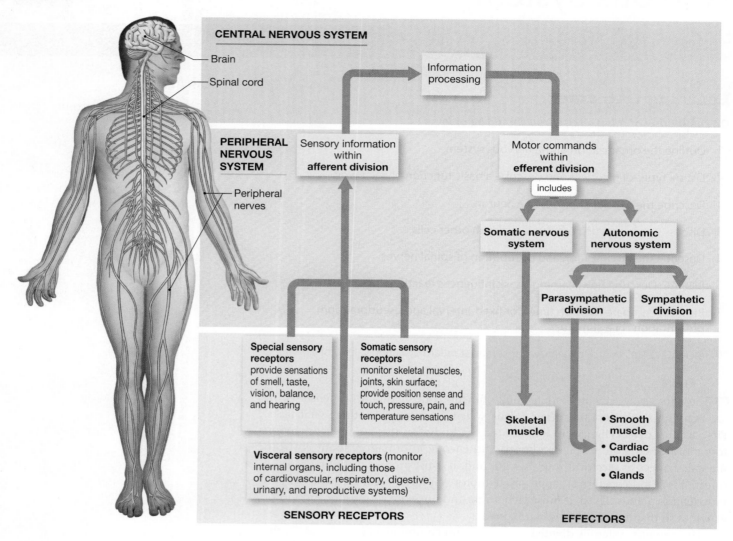

the **peripheral nervous system (PNS),** which communicates with the CNS by way of cranial and spinal nerves, collectively called *peripheral nerves.* A **nerve** is a bundle of neurons plus any associated blood vessels and connective tissue. The PNS is responsible for providing the CNS with information concerning changes both inside the body and in the surrounding environment. Sensory information is sent along PNS nerves that join the CNS in either the spinal cord or the brain. The CNS evaluates the sensory data and determines whether muscle and gland activities should be modified in response to the changes. Motor commands from the CNS are then relayed to PNS nerves that carry the commands to specific muscles and glands.

The PNS is divided into afferent and efferent divisions. The **afferent division** receives sensory information from **sensory receptors,** which are the cells and organs that detect changes in the body and the surrounding environment, and then sends that information to the CNS for interpretation. The CNS decides the appropriate response to the sensory information and sends motor commands to the PNS **efferent division,** which controls the activities of **effectors,** the general term for all the muscles and glands of the body. **Somatic effectors** are skeletal muscles, and **visceral effectors** are cardiac muscle, smooth muscle, and glands.

The efferent division is divided into two parts. One part, the **somatic nervous system,** conducts motor responses to skeletal muscles. The other part, the **autonomic nervous system,** consists of the **sympathetic** and **parasympathetic divisions,** both of which send commands to smooth muscles, cardiac muscles, and glands.

The nerves of the PNS are divided into two groups according to which part of the CNS they communicate with. **Cranial nerves** communicate with the brain and pass into the face and neck through foramina in the skull. **Spinal nerves** join the spinal cord at intervertebral foramina and pass either into the upper and lower limbs or into the body wall. There are 12 pairs of cranial nerves and 31 pairs of spinal nerves, and each pair transmits specific information between the CNS and the PNS. Functionally, all spinal nerves are **mixed nerves,** which means they carry both sensory signals and motor signals. Cranial nerves are either entirely sensory or mixed. Although a cranial or spinal nerve may transmit both sensory and motor impulses, a single neuron within the nerve transmits only one type of signal.

Lab Activity 1   Histology of the Nervous System

Two types of cells populate the nervous system: glial cells and neurons (discussed in Exercise 10). Glial cells have a supportive role in protecting and maintaining nerve tissue. Neurons are the communication cells of the nervous system and are capable of propagating and transmitting electrical impulses to respond to the ever-changing needs of the body.

Glial Cells

Glial cells, which collectively make up a network called the **neuroglia** (noo-RŌG-lē-uh), are the most abundant cells in the nervous system. They protect, support, and anchor neurons in place. In the CNS, glial cells are involved in the production and circulation of the cerebrospinal fluid that circulates in the ventricles of the brain and in the central canal of the spinal cord. In both the CNS and the PNS, glial cells isolate and support neurons with myelin.

The CNS has four types of glial cells (Figure 23.2). The glial cells known as **oligodendrocytes** (o-li-gō-DEN-drō-sīts) wrap around axons of neurons in the CNS and form a fatty **myelin sheath. Microglia** (mī-KROG-lē-uh) are phagocytic glial cells that remove microbes and cellular debris from CNS tissue. **Ependymal** (ep-EN-dī-mul) cells line the ventricles of the brain and the central canal of the spinal cord; these glial cells contribute to the production of the cerebrospinal fluid. **Astrocytes** (AS-trō-sīts), shown in Figure 23.3, hold neurons in place and

Figure 23.2 The Classification of Glial Cells The categories and functions of the various glial cell types in the CNS and the PNS.

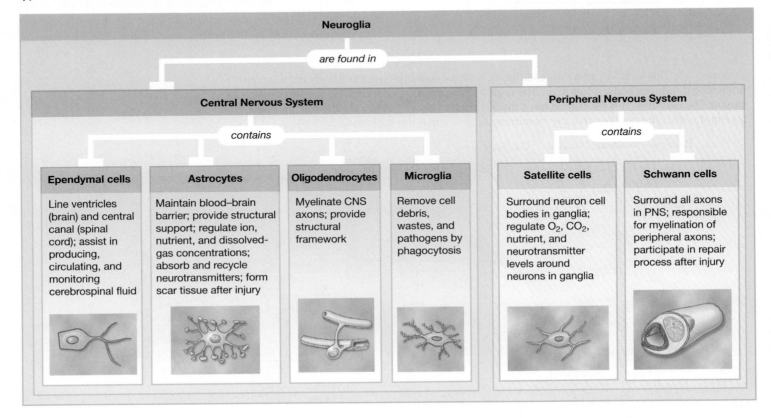

	Neuroglia	
	are found in	

Central Nervous System				Peripheral Nervous System	
	contains				contains
Ependymal cells	**Astrocytes**	**Oligodendrocytes**	**Microglia**	**Satellite cells**	**Schwann cells**
Line ventricles (brain) and central canal (spinal cord); assist in producing, circulating, and monitoring cerebrospinal fluid	Maintain blood–brain barrier; provide structural support; regulate ion, nutrient, and dissolved-gas concentrations; absorb and recycle neurotransmitters; form scar tissue after injury	Myelinate CNS axons; provide structural framework	Remove cell debris, wastes, and pathogens by phagocytosis	Surround neuron cell bodies in ganglia; regulate O_2, CO_2, nutrient, and neurotransmitter levels around neurons in ganglia	Surround all axons in PNS; responsible for myelination of peripheral axons; participate in repair process after injury

isolate one neuron from another. They also wrap footlike extensions around blood vessels, creating a blood-brain barrier that prevents certain materials from passing out of the blood and into nerve tissue.

Figure 23.3 Astrocytes Micrograph of astrocytes showing the many cellular extensions of this type of glial cell.

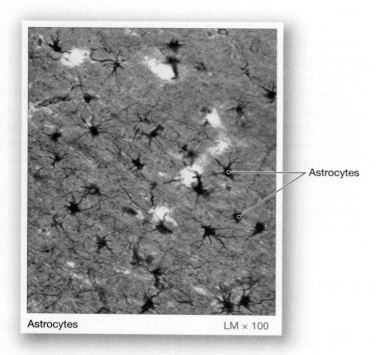

Astrocytes   LM × 100

The PNS has two types of glial cell. Where neuron cell bodies cluster in groups called *ganglia*, the glial cells called **satellite cells** encase each cell body and isolate it from the interstitial fluid to regulate the neuron's chemical environment. **Schwann cells** surround and myelinate PNS axons in spinal and cranial nerves.

Neurons

A neuron has three distinguishable features: dendrites, a cell body (soma), and an axon (**Figures 23.4** and **23.5**). The numerous dendrites carry information into the large, rounded cell body, which contains the nucleus and organelles of the cell. The **perikaryon** (per-i-KAR-ē-on), which is the entire area of the cell body surrounding the nucleus, contains such organelles as mitochondria, free ribosomes, and fixed ribosomes. Also found in the perikaryon are **Nissl bodies,** which are groups of free ribosomes and rough endoplasmic reticulum. Nissl bodies account for the dark regions that are clearly visible in a sagittal section of the brain.

The first part of the axon of a neuron, called the **initial segment,** extends from a narrow part of the cell body referred to as the **axon hillock.** The axon may divide into several **collateral branches** that subdivide into smaller branches called **telodendria** (tel-ō-DEN-drē-uh). At the distal tip of each telodendrion of the neuron is a **synaptic terminal** (also called *synaptic knob* or *end bulb*) that houses **synaptic vesicles** full of **neurotransmitter** molecules. These molecules are released by the neuron and are the means by which it communicates with another cell, either another neuron

Figure 23.4 The Anatomy of a Representative Neuron A neuron has a cell body (soma), some branching dendrites, and a single axon. The region of the cytoplasm around the nucleus is the perikaryon. The neuron in this illustration has a myelin sheath covering the axon. The outer layer of the myelin sheath is the neurilemma, the membrane of the axon is the axolemma.

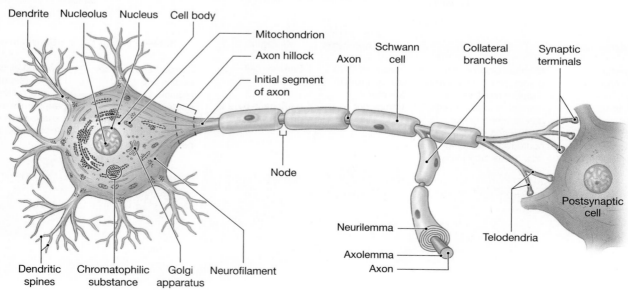

Figure 23.5 A Representative Motor Neuron Micrograph of a neuron having multiple dendrites and a single axon.

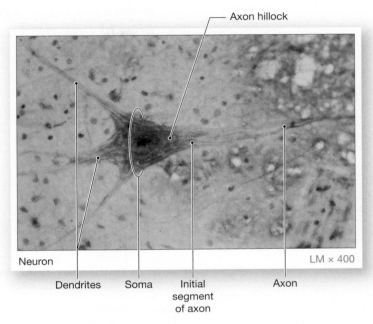

Axon hillock

Neuron

LM × 400

Dendrites   Soma   Initial segment of axon   Axon

or a muscle or gland effector cell. The synaptic terminal is the transmitting part of the **synapse,** which is the general term for the neural communication site. At any given synapse, the neuron-releasing neurotransmitter is called the *presynaptic neuron*. If this neuron communicates with another neuron, the latter is called the *postsynaptic neuron*. If the presynaptic neuron communicates with a muscle or gland effector cell, that cell is called the *postsynaptic cell*. A small gap called the **synaptic cleft** separates the presynaptic neuron from the postsynaptic neuron or postsynaptic cell.

As described previously, axons are myelinated by glial cells. In the PNS, a Schwann cell wraps around and encases a small section of axon in multiple layers of the Schwann cell's membrane. Any region of an axon covered in this membrane is called a **myelinated internode.** Between the internodes are gaps in the sheath, called **nodes** (also called *nodes of Ranvier*) as shown in **Figure 23.6**. The membrane of the axon, the **axolemma,** is exposed at the nodes, and this exposure permits a nerve impulse to arc rapidly from node to node. The **neurilemma** (noo-ri-LEM-uh), or outer layer of the Schwann cell, covers the axolemma at the myelinated internodes.

Any regions of the PNS and CNS containing large numbers of myelinated neurons are called *white matter* because of the white color of the myelin. Regions containing mostly unmyelinated neurons are called *gray matter* because without any myelin present, gray is the predominant color due

Figure 23.6 Myelinated Neuron Micrograph of a myelinated neuron stained to show the myelin sheath and nodes.

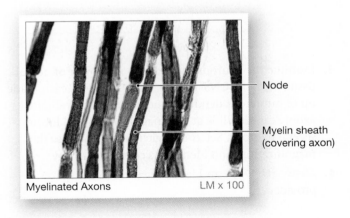

Node

Myelin sheath (covering axon)

Myelinated Axons

LM × 100

to the dark color of the neuron's organelles. White and gray matter are clearly visible in sections of the brain and spinal cord.

QuickCheck Questions

1.1 What are the two major types of cells in the nervous system?

1.2 What are the three main regions of a neuron?

1.3 What is a node on an axon?

In the Lab 1

Materials

☐ Compound microscope

☐ Prepared slides of
- Astrocytes
- Neurons
- Myelinated nerve tissue (teased)

Procedures

1. Scan the astrocytes slide at low magnification and locate a group of glial cells. Examine a single astrocyte at high magnification and note the numerous cellular extensions.

2. *Draw It!* Draw an astrocyte in the space provided.

Astrocyte

3. Examine the neurons slide, which is a smear of neural tissue from the CNS and has many neurons, each made up of numerous dendrites and a single unmyelinated axon. Scan the slide at low magnification and locate several neurons. Select a single neuron, increase the magnification, and identify its cellular anatomy.

4. *Draw It!* Draw and label a neuron in the space provided.

Neuron

5. The myelinated nerve slide is a preparation from a nerve that has been teased apart to separate the individual

myelinated axons. Use Figure 23.6 as a reference and examine the slide at each magnification; identify the myelin sheath and the nodes.

6. *Draw It!* Draw and label a sketch of your observations in the space provided. ∎

Myelinated axon

Lab Activity 2 Anatomy of a Nerve

Cranial and spinal nerves are protected and organized by three layers of connective tissue in much the same way a skeletal muscle is organized (**Figure 23.7**). The nerve is wrapped in an outer covering called the **epineurium.** Beneath this layer is the **perineurium,** which separates the axons into bundles called *fascicles*. Inside a fascicle, the **endoneurium** surrounds each axon and isolates it from neighboring axons.

QuickCheck Questions

2.1 Name the three connective tissue layers that organize a nerve.

2.2 Describe how these connective tissue layers are arranged in a typical spinal nerve.

In the Lab 2

Materials

☐ Spinal cord laboratory model

☐ Spinal cord chart

☐ Compound microscope

☐ Prepared slide of spinal nerve

Procedures

1. Examine the spinal nerve slide at low magnification and locate the nerve section. Identify the epineurium and note how it encases the nerve.

2. Examine a single fascicle and distinguish between the perineurium and the epineurium. Locate the individual axons inside a fascicle.

Figure 23.7 Anatomy of a Spinal Nerve A spinal nerve consists of an outer epineurium enclosing a variable number of fascicles (bundles of neurons). The fascicles are wrapped by the perineurium, and within each fascicle the individual axons are encased by the endoneurium. Schwann cells encompass the axons and create a myelin sheath over them.

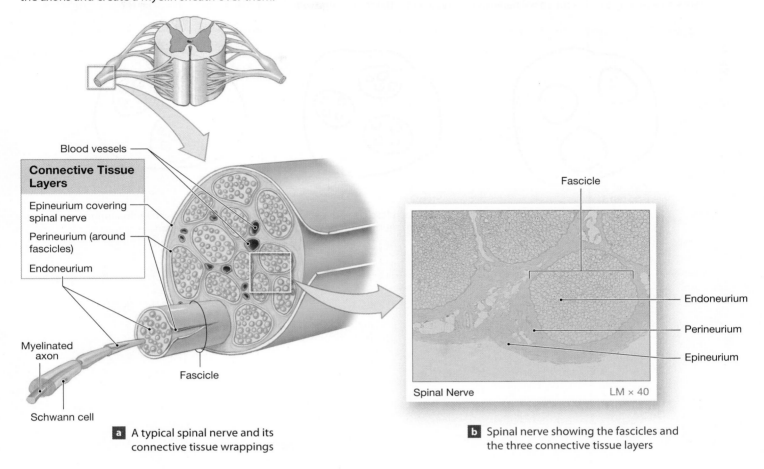

Connective Tissue Layers

Blood vessels

Epineurium covering spinal nerve

Perineurium (around fascicles)

Endoneurium

Myelinated axon

Schwann cell

Fascicle

a A typical spinal nerve and its connective tissue wrappings

Fascicle

Endoneurium

Perineurium

Epineurium

Spinal Nerve   LM × 40

b Spinal nerve showing the fascicles and the three connective tissue layers

3. *Draw It!* Draw and label the nerve in the space provided.

Spinal nerve

For additional practice, complete the ***Sketch to Learn*** activity (on p. 320) ■

Lab Activity 3   **BIOPAC**
Reaction Time

The beginning of a race is a classic example of a **stimulus-response** situation, where people hear a *stimulus* (the starter's pistol) and react to it in some way (*response*). There are two key factors in stimulus-response: **reaction time** and **learning.** *Reaction time* is the delay between when the stimulus is presented and when you do something about it. *Learning* is the acquisition of knowledge or skills as a result of experience and/or instruction.

The delay between hearing the signal and responding is a function of the length of time needed for the afferent signal

Sketch to Learn

Here's an easy way to draw the spinal nerve as observed under the microscope.

Sample Sketch

  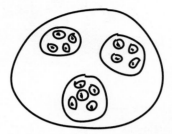

Step 1
- Draw an oval to represent the spinal nerve.
- Add 3–4 smaller ovals inside for fascicles.

Step 2
- Fill each fascicle with ovals to represent the endoneurium.

Step 3
- Add a dot to each endoneurium to show the axon.
- Label your sketch.

Your Sketch

to reach the brain and for the brain to send an efferent signal to the muscles. With learning, the time for the various steps in the process can be shortened. As people learn what to expect, reaction time typically decreases. Reaction time varies from person to person and from situation to situation, and most people have delayed reaction times late at night and early in the morning. Longer reaction times are also a sign that people are paying less attention to the stimulus and/or are processing information.

This lesson shows how easily and quickly people learn, as demonstrated by their ability to anticipate when to press a button. The lesson uses a relatively simple variation of stimuli (pseudorandom versus fixed-interval) to determine what results in the shortest reaction times. In the **pseudorandom stimuli** segments, the computer generates a click once every one to ten seconds. For the **fixed-interval stimuli** segments, the computer generates a click every four seconds.

With the pseudorandom presentation, the subject cannot predict when the next click will occur, and the result is that both learning and decrease in reaction time are minimal. When fixed-interval trials are performed repeatedly, average reaction time typically decreases each time new data are recorded, up to a point. Eventually, the minimal reaction time required to process information is reached, and then reaction time becomes constant.

This lesson takes a relatively simple look at reaction time and demonstrates how changing one small aspect of procedure can result in differences in reaction times. You will probably notice a difference in average reaction times between the pseudorandom and fixed-interval presentation trials, and this difference will most likely favor the segments with fixed-interval presentation stimuli. Part of the difference is probably due to the random versus nonrandom presentation of the stimuli. However, you might also notice that reaction time decreases when you change from pseudorandom presentation to fixed-interval presentation, suggesting that maybe you just got better with practice.

QuickCheck Questions

3.1 What is a stimulus-response situation?

3.2 What is reaction time?

In the Lab 3

Materials

- ☐ BIOPAC acquisition unit (MP36/35/30)
- ☐ BIOPAC software: Biopac Student Lab (BSL) v3.7.5 or better
- ☐ BIOPAC hand switch (SS10L)
- ☐ BIOPAC headphones (OUT1)
- ☐ Computer: PC Windows 7, Vista, or XP; Mac OS X 10.4–10.6

Procedures

The reaction time investigation is divided into four sections: setup, calibration, data recording, and data analysis. Read each section completely before attempting a recording. If you encounter a problem or need further explanation of a concept, ask your instructor.

Data collected in the recording segments must be recorded in the laboratory. You may record the data by hand or choose Edit > Journal > Paste Measurements to paste the data into your electronic journal for future reference.

While you record a segment, markers are inserted for each response with the hand switch. In this exercise, all of the markers and labels are inserted automatically. Markers appear at the top of the computer windows as inverted triangles.

Section 1: Setup

1. Turn on your computer but keep the BIOPAC acquisition unit off.
2. Plug the equipment in as shown in **Figure 23.8**: hand switch (SS10L) in CH 1 of the acquisition unit and headphones (OUT1) in back of the unit. Turn the acquisition unit on.
3. Start the Biopac Student Lab program. Choose lesson "L11-React-1," and type in a unique filename when prompted.

Section 2: Calibration

1. To prepare for the calibration recording, the subject should be seated and relaxed, with headphones on and eyes closed. Hold the hand switch with your dominant hand, so that the thumb is ready to press the button.

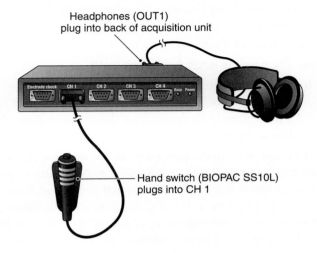

Figure 23.8 Equipment Setup Setup of the BIOPAC hardware for the reaction time lab activity.

Headphones (OUT1) plug into back of acquisition unit

Hand switch (BIOPAC SS10L) plugs into CH 1

Note: When the Calibrate button is clicked in the next step, system feedback may cause the volume through the headphones to be very loud. Use "Lesson Preferences" under the file menu to adjust headphone volume, or position the headphones slightly off the ears to reduce the sound.

2. Click on Calibrate. Before the calibration begins, a pop-up window may appear, reminding you to press the button when you hear a click. Click OK to begin the calibration recording.
3. Press the hand switch when you hear a click, approximately four seconds into the recording. Briefly depress the button, then release it. Do not hold the button down and do not press it more than once.
4. Wait for the calibration to end. The calibration will run for eight seconds and then stop automatically.
5. Review the data on the screen. If your screen is similar to **Figure 23.9**, then proceed to the Data Recording

Figure 23.9 Sample Calibration Data Repeat the calibration procedures if your calibration graph is not similar to this graph.

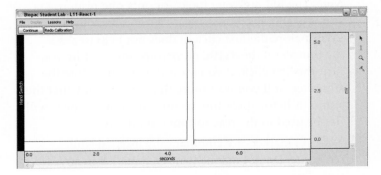

section. If your calibration screen does not resemble Figure 23.9, then repeat the calibration to obtain a similar screen. Click Redo Calibration and repeat the calibration procedure.

Two reasons for incorrect data are:

a. The baseline is not 0 millivolt (mV).

b. The data are excessively noisy, meaning more than approximately 5 mV peak-to-peak. Your data may be a little more or less noisy than the example shown in Figure 23.9.

If the Calibrate button reappears in the window, check the connections and repeat the calibration, making sure to press the button firmly but briefly. If no signal is detected from the hand switch (flat line at 0 mV), the program will automatically return you to the beginning of the calibration procedure. If this happens, check the connections to the hand switch and make sure you are pressing the button firmly. Click Redo Calibration and repeat the calibration.

Section 3: Data Recording

Prepare for the recording. You will record four segments, each requiring you to press a button (response) as soon as possible after hearing a click (stimulus).

a. Segments 1 and 3 present the stimuli at pseudorandom intervals every one to ten seconds.

b. Segments 2 and 4 present the stimuli at fixed intervals four seconds apart.

To work efficiently, read this entire section before beginning the recording step. Screen prompts are similar in the recent versions of the Biopac Student Lab (BSL) software. BSL 4 uses 'Continue' and 'Record' buttons to allow review/prep between segments; BSL 3.7.5–3.7.7 uses 'Record' and 'Resume' buttons.

From this point on, two persons are involved: a subject and a director. The subject should be seated and relaxed, with headphones on and eyes closed. The subject should hold the hand switch with the dominant hand, so that the thumb is ready to press the button. The director watches the screen and presses the Record and Resume buttons as required.

Note: The BSL software looks for only one response per stimulus. Because the software ignores responses that occur before the first click, it does not help to press the button on the SS10L numerous times before you hear the first click. If you press the button before the stimulus, or if you wait more than one second after the stimulus before pressing the button, your response will not be used in the reaction time summary.

Note: All markers are automatically inserted while recording. Do not manually insert a marker in any recording segment of this lesson.

Segment 1: Pseudorandom Dominant

1. Click on Continue and when ready click on Record. Once the director clicks on Record, a pseudorandom presentation trial will begin, with a click produced randomly every one to ten seconds. The recording will suspend automatically after 10 clicks.

2. As soon as the subject hears a click through the headphones, he or she should press and release the hand switch button. A marker will automatically be inserted each time a click is generated, and an upward-pointing "pulse" will be displayed on the screen each time the button is pressed.

3. Review the data on the screen. After 10 clicks, the resulting graph should resemble **Figure 23.10**. If the subject pressed the button correctly, a pulse will be displayed after each marker. If the screen is not similar to Figure 23.10 then click Redo, otherwise continue to Segment 2. Three probable reasons for incorrect data are:

    a. The recording did not capture a pulse for each click.

    b. The pulse occurs before the marker, indicating that the subject responded prematurely.

    c. The duration of the pulse extends into the next marker, indicating that the subject held the button down too long.

    Note: The subject is allowed to miss some responses, but if more than two are missed, the recording should be redone.

Segment 2: Fixed-Interval Dominant

In this segment and in segment 4, the subject will respond to a stimulus sounded every four seconds. Data recording will continue from the point it last stopped and a marker labeled "fixed-interval" will automatically be inserted when recording.

4. Click on Continue and when ready click on Record. The subject should press and release the hand switch button

Figure 23.10   Representation of Response Pulses to the Click Stimuli Each spike on the graph represents a push of the hand switch button.

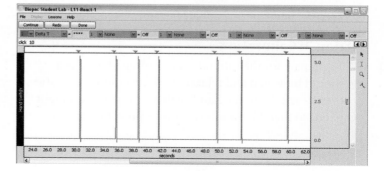

at the sound of each click. The recording will suspend automatically after 10 clicks.

5. Review the data on the screen. After 10 clicks, the resulting graph on your screen should resemble Figure 23.10. If it does, go to Segment 4; if it doesn't then click Redo and repeat the recording.

Segment 3: Pseudorandom—Non-dominant

In this segment, the subject will respond to a pseudorandom presentation trial for the nondominant hand. The recording will continue from the point it last stopped, and a marker labeled "pseudorandom, non-dominant" will be inserted when Record is pressed.

6. Click on Continue and when ready click on Record. The subject should press and release the hand switch button at the sound of each click. The recording will suspend automatically after 10 clicks.

7. Review the data on the screen. After 10 clicks, the resulting graph on your screen should resemble Figure 23.10. If it does, go to Segment 3; if it doesn't, click Redo and repeat the recording. Data could be incorrect for the reasons listed in step 3.

Segment 4: Fixed-Interval—Non-dominant

In this segment the subject will respond to a fixed-interval stimulus with the non-dominant hand. The recording will continue from the point it last stopped, and a marker labeled "fixed-interval, non-dominant" will automatically be inserted.

8. Click on Continue and when ready click on Record. The subject should press and release the hand switch button at the sound of each click. The recording will suspend automatically after 10 clicks.

9. Review the data on the screen. After 10 clicks, the resulting graph on your screen should resemble Figure 23.10. If it does, go to step 10; if it doesn't then click Redo and repeat the recording.

10. The director clicks Done. A pop-up window with options will appear. Make a choice and continue as directed. If choosing the "Record from another Subject" option, remember that each subject will need to use a unique filename.

Section 4: Data Analysis

To compare the reaction times from the two types of presentation schedules, you can summarize the results as statistics, or measures of a population. Certain statistics are usually reported for the results of a study: mean, range, variance, and standard deviation. Mean is a measure of a central tendency. Range, variance, and standard deviation are measures of distribution—in other words, the "spread" of data. Using mean and distribution, investigators can compare the performance

of groups. You will calculate your group statistics, but you will not do formal comparisons between groups.

The **mean** is the average of the sum of the reaction times divided by the number of subjects (n).

The **range** is the highest score minus the lowest score. Because range is affected by extremely high and low reaction times, investigators also describe the *spread*, or distribution, of reaction times with two related statistics: variance and standard deviation.

Variance is the average squared deviation of each number from its mean.

Standard deviation is the square root of the variance.

1. Enter the Review Saved Data mode from the Lessons menu. After the Done button was pressed in the previous section, the program automatically took all 10 reaction times and calculated average reaction times for each trial and placed them in the journal. Use this journal information to fill in your data report.

2. Set up your display window for optimal viewing of the first marker and pulse of the first segment. **Figure 23.11** shows the "selected area" of the first stimulus-response marker. Use the horizontal scroll bar to adjust the width of the waveforms and the vertical scroll bar to adjust their height. You may also find the Zoom tool useful for examining a segment of a graph with the X and Y axes expanded.

3. The measurement boxes are above the marker region in the data window. Set the boxes up as follows:

Channel	Measurement
CH1	Delta T
CH1	None
CH1	None
CH1	None

Figure 23.11 Highlighting a Stimulus-Response Pass Analysis of response time to click stimulus.

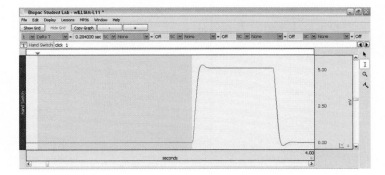

The Delta T (read as "delta time") measurement is the difference between the time at the end of the area selected by the I-beam tool and the time at the beginning of that area, including endpoints. The "none" measurement turns the measurement channel off.

You can record this and all other measurement data by hand or choose Edit > Journal > Paste Measurements to paste the data to your journal for future reference.

4. Select an area from the first marker to the leading edge of the first pulse (Figure 23.11), and note the Delta T measurement. The marker indicates the start of the stimulus click. The leading edge of the pulse (the point where the pulse first reaches its peak) indicates when the button was first pressed. The threshold the program uses to calculate reaction time is 1.5 mV. The reaction time for the event shown is 0.265 second (see Delta T).

5. Look at the first reaction time result in your journal, and compare this with the Delta T measurement you found

above. The two measurements should be approximately the same. Repeat the comparison on other pulses until you are convinced that your journal readings are accurate. You can move around using the marker tools.

6. Transfer your data from the journal to **Table 23.1** in section A of the Reaction Time Review & Practice Sheet at the end of this exercise. This step may not be necessary if your instructor allows you to print out your journal and include it with the review sheet.

7. Collect data from at least four other students in your class as needed to complete **Tables 23.2, 23.3, 23.4,** and **23.5** of the Reaction Time Review & Practice Sheet.

8. Either save the data or journal notes to a computer drive, or print the data file. Exit the program. ∎

Review & Practice Sheet

EXERCISE 23

Name ______________________________

Date ______________________________

Section ______________________________

Organization of the Nervous System

A. Description

Write a description for each of the following structures.

1. dendrite

2. collateral branches

3. synaptic terminal

4. axon hillock

5. telodendria

6. myelinated internode

7. neurilemma

8. synaptic vesicles

9. axolemma

10. cell body

B. Labeling

1. Label **Figure 23.12**.

Figure 23.12 **Myelinated Axon**

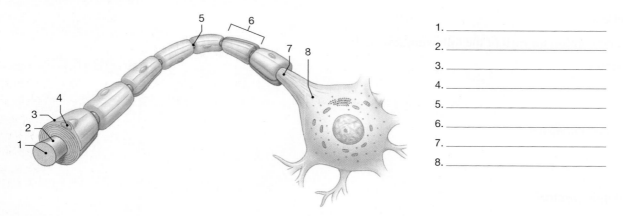

1. _____
2. _____
3. _____
4. _____
5. _____
6. _____
7. _____
8. _____

2. Label **Figure 23.13**.

Figure 23.13 **Nervous System Overview**

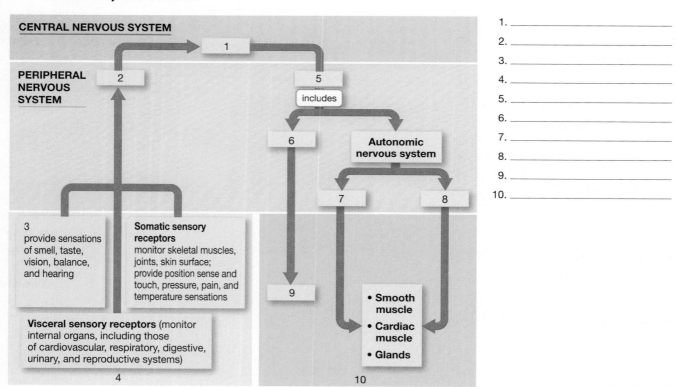

1. _____
2. _____
3. _____
4. _____
5. _____
6. _____
7. _____
8. _____
9. _____
10. _____

C. Short-Answer Questions

1. Compare the CNS and the PNS.

2. Describe the microscopic appearance of an astrocyte.

3. Molecules of what substances are stored in synaptic terminals?

4. Name two functions of Schwann cells.

5. List six types of glial cells, and indicate which are found in the CNS and which are found in the PNS.

D. Analysis and Application

1. Which type of glial cell in the CNS is found in the white matter of the brain and spinal cord?

2. While observing a microscopic specimen of nerve tissue from the brain, you notice an axon encased by a different cell. Describe the covering over the axon and identify the cell that has surrounded the axon.

E. **Clinical Challenge**

How would an injury to the afferent neurons in the left leg affect the victim's sensory and motor functions?

Name _____

Date _____

Section _____

BIOPAC:
Reaction Time

A. Data and Calculations

Subject Profile

Name _____   Height _____

Age _____   Weight _____

Gender _____

1. Manual Calculation of Reaction Time

   Transfer your data from the journal to **Table 23.1**. Calculate the reaction time for the first click in segment 1:

   Delta T = time at end of selected area – time at beginning of selected area = _____ seconds

2. Summary of Subject's Results (from software journal)

Table 23.1	Reaction Data			
Stimulus Number	*Pseudorandom*		*Fixed-Interval*	
	Segment 1 (Dominant)	**Segment 3 (Non-dominant)**	**Segment 2 (Dominant)**	**Segment 4 (Non-dominant)**
1.				
2.				
3.				
4.				
5.				
6.				
7.				
8.				
9.				
10.				
Mean				

3. Comparison of Reaction Time and Number of Presentations
Complete **Table 23.2** with data from the first fixed-interval trial (data segment 2), and calculate the mean for each presentation to determine if reaction times vary as a subject progresses through the series of stimulus events.

Table 23.2	Comparison of Reaction Times					
	Pseudorandom Trial Data (Segment 1, Dominant)			Fixed-Interval Trial Data (Segment 2, Dominant)		
Student's Name	Stimulus 1	Stimulus 5	Stimulus 10	Stimulus 1	Stimulus 5	Stimulus 10
1.						
2.						
3.						
4.						
5.						
Mean						

4. Group Summary

Complete **Table 23.3** with the means for five students, and then calculate the group mean.

Table 23.3	Reaction Time Means			
	Pseudorandom Trials		Fixed-Interval Trials	
Class Data Student Means	Dominant	Non-dominant	Dominant	Non-dominant
1.				
2.				
3.				
4.				
5.				
Group mean				

5. Variance and Standard Deviation

$$\text{Variance} = \frac{1}{n-1}\sum_{j=1}^{n}(x_j - x)^2$$

$$\text{Standard Deviation} = \sqrt{\text{variance}}$$

where

n = number of students

x_j = mean reaction time for each student

$\bar{x}$ = group mean (constant for all students)

$\sum_{j=1}^{n}$ = sum of all student data

Calculate the variance and standard deviation for five students with data from Segment 3: Pseudorandom Non-dominant (**Table 23.4**) and from Segment 4: Fixed-Interval Non-dominant (**Table 23.5**).

Table 23.4	BIOPAC Segment 3: Pseudorandom Non-dominant Data			
Student	Enter Mean Reaction Time for Student (x_j)	Enter Group Mean ($\bar{x}$)	Calculate Deviation ($\bar{x}_j - \bar{x}$)	Calculate Deviation^2 ($\bar{x}_j - \bar{x}$)
1.				
2.				
3.				
4.				
5.				

Sum the data for all students in the Deviation^2 column $= \displaystyle\sum_{j=1}^{n} (x_j - \bar{x})^2$ $=$

Variance (σ^2) $=$ Multiply by 0.25 $= \dfrac{1}{n-1}$ $=$

Standard Deviation $=$ Square root of variance $= \sqrt{\text{variance}}$ $=$

Table 23.5	BIOPAC Segment 4: Fixed-Interval Non-dominant Data			
Student	Enter Mean Reaction Time for Student ($\bar{x}$)	Enter Group Mean ($\bar{x}$)	Calculate Deviation ($\bar{x}_j - \bar{x}$)	Calculate Deviation^2 ($\bar{x}_j - \bar{x}$)
1.				
2.				
3.				
4.				
5.				

Sum the data for all students in the Deviation^2 column $= \displaystyle\sum_{j=1}^{n} (x_j - \bar{x})^2$ $=$

Variance (σ^2) $=$ Multiply by 0.25 $= \dfrac{1}{n-1}$ $=$

Standard Deviation $=$ Square root of variance $= \sqrt{\text{variance}}$ $=$

B. Short-Answer Questions

1. Describe the changes in mean reaction time between the first and tenth stimuli presentation.

   Segment 1:_____

   _____

   Segment 3: _____

   _____

   Which segment showed the greater change in mean reaction time, segment 1 or segment 3?

2. From Tables 23.2 and 23.3, estimate the minimum reaction time at which reaction time becomes constant: _____ seconds. What physiological processes occur between the time a stimulus is presented and the time the button is pressed?

   _____

3. From Table 23.2, which presentation schedule had the lower group mean, the pseudorandom schedule or the fixed-interval schedule? _____

4. From Tables 23.4 and 23.5, which presentation schedules seem to have less variation (lower variance and lower standard deviation), the pseudorandom schedules or the fixed-interval schedules?

   _____ Pseudorandom _____ Fixed-interval

5. Based on what you see in Tables 23.4 and 23.5, state a plausible relationship between the difficulty of a task and the reaction time statistics for the task: mean, variance, and standard deviation.

   _____

   _____

6. What differences would you expect between the effect of learning a task on your reaction time when you perform the task with your dominant hand and the effect of learning on reaction time when you perform the task with your non-dominant hand?

   _____

   _____

   _____

Spinal Cord, Spinal Nerves, and Reflexes

Learning Outcomes

On completion of this exercise, you should be able to:

1. Identify the major surface features of the spinal cord, including the spinal meninges.

2. Identify the sectional anatomy of the spinal cord.

3. Describe the organization and distribution of spinal and peripheral nerves.

4. List the events of a typical reflex arc.

5. Describe how to perform and interpret the stretch reflex, and the biceps and triceps reflexes.

The **spinal cord** is the long, cylindrical portion of the central nervous system located in the spinal cavity of the vertebral column. It connects the peripheral nervous system (PNS) with the brain. Sensory information from the PNS enters the spinal cord and ascends to the brain. Motor signals from the brain descend the spinal cord and exit the spinal cord to reach the effectors. The spinal cord is more than just a conduit to and from the brain, however. It also processes information and produces **spinal reflexes.** A classic example of a spinal reflex is the stretch reflex that occurs when the tendon over the patella is struck; the spinal cord responds to the tap by stimulating the extensor muscles of the leg in the well-known "knee-jerk" reflex.

Need More Practice and Review?

Build your knowledge—and confidence!—in the Study Area of MasteringA&P® at www.masteringaandp.com with Pre-lab Quizzes, Post-lab Quizzes, Practice Anatomy Lab™ (PAL™) 3.0 virtual anatomy practice tool, PhysioEx™ 9.0 laboratory simulations, and A&P Flix™ with Quizzes.

PAL | practice anatomy lab | For this lab exercise, follow these navigation paths:
- PAL>Human Cadaver>Nervous System
- PAL> Anatomical Models>Nervous System
- PAL>Histology>Nervous Tissue

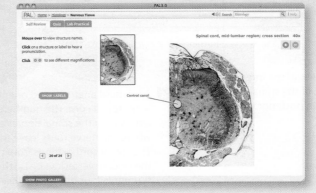

MasteringA&P®

Lab Activity 1 Gross Anatomy of the Spinal Cord

The spinal cord is continuous with the inferior portion of the brain stem (**Figure 24.1**). It passes through the foramen magnum, descends approximately 45 cm (18 in.) down the spinal canal of the vertebral column, and terminates between lumbar vertebrae L_1 and L_2. In young children, the spinal cord extends through most of the spine. After the age of four, the spinal cord stops lengthening, but the spine continues to grow. By adulthood, therefore, the spinal cord is shorter than the spine and descends only to the level of the upper lumbar vertebrae.

Figure 24.1 Gross Anatomy of the Adult Spinal Cord

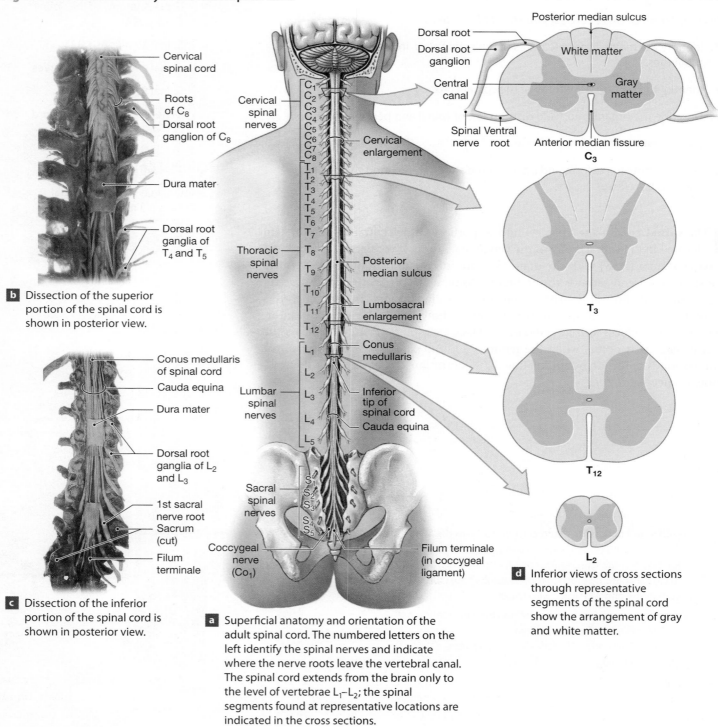

b Dissection of the superior portion of the spinal cord is shown in posterior view.

c Dissection of the inferior portion of the spinal cord is shown in posterior view.

a Superficial anatomy and orientation of the adult spinal cord. The numbered letters on the left identify the spinal nerves and indicate where the nerve roots leave the vertebral canal. The spinal cord extends from the brain only to the level of vertebrae L_1–L_2; the spinal segments found at representative locations are indicated in the cross sections.

d Inferior views of cross sections through representative segments of the spinal cord show the arrangement of gray and white matter.

The diameter of the spinal cord is not constant along its length. Two enlarged regions occur where the spinal nerves of the limbs join the spinal cord. The **cervical enlargement** in the neck supplies nerves to the upper limbs. The **lumbar enlargement** occurs near the distal end of the cord, where nerves supply the pelvis and lower limbs. Inferior to the lumbar enlargement, the spinal cord narrows and terminates at the **conus medullaris.** Spinal nerves fan out from the conus medullaris in a group called the **cauda equina** (KAW-duh ek-WI-nuh), the "horse's tail." A thin thread of fibrous tissue, the **filum terminale,** extends past the conus medullaris to anchor the spinal cord in the sacrum.

The spinal cord is organized into 31 segments. Each segment is attached to two spinal nerves, one on each side of the segment (Figure 24.2). Each of the two spinal nerves on a given cord segment is formed by the joining of two lateral extensions of the segment. One of these extensions, called the **dorsal root,** contains sensory neurons entering the spinal cord from sensory receptors. The dorsal root swells at the **dorsal root ganglion,** which is where cell bodies of sensory neurons cluster. The other extension, the **ventral root,** consists of motor neurons exiting the CNS and leading to effectors. The two roots join to form the spinal nerve. Each spinal nerve is therefore a *mixed nerve*, carrying both sensory and motor information. (The first spinal nerve does not have a dorsal root and is therefore a motor nerve.)

Figure 24.2 illustrates the spinal cord in transverse section to show the internal anatomy, also called the *sectional anatomy*. The cord is divided by the deep and conspicuous **anterior median fissure** and by the shallow **posterior median sulcus.** The periphery of the cord consist of myelinated neurons grouped into three masses called **columns** (*funiculi*). Deep to the white columns is gray matter organized into horns. The **gray horns** contain many glial cells and neuron cell bodies. Each horn contains a specific type of neuron. The **posterior gray horns** carry sensory neurons into the spinal cord, and the **anterior gray horns** carry somatic motor neurons out of the cord and to skeletal muscles. In the sacral region, the anterior gray horns have preganglionic neurons of the parasympathetic nervous system. The **lateral gray horns** occur in spinal segments T_1 through L_2 and consist of visceral motor neurons. Axons may cross to the opposite side of the spinal cord at the crossbars of the horns, called the **anterior** and **posterior gray commissures.** Between the gray commissures is a hole, called the **central canal,** which contains cerebrospinal fluid. The central canal is continuous with the fluid-filled ventricles of the brain. Collectively, all these structures are sometimes referred to as the spinal cord's **gray matter.**

The columns are organized on each side of the cord into the **posterior, lateral,** and **anterior white columns.** The two anterior white columns are connected by the **anterior white commissure.** Within each white column, myelinated axons form distinct bundles of neurons, called either **tracts** (*fasciculi*). (Recall that a bundle of neurons in the PNS is called a *nerve*.)

Make a Prediction

Consider what you know about myelinated neurons. Which anatomical structure's sensory information in the spinal cord ascends to the brain?

QuickCheck Questions

1.1 How is the white and gray matter of the spinal cord organized?

1.2 Which structure is useful in determining which portion of a spinal cord cross section is the anterior region?

1.3 Why is the spinal cord shorter than the vertebral column?

In the Lab 1

Materials

☐ Spinal cord model

☐ Spinal cord chart

☐ Dissection microscope

☐ Compound microscope

☐ Prepared slide of transverse section of spinal cord

Procedures

1. Review Figures 24.1 and 24.2.
2. Locate each surface feature of the spinal cord on the spinal cord model and chart.
3. Review the internal anatomy of the spinal cord on the spinal cord model.
4. Examine the microscopic features of the spinal cord in transverse section by following this sequence:
   - View the slide at low magnification with the dissection microscope. Identify the anterior and posterior regions.
   - Transfer the slide to a compound microscope. Move the slide around to survey the preparation at low magnification, again identifying the posterior and anterior aspects.
   - Examine the central canal and gray horns. Can you distinguish among the posterior, lateral, and anterior gray horns? Locate the gray commissures.
   - Examine the white columns. What is the difference between gray and white matter in the CNS?

For additional practice, complete the *Sketch to Learn* activity (on p. 338). ■

Lab Activity 2   Spinal Meninges

The spinal cord is protected within three layers of **spinal meninges** (me-NIN-jēz). The outer layer, the **dura mater** (DOO-ruh MĀ-ter), is composed of tough, fibrous connective tissue (Figure 24.3). The fibrous tissue attaches to the bony

Figure 24.2 **The Sectional Organization of the Spinal Cord**

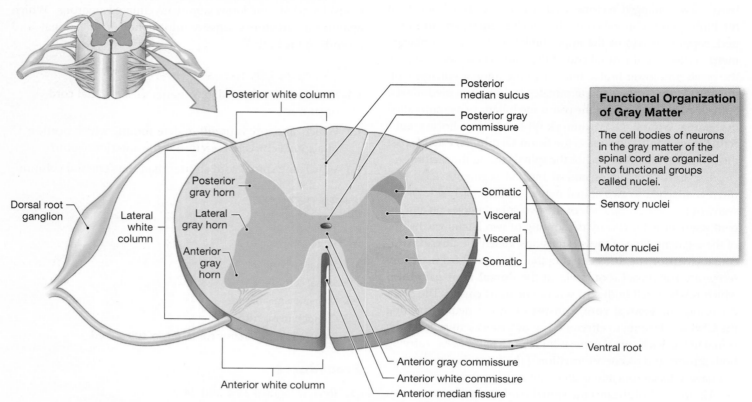

Posterior white column

Posterior median sulcus

Posterior gray commissure

Dorsal root ganglion

Lateral white column

Posterior gray horn

Lateral gray horn

Anterior gray horn

Functional Organization of Gray Matter

The cell bodies of neurons in the gray matter of the spinal cord are organized into functional groups called nuclei.

Somatic
Visceral

Sensory nuclei

Visceral
Somatic

Motor nuclei

Ventral root

Anterior gray commissure

Anterior white commissure

Anterior median fissure

Anterior white column

a The left half of this sectional view shows important anatomical landmarks, including the three columns of white matter. The right half indicates the functional organization of the nuclei in the anterior, lateral, and posterior gray horns.

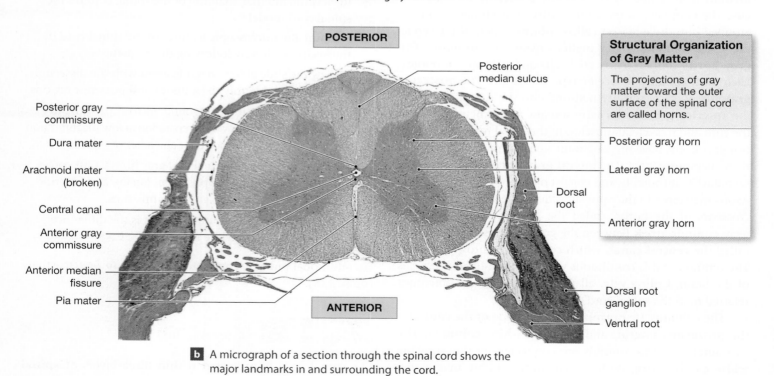

POSTERIOR

Posterior median sulcus

Posterior gray commissure

Dura mater

Arachnoid mater (broken)

Central canal

Anterior gray commissure

Anterior median fissure

Pia mater

ANTERIOR

Structural Organization of Gray Matter

The projections of gray matter toward the outer surface of the spinal cord are called horns.

Posterior gray horn

Lateral gray horn

Dorsal root

Anterior gray horn

Dorsal root ganglion

Ventral root

b A micrograph of a section through the spinal cord shows the major landmarks in and surrounding the cord.

Figure 24.3 **The Spinal Cord and Spinal Meninges**

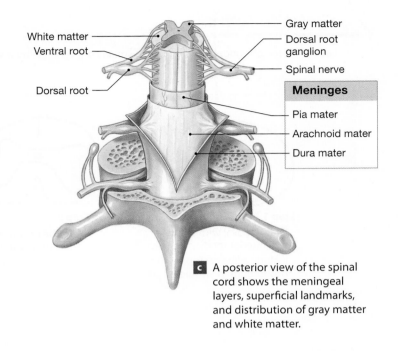

c A posterior view of the spinal cord shows the meningeal layers, superficial landmarks, and distribution of gray matter and white matter.

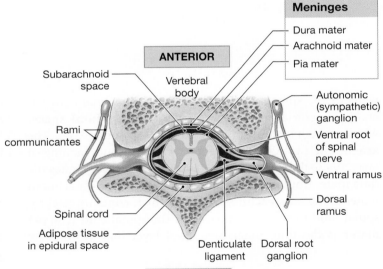

d A sectional view through the spinal cord and meninges shows the peripheral distribution of spinal nerves.

a Anterior view of the spinal cord and spinal nerve roots in the vertebral canal. The dura mater and arachnoid mater have been reflected.

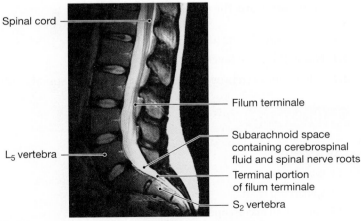

b MRI of inferior portion of spinal cord is shown in sagittal view.

Sketch to Learn

Let's reinforce our understanding of the spinal cord by drawing a section of one. It's easier than it might seem.

Sample Sketch

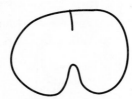

Step 1
- Draw an oval with a fold for the anterior median fissure.
- Include a line on top for the posterior median sulcus.

Step 2
- Add the gray horns and central canal.

Step 3
- Draw the dorsal and ventral roots by starting at the spinal cord and sketching outward.
- Label your sketch.

Your Sketch

walls of the spinal canal and supports the spinal cord laterally. Superficial to the dura mater is the **epidural space,** which contains adipose tissue to pad the spinal cord. The **arachnoid** (a-RAK-noyd) **mater** is the second meningeal layer. A small cavity called the **subdural space** separates the dura mater from the arachnoid mater. Deep to the arachnoid mater is the **subarachnoid space,** which contains cerebrospinal fluid to protect and cushion the spinal cord. The **pia mater** is the thin, inner meningeal layer that lies directly over the spinal cord. Blood vessels supplying the spinal cord are held in place by the pia mater. The pia mater extends laterally on each side of the spinal cord as the **denticulate ligament,** which joins the dura mater for lateral support to the spinal cord. Another extension of the pia mater, the filum terminale, supports the spinal cord inferiorly.

QuickCheck Questions

2.1 Name the three layers of spinal meninges.

2.2 Where does cerebrospinal fluid circulate in the spinal cord?

Clinical Application Epidural Injections and Spinal Taps

During childbirth, the expectant mother may receive an **epidural block,** a procedure that introduces anesthesia in the epidural space. A thin needle is inserted between two lumbar vertebrae, and the drug is injected into the epidural space. The anesthetic numbs only the spinal nerves of the pelvis and lower limbs and reduces the discomfort the woman feels during the powerful labor contractions of her uterus.

A **spinal tap** is a procedure in which a needle is inserted into the subarachnoid space to withdraw a sample of cerebrospinal fluid. The fluid is then analyzed for the presence of microbes, wastes, and metabolites. To prevent injury to the spinal cord, the needle is inserted into the lower lumbar region inferior to the cord. ■

In the Lab 2

Materials

☐ Spinal cord model
☐ Spinal cord chart
☐ Compound microscope
☐ Prepared slide of transverse section of spinal cord

Procedures

1. Review the spinal meninges in Figure 24.3.
2. Locate the spinal meninges on the spinal cord model and chart.

3. Use the compound microscope to examine the spinal meninges in transverse section. Move the slide around to survey the preparation. Locate the dura mater, arachnoid mater, and pia mater and the associated spaces between the meninges.

4. Add the spinal meninges to the Sketch to Learn activity (on p. 338) you began in Lab Activity 1. ∎

Lab Activity 3   Spinal and Peripheral Nerves

Two types of nerves connect PNS sensory receptors and effectors to the CNS: 12 pairs of cranial nerves and 31 pairs of spinal nerves. As their names indicate, cranial nerves connect with the brain and spinal nerves communicate with the spinal cord. As noted at the opening of this exercise, spinal nerves branch into PNS nerves, and it is spinal nerves that make up the axons of PNS sensory and motor neurons.

The two spinal nerves on a given spine segment exit the vertebral canal by passing through an intervertebral foramen between two adjacent vertebrae (shown in Figure 24.3). Each spinal nerve divides into a series of **peripheral nerves.** The posterior branch is called the **dorsal ramus** and supplies the skin and muscles of the back, and the anterior branch, called the **ventral ramus,** innervates the anterior and lateral skin and muscles. The ventral ramus has additional branches, called the **rami communicantes,** that innervate autonomic ganglions. The rami communicantes consists of two branches: a **white ramus,** which passes ANS preganglionic neurons from the spinal nerve into the ganglion, and a **gray ramus,** which carries ganglionic neurons back into the spinal nerve. Once in the spinal nerve, the ganglionic neurons travel in the ventral or dorsal ramus to their target effector. As their names imply, the white ramus has *myelinated* preganglionic neurons and the gray ramus has *unmyelinated* ganglionic neurons.

The 31 pairs of spinal nerves are named after the vertebral region in which they are associated (**Figure 24.4**). There are 8 **cervical nerves** (C_1 through C_8), 12 **thoracic nerves** (T_1 through T_{12}), 5 **lumbar nerves** (L_1 through L_5), 5 **sacral nerves** (S_1 through S_5), and a single **coccygeal nerve** (Co_1). The cervical nerves exit superior to their corresponding vertebrae, except for C_8, which exits inferior to vertebra C_7. The thoracic and lumbar spinal nerves are named after the vertebra immediately above each nerve, which means that thoracic nerve T_1 is inferior to vertebra T_1. Only the spinal nerves that have autonomic neurons carry visceral motor information. Cervical, some lumbar, and coccygeal spinal nerves do not have ANS neurons.

Groups of spinal nerves join in a network called a **plexus.** As muscles form during fetal development, the spinal nerves that supplied the individual muscles interconnect and create a plexus. There are four of these regions: the cervical, brachial, lumbar, and sacral plexuses (**Figure 24.5**). Note that thoracic

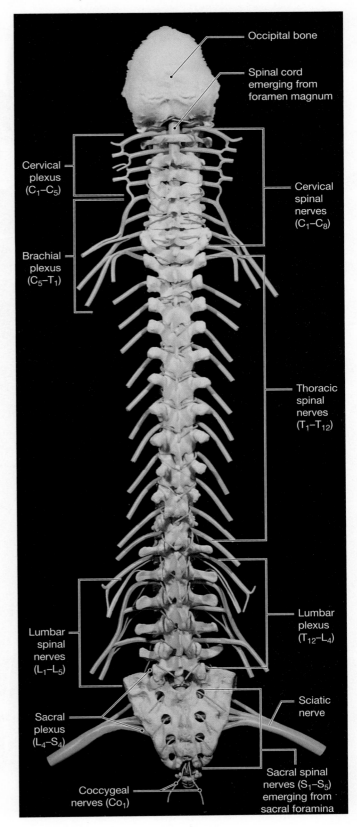

Figure 24.4 Posterior View of Spinal Nerves Exiting Vertebral Column The yellow wires represent the 31 pairs of spinal nerves. Groups of nerves are interwoven into a network called a plexus. There are four plexuses: cervical, brachial, lumbar, and sacral.

Occipital bone

Spinal cord emerging from foramen magnum

Cervical plexus (C_1–C_5)

Cervical spinal nerves (C_1–C_8)

Brachial plexus (C_5–T_1)

Thoracic spinal nerves (T_1–T_{12})

Lumbar spinal nerves (L_1–L_5)

Lumbar plexus (T_{12}–L_4)

Sciatic nerve

Sacral plexus (L_4–S_4)

Coccygeal nerves (Co_1)

Sacral spinal nerves (S_1–S_5) emerging from sacral foramina

Figure 24.5 **Peripheral Nerves and Plexuses** Spinal nerves branch as peripheral nerves that spread to specific regions of the body. The major peripheral nerves are illustrated in this figure. Groups of peripheral nerves may intertwine into a network called a plexus. There are four major nerve plexuses: cervical, brachial, lumbar, and sacral.

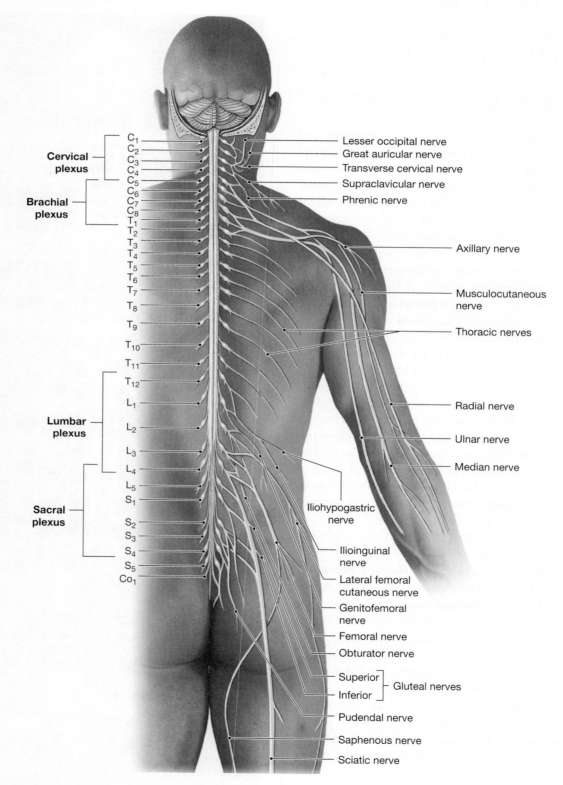

spinal nerves T_2 through T_{11} are not part of any plexus but instead constitute **intercostal nerves** that enter the spaces between the ribs. The intercostal nerves innervate the intercostal muscles and abdominal muscles and receive sensations from the lateral and anterior trunk.

Cervical Plexus

The eight cervical spinal nerves supply the neck, shoulder, upper limb, and diaphragm. The various branches of the **cervical plexus** contain nerves C_1 through C_4 and parts of C_5. This plexus innervates muscles of the larynx plus the sternocleidomastoid, trapezius, and the skin of the upper chest, shoulder, and ear. Nerves C_3 through C_5, called the phrenic nerves, control the diaphragm, the muscle used for breathing.

Brachial Plexus

The **brachial plexus** includes the parts of spinal nerve C_5 not involved with the cervical plexus plus nerves C_6, C_7, C_8, and T_1. This plexus is more complex than the cervical plexus and branches to innervate the shoulder, the upper limb, and some muscles on the trunk. The major branches of this plexus are the axillary, radial, musculocutaneous, median, and ulnar nerves. The **axillary nerve** (C_5 and C_6) supplies the deltoid and teres minor muscles and the skin of the shoulder. The **radial nerve** (C_5 through T_1) controls the extensor muscles of the upper limb as well as the skin over the posterior and lateral margins of the arm. The **musculocutaneous nerve** (C_5 through C_7) supplies the flexor muscles of the upper limb and the skin of the lateral forearm. The **median nerve** (C_6 through T_1) innervates the flexor muscles of the forearm and digits, the pronator muscles, and the lateral skin of the hand. The **ulnar nerve** (C_8 and T_1) controls the flexor carpi ulnaris muscle of the forearm, other muscles of the hand, and the medial skin of the hand. Notice how overlap occurs in the brachial plexus. For example, spinal nerve C_6 innervates both flexor and extensor muscles.

Lumbar and Sacral Plexuses

The largest nerve network is called the **lumbosacral plexus.** It is a combination of the **lumbar plexus** (T_{12}, L_1 through L_4) and the **sacral plexus** (L_4, L_5, S_1 through S_4). The major nerves of the lumbar plexus innervate the skin and muscles of the abdominal wall, genitalia, and thigh. The **genitofemoral nerve** supplies some of the external genitalia and the anterior and lateral skin of the thigh. The **lateral femoral cutaneous nerve** innervates the skin of the thigh from all aspects except the medial region. The **femoral nerve** controls the muscles of the anterior thigh and the adductor muscles and medial skin of the thigh.

The sacral plexus consists of two major nerves, the sciatic and the pudendal. The **sciatic nerve** descends the posterior lower limb and sends branches into the posterior thigh muscles and the musculature and skin of the leg. The **pudendal nerve** supplies the muscular floor of the pelvis, the perineum, and parts of the skin of the external genitalia.

QuickCheck Questions

3.1 What are the two groups of nerves in the PNS?

3.2 Which branch of a peripheral nerve innervates the limbs?

3.3 Name the four plexuses in the body.

In the Lab 3

Materials

☐ Spinal cord model
☐ Spinal cord chart

Procedures

1. Review the major spinal nerves and plexuses shown in Figures 24.4 and 24.5.

2. Locate each nerve plexus on the spinal cord model and chart.

3. Locate the spinal nerves assigned by your instructor on the spinal cord model. ■

Lab Activity 4   Spinal Reflexes

Reflexes are automatic neural responses to specific stimuli. Most reflexes have a protective function. Touch something hot, and the withdrawal reflex removes your hand to prevent tissue damage. Shine a bright light into someone's eyes, and the pupils constrict to protect the retina from excessive light. Reflexes cause rapid adjustments to maintain homeostasis. The CNS does minimal processing to respond to the stimulus. The sensory and motor components of a reflex are "prewired" and initiate the reflex upon stimulation.

Figure 24.6 depicts the five steps involved in a typical reflex pathway, called a **reflex arc.** First, a receptor is activated by a stimulus. The receptor in turn activates a sensory neuron that enters the CNS, where the third step, information processing, occurs. The processing is performed at the synapse between the sensory and motor neurons. A conscious thought or recognition of the stimulus is not required to evaluate the sensory input of the reflex. The processing results in activation of a motor neuron that elicits the appropriate action, a response by the effectors. In this basic reflex arc, only two neurons are involved: one sensory and one motor. Complex reflex arcs include **interneurons** between the sensory and motor neurons.

There are many types of reflexes. **Innate** reflexes are the inborn responses of a newborn baby, such as grasping an

Figure 24.6 Components of a Reflex Arc in the Patellar Reflex A reflex has five main components: a sensory receptor, a sensory neuron, the CNS, a motor neuron, and an effector.

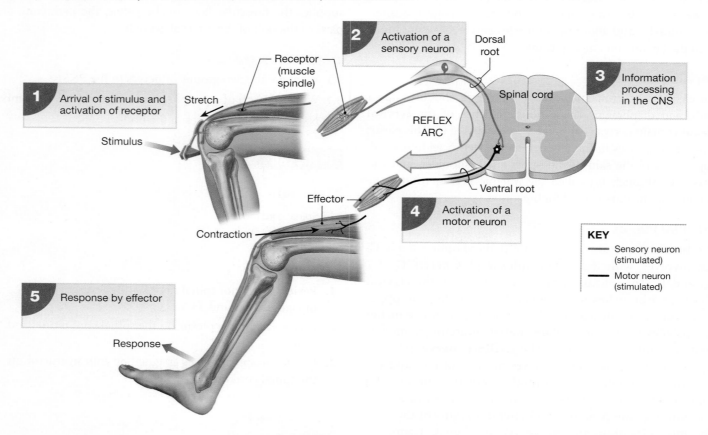

object and suckling the breast for milk. **Cranial** reflexes have pathways in cranial nerves. **Visceral** reflexes pertain to the internal organs. **Spinal** reflexes process information in the spinal cord rather than the brain. **Somatic** reflexes involve skeletal muscles. The number of synapses in a reflex can also be used to classify reflexes. In the arc of a **monosynaptic reflex,** there is only one synapse between the sensory and motor neurons. In the arc of a **polysynaptic reflex,** there are numerous interneurons between sensory and motor neurons. The response of a polysynaptic reflex is more complex and may include both stimulation and inhibition of muscles. Reflexes are used as a diagnostic tool to evaluate the function of specific regions of the brain and spinal cord. An abnormal reflex or the lack of a reflex indicates a loss of neural function resulting from disease or injury.

You are probably familiar with the "knee jerk," a type of **stretch reflex** called the **patellar reflex,** shown in Figure 24.6. This reflex occurs when the tendon over the patella is hit with a rubber percussion hammer. Tapping on the patellar tendon stretches receptors called **muscle spindles** in the quadriceps muscle group of the anterior thigh. This stimulus evokes a rapid motor reflex to contract the quadriceps and shorten the muscles. **Figure 24.7** shows other tendons that may be gently struck to study additional somatic reflexes.

QuickCheck Questions

4.1 What are the components of a reflex arc?

4.2 How can reflexes be used diagnostically?

In the Lab   4

Materials

☐ Reflex (percussion) hammer (with rubber head)

☐ Lab partner

Procedures

1. Patellar reflex (Figure 24.6):
   - Have your partner sit and cross the legs at the knee.
   - On the partner's top leg, gently tap below the patella with the percussion hammer to stimulate the tendon of the rectus femoris muscle.
   - What is the response?
   - How might this reflex help maintain upright posture?

2. Biceps reflex (Figure 24.7b):
   - This reflex tests the response of the biceps brachii muscle.
   - Have your partner rest an arm on the laboratory benchtop.

Figure 24.7 Somatic Reflexes Effectors for somatic reflexes are skeletal muscles. Stretch reflexes involve tapping a tendon with a percussion hammer and stimulating the attached muscle.

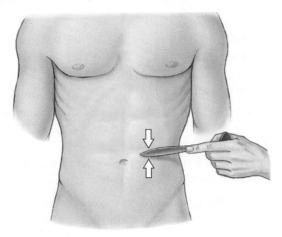

a **Abdominal reflex.** Gently stroking the sides of the abdomen causes the abdominal reflex, an example of a superficial reflex.

b **Biceps reflex.** Tapping the tendon initiates a stretch reflex.

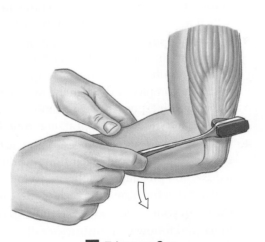

c **Triceps reflex.**

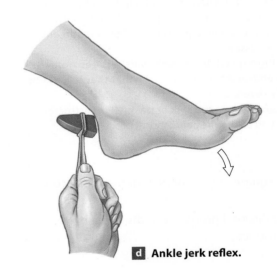

d **Ankle jerk reflex.**

- Place a finger over the tendon of the biceps brachii and gently tap your finger with the percussion hammer.
- What is the response?

3. Triceps reflex (Figure 24.7c):
- This reflex tests the response of the triceps brachii muscle.
- Loosely support one of your partner's forearms.
- Gently tap the tendon of the triceps brachii at the posterior elbow.
- What is the response?

4. Ankle calcanean reflex (Figure 24.7d):
- This reflex tests the response of the gastrocnemius muscle when the calcanean (Achilles) tendon is stretched.
- Have your partner sit in a chair and extend one leg forward so that the foot is off the floor.
- Gently tap the calcanean tendon with the percussion hammer.
- What is the response? ▪

Lab Activity 5   Dissection of the Spinal Cord

Dissecting a preserved sheep or cow spinal cord provides you the opportunity to examine the meningeal layers and the internal anatomy.

⚠ **Safety Alert:** Dissecting the Spinal Cord

You *must* practice the highest level of laboratory safety while handling and dissecting the spinal cord. Keep the following guidelines in mind during the dissection.

1. Be sure to use only a *preserved* spinal cord for dissection because fresh spinal cords can carry disease.
2. Wear gloves and safety glasses to protect yourself from the fixatives used to preserve the specimen.
3. Do not dispose of the fixative from your specimen. You will later store the specimen in the fixative to keep the specimen moist and to prevent it from decaying.
4. Be extremely careful when using a scalpel or other sharp instrument. Always direct cutting and scissor motions away from you to prevent an accident if the instrument slips on moist tissue.
5. Before cutting a given tissue, make sure it is free from underlying and/or adjacent tissues so that they will not be accidentally severed.
6. Never discard tissue in the sink or trash. Your instructor will inform you of the proper disposal procedure. ▲

QuickCheck Questions

5.1 What safety equipment is required for the spinal cord dissection?

5.2 Describe the disposal procedure as discussed by your laboratory instructor.

In the Lab   5

Materials

☐ Gloves
☐ Safety glasses
☐ Segment of preserved sheep or cow spinal cord
☐ Dissection pan
☐ Dissection pins
☐ Scissors
☐ Scalpel
☐ Forceps
☐ Blunt probe

Procedures

Put on gloves and safety glasses before opening the container of preserved spinal cord segments or handling one of the segments.

1. Lay the spinal cord on the dissection pan and cut a thin cross section about 2 cm (0.75 in.) thick. Lay this cross section flat on the dissection pan and observe the internal anatomy. Use **Figure 24.8** as a guide to help locate the various anatomical features of the spinal cord.
2. Identify the gray horns, central canal, and white columns. What type of tissue is found in the gray horns? What type is found in the white columns? How can you determine which margin of the cord is the posterior margin?
3. Locate the spinal meninges by pulling the outer tissues away from the spinal cord with a forceps and blunt probe. Slip your probe between the meninges on the lateral spinal cord. Cut completely through the meninges and gently peel them back to expose the ventral and dorsal roots. How does the dorsal root differ in appearance from the ventral root?
4. Closely examine the meninges. With your probe, separate the arachnoid mater from the dura mater. With a dissection pin, attempt to loosen a free edge of the pia mater. What function does each of these membranes serve?
5. Clean up your work area, wash the dissection pan and tools, and follow your instructor's directions for proper disposal of the specimen. ▪

Figure 24.8 Sheep Spinal Cord Dissection The sheep spinal cord in this transverse section shows its internal organization and the three spinal meninges.

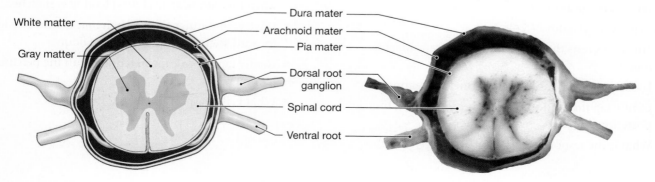

Name _____

Date _____

Section _____

Spinal Cord, Spinal Nerves, and Reflexes

A. Matching

Match each term in the left column with its correct description from the right column.

_____ 1. lateral gray horn	**A.** site of cerebrospinal fluid circulation
_____ 2. bundle of axons	**B.** sensory branch entering spinal cord
_____ 3. rami communicantes	**C.** surrounds axons of peripheral nerve
_____ 4. subarachnoid space	**D.** contains visceral motor cell bodies
_____ 5. ventral root	**E.** tapered end of spinal cord
_____ 6. dorsal ramus	**F.** fascicle
_____ 7. dorsal root ganglion	**G.** posterior branch of a spinal nerve
_____ 8. conus medullaris	**H.** motor branch exiting spinal cord
_____ 9. endoneurium	**I.** leads to autonomic ganglion
_____ 10. dorsal root	**J.** contains sensory cell bodies

B. Labeling

1. Label the sectional anatomy of the spinal cord in **Figure 24.9**.

Figure 24.9 Anatomy of the Spinal Cord

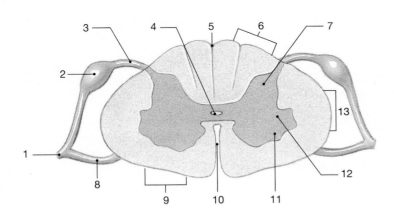

1. _____
2. _____
3. _____
4. _____
5. _____
6. _____
7. _____
8. _____
9. _____
10. _____
11. _____
12. _____
13. _____

C. Short-Answer Questions

1. Describe the organization of white and gray matter in the spinal cord.

2. Describe the spinal meninges.

3. Discuss the major nerves of the brachial plexus.

4. List the five basic steps of a reflex.

D. Analysis and Application

1. Starting in the spinal cord, trace a motor pathway to the adductor muscles of the thigh. Include the spinal cord root, spinal nerve, nerve plexus, and specific peripheral nerve involved in the pathway.

2. Compare the types of neurons that synapse in the posterior, lateral, and anterior gray horns of the spinal cord.

E. Clinical Challenge

1. How can an injury to a peripheral nerve cause loss of both sensory and motor functions?

2. How does the stretch reflex cause the quadriceps femoris muscle group to contract?

3. A woman injures her neck in a car accident and now has difficulty breathing. Which spinal nerves may be involved in this case?

Anatomy of the Brain

Learning Outcomes

On completion of this exercise, you should be able to:

1. Name the three meninges that cover the brain.

2. Describe the extensions of the dura mater.

3. Identify the six major regions of the brain and a basic function of each.

4. Identify the surface features of each region of the brain.

5. Identify the 12 pairs of cranial nerves.

6. Identify the anatomy of a dissected sheep brain.

The brain, which occupies the cranial cavity, is one of the largest organs in the body. The adult brain weighs approximately 1.4 kg (3 pounds) and has an average volume of 1,200 cc; the brain of a newborn weighs only 350 to 400 g. Adult males tend to have larger bodies and therefore have larger brains than females, but of course, this size difference offers no intellectual advantage to the males.

Approximately 100 billion neurons in the brain interconnect with over 1 trillion synapses as vast biological circuitry that no electronic computer has yet to surpass. Every second, the brain performs a huge number of calculations, interpretations, and visceral-activity adjustments and coordinations to maintain homeostasis.

The brain is organized into six major regions: cerebrum (se-RĒ-brum or SER-e-brum), diencephalon (dī-en-SEF-a-lon), mesencephalon, pons, medulla oblongata, and cerebellum (**Figure 25.1**). The medulla oblongata, pons, and mesencephalon (midbrain) are collectively called the **brain stem.** Some anatomists also include the diencephalon as part of the brain stem.

Need More Practice and Review?

Build your knowledge—and confidence!—in the Study Area of MasteringA&P® at www.masteringaandp.com with Pre-lab Quizzes, Post-lab Quizzes, Practice Anatomy Lab™ (PAL™) 3.0 virtual anatomy practice tool, PhysioEx™ 9.0 laboratory simulations, and A&P Flix™ with Quizzes.

PAL | practice anatomy lab™ For this lab exercise, follow these navigation paths:
- PAL>Human Cadaver>Nervous System>Central Nervous System
- PAL>Anatomical Models>Nervous System>Central Nervous System

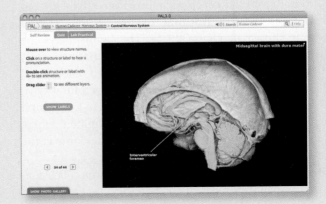

MasteringA&P®

Figure 25.1 The Human Brain The major regions of the human brain.

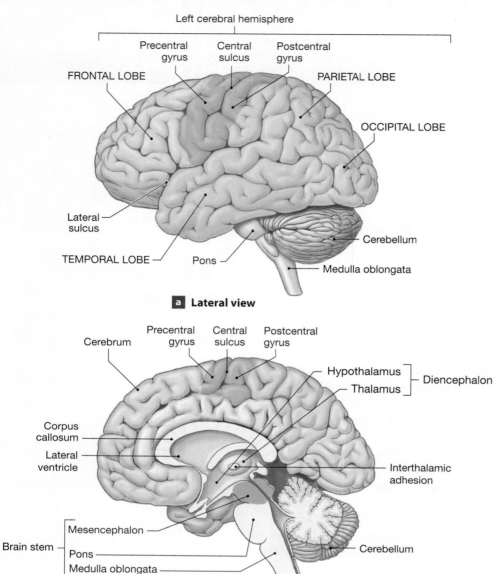

a **Lateral view**

b **Midsaggital section**

diencephalon (Figure 25.1b). Inferior to the diencephalon is the **mesencephalon** (midbrain) of the brain stem. The **pons** is the large, swollen region of the brain stem just inferior to the mesencephalon, and the **medulla oblongata** is the most inferior part of the brain stem, connecting the brain to the spinal cord. The **cerebellum** is the oval mass posterior to the brain stem.

Lab Activity 1 Cranial Meninges and Ventricles of the Brain

Cranial Meninges

The brain is encased in layers of tough, protective **cranial meninges.** Circulating between certain meningeal layers, **cerebrospinal fluid (CSF)** cushions the brain and prevents it from contacting the cranial bones during a head injury, much like a car's airbag prevents a passenger from hitting the dashboard. The cranial meninges are anatomically similar to, and continuous with, the spinal meninges of the spinal cord. Like their spinal counterparts, the cranial meninges consist of three layers: dura mater, arachnoid mater, and pia mater (**Figure 25.2**).

Make a Prediction

Which body fluid is filtered to produce CSF?

The **cerebrum** is the largest region of the brain. It is divided into right and left **cerebral hemispheres** by the deep groove known as the **longitudinal fissure.** A left cerebral hemisphere is shown in Figure 25.1a. The hemispheres are covered with a folded **cerebral cortex** (*cortex*, bark or rind) of gray matter, where neurons are not myelinated. Each small fold of the cerebral cortex is called a **gyrus** (JĪ-rus; plural *gyri*), and each shallow groove is called a **sulcus** (SUL-kus; plural *sulci*). Deep in the cerebrum is the brain's white matter, where myelinated neurons that occur in thick bands interconnect the various regions of the brain.

Inferior to the cerebrum are the **thalamus** (THAL-a-mus) and **hypothalamus,** which together make up the

The **dura mater** (DOO-ruh MĀ-ter; *dura*, tough + *mater*, mother), the outer meningeal covering, consists of an **endosteal layer** fused with the periosteum of the cranial bones and a **meningeal layer** that faces the arachnoid mater. The endosteal layer is referred to as the *outer dural layer*, and the meningeal layer is

Study Tip A Sea Horse's Guide to the Brain

When examining the brain in sagittal section, most people notice how the brain stem and diencephalon form the shape of a sea horse. The pons is the horse's belly, the mesencephalon the neck, the diencephalon the head, and the medulla oblongata the tail. Imagine the sea horse is wearing the cerebellum as a backpack and the cerebrum as a very large hat. ■

Figure 25.2 Brain, Cranium, and Meninges The brain is protected by the cranium and by a three-layered covering called the cranial meninges.

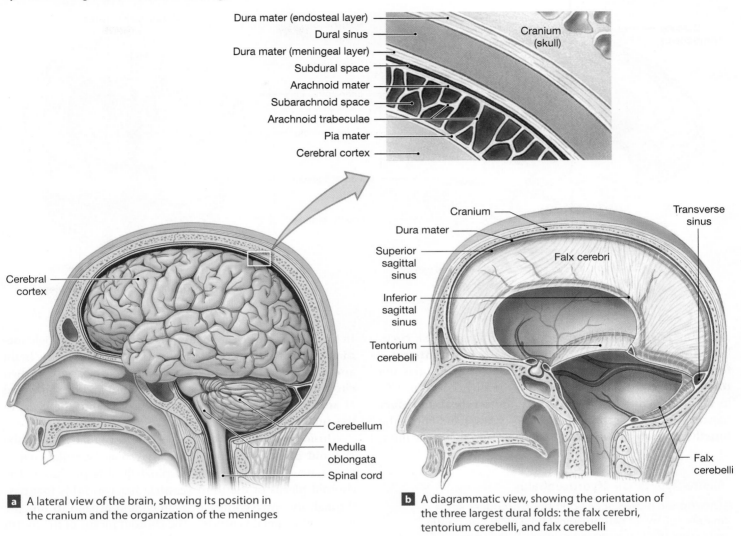

Dura mater (endosteal layer)
Dural sinus
Dura mater (meningeal layer)
Subdural space
Arachnoid mater
Subarachnoid space
Arachnoid trabeculae
Pia mater
Cerebral cortex

Cranium (skull)

Cerebral cortex

Cerebellum
Medulla oblongata
Spinal cord

Cranium
Dura mater
Superior sagittal sinus
Inferior sagittal sinus
Tentorium cerebelli

Transverse sinus
Falx cerebri
Falx cerebelli

a A lateral view of the brain, showing its position in the cranium and the organization of the meninges

b A diagrammatic view, showing the orientation of the three largest dural folds: the falx cerebri, tentorium cerebelli, and falx cerebelli

referred to as the *inner dural layer*. Between the two layers are large blood sinuses, collectively called **dural sinuses,** that drain blood from cranial veins into the jugular veins. The **superior** and **inferior sagittal sinuses** are large veins in the dura mater between the two hemispheres of the cerebrum. The **transverse sinus** is in the dura mater between the cerebrum and the cerebellum. Between the dura mater and the underlying arachnoid mater is the **subdural space.**

Deep to the dura mater is the **arachnoid** (a-RAK-noyd; *arachno,* spider) **mater,** named after the weblike connection this membrane has with the underlying pia mater. The arachnoid mater forms a smooth covering over the brain.

On the surface of the brain is the **pia** (PĒ-uh; *pia,* delicate) **mater,** which contains many blood vessels supplying the brain. Between the arachnoid mater and pia mater is the **subarachnoid space,** where the CSF circulates.

The dura mater has extensions that help stabilize the brain (Figure 25.2b). A midsagittal fold in the dura mater forms the **falx cerebri** (FALKS SER-e-brī; *falx,* sickle shaped) and separates the right and left hemispheres of the cerebrum. Posteriorly, the dura mater folds again as the **tentorium cerebelli** (ten-TŌ-rē-um ser-e-BEL-ē; *tentorium,* a covering) and separates the cerebellum from the cerebrum. The **falx cerebelli** is a dural fold between the hemispheres of the cerebellum.

Ventricles

Deep in the brain are four chambers called **ventricles** (Figure 25.3). Two **lateral ventricles,** one in each cerebral hemisphere, extend deep into the cerebrum as horseshoe-shaped chambers. At the midline of the brain, the lateral ventricles are separated by a thin membrane called the **septum pellucidum** (pe-LOO-si-dum; *pellucid,* transparent). A brain sectioned

Figure 25.3 Ventricles of the Brain The orientation and extent of the ventricles as they would appear if the brain were transparent.

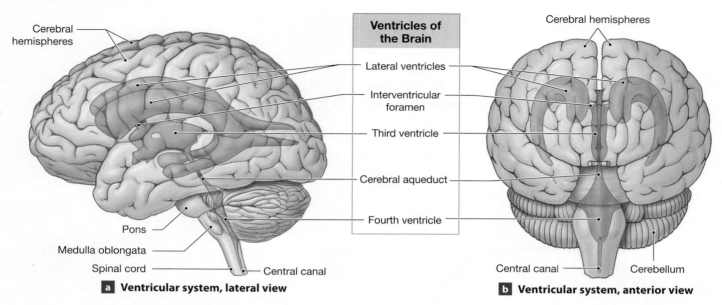

Ventricles of the Brain

a Ventricular system, lateral view

Labels: Cerebral hemispheres, Lateral ventricles, Interventricular foramen, Third ventricle, Cerebral aqueduct, Fourth ventricle, Pons, Medulla oblongata, Spinal cord, Central canal

b Ventricular system, anterior view

Labels: Cerebral hemispheres, Central canal, Cerebellum

at the midsagittal plane exposes this membrane. CSF circulates from the lateral ventricles through the **interventricular foramen** (also called the *foramen of Monro*) and enters the **third ventricle,** a small chamber in the diencephalon. CSF in the third ventricle passes through the **cerebral aqueduct** (*aqueduct of the midbrain* or *aqueduct of Sylvius*) and enters the **fourth ventricle** between the brain stem and the cerebellum.

In the fourth ventricle, two **lateral apertures** and a single **median aperture** direct CSF laterally to the exterior of the brain and spinal cord and into the subarachnoid space. CSF then circulates around the brain and spinal cord and is reabsorbed at **arachnoid granulations,** which project into the veins of the dural sinuses (**Figure 25.4**).

Inside each ventricle is a specialized capillary called the **choroid plexus** (KŌ-royd PLEK-sus; *choroid,* vascular coat + *plexus,* network) where cerebrospinal fluid is produced. The choroid plexus of the third ventricle has two folds that pass through the interventricular foramen and expand to line the floor of the lateral ventricles. The choroid plexus of the fourth ventricle lies on the posterior wall of the ventricle.

Clinical Application Hydrocephalus

The choroid plexus of an adult brain produces approximately 500 mL of cerebrospinal fluid daily, constantly replacing the 150 mL that circulates in the ventricles and subarachnoid space. Because CSF is constantly being made, a volume equal to that produced must be removed from the central nervous system to prevent a buildup of fluid pressure in the ventricles. In an infant, if CSF production exceeds CSF reabsorption, the increase in cranial pressure expands the unfused skull, creating a condition called *hydrocephalus.* There are two types of hydrocephalus; internal and external. Internal hydrocephalus occurs when CSF accumulates in the ventricles inside the brain. This form of hydrocephalus is almost always fatal because of damaging distortion of the brain tissue. External hydrocephalus is the buildup of CSF in the subdural space, resulting in an enlarged skull and possible brain damage caused by high fluid pressure on the delicate neural tissues. Surgical treatment of external hydrocephalus involves installation of small tubes called shunts to drain the excess CSF and reduce intracranial pressure. ■

QuickCheck Questions

1.1 What are the functions of the cranial meninges?

1.2 Between which meningeal layers does CSF circulate?

1.3 Where does CSF circulate, and where is it returned to the blood?

In the Lab   1

Materials

☐ Brain model

☐ Brain chart

☐ Ventricular system model

☐ Preserved and sectioned human brain (if available)

Figure 25.4 Formation and Circulation of Cerebrospinal Fluid

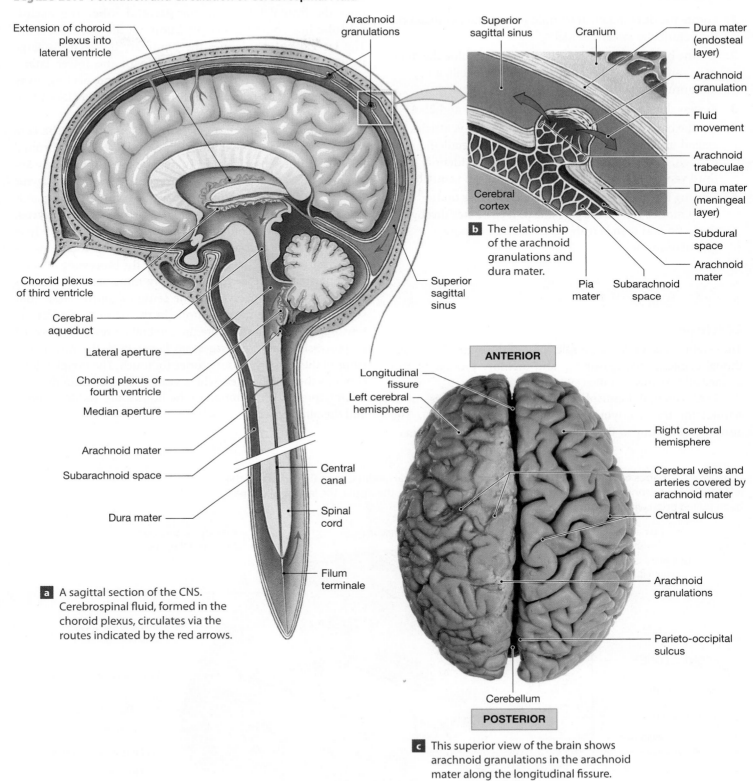

Extension of choroid plexus into lateral ventricle

Arachnoid granulations

Choroid plexus of third ventricle

Cerebral aqueduct

Lateral aperture

Choroid plexus of fourth ventricle

Median aperture

Arachnoid mater

Subarachnoid space

Dura mater

Superior sagittal sinus

Central canal

Spinal cord

Filum terminale

a A sagittal section of the CNS. Cerebrospinal fluid, formed in the choroid plexus, circulates via the routes indicated by the red arrows.

Superior sagittal sinus

Cranium

Dura mater (endosteal layer)

Arachnoid granulation

Fluid movement

Arachnoid trabeculae

Dura mater (meningeal layer)

Subdural space

Arachnoid mater

Cerebral cortex

Pia mater

Subarachnoid space

b The relationship of the arachnoid granulations and dura mater.

ANTERIOR

Longitudinal fissure

Left cerebral hemisphere

Right cerebral hemisphere

Cerebral veins and arteries covered by arachnoid mater

Central sulcus

Arachnoid granulations

Parieto-occipital sulcus

Cerebellum

POSTERIOR

c This superior view of the brain shows arachnoid granulations in the arachnoid mater along the longitudinal fissure.

Procedures

1. Locate the dura mater, arachnoid mater, and pia mater on the ventricular system model.

2. On the brain model or preserved brain, examine the dura mater and identify the falx cerebri, falx cerebelli, and tentorium cerebelli.

3. Review the ventricular system in Figures 25.3 and 25.4.

4. On the brain model, observe how the lateral ventricles extend into the cerebrum. If your model is detailed enough, locate the interventricular foramen. Identify the third ventricle, cerebral aqueduct, and fourth ventricle.

5. Starting from one of the two lateral ventricles on the brain model, trace a drop of CSF as it circulates through the brain and then is reabsorbed at an arachnoid granulation. ■

Lab Activity 2   Regions of the Brain

Cerebrum

The cerebrum is the most complex part of the brain. Conscious thought, intellectual reasoning, and memory processing and storage all take place in the cerebrum.

Each cerebral hemisphere consists of five lobes, most named for the overlying cranial bone (**Figure 25.5**). The anterior cerebrum is the **frontal lobe,** and the prominent

central sulcus, located approximately midposterior, separates the frontal lobe from the **parietal lobe.** The **occipital lobe** lies under the occipital bone of the posterior skull. The **lateral sulcus** defines the boundary between the large frontal lobe and the **temporal lobe** of the lower lateral cerebrum. Cutting into the lateral sulcus and peeling away the temporal lobe reveals a fifth lobe, the **insula** (IN-sū-luh; *insula*, island).

Regional specializations occur in the cerebrum. The central sulcus separates the motor region of the cerebrum (frontal lobe) from the sensory region (parietal lobe). Immediately anterior to the central sulcus is the **precentral gyrus.** This gyrus contains the primary motor cortex, where voluntary commands to skeletal muscles are generated. The **postcentral gyrus,** on the parietal lobe, contains the primary sensory cortex, where the general sense of touch is perceived. The other four senses—sight, hearing, smell, and taste—involve the processing of complex information received from many more sensory neurons than the number involved in the sense of touch. These four senses thus require more neurons in the brain to process the sensory signals, and therefore the cerebral cortex areas devoted to processing these messages are larger than the postcentral gyrus of the primary sensory cortex for touch. The occipital lobe contains the visual cortex, where visual impulses from the eyes are interpreted. The temporal lobe houses the auditory cortex and the olfactory cortex.

Figure 25.5 Lobes of a Cerebral Hemisphere Major anatomical landmarks on the surface of the left cerebral hemisphere. Association areas are colored. To expose the insula, the lateral sulcus has been pulled open with two retractors.

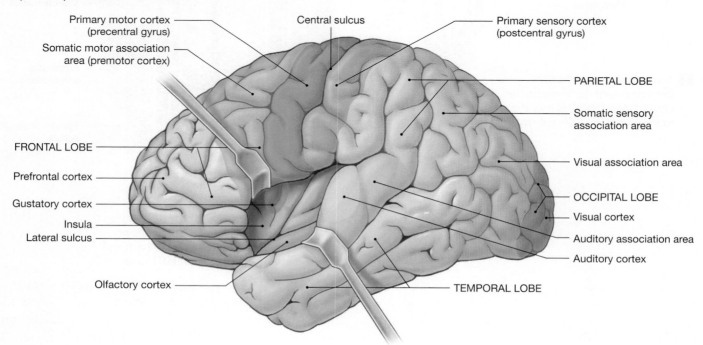

Figure 25.5 also shows numerous **association areas,** regions that either interpret sensory information from more than one sensory cortex or integrate motor commands into an appropriate response. The **premotor cortex** is the somatic motor association area of the anterior frontal lobe. Auditory and visual association areas occur near the corresponding sensory cortex in the occipital lobe.

Deep structures of the cerebrum are visible when the brain is sectioned, as in Figures 25.6 and 25.7. The cerebral hemispheres are connected by a deep, thick tract of white matter called the **corpus callosum** (kōr-pus ka-LŌ-sum; *corpus*, body + *callosum*, hard). This structure, which bridges the two hemispheres at the base of the longitudinal fissure, is easily identified as the curved white structure at the base of the cerebrum. The inferior portion of the corpus callosum is the **fornix** (FOR-niks), a white tract connecting deep structures of the limbic system, the "emotional" brain. The fornix narrows anteriorly and meets the **anterior commissure** (kom-MIS-sur), another tract of white matter connecting the cerebral hemispheres.

In each cerebral hemisphere, paired masses of gray matter called **basal nuclei** are involved in automating voluntary muscle contractions. Each basal nucleus consists of a medial **caudate nucleus** and a lateral **lentiform nucleus** (see Figures 25.6 and 25.7). The latter is made up of two parts: a **putamen** (pū-TĀ-men; shell) and a **globus pallidus** (glō-bus PAL-i-dus; *globus*, ball + *pallidus*, pale). At the tip of the caudate nucleus is the **amygdaloid** (ah-MIG-da-loyd; almond) **body.** Between the caudate nucleus and the lentiform nucleus lies the **internal capsule,** a band of white matter that connects the cerebrum to the diencephalon, brain stem, and cerebellum.

Diencephalon: The Thalamus and Hypothalamus

The diencephalon is embedded in the cerebrum and is exposed only at the inferior aspect of the brain. The thalamus region of the diencephalon maintains a crude sense of awareness. All sensory impulses except smell and proprioception (the sense of muscle, bone, and joint position) pass into the thalamus and are relayed to the proper sensory cortex for interpretation. Nonessential sensory data are filtered out by the thalamus and do not reach the sensory cortex. In sagittal section, the **interthalamic adhesion,** also called the *massa intermedia*, is an oval structure in the diencephalon that connects the right and left sides of the thalamus (Figure 25.7). The **pineal** (PIN-ē-ul) **gland** is the cone-shaped structure superior to the mesencephalon positioned between the cerebrum and the cerebellum.

The hypothalamus is the floor of the diencephalon. On the inferior surface of the brain, a pair of rounded **mamillary** (MAM-i-lar-ē; *mammilla*, little breast) **bodies** are visible inferior to the hypothalamus (Figure 25.7). These bodies are hypothalamic nuclei that control eating reflexes for licking, chewing, sucking, and swallowing. Anterior to the mamillary bodies is the **infundibulum** (in-fun-DIB-ū-lum; *infundibulum*, funnel), the stalk that attaches the **pituitary gland** to the hypothalamus.

Mesencephalon (Midbrain)

The mesencephalon (Figure 25.8; also see Figure 25.7) is posteriorly covered by the cerebrum. Posterior to the cerebral aqueduct is the **corpora quadrigemina** (KOR-pōr-uh qui-dri-JEM-i-nuh), a series of four bulges next to the pineal gland of the diencephalon. The two members of the superior pair of bulges are the **superior colliculi** (ko-LĪK-u-lē; *colliculus*, small hill), which function as a visual reflex center to move the eyeballs and the head, to keep an object centered on the retina of the eye. The two members of the inferior pair of bulges are the **inferior colliculi,** which function as an auditory reflex center to move the head, to locate and follow sounds. The anterior mesencephalon between the pons and the hypothalamus consists of the **cerebral peduncles** (*peduncles*, little feet), a group of white fibers connecting the cerebral cortex with other parts of the brain.

Pons

The pons is located inferior to the mesencephalon (Figures 25.7 and 25.8). The pons functions as a relay station to direct sensory information to the thalamus and cerebellum. It also contains certain sensory, somatic motor, and autonomic cranial nerve nuclei.

Medulla Oblongata

The medulla oblongata is the inferior part of the brain stem and is continuous with the spinal cord (Figures 25.7 and 25.8). Sensory information in ascending tracts in the spinal cord enter the brain at the medulla oblongata, and motor commands in descending tracts pass through the medulla oblongata and into the spinal cord. The anterior surface of the medulla oblongata has two prominent folds called **pyramids** where some motor tracts cross over, or *decussate*, to the opposite side of the body. The medulla oblongata also functions as an autonomic center for visceral functions. Nuclei in this portion of the brain are vital reflex centers for the regulation of cardiovascular, respiratory, and digestive activities.

Cerebellum

The cerebellum (Figure 25.9) is inferior to the occipital lobe of the cerebrum and is covered by a layer called the **cerebellar cortex.** Small folds on the cerebellar cortex are called **folia** (FŌ-lē-uh; *folia*, leaves; singular *folium*). The cerebellum is divided into right and left **cerebellar hemispheres,** which are separated by a narrow **vermis** (VER-mis; *vermis*, worm). Each cerebellar hemisphere consists of two lobes: a smaller **anterior lobe,** which is directly inferior to the cerebrum, and a

Figure 25.6 **The Basal Nuclei** The basal nuclei are masses of gray matter deep in the cerebrum.

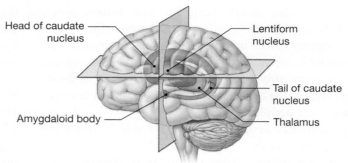

a The relative positions of the basal nuclei in the brain

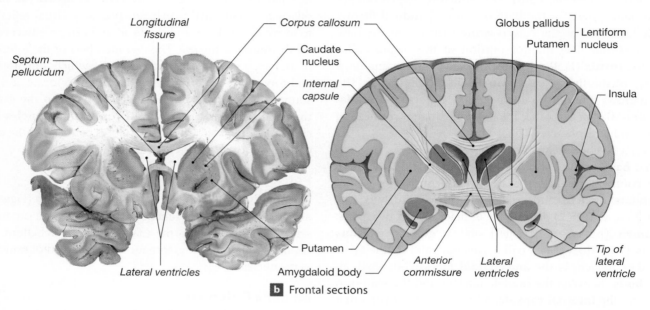

b Frontal sections

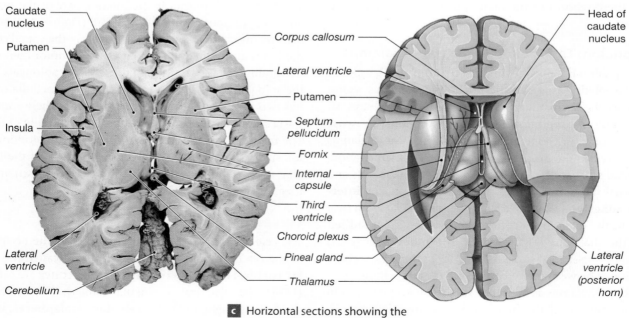

c Horizontal sections showing the caudate nucleus immediately lateral to the ventricles

Figure 25.7 **The Brain in Midsagittal and Frontal Sections** Midsagittal and frontal sections show the relationship among internal structures of the brain.

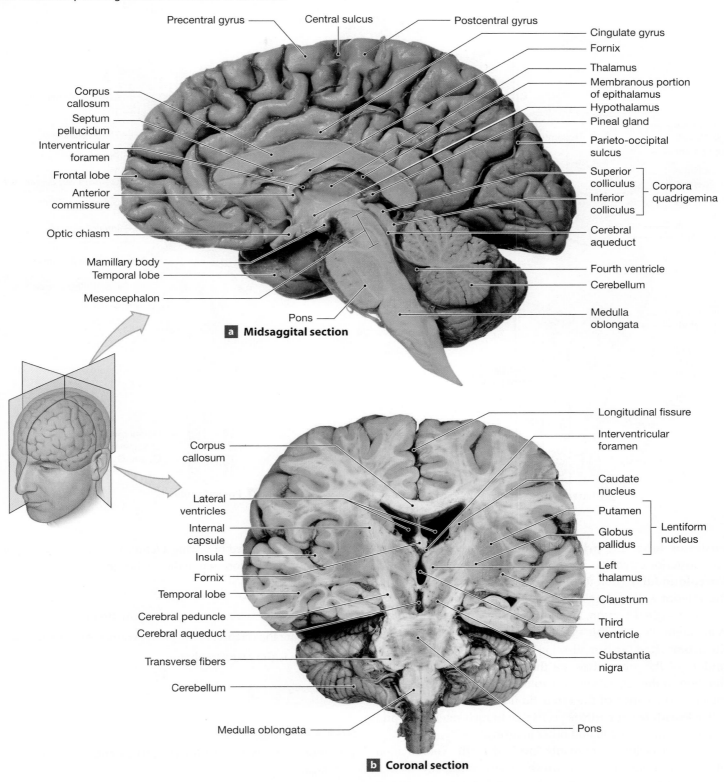

a Midsaggital section

b Coronal section

Figure 25.8 Brain Stem and Diencephalon The medulla, pons, and mesencephalon constitute the brain stem.

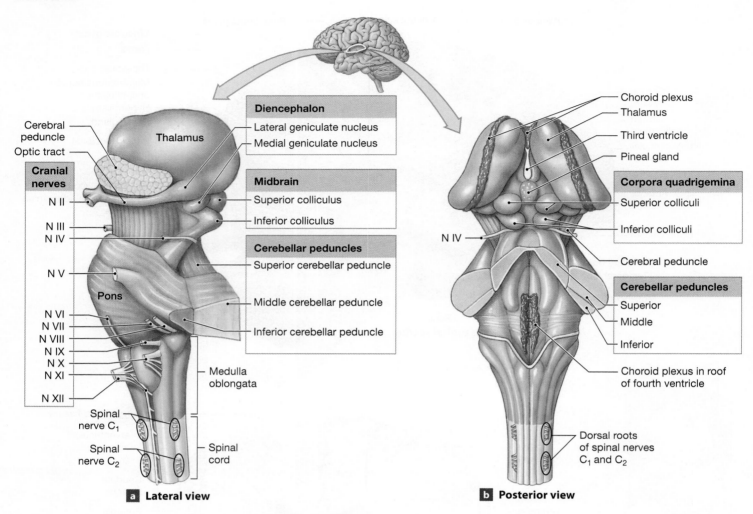

a **Lateral view**

b **Posterior view**

posterior lobe. The **primary fissure** separates the anterior and posterior cerebellar lobes. In a sagittal section, a smaller **flocculonodular** (flok-ū-lō-NOD-ū-lar) **lobe** is visible where the anterior wall of the cerebellum faces the pons.

In a sagittal section, the white matter of the cerebellum is apparent. Because this tissue is highly branched, it is called the **arbor vitae** (ar-bor VĪ-tē; *arbor*, tree + *vitae*, life). In the middle of the arbor vitae are the **cerebellar nuclei,** which function in the regulation of involuntary skeletal muscle contraction. The cortex of the cerebellum contains large neurons called **Purkinje** (pur-KIN-jē) cells that branch extensively and synapse with up to 200,000 other neurons.

The cerebellum is primarily involved in the coordination of somatic motor functions, which means principally skeletal muscle contractions. Adjustments to postural muscles occur when impulses from the cranial nerve of the inner ear pass into the flocculonodular lobe, the part of the cerebellum where information concerning equilibrium is processed. Learned muscle patterns,

such as those involved in serving a tennis ball or playing the guitar, are stored and processed in the cerebellum.

QuickCheck Questions

2.1 What are the six major regions of the brain?

2.2 How are the cerebral hemispheres connected to each other?

2.3 Where is the mesencephalon?

In the Lab 2

Materials

☐ Brain model (midsagittal, frontal, and horizontal sections)
☐ Brain chart
☐ Preserved and sectioned human brain (if available)
☐ Compound microscope
☐ Microscope slide of cerebellar cortex

Figure 25.9 **Cerebellum** The cerebellum is posterior to the brain stem.

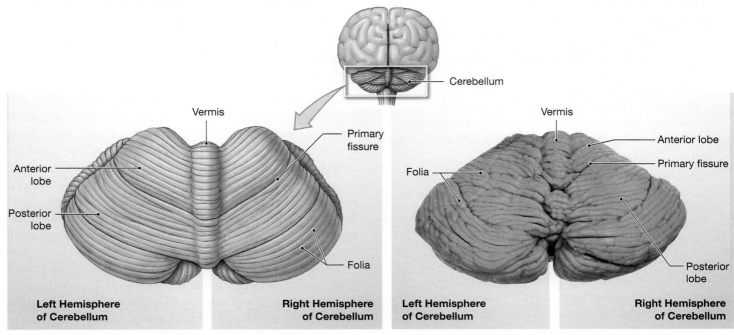

Vermis

Primary fissure

Anterior lobe

Posterior lobe

Folia

Left Hemisphere of Cerebellum

Right Hemisphere of Cerebellum

Vermis

Anterior lobe

Primary fissure

Folia

Posterior lobe

Left Hemisphere of Cerebellum

Right Hemisphere of Cerebellum

a The posterior, superior surface of the cerebellum, showing major anatomical landmarks and regions

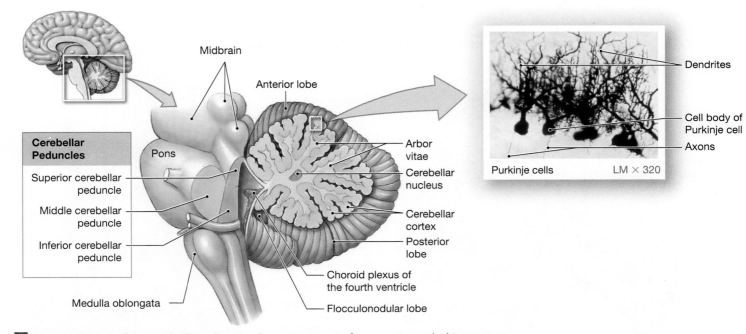

Midbrain

Anterior lobe

Dendrites

Cell body of Purkinje cell

Axons

Purkinje cells   LM × 320

Cerebellar Peduncles

Pons

Superior cerebellar peduncle

Middle cerebellar peduncle

Inferior cerebellar peduncle

Medulla oblongata

Arbor vitae

Cerebellar nucleus

Cerebellar cortex

Posterior lobe

Choroid plexus of the fourth ventricle

Flocculonodular lobe

b A sectional view of the cerebellum, showing the arrangement of gray matter and white matter

Procedures

1. Review the brain anatomy in Figures 25.5 through 25.9.
2. On the brain model, identify the following:
   - *Cerebrum* Note how the longitudinal fissure separates it into two cerebral hemispheres. Identify the five lobes of each hemisphere, along with the central sulcus, precentral gyri, and postcentral gyri. View the brain model in midsagittal section and identify the corpus callosum, fornix, and anterior commissure. In a frontal section and a horizontal section, locate the internal capsule, lentiform nucleus, and caudate nucleus.

Distinguish between the putamen and the globus pallidus of the lentiform nucleus.

- *Diencephalon* In a midsagittal section of the brain model, identify the thalamus, recognizable as the lateral wall around the diencephalon, and the wedge-shaped hypothalamus, inferior to the thalamus. Observe the third ventricle around the thalamus and the interthalamic adhesion. Identify the infundibulum, which attaches the pituitary gland to the hypothalamus. Locate the mamillary bodies and pineal gland.

- *Brain stem* Identify the medulla oblongata, pons, and mesencephalon. Locate the two pyramids on the medulla's anterior surface and the cerebral peduncles on the lateral sides of the mesencephalon. Identify the corpora quadrigemina of the mesencephalon, distinguishing between the superior and inferior colliculi.

- *Cerebellum* Locate the right and left hemispheres and the vermis separating them. In each hemisphere, identify the primary fissure and the anterior and posterior lobes. In a midsagittal section, locate the arbor vitae and the cerebellar nuclei.

3. Observe the cerebellar cortex slide and identify the Purkinje cells. Note the large size and many branched dendrites and a single thin axon. ■

Lab Activity 3 Cranial Nerves

Cranial nerves emerge from the brain at specific locations and pass through various foramina of the skull to reach the peripheral structures they innervate. Like spinal nerves, cranial nerves occur in pairs, 12 pairs in the case of the cranial nerves. The nerves are identified by name and are numbered with Roman numerals from N I to N XII. The numbers are assigned according to the locations at which the nerves contact the brain, with N I being most anterior and N XII most posterior. Some cranial nerves are entirely sensory nerves, but most are mixed. However, those mixed nerves that conduct primarily motor commands are considered motor nerves even though they have a few sensory fibers to inform the brain about muscle tension and position. **Figure 25.10** shows the position of each cranial nerve on the inferior surface of the brain. **Table 25.1** summarizes the cranial nerves and includes the foramen through which each nerve passes.

Figure 25.10 Origins of the Cranial Nerves Twelve pairs of cranial nerves connect the brain to organs mostly in the head and neck.

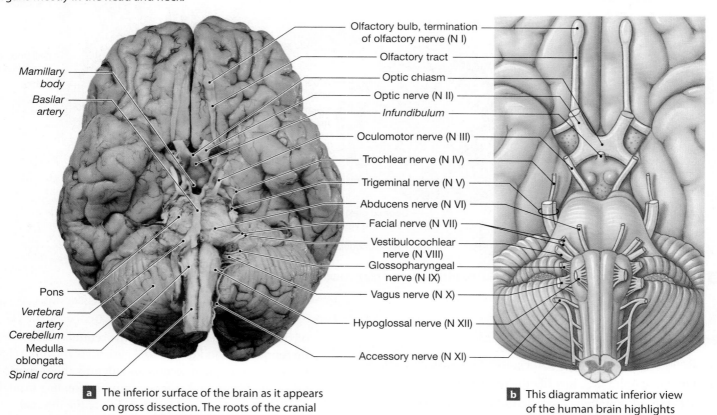

Mamillary body
Basilar artery
Pons
Vertebral artery
Cerebellum
Medulla oblongata
Spinal cord

Olfactory bulb, termination of olfactory nerve (N I)
Olfactory tract
Optic chiasm
Optic nerve (N II)
Infundibulum
Oculomotor nerve (N III)
Trochlear nerve (N IV)
Trigeminal nerve (N V)
Abducens nerve (N VI)
Facial nerve (N VII)
Vestibulocochlear nerve (N VIII)
Glossopharyngeal nerve (N IX)
Vagus nerve (N X)
Hypoglossal nerve (N XII)
Accessory nerve (N XI)

a The inferior surface of the brain as it appears on gross dissection. The roots of the cranial nerves are clearly visible.

b This diagrammatic inferior view of the human brain highlights the origins of the cranial nerves.

Table 25.1	**Cranial Nerve Branches and Functions**				
Cranial Nerve (Number)	**Sensory Ganglion**	**Branch**	**Primary Function**	**Foramen**	**Innervation**
Olfactory (I)			Special sensory	Olfactory foramina of ethmoid	Olfactory epithelium
Optic (II)			Special sensory	Optic canal	Retina of eye
Oculomotor (III)			Motor	Superior orbital fissure	Inferior, medial, superior rectus, inferior oblique, and levator palpebrae superioris muscles; intrinsic eye muscles
Trochlear (IV)			Motor	Superior orbital fissure	Superior oblique muscle
Trigeminal (V)	Semilunar		Mixed	Superior orbital fissure	Areas associated with the jaws
		Ophthalmic	Sensory	Superior orbital fissure	Orbital structures, nasal cavity, skin of forehead, upper eyelid, eyebrows, and nose (part)
		Maxillary		Foramen rotundum	Lower eyelid; superior lip, gums, and teeth; cheek, nose (part) palate, and pharynx (part)
		Mandibular		Foramen ovale	*Sensory:* inferior gums, teeth, lips, palate (part), and tongue (part) *Motor:* muscles of mastication
Abducens (VI)			Motor	Superior orbital fissure	Lateral rectus muscle
Facial (VII)	Geniculate		Mixed	Internal acoustic canal to facial canal; exits at stylomastoid foramen	*Sensory:* taste receptors on anterior 2/3 of tongue *Motor:* muscles of facial expression, lacrimal gland, submandibular gland, and sublingual salivary glands
Vestibulocochlear (Acoustic) (VIII)		Cochlear Vestibular	Special sensory	Internal acoustic canal	Cochlea (receptors for hearing) Vestibule (receptors for motion and balance)
Glossopharyngeal (IX)	Superior (jugular) and inferior (petrosal)		Mixed	Jugular foramen	*Sensory:* posterior 1/3 of tongue; pharynx and palate (part); receptors for blood pressure, pH, oxygen, and carbon dioxide concentrations *Motor:* pharyngeal muscles and parotid salivary gland
Vagus (X)	Jugular and nodose		Mixed	Jugular foramen	*Sensory:* pharynx; auricle and external acoustic canal; diaphragm; visceral organs in thoracic and abdominopelvic cavities *Motor:* palatal and pharyngeal muscles and visceral organs in thoracic and abdominopelvic cavities
Accessory (XI)		Internal	Motor	Jugular foramen	Skeletal muscles of palate, pharynx, and larynx (with vagus nerve)
		External	Motor	Jugular foramen	Sternocleidomastoid and trapezius muscles
Hypoglossal (XII)			Motor	Hypoglossal canal	Tongue musculature

Make a Prediction

Consider the major sensory organs of the head and predict how many cranial nerves are sensory nerves.

Olfactory Nerve (N I)

The **olfactory nerve** is composed of bundles of sensory fibers for the sense of smell and is located in the roof of the nasal cavity. The nerve passes through the cribriform plate of the ethmoid bone and enters an enlarged **olfactory bulb,** which then extends into the cerebrum as the **olfactory tract.**

Optic Nerve (N II)

The **optic nerve** carries visual information. This nerve originates in the retina, the neural part of the eye that is sensitive to changes in the amount of light entering the eye. The nerve is easy to identify as the X-shaped structure at the **optic chiasm** inferior to the hypothalamus. It is at this point that some of the sensory fibers cross to the nerve on the opposite side of the brain. The optic nerve enters the thalamus, which relays the visual signal to the occipital lobe. Some of the fibers enter the superior colliculus for visual reflexes.

Oculomotor Nerve (N III)

The **oculomotor nerve** innervates four extraocular eye muscles—the superior, medial, and inferior rectus muscles, and the inferior oblique muscle—and the levator palpebrae muscle of the eyelid. Autonomic motor fibers also control the intrinsic muscles of the iris and the ciliary body. The oculomotor nerve is located on the ventral mesencephalon just posterior to the optic nerve.

Trochlear Nerve (N IV)

The **trochlear** (TRŌK-lē-ar) **nerve** supplies motor fibers to the superior oblique muscle of the eye and originates where the mesencephalon joins the pons. The root of the nerve exits the mesencephalon on the lateral surface. Because it is easily cut or twisted off during removal of the dura mater, many dissection specimens do not have this nerve intact. The superior oblique eye muscle passes through a trochlea, or "pulley"; hence the name of the nerve.

Trigeminal Nerve (N V)

The **trigeminal** (trī-JEM-i-nal) **nerve** is the largest of the cranial nerves. It is located on the lateral pons near the medulla oblongata and services much of the face. In life, the nerve has three branches: *ophthalmic, maxillary,* and *mandibular.* The ophthalmic branch innervates sensory structures of the forehead, eye orbit, and nose. The maxillary branch contains sensory fibers for structures in the roof of the mouth, including half of the maxillary teeth. The mandibular branch carries the motor portion of the nerve to the muscles of mastication. Sensory signals from the lower lip, gum, muscles of the tongue, and one-third of the mandibular teeth are also part of the mandibular branch.

Abducens Nerve (N VI)

The **abducens** (ab-DŪ-senz) **nerve** controls the lateral rectus extraocular muscle. When this muscle contracts, the eyeball is abducted; hence the name. The nerve originates on the medulla oblongata and is positioned posterior and medial to the trigeminal nerve.

Facial Nerve (N VII)

The **facial nerve** is located on the medulla oblongata, posterior and lateral to the abducens nerve. It is a mixed nerve, with sensory fibers for the anterior two-thirds of the taste buds and somatic and autonomic motor fibers. The somatic motor neurons innervate the muscles of facial expression, such as the zygomaticus muscle. Visceral motor neurons control the activity of the salivary glands, lacrimal (tear) glands, and nasal mucous glands.

Vestibulocochlear Nerve (N VIII)

The **vestibulocochlear nerve,** also called the *auditory nerve,* is a sensory nerve of the inner ear located on the medulla oblongata near the facial nerve. The vestibulocochlear nerve has two branches. The vestibular branch gathers information regarding the sense of balance from the vestibule and semicircular canals of the inner ear. The cochlear branch conducts auditory sensations from the cochlea, the organ of hearing in the inner ear.

Glossopharyngeal Nerve (N IX)

The **glossopharyngeal** (glos-ō-fah-RIN-jē-al) **nerve** is a mixed nerve of the tongue and throat. It supplies the medulla oblongata with sensory information from the posterior third of the tongue (remember, the facial nerve innervates the anterior two-thirds of the taste buds) and from the palate and pharynx. The glossopharyngeal nerve also conveys barosensory and chemosensory information from the carotid sinus and the carotid body, where blood pressure and dissolved blood gases are monitored, respectively. Motor innervation by the glossopharyngeal nerve controls the pharyngeal muscles involved in swallowing and in the activity of the salivary glands.

Vagus Nerve (N X)

The **vagus** (VĀ-gus) **nerve** is a complex nerve on the medulla oblongata that has mixed sensory and motor functions. Sensory neurons from the pharynx, diaphragm, and most of the internal organs of the thoracic and abdominal cavities ascend along the vagus nerve and synapse with autonomic nuclei in the medulla. The motor portion controls the involuntary muscles of the respiratory, digestive, and cardiovascular systems. The vagus is the only cranial nerve to descend below the neck. It enters the ventral body cavity, but it does not pass to the thorax via the spinal cord; rather, it follows the musculature of the neck. Because this nerve regulates the activities of the organs of the thoracic and abdominal cavities, disorders of the nerve result in systemic disruption of homeostasis. Parasympathetic fibers in the vagus nerve control swallowing, digestion, heart rate, and respiratory patterns. If this control is compromised, sympathetic stimulation goes unchecked, and the organs respond as during exercise or stress. The cardiovascular and respiratory systems increase their activities, and the digestive system shuts down.

Accessory Nerve (N XI)

The **accessory nerve** is a motor nerve controlling the skeletal muscles involved in swallowing and the sternocleidomastoid and trapezius muscles of the neck. It is the only cranial nerve with fibers originating from both the medulla oblongata and the spinal cord. Numerous threadlike branches from these two regions unite in the spinal accessory nerve.

Hypoglossal Nerve (N XII)

The **hypoglossal** (hī-pō-GLOS-al) **nerve** is located on the medulla oblongata medial to the vagus nerve. This motor nerve supplies motor fibers that control tongue movements for speech and swallowing.

QuickCheck Questions

3.1 List three cranial nerves that are sensory nerves.

3.2 Which cranial nerve enters the ventral body cavity?

In the Lab 3

Materials

☐ Brain model

☐ Brain chart

☐ Isopropyl (rubbing) alcohol

☐ Wintergreen oil

☐ Eye chart

☐ Sugar solution

☐ Quinine solution

☐ Tuning fork

☐ Beaker of ice and cold probes

☐ Beaker of warm water and warm probes

Procedures

1. Review the cranial nerves in Figure 25.10.

2. Locate each cranial nerve on the brain model and chart.

3. Your instructor may ask you to test the function of selected cranial nerves. **Table 25.2** lists the basic tests used to assess the general function of each nerve.

For additional practice, complete the *Sketch to Learn* activity. ∎

 Sketch to Learn

To help you remember the cranial nerves, let's draw each nerve on the provided picture of the brain. Pay close attention to the location of the base of each nerve and identify that part of the brain where the nerve emerges. Refer to Figure 25.10 for each nerve's exact location.

Sample Sketch

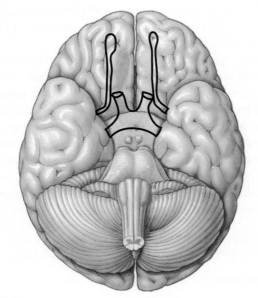

Your Sketch

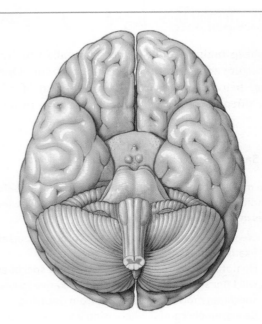

Step 1
· Draw the olfactory nerve lateral to the longitudinal fissure.
· Make a large x-shape for the optic nerve and chiasma.

Step 2
· Use the sample as reference for nerve placement and add the remaining nerves to the brain template.
· Label each nerve.

Table 25.2	Cranial Nerve Tests
Cranial Nerve	**Nerve Function Test**
I. Olfactory	Hold open container of rubbing alcohol under subject's nose and have subject identify odor. Repeat with open container of wintergreen oil.
II. Optic	Test subject's visual field by moving a finger back and forth in front of subject's eyes. Use eye chart to test visual acuity.
III. Oculomotor	Examine subject's pupils for equal size. Have subject follow an object with eyes.
IV. Trochlear	Tested with oculomotor nerve. Have subject roll eyes downward.
V. Trigeminal	Check motor functions of nerve by having subject move mandible in various directions. Check sensory functions with warm and cold probes on forehead, upper lip, and lower jaw.
VI. Abducens	Tested with oculomotor nerve. Have subject move eyes laterally.
VII. Facial	Use sugar solution to test anterior of tongue for sweet taste reception. Observe facial muscle contractions for even muscle tone on each side of face while subject smiles, frowns, and purses lips.
VIII. Vestibulocochlear	Cochlear branch—Hold vibrating tuning fork in air next to ear, and then touch fork to mastoid process for bone-conduction test. Vestibular branch—Have subject close eyes and maintain balance.
IX. Glossopharyngeal	While subject coughs, check position of uvula on posterior of soft palate. Use quinine solution to test posterior of tongue for bitter taste reception.
X. Vagus	While subject coughs, check position of uvula on posterior of soft palate.
XI. Spinal accessory	Hold subject's shoulder while the subject rotates it to test the strength of sternocleidomastoid muscle. Hold head while subject rotates it to test trapezius strength.
XII. Hypoglossal	Observe subject protract and retract tongue from mouth, and check for even movement on two lateral edges of tongue.

Lab Activity 4 Sheep Brain Dissection

The sheep brain, like all other mammalian brains, is similar in structure and function to the human brain. One major difference between the human brain and that of other animals is the orientation of the brain stem relative to the body axis.

 Safety Alert: Brain Dissection

You *must* practice the highest level of laboratory safety while handling and dissecting the brain. Keep the following guidelines in mind during the dissection.

1. Wear gloves and safety glasses to protect yourself from the fixatives used to preserve the specimen.
2. Do not dispose of the fixative from your specimen. You will later store the specimen in the fixative to keep the specimen moist and prevent it from decaying.
3. Be extremely careful when using a scalpel or other sharp instrument. Always direct cutting and scissor motions away from you to prevent an accident if the instrument slips on moist tissue.
4. Before cutting a given tissue, make sure it is free from underlying and/or adjacent tissues so that they will not be accidentally severed.
5. Never discard tissue in the sink or trash. Your instructor will inform you of the proper disposal procedure. ▲

The human body has a vertical axis, and the brain stem and spinal cord are positioned vertically. In four-legged animals, the body axis is horizontal and the brain stem and spinal cord are also horizontal.

All vertebrate animals—sharks, fish, amphibians, reptiles, birds, and mammals—have a brain stem for basic body functions. These animals can learn through experience, a complex neurological process that requires higher-level processing and memory storage, as occurs in the human cerebrum. Imagine the complex motor activity necessary for locomotion in these animals.

Dissecting a sheep brain enhances your study of models and charts of the human brain. Take your time during the dissection and follow the directions carefully. Refer to this manual and its illustrations often during the procedures.

In the Lab 4

Materials

☐ Gloves
☐ Safety glasses
☐ Preserved sheep brain (preferably with dura mater intact)
☐ Dissection pan
☐ Scissors
☐ Blunt probe
☐ Large dissection knife

Procedures

Put on gloves and safety glasses before handling the brain.

I. The Meninges

If your sheep brain does not have the dura mater, skip to part II.

1. On the intact dura mater, locate the falx cerebri and the tentorium cerebelli on the overlying dorsal surface of the dura mater. How does the tissue of the falx cerebri compare with the dura mater covering the hemispheres?

2. If your specimen still has the ethmoid, a mass of bone on the anterior frontal lobe, slip a probe between the bone and the dura mater. Carefully pull the bone off the specimen, using scissors to snip away any attached dura mater. Examine the removed ethmoid and identify the crista galli, which is the crest of bone where the meninges attach.

3. Gently insert a probe between the dura mater and the brain and gently work the probe back and forth to separate the two. With scissors, cut completely around the base of the dura mater, leaving the inferior portion intact over the cranial nerves. Make small cuts with the scissors and be careful not to cut or remove any of the cranial nerves. Do not lift the dura too high or the cranial nerves will detach from the brain.

4. Cut completely through the lateral sides of the tentorium cerebelli and then remove the dura mater in one piece by grasping it with your (gloved) hand and peeling it off the brain.

5. Open the detached dura mater and identify the falx cerebri and tentorium cerebelli. (One difference between the sheep brain and the human brain is that the sheep brain does not have a falx cerebelli.)

II. External Brain Anatomy

1. Examine the cerebrum, identifying the frontal, parietal, occipital, and temporal lobes. The insula is a deep lobe and is not visible externally. Note the longitudinal fissure separating the right and left cerebral hemispheres. Observe the gyri and sulci on the cortical surface. Examine the surface between sulci for the arachnoid mater and pia mater.

2. Identify the cerebellum and compare the size of the folia with the size of the cerebral gyri. Unlike the human brain, the sheep cerebellum is not divided medially into two lateral hemispheres.

3. To examine the dorsal anatomy of the mesencephalon, position the sheep brain as in Figure 25.11 and use your fingers to gently depress the cerebellum. The mesencephalon will then be visible between the cerebrum and cerebellum. Now identify the four elevated masses of the corpora quadrigemina and distinguish between the superior colliculi and the inferior colliculi. The pineal gland of the diencephalon is superior to the mesencephalon.

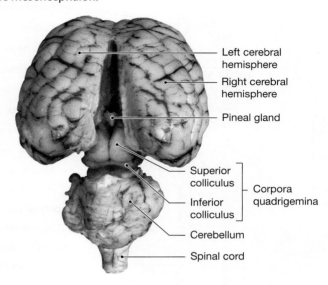

Figure 25.11 Dorsal View of the Sheep Brain The cerebellum is pushed down to show the location of the corpora quadrigemina of the mesencephalon.

Left cerebral hemisphere
Right cerebral hemisphere
Pineal gland
Superior colliculus
Inferior colliculus
Corpora quadrigemina
Cerebellum
Spinal cord

4. Turn the brain over to view the ventral surface, as in Figure 25.12. Note how the spinal cord joins the medulla oblongata. Identify the pons and the cerebral peduncles of the mesencephalon. Locate the single mamillary body on the hypothalamus. (Remember that the mamillary body of the human brain is a *paired* mass.) The pituitary gland has most likely been removed from your specimen; however, you can still identify the stub of the infundibulum that attaches the pituitary to the hypothalamus.

5. Using Figure 25.12 as a guide, identify as many cranial nerves on your sheep brain as possible. Nerves I through III and nerve V are usually intact and easy to identify. Your laboratory instructor may ask you to observe several sheep brains in order to study all the cranial nerves. The three branches of the trigeminal nerve were cut when the brain was removed from the sheep and therefore are not present on any specimen. The glossopharyngeal nerve may have been removed inadvertently when the specimen was being prepared. Even if this nerve is present in your specimen, however, it is difficult to identify on the sheep brain.

III. Internal Brain Anatomy—Sagittal and Frontal Sections

Sagittal Section

1. To study the interal organization of the brain, make a midsagittal section to expose the deep structures. Lay the sheep brain in the dissection pan so that the superior surface faces you, as in Figure 25.11. Place the blade of a large dissection knife in the anterior region of the longitudinal fissure and section the brain by cutting it

Figure 25.12 Ventral View of the Sheep Brain Cranial nerves are clearly visible in the ventral view of the sheep brain.

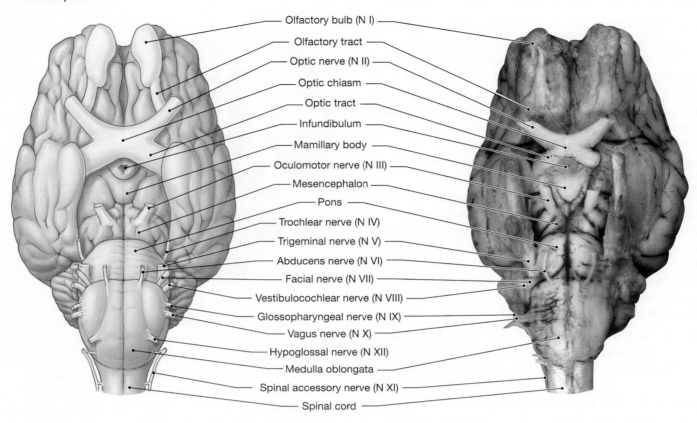

Olfactory bulb (N I)
Olfactory tract
Optic nerve (N II)
Optic chiasm
Optic tract
Infundibulum
Mamillary body
Oculomotor nerve (N III)
Mesencephalon
Pons
Trochlear nerve (N IV)
Trigeminal nerve (N V)
Abducens nerve (N VI)
Facial nerve (N VII)
Vestibulocochlear nerve (N VIII)
Glossopharyngeal nerve (N IX)
Vagus nerve (N X)
Hypoglossal nerve (N XII)
Medulla oblongata
Spinal accessory nerve (N XI)
Spinal cord

in half along the fissure. Use as few cutting strokes as possible to prevent damage to the brain tissue.

2. Using **Figure 25.13** as a guide, identify the internal anatomical features of the sheep brain. Gently slide a blunt probe between the corpus callosum and fornix and into the lateral ventricle to determine how deep the ventricle extends into the cerebrum. Inside the lateral ventricle, locate the choroid plexus, which appears as a granular mass of tissue.

Frontal Section

1. To view deep structures of the cerebrum and diencephalon, put the two halves of the brain together and use a large dissection knife to cut a frontal section through the infundibulum. Make another frontal section

just posterior to the first to slice off a thin slab of brain. Lay the slab in the dissection pan with the anterior side up. (The anterior side is the surface where you made your first cut.)

2. Using **Figure 25.14** as a guide, notice the distribution of gray matter and white matter. Observe how the corpus callosum joins each cerebral hemisphere. Lateral to the lateral ventricles is the gray matter of the basal nuclei.

IV. Cleanup and Disposal of Brain

When finished, store or discard the sheep brain as directed by your laboratory instructor. Proper disposal of all biological waste protects the local environment and is mandated by local, state, and federal regulations. ■

Figure 25.13 Midsagittal Section of the Sheep Brain Internal anatomy of the sheep brain in sagittal section.

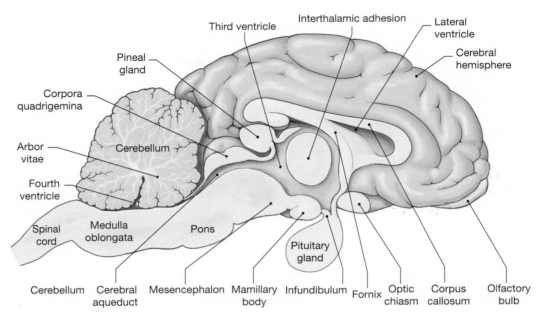

Third ventricle

Interthalamic adhesion

Lateral ventricle

Pineal gland

Cerebral hemisphere

Corpora quadrigemina

Arbor vitae

Cerebellum

Fourth ventricle

Spinal cord

Medulla oblongata

Pons

Pituitary gland

Cerebellum   Cerebral aqueduct   Mesencephalon   Mamillary body   Infundibulum   Fornix   Optic chiasm   Corpus callosum   Olfactory bulb

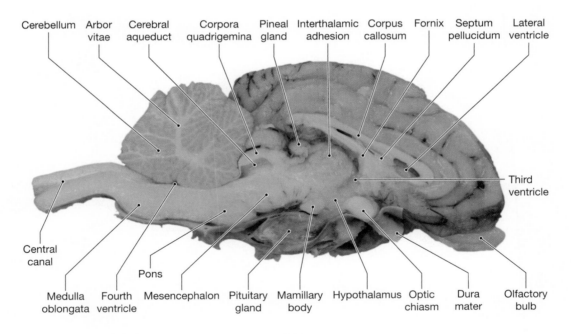

Cerebellum   Arbor vitae   Cerebral aqueduct   Corpora quadrigemina   Pineal gland   Interthalamic adhesion   Corpus callosum   Fornix   Septum pellucidum   Lateral ventricle

Third ventricle

Central canal

Pons

Medulla oblongata   Fourth ventricle   Mesencephalon   Pituitary gland   Mamillary body   Hypothalamus   Optic chiasm   Dura mater   Olfactory bulb

Figure 25.14 Frontal Section of the Sheep Brain Internal anatomy of the sheep brain in frontal section.

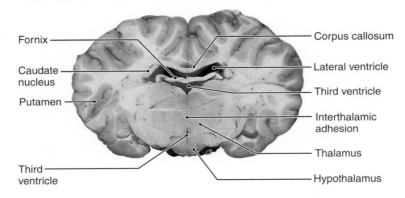

Fornix

Corpus callosum

Caudate nucleus

Lateral ventricle

Putamen

Third ventricle

Interthalamic adhesion

Thalamus

Third ventricle

Hypothalamus

Review & Practice Sheet

EXERCISE 25

Name ______________________

Date ______________________

Section ______________________

Anatomy of the Brain

A. Description

Describe each of the following structures.

1. cerebrum

2. mamillary body

3. longitudinal fissure

4. optic chiasm

5. falx cerebri

6. hypothalamus

7. dura mater

8. vermis

9. subarachnoid space

10. septum pellucidum

11. thalamus

12. corpus callosum

13. pineal gland

14. superior colliculus

15. arbor vitae

B. Labeling

1. Label **Figure 25.15**, which shows the inferior surface of the brain.

Figure 25.15 Inferior Surface of the Human Brain

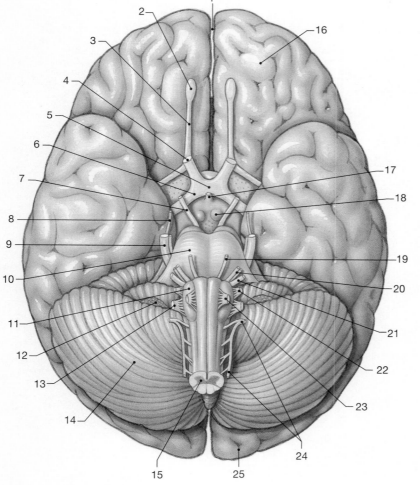

1. _____
2. _____
3. _____
4. _____
5. _____
6. _____
7. _____
8. _____
9. _____
10. _____
11. _____
12. _____
13. _____
14. _____
15. _____
16. _____
17. _____
18. _____
19. _____
20. _____
21. _____
22. _____
23. _____
24. _____
25. _____

2. Label Figure 25.16, a midsagittal close-up view of the human brain.

Figure 25.16 **Detail of Sagittal Section of the Human Brain**

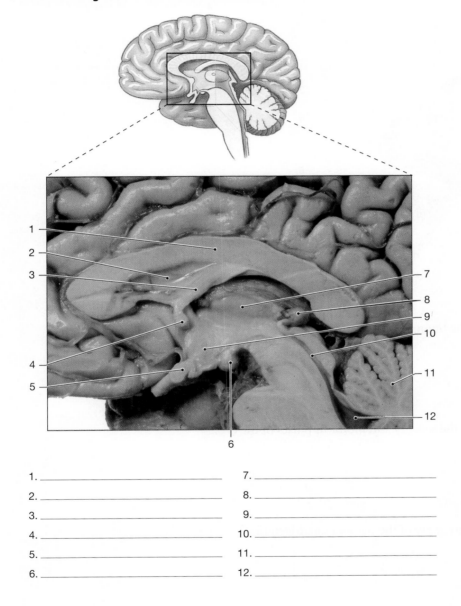

1. _____   7. _____

2. _____   8. _____

3. _____   9. _____

4. _____   10. _____

5. _____   11. _____

6. _____   12. _____

3. Label **Figure 25.17**, a cast of the ventricles of the brain.

Figure 25.17 Ventricles of the Brain A cast of the ventricles.

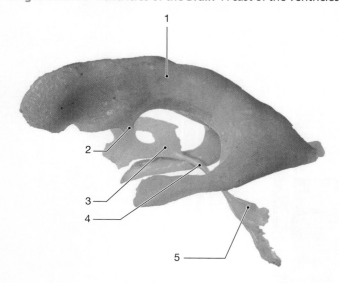

1. _____

2. _____

3. _____

4. _____

5. _____

C. Short-Answer Questions

1. List the six major regions of the brain.

2. Which cranial nerves conduct the sensory and motor impulses of the eye?

3. List the location and specific anatomy of the corpora quadrigemina.

4. Describe the extensions of the dura mater.

5. What is the function of the precentral gyrus?

6. Trace a drop of CSF from a lateral ventricle to reabsorption at an arachnoid granulation.

D. Analysis and Application

1. Imagine watching a bird fly across your line of vision. What part of your brain is active in keeping an image of the moving bird on your retina?

2. You have just eaten a medium-sized pepperoni pizza and are now laying down to digest it. Which cranial nerve stimulates the muscular activity of your digestive tract?

3. A child is preoccupied with a large cherry lollipop. What part of the child's brain is responsible for the licking and eating reflexes?

4. Your favorite movie has made you cry yet again. Which cranial nerve is responsible for your tears?

E. Clinical Challenge

1. A patient is brought into the emergency room with severe whiplash. He is not breathing and has lost cardiac function. What part of the brain has most likely been damaged?

2. A woman is admitted to the hospital with Bell's palsy caused by an inflamed facial nerve. What symptoms will you, as the attending physician, observe, and how would you test her facial nerve?

Autonomic Nervous System

Learning Outcomes

On completion of this exercise, you should be able to:

1. Compare the location of the preganglionic outflow from the CNS in the sympathetic and parasympathetic divisions.

2. Compare the lengths of and the neurotransmitters released by each fiber in the sympathetic and parasympathetic divisions.

3. Trace the sympathetic pathways into a chain ganglion, into a collateral ganglion, and into the adrenal medulla.

4. Trace the parasympathetic pathways into cranial nerves III, VII, IX, and X, and into the pelvic nerves.

5. Compare the responses to sympathetic and parasympathetic innervation.

Lab Activities

1 The Sympathetic (Thoracolumbar) Division 375

2 The Parasympathetic (Craniosacral) Division 380

Clinical Application

Stress and the ANS 375

The autonomic nervous system (ANS) controls the motor and glandular activity of the visceral effectors. Most internal organs have **dual innervation** in that they are innervated by both sympathetic and parasympathetic nerves of the ANS. Thus, the two divisions of the ANS share the role of regulating autonomic function. Typically, one division stimulates a given effector, and the other division inhibits that same effector. Autonomic motor pathways originate in the brain and enter the cranial and spinal nerves.

An autonomic pathway consists of two groups of neurons, both of which have names that reflect the fact that they synapse with one another in bulblike PNS structures called **ganglia** (GANG-lē-uh). An autonomic neuron between the CNS and a sympathetic or parasympathetic ganglion is called a **preganglionic neuron;** an autonomic neuron between the ganglion and the target muscle or gland is a

Need More Practice and Review?

Build your knowledge—and confidence!—in the Study Area of MasteringA&P® at www.masteringaandp.com with Pre-lab Quizzes, Post-lab Quizzes, Practice Anatomy Lab™ (PAL™) 3.0 virtual anatomy practice tool, PhysioEx™ 9.0 laboratory simulations, and A&P Flix™ with Quizzes.

PAL | practice anatomy lab For this lab exercise, follow these navigation paths:
- PAL>Human Cadaver>Nervous System>Autonomic Nervous System
- PAL>Anatomical Models>Nervous System>Autonomic Nervous System

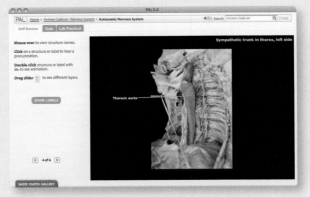

---

ganglionic neuron (Figure 26.1). The axons of autonomic neurons are called *fibers*. **Preganglionic fibers** are axons that synapse with ganglionic neurons in the ganglion, whereas **ganglionic fibers** synapse with the effectors: smooth muscles, the heart, and glands.

The preganglionic neurons of both divisions release acetylcholine (ACh) into a ganglion, but the ganglionic neurons of the two divisions release different neurotransmitters to the target effector cells. During times of excitement, emotional stress, and emergencies, sympathetic ganglionic neurons release norepinephrine (NE) to effectors and cause a sympathetic **fight-or-flight response** that increases overall alertness. Heart rate, blood pressure, and respiratory rate all increase, sweat glands secrete, and digestive and urinary functions cease. Parasympathetic ganglionic neurons release ACh, which slows the body for normal, energy-conserving homeostasis. This parasympathetic **rest-and-digest response** decreases cardiovascular and respiratory activity and increases the rate at which food and wastes are processed and eliminated.

Make a Prediction

A 65-year-old male has outpatient surgery. While in the recovery room, he is told that he may go home once he urinates. Why is urination a good indicator that it is safe to allow the patient to go home?

The two major anatomical differences between the sympathetic and parasympathetic subdivisions of the ANS are the location of preganglionic exit points from the CNS and the location of autonomic ganglia in the PNS.

Figure 26.1 An Overview of ANS Pathways Sympathetic pathways consist of short preganglionic neurons that release acetylcholine (ACh) in sympathetic ganglia. They synapse with long ganglionic neurons that release norepinephrine (NE) at an effector. The sympathetic response is generalized as a fight-or-flight response. Parasympathetic pathways have long preganglionic neurons that exit the CNS either directly from the brain (shown) or by passing down the spinal cord to the sacral region (not shown). They release ACh in terminal and intramural ganglia located in or near the effector organ. Preganglionic parasympathetic neurons synapse with short ganglionic neurons that also release ACh. The general parasympathetic response is a rest-and-digest response.

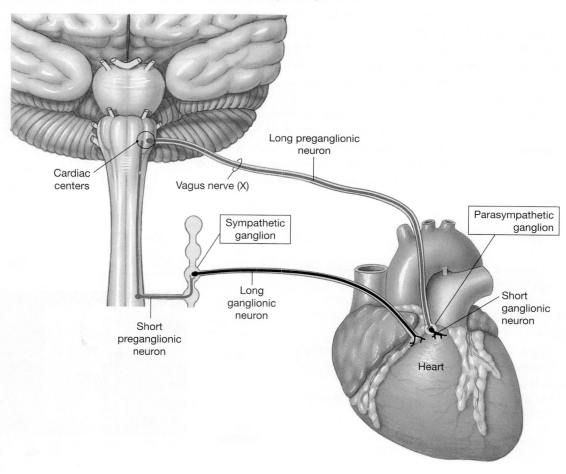

Cardiac centers
Vagus nerve (X)
Long preganglionic neuron
Sympathetic ganglion
Parasympathetic ganglion
Short preganglionic neuron
Long ganglionic neuron
Short ganglionic neuron
Heart

Location of Preganglionic Exit Points from CNS

Sympathetic preganglionic neurons exit the spinal cord at segments T_1 through L_2 and enter the thoracic and first two lumbar spinal nerves. Because of this nerve distribution, the sympathetic division is also called the **thoracolumbar** (tho-ra-kō-LUM-bar) **division.** In the parasympathetic division, the efferent neurons originating in the brain either exit the cranium in certain cranial nerves or descend the spinal cord and exit at the sacral level. The parasympathetic division is also called the **craniosacral** (krā-nē-ō-SĀ-krul) **division** (Figure 26.1).

Location of Autonomic Ganglia in PNS

All autonomic ganglia are in the PNS, but their proximity to the CNS provides another difference between the sympathetic and parasympathetic divisions. Sympathetic ganglia are located close to the spinal cord. This location results in short sympathetic preganglionic neurons and long sympathetic ganglionic neurons. Parasympathetic ganglia are located either near or within the visceral effectors. With the ganglia farther away from the CNS, parasympathetic preganglionic neurons are long and parasympathetic ganglionic neurons are short. In Figure 26.1, notice both the difference in the locations of the sympathetic and parasympathetic ganglia and the difference in the preganglionic and ganglionic lengths.

Lab Activity 1 The Sympathetic (Thoracolumbar) Division

The organization of the sympathetic division of the ANS is diagrammed in **Figure 26.2**. Preganglionic neurons originate in the pons and the medulla of the brain stem. These autonomic

motor neurons descend in the spinal cord to the thoracic and lumbar segments, where their somae are located in the lateral gray horns. Preganglionic axons exit the spinal cord in ventral roots and pass into a spinal nerve which branches into a sympathetic chain ganglion. In the chain ganglion, the preganglionic neuron will either synapse with a ganglionic neuron or pass through the chain ganglion and synapse in a collateral ganglion or in the adrenal medulla.

In a typical sympathetic pathway, the short sympathetic preganglionic fibers release ACh at the synapse where ACh is excitatory to the ganglionic fiber. The long ganglionic axon then releases norepinephrine at its synapse with the effector. How the NE affects the effector depends on the type of NE receptors present in the effector's cell membrane. Generally, the sympathetic response is to prepare the body for increased activity or a crisis situation; this is the fight-or-flight response that occurs during exercise, excitement, and emergencies.

Figure 26.3 outlines the general distribution of the sympathetic pathways, showing the *sympathetic chain ganglia* positioned lateral to the lower portion of the spinal cord. For simplicity, the left side of the figure shows sympathetic nerves to structures of the skin, blood vessels, and adipose tissue; the right side of Figure 26.3 details sympathetic distribution to organs in the head and ventral body cavity. In real life, sympathetic distribution is the same on both sides of the spinal cord.

Three types of sympathetic ganglia occur in the body: sympathetic chain ganglia, collateral ganglia, and modified ganglia in the adrenal medulla. Preganglionic neurons extend from the thoracic and lumbar segments of the spinal cord and pass into sympathetic ganglia where they synapse with ganglionic neurons that exit the ganglia and innervate the effectors of the thoracic cavity, head, body wall, and limbs.

Sympathetic chain ganglia (Figure 26.3) are located lateral to the spinal cord and are also called **paravertebral ganglia.** All sympathetic preganglionic neurons pass through a sympathetic chain ganglion but only the ones that supply the head, body wall, and limbs will synapse in the chain with a ganglionic neuron. Neurons that supply the abdominopelvic cavity do not synapse in the chain ganglia; instead, they pass through the chain ganglia and synapse in collateral ganglia.

Collateral ganglia are located anterior to the vertebral column and contain ganglionic neurons that lead to organs in

Figure 26.2 Organization of the Sympathetic Division of the ANS Preganglionic neurons of the sympathetic division exit thoracic and lumbar segments of the spinal cord and enter sympathetic ganglia where they synapse with ganglionic neurons that supply the target organs with sympathetic control.

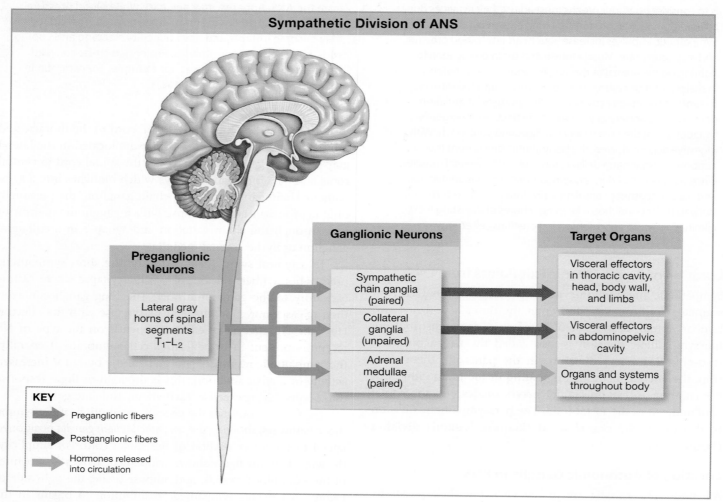

the abdominopelvic cavity. The preganglionic fibers associated with collateral ganglia pass through the sympathetic chain ganglia without synapsing and join to form a network called the **splanchnic** (SPLANK-nik) **nerves.** This network divides and sends branches into the collateral ganglia, where the preganglionic fibers synapse with ganglionic neurons. The ganglionic fibers then synapse with abdominopelvic effectors. The collateral ganglia are named after the adjacent blood vessels. The **celiac** (SĒ-lē-ak) **ganglion** supplies the liver, gallbladder, stomach, pancreas, and spleen. The **superior mesenteric** (mez-en-TER-ik) **ganglion** innervates the small intestine and parts of the large intestine. The **inferior mesenteric ganglion** controls most of the large intestine, the kidneys, the bladder, and the sex organs.

The third type of sympathetic ganglion is associated with the adrenal glands, also called *suprarenal glands*, which are positioned on top of the kidneys. Each adrenal gland has an outer cortex layer that produces hormones and an inner region called the **adrenal medulla.** It is this region that contains sympathetic ganglia and ganglionic neurons. During sympathetic stimulation, the ganglionic neurons in the medulla, like other sympathetic ganglionic neurons, release epinephrine into the bloodstream and contribute to the fight-or-flight response.

Sympathetic Pathways

Figure 26.4 shows the sympathetic pathways in more detail. The pathway utilizing the sympathetic chain ganglia passes through areas called the **white ramus** and the **gray ramus.** (Collectively, these two regions are known as the **rami communicantes.**) Once a preganglionic fiber enters a sympathetic chain ganglion via the white ramus, the fiber usually synapses with a ganglionic neuron, as shown in Figure 26.4a. The ganglionic fiber exits the sympathetic chain ganglion via either the gray ramus or an autonomic nerve. The gray ramus directs

Figure 26.3 Distribution of Sympathetic Innervation The distribution of sympathetic fibers is the same on both sides of the body. For clarity, the innervation of somatic structures is shown here on the left, and the innervation of visceral structures on the right.

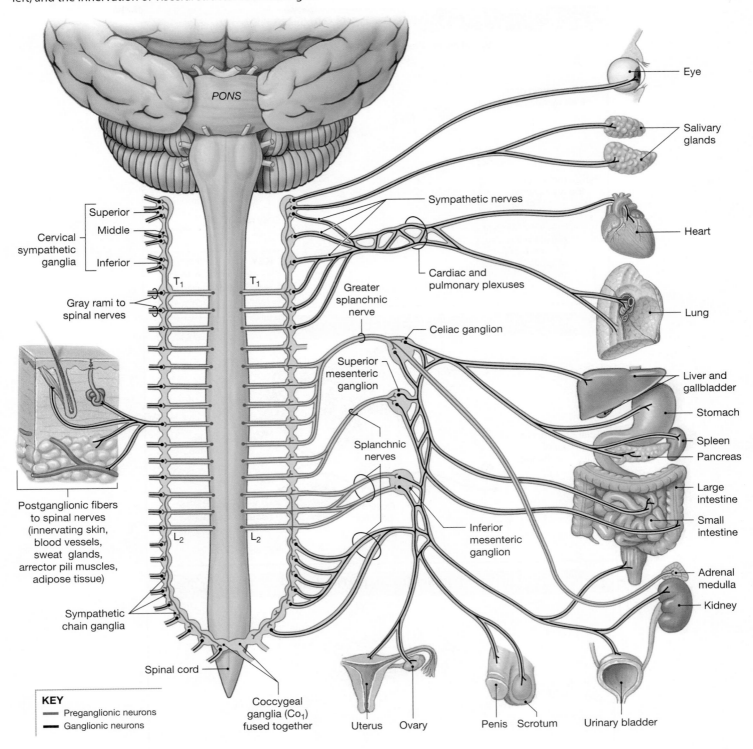

Figure 26.4 **Sympathetic Ganglia and Pathways** Sympathetic ganglia are located in three regions: Sympathetic chain ganglia are lateral to the spinal cord; collateral ganglia supply the abdominal organs and are anterior to the spinal cord; and the adrenal medullae are the middle portions of the adrenal (suprarenal) glands. Sympathetic preganglionic neurons synapse in the medullae and ganglionic neurons release NE into the blood.

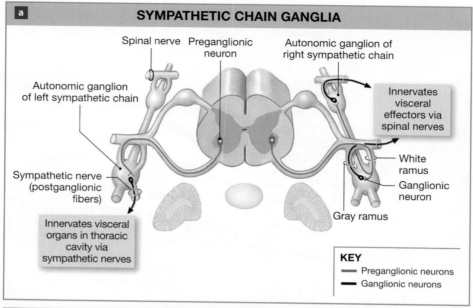

a **SYMPATHETIC CHAIN GANGLIA**

Spinal nerve
Preganglionic neuron
Autonomic ganglion of right sympathetic chain

Autonomic ganglion of left sympathetic chain

Innervates visceral effectors via spinal nerves

Sympathetic nerve (postganglionic fibers)

White ramus

Ganglionic neuron

Gray ramus

Innervates visceral organs in thoracic cavity via sympathetic nerves

KEY
— Preganglionic neurons
— Ganglionic neurons

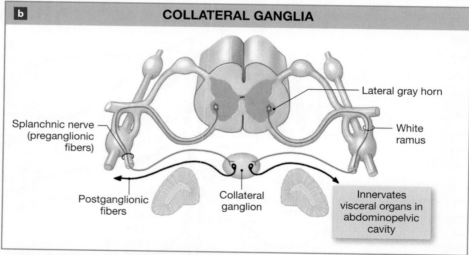

b **COLLATERAL GANGLIA**

Splanchnic nerve (preganglionic fibers)

Lateral gray horn

White ramus

Postganglionic fibers

Collateral ganglion

Innervates visceral organs in abdominopelvic cavity

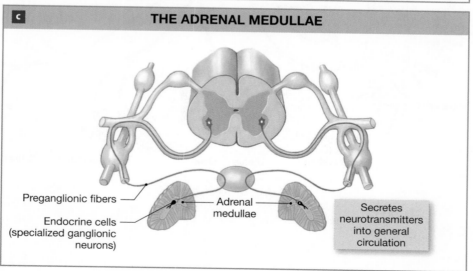

c **THE ADRENAL MEDULLAE**

Preganglionic fibers

Adrenal medullae

Endocrine cells (specialized ganglionic neurons)

Secretes neurotransmitters into general circulation

the ganglionic fiber into a spinal nerve leading to a general somatic structure, such as blood vessels supplying skeletal muscles. A ganglionic fiber in an autonomic nerve passes into the thoracic cavity to innervate the thoracic viscera.

Notice in Figure 26.4a that all the sympathetic chain ganglia on the same side of the spinal cord are interconnected. A single preganglionic neuron may enter one sympathetic chain ganglion and branch into many different chain ganglia, to synapse with up to 32 ganglionic neurons. This fanning out of preganglionic neurons within the sympathetic chain ganglia contributes to the widespread effect that sympathetic stimulation has on the body.

Figure 26.4b details the pathway involving collateral ganglia. Note how the preganglionic axons pass through the chain ganglia and enter the splanchnic nerve (described earlier) before entering the collateral ganglia. Sympathetic neurons supplying the adrenal gland do not synapse in a sympathetic chain ganglion or a collateral ganglion. Instead, the preganglionic fibers penetrate deep into the adrenal gland and synapse with ganglionic neurons in the adrenal medulla, as noted earlier.

QuickCheck Questions

1.1 Why is the sympathetic division of the ANS also called the thoracolumbar division?

1.2 What is the body's general response to sympathetic stimulation?

1.3 How do the heart, lungs, and digestive tract respond to sympathetic stimulation?

Study Tip Understanding Sympathetic Ganglia

Following is a brief summary of sympathetic ganglia:

- First, remember that all sympathetic ganglionic neurons enter chain ganglia.
- *Chain ganglia.* Sympathetic preganglionic neurons pass into chain ganglia, synapse with ganglionic neurons that exit the chain to innervate thoracic and integumentary organs.
- *Collateral ganglia.* Ganglionic neurons pass through chain ganglia and synapse with ganglionic neurons in collateral ganglia which supply organs in the abdomino-pelvic cavity.
- *Adrenal medullae.* Sympathetic ganglionic neurons pass through chain and collateral ganglia and enter the medulla of the adrenal glands where ganglionic neurons release adrenaline into the bloodstream. ■

In the Lab   1

Materials

- ☐ Nervous system chart
- ☐ Spinal cord model

Procedures

1. Review the anatomy and sympathetic pathways presented in Figures 26.1 through 26.4. Complete the *Sketch to Learn* activity below.

 Sketch to Learn

Use the provided template and practice drawing sympathetic pathways.

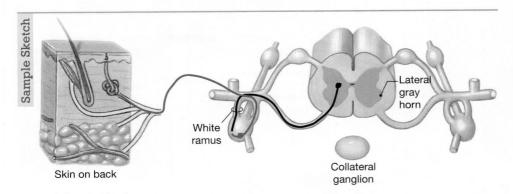

Sample Sketch

White ramus

Skin on back

Lateral gray horn

Collateral ganglion

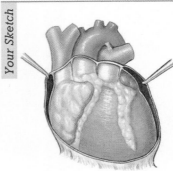

Your Sketch

Step 1
- Draw a black line for the preganglionic neuron, which passes from the lateral gray horn to the sympathetic ganglion.

Step 2
- Draw a red line for the ganglionic neuron, which passes from the ganglion, to the gray ramus, and out the dorsal ramus to enter the skin.

Step 3
- On the right side of the template, draw a sympathetic pathway from the spinal cord to the heart, using black and red lines for the preganglionic and ganglionic neurons.

2. On the spinal cord model, locate the lateral gray horns, ventral roots, and the components of the rami communicantes.

3. On a chart of the nervous system, or in Figure 26.3, locate a sympathetic chain ganglion, a collateral ganglion, and the medulla of the adrenal gland.

4. On the nervous system chart, trace the following sympathetic pathways:

   a. Preganglionic fiber synapsing in a collateral ganglion

   b. Ganglionic fiber exiting a chain ganglion and passing into a spinal nerve

   c. Preganglionic fiber synapsing in the adrenal medulla ■

Lab Activity 2 The Parasympathetic (Craniosacral) Division

The organization of the parasympathetic division of the ANS is diagrammed in **Figure 26.5**. In this division, the preganglionic neurons leave the CNS either via cranial nerves III, VII, IX, and X, or via the sacral level of the spinal cord. Parasympathetic preganglionic neurons release acetylcholine, which is always excitatory to a ganglionic fiber. The parasympathetic ganglionic fibers also release ACh to their visceral effectors. How the ACh affects the effectors depends on the type of ACh receptors present in the cell membrane of the effector cells. Generally, the parasympathetic

Figure 26.5 Organization of the Parasympathetic Division of the ANS Preganglionic neurons of the parasympathetic division are in cranial nerves III, VII, IX, and X and in pelvic nerves in sacral spinal cord segments. Preganglionic neurons in these cranial and sacral nerves synapse in parasympathetic ganglia located near or within the effectors where they synapse with ganglionic neurons that supply the target organs with parasympathetic control.

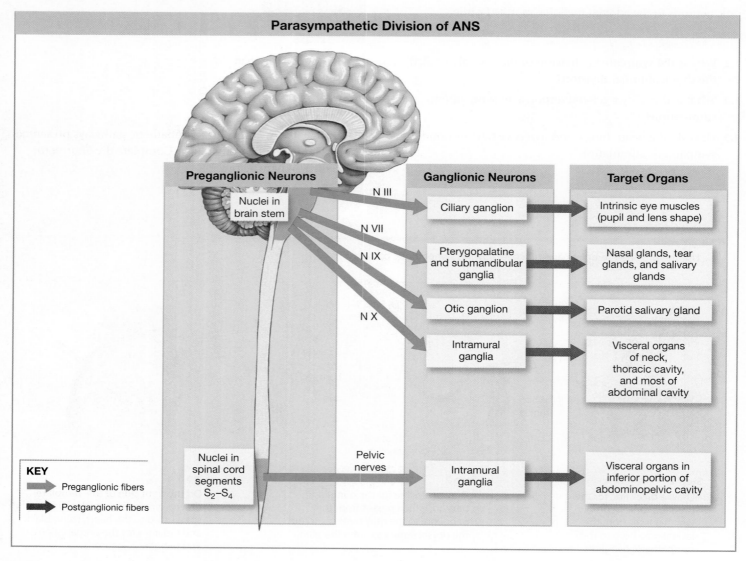

Parasympathetic Division of ANS

Preganglionic Neurons		Ganglionic Neurons	Target Organs
Nuclei in brain stem	N III	Ciliary ganglion	Intrinsic eye muscles (pupil and lens shape)
	N VII	Pterygopalatine and submandibular ganglia	Nasal glands, tear glands, and salivary glands
	N IX	Otic ganglion	Parotid salivary gland
	N X	Intramural ganglia	Visceral organs of neck, thoracic cavity, and most of abdominal cavity
Nuclei in spinal cord segments S_2–S_4	Pelvic nerves	Intramural ganglia	Visceral organs in inferior portion of abdominopelvic cavity

KEY
➡ Preganglionic fibers
➡ Postganglionic fibers

response is a rest-and-digest response that slows body functions and promotes digestion and waste elimination.

There are two main types of parasympathetic ganglia: terminal and intramural. **Terminal ganglia** are located near the eye and salivary glands; **intramural** (within walls) **ganglia** are embedded in the walls of effector organs. In the brain, parasympathetic preganglionic neurons branch into four cranial nerves: oculomotor, facial, glossopharyngeal, and vagus (**Figure 26.6**). For the first three of these nerves, there is a separate terminal ganglion for each one. The oculomotor nerve

Figure 26.6 Distribution of Parasympathetic Innervation Parasympathetic nerves are in cranial nerves III, VII, IX, and X and in sacral nerves of the sacral part of the spinal cord. For clarity, only the right side of the figure shows nerves but in real life each nerve is paired.

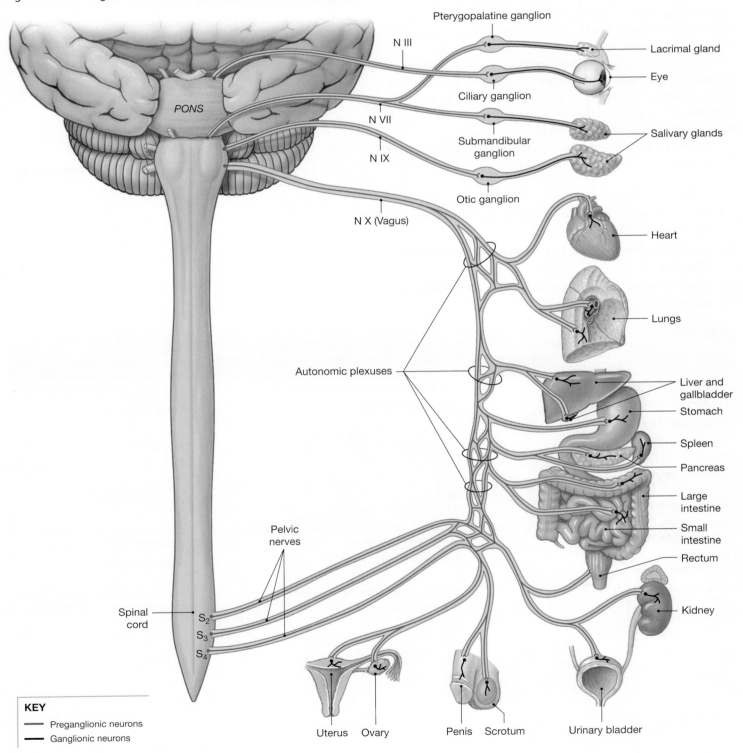

KEY
— Preganglionic neurons
— Ganglionic neurons

(N III) to the eyes enters the **ciliary ganglion,** the facial nerve (N VII) passes into the **pterygopalatine** (TER-i-gō-PAL-a-tin) and **submandibular ganglia,** and the glossopharyngeal nerve (N IX) includes the **otic ganglion.** Intramural ganglia receive preganglionic neurons in the vagus nerve (N X), which exits the brain, travels down the musculature of the neck, enters the ventral body cavity, and spreads into the intramural ganglia of the internal organs. The sacral portion of the parasympathetic division contains preganglionic neurons in sacral segments S_2, S_3, and S_4. The preganglionic fibers remain separate from spinal nerves and exit from spinal segments S_2 through S_4 as **pelvic nerves.**

Networks of preganglionic neurons, called **autonomic plexuses,** occur between the vagus nerve and the pelvic nerves. In these plexuses, sympathetic preganglionic neurons and parasympathetic preganglionic neurons intermingle as they pass to their respective autonomic ganglia.

Table 26.1 compares the sympathetic and parasympathetic divisions.

QuickCheck Questions

2.1 Why is the parasympathetic division also called the craniosacral division?

2.2 What is the body's general response to parasympathetic stimulation?

2.3 How do the heart, lungs, and digestive tract respond to parasympathetic stimulation?

In the Lab 2

Material

☐ Nervous system chart

Procedures

1. Review the anatomy and parasympathetic pathways presented in Figures 26.5 and 26.6.

2. On a chart of the nervous system, identify the oculomotor, facial, glossopharyngeal, and vagus cranial nerves. In which part of the brain are these nerves located?

3. On the nervous system chart, trace the following parasympathetic pathways:

   a. Preganglionic fiber entering a pelvic nerve and traveling to the urinary bladder

   b. Vagus nerve from the brain to the heart

   c. Preganglionic fiber synapsing in a ciliary ganglion ■

Table 26.1	Comparison of Sympathetic and Parasympathetic Divisions of Autonomic Nervous System	
Characteristic	**Sympathetic Division**	**Parasympathetic Division**
Location of CNS visceral motor neurons	Lateral gray horns of spinal segments T_1–L_2	Brain stem and spinal segments S_2–S_4
Location of PNS ganglia	Near vertebral column	Typically intramural
Preganglionic fibers		
Length Neurotransmitter released	Relatively short ACh	Relatively long ACh
Ganglionic fibers		
Length Neurotransmitter released	Relatively long Normally NE; sometimes ACh	Relatively short ACh
Neuromuscular or neuroglandular junction	Varicosities and enlarged terminal knobs that release transmitter near target cells	Junctions that release transmitter to special receptor surface
Degree of divergence from CNS to ganglion cells	Approximately 1:32	Approximately 1:6
General function(s)	Stimulates metabolism; increases alertness; prepares for emergency ("fight or flight")	Promotes relaxation, nutrient uptake, energy storage ("rest and digest")

Name _____

Date _____

Section _____

Autonomic Nervous System

A. Description

Write a brief description of each ANS structure listed.

1. preganglionic neuron

2. gray ramus

3. adrenal medulla

4. rami communicantes

5. thoracolumbar division of ANS

6. collateral ganglion

7. intramural ganglion

8. white ramus

9. ganglionic neuron

10. craniosacral division of ANS

B. Short-Answer Questions

1. Discuss the anatomy of the sympathetic chain ganglia. How do fibers enter and exit these ganglia?

2. Which cranial nerves are involved in the parasympathetic division of the ANS?

3. Compare the lengths of preganglionic and ganglionic neurons in the sympathetic and parasympathetic divisions of the ANS.

C. Drawing

1. ***Draw It!*** In Figure 26.7, draw the preganglionic and ganglionic neurons for a sympathetic pathway from the CNS to visceral effectors in the skin.

2. ***Draw It!*** In Figure 26.7, draw the preganglionic and ganglionic neurons for a sympathetic pathway from the CNS to the stomach.

Figure 26.7 **Sympathetic Pathways**

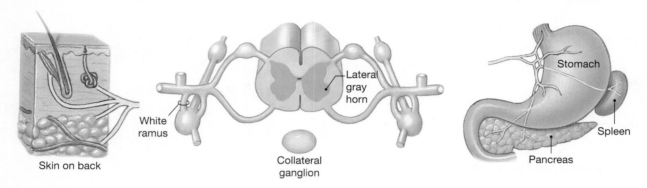

Skin on back

White ramus

Lateral gray horn

Collateral ganglion

Stomach

Spleen

Pancreas

D. Analysis and Application

1. As a child, you might have been told to wait for up to an hour after eating before going swimming. Explain the rationale for this statement.

2. Compare the outflow of preganglionic neurons from the CNS in the sympathetic and parasympathetic divisions of the ANS.

E. Clinical Challenge

1. List four responses to sympathetic stimulation and four responses to parasympathetic stimulation.

2. Compare the effect that neurotransmitters from sympathetic and parasympathetic ganglionic fibers have on smooth muscle in the digestive tract, on cardiac muscle, and on arterioles in skeletal muscles.

General Senses

Learning Outcomes

On completion of this exercise, you should be able to:

1. List the receptors for the general senses and those for the special senses.

2. Discuss the distribution of cutaneous receptors.

3. Describe the two-point discrimination test.

4. Describe and give examples of adaptation in sensory receptors.

5. Explain how referred pain can occur.

C hanges in the body's internal and external environments are detected by special cells called *sensory receptors.* Most of these receptors are sensitive to a specific stimulus. The taste buds of the tongue, for example, are stimulated by chemicals dissolved in saliva and not by sound waves or light rays.

The human senses may be grouped into two broad categories: general senses and special senses. The **general senses,** which have simple neural pathways, are touch, temperature, pain, chemical and pressure detection, and body position (proprioception). The **special senses** have complex pathways, and the receptors for these senses are housed in specialized organs. The special senses include gustation (taste), olfaction (smell), vision, audition (hearing), and equilibrium. In this exercise you will study the receptors of the general senses.

A sensory neuron monitors a specific region called a **receptive field.** Overlap in adjacent receptive fields enables the brain to detect where a stimulus was applied to the body. The neuron of a given receptive field is connected to a specific area of the sensory cortex. This neural connection is called a **labeled line,** and the

Clinical Application

Need More Practice and Review?

Build your knowledge—and confidence!—in the Study Area of MasteringA&P® at www.masteringaandp.com with Pre-lab Quizzes, Post-lab Quizzes, Practice Anatomy Lab™ (PAL™) 3.0 virtual anatomy practice tool, PhysioEx™ 9.0 laboratory simulations, and A&P Flix™ with Quizzes.

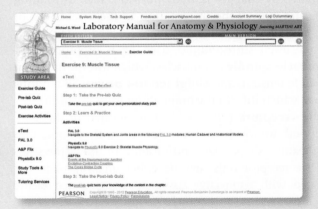

CNS interprets sensory information entirely on the basis of the labeled line over which the information arrives.

Nerve impulses are similar to bursts of messages over a telegraph wire. The pattern of action potentials is called **sensory coding** and provides the CNS with such information as intensity, duration, variation, and movement of the stimulus. The cerebral cortex cannot tell the difference between true and false sensations, however. For example, when you rub your eyes, you sometimes see flashes of light. The eye rubbing activates the optic nerve, and the sensory cortex interprets this false impulse as a visual signal. Sometimes the body projects a sensation, usually pain, to another part of the body. This phenomenon is called **referred pain.**

A sensory receptor is either a tonic or a phasic receptor. **Tonic receptors** are always active; pain receptors are one example. **Phasic receptors** are usually inactive and are "turned on" with stimulation. These receptors provide information on the rate of change of a stimulus. Some examples are root-hair plexuses, tactile corpuscles, and lamellated corpuscles, which are all phasic receptors for touch.

Lab Activity 1   General-Sense Receptors

Many kinds of sensory receptors transmit information to the CNS. **Thermoreceptors** are sensors for changes in temperature and have wide distribution in the body, being found in the dermis, skeletal muscles, and hypothalamus. (The hypothalamus is the body's internal thermostat.) **Chemoreceptors,** which are found in the medulla, in arteries near the heart, and in the heart, monitor changes in the concentrations of various chemicals present in body fluids. **Nociceptors** (nō-sē-SEP-turz) are pain receptors in the epidermis.

Mechanoreceptors, which are touch receptors that bend when stimulated, come in three types: baroreceptors, proprioceptors, and tactile receptors. **Baroreceptors** monitor pressure changes in liquids and gases. These receptors are typically the tips of sensory neuron dendrites in blood vessels and the lungs. **Proprioceptors** (prō-prē-ō-SEP-turz) are stimulated by changes in body position, such as rotating the head, and convey the information to the cerebellum of the brain so that the CNS knows where we are located in our three-dimensional surroundings. Two types of proprioceptors are **muscle spindles** in muscles, which inform the brain about muscle tension, and **Golgi tendon organs** in tendons near joints, which inform the brain about joint position.

Tactile receptors (**Figure 27.1**) are located in the skin. They respond to touch and provide us with information regarding texture, shape, size, and location of the tactile stimulation. The receptor cells may be either unencapsulated or encapsulated with connective tissue. Unencapsulated tactile receptors are very sensitive to touch. The unencapsulated tactile

receptors known as **free nerve endings** are simple receptors that are the exposed tips of dendrites in tissues. **Merkel cells** are unencapsulated tactile receptors that are in direct contact with the epidermis and respond to sensory neurons at swollen synapses called **tactile discs. Root-hair plexuses** are unencapsulated tactile receptors composed of sensory neuron dendrites wrapped around hair roots. These receptors are stimulated when an insect, for example, lands on your bare arm and moves one of the hairs there.

Encapsulated tactile receptors have branched dendrites that are covered by specialized cells. **Tactile corpuscles,** also called **Meissner's** (MĪS-nerz) **corpuscles,** are nerve endings located in the dermal papillae of the skin. The dendrites are wrapped in special Schwann cells and are highly sensitive to pressure, change in shape, and touch. Tactile corpuscles are phasic receptors. Deeper in the dermis are encapsulated tactile receptors called either **lamellated** (LAM-e-lā-ted; *lamella,* thin plate) **corpuscles** or **Pacinian** (pa-SIN-ē-an) **corpuscles.** These receptors have a large dendrite encased in concentric layers of connective tissue and respond to deep pressure and vibrations. **Ruffini** (roo-FĒ-nē) **corpuscles** are surrounded by collagen fibers embedded in the dermis. Changes in either the tension or shape of the skin tug on the collagen fibers, and the tugging stimulates Ruffini corpuscles.

QuickCheck Questions

1.1 What is chemoreception?

1.2 What is baroreception?

1.3 What is the stimulus for mechanoreceptors?

In the Lab   1

Materials

☐ Compound microscope
☐ Microscope slide of tactile corpuscles
☐ Microscope slide of lamellated corpuscles

Procedures

1. Gently touch a hair on your forearm and notice how you suddenly sense the touch. Is the root-hair plexus a phasic receptor or a tonic receptor?

2. Examine the tactile corpuscle slide at low magnification and locate the junction between the epidermis and dermis. Locate the tactile corpuscles in the papillary region of the dermis, where the dermis folds to attach the epidermis (**Figure 27.2**). Observe the tactile corpuscles at medium magnification.

3. Using **Figure 27.3** for reference, examine the slide of the lamellated corpuscles at low magnification. Note the multiple layers wrapped around the dendritic process of the receptor.

Figure 27.1 Tactile Receptors in the Skin The location and general appearance of six important tactile receptors.

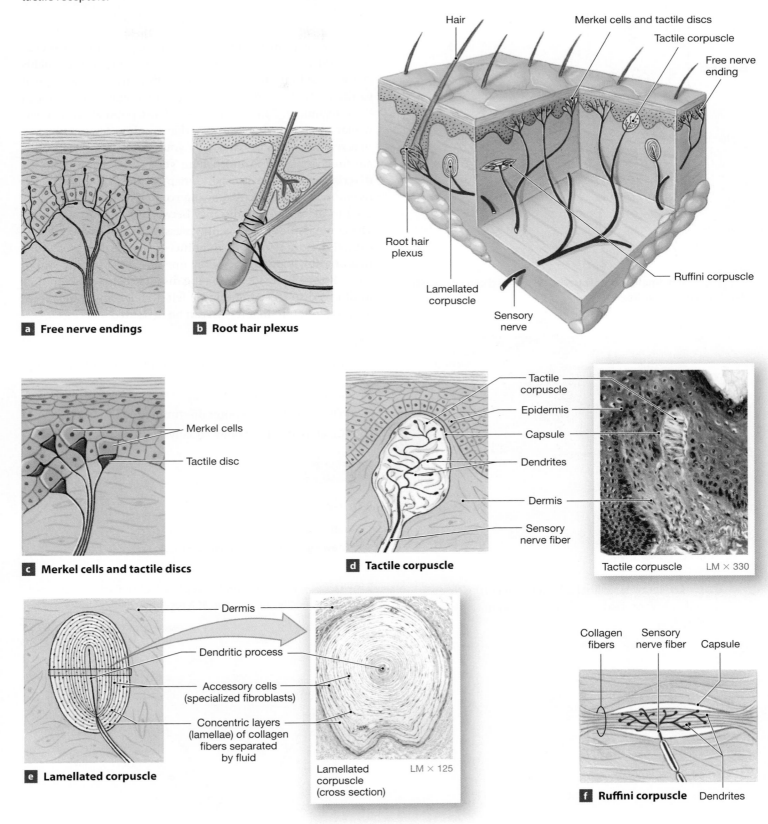

a Free nerve endings

b Root hair plexus

Hair

Merkel cells and tactile discs

Tactile corpuscle

Free nerve ending

Root hair plexus

Lamellated corpuscle

Sensory nerve

Ruffini corpuscle

Merkel cells

Tactile disc

c Merkel cells and tactile discs

Tactile corpuscle

Epidermis

Capsule

Dendrites

Dermis

Sensory nerve fiber

d Tactile corpuscle

Tactile corpuscle   LM × 330

Dermis

Dendritic process

Accessory cells (specialized fibroblasts)

Concentric layers (lamellae) of collagen fibers separated by fluid

e Lamellated corpuscle

Lamellated corpuscle (cross section)   LM × 125

Collagen fibers

Sensory nerve fiber

Capsule

f Ruffini corpuscle   Dendrites

Figure 27.2 **Tactile Corpuscle** Tactile corpuscles in the dermal papillae.

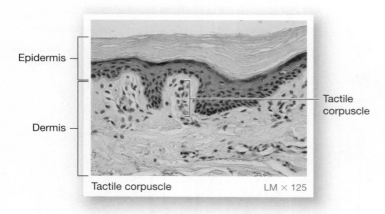

Epidermis

Dermis

Tactile corpuscle

Tactile corpuscle

LM × 125

Figure 27.3 **Lamellated Corpuscle** Lamellated corpuscle with dendritic process visible.

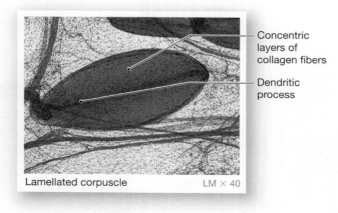

Concentric layers of collagen fibers

Dendritic process

Lamellated corpuscle

LM × 40

4. ***Draw It!*** Draw and label a lamellated corpuscle as viewed at low or medium magnification in the space provided. ■

Lamellated corpuscle

Lab Activity 2   Two-Point Discrimination Test

A sensory neuron monitors a specific region called a **receptive field** (Figure 27.4). Overlap in adjacent receptive fields enables the brain to detect where a stimulus is being applied to the body. Sensory receptors are not evenly distributed in the integument. Some areas have a dense population of a particular receptor, whereas others have only a few or none of that receptor. This explains why your fingertips, for example, are more sensitive to touch than your scalp. The **two-point discrimination test** is used to map the distribution of touch receptors on the skin. A drawing compass with two points is used to determine the distance between cutaneous receptors. The compass points are gently pressed into the skin, and the subject decides if one or two points are felt. If the sensation is that of a single point, then only one receptor has been stimulated. By gradually increasing the distance between the points until two distinct sensations are felt, the density of the receptor population in that region can be measured.

Make a Prediction

Predict the approximate size of the receptive field of a fingertip.

QuickCheck Questions

2.1 What does the two-point discrimination test measure?

2.2 Are all parts of the body equally sensitive to touch?

In the Lab   2

Materials

☐ Lab partner

☐ Drawing compass with millimeter scale

Figure 27.4 **Receptors and Receptive Fields** Each receptor cell monitors a specific area known as a receptive field.

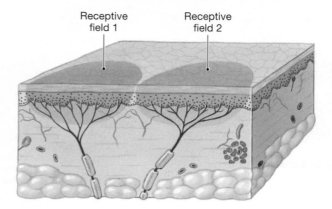

Receptive field 1

Receptive field 2

Procedures

1. Push the two points of the compass as close together as they will go. Now read from the millimeter scale how far apart the points are, and record this distance in the space provided in the leftmost column of Table 27.1, under "Index Finger." Gently place the points on the tip of an index finger of your laboratory partner, and then record whether one point or two points are felt. Slightly spread the compass points, record the distance apart as read from the millimeter scale, and place them again on the same area of the fingertip; again record whether one point or two points are felt. Repeat this procedure until the subject feels two distinct points. Record this distance in Table 27.1.

2. Reset the compass so that the two points are as close to each other as possible, and repeat the test on the back of the hand, the back of the neck, and one side of the nose. Record the data in Table 27.1. ∎

Lab Activity 3 | Distribution of Tactile Receptors

An experiment similar to the two-point discrimination test is to test whether a subject can feel a single touch from a stiff bristle of hair. In this activity, you will compare the sensitivity of two sites: the anterior and posterior of the forearm. The bristle used to measure this sensitivity is called a Von Frey hair.

QuickCheck Question

3.1 What do the Von Frey hairs measure?

In the Lab 3

Materials

☐ Lab partner
☐ Von Frey hairs (stiff boar bristles from hair brush)
☐ Water-soluble felt-tipped marker

Grid 1: Data for posterior forearm

Grid 2: Data for anterior forearm

Procedures

1. With the felt-tipped marker, draw a small box, approximately 1-inch square, on your partner's posterior forearm and another box on the anterior forearm. Divide the box into 16 smaller squares.

Table 27.1	Two-Point Discrimination Test Data						
Index Finger		*Back of Hand*		*Back of Neck*		*Side of Nose*	
Distance between points (mm)	**1 point or 2 points felt?**	**Distance between points (mm)**	**1 point or 2 points felt?**	**Distance between points (mm)**	**1 point or 2 points felt?**	**Distance between points (mm)**	**1 point or 2 points felt?**

Table 27.2	Tactile Density Tests
Region	**Number of Touches Felt**
Posterior forearm	
Anterior forearm	

Table 27.3	Thermoreceptor Density Tests
Temperature	**Number of Touches Felt**
Cold probe	
Warm probe	

2. *Note: Throughout this activity, the subject must look away and not watch as each test is run.* To test the posterior forearm's sensitivity to touch, use a Von Frey hair to gently touch the skin inside one small square. Be careful to touch only one point inside the square, and apply only enough pressure to slightly bend the bristle and stimulate the superficial tactile corpuscles. (Too much pressure will stimulate the underlying deep-touch receptors, the lamellated corpuscles.) In Grid 1: Data for posterior forearm, mark the corresponding small square. Draw a dot if the subject felt the bristle and an X if the subject did not feel the bristle.

3. Repeat this procedure in each of the other 15 small squares you have drawn on the subject's posterior forearm. Remember that the subject must be looking away as you administer the test. When finished, the grid will contain either a dot or an X in each square.

4. Repeat steps 2 and 3 on the subject's anterior forearm, touching the skin and marking the 16 small squares of Grid 2: Data for anterior forearm.

5. Repeat the experiment with you as the subject and your partner administering the tests.

6. Compare the results (a) between the two regions of your forearm, (b) between the two regions of your partner's forearm, and (c) between your forearm and your partner's. Record your final count in **Table 27.2**. ∎

Lab Activity 4   Distribution of Thermoreceptors

This experiment is a simple process of mapping the general distribution of thermoreceptors in the skin.

QuickCheck Question

4.1 Where are thermoreceptors located?

In the Lab   4

Materials

☐ Lab partner
☐ Small probes (small, blunt metal rods or straightened paper clips)

☐ Water-soluble felt-tipped marker
☐ Beaker filled with ice water
☐ Beaker filled with 45°C water

Grid 3: Data for thermoreception

Procedures

1. With the felt-tipped marker, draw a small box, approximately 1-inch square, on your partner's anterior forearm just above the wrist. Divide the box into 16 smaller squares.

2. Place a probe in the cold water for several minutes. Remove the probe from the water, dry it, and—with the subject not watching—touch the probe lightly to one of the small squares on the subject's arm. If the subject feels the cold, mark a C in the appropriate square of Grid 3: Data for thermoreception. If no cold is felt, mark an X in the square. (Write small, because during this activity you need to make two marks in each small square.)

3. Repeat in each of the other 15 squares, returning the probe to the cold water for a minute or so after each test.

4. Place a probe in the warm water for several minutes. *Important: The probe should be only warm to the touch, not hot enough to burn or cause pain.* Repeat steps 2 and 3 for each square on the subject's arm, placing an H in each appropriate square of the thermoreception data grid when heat is felt.

5. Have your partner repeat the experiment on your anterior wrist, and then compare the two sets of data. Record your final data in **Table 27.3**. ∎

Lab Activity 5   Receptor Adaptation

Receptors display **adaptation,** which means a reduction in sensitivity to repeated stimulus. When a receptor is stimulated, it first responds strongly, but then the response declines as the stimulus is repeated. **Peripheral adaptation** is the decline in response to stimuli at receptors. This type of adaptation reduces the amount of sensory information the CNS must process. **Central adaptation** occurs in the CNS. Inhibition along a sensory pathway reduces sensory information. Phasic receptors are fast-adapting receptors, and tonic receptors are slow adapting. In this experiment you will investigate the adaptation of thermoreceptors.

QuickCheck Question

5.1 Define the term *adaptation.*

In the Lab 5

Materials

- ☐ Lab partner
- ☐ Bowl filled with ice water
- ☐ Bowl filled with 45°C water
- ☐ Bowl filled with room-temperature water

Procedures
Test 1

1. Immerse your partner's left hand in the ice water. Note in the "Initial Sensation" column of Table 27.4 that the subject felt the cold. After two minutes, record in the table your partner's description of what temperature sensation is being felt in the left hand.

2. Leaving the left hand immersed, have your partner immerse her or his right hand in the ice water and describe whether the water feels colder to the left hand or to the right hand. Note this description in the "Initial Sensation" column of Table 27.4.

Test 2

Wait five minutes before proceeding to the next test, to allow blood to flow in the subject's hands and restore normal temperature sensitivity.

1. Place your partner's left hand in the ice water and his or her right hand in the 45°C water.
2. Keep hands immersed for two minutes.
3. Remove the left hand from the ice water and immerse it in the room-temperature water. Record in Table 27.4 whether your partner senses the room-temperature water as feeling hotter or colder than the ice water.
4. Now remove the right hand from the 45°C water, and immerse it in the room-temperature water. Record in Table 27.4 whether your partner senses the room-temperature water as being hotter or colder than the 45°C water. ∎

Lab Activity 6 Referred Pain

Referred pain means that the part of the body where pain is felt is different from the part of the body at which the painful stimulus is applied. A well-documented example of referred pain is the sensation felt in the left medial arm during a heart attack. The arm and heart are supplied by nerves from the same segments of the spinal cord and interneurons in these segments spread the incoming pain signals from the heart to areas innervated by that segment. In this activity, you will immerse your elbow in ice water and note any referred pain.

Figure 27.5 shows referred pain from the heart, along with three other common types. Liver and gallbladder pain is referred to the superior margin of the right shoulder. Pain from the appendix, located in the lower right quadrant, is referred to the medial area of the abdomen. Stomach, small intestine, and colon pain are also referred to the medial abdomen. The ureters are ducts in the medial aspect of the abdomen that transport urine from the kidneys to the urinary bladder; ureter pain is referred to the lateral abdomen.

QuickCheck Questions

6.1 Define the term *referred pain.*

6.2 Explain how referred pain occurs.

Table 27.4	Adaptation Tests	
Test 1	**Initial Sensation**	**Sensation After 2 Minutes**
Left hand in ice water		
Both hands in ice water		—
Test 2	**Movement After 2 Minutes**	**Sensation in Room-Temperature Water**
Left hand ice water, right hand 45°C water	Left hand to room-temperature water	Hotter or colder than ice water? (circle one)
Left hand ice water, right hand in 45°C water	Right hand to room-temperature water	Hotter or colder than 45°C water? (circle one)

Clinical Application Brain Freeze

An excellent example of referred pain is the pain in the forehead some people feel after quickly consuming a cold drink or a bowl of ice cream. The resulting "brain freeze," as it is commonly called, is pain referred from the nerves of the throat because these nerves also innervate the forehead. ∎

Figure 27.5 Referred Pain Pain sensations from visceral organs are often perceived as involving specific regions of the body surface innervated by the same spinal segments. Each region of perceived pain is labeled according to the organ at which the pain originates.

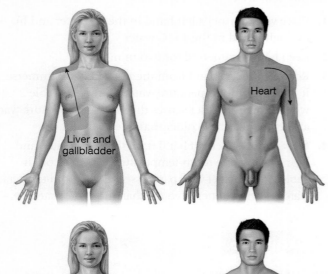

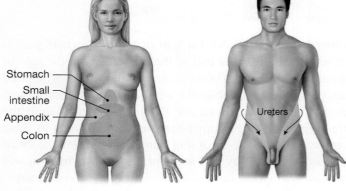

In the Lab 6

Materials

☐ Lab partner
☐ Bowl filled with ice water

Procedures

1. Place your elbow in the bowl of ice water and immediately note the location and type of your initial sensation. Your lab partner will record your observations and comments in **Table 27.5**.

2. Keep your elbow submerged and describe the location and type of sensation at 30, 60, 90, and 120 seconds. Record the information in Table 27.5. ■

Lab Activity 7   Proprioception

Proprioception is the sense of body position. It is the sense that gives us ownership of our bodies, enabling us to walk without having to watch our feet, say, or to reach up and scratch an ear without looking in a mirror.

QuickCheck Question

7.1 Define the term *proprioception*.

In the Lab 7

Materials

☐ Sheet of paper
☐ Red and black felt-tipped pens
☐ Ruler

Procedures

1. Using the dominant hand and the black pen, have your partner make a small circle (1/4-inch diameter) in the middle of the sheet of paper.

2. Instruct your partner to place the tip of the pen in the middle of the circle and then remain in that position with eyes closed for a few moments.

3. Keeping the eyes closed, have your partner lift the pen 3 to 4 inches off the paper and then try to make a mark within the circle. Be sure the person's arm is not resting on the table during the procedure.

4. Repeat step 3 until 10 marks have been made on the paper.

5. Repeat the experiment with your partner using his or her nondominant hand and the red pen.

6. Observe the pattern of the two sets of marks. Measure the farthest distance between two marks from the same hand.

7. Record the marking accuracy results in **Table 27.6**. ■

Table 27.5	Referred Pain from Elbow Tests		
Time		**Location**	**Type of Sensation**
Upon immersion in ice water			
30 seconds after immersion			
60 seconds after immersion			
90 seconds after immersion			
120 seconds after immersion			

Table 27.6	Proprioception Tests		
Hand	**Number of Marks Within Circle**	**Cluster Shape of Marks**	**Farthest Distance Between Marks**
Dominant			
Nondominant			

Name _____

Date _____

Section _____

General Senses

A. Description

Describe each of the following terms.

1. Pacinian corpuscle

2. Meissner's corpuscle

3. muscle spindles

4. receptive field

5. phasic receptor

6. tonic receptor

B. Labeling

Label the tactile receptors in **Figure 27.6**.

Figure 27.6 **Tactile Receptors in the Skin**

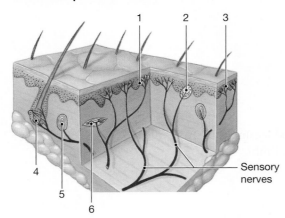

1. _____

2. _____

3. _____

4. _____

5. _____

6. _____

C. Short-Answer Questions

1. Describe the two-point discrimination test, and explain how it demonstrates receptor density in the skin.

2. Based on your results from Lab Activity 5, which type of thermoreceptor adapts more quickly, warm or cold?

D. Analysis and Application

1. In Lab Activities 3 and 4, which type of receptor was most abundant: touch, cold, or warm?

2. Beth wears her hair in a tight ponytail. At first her tactile receptors are sensitive and she feels the pull from her hair, but after a while she no longer feels it. Explain the loss of sensation.

3. In Lab Activity 6, where did you feel the pain when you had your elbow in the cold water? Explain why pain is felt in this location.

4. Explain the differences in the proprioception marking accuracy test between your dominant and nondominant hands.

E. Clinical Challenge

A patient describes the sensation that he still feels an amputated leg and foot. How might this "good phantom" assist the individual in using a prosthetic limb?

Special Senses: Olfaction and Gustation

EXERCISE

28

Learning Outcomes

On completion of this exercise, you should be able to:

1. Describe the location and structure of the olfactory receptors.

2. Identify the microscopic features of the olfactory epithelium.

3. Describe the location and structure of taste buds and papillae.

4. Identify the microscopic features of taste buds.

5. Explain why olfaction accentuates gustation.

Lab Activities

The special senses are gustation (taste), olfaction (smell), vision, audition (hearing), and equilibrium. In this exercise you will study the receptors for olfaction and gustation. The receptors for all the special senses are housed in specialized organs, and information from the receptors is processed in dedicated areas of the cerebral cortex. Neural pathways for the special senses are complex, often branching out to different regions of the brain for integration with other sensory input.

Lab Activity 1 Olfaction

Olfactory receptor cells are located in the **olfactory epithelium** lining the roof of the nasal cavity (**Figure 28.1**). These receptors are bipolar neurons with many cilia that are sensitive to airborne molecules. Most of the air we inhale passes through the nasal cavity and into the pharynx. Sniffing increases our sense of smell by pulling more air across the olfactory receptor cells.

Need More Practice and Review?

Build your knowledge—and confidence!—in the Study Area of MasteringA&P® at www.masteringaandp.com with Pre-lab Quizzes, Post-lab Quizzes, Practice Anatomy Lab™ (PAL™) 3.0 virtual anatomy practice tool, PhysioEx™ 9.0 laboratory simulations, and A&P Flix™ with Quizzes.

PAL For this lab exercise, follow these navigation paths:
- PAL>Human Cadaver>Nervous System>Special Senses
- PAL>Anatomical Models>Nervous System>Special Senses
- PAL>Histology>Special Senses

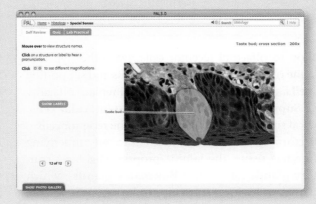

Figure 28.1 The Nose and Olfactory Epithelium

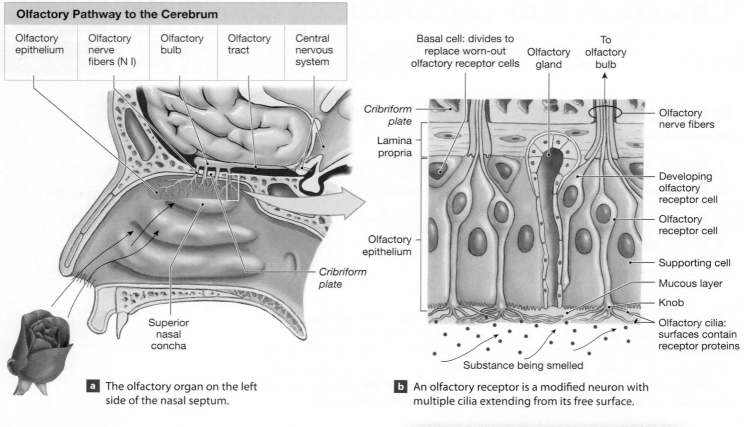

Olfactory Pathway to the Cerebrum				
Olfactory epithelium	Olfactory nerve fibers (N I)	Olfactory bulb	Olfactory tract	Central nervous system

Cribriform plate

Cribriform plate

Superior nasal concha

a The olfactory organ on the left side of the nasal septum.

Basal cell: divides to replace worn-out olfactory receptor cells

Olfactory gland

To olfactory bulb

Cribriform plate

Lamina propria

Olfactory epithelium

Olfactory nerve fibers

Developing olfactory receptor cell

Olfactory receptor cell

Supporting cell

Mucous layer

Knob

Olfactory cilia: surfaces contain receptor proteins

Substance being smelled

b An olfactory receptor is a modified neuron with multiple cilia extending from its free surface.

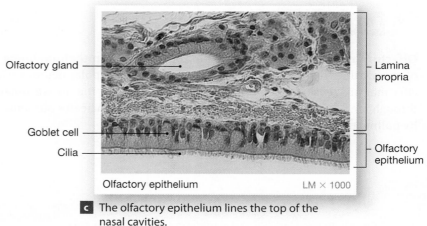

Olfactory gland

Goblet cell

Cilia

Olfactory epithelium

LM × 1000

Lamina propria

Olfactory epithelium

c The olfactory epithelium lines the top of the nasal cavities.

In addition to olfactory receptor cells, two other types of cells occur in the olfactory epithelium: basal cells and supporting cells. **Basal cells** are stem cells that divide and replace olfactory receptor cells. **Supporting cells,** also called **sustentacular cells,** provide physical support and nourishment to the receptor cells.

The olfactory epithelium is attached to an underlying layer of connective tissue, the lamina propria. This layer contains **olfactory glands** (also called **Bowman's glands**), which secrete mucus. In order for a substance to be smelled, volatile molecules from the substance must diffuse through the air from the substance to your nose. Once in the nose, the molecules must diffuse through the mucus secreted by the olfactory glands before they can stimulate the cilia of the olfactory receptor cells. The sense of smell is drastically reduced by colds and allergies because mucus production increases in the nasal cavity and keeps molecules from reaching the olfactory receptor cells.

The olfactory nerve, cranial nerve N I, passes through the cribriform plate of the ethmoid and enters the brain at the olfactory bulb. The bulb continues as the olfactory tract to the temporal lobe, where the olfactory cortex is located. (Unlike the pathways for many other senses, the olfactory pathway does not have a synapse in the thalamus.) Some branches of the olfactory tract synapse in the hypothalamus and limbic system of the brain, which explains the strong emotional responses associated with olfaction.

QuickCheck Questions

1.1 Where are the olfactory receptor cells located?

1.2 What is the function of the olfactory glands?

In the Lab 1

Materials

☐ Compound microscope
☐ Prepared slide of olfactory epithelium

Procedures

1. At low magnification, focus on the olfactory epithelium. Identify the supporting cells and olfactory receptor cells, both shown in Figure 28.1.

2. Locate the lamina propria and the olfactory glands.

3. ***Draw It!*** In the space provided, sketch the olfactory epithelium as viewed at medium magnification. ∎

Olfactory epithelium

Lab Activity 2 Olfactory Adaptation

Adaptation is defined as a reduction in sensitivity to a repeated stimulus. When a receptor is first stimulated, it responds strongly, but then the response declines as the stimulus is repeated. We say the receptor has *adapted* to the stimulus. **Peripheral adaptation** is adaptation that happens because the receptors become desensitized to the stimulus. This type of adaptation occurs rapidly in phasic receptors, and the resulting adaptation reduces the amount of sensory information the CNS must process. **Central adaptation** occurs in the CNS, due to inhibition of sensory neurons along a sensory pathway. It is central adaptation that allows us to smell a new odor while we have adapted to reduce our awareness of an initial odor.

The olfactory pathway is quick to adapt to a repeated stimulus. A few minutes after you apply cologne or perfume, for instance, you do not smell it as much as you did initially. However, if a new odor is present, the nose is immediately capable of sensing the new scent, proof that what is going on is central adaptation and not receptor fatigue. (If the receptors were fatigued instead of adapted, you would not sense new stimuli once the receptors reached exhaustion.) Olfactory adaptation occurs along the olfactory pathway in the brain, not in the receptors.

In this activity, you determine the length of time it takes for your olfactory epithelium to adapt to a particular odor. Use care when smelling the vials. Do not put the vial directly under your nose and inhale. Instead, hold the open vial about 6 inches in front of your nose and wave your hand over the opening to waft the odor toward your nose.

Make a Prediction

How much time do you think it will take until you adapt to the wintergreen oil in the following activity?

QuickCheck Questions

2.1 Where does olfactory adaptation occur?

2.2 When does olfactory adaptation occur?

In the Lab 2

Materials

☐ Vial containing oil of wintergreen
☐ Vial containing isopropyl alcohol
☐ Stop watch
☐ Lab partner

Procedures

1. Hold the vial of wintergreen oil near your face and waft the fumes toward your nose. Ask your partner to start the stop watch.

2. Breathe through your nose to smell the oil. Continue wafting and smelling until you no longer sense the odor. Have your partner stop the stop watch at that instant, and record in **Table 28.1** the time it took for adaptation to occur.

Table 28.1	Olfactory Adaptation Tests	
Student	**Olfactory Adaptation Time for Wintergreen Oil(s)**	**Olfactory Adaptation Time for Isopropyl Alcohol(s)**

3. Immediately following loss of sensitivity to the wintergreen oil, smell the vial of alcohol. Explain how you can smell the alcohol but can no longer smell the wintergreen oil.

4. Wait about three minutes and then repeat steps 1 through 3 using the alcohol as the first vial. Is there a difference in adaptation time for the two substances?

5. Repeat steps 1 through 4 with your partner doing the smelling and you doing the timing. Record your partner's olfactory adaptation time in Table 28.1.

6. Repeat steps 1 through 4 with several other classmates and record the times in Table 28.1. ■

Lab Activity 3   Gustation

Gustation (gus-TĀ-shun) is the sense of taste. The receptors for gustation are **gustatory cells** located in **taste buds** that cover the surface of the tongue, the pharynx, and the soft palate (**Figure 28.2**). A taste bud can contain up to 100 gustatory cells. The gustatory cells are replaced every 10 to 12 days by basal cells, which divide and produce transitional cells that mature into the gustatory cells. Each gustatory cell has a small hair, or **microvillus,** that projects through a small **taste pore.** Contact with food dissolved in saliva stimulates the microvilli to produce gustatory impulses.

Figure 28.2 Gustatory Reception

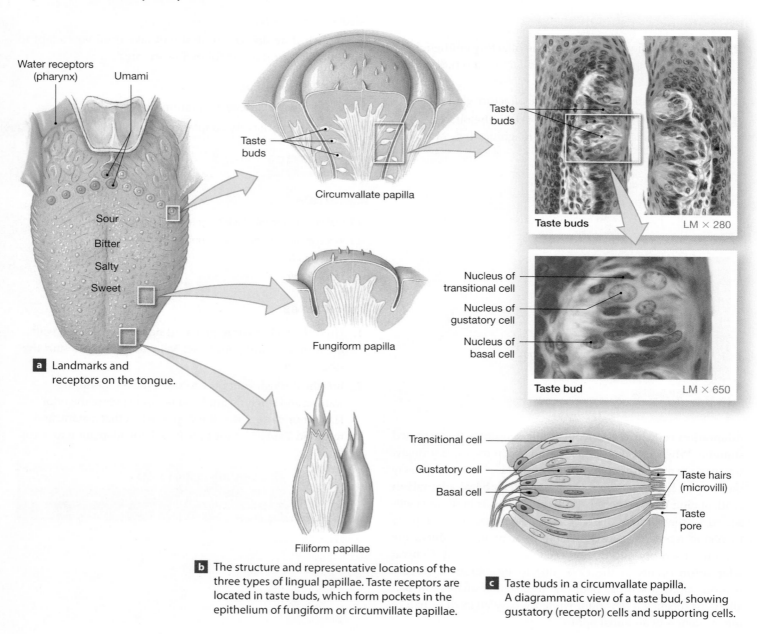

a Landmarks and receptors on the tongue.

Circumvallate papilla

Taste buds

Taste buds   LM × 280

Nucleus of transitional cell

Nucleus of gustatory cell

Nucleus of basal cell

Taste bud   LM × 650

Fungiform papilla

Filiform papillae

Transitional cell

Gustatory cell

Basal cell

Taste hairs (microvilli)

Taste pore

b The structure and representative locations of the three types of lingual papillae. Taste receptors are located in taste buds, which form pockets in the epithelium of fungiform or circumvallate papillae.

c Taste buds in a circumvallate papilla. A diagrammatic view of a taste bud, showing gustatory (receptor) cells and supporting cells.

An adult has approximately 10,000 taste buds located inside elevations called **papillae** (pa-PIL-lē), detailed in Figure 28.3. The base of the tongue has a number of circular papillae, called **circumvallate** (sir-kum-VAL-āt) **papillae,** arranged in the shape of an inverted V across the width of the tongue. The tip and sides of the tongue contain button-like **fungiform** (FUN-ji-form) **papillae.** Approximately two-thirds of the anterior portion of the tongue is covered with **filiform** (FIL-i-form) **papillae,** which do not contain taste buds but instead provide a rough surface for the movement of food.

Tastes can be grouped into five categories: sweet, salty, sour, bitter, and umami (oo-MAH-mē), this last one being the comforting taste of proteins in soup broth. Although water is tasteless, water receptors occur in taste buds of the pharynx. Sensory information from the taste buds is carried to the brain by parts of three cranial nerves: the vagus nerve (N X) serving the pharynx, the facial nerve (N VII) serving the anterior two-thirds of the tongue, and the glossopharyngeal nerve (N IX) serving the posterior one-third of the tongue. Children have more taste buds than adults, and at around the age of 50 the number of taste buds begins to rapidly decline. This difference in taste bud density helps to explain why children might complain that a food is too spicy whereas a grandparent responds that it tastes bland.

The sense of taste is a genetically inherited trait, and consequently two individuals can perceive the same substance in different ways. The chemical phenylthiocarbamide (PTC), for example, tastes bitter to some individuals, sweet to others,

Figure 28.3 Taste Buds Taste buds are visible along the walls of papillae on the tongue surface.

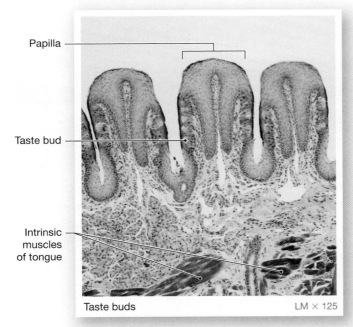

Papilla

Taste bud

Intrinsic muscles of tongue

Taste buds   LM × 125

Table 28.2	PTC Taste Tests		
Student		Taster	Nontaster
Percentage of the class:			

and tasteless to still others. Approximately 30 percent of the population are nontasters of PTC.

QuickCheck Questions

3.1 Where are taste buds located in the tongue?

3.2 What are the taste receptor cells called?

In the Lab 3

Materials

☐ Compound microscope

☐ Prepared slide of taste buds

☐ PTC taste paper

Procedures

1. Observing the taste bud slide at low and medium magnifications, use Figure 28.2 as a guide as you identify the three types of papillae and the taste buds.

2. Place a strip of PTC paper on your tongue and chew it several times. Are you a taster or a nontaster? If your instructor has each student in the class record her or his taster-or-nontaster results on the chalkboard, calculate the percentage of tasters versus nontasters and complete Table 28.2.

For additional practice, complete the **Sketch to Learn** activity (on p. 400). ∎

Lab Activity 4   Relationship Between Olfaction and Gustation

The sense of taste is thousands of times more sensitive when gustatory and olfactory receptor cells are stimulated simultaneously. If the sense of smell is decreased, as during nasal congestion or the flu, food can taste bland. In this activity, pieces of apple and onion will be placed on your partner's tongue while his or her nose is closed. Your partner will then attempt to identify which food was placed on the tongue without smelling it.

QuickCheck Question

4.1 What is this experiment designed to demonstrate?

Sketch to Learn

Let's draw the taste bud slide. Imagine the papillae as cone-shaped hills and the taste buds as caves in the sides of the hills. Mental images not only help you recall information, but they also show the anatomical associations of the particular region.

Sample Sketch

Step 1
- Draw 2 papillae of approximately equal size.

Step 2
- Add several ovals to each side of the papillae.
- Place dots inside small ovals for gustatory cells.

Step 3
- Use lines under the papillae to show skeletal muscle of the tongue.
- Label your sketch.

Your Sketch

In the Lab 4

Materials

- ☐ Diced onion
- ☐ Diced apple
- ☐ Paper towels
- ☐ Lab partner

Procedures

1. Dry the surface of your partner's tongue with a clean paper towel.

2. Have your partner stand with eyes closed and nose pinched shut with the thumb and index finger.

3. Place a piece of either onion or apple on the dried tongue, and ask if the food can be identified. Record the reply, yes or no, in **Table 28.3**.

4. Have your partner, still with eyes closed and nose pinched shut, chew the piece of food, and again ask if it can be identified. Record the reply in Table 28.3.

5. Have your partner release the nose pinch, and ask one last time if the food can be identified. Record the reply in Table 28.3. ■

Table 28.3	Gustatory and Olfactory Sensation Tests		
Food	**Dry Tongue**	**After Chewing**	**Open Nostrils**
Apple	_____	_____	_____
Onion	_____	_____	_____

Name _____

Date _____

Section _____

Special Senses: Olfaction and Gustation

A. Definitions

Define or describe each term.

1. central adaptation

2. peripheral adaptation

3. filiform papilla

4. gustation

5. Bowman's gland

6. basal cell

7. circumvallate papilla

8. sustentacular cell

9. olfaction

10. taste bud

B. Drawing

Draw It! Sketch the tongue, and identify where the taste buds for sweet, bitter, sour, and salty compounds are concentrated.

C. Labeling

Label the micrographs in **Figure 28.4**.

Figure 28.4   Gustatory and Olfactory Receptors

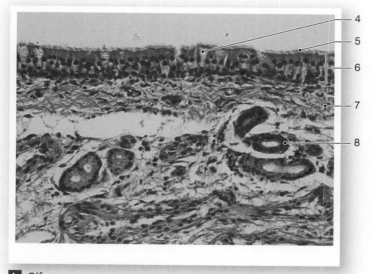

a Gustatory receptors

b Olfactory receptors

1. _____

2. _____

3. _____

4. _____

5. _____

6. _____

7. _____

8. _____

D. Short-Answer Questions

1. What type of glands occur in the olfactory epithelium?

2. Where are the receptors for taste located?

3. Which cranial nerve innervates the nose?

4. Which cranial nerves relay sensory information from the tongue?

E. Analysis and Application

1. Describe an experiment that demonstrates how the senses of taste and smell are linked.

2. Several minutes after applying cologne, you notice that you cannot smell the fragrance. Explain this response by your olfactory receptors.

3. Examine the data in Table 28.1. Did adaptation time differ greatly from one individual to another?

4. Does the class data in Table 28.2 clearly demonstrate which trait, taster or nontaster, is genetically dominant?

F. Clinical Challenge

Why does a cold affect your sense of smell?

Anatomy of the Eye

Learning Outcomes

On completion of this exercise, you should be able to:

1. Identify and describe the accessory structures of the eye.

2. Explain the actions of the six extraocular eye muscles.

3. Describe the external and internal anatomy of the eye.

4. Describe the cellular organization of the retina.

5. Identify the structures of a dissected cow or sheep eye.

The eyes are complex and highly specialized sensory organs, allowing us to view everything from the pale light of stars to the intense, bright blue of the sky. To function in such a wide range of light conditions, the retina of the eye contains two types of receptors, one for night vision and another for bright light and color vision. Because the level and intensity of light are always changing, the eye must regulate the size of the pupil, which allows light to enter the eye. Six oculomotor muscles surrounding the eyeball allow it to move. Four of the 12 cranial nerves control the muscular activity of the eyeball and transmit sensory signals to the brain. A sophisticated system of tear production and drainage keeps the surface of the eyeball clean and moist. In this exercise you examine the anatomy of the eye and dissect the eye of a sheep or a cow.

Lab Activity 1   External Anatomy of the Eye

The human eyeball is a spherical organ measuring about 2.5 cm (1 in.) in diameter. Only about one-sixth of the eyeball is visible between the eyelids; the rest is recessed

Need More Practice and Review?

Build your knowledge—and confidence!—in the Study Area of MasteringA&P® at www.masteringaandp.com with Pre-lab Quizzes, Post-lab Quizzes, Practice Anatomy Lab™ (PAL™) 3.0 virtual anatomy practice tool, PhysioEx™ 9.0 laboratory simulations, and A&P Flix™ with Quizzes.

PAL practice anatomy lab   For this lab exercise, follow these navigation paths:
- PAL>Human Cadaver>Nervous System>Special Senses
- PAL>Anatomical Models>Nervous System>Special Senses
- PAL>Histology>Special Senses

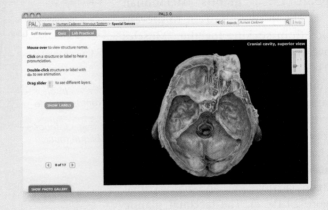

in the bony orbit of the skull. Most of the external features of the eye are accessory structures of the eyeball and not a physical part of the eyeball itself. The accessory structures of the eye are the upper eyelid, lower eyelid, eyebrow, eyelashes, lacrimal apparatus, and six extraocular (external) eye muscles.

The eyelids, called **palpebrae** (pal-PĒ-brē), distribute tears across the surface of the eye to keep it moist. The anterior surface of the eyelid is covered with skin and the edge has short hairs, called the **eyelashes** (Figure 29.1). Eyelashes and the eyebrows protect the eyeball from foreign objects, such as perspiration and dust, and partially shade the eyeball from the sun. Modified sweat glands called **ciliary** (SIL-ē-ar-ē) **glands,** located at the base of the eyelashes, help to lubricate the eyeball. The cleft between the eyelids is the **palpebral fissure.** The two points where the upper and lower lids meet are the **lateral canthus** (KAN-thus; corner) and **medial canthus.** A red, fleshy structure in the medial canthus, the **lacrimal caruncle** (KAR-ung-kul; small soft mass), contains modified sebaceous and sweat glands. Secretions from the lacrimal caruncle accumulate in the medial canthus during long periods of sleep. A thin mucous membrane called the **palpebral conjunctiva** (kon-junk-TĪ-vuh) covers the underside of the eyelids and reflects over most of the anterior surface of the eyeball as the **ocular conjunctiva.** The conjunctiva has glands that secrete mucus to reduce friction and moisten the eyeball surface. The blood vessels you see over the white parts of your eyeballs are vessels in the ocular conjunctiva.

Eyelids have internal **tarsal plates** of fibrous tissue that give the lids their shape and support. **Tarsal glands** (Meibomian glands) secrete an oily lubricant to prevent the eyelids from sticking together. Muscles that move the eyelids insert on the tarsal plates. The **levator palpebrae superioris** muscle raises the upper eyelid, and the **orbicularis oculi** muscle closes the eyelids. Blinking the eyelids keeps the eyeball surface lubricated and clean.

The **lacrimal** (LAK-ri-mal; *lacrima,* a tear) **apparatus** consists of the lacrimal glands, lacrimal canals, lacrimal sac, and nasolacrimal duct (Figure 29.1b, c). The **lacrimal glands**

Figure 29.1 Accessory Structures of the Eye External to the eyeball are accessory structures that protect and support the eyeball.

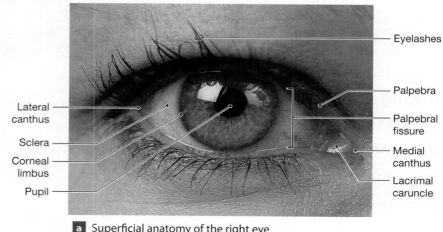

a Superficial anatomy of the right eye and its accessory structures.

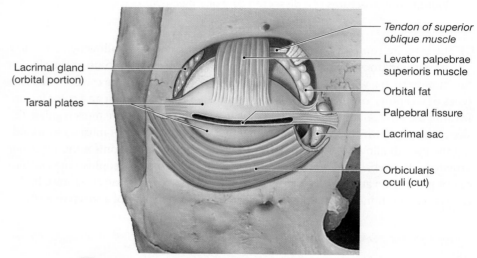

b Diagrammatic representation of superficial dissection of the right eye.

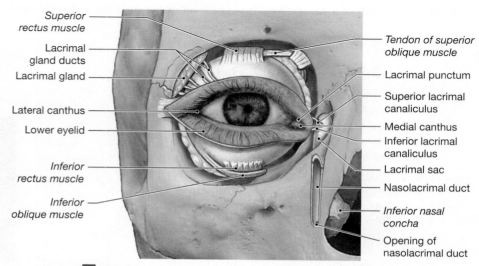

c This diagrammatic representation of a deeper dissection of the right eye shows its position in the eye orbit and its relationship to accessory structures, especially the lacrimal apparatus.

Clinical Application Infections of the Eye

When a ciliary gland is blocked, it becomes inflamed as a **sty.** Because it is on the tip of the eyelid, the sty irritates the eyeball. **Conjunctivitis** is an inflammation of the conjunctiva that can be caused by bacteria, dust, smoke, or air pollutants; the infected eyeball usually appears red and irritated. The bacterial form of this infection is contagious and spreads easily among young children and individuals sharing such objects as tools and office equipment. ■

are superior and lateral to each eyeball. Each gland contains 6 to 12 excretory **lacrimal ducts** that deliver to the anterior surface of the eyeball a slightly alkaline solution, called either *lacrimal fluid* or *tears*, which cleans, moistens, and lubricates the surface. The lacrimal fluid also contains an antibacterial enzyme called *lysozyme* that attacks any bacteria that may be on the surface of the eyeball. The fluid moves medially across the eyeball surface and enters two small openings of the medial canthus, the **superior** and **inferior lacrimal puncta** (PUNGK-tă). From there, the lacrimal fluid passes into two ducts, the **lacrimal canaliculi,** which lead to an expanded portion of the nasolacrimal duct called the **lacrimal sac.** The **nasolacrimal duct** drains the tears into the nasal cavity.

Six extraocular muscles control the movements of the eyeball (**Figure 29.2**). The **superior rectus, inferior rectus, medial rectus,** and **lateral rectus** are straight muscles that move the eyeball up and down and side to side. The **superior** and **inferior oblique** muscles attach diagonally on the eyeball. The superior oblique has a tendon passing through the **trochlea** (*trochlea*, pully) located on the upper orbit. This muscle rolls the eyeball downward, and the inferior oblique rolls it upward.

QuickCheck Questions

1.1 How are lacrimal secretions drained from the surface of the eyeball?

1.2 Where are the two parts of the conjunctiva located?

In the Lab 1

Materials

☐ Dissectible eye model
☐ Eyeball chart

Procedures

1. Review the structures of the eye in Figure 29.1 and the extraocular muscles in Figure 29.2.
2. On the eye model and chart, locate the four structures of the lacrimal apparatus and the six accessory structures (count the six extraocular muscles as one accessory structure).
3. On the model, identify the six extraocular muscles. Describe how each one moves the eye. ■

Figure 29.2 Extraocular Eye Muscles Lateral surface of the right eye, illustrating the muscles that move the eyeball. The medial rectus muscle is not visible in this view. The levator palpebrae superioris muscle, which raises the upper eyelid, is not classified as one of the six extraocular muscles that move the eye.

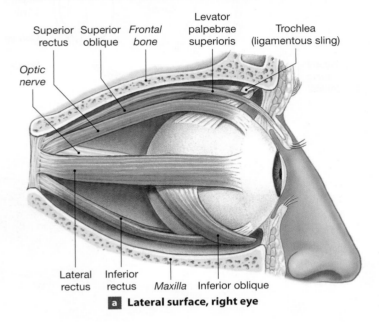

a Lateral surface, right eye

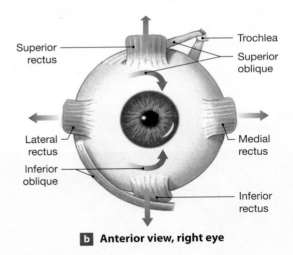

b Anterior view, right eye

Lab Activity 2 Internal Anatomy of the Eye

The wall of the eyeball is anatomically complex with three layers to serve a variety of functions including support and protection, adjustment of the lens to focus light, and reception of light for the sense of vision (**Figure 29.3**). The outermost layer is called the **fibrous tunic** because of the abundance of dense connective tissue. The **sclera** (SKLER-uh; hard) is

Figure 29.3 **Anatomy of the Eye** The eyeball wall and internal features of the eye.

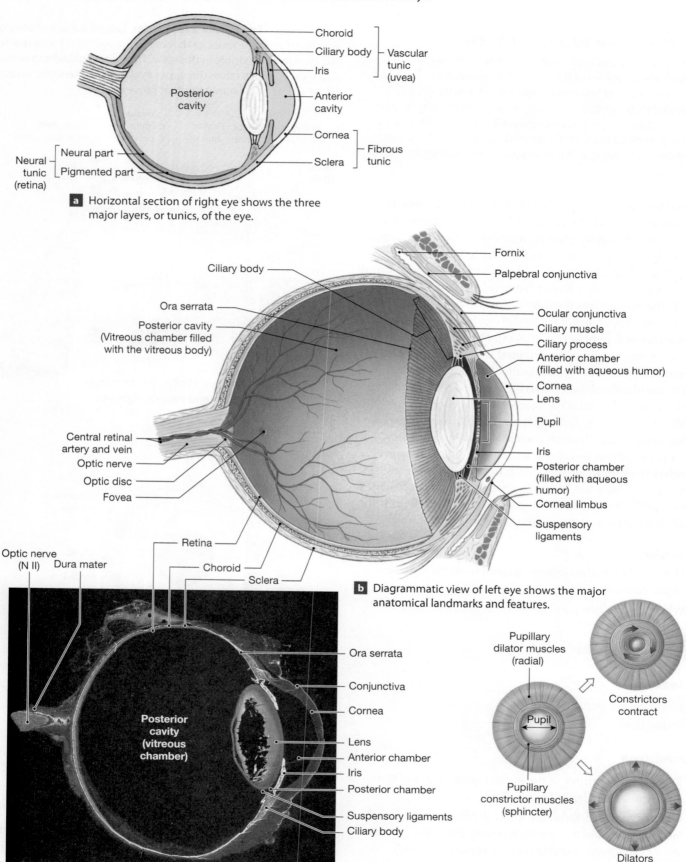

Choroid ⎤
Ciliary body ⎬ Vascular tunic (uvea)
Iris ⎦

Posterior cavity

Anterior cavity

Cornea ⎤
Sclera ⎬ Fibrous tunic

Neural tunic (retina) ⎡ Neural part
⎣ Pigmented part

a Horizontal section of right eye shows the three major layers, or tunics, of the eye.

Ciliary body

Ora serrata

Posterior cavity (Vitreous chamber filled with the vitreous body)

Fornix

Palpebral conjunctiva

Ocular conjunctiva

Ciliary muscle

Ciliary process

Anterior chamber (filled with aqueous humor)

Cornea

Lens

Pupil

Iris

Posterior chamber (filled with aqueous humor)

Corneal limbus

Suspensory ligaments

Central retinal artery and vein

Optic nerve

Optic disc

Fovea

Retina

Choroid

Sclera

b Diagrammatic view of left eye shows the major anatomical landmarks and features.

Optic nerve (N II) Dura mater

Ora serrata

Conjunctiva

Cornea

Posterior cavity (vitreous chamber)

Lens

Anterior chamber

Iris

Posterior chamber

Suspensory ligaments

Ciliary body

d Sagittal section of the eye.

Pupillary dilator muscles (radial)

Pupil

Constrictors contract

Pupillary constrictor muscles (sphincter)

Dilators contract

c The action of the pupillary muscles and changes in pupil diameter.

the white part of the fibrous tunic that resists punctures and maintains the shape of the eyeball. It covers the eyeball except at the transparent **cornea** (KOR-nĕ-uh), which is the region of the fibrous tunic where light enters the eye. The cornea consists primarily of many layers of densely packed collagen fibers. The **corneal limbus** (LIM-bus; border) is the border between the sclera and the cornea. Around the corneal limbus is the **canal of Schlemm,** also called the **scleral venous sinus,** a small passageway that drains lacrimal fluid into veins in the sclera.

The second of the eyeball's three layers is the **vascular tunic (uvea)** and is organized into the *iris, ciliary body,* and the *choroid.* The most posterior portion of this layer, the **choroid,** is highly vascularized and contains a dark pigment (melanin) that absorbs light to prevent reflection.

The anterior part of the uvea is the pigmented **iris.** It has a central aperture called the **pupil.** Posterior to the iris is the transparent **lens,** the part of the eye that focuses light. In the iris, **pupillary sphincter muscles** and **pupillary dilator muscles** change the diameter of the pupil to regulate the amount of light entering the lens. In bright light and for close vision, the pupil constricts as a result of parasympathetic activation that causes the pupillary sphincter muscles to contract and the pupillary dilator muscles to relax. In low light and for distant vision, sympathetic stimulation causes the dilator muscles to contract and the sphincter muscles to relax; as a result, the pupil expands and more light enters the eye. Around the lens the uvea is the wedge-shaped enlarged **ciliary body** where the iris attaches. In the ciliary body is the **ciliary muscle** that adjusts the shape of the lens for near and far vision. The **ciliary process** is a series of folds at the edge of the ciliary body and has thin **suspensory ligaments** that extend to the lens.

The innermost of the eyeball's three layers is the **neural tunic,** usually referred to as the **retina.** This layer contains an outer **pigmented part** covering the choroid and a **neural part** containing light-sensitive photoreceptors. The anterior margin of the retina, where the choroid of the vascular tunic is exposed, is the **ora serrata** (Ō-ra ser-RA-tuh; serrated mouth) and appears as a jagged edge, much like a serrated knife.

The lens divides the eyeball into an **anterior cavity,** the area between the lens and the cornea; and a **posterior cavity** (also called *vitreous chamber*), the area between the lens and the retina (**Figure 29.4**). The anterior cavity is further subdivided into an **anterior chamber** between the iris and the cornea and a **posterior chamber** between the iris and the lens. Capillaries of the ciliary processes form a watery fluid called

aqueous humor (AK-wē-us, watery; HŪ-mor, fluid) that is secreted into the posterior chamber and circulates through the pupil and into the anterior chamber. Around the corneal limbus is the **scleral venous sinus (canal of Schlemm),** a series of small veins that reabsorb the aqueous humor. The aqueous humor helps maintain the intraocular pressure of the eyeball and supplies nutrients to the lens and cornea. The posterior cavity, larger than the anterior cavity, contains the **vitreous** (VIT-rē-us; *vitreo-,* glassy) **body,** a clear, jellylike substance that

---

Clinical Application Diseases of the Eye

Glaucoma is a disease in which the intraocular pressure of the eye is elevated. The increased pressure damages the optic nerve and may eventually result in blindness. If the canal of Schlemm becomes blocked, fluid accumulates in the anterior cavity and intraocular pressure rises. Diabetic individuals are at risk of developing **diabetic retinopathy,** a proliferation and rupturing of blood vessels over the retina. These vascular changes occur gradually, but eventually vision declines as photoreceptors are damaged. ∎

Figure 29.4 Chambers of the Eye The eyeball is a hollow organ filled with aqueous humor and vitreous body.

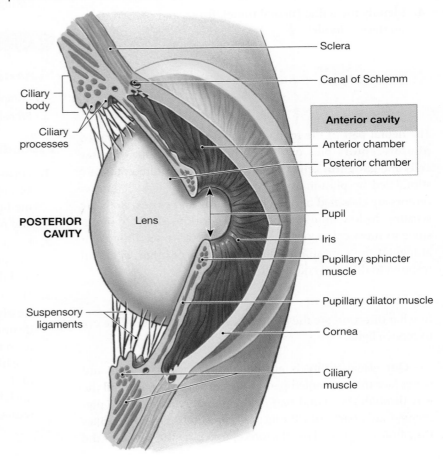

holds the retina against the choroid and prevents the eyeball from collapsing.

QuickCheck Questions

2.1 List the three major layers that form the wall of the eye.

2.2 Trace a drop of aqueous humor circulating in the eye.

2.3 How does the pupil regulate the amount of light that enters the lens?

In the Lab 2

Materials

☐ Dissectible eye model
☐ Eyeball chart

Procedures

1. On the eye model and chart, identify the cavities and chambers of the eye and the three major layers of the eyeball wall.

2. Identify the sclera and cornea on the eye model. Also locate the corneal limbus and canal of Schlemm.

3. On the model, locate the choroid, and identify the ciliary body and associated structures.

4. Identify the retina (neural tunic), fovea, and ora serrata on the eye model. ■

Lab Activity 3   Cellular Organization of the Retina

The neural part of the retina contains sensory receptors called **photoreceptors** plus two types of sensory neurons: **bipolar cells** and **ganglion cells** (Figure 29.5). The photoreceptors are stimulated by photons, which are particles of light. Photoreceptors are classified into two types: **rods** and **cones.** Rods are sensitive to low illumination and to motion. They are insensitive to most colors of light, and therefore we see little color at night. Cones are stimulated by moderate or bright light and respond to different colors of light.

Make a Prediction

In what direction are the photoreceptors positioned in the eye to receive light?

Our visual acuity is attributed to cones. The rods and cones face the pigmented part of the retina. Light passes all the way through the neural part of the retina, reflects off the pigmented area back into the neural part, and only then strikes the photoreceptors. The photoreceptors pass the signal to the

bipolar cells, which in turn pass the signal to the ganglion cells. The axons of the ganglion cells converge at an area of the neural part of the retina called the **optic disc,** where the optic nerve enters the eyeball. Cells called **horizontal cells** form a network that either inhibits or facilitates communication between the photoreceptors and the bipolar cells. **Amacrine** (AM-a-krin) **cells** enhance communication between bipolar and ganglion cells.

The optic disc lacks photoreceptors and is a "blind spot" in your field of vision. Because the visual fields of your two eyes overlap, however, the blind spot is filled in and not noticeable. Lateral to the optic disc is an area of high cone density called the **macula lutea** (LOO-tē-uh; yellow spot). In the center of the macula lutea is a small depression called the **fovea** (FŌ-vē-uh, shallow depression). The fovea is the area of sharpest vision because of the abundance of cones. Rods are most numerous at the periphery of the neural part of the retina, and we see best at night by looking out of the corners of our eyes. There are no rods in the fovea.

QuickCheck Questions

3.1 What is the optic disc, and why don't you see a blind spot in your field of vision?

3.2 Name the different types of cells in the neural part of the retina and describe how they are organized.

In the Lab 3

Materials

☐ Compound microscope
☐ Prepared slide of retina

Procedures

1. Focus the slide at low magnification, and use Figure 29.5 as a reference while observing the specimen. Change to medium or high magnification as you examine the neural and pigmented parts of the retina.

2. Locate the thick vascular tunic on the edge of the specimen. Next to the choroid part of the vascular tunic, find the pigmented part of the retina.

3. The three types of cells in the neural part of the retina are clearly visible where the nuclei are grouped into three distinct bands. The photoreceptors—the rods and cones—form the dense band of nuclei next to the pigmented part. The bipolar cells form a thinner cluster of nuclei next to the photoreceptors. The ganglion cells have scattered nuclei and appear on the edge of the neural part of the retina. Locate each layer of cells on the slide.

Figure 29.5 Organization of the Retina The retina has three major layers of cells.

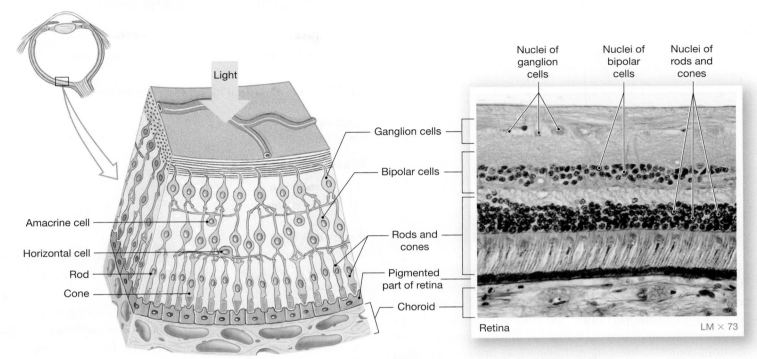

Nuclei of ganglion cells

Nuclei of bipolar cells

Nuclei of rods and cones

Light

Ganglion cells

Bipolar cells

Amacrine cell

Rods and cones

Horizontal cell

Rod

Pigmented part of retina

Cone

Choroid

Retina

LM × 73

a The cellular organization of the retina. The photoreceptors are closest to the choroid, rather than near the posterior cavity (vitreous chamber).

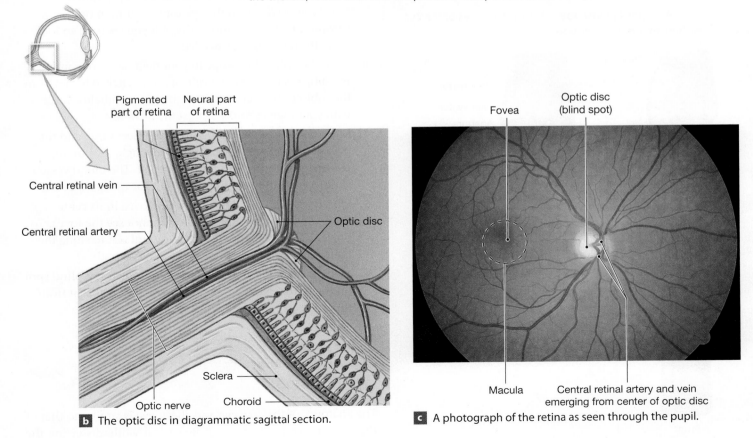

Pigmented part of retina

Neural part of retina

Central retinal vein

Central retinal artery

Optic disc

Sclera

Choroid

Optic nerve

b The optic disc in diagrammatic sagittal section.

Fovea

Optic disc (blind spot)

Macula

Central retinal artery and vein emerging from center of optic disc

c A photograph of the retina as seen through the pupil.

4. ***Draw It!*** Draw the retina slide at high magnification. ■

Retina

Lab Activity 4   Observation of the Retina

The retina is the only location in the body where blood vessels may be directly observed. The retinal blood vessels enter the eyeball by passing through the optic disc and then spread out into the neural part of the retina to provide blood to the photoreceptors and sensory neurons. To observe this vascularization, clinicians use a lighted magnifying instrument called an **ophthalmoscope** (**Figure 29.6**). The instrument shines a

Figure 29.6 An Ophthalmoscope Internal structures of the eye are visible using an ophthalmoscope.

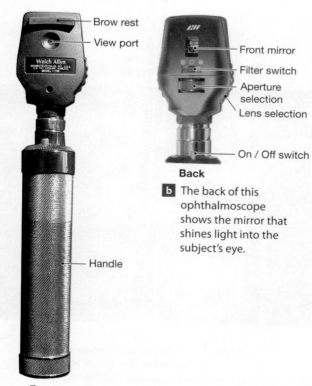

Brow rest
View port
Welch Allyn
Handle
Front
a The front of this ophthalmoscope shows the operator's view port.

Front mirror
Filter switch
Aperture selection
Lens selection
On / Off switch
Back
b The back of this ophthalmoscope shows the mirror that shines light into the subject's eye.

beam of light into the eye while the examiner looks through a lens called a viewing port to observe the retina.

QuickCheck Questions

4.1 An ophthalmoscope is used to observe what part of the eyeball?

4.2 Where do the retinal blood vessels enter the eyeball?

In the Lab 4

Materials

☐ Ophthalmoscope
☐ Laboratory partner

Procedures

Important: To protect the subject's eye, make only quick observations with the ophthalmoscope, moving the light beam away from the eye after about two seconds.

1. Before observing the retina, familiarize yourself with the parts of the ophthalmoscope, using Figure 29.6 as a reference.

2. The examination is best performed in a darkened room. Sit face to face with your partner, the *subject*, who should be relaxed. Be careful not to shine the light from the ophthalmoscope into the eye for longer than one to two seconds at a time. Additionally, ask your partner to look away from the light as needed.

3. Hold the ophthalmoscope in your right hand to examine the subject's right retina. Begin approximately 6 inches from the subject's right eye and look into the ophthalmoscope with your right eyebrow against the brow rest.

4. Move the instrument closer to the subject's eye, and tilt it so that light enters the pupil at an angle. The orange-red image is the interior of the eyeball. The blood vessels should be visible as a branched line, as in Figure 29.5c.

5. Observe the macula lutea, with the fovea in its center. Move closer to the subject if you cannot see the fovea. To prevent damage to the fovea, be careful not to shine the light on the fovea for longer than one second.

6. Medial to the macula lutea is the optic disc, the blind spot on the retina. Notice how blood vessels are absent from this area. ■

Lab Activity 5   Dissection of the Cow or Sheep Eye

The anatomy of the cow eye and sheep eye is similar to that of the human eye (**Figure 29.7**). Be careful while dissecting the eyeball because the sclera is fibrous and difficult to cut. Use

Figure 29.7 Anatomy of the Sheep Eye Fresh sheep eye with extraocular muscles intact.

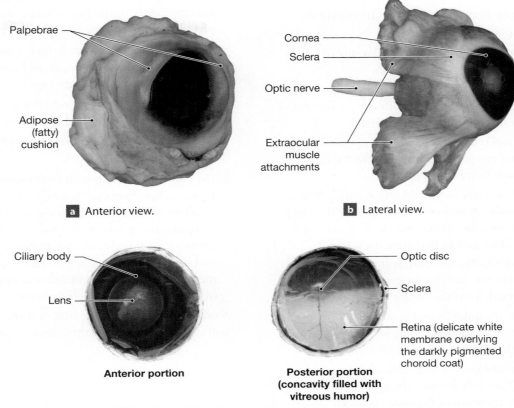

Palpebrae

Adipose (fatty) cushion

a Anterior view.

Cornea

Sclera

Optic nerve

Extraocular muscle attachments

b Lateral view.

Ciliary body

Lens

Anterior portion

Optic disc

Sclera

Retina (delicate white membrane overlying the darkly pigmented choroid coat)

Posterior portion (concavity filled with vitreous humor)

c Internal view of frontal sections. The tapetum lucidum is the greenish-blue membrane of the choroid.

small strokes with the scalpel, and cut away from your fingers. Do not allow your laboratory partner to hold the eyeball while you dissect.

⚠ Safety Alert: Dissecting the Eyeball

You must practice the highest level of laboratory safety while handling and dissecting the eyeball. Keep the following guidelines in mind during the dissection.

1. Wear gloves and safety glasses to protect yourself from the fixatives used to preserve the specimen.
2. Be extremely careful when using a scalpel or other sharp instrument. Always direct cutting and scissor motions away from you to prevent an accident if the instrument slips on moist tissue.
3. Before cutting a given tissue, make sure it is free from underlying and/or adjacent tissues so that they are not accidentally severed.
4. Never discard tissue in the sink or trash. Your instructor will inform you of the proper disposal procedure. ▲

In the Lab 5

Materials

☐ Gloves
☐ Safety glasses
☐ Fresh or preserved cow or sheep eye
☐ Dissection pan
☐ Scissors
☐ Scalpel
☐ Blunt probe
☐ Newspaper

Procedures

1. Examine the external features of the cow or sheep eye. Depending on how the eye was removed from the animal, your specimen may have, around the eyeball, adipose tissue, portions of the extraocular muscles, and the palpebrae. If so, note the amount of adipose tissue,

which cushions the eyeball. If your specimen lacks these structures, observe them in Figure 29.7.

- Identify the optic nerve (cranial nerve II) exiting the eyeball at the posterior wall.

- Examine the remnants of the extraocular muscles and, if present, the palpebrae and eyelashes.

- Locate the corneal limbus, where the white sclera and the cornea join. The cornea, which is normally transparent, will be opaque if the eye has been preserved.

2. Holding the eyeball securely, use scissors to remove any adipose tissue and extraocular muscles from the surface, taking care not to remove the optic nerve.

3. Hold the eyeball securely in the dissection pan, and with a sharp scalpel make an incision about 0.6 cm (0.25 in.) back from the cornea. Use numerous small, downward strokes over the same area to penetrate the sclera.

4. Insert the scissors into the incision, and cut around the circumference of the eyeball, being sure to maintain the 0.6-cm distance back from the cornea.

5. Carefully separate the anterior and posterior cavities of the eyeball. The vitreous body should stay with the posterior cavity. Examine the anterior portion of the eyeball.

- Place a blunt probe between the lens and the ciliary processes, and carefully lift the lens up a little. The halo of delicate transparent filaments between the lens and the ciliary processes is formed by the suspensory ligaments. Notice the ciliary body, where the suspensory ligaments originate, and the heavily pigmented iris with the pupil in its center.

- Remove the vitreous body from the posterior cavity, set it on a piece of newspaper, and notice how it causes refraction (bending) of light rays.

- The retina is the tan membrane that is easily separated from the heavily pigmented choroid of the vascular tunic.

- Examine the optic disc, where the retina attaches to the posterior of the eyeball.

- The choroid has a greenish-blue membrane, the **tapetum lucidum** (ta-PĒ-tum, a carpet; LU-sid-um, clear) which improves night vision in many animals, including sheep and cows. When headlights shine in a cow's eyes at night, this membrane reflects the light and makes the eyes glow. Humans do not have this membrane, and our night vision is not as good as that of animals that have the membrane.

6. Never discard tissue in the sink or trash. Your instructor will inform you of the proper disposal procedure. ■

Name _____

Date _____

Section _____

Anatomy of the Eye

A. Definitions

Define or describe each structure of the eye.

1. corneal limbus

2. iris

3. vitreous body

4. optic disc

5. ciliary muscle

6. canthus

7. cornea

8. sclera

9. lacrimal gland

10. suspensory ligaments

11. fovea

12. nasolacrimal duct

B. Short-Answer Questions

1. Describe the ciliary body region of the vascular tunic.

2. Describe the two cavities and two chambers of the eye and the circulation of aqueous humor through them.

3. How do the two kinds of muscles in the iris respond to high levels and low levels of light entering the eye?

C. Labeling

1. Label the structures of the eye in **Figure 29.8**.

Figure 29.8 Anatomy of the Eye A horizontal section of the right eye.

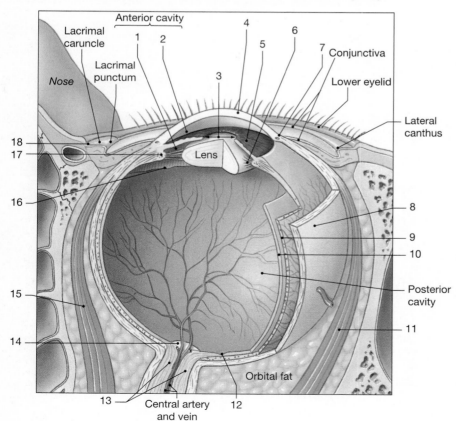

1. _____

2. _____

3. _____

4. _____

5. _____

6. _____

7. _____

8. _____

9. _____

10. _____

11. _____

12. _____

13. _____

14. _____

15. _____

16. _____

17. _____

18. _____

2. Label the cells of the retina in Figure 29.9.

Figure 29.9 **The Retina** Cellular organization of the retina.

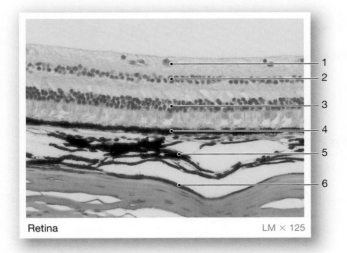

Retina   LM × 125

1. _____

2. _____

3. _____

4. _____

5. _____

6. _____

3. Label the structures of the lacrimal apparatus in Figure 29.10.

Figure 29.10 **Organization of the Lacrimal Apparatus**

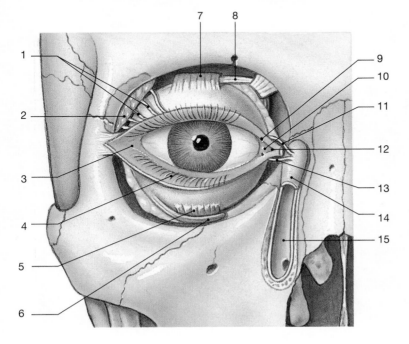

1. _____

2. _____

3. _____

4. _____

5. _____

6. _____

7. _____

8. _____

9. _____

10. _____

11. _____

12. _____

13. _____

14. _____

15. _____

D. Analysis and Application

1. Which extraocular muscles move the eyes to the right to look at an object in the far right of your visual field?

2. Name the structures of the eye through which light passes, starting with the pupil and including the three cell layers of the retina.

3. What causes the pupils to dilate in someone who is excited or frightened?

E. Clinical Challenge

Explain how a blocked lacrimal punctum would affect drainage of lacrimal secretions from the surface of the eye.

Physiology of the Eye

Learning Outcomes

On completion of this exercise, you should be able to:

1. Demonstrate the use of a Snellen eye chart and an astigmatism chart.

2. Explain the terms *myopia, hyperopia, presbyopia,* and *astigmatism*.

3. Explain why a blind spot exists in the eye and describe how it is mapped.

4. Describe how to measure accommodation.

5. Discuss the role of convergence in near vision.

6. Describe how to record an electrooculogram.

7. Compare eye movement when the eye is fixated on a stationary object with movement when the eye is tracking an object.

8. Measure duration of saccades and fixation during reading.

S ight is the only sense that requires alteration and adjustment of the stimulus before it strikes the receptors. The eye must make rapid adjustments with the iris and lens to bring light rays into focus on the retina. The iris regulates how much light can pass into the lens; the lens can change shape to focus close and distant light rays.

In this exercise you will perform several eye tests to measure how your eyes focus compared to the physiological normal eye and you will map your blind spot where the optic nerve exits the eyeball. Using the BIOPAC system you will measure and record an electrooculogram.

Need More Practice and Review?

Build your knowledge—and confidence!—in the Study Area of MasteringA&P® at www.masteringaandp.com with Pre-lab Quizzes, Post-lab Quizzes, Practice Anatomy Lab™ (PAL™) 3.0 virtual anatomy practice tool, PhysioEx™ 9.0 laboratory simulations, and A&P Flix™ with Quizzes.

Lab Activity 1 Visual Acuity

Sharpness of vision, or **visual acuity,** is tested with a Snellen eye chart, which consists of black letters of various sizes printed on white cardboard (**Figure 30.1**). A person with a visual acuity of 20/20 is considered to have normal, or **emmetropic** (em-ē-TRŌ-pik), vision. If your visual acuity is 20/30, for example, you can see at 20 feet what an emmetropic eye can see at 30 feet; 20/30 vision is not as sharp as 20/20 vision.

An eye that focuses an image in front of the retina is **myopic** (mi-Ō-pik), or nearsighted, and can clearly see close objects but not distant ones. An eye that focuses an image behind the retina is **hyperopic** (hī-per-Ō-pik), or farsighted, and can only see distant objects clearly. Corrective lenses are used to adjust for both conditions.

Make a Prediction

Consider your eyesight. Can you predict your own visual acuity?

QuickCheck Questions

1.1 What is visual acuity, and how can it be measured?

1.2 How is the myopic eye different from the emmetropic eye?

In the Lab 1

Materials

- ☐ Snellen eye chart
- ☐ 25-foot tape measure
- ☐ Masking tape
- ☐ Laboratory partner

Figure 30.1 Snellen Eye Chart

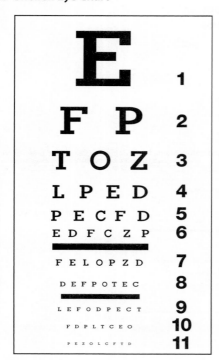

Procedures

1. Mount the eye chart on a wall at eye level. Along the floor, measure off a distance of 20 feet in front of the chart, and mark that spot on the floor with a piece of tape.

2. If you wear glasses or contact lenses, remove them before performing the vision test.

3. Stand at the 20-foot mark, and cover your left eye with either a cupped hand or an index card. While your partner stands next to the eye chart, read a line where you can easily make out all the letters. Continue to view progressively smaller letters until your partner announces that you have not read the letters correctly. Record in **Table 30.1** the visual-acuity value of this line.

4. Repeat this process with your right eye. Record your data in Table 30.1.

5. Now, using both eyes, read the smallest line you can see clearly. Record your data in Table 30.1.

6. If you wear glasses or contact lenses, repeat the test while wearing your corrective lenses. Record your data in Table 30.1. ■

Lab Activity 2 Astigmatism Test

Astigmatism (ah-STIG-mah-tizm) is a reduction in sharpness of vision due to an irregularly shaped cornea or lens. When either of these surfaces is misshapen, it bends, or **refracts,** light rays incorrectly, resulting in blurred vision. The chart used to test for astigmatism has 12 sets of three lines laid out in a circular arrangement resembling a clock face (**Figure 30.2**).

QuickCheck Questions

2.1 What is astigmatism?

2.2 Describe the chart used to test for astigmatism.

In the Lab 2

Materials

- ☐ Astigmatism chart
- ☐ 25-foot tape measure

Table 30.1	Visual Acuity	
	Acuity (without corrective lenses)	**Acuity (with corrective lenses)**
Left eye	_____	_____
Right eye	_____	_____
Both eyes	_____	_____

Figure 30.2 Astigmatism Test Chart

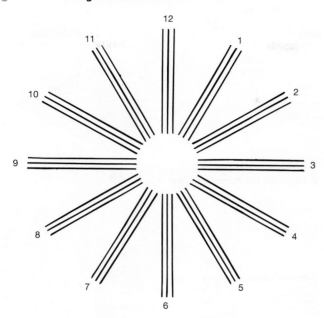

Procedures

1. Mount the astigmatism chart on a wall at eye level. Along the floor, measure off a distance of 20 feet in front of the chart, and mark that spot on the floor with a piece of tape.

2. If you wear glasses or contact lenses, remove them before performing this astigmatism test.

3. Stand at the 20-foot mark, and look at the white circle in the center of the chart. If all the radiating lines appear equally sharp and equally black, you do not have astigmatism. If some lines appear blurred or are not consistently dark, you have astigmatism. ■

Lab Activity 3 Blind-Spot Mapping

The optic disk, or **blind spot,** is an area of the retina lacking photoreceptors. Normally you do not see your blind spot because the visual fields of your two eyes overlap and "fill in" the information missing from the blind spot.

QuickCheck Question

3.1 What is the blind spot?

In the Lab 3

Material

☐ Figure 30.3

Figure 30.3 The Optic Disk Close your left eye and stare at the cross with your right eye, keeping the cross in the center of your field of vision. Begin with the page a few inches away from your eye, and gradually increase the distance. The dot will disappear when its image falls on the blind spot, at your optic disk. To check the blind spot in your left eye, close your right eye and repeat this sequence while you stare at the dot.

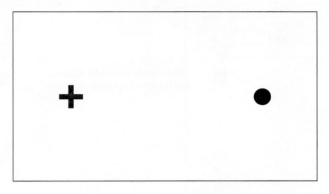

Procedures

1. If you wear glasses, try the mapping procedures both with and without your glasses. Hold Figure 30.3 about 2 inches from your face with the cross in front of your right eye. Close your left eye, and stare at the cross with your right eye.

2. Slowly move the page away from your face. The dot disappears when its image falls on your blind spot.

3. If you have difficulty mapping your blind spot, remember not to move your eyes as the page moves. ■

Lab Activity 4 Accommodation

The process called **accommodation,** by which the eye lens changes shape to focus light on the retina, is detailed in Figure 30.4.

For objects 20 feet or further from the eye, the light rays leaving the objects and entering the eye are parallel to one another, as shown in Figure 30.4a. The ciliary muscle is relaxed, and the ciliary body is behind the lens. This causes the suspensory ligaments to pull the lens flat for proper refraction of the parallel light rays. Objects closer than 20 feet to the eye have divergent, or spreading, light rays that require more refraction to be focused on the retina. The ciliary muscle contracts, and the ciliary body shifts forward, releasing the tension on the suspensory ligaments. This release of tension causes the lens to bulge and become spherical rather than flat. The spherical lens increases refraction, and the divergent rays are bent into focus on the retina.

Figure 30.4 Image Formation and Visual Accommodations A lens refracts light toward a specific point. The distance from the center of the lens to that point is the focal distance of the lens.

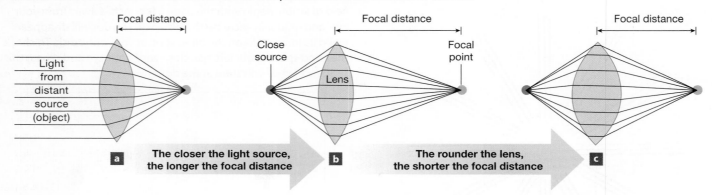

a — The closer the light source, the longer the focal distance

b — The rounder the lens, the shorter the focal distance

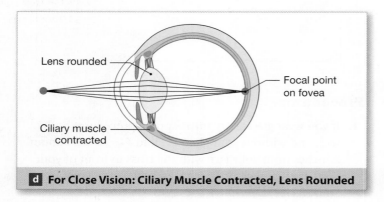

d For Close Vision: Ciliary Muscle Contracted, Lens Rounded

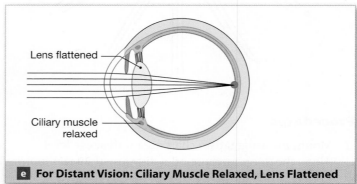

e For Distant Vision: Ciliary Muscle Relaxed, Lens Flattened

Reading and other activities requiring near vision cause eyestrain and fatigue because of the contraction of the ciliary muscle for accommodation. The lens gradually loses its elasticity as we age and causes a form of farsightedness called **presbyopia** (prez-bē-Ō-pē-uh). Many individuals have difficulty reading small type by the age of 40 and may require reading glasses to correct for the reduction in accommodation.

Accommodation is determined by measuring the closest distance from which one can see an object in sharp focus. This distance is called the **near point** of vision. A simple test for near-point vision involves moving an object toward the eye until it becomes blurred.

QuickCheck Questions

4.1 What is accommodation?

4.2 Why does presbyopia occur with aging?

Table 30.2	Near-Point Determination		
Name and Age	**Right Eye**	**Left Eye**	**Both Eyes**
_____	_____	_____	_____
_____	_____	_____	_____
_____	_____	_____	_____

In the Lab 4

Materials

☐ Pencil ☐ Ruler

Procedures

1. Hold a pencil with the eraser up approximately 2 feet from your eyes and look at the ribs in the metal eraser casing.

2. Close your left eye, and slowly bring the pencil toward your open right eye.

3. Measure the distance from the eye just before the metal casing blurs. Record your measurements in **Table 30.2**.

4. Repeat the procedure with your left eye open.

5. Repeat the procedure with both eyes open.

6. Compare near-point distances with classmates of various ages. Record this comparative data in Table 30.2.

7. To focus on near objects, the eyes must rotate medially, a process called **convergence.** To observe convergence, hold the pencil at arm's length, stare at it with both eyes open, and slowly move it closer to your eyes. Which way did your eyes move?

For additional practice, complete the *Sketch to Learn* activity. ■

Sketch to Learn

This sketch will help us understand the difference in near and distant vision. We will draw a lens, suspensory ligaments, an arrow for the ciliary muscle, and the retina.

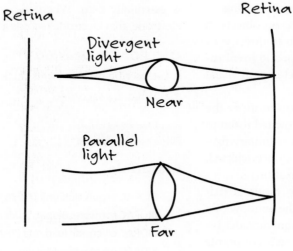

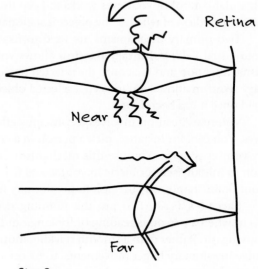

Step 1

- Draw a vertical line for the retina.
- Draw an oval lens and a flat lens and label them "near" and "far."

Step 2

- Trace light rays from a near and far object and label.

Step 3

- Near: use arrow to show ciliary muscle contracting and suspensory ligaments loose.
- Far: use a wavy arrow for a relaxed muscle and lines for tight suspensory ligaments.

One of the most important functions your eyes can perform is to "fix," or "lock," on a specific object in such a way that the image is projected onto your retina at the area of greatest acuity, the fovea. Muscular control of your eyes works to keep the image on your fovea, regardless of whether the object is stationary or moving.

Two primary mechanisms are used to fixate on objects in your visual field: **voluntary fixation** allows you to direct your visual attention and lock onto the selected object, and **involuntary fixation** allows you to keep a selected object in your visual field once it has been found.

Voluntary fixation involves a conscious effort to move the eyes. You can, for instance, pick a person in a crowded room to fix your eyes upon. You use this mechanism of voluntary fixation to initially select objects in your visual field; once selected, your brain "hands off" the task to involuntary fixation.

Saccades (sa-KĀDS) are the jumping eye motions that occur when a person is reading or looking out the window of a moving car. Rather than a smooth tracking motion, saccades involve involuntary larger movements, to fix on a series of points in rapid succession. When this happens, your eye jumps from point to point at a rate of about three jumps per second. During saccades, the brain suppresses visual images so that you do not "see" (are not aware of) the transitional images between the fixation points. When you are reading, your eye typically spends about 10 percent of the time in saccades, moving from fixation point to fixation point, with the other 90 percent of the time fixating on words. Even when you fixate on a stationary object, your eyes are not still but exhibit tiny, involuntary movements called microsaccades. These small, jerky movements keeps the visual image moving on the fovea to prevent the image from fading because of sensory adaptation of photoreceptors.

When you wish to follow a moving object, you use large, slow movements called **tracking movements.** As you watch a bird fly across your visual field, your eyes are following an apparently smooth motion and tracking the moving bird. Although you have voluntarily directed your eyes to the bird, tracking movements are involuntary.

Eye movement can be recorded as an **electrooculogram (EOG),** a recording of voltage changes that occur as eye position changes. In patients with an eye movement impairment, the EOG is used to measure neuromuscular signals, not the muscles of the eye. Electrically, the eye is a spherical battery, with the positive terminal in front at the cornea and the negative terminal behind at the retina. The potential between the front and back of the eyeball is between 0.4 and 1 mV. By placing electrodes on either side of the eye, you can measure eye movement up to 70 degrees either left or right or up and down, where 0 degrees represents the eye pointed straight ahead and 90 degrees is directly lateral or vertical to the eyes. The electrodes measure the changes in potential as the cornea moves nearer or farther from the electrodes. When the eye is looking straight ahead, it is about the same distance from either electrode, and so the signal is essentially zero. When the cornea is closer to the positive electrode, that electrode records a positive difference in voltage.

QuickCheck Questions

5.1 Describe how the eyes track an object.

5.2 What are saccades?

In the Lab 5

Materials

- ☐ BIOPAC acquisition unit (MP36/35/30)
- ☐ BIOPAC software: Biopac Student Lab (BSL) v3.7.5 or better
- ☐ BIOPAC electrode lead sets (SS2L)—2 lead sets per subject
- ☐ BIOPAC disposable vinyl electrodes (EL503)—6 electrodes per subject
- ☐ BIOPAC electrode gel (GEL1) and abrasive pad (ELPAD)
- ☐ Computer: PC Windows 7, Vista, or XP; Mac OS X 10.4–10.6
- ☐ Skin cleanser or rubbing alcohol
- ☐ Cotton balls
- ☐ Tape measure
- ☐ Pendulum—any object attached to approximately 61 cm (24 inches) of string
- ☐ Reading passages—easy: entertainment article, hard: scientific article
- ☐ Two laboratory partners

Procedures

The EOG investigation is divided into four sections: setup, calibration, recording, and data analysis. Read each section completely before attempting a recording. If you encounter a problem or need further explanation of a concept, ask your laboratory instructor.

This experiment requires three persons. **You** will perform the test movements, one of your partners will be the **subject,** and the other partner will be the **recorder.** The recorder may either record the data by hand or choose Edit > Journal > Paste Measurements to paste the data to the electronic journal for future reference.

Most response markers and labels are inserted automatically as a segment of the test is recorded. Markers appear at the top of the window as inverted triangles. The recorder may insert and label the marker either during or after the data are collected. To insert markers, press ESC on a Mac or F9 on a PC.

Section 1: Setup

1. Make sure the acquisition unit is turned off.

2. Plug the equipment in as shown in Figure 30.5, with one lead set (SS2L) into CH 1 for horizontal and the other lead set (SS2L) into CH 2 for vertical. Note that each lead set has a pinch connector at the point where the red, white, and black leads attach to the main lead; you will use these pinch connectors in a moment when you attach the SS2L lead sets to the subject. Turn on the acquisition unit.

3. Have the subject remove all jewelry, especially rings, bracelets, and studs. Also, *be sure the subject is not in contact with any metal objects (faucets, pipes, and so forth).*

4. Place six electrodes (EL503) on the subject as shown in Figure 30.6. *Important:* For accurate recordings, the electrodes must be horizontally and vertically aligned as described in the following steps. Before positioning each electrode, clean the subject's skin with a cotton ball dipped in skin cleanser or rubbing alcohol, and then apply a small dab of electrode gel (GEL1).

   For optimal electrode adhesion, the electrodes should be placed on the skin at least five minutes before the start of the calibration procedure. *Note:* Because these electrodes are attached near the eye, be very careful if using alcohol to clean the skin.

   ■ Attach one electrode above the right eyebrow and one below the right eye, with the two aligned vertically.

   ■ Attach one electrode to the lateral side of the right eye and one to the lateral side of the left eye, with the two aligned horizontally.

   ■ Attach the fifth electrode above the nose and the sixth above the left eyebrow. These two electrodes serve as electrical grounds, and it is not critical that they be aligned.

5. Attach the pinch connector of the vertical SS2L lead set from CH 2 to the subject's shirt to relieve strain on the cable. Then attach the three leads to three of the EL503 electrodes on the subject's face, following the arrangement shown in Figure 30.6: the black lead to the electrode above the nose, the red lead to the electrode above the right eye, and the white lead to the electrode below the right eye. It is recommended that the electrode leads run behind the ears, as shown, to give proper cable strain relief.

6. Attach the pinch connector of the horizontal SS2L lead set from CH 1 to the subject's shirt to relieve strain on the cable. Then attach the three leads to the other three EL503 electrodes on the subject's face, following the arrangement shown in Figure 30.7: the black lead to the electrode above the left eye, the red lead to the electrode on the lateral side of the right eye, and the white lead to the electrode to the lateral side of the left eye. Again, it is recommended that the electrode leads run behind the ears, as shown, to give proper cable strain relief.

Figure 30.6 Lead Placement for CH 2 (Vertical) Electrodes

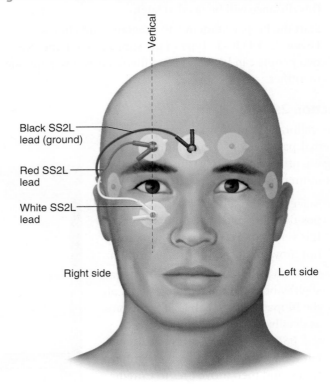

Vertical

Black SS2L lead (ground)

Red SS2L lead

White SS2L lead

Right side   Left side

Figure 30.5 Electrooculogram Setup

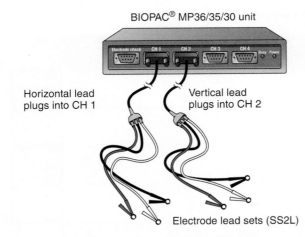

BIOPAC® MP36/35/30 unit

Horizontal lead plugs into CH 1

Vertical lead plugs into CH 2

Electrode lead sets (SS2L)

Figure 30.7 Lead Placement for CH 1 (Horizontal) Electrodes

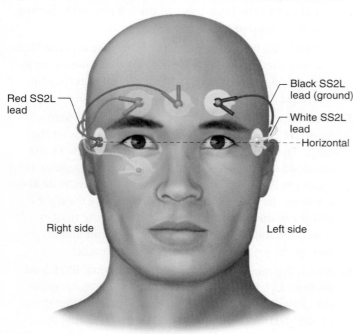

Red SS2L lead

Black SS2L lead (ground)

White SS2L lead

Horizontal

Right side   Left side

7. Have the subject sit so that his or her eyes are in line with the center of the computer screen and he or she can see the screen easily with no head movement. Supporting the head to minimize movement is recommended.

8. Note the distance from the eyes to the computer screen; this distance will be needed later.

9. Start the Biopac Student Lab program, and choose Lesson L10-EOG-1. Type in the subject's filename. No two people can have the same filename, so use a unique identifier.

Section 2: Calibration

The calibration procedure establishes the internal parameters for the hardware (such as gain, offset, and scaling) and is critical for optimum performance.

1. Make sure the subject is seated in the position described in Setup: Step 7. It is very important that the subject not move the head during calibration.

2. Click on Calibrate. Screen prompts vary slightly between the recent versions of the Biopac Student Lab (BSL) software, as detailed below.

   BSL 4 users: Move eyes to extreme left then extreme right and repeat four times, and then move eyes to extreme

up then extreme down and repeat four times. Calibration will continue for about 20 seconds.

   BSL 3.7.5–3.7.7 users: A new window will be established, and a dialog box will pop up. The journal will be hidden from view during calibration. Click on OK to begin the calibration. A dot will go in a counterclockwise rotation around the screen, and the subject will need to *track the dot with the eyes while keeping the head perfectly still.* Calibration will continue for about 10 seconds.

3. Calibration will stop automatically.

4. Check the calibration data at the end of the recording. There should be fluctuation in the data for each channel. If your data recording is similar to **Figure 30.8**, proceed to Section 3, or click Redo and repeat the calibration sequence.

Possible reasons for the differences: If the subject did not follow the dot with the eyes or if the subject blinked, your recording will have large spikes. If one of the electrodes peeled up from the subject's face, your recording will have a too-large baseline drift.

Section 3: Recording Lesson Data

Prepare for the recording. You will record up to eight segments: real and simulated pendulum; real and simulated vertical tracking; read silently, easy and hard text; read aloud; and *optional* dot plot. In order to work efficiently, read this entire section so you will know what to do before you begin recording. Screen prompts are similar in the recent versions of the Biopac Student Lab (BSL) software. BSL 4 uses 'Continue' and 'Record' buttons to allow review/prep between segments; BSL 3.7.5–3.7.7 uses 'Record' and 'Resume' buttons.

Hints for Obtaining Optimal Data

a. The object should always be tracked with the eyes, *not the head*. The subject needs to sit so that head movement is minimized during recording.

Figure 30.8 Calibration Recording

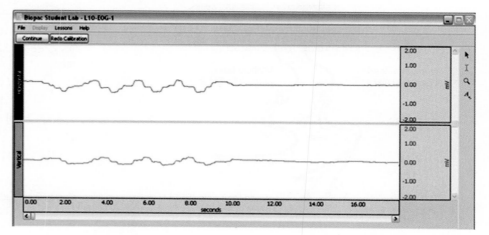

b. The subject should focus on one point of the object and should maintain that focus consistently.

c. There should be enough space near the subject so that you are able to move an object around the head at a distance of about 25 cm (10 in.). When you are moving the object, try to keep it always at the same distance from the subject's head.

d. During recording, the subject should not blink. If unavoidable, the recorder should mark the blink on the recording.

e. Make sure the six electrodes stay firmly attached to the subject's face.

f. The larger the monitor, the better the eye-tracking portion of this lesson will work.

Segment 1—Pendulum Horizontal Tracking

1. You and the subject should face each other in such a way that the subject is not looking at the computer screen.

2. Hold the pendulum in front of the subject's head at a distance of about 25 cm (10 in.) and lift it up approximately 45 degrees while maintaining a taut string. Pendulum should be centered with subject's eyes when it is swinging and subject should be able to see full swing range without moving head.

3. Click on Continue and when ready click on Record and set pendulum in motion to begin recording Segment 1 data. Record until pendulum stops swinging.

4. Click on Suspend to halt the recording and review the data. Look for diminishing amplitudes, especially for the horizontal EOG (CH 1). Repeat the recording if necessary. A few blinks may be unavoidable; they will show in the data, but would not necessitate redoing the recording. The data will be incorrect if any of the following occur.

   a. Channel connections are incorrect.

   b. Lead connections are incorrect (for instance, if the red lead is not connected to the electrode at the subject's right temple).

   c. The suspend button is pressed prematurely.

   d. The electrode peels up, giving a large baseline drift.

   e. The subject looks away or moves the head.

Segment 2—Simulate Pendulum

5. Subject remains facing Director, away from computer screen, and tracks an imaginary pendulum.

6. Click on Continue and when ready click on Record.

7. Subject imagines the pendulum's arc decreasing with each swing cycle until it is stationary.

8. Click on Suspend when the graph shows little or no eye movement on CH 1 Horizontal.

Segment 3—Vertical Tracking

9. You and the subject should face each other in such a way that the subject is not looking at the computer screen.

10. Hold a pen in front of the subject's head at a distance of about 25 cm (10 in.). Center the pen relative to the head so that the subject's eyes are looking straight ahead. The subject may need to blink before resuming recording. Instruct the subject to pick a point on the pen that is directly in his or her line of vision.

11. Click on Continue and when ready click on Record. The Segment 3 data will be recorded at the point where Segment 2 data stopped. Record for about 30 seconds.

    ■ Hold the pen in front of the subject for about three seconds, then move it vertically up to the top edge of the subject's field of vision, then back past the center point and vertically down to the bottom edge of the field of vision, then back to center. From the time you start moving up till you return to center from the maximum down position, about 10 seconds should elapse. Say the movement directions aloud so that the recorder will know where to place markers to denote direction.

    ■ The recorder inserts a marker with each change of direction you call out, and then labels markers "U" for up and "D" for down. Markers may also be entered or edited after the data are recorded.

12. Click on Suspend to stop the recording and review the data on the screen. Look for large deflections for the vertical EOG (CH 2) and very little deflection for the horizontal EOG (CH 1). There should be a positive peak where the subject's eyes reached their maximum displacement upward and a negative peak where the eyes reached their maximum displacement downward. Repeat the recording if necessary.

Segment 4—Simulate Vertical Tracking

13. Subject remains facing Director, away from computer screen, and tracks an imaginary pen moving vertically.

14. Click on Continue and when ready click on Record.

15. Subject tracks an imaginary pen starting at center position, then moving to the upper and then the lower edges of the visual field, and finally returning to center. Complete *five* upper/lower cycles.

16. Click on Suspend when the graph shows little or no eye movement on CH 2 Vertical.

Segment 5—Read Silently–Easy Text

17. Turn to the end of the Review & Practice Sheet, where the reading sample appears, and hold the *easy text* page in front of the subject about 25 cm (10 in.) in front of his or her eyes.

18. Click on Continue and when ready click on Record. The subject should read for about 20 seconds, reading silently to reduce any electromyogram (EMG) artifacts from facial muscle contraction. The recorder will insert a marker (ESC on Mac or F9 on PC) when the subject starts each new line of the reading sample. The recorder should watch the subject's eyes for vertical movements that indicate when the subject has finished reading one line and has begun to read the next line.

19. Click on Suspend to stop the recording and review the data on the screen. Repeat the recording if necessary.

Segment 6—Read Silently–Hard Text

20. Turn to the end of the Review & Practice Sheet, where the reading sample appears, and hold the *hard text* page in front of the subject about 25 cm (10 in.) in front of his or her eyes.

21. Click on Continue and when ready click on Record. The subject should read for about 20 seconds, reading silently to reduce any electromyogram (EMG) artifacts from facial muscle contraction. The recorder will insert a marker (ESC on Mac or F9 on PC) when the subject starts each new line of the reading sample. The recorder should watch the subject's eyes for vertical movements that indicate when the subject has finished reading one line and has begun to read the next line.

Segment 7—Read Aloud

22. Turn to the end of the Review & Practice Sheet, where the reading sample appears, and hold the *hard text* page in front of the subject about 25 cm (10 in.) in front of his or her eyes.

23. Click on Continue and when ready click on Record. The subject should read aloud for about 20 seconds. Subject should remain relaxed and should try not to blink during the recording. The recorder will insert a marker (ESC on Mac or F9 on PC) when the subject starts each new line of the reading sample. The recorder should watch the subject's eyes for vertical movements that indicate when the subject has finished reading one line and has begun to read the next line.

24. Click on Suspend to stop the recording and review the data on the screen. Repeat the recording if necessary.

Segment 8—Optional: Dot Plot

25. Click on Stop and Yes, and then click on Dot Plot if available. The subject should focus on the center of the cross. A fixed focus would hold the colored dot in the center of the cross. Dot movement indicates microsaccadic eye movement.

26. Click on Stop to end the Dot Plot. Review the data on the screen and click Redo to repeat the Dot Plot if desired.

27. Click on Done. A dialog box comes up, asking if you are sure you want to stop the recording. Clicking Yes will end the data-recording segment and automatically save the data. Clicking No will take you back to the Resume or Stop options.

28. Disconnect the SS2L pinch connectors from the subject's face and clothing, and peel the electrodes off the subject's face. Throw out the electrodes. Wash the electrode gel residue from the skin, using soap and water. The electrodes may leave a slight ring on the skin for a few hours. This is normal and does not indicate that anything is wrong.

Section 4: Data Analysis

1. Enter the Review Saved Data mode from the Lessons menu. A window that looks like **Figure 30.9** should open. Note the CH designations: 40 for horizontal and 41 for vertical.

2. Set up your display window for optimal viewing of the first data segment, which is the one between time 0 and the first marker. The following tools help you adjust the data window.

Autoscale horizontal	Horizontal (time) scroll bar
Autoscale waveforms	Vertical (amplitude) scroll bar
Zoom tool	Zoom previous

Grids—Turn grids on and off by choosing Preferences from the File menu.

3. The measurement boxes are above the marker region in the data window. Each measurement has three sections: channel number, measurement type, and result. Channel number and measurement type are pull-down menus that

Figure 30.9 **Recording While Looking Left and Right**

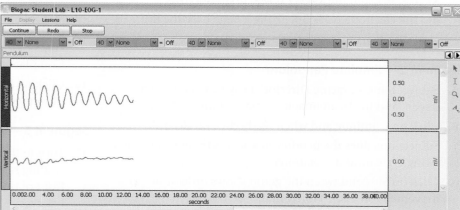

are activated when you click on them. Set them up as follows:

CH 40   **Delta T** (difference between time at end of selected area and time at beginning of area)

CH 40   **P-P** (peak-to-peak difference between maximum and minimum amplitude values in selected range)

CH 40   **Slope** (difference in magnitude of two endpoints of selected area divided by time interval between endpoints; indicates relative speed of eye movement)

Note: The selected area is the area selected by the I-beam tool, including the endpoints.

Remember, you can either record this and all other measurement information individually by hand, or choose Edit > Journal > Paste Measurements to paste the data to your journal for future reference.

When interpreting the different bumps in the data, in general, remember the following:

- Large vertical bumps represent either blinks or eye movement from one line to another.

- Large horizontal bumps represent the eyes moving left to start the next line.

- Small bumps are saccades.

4. Measure the amplitude change for each pendulum tracking cycle (Figure 30.9); use the horizontal scroll bar to move through the recording.

Figure 30.10 Recording of Saccades

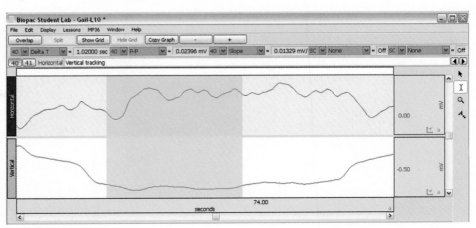

5. Repeat measurements for each simulated pendulum cycle.

6. Set up the measurement boxes as follows:

CH 41   **Delta T**

CH 41   **P-P**

CH 41   **Slope**

7. Measure the amplitude change for each object tracking cycle.

8. Repeat measurements for each simulated tracking cycle.

9. Scroll to the Read silently 1 data.

10. Measure the duration (Delta T) for each saccade in the data (**Figure 30.10**).

11. Repeat measurements on Read Silently 2 saccades.

12. Repeat measurements on Read Aloud saccades.

13. Save or print the data file. You may save the data to a storage device, save notes that are in the journal, or print the data file.

14. Exit the program. ■

Name _____

Date _____

Section _____

Physiology of the Eye

A. Definitions

Define each term.

1. emmetropic

2. astigmatism

3. hyperopic

4. myopic

5. presbyopia

6. visual acuity

7. refraction

8. accommodation

B. Short-Answer Questions

1. Explain the difference between viewing a distant object and viewing one close to the eye.

2. How is the Snellen eye chart used to measure visual acuity?

3. Describe how the lens changes shape to view a close object.

C. Analysis and Application

1. Describe an experiment that demonstrates the presence of a blind spot on the retina.

2. Explain how the nearsighted eye cannot view distant objects in focus but can focus on near objects.

D. Clinical Challenge

Explain why many people 40 years and older need to wear glasses when they read.

Name ____________________

Date ____________________

Section ____________________

BIOPAC: Electrooculogram

Review & Practice Sheet

EXERCISE 30

A. Data and Calculations

Subject Profile

Name ____________________ Height ____________

Age ____________________ Weight ____________

Gender ____________________

1. Complete **Table 30.3** using Segments 1–2 data. Be careful to be consistent with units (milliseconds versus seconds).

Table 30.3 Segments 1–2—Pendulum Tracking vs. Simulation Tracking (using Horizontal data)

Cycle	Pendulum			Simulation		
	Delta T (CH 40)	P-P (CH 40)	Slope (CH 40)	Delta T (CH 40)	P-P (CH 40)	Slope (CH 40)
1						
2						
3						
4						
5						
6						
7						

Note: Horizontal data (CH 40) is used for analysis.

2. Complete **Table 30.4** using Segments 3–4 data. Be careful to be consistent with units (milliseconds versus seconds). Use Vertical data (CH 41) for analysis.

Table 30.4 Segments 3–4—Vertical Tracking vs. Simulation Tracking

Cycle	Real Object			Simulation		
	Delta T (CH 41)	P-P (CH 41)	Slope (CH 41)	Delta T (CH 41)	P-P (CH 41)	Slope (CH 41)
1						
2						
3						
4						
5						
6						
7						

3. Complete **Table 30.5** with Segments 5–7 data. (You may not have seven saccades per line.) Use Vertical data (CH 41) for analysis.

Table 30.5	Segments 5–7—Saccades		Read Silently 1		Read Silently 2		Read Aloud	
Measurement			1st line	2nd line	1st line	2nd line	1st line	2nd line
Number of saccades								
Saccade duration (Delta T)		#1						
		#2						
		#3						
		#4						
		#5						
		#6						
		#7						
Total duration of saccades/line								
Total reading time/line								
Saccade % of reading time								

B. Short-Answer Questions

1. Explain how an electrooculogram is recorded.

2. What is the difference between a voluntary fixation and involuntary fixation?

C. Analysis and Application

1. Examine your data in Table 30.3 and answer the following questions:

a. What are the differences in amplitude during the pendulum movement recording and the simulated movement recording?

b. What are the differences in period frequency during the pendulum movement recording and the simulated movement recording?

c. Did the relative speed of the eye movements (the slope in the recording waves) change during the pendulum movement recording?

2. Examine your data in Table 30.4 and answer the following questions:

    a. What are the differences in amplitude during the object movement recording and the simulated movement recording?

    b. What are the differences in period frequency during the object movement recording and the simulated movement recording?

    c. Did the relative speed of the eye movements (the slope in the recording waves) change during the object movement recording?

3. Examine the data in Table 30.5 and answer the following questions:

    a. How is the speed of eye movement different while reading a challenging passage as compared to reading an easy passage?

    b. How does eye movement compare between reading silently and aloud?

Easy Reading Sample for Segment 5

> Row, row, row your boat, gently down
> the stream, merrily, merrily, merrily,
> merrily, life is but a dream.

Difficult Reading Sample for Segments 6–7

> Alas, poor Yorick! I knew him, Horatio,
> a fellow of infinite jest, of most
> excellent fancy. He hath borne me on
> his back a thousand times, and now
> how abhorr'd in my imagination it is!
> My gorge rises at it. Here hung those
> lips hat I have kissed I know
> not how oft. Where be your gibes
> now? Your gambol? Your songs?

Anatomy of the Ear

Learning Outcomes

On completion of this exercise, you should be able to:

1. Identify and describe components of the external, middle, and inner ear.

2. Describe the anatomy of the cochlea.

3. Describe components of the semicircular canals and the vestibule and explain their role in static and dynamic equilibrium.

T he ear is divided into three regions: the **external,** or outer, **ear;** the **middle ear;** and the **inner ear.** The external ear and middle ear direct sound waves to the inner ear for hearing. The inner ear serves two unique functions: balance and hearing. Without a sense of balance, you would not know, at any given moment, where your body is relative to the ground and in three-dimensional space. You would be unable to stand, let alone walk, or even drive a car. Your sense of hearing enables you to enjoy your favorite song while simultaneously conversing with a friend.

In this exercise you identify the anatomical features of the ear and look at the sensory receptors for equilibrium and hearing.

Make a Prediction

Receptors for balance and hearing are hair cells. How are these cells stimulated?

Lab Activity 1   External and Middle Ear

The pinna, or **auricle,** is the flap of the outer ear that funnels sound waves into the **external acoustic meatus,** a tubular chamber that delivers sound waves to the

Need More Practice and Review?

Build your knowledge—and confidence!—in the Study Area of MasteringA&P® at www.masteringaandp.com with Pre-lab Quizzes, Post-lab Quizzes, Practice Anatomy Lab™ (PAL™) 3.0 virtual anatomy practice tool, PhysioEx™ 9.0 laboratory simulations, and A&P Flix™ with Quizzes.

PAL ⌷ practice anatomy lab   For this lab exercise, follow these navigation paths:
- PAL>Human Cadaver>Nervous System>Special Senses
- PAL>Anatomical Models>Nervous System>Special Senses
- PAL>Histology>Special Senses

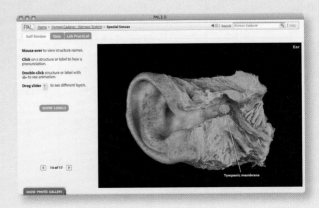

tympanic (tim-PAN-ik) **membrane** (*tympanum*), commonly called the *eardrum* (**Figure 31.1**). The auricle has an inner foundation of elastic cartilage covered with adipose tissue and skin. The tympanic membrane is a thin sheet of fibrous connective tissue stretched across the distal end of the external acoustic meatus and separating the external ear from the middle ear. The meatus contains wax-secreting cells in **ceruminous glands** plus many hairs that prevent dust and debris from entering the middle ear.

The middle ear (**Figure 31.2**) is the **tympanic cavity** inside the petrous part of the temporal bone. It is connected to the back of the upper throat (the nasopharynx) by the **auditory tube,** also called either the *pharyngotympanic* or *Eustachian tube*. This tube equalizes pressure between the external air and the cavity of the middle ear. Three small bones of the middle ear, called **auditory ossicles,** transfer vibrations from the external ear to the inner ear. The **malleus** (*malleus*, hammer) is connected on one side to the tympanic membrane and on the other side to the **incus** (*incus,* anvil), which is in contact with the third auditory ossicle, the **stapes** (*stapes*, stirrup). Vibrations of the tympanic membrane are transferred to

Clinical Application Otitis Media

Infection of the middle ear is called **otitis media.** It is most common among infants and children but occurs infrequently in adults. The infection source is typically a bacterial invasion of the throat that has migrated to the middle ear by way of the auditory tube. In children, the auditory tube is narrow and more horizontal than in adults. This orientation permits pathogens originally present in the throat to infect the middle ear. Children who frequently get middle ear infections may have small tubes implanted through the tympanic membrane to drain liquid from the middle ear into the external acoustic meatus.

In severe cases, the microbes infect the air cells of the mastoid process (Exercise 14), causing a condition known as **mastoiditis.** The passageways between the air cells become congested, and swelling occurs behind the auricle. This condition is serious, as it may spread to the brain. Powerful antibiotic therapy is necessary to treat the infection. Otherwise, a mastoidectomy may be necessary, a procedure that involves opening and draining the mastoid air cells. ∎

Figure 31.1 **Anatomy of the Ear** The boundaries separating the three main anatomical regions of the ear (external, middle, and inner) are indicated by the dashed lines.

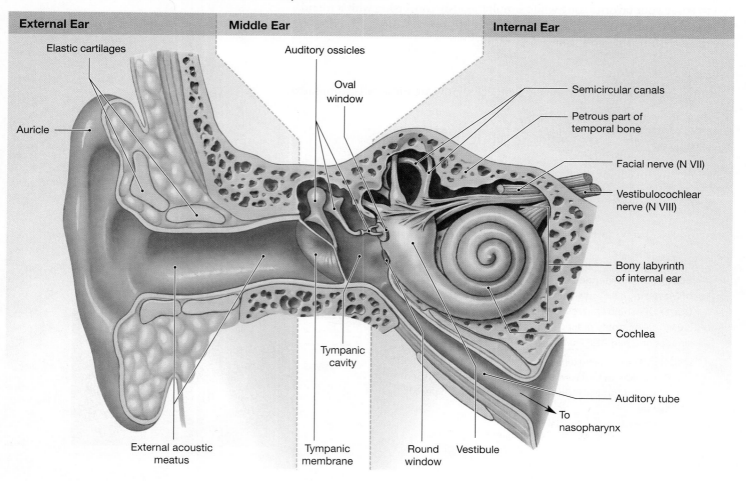

External Ear
Middle Ear
Internal Ear

Elastic cartilages
Auditory ossicles
Oval window
Semicircular canals
Petrous part of temporal bone
Auricle
Facial nerve (N VII)
Vestibulocochlear nerve (N VIII)
Bony labyrinth of internal ear
Cochlea
Auditory tube
To nasopharynx
External acoustic meatus
Tympanic cavity
Tympanic membrane
Round window
Vestibule

Figure 31.2 The Middle Ear

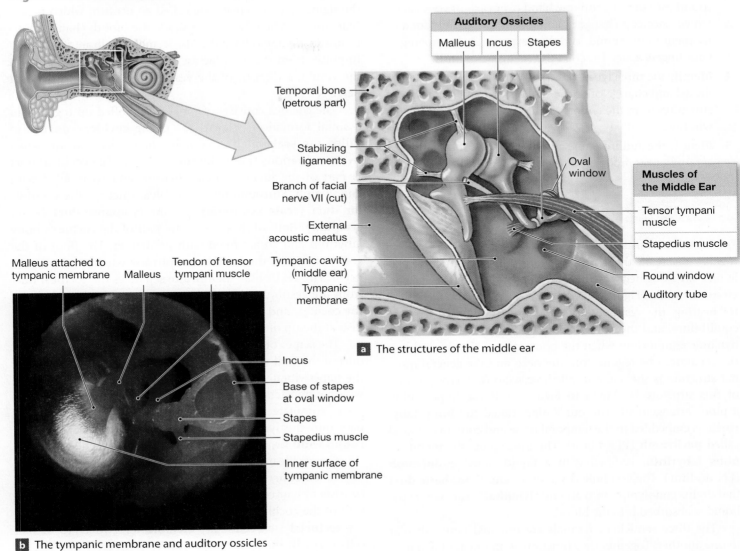

a The structures of the middle ear

b The tympanic membrane and auditory ossicles

the malleus, which then conducts the vibrations to the incus and stapes. The stapes in turn pushes on the **oval window** of the inner ear to stimulate the auditory receptors.

The smallest skeletal muscles of the body are attached to the auditory ossicles. The **tensor tympani** (TEN-sor tim-PAN-ē) **muscle** attaches to the malleus, and the **stapedius** (sta-PĒ-dē-us) **muscle** inserts on the stapes.

QuickCheck Questions

1.1 What is the function of the auricle?

1.2 Which two regions of the ear does the tympanic membrane separate?

1.3 What is the function of the auditory tube?

In the Lab 1

Materials

☐ Ear model

☐ Ear chart

Procedures

1. Review the three major regions of the ear in Figure 31.1 and the features of the middle ear in Figure 31.2.

2. Identify the auricle, external acoustic meatus, and tympanic membrane on the ear model and chart.

3. To appreciate how important the auricle is in directing sound into the ear, cup one hand over each of your ears. Do you notice a change in sound? Listen carefully for a moment to the sound, and then experiment by moving your fingers apart. Describe the change in sounds.

4. Identify the three types of auditory ossicles on the ear model and chart. Notice the sequence of articulated structures from the tympanic membrane to the oval window.

5. Identify the auditory tube and the muscles of the middle ear on the ear model and chart. ∎

Lab Activity 2   Inner Ear

The inner ear consists of three regions (Figure 31.3). Moving medial to lateral, these regions are a helical **cochlea** (KOK-lē-uh), an elongated **vestibule** (VES-ti-bū-l), and three **semicircular canals** (Figure 31.3a). The cochlea contains receptors for hearing; the vestibule is receptive to stationary, or static, equilibrium; and the semicircular canals contain receptors for dynamic equilibrium when the body moves. No physical barrier separates one region from the next, and the general internal structure is the same in all three regions. A cross section of this structure is shown in Figure 31.3b—a "pipe within a pipe" arrangement. The outer pipe, called the **bony labyrinth,** is embedded in the temporal bone and contains a liquid called **perilymph** (PER-i-limf). The inner pipe, the **membranous labyrinth,** is filled with a liquid called **endolymph** (EN-dō-limf). The vestibule contains an **endolymphatic duct** that drains endolymph into an **endolymphatic sac,** where the liquid is absorbed into the blood.

The three semicircular canals are oriented perpendicular to one another. Together they function as an organ of dynamic equilibrium and work to maintain equilibrium when the body is in motion. Inside the canals, the membranous labyrinth is called the **semicircular ducts.** At one end of each semicircular duct is a swollen **ampulla** (am-PŪL-luh) that houses the balance receptors called **cristae** (Figure 31.3c). Each crista is composed of hair cells and supporting cells, with the cilia of the hair cells extending upward from the crista into a gelatinous material called the **cupula** (KŪ-pū-la). Movement of the head causes the endolymph inside the semicircular ducts to either push or pull on the cupula, so that the embedded hair cells are either bent or stretched.

In the vestibule, the membranous labyrinth contains two sacs, the **utricle** (Ū-tri-kul) and the **saccule** (SAK-ū-l), which contain **maculae** (MAK-ū-lē), receptors that work to maintain static equilibrium. Like the cristae of the ampullae, the maculae of the utricle and saccule have hair cells and a **gelatinous material.** Embedded in the gel are calcium carbonate crystals

called **statoconia** (Figure 31.3e). The gelatinous material and the statoconia collectively are called an **otolith,** which means "ear stone." When the head is tilted, the otolith changes position, and the hair cells in the utricle and saccule are stimulated. Impulses from the maculae are passed to sensory neurons in the vestibular branch of the vestibulocochlear nerve (cranial nerve N VIII).

The cochlea consists of three ducts rolled up together in a spiral formation (Figure 31.4). The **cochlear duct,** also called the **scala media,** contains hair cells that are sensitive to vibrations caused by sound waves. The cochlear duct is part of the membranous labyrinth and so is filled with endolymph. Surrounding the cochlear duct are the **vestibular duct (scala vestibule)** and the **tympanic duct (scala tympani).** Both of these ducts are part of the cochlea's bony labyrinth and so are filled with perilymph. The floor of the cochlear duct is the **basilar membrane** where the hair cells occur. The **vestibular membrane** separates the cochlear duct from the vestibular duct. These two ducts follow the helix of the cochlea, and the vestibular and tympanic ducts interconnect at the tip of the spiral.

The stapes of the middle ear is connected to the vestibular duct at the oval window. When incoming sound waves make the stapes vibrate against the oval window, the pressure on the window transfers the waves to the ducts of the cochlea. The waves stimulate the hair cells in the cochlear duct and then pass into the tympanic duct, where a second window, the **round window,** stretches to dissipate the wave energy.

The cochlear duct contains the sensory receptor for hearing, called either the **spiral organ** or the **organ of Corti.** It consists of hair cells and supporting cells. Extending from the wall of the cochlear duct and projecting over the hair cells is the **tectorial** (tek-TOR-ē-al) **membrane.** Two types of hair cells occur in the spiral organ: **inner hair cells** that rest on the basilar membrane near the proximal portion of the tectorial membrane and **outer hair cells** at the tip of the membrane. The long stereocilia of the hair cells extend into the endolymph and contact the tectorial membrane. Sound waves cause liquid movement in the cochlea, and the hair cells are bent and stimulated as they are pushed against the tectorial membrane. The hair cells synapse with sensory neurons in the cochlear branch of the vestibulocochlear nerve (N VIII), which transmits the impulses to the auditory cortex of the brain. The **spiral ganglia** contain cell bodies of sensory neurons in the cochlear branch of the vestibulocochlear nerve.

QuickCheck Questions

2.1 What are the three kinds of sensory receptors of the inner ear, and what is the function of each?

2.2 What is the function of the semicircular canals?

2.3 What is the function of the vestibule?

Figure 31.3 The Inner Ear The inner ear is located in the petrous part of each temporal bone.

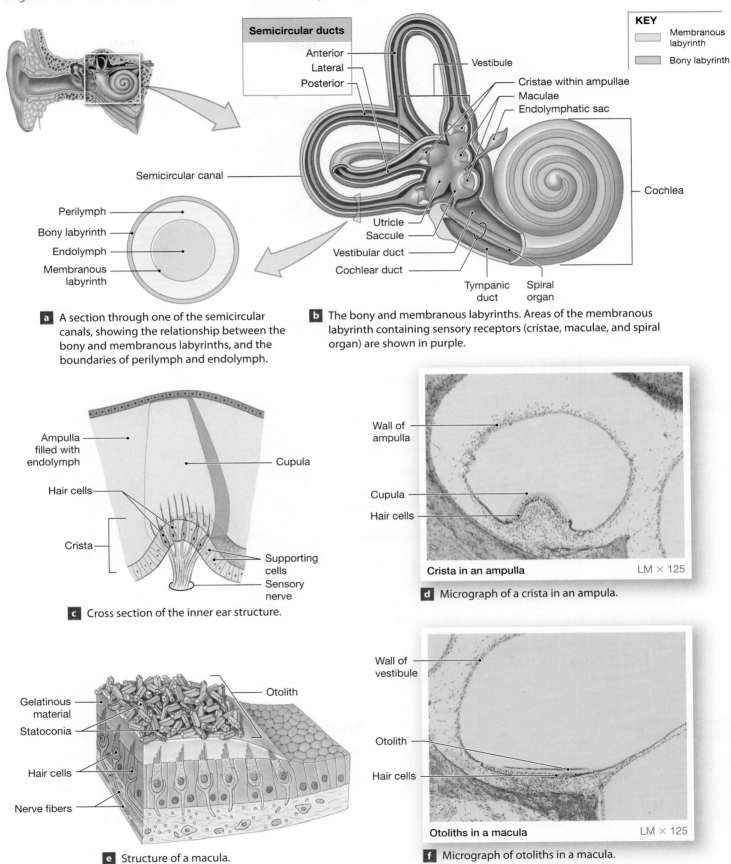

KEY

- Membranous labyrinth
- Bony labyrinth

Semicircular ducts
- Anterior
- Lateral
- Posterior

Vestibule
Cristae within ampullae
Maculae
Endolymphatic sac

Semicircular canal

Cochlea

Perilymph
Bony labyrinth
Endolymph
Membranous labyrinth

Utricle
Saccule
Vestibular duct
Cochlear duct

Tympanic duct
Spiral organ

a A section through one of the semicircular canals, showing the relationship between the bony and membranous labyrinths, and the boundaries of perilymph and endolymph.

b The bony and membranous labyrinths. Areas of the membranous labyrinth containing sensory receptors (cristae, maculae, and spiral organ) are shown in purple.

Ampulla filled with endolymph

Cupula

Hair cells

Crista

Supporting cells

Sensory nerve

c Cross section of the inner ear structure.

Wall of ampulla

Cupula

Hair cells

Crista in an ampulla   LM × 125

d Micrograph of a crista in an ampula.

Gelatinous material
Statoconia

Otolith

Hair cells

Nerve fibers

e Structure of a macula.

Wall of vestibule

Otolith

Hair cells

Otoliths in a macula   LM × 125

f Micrograph of otoliths in a macula.

Figure 31.4 The Cochlea The ducts of the cochlea are coiled approximately 2.5 times.

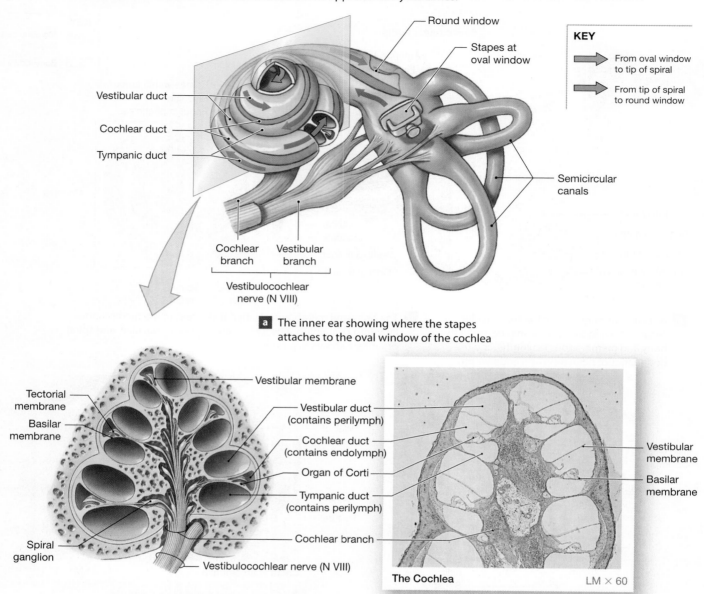

KEY

➜ From oval window to tip of spiral

➜ From tip of spiral to round window

Round window

Stapes at oval window

Vestibular duct

Cochlear duct

Tympanic duct

Semicircular canals

Cochlear branch — Vestibular branch

Vestibulocochlear nerve (N VIII)

a The inner ear showing where the stapes attaches to the oval window of the cochlea

Tectorial membrane

Basilar membrane

Vestibular membrane

Vestibular duct (contains perilymph)

Cochlear duct (contains endolymph)

Organ of Corti

Tympanic duct (contains perilymph)

Cochlear branch

Spiral ganglion

Vestibulocochlear nerve (N VIII)

Vestibular membrane

Basilar membrane

The Cochlea   LM × 60

b Structure of the cochlea in the temporal bone as seen in section, showing the turns of the vestibular duct, cochlear duct, and tympanic duct

In the Lab 2

Materials

- ☐ Ear model
- ☐ Ear chart
- ☐ Compound microscope
- ☐ Prepared slide of crista
- ☐ Prepared slide of macula
- ☐ Prepared slide of cochlea

Procedures

1. Review the anatomy of the inner ear in Figures 31.3 and 31.4.

2. On the ear model and/or chart, distinguish among the anterior, posterior, and lateral semicircular canals, and then locate the ampulla at the base of each canal.

Figure 31.4 The Cochlea *(continued)*

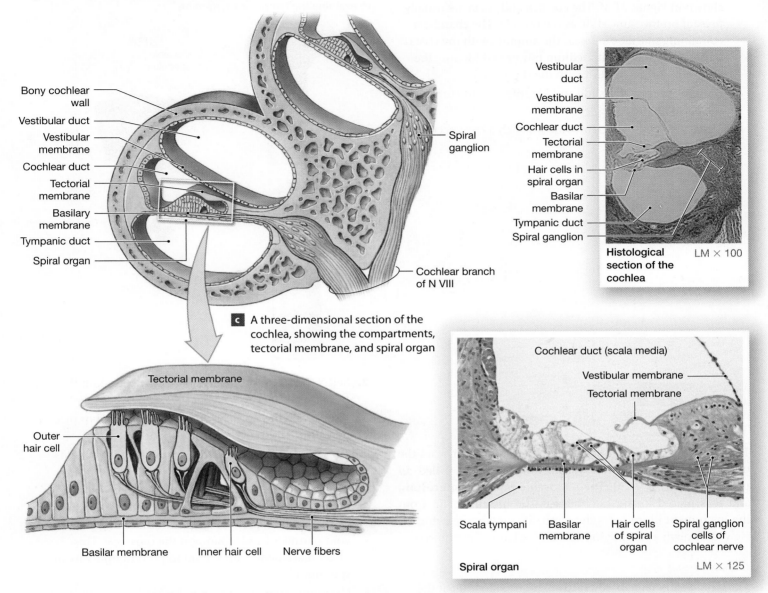

Bony cochlear wall
Vestibular duct
Vestibular membrane
Cochlear duct
Tectorial membrane
Basilary membrane
Tympanic duct
Spiral organ
Spiral ganglion
Cochlear branch of N VIII

c A three-dimensional section of the cochlea, showing the compartments, tectorial membrane, and spiral organ

Tectorial membrane
Outer hair cell
Basilar membrane
Inner hair cell
Nerve fibers

Vestibular duct
Vestibular membrane
Cochlear duct
Tectorial membrane
Hair cells in spiral organ
Basilar membrane
Tympanic duct
Spiral ganglion
Histological section of the cochlea   LM × 100

Cochlear duct (scala media)
Vestibular membrane
Tectorial membrane
Scala tympani   Basilar membrane   Hair cells of spiral organ   Spiral ganglion cells of cochlear nerve
Spiral organ   LM × 125

d Diagrammatic and sectional views of the receptor hair cell complex of the spiral organ

3. On the ear model and/or chart, identify the utricle, saccule, endolymphatic duct, and endolymphatic sac. Note the vestibular branch of the vestibulocochlear nerve (N VIII).

4. Observe the cochlea on the model and/or chart and identify the various ducts and membranes. Examine the organ of Corti and locate the inner and outer hair cells and the tectorial membrane.

5. Examine the crista slide and identify the structures shown in Figure 31.3d. The cochlea is usually present on the same slide as the crista. Search for the cone-shaped crista at the base of the cupula. Observe the crista at medium power and identify the hair cells and the cupula.

6. *Draw It!* In the space provided, sketch a cross section of the crista.

Crista

7. Examine the macula slide and identify the structures shown in Figure 31.3f. The cochlea and crista are usually present on the same slide as the macula. The chambers containing maculae are near the ampulla with the crista. Observe the macula at medium power and identify the hair cells and the otolith with the statoconia.

8. Examine the cochlea slide and identify the structures shown in Figure 31.4.

9. *Draw It!* In the space provided, sketch a cross section of the cochlea. ■

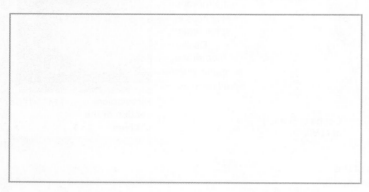

Cochlea

Lab Activity 3   Examination of the Tympanic Membrane

The tympanic membrane separating the external ear from the middle ear can be examined with an instrument called an **otoscope** (Figure 31.5). The removable tip is the **speculum,** and it is placed in the external acoustic meatus. Light from the instrument illuminates the tympanic membrane, which is viewed through a magnifying lens on the back of the otoscope.

QuickCheck Questions

3.1 What is the name of the instrument used to look at the tympanic membrane?

3.2 What is the removable tip of the instrument called?

In the Lab 3

Materials

☐ Otoscope
☐ Alcohol wipes
☐ Laboratory partner

Procedures

1. Using Figure 31.5 as a reference, identify the parts of the otoscope.

Figure 31.5 Otoscope The speculum is placed in the ear canal to examine the tympanic membrane.

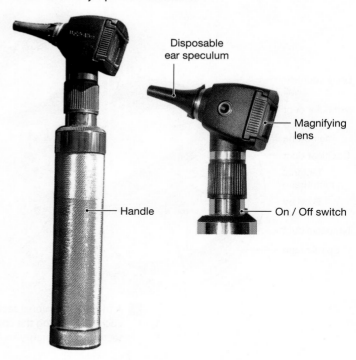

Disposable ear speculum

Magnifying lens

Handle

On / Off switch

2. Select the shortest but *largest-diameter* speculum that will fit into your partner's ear.

3. Either wipe the tip clean with an alcohol pad or place a new disposable cover over the speculum.

4. Turn on the otoscope light. Be sure the light beam is strong.

5. Hold the otoscope between your thumb and index finger, and either sit or stand facing one of your partner's ears. Place the tip of your extended little finger against your partner's head to support the otoscope. This finger placement is important to prevent injury by the speculum.

6. Carefully insert the speculum into the external acoustic meatus while gently pulling the auricle up and posterolaterally. Neither the otoscope nor the pulling should hurt your partner. If your partner experiences pain, stop the examination.

7. Looking into the magnifying lens, observe the walls of the external acoustic meatus. Note if there is any redness in the walls or any buildup of wax.

8. Manipulate the auricle and speculum until you see the tympanic membrane. A healthy membrane appears white. Also notice the vascularization of the region.

9. After the examination, either clean the speculum with a new alcohol wipe or remove the disposable cover. Dispose of used wipes and covers in a biohazard container. ■

Name _____

Date _____

Section _____

Anatomy of the Ear

A. Definition

Define each of the following terms.

1. statoconia

2. basilar membrane

3. tectorial membrane

4. membranous labyrinth

5. cupula

6. tympanic membrane

7. cochlea

8. semicircular canal

9. round window

10. oval window

11. maculae

12. scala tympani

B. Short-Answer Questions

1. Describe the components of the middle ear.

2. Describe the components of the inner ear.

3. Describe the receptors for hearing.

C. Labeling

1. Label the structures of the ear in **Figure 31.6**.

Figure 31.6 **Anatomy of the Ear**

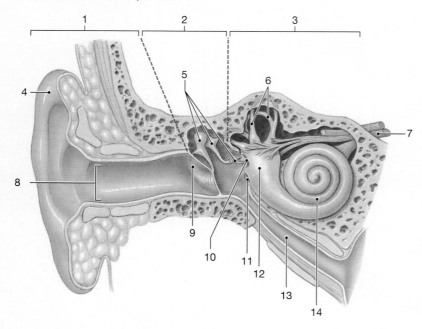

1. _____
2. _____
3. _____
4. _____
5. _____
6. _____
7. _____
8. _____
9. _____
10. _____
11. _____
12. _____
13. _____
14. _____

2. Label the structures of the ear in **Figure 31.7**.

Figure 31.7 The Cochlea

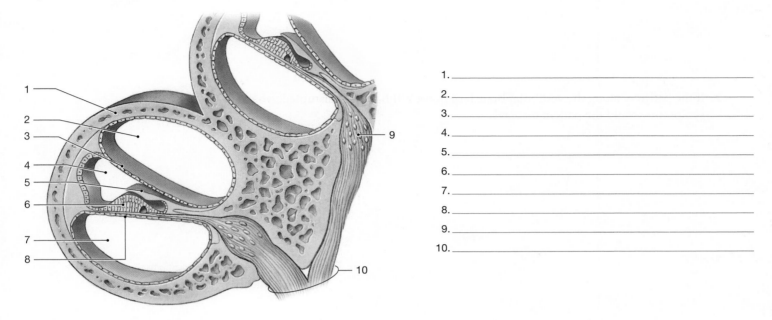

1. _____
2. _____
3. _____
4. _____
5. _____
6. _____
7. _____
8. _____
9. _____
10. _____

D. Analysis and Application

1. Which parts of the inner ear are organized in a "pipe within a pipe" arrangement?

2. How are sound waves transmitted to the inner ear?

3. Which structures in the middle ear reduce the ear's sensitivity to sound?

E. Clinical Challenge

1. Explain how children have more middle ear infections than adults.

2. If the pathway along the vestibular branch of nerve VIII has been disrupted, what symptoms would the patient display?

Physiology of the Ear

Learning Outcomes

On completion of this exercise, you should be able to:

1. Explain the difference between static and dynamic equilibrium by performing various comparative tests.

2. Describe the role of nystagmus in equilibrium.

3. Test your range of hearing by using tuning forks.

4. Compare the conduction of sound through air versus bone (Rinne test).

The inner ear serves two unique functions: balance and hearing. Imagine life without a sense of balance. We would not be able to stand, let alone play sports or enjoy a brisk walk. Balance, called **equilibrium,** feeds the brain a constant stream of information detailing the body's position relative to the ground. Although we live in a three-dimensional world, our bodies function in only two dimensions: front to back and side to side. Because our sense of the third dimension, top to bottom, is ground-based, we can get disoriented in deep water or while piloting an airplane. **Vertigo,** or motion sickness, may occur when the CNS receives conflicting sensory information from the inner ear, the eyes, and other receptors. For example, when you read in a moving car, your eyes are concentrating on the steady book, but your inner ear is responding to the motion of the car. The CNS receives the opposing sensory signals and may respond with vomiting, dizziness, sweating, and other symptoms of motion sickness.

Have you ever stood at your microwave oven and listened to your popcorn pop, waiting to push the stop button for a perfect batch? The ear is a dynamic sense organ capable of hearing multiple sound waves simultaneously. We can,

Need More Practice and Review?

Build your knowledge—and confidence!—in the Study Area of MasteringA&P® at www.masteringaandp.com with Pre-lab Quizzes, Post-lab Quizzes, Practice Anatomy Lab™ (PAL™) 3.0 virtual anatomy practice tool, PhysioEx™ 9.0 laboratory simulations, and A&P Flix™ with Quizzes.

MasteringA&P®

for example, talk with a friend while listening to music and still hear the phone ring. The receptors for hearing, the **auditory** receptors, are located in the cochlea of the inner ear (as discussed in Exercise 31).

Make a Prediction

What must an object be able to do to produce sound**?**

Lab Activity 1   **Equilibrium**

The receptors for equilibrium are in the membranous labyrinth of the vestibule of the inner ear. The crista in the ampulla of each semicircular duct is the receptor for **dynamic equilibrium** and responds to such movements as tilting of the head (Figure 32.1). When the head and body move suddenly,

Figure 32.1 The Vestibular Complex

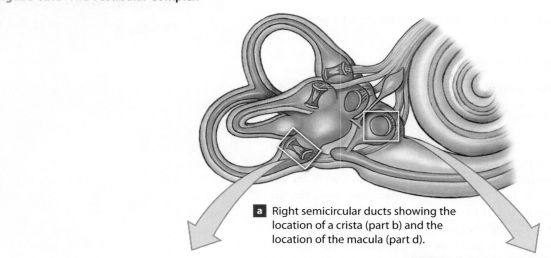

a Right semicircular ducts showing the location of a crista (part b) and the location of the macula (part d).

Anterior semicircular duct for "yes"

Lateral semicircular duct for "no"

Posterior semicircular duct for "tilting head"

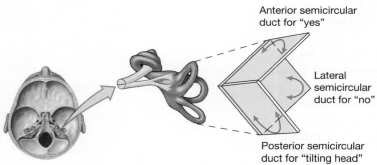

b This superior view shows the planes of sensitivity for the semicircular ducts.

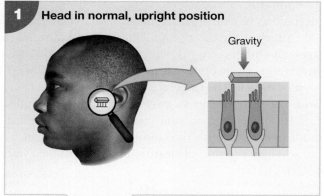

1 **Head in normal, upright position**

Gravity

Direction of duct rotation

Direction of relative endolymph movement

Direction of duct rotation

Semicircular duct

Ampulla

At rest

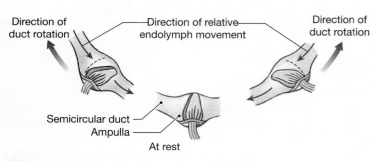

c Endolymph movement along the length of the duct moves the cupula and stimulates the hair cells.

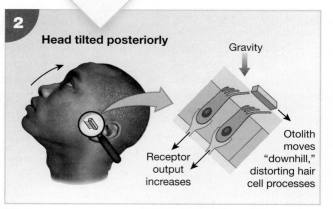

2 **Head tilted posteriorly**

Gravity

Receptor output increases

Otolith moves "downhill," distorting hair cell processes

d A diagrammatic view of macular function when the head is held horizontally **1** and then tilted back **2**.

the otoliths, because they are heavier than the hair cells, lag behind and distort the hair cells, causing them to produce sensory impulses. The maculae, receptors for **static equilibrium,** sense changes in body position relative to the direction of the pull exerted by gravity, such as when you stand on your head. Visual awareness of your surroundings enhances your sense of equilibrium as your brain continuously compares body position relative to the positions of surrounding stationary objects. A loss of visual references usually results in a loss of balance.

QuickCheck Questions

1.1 What is the function of the cristae in the semicircular ducts?

1.2 Which type of equilibrium do the maculae sense?

In the Lab   1

Material

☐ Laboratory partner

Procedures

1. With your partner standing by to help if you lose your balance any time during this activity, stand on both feet in a clear area of the room.

2. With your arms at your sides and your eyes open, raise one foot and try to stand perfectly still for 45 seconds. Record your observations in **Table 32.1**.

3. While still standing on one foot, close both eyes and again try to stand perfectly still for 45 seconds. Record your observations in Table 32.1. ■

Lab Activity 2   Nystagmus

Nystagmus (nis-TAG-mus) is a reflex movement of the eyes in an attempt to maintain balance. It occurs as the visual system attempts to provide the brain with stationary references. When the head moves to the right, the eyes first move slowly to the left then quickly jump to the right to fix on some stationary object. The brain uses this object as a reference point by comparing the object's position with the body's position. This cycle of fast and slow eye movements provides the brain with brief "snapshots" of the stationary object rather than with a visual

Table 32.1	Equilibrium Tests
Standing Position	**Observation**
On one foot, eyes open	
On one foot, eyes closed	

signal that is blurred because of the movement. Nystagmus can be used to evaluate how well receptors in the semicircular canals function. Immediately following any rotational motion of the head, endolymph in the semicircular ducts sloshes back and forth and continues to stimulate the receptors as if the head were still moving.

In this activity, you will look for nystagmus as a subject spins in a swivel chair. If a person spins in a chair with the head tilted forward, receptors in the lateral semicircular canals are stimulated, and the eyes move laterally. Spinning with the head leaning toward the shoulder stimulates the anterior semicircular canal. Nystagmus is a vertical eye movement.

The eye movement called *saccades,* which is similar to nystagmus, is a tracking mechanism that occurs when a person is reading or looking out the window of a moving car. In saccades, the eyes first fix on one object, then rapidly jump forward to another object.

QuickCheck Questions

2.1 Define *nystagmus.*

2.2 How does nystagmus help you maintain equilibrium?

In the Lab   2

Materials

☐ Swivel chair
☐ Four subjects
☐ Four spotters

Procedures

1. Seat the first subject in a swivel chair; have the spotters use their feet to support the chair base.

2. Instruct the subject to stare straight ahead with both eyes open.

3. Spin the chair clockwise for 10 rotations (approximately one rotation per second).

4. Quickly and carefully stop the chair, and observe the subject's eye movements. Record your observations in **Table 32.2**.

⚠ **Safety Alert:** Loss of Balance

This experiment involves spinning a subject in a chair and observing eye movements. *Do not use anyone who is prone to motion sickness.* The observer will spin the subject and record eye movement. To prevent accidents, four other individuals must hold the base of the chair steady and serve as spotters to help the subject stay in the chair during the experiment. The subject should remain in the chair for about two minutes after the experiment to regain normal equilibrium. ▲

Table 32.2	Nystagmus Tests
Rotation and Head Position	**Eye Movements**
Clockwise, head facing straight forward	
Clockwise, head flexed	
Clockwise, head hyperextended	
Clockwise, head tilted to side	

5. Keep the subject seated for approximately two minutes after spinning to regain balance.

6. Using a second subject, repeat the clockwise rotation but with the subject's head flexed. Record your observations of eye movements in Table 32.2.

7. Using a third subject, repeat the clockwise rotation but with the subject's head hyperextended. Record your observations of eye movements in Table 32.2.

8. Using a fourth subject, repeat the clockwise rotation but with the subject's head tilted toward either shoulder. Record your observations of eye movements in Table 32.2. ■

Lab Activity 3 Hearing

We rely on the sense of hearing for communication and for an awareness of events in our immediate surroundings. Sounds are produced by vibrating objects as the vibrations cause the air around the objects first to compress and then to decompress, sending a wave of compressed and decompressed regions outward from the objects (**Figure 32.2**).

The number of compressed regions that pass a given point in one second is the **frequency,** or pitch, of the sound. An object vibrating rapidly produces a higher pitch than an object vibrating more slowly. The unit **hertz (Hz)** is used for the frequency of the compressed waves. (An alternative expression for this unit is cycles per second, or cps.) Humans can hear sounds from about 20 to 20,000 Hz (20 to 20,000 cps). The *intensity,* or **amplitude,** of a sound is measured in **decibels (dB).** The higher, or "taller," a sound wave is, the higher the amplitude of the sound and the greater its decibel rating.

Hearing occurs when sound waves enter the external auditory canal and strike the tympanic membrane, causing the membrane to vibrate at the frequency of the sound waves (**Figure 32.3**). The vibrations in the tympanic membrane are passed along to the auditory ossicles, causing them to vibrate. The malleus, which is connected to the tympanic membrane, vibrates and moves the incus, which moves the stapes. Vibrations of the stapes move the oval window, creating pressure waves in the perilymph of the vestibular duct. The pressure waves correspond to the sound waves that initially hit the tympanic membrane. The pressure

Figure 32.2 The Nature of Sound

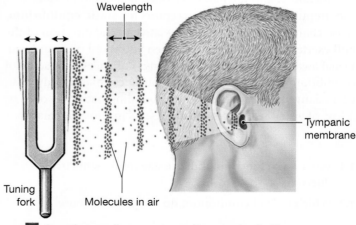

a Sound waves (here, generated by a tuning fork) travel through the air as pressure waves.

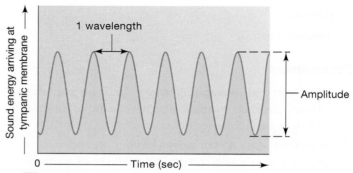

b A graph showing the sound energy arriving at the tympanic membrane. The distance between wave peaks is the wavelength. The number of waves arriving each second is the frequency, which we perceive as pitch. Frequencies are reported in cycles per second (cps), or hertz (Hz). The amount of energy in each wave determines the wave's amplitude, or intensity, which we perceive as the loudness of the sound.

waves pass through the cochlear duct and into the tympanic duct, causing the basilar membrane to vibrate. The pressure waves are dissipated by the stretching of the round window. If the pressure waves are not dissipated, they will cause interference with the next set of pressure waves, much like an ocean wave bouncing off a sea wall and slapping into the next incoming wave. In the ear, this interference would cause a loss of auditory acuity and loss of the ability to correctly discriminate between similar sounds.

The vibrations in the basilar membrane push specific hair cells into the tectorial membrane. The cells fire sensory impulses when their stereocilia touch the tectorial membrane and bend. The auditory information is passed over the cochlear branch of the vestibulocochlear nerve.

Deep (low-frequency) sounds have long sound waves that stimulate the distal portion of the basilar membrane. High-pitch

Figure 32.3 Sound and Hearing Steps in the reception and transduction of sound energy.

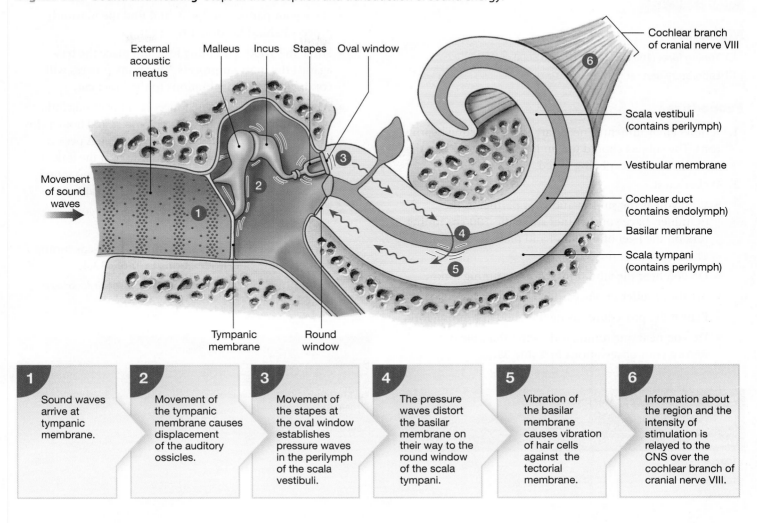

External acoustic meatus

Malleus Incus Stapes Oval window

Cochlear branch of cranial nerve VIII

Movement of sound waves

Scala vestibuli (contains perilymph)

Vestibular membrane

Cochlear duct (contains endolymph)

Basilar membrane

Scala tympani (contains perilymph)

Tympanic membrane

Round window

1	2	3	4	5	6
Sound waves arrive at tympanic membrane.	Movement of the tympanic membrane causes displacement of the auditory ossicles.	Movement of the stapes at the oval window establishes pressure waves in the perilymph of the scala vestibuli.	The pressure waves distort the basilar membrane on their way to the round window of the scala tympani.	Vibration of the basilar membrane causes vibration of hair cells against the tectorial membrane.	Information about the region and the intensity of stimulation is relayed to the CNS over the cochlear branch of cranial nerve VIII.

(high-frequency) sounds have short waves that stimulate the basilar membrane close to the oval window.

Deafness can be the result of many factors, not all of which are permanent. The two main categories of deafness are conduction deafness and nerve deafness. **Conduction deafness** involves damage either to the tympanic membrane or to one or more of the auditory ossicles. Proper conduction produces vibrations heard equally in both ears. If a conduction problem exists, sounds are normally heard best in the unaffected ear. Conduction tests with tuning forks, however, cause *the ear with the deafness to hear the sound louder than the normal ear.* This is due to an increased sensitivity to sounds in the ear with conduction deafness. Hearing aids are often used to correct for conduction deafness.

Nerve deafness is a result of damage to either the cochlea or the cochlear nerve. Repetitive exposure to excessively loud noises, such as music and machinery, can damage the delicate spiral organ. Nerve deafness cannot be corrected and results in a permanent loss of hearing, usually within a specific range of frequencies.

In the following tests, sound vibrations from tuning forks are conducted through the bones of the skull, bypassing normal conduction by the external and middle ear. Because the inner ear is surrounded by the temporal bone, vibrations are transmitted directly from the bone into the cochlea.

QuickCheck Questions

3.1 Where in the ear are the receptors for hearing?

3.2 What is the difference between conduction deafness and nerve deafness?

! **Safety Alert:** Use of Tuning Forks

To prevent damage to the tympanic membrane, *never insert a tuning fork into the external auditory canal.*

Tuning forks are designed to vibrate at a certain frequency. Gently tapping the fork on the side of your palm is sufficient to cause it to vibrate. ▲

In the Lab 3

Materials

☐ Tuning forks (100 cps, 1000 cps, and 5000 cps)
☐ Laboratory partner

Procedures

1. A quiet environment is necessary for conducting hearing tests. The subject should use an index finger to close the ear opposite the ear being tested.

2. Weber's test:
   - With you as the subject and your partner as the examiner, have your partner strike the 100-cps tuning fork on the heel of the hand and place its base on top of your head, at the center.
   - Do you hear the vibrations from the tuning fork?
   - Are they louder in one ear than in the other?
   - Repeat the procedures using the 1000-cps tuning fork.
   - Do you hear one tuning fork better than the other? Record your observations in **Table 32.3**.

3. Rinne test:
   - Have your partner sit down, and find the mastoid process behind his or her right ear.
   - Strike the 1000-cps tuning fork and place the base against the mastoid process. This bony process will conduct the sound vibrations to the inner ear.
   - When your partner can no longer hear the tuning fork, quickly move it from the mastoid process and hold it close to—*but not touching*—her or his external ear. A person with normal hearing should be able to hear the fork. If conduction deafness exists in the middle ear, no sound will be heard. Record your observations in **Table 32.4**.
   - Repeat this test on the left ear, and record your observations in Table 32.4.
   - Repeat this test on each ear with the 5000-cps tuning fork. Record your observations in Table 32.4.

4. For additional practice, complete the **Sketch to Learn** activity below. ■

Table 32.3	Weber's Hearing Tests	
	Observations	
Frequency	**Right Ear**	**Left Ear**
100 cps		
1000 cps		

Table 32.4	Rinne Hearing Tests	
	Observations	
Frequency	**Right Ear**	**Left Ear**
1000 cps		
5000 cps		

 Sketch to Learn

Draw sound waves that are of the same frequency but differ in amplitude.

Sample Sketch

Step 1
- Draw 3–4 "waves" that are the same height and distance apart. These are the low amplitude sound waves of a quiet noise.

Your Sketch

Step 2
- Above the quiet waves draw higher waves with the same spacing as the first set.
- Label the loud and quiet sound waves.

Name _____

Date _____

Section _____

Physiology of the Ear

A. Definition

Define or describe each of the following terms.

1. nerve deafness

2. motion sickness

3. nystagmus

4. frequency of sound

5. amplitude of sound

6. conduction deafness

B. Drawing

Draw It! Draw a sound wave for a high pitch sound and a low pitch sound.

C. **Short-Answer Questions**

    1. Describe the process of hearing.

    2. Explain how sound waves striking the tympanic membrane result in movement of fluids in the inner ear.

    3. Describe the receptors for dynamic and static equilibrium.

    4. What is the range of sound frequencies that humans can hear?

D. **Analysis and Application**

    1. How could the stereocilia of hair cells in the organ of Corti become damaged?

    2. Explain the phenomenon of nystagmus.

E. **Clinical Challenge**

If conduction deafness exists, why is the sound of a tuning fork heard better in the deaf ear?

Endocrine System

Learning Outcomes

On completion of this exercise, you should be able to:

1. Compare the two regulatory systems of the body: the nervous and endocrine systems.

2. Identify each endocrine gland on laboratory models.

3. Describe the histological appearance of each endocrine gland.

4. Identify each endocrine gland when viewed microscopically.

G lands of the body are classified into two major groups. **Endocrine glands** produce regulatory molecules called **hormones** that slowly cause changes in the metabolic activities of **target cells,** which are any cells that contain membrane receptors for the hormones. Endocrine glands are commonly called *ductless glands* because they secrete their hormones into the surrounding extracellular fluid instead of secreting into a duct. The other kind of glands, **exocrine glands,** secrete substances into a duct for transport and release onto a free surface of the body. Examples of exocrine glands are the sweat glands and sebaceous glands of the skin.

Two systems regulate homeostasis: the nervous system and the endocrine system. These systems must coordinate their activities to maintain control of internal functions. The nervous system responds rapidly to environmental changes, sending electrical commands that can produce an immediate response in any part of the body. The duration of each electrical impulse is brief, measured in milliseconds. In contrast, the endocrine system maintains long-term control. In response to stimuli, endocrine glands release their hormones, and the hormones then slowly

Need More Practice and Review?

Build your knowledge—and confidence!—in the Study Area of MasteringA&P® at www.masteringaandp.com with Pre-lab Quizzes, Post-lab Quizzes, Practice Anatomy Lab™ (PAL™) 3.0 virtual anatomy practice tool, PhysioEx™ 9.0 laboratory simulations, and A&P Flix™ with Quizzes.

PAL | practice anatomy lab   For this lab exercise, follow these navigation paths:
- PAL>Human Cadaver>Endocrine System
- PAL>Anatomical Models>Endocrine System
- PAL>Histology>Endocrine System

PhysioEx™ 9.0 For this lab exercise, go to this topic:
- PhysioEx Exercise 4: Endocrine System Physiology

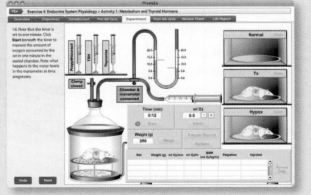

MasteringA&P®

cause changes in the metabolic activities of their target cells. Typically, the effect of a hormone is prolonged and lasts several hours.

The secretion of many hormones is regulated by negative feedback mechanisms. In **negative feedback,** a stimulus causes a response that either reduces or removes the stimulus. An excellent analogy is the operation of an air conditioner. When a room heats up, the warm air activates a thermostat that then turns on the compressor of the air conditioner. Cooled air flowing in cools the room and removes the stimulus (the warm air). Once the stimulus is removed, the unit shuts off. Negative feedback is therefore a self-limiting mechanism.

An example of negative feedback control of hormonal secretion is the regulation of insulin, a hormone from the pancreas that lowers the concentration of glucose in the blood. When blood glucose levels are high, as they are after a meal, the pancreas secretes insulin. The secreted insulin stimulates the body's cells to increase their glucose consumption and storage, thus lowering the concentration of glucose in the blood. As this concentration returns to normal, insulin secretion stops.

In this exercise, you study the following glands: pituitary, thyroid, parathyroid, thymus, pancreas, adrenal, testes, and ovaries. (The ovaries and testes are also important reproductive organs and their anatomy is presented in more detail in Exercise 45.)

Lab Activity 1   Pituitary Gland

Anatomy

The **pituitary gland,** or **hypophysis** (hī-POF-i-sis), is located in the sella turcica of the sphenoid of the skull, immediately inferior to the hypothalamus of the brain (Figure 33.1). A stalk called the **infundibulum** attaches the pituitary to the brain at the hypothalamus. The pituitary gland is organized into two lobes, an **anterior lobe,** also called the **adenohypophysis** (ad-e-nō-hī-POF-i-sis), and a **posterior lobe,** also called either the **neurohypophysis** (noo-rō-hī-POF-i-sis) or the **pars nervosa.** The main portion of the anterior lobe is the **pars distalis** (dis-TAL-is); the **pars tuberalis** is a narrow portion that wraps round the infundibulum; the **pars intermedia** (in-ter-MĒ-dē-uh) is found in the interior of the gland, forming the boundary between the anterior and posterior lobes.

The two pituitary lobes are easily distinguished from each other by how they accept stain. The posterior lobe consists mostly of lightly stained unmyelinated axons from hypothalamic neurons. Darker-stained cells called **pituicytes** are scattered in the lobe and are similar to glial cells in function.

The darker-staining anterior lobe is populated by a variety of cell types that are classified into two main groups determined by their histological staining qualities. **Chromophobes** are light-colored cells that do not react to most stains. **Chromophils**

Figure 33.1 **The Anatomy and Orientation of the Pituitary Gland**

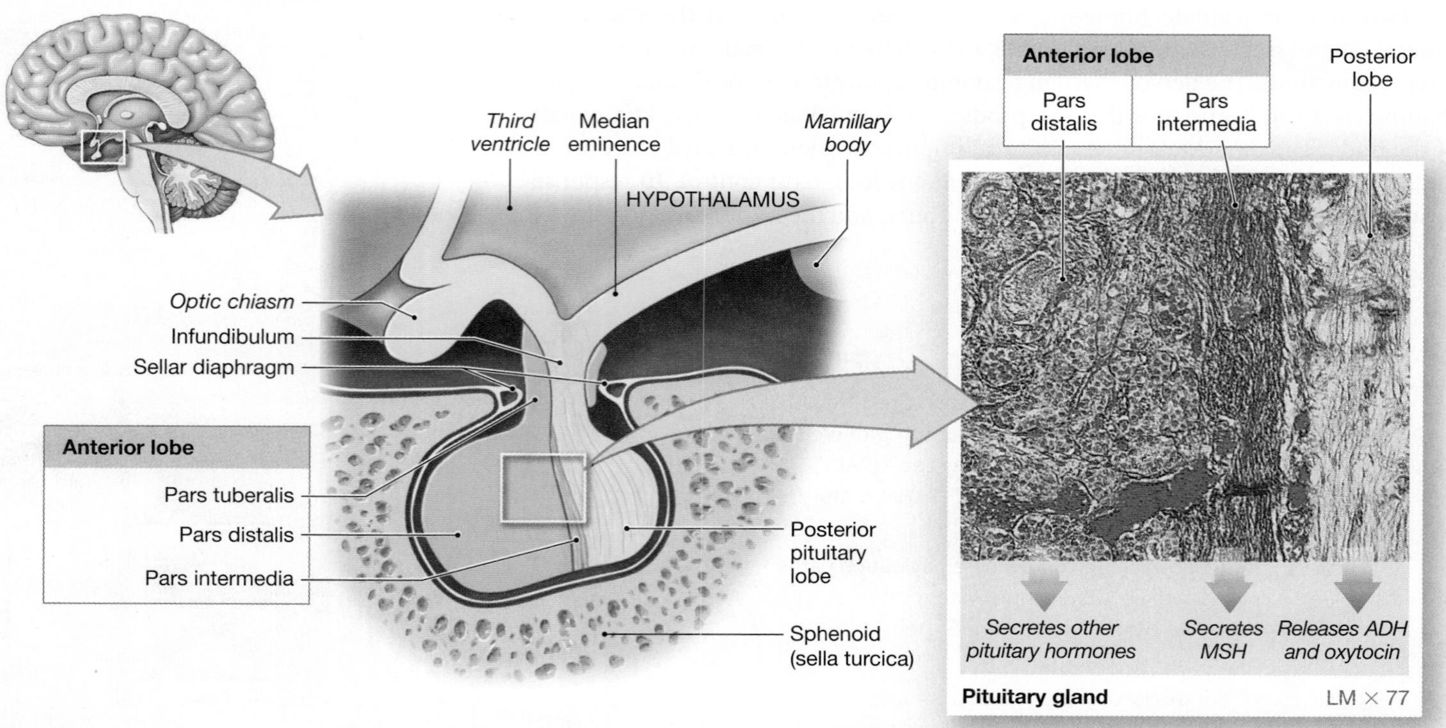

a Relationship of the pituitary gland to the hypothalamus

b Histological organization of pituitary gland showing the anterior and posterior lobes of the pituitary gland

react to histological stains and are darker than chromophobes. Chromophils are subdivided into **acidophils,** which react with acidic stains, and **basophils,** which react with basic stains. In most slide preparations, basophils are stained darker than the more numerous reddish acidophils.

Hormones

The pituitary gland is commonly called the *master gland* because it has a critical role in regulating endocrine function and produces hormones that control the activity of many other endocrine glands. **Regulatory hormones** from the hypothalamus travel down a plexus of blood vessels in the infundibulum and signal the pars distalis to secrete **tropic hormones** that target other endocrine glands, inducing them to produce and secrete their own hormones. The pars intermedia produces a single hormone, melanocyte-stimulating hormone (MSH).

The posterior lobe does not produce hormones. Instead, its function is to store and release antidiuretic hormone (ADH) and oxytocin (OT), which are both produced in the hypothalamus and then passed down the infundibulum to the pituitary gland.

QuickCheck Questions

1.1 Where is the pituitary gland located?

1.2 What is the main staining difference between the anterior and posterior pituitary lobes?

In the Lab 1

Materials

☐ Torso model
☐ Endocrine chart
☐ Dissecting microscope
☐ Compound microscope
☐ Prepared slide of pituitary gland

Procedures

1. Locate the pituitary gland on the torso model and endocrine chart.

2. Use the dissecting microscope to survey the pituitary gland slide at low magnification. Distinguish between the two lobes of the gland.

3. Examine the slide at low and medium powers with the compound microscope. Identify the anterior and posterior lobes, noting the different cell arrangements in each.

For additional practice, complete the *Sketch to Learn* activity. ■

 Sketch to Learn

Your endocrine lab studies involve a histological examination of the major endocrine glands. Drawing each slide helps you identify and recall the important features of each gland. Let's start by sketching the pituitary gland as seen at low magnification.

Sample Sketch

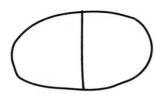 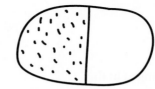 

Step 1
• Draw an oval and divide it with a line.

Step 2
• Detail the anterior pituitary with dots to represent the many endocrine cells of this part of the gland.

Step 3
• Use short lines to illustrate the histology of the neurohypophysis.
• Label your sketch.

Your Sketch

Lab Activity 2 Thyroid Gland

Anatomy

The **thyroid gland** is located in the anterior aspect of the neck, directly inferior to the thyroid cartilage (Adam's apple) of the larynx and just superior to the trachea (Figure 33.2). This gland consists of two lateral lobes connected by a central mass, the **isthmus** (IS-mus). The thyroid produces two groups of hormones associated with the regulation of cellular metabolism and calcium homeostasis.

The thyroid gland is very distinctive. It is composed of spherical **follicles** embedded in connective tissue. Each follicle is composed of a single layer of simple cuboidal epithelial cells called **follicle cells,** or *follicular cells*. The lumen of each follicle is filled with a glycoprotein called **thyroglobulin** (thī-rō-GLOB-ū-lin) that stores thyroid hormones. On the superficial margins of the follicles are **C cells,** also called *parafollicular cells*, which are larger and less abundant than the follicle cells. On most slides, the C cells have a light-stained nucleus.

Hormones

Follicle cells produce the hormones **thyroxine (T_4)** and **triiodothyronine (T_3),** both of which regulate metabolic rate. These hormones are synthesized in the form of the glycoprotein thyroglobulin. It is secreted into the lumen of the follicles and stored there until needed by the body, at which time it is reabsorbed by the follicle cells and released into the blood.

C cells produce the hormone **calcitonin (CT),** which decreases blood calcium levels. Calcitonin stimulates osteoblasts in bone tissue to store calcium in bone matrix and lower fluid calcium levels. It also inhibits osteoclasts in bone from dissolving bone matrix and releasing calcium. Calcitonin has a minor role in calcium regulation in humans but is more active in lowering blood calcium in other animals.

QuickCheck Questions

2.1 Where is the thyroid gland located?

2.2 How are the various types of thyroid cells arranged in the gland?

In the Lab 2

Materials

- ☐ Torso model
- ☐ Endocrine chart
- ☐ Dissecting microscope
- ☐ Compound microscope
- ☐ Prepared slide of thyroid gland

Clinical Application Hyperthyroidism

Hyperthyroidism occurs when the thyroid gland produces too much T_4 and T_3. Because these hormones increase mitochondrial ATP production and increase metabolic rate, individuals with this endocrine disorder are often thin, restless, and emotionally unstable. They fatigue easily because the cells are consuming rather than storing high-energy ATP molecules. *Graves' disease* is a form of hyperthyroidism that occurs when the body has an autoimmune response and produces antibodies that attack the thyroid gland. The gland enlarges to the point that it protrudes from the throat; the enlarged mass is called a *goiter*. Fat tissue is also deposited deep in the eye orbits, causing the eyeballs to protrude, a condition called *exophthalmos*. Treatment for hyperthyroidism may include partial removal of the gland or destruction of parts of it with radioactive iodine. ∎

Procedures

1. Review the features of the thyroid gland in Figure 33.2.
2. Locate the thyroid gland on the torso model and endocrine chart.
3. Scan the thyroid slide with the dissecting microscope and observe the many thyroid follicles.
4. Use the compound microscope to view the thyroid slide at low and medium powers. Locate a follicle, some follicle cells, thyroglobulin, and C cells.
5. *Draw It!* In the space provided, sketch several follicles as observed at medium magnification. ∎

Follicles of a thyroid gland

Figure 33.2 **The Thyroid Gland**

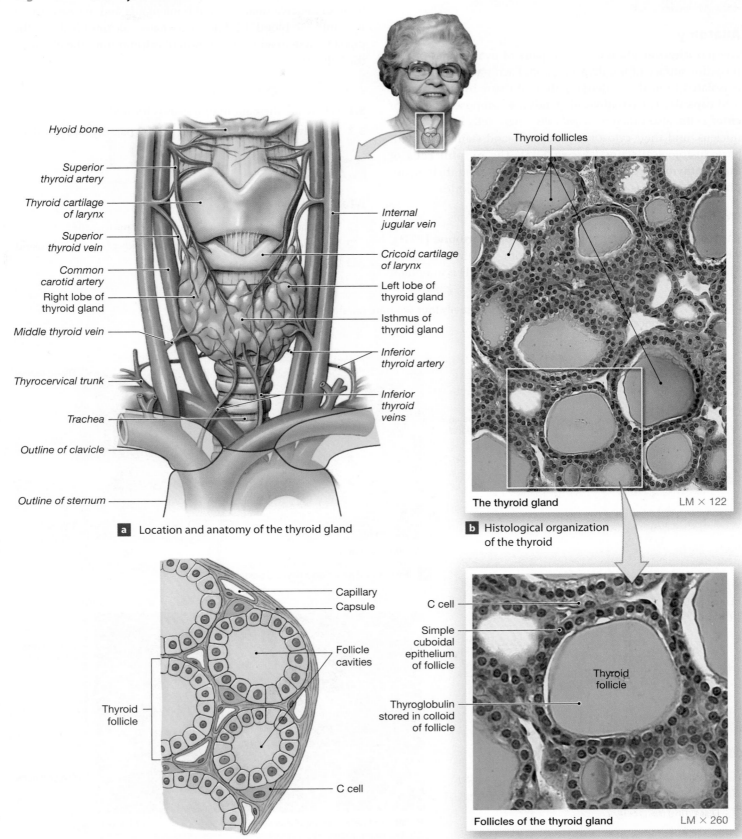

a Location and anatomy of the thyroid gland

b Histological organization of the thyroid

The thyroid gland — LM × 122

Follicles of the thyroid gland — LM × 260

c Histological details of the thyroid gland showing thyroid follicles and both of the cell types in the follicular epithelium

Lab Activity 3   Parathyroid Glands

Anatomy

The **parathyroid glands** are two pairs of oval masses on the posterior surface of the thyroid gland. Each parathyroid gland is isolated from the underlying thyroid tissue by the parathyroid **capsule.** The parathyroid glands are composed mostly of **chief cells,** also called *principal cells.* These cells have a round nucleus, and their cytosol is basophilic and stains pale with basic histological stains (Figure 33.3). The **oxyphil cells** of the parathyroid are larger than the chief cells, and their acidophilic cytosol reacts to acidic stains and turns colorless.

Hormone

The parathyroid glands produce **parathyroid hormone (PTH),** which is antagonistic to calcitonin from the thyroid gland. Although CT is relatively ineffective in humans, PTH is important in maintaining blood calcium level by stimulating osteoclasts in bone to dissolve small areas of bone matrix and release calcium ions into the blood. PTH also stimulates calcium uptake in the digestive system and reabsorption of calcium from the filtrate in the kidneys.

QuickCheck Questions

3.1 Where are the parathyroid glands located?

3.2 What two types of cells make up the parathyroid glands?

In the Lab 3

Materials

- ☐ Torso model
- ☐ Endocrine chart
- ☐ Dissecting microscope
- ☐ Compound microscope
- ☐ Prepared slide of parathyroid gland

Figure 33.3 The Parathyroid Glands There are usually four separate parathyroid glands bound to the posterior surface of the thyroid gland.

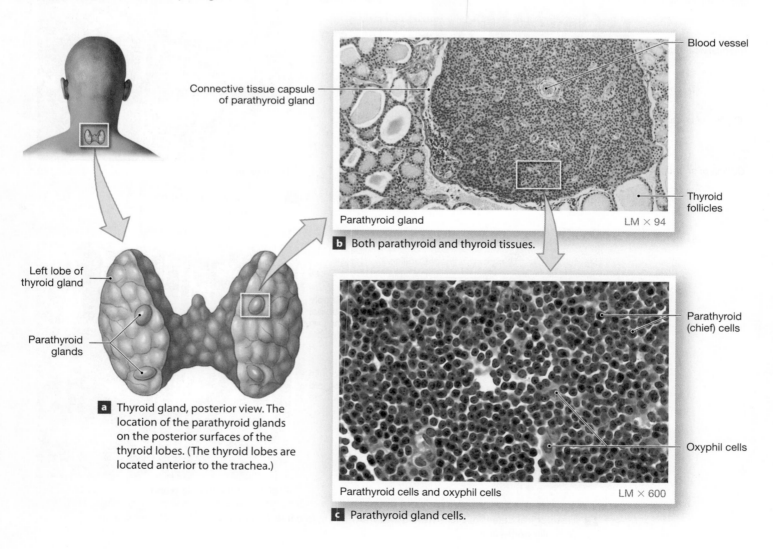

Connective tissue capsule of parathyroid gland

Blood vessel

Thyroid follicles

Parathyroid gland   LM × 94

b Both parathyroid and thyroid tissues.

Left lobe of thyroid gland

Parathyroid glands

a Thyroid gland, posterior view. The location of the parathyroid glands on the posterior surfaces of the thyroid lobes. (The thyroid lobes are located anterior to the trachea.)

Parathyroid (chief) cells

Oxyphil cells

Parathyroid cells and oxyphil cells   LM × 600

c Parathyroid gland cells.

Procedures

1. Review the parathyroid glands in Figure 33.3.
2. Locate the parathyroid glands on the torso model and endocrine chart.
3. Examine the parathyroid slide with the dissecting microscope. Scan the gland for thyroid follicles that may be on the slide near the parathyroid tissue.
4. Observe the parathyroid slide at low and medium powers with the compound microscope. Locate the dark-stained chief cells and the light-stained oxyphil cells.
5. *Draw It!* In the space provided, sketch the parathyroid gland as observed at medium magnification. ■

Parathyroid gland

Lab Activity 4 Thymus Gland

Anatomy

The **thymus gland** is located inferior to the thyroid gland, in the thoracic cavity posterior to the sternum (Figure 33.4). Because hormones secreted by the thymus gland facilitate development of the immune system, the gland is larger and more active in youngsters than in adults.

The thymus gland is organized into two main lobes, with each lobe made up of many **lobules,** which are very small lobes. The lobules in each lobe of the thymus gland are separated from one another by septae made up of fibrous connective tissue. Each lobule consists of a dense outer **cortex** and a light-staining central **medulla.** The cortex is populated by reticular cells that secrete the thymic hormones. In the medulla, other reticular cells cluster together into distinct oval masses called **thymic corpuscles** (Hassall's corpuscles). Surrounding the corpuscles are developing white blood cells called **lymphocytes** that eventually enter the blood. Adipose and other connective tissues are abundant in an adult thymus because the function and size of the gland decrease after puberty.

Hormone

Although the reticular cells of the thymus gland produce several hormones, the function of only one, **tymosin,** is understood. Tymosin is essential in the development and maturation of the immune system. Removal of the gland during early childhood usually results in a greater susceptibility to acute infections.

QuickCheck Questions

4.1 Where is the thymus gland located?
4.2 What are the main histological features of the thymus gland?

In the Lab 4

Materials

- ☐ Torso model
- ☐ Endocrine chart
- ☐ Dissecting microscope
- ☐ Compound microscope
- ☐ Prepared slide of thymus gland

Procedures

1. Review the anatomy of the thymus gland in Figure 33.4.
2. Locate the thymus gland on the torso model and endocrine chart.
3. Scan the slide of the thymus gland with the dissecting microscope and distinguish between the cortex and the medulla.
4. Examine the thymus slide with the compound microscope at low magnification to locate a stained thymic corpuscle. Increase the magnification and examine the corpuscle. The cells surrounding the corpuscles are lymphocytes.
5. *Draw It!* In the space provided, sketch the thymus gland as observed at medium magnification. ■

Thymus gland

Figure 33.4 The Thymus

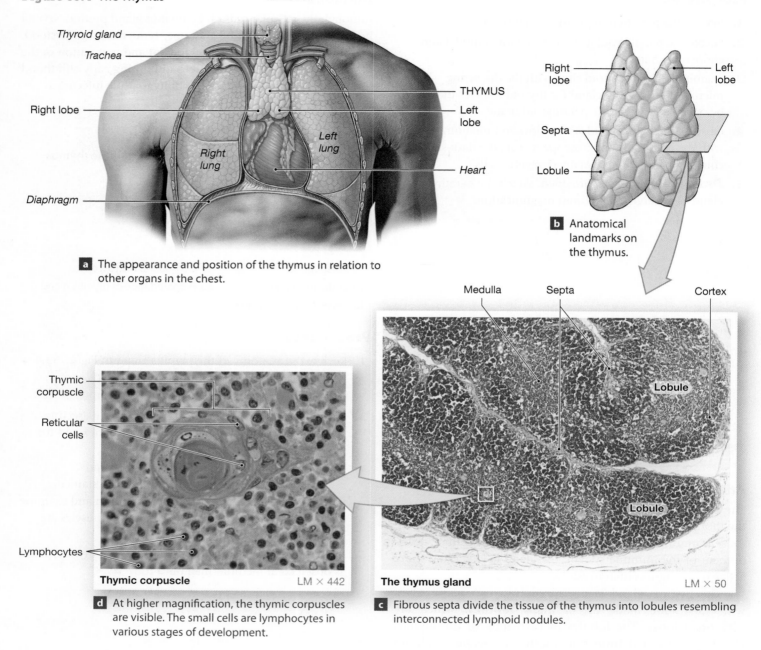

Thyroid gland

Trachea

Right lobe

Right lung

Left lung

Heart

Diaphragm

THYMUS

Left lobe

a The appearance and position of the thymus in relation to other organs in the chest.

Right lobe

Left lobe

Septa

Lobule

b Anatomical landmarks on the thymus.

Thymic corpuscle

Reticular cells

Lymphocytes

Thymic corpuscle LM × 442

d At higher magnification, the thymic corpuscles are visible. The small cells are lymphocytes in various stages of development.

Medulla

Septa

Cortex

Lobule

Lobule

The thymus gland LM × 50

c Fibrous septa divide the tissue of the thymus into lobules resembling interconnected lymphoid nodules.

Lab Activity 5 Adrenal Glands

Anatomy

Superior to the kidneys are **adrenal glands** (a-DRĒ-nal), so-called because of the adrenaline they secrete (**Figure 33.5**). A protective **adrenal capsule** encompasses the gland and attaches it to the kidney. The gland is organized into two major regions: the outer **adrenal cortex** and the inner **adrenal medulla.**

The adrenal cortex is differentiated into three distinct regions, each producing specific hormones. The **zona glomerulosa** (glō-mer-ū-LŌ-suh) is the outermost cortical region. Cells in this area are stained dark and arranged in oval clusters. The next layer, the **zona fasciculata** (fa-sik-ū-LA-tuh), is made up of larger cells organized in tight columns. These cells contain large amounts of lipid, making them appear lighter than the surrounding cortical layers (Figure 33.5). The deepest layer of the cortex, next to the medulla, is the **zona reticularis** (re-tik-ū-LAR-is). Cells in this area are small and loosely linked together in chainlike structures. The many blood vessels in the adrenal medulla give this tissue a dark red color.

Figure 33.5 The Adrenal Gland

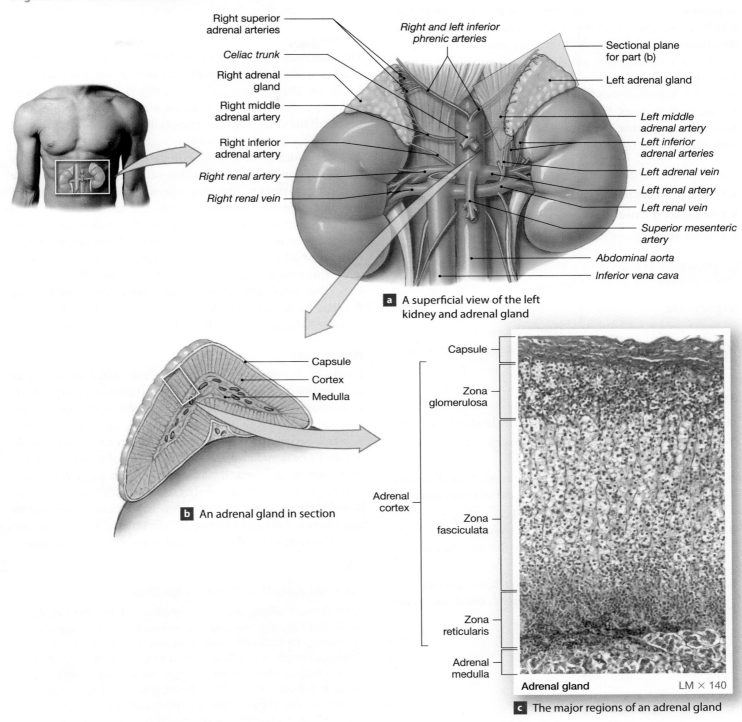

Right superior adrenal arteries

Right and left inferior phrenic arteries

Celiac trunk

Right adrenal gland

Right middle adrenal artery

Right inferior adrenal artery

Right renal artery

Right renal vein

Sectional plane for part (b)

Left adrenal gland

Left middle adrenal artery

Left inferior adrenal arteries

Left adrenal vein

Left renal artery

Left renal vein

Superior mesenteric artery

Abdominal aorta

Inferior vena cava

a A superficial view of the left kidney and adrenal gland

Capsule

Cortex

Medulla

b An adrenal gland in section

Capsule

Zona glomerulosa

Adrenal cortex

Zona fasciculata

Zona reticularis

Adrenal medulla

Adrenal gland   LM × 140

c The major regions of an adrenal gland

Hormones

The adrenal cortex secretes hormones collectively called *adrenocortical steroids*, or simply *corticosteroids*. These hormones are lipid-based steroids. The zona glomerulosa secretes a group of hormones called **mineralocorticoids** that regulate, as their name implies, mineral or electrolyte concentrations of body fluids. A good example is **aldosterone,** which stimulates the kidneys to reabsorb sodium from the liquid being processed into urine. The zona fasciculata produces a group of hormones called **glucocorticoids** that are involved in fighting stress, increasing glucose metabolism, and preventing inflammation. Two of the glucocorticoids, **cortisol** and **corticosterone** (kor-ti-KOS-te-rōn), are commonly found in creams used to treat rashes and allergic responses of the skin. The zona reticularis produces **androgens,** which are male sex hormones. Both males and females produce small quantities of androgens in the zona reticularis.

The adrenal medulla is regulated by sympathetic neurons from the hypothalamus. In times of stress, exercise, or emotion, the hypothalamus stimulates the adrenal medulla to release its hormones, the neurotransmitters **epinephrine (E)** and **norepinephrine (NE),** into the blood, resulting in a body-wide sympathetic fight-or-flight response.

QuickCheck Questions

5.1 Where are the adrenal glands located?

5.2 What are the two major regions of the adrenal gland?

5.3 What are the three layers of the adrenal cortex?

In the Lab 5

Materials

- ☐ Torso model
- ☐ Endocrine chart
- ☐ Dissecting microscope
- ☐ Compound microscope
- ☐ Prepared slide of adrenal gland

Procedures

1. Review the components of the adrenal gland in Figure 33.5.

2. Locate the adrenal gland on the torso model and endocrine chart.

3. Examine the slide of the adrenal gland with the dissecting microscope and distinguish among the adrenal capsule, adrenal cortex, and adrenal medulla.

4. Observe the adrenal gland with the compound microscope and differentiate among the three layers of the adrenal cortex.

5. *Draw It!* In the space provided, sketch the adrenal gland, showing the details of the three cortical layers and the medulla. ∎

Adrenal gland

Lab Activity 6 Pancreas

Anatomy

The **pancreas,** a glandular organ that lies posterior to the stomach (Figure 33.6), performs important exocrine and endocrine functions. The exocrine cells secrete digestive enzymes, buffers, and other molecules into a pancreatic duct that empties into the small intestine. The endocrine cells produce hormones that regulate blood sugar metabolism.

The pancreas is densely populated by dark-stained cells called the **pancreatic acini.** These cells make up the exocrine part of the pancreas, and they secrete pancreatic juice, which contains digestive enzymes. Connective tissues and pancreatic ducts are dispersed in the tissue. The endocrine cells of the pancreas occur in isolated clusters of **pancreatic islets** (ī-letz), or *islets of Langerhans* (LAN-ger-hanz), that are scattered throughout the gland. Each islet houses four types of endocrine cells: **alpha cells, beta cells, delta cells,** and **F cells.** These cells are difficult to distinguish with routine staining techniques and will not be individually examined.

Hormones

Pancreatic hormones affect carbohydrate metabolism. Alpha cells secrete the hormone **glucagon** (GLOO-ka-gon), which raises blood sugar concentration by catabolizing glycogen to glucose for cellular respiration. This process is called *glycogenolysis.* Beta cells secrete **insulin** (IN-suh-lin), which accelerates glucose uptake by cells and also accelerates the rate of glycogenesis, the formation of glycogen. Insulin lowers blood sugar concentration by promoting the removal of sugar from the blood. Normal blood plasma glucose concentration is generally considered to range between 70 and 110 mg/dL. The interaction of pancreatic and other hormones plays a key role in regulating blood sugar.

Clinical Application Diabetes Mellitus

In **diabetes mellitus,** glucose in the blood cannot enter cells, and blood glucose levels rise above normal levels. In **type I diabetes,** the beta cells in the pancreas do not produce enough insulin, and cells are not stimulated to take in glucose. Type I diabetics take insulin to regulate their blood sugar. **Type II diabetes** occurs when the body becomes less responsive to insulin. The pancreas produces adequate amounts of insulin, but the body is not responsive to it. Type II diabetics take oral medication and may eventually begin to take insulin.

Diabetes is a self-aggravating disease. Because they are glucose-starved, the pancreatic alpha cells respond as during hypoglycemia and secrete glucagon to signal cells to break down glycogen into glucose. As cells release sugar, blood glucose concentration increases. ∎

Figure 33.6 The Pancreas The pancreas, which is dominated by exocrine pancreatic acini cells, contains endocrine cells in clusters known as the pancreatic islets.

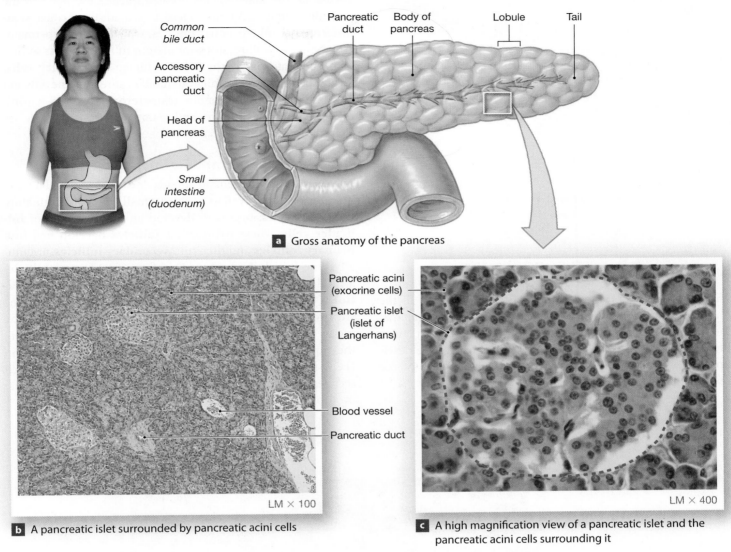

a Gross anatomy of the pancreas

b A pancreatic islet surrounded by pancreatic acini cells

LM × 100

c A high magnification view of a pancreatic islet and the pancreatic acini cells surrounding it

LM × 400

QuickCheck Questions

6.1 Where is the pancreas located?

6.2 What is the exocrine function of the pancreas?

6.3 Where are the endocrine cells located in the pancreas?

In the Lab 6

Materials

☐ Torso model   ☐ Compound microscope
☐ Endocrine chart   ☐ Prepared slide of pancreas

Procedures

1. Review the histology of the pancreas in Figure 33.6.

2. Locate the pancreas on the torso model and endocrine chart.

3. Use the compound microscope to locate the dark-stained pancreatic acini cells and the oval pancreatic ducts. Identify the clusters of pancreatic islets, the endocrine portion of the gland.

4. *Draw It!* In the space provided, sketch the pancreas, labeling the pancreatic islets and the pancreatic acini cells. ■

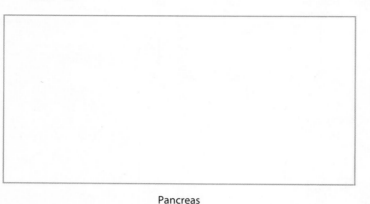

Pancreas

Lab Activity 7   Testes and Ovaries

The testes and ovaries are **gonads,** specialized organs of the male and female reproductive organs that produce *gametes,* the spermatozoa and ova that fuse at fertilization to start a new life. The testes and ovaries secrete hormones to regulate development of the reproductive system and maintenance of the sexually mature adult.

Testes—Anatomy and Hormones

Testes are the male gonads, located outside the body in the pouchlike scrotum. A testis, or testicle, is made up of coiled **seminiferous** (se-mi-NIF-er-us) **tubules,** which produce spermatozoa. **Figure 33.7** illustrates seminiferous tubules in cross section with spermatozoa in the tubular lumen. **Interstitial cells,** located between the seminiferous tubules, are endocrine cells and secrete the male sex hormone **testosterone** (tes-TOS-ter-ōn), the hormone that produces and maintains secondary male sex characteristics, such as facial hair.

Ovaries—Anatomy and Hormones

The **ovaries** are the female gonads, located in the pelvic cavity. During each ovarian cycle, a small group of immature eggs, or oocytes, begins to develop an outer capsule of **follicular cells.** These **primordial follicles** develop first into **primary follicles** and then into **secondary follicles.** Eventually, one follicle becomes a **Graafian** (GRAF-ē-an) **follicle,** a fluid-filled bag containing an oocyte for release at **ovulation** (**Figure 33.8**). The follicles are temporary endocrine structures and secrete the hormone **estrogen** to prepare the uterus for implantation of a fertilized egg. After ovulation, the ruptured Graafian follicle becomes the **corpus luteum** (LOO-tē-um), another temporary endocrine structure, which secretes **progesterone** (pro-JES-ter-ōn), the hormone that promotes further thickening of the uterine wall.

Figure 33.7 The Testes

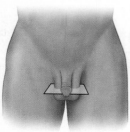

a Location of the testes in the scrotum

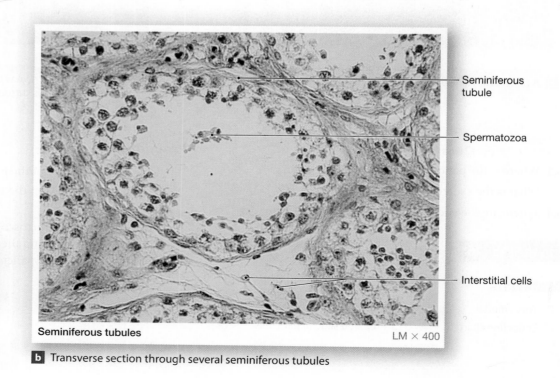

Seminiferous tubule

Spermatozoa

Interstitial cells

Seminiferous tubules

LM × 400

b Transverse section through several seminiferous tubules

Figure 33.8 The Ovary

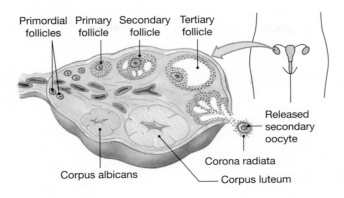

Primordial follicles   Primary follicle   Secondary follicle   Tertiary follicle

Released secondary oocyte

Corpus albicans

Corona radiata

Corpus luteum

a Follicular development during the ovarian cycle

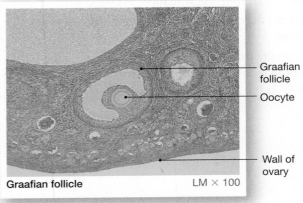

Graafian follicle

Oocyte

Wall of ovary

Graafian follicle   LM × 100

b Section of an ovary showing a Graafian follicle

Clinical Application Steroid Abuse

Anabolic steroids are androgens, precursors to the male sex hormone testosterone. Because testosterone stimulates muscle development and enhances the competitiveness of males, a surprising number of high school, college, and professional athletes use anabolic steroids to increase their strength, endurance, and athletic drive. The most commonly used steroid is *androstedione,* which is converted by the body to testosterone. Elevated blood testosterone levels inhibit secretion of the regulatory hormone GnRH from the hypothalamus. Decreased GnRH secretion keeps the anterior pituitary lobe from secreting FSH and LH, resulting in a decrease in sperm and testosterone production. Although steroids are used to enhance athletic performance, abuse of the steroids actually harms males and causes sterility and a decrease in testosterone secretion.

Steroid abuse by female athletes is just as dangerous as abuse by males. Females produce small amounts of androgen in the adrenal cortex that is converted into estrogen, the main female sex hormone. Steroid use by female athletes also increases muscle mass but can cause irregular menstrual cycles and increased body hair and other secondary sex characteristics and, in some cases, baldness. In both sexes, steroid use can lead to liver failure, premature closure of epiphyseal plates, cardiovascular problems, and infertility. ∎

QuickCheck Questions

7.1 Where are the testes and ovaries located?

7.2 Where are the endocrine cells in the male gonad?

7.3 What are the endocrine structures in the ovaries?

In the Lab 7

Materials

☐ Torso models
☐ Endocrine charts
☐ Compound microscope
☐ Prepared slides of testis and ovary

Procedures

Testis

1. Review the structures of the testis in Figure 33.7.
2. Locate the testes on a torso model and an endocrine chart.
3. Scan the testis slide at low magnification, and identify the seminiferous tubules. Increase the magnification to locate interstitial cells between the seminiferous tubules.
4. *Draw It!* In the space provided, sketch a cross section of a testis, detailing the seminiferous tubules and interstitial cells.

Testis

Ovary

1. Review the ovarian structures in Figure 33.8.
2. Locate the ovaries on a torso model and an endocrine chart.
3. Scan the ovary slide at low power to locate the large Graafian follicle. Identify the developing ovum inside the follicle.

4. ***Draw It!*** In the space provided, sketch the Graafian follicle. ■

Graafian follicle

Name _____

Date _____

Section _____

Endocrine System

A. Definition

Define or describe each of the following terms.

1. thyroid follicle

2. adrenal medulla

3. thymic corpuscle

4. seminiferous tubules

5. zona glomerulosa

6. parathyroid gland

7. Graafian follicle

8. adenohypophysis

9. target cell

10. pancreatic islets

11. interstitial cells

12. zona reticularis

B. Drawing

Draw It! Sketch a section of the thyroid gland and detail several thyroid follicles.

C. Short-Answer Questions

1. Describe how negative feedback regulates the secretion of most hormones.

2. Explain how the pituitary gland functions as the master gland of the body.

3. Describe the hormones produced by the three layers of the adrenal cortex.

4. Describe what structures are involved in the ovulation of an egg.

5. What are the endocrine functions of the pancreas?

D. Labeling

Label the features of each endocrine gland in **Figure 33.9**.

1. _____

2. _____

3. _____

4. _____

5. _____

6. _____

7. _____

8. _____

9. _____

10. _____

11. _____

12. _____

13. _____

14. _____

15. _____

Figure 33.9 **Endocrine Organs**

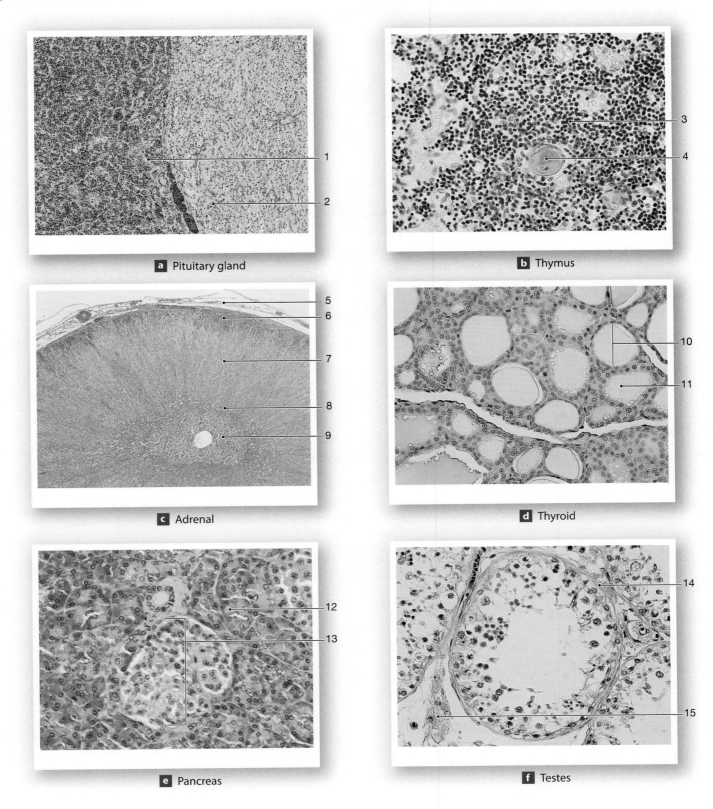

a Pituitary gland

b Thymus

c Adrenal

d Thyroid

e Pancreas

f Testes

E.   Analysis and Application

1. What is the difference between type I and type II diabetes?

2. Why is steroid use among athletes dangerous to their health?

3. Compare the histology of an adult thymus with that of an infant.

4. How is blood calcium regulated by the endocrine system?

F.   Clinical Challenge

1. How do the pancreatic alpha cells of a type II diabetic contribute to high blood sugar concentration?

2. What symptoms would someone with hyperthyroidism exhibit?

Blood

Learning Outcomes

On completion of this exercise, you should be able to:

1. List the functions of blood.

2. Describe each component of blood.

3. Distinguish each type of blood cell on a blood-smear slide.

4. Describe the antigen-antibody reactions of the ABO and Rh blood groups.

5. Safely collect a blood sample using the blood lancet puncture technique.

6. Safely type a sample of blood to determine the ABO and Rh blood types.

7. Correctly perform a hematocrit test.

8. Correctly perform a coagulation test.

9. Describe how to discard blood-contaminated wastes properly.

Lab Activities

Clinical Application

B lood is a fluid connective tissue that flows through blood vessels of the cardiovascular system. Blood consists of cells and cellular pieces, collectively called the **formed elements,** carried in an extracellular fluid called blood **plasma** (PLAZ-muh). Blood has many functions. It controls the chemical composition of all interstitial fluid by regulating pH and electrolyte levels. It supplies trillions of cells with life-giving oxygen, nutrients, and regulating molecules. Some of its formed elements protect the body from invasion by foreign organisms, such as bacteria, and other formed elements manufacture substances needed for defense against specific biological and chemical threats. In response to injury, blood has the ability to change from a liquid to a gel so as to clot and stop bleeding.

Need More Practice and Review?

Build your knowledge—and confidence!—in the Study Area of MasteringA&P® at www.masteringaandp.com with Pre-lab Quizzes, Post-lab Quizzes, Practice Anatomy Lab™ (PAL™) 3.0 virtual anatomy practice tool, PhysioEx™ 9.0 laboratory simulations, and A&P Flix™ with Quizzes.

PAL | practice anatomy lab   For this lab exercise, follow this navigation path:
- PAL>Histology>Cardiovascular System

PhysioEx™ 9.0 For this lab exercise, go to this topic:
- PhysioEx Exercise 11: Blood Analysis

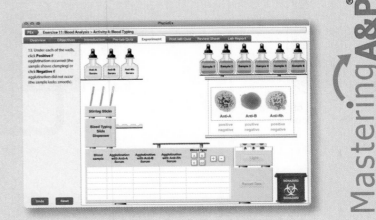

MasteringA&P®

Lab Activity 1 Composition of Whole Blood

A sample of blood is approximately 55 percent plasma and 45 percent formed elements (Figure 34.1). Plasma is 92 percent water and contains proteins that regulate the osmotic pressure of blood, proteins for clotting, and **antibodies,** the immune system proteins that protect the body from invading pathogens and molecules, collectively referred to as **antigens.** Electrolytes, hormones, nutrients, and some blood gases are transported in the blood plasma. The formed elements are organized into three groups of cells and pieces of cells: red blood cells, white blood cells, and platelets. When stained, each group is easy to identify with a microscope. The reddish cells are erythrocytes, the cells that have visible nuclei are leukocytes, and the small cell fragments between the erythrocytes and leukocytes are platelets.

Red blood cells (RBCs), also called **erythrocytes** (e-RITH-rō-sīts), are red and lack a nucleus. The most abundant of all blood cells, RBCs are biconcave discs that are noticeably thin in the center (Figure 34.2). On a laboratory slide, the thin central section of each disc is not as deeply stained as the surrounding rim. The biconcave shape gives each RBC more surface area than a flat-faced disc would have, an important feature that allows rapid gas exchange between the blood and the tissues of the body. Their shape also allows RBCs to flex and squeeze through narrow capillaries.

The major function of RBCs is to transport blood gases. They pick up oxygen in the lungs and carry it to the cells of the body. While supplying the cells with oxygen, the blood acquires carbon dioxide from the cells. The plasma and RBCs convey the carbon dioxide to the lungs for removal during exhalation. To accomplish the task of gas transport, each RBC contains millions of hemoglobin (Hb) molecules. **Hemoglobin** (HĒ-mō-glō-bin) is a complex protein molecule containing as part of its structure four iron atoms that bind loosely to oxygen and carbon dioxide molecules.

The second type of formed element is **white blood cells (WBCs),** also called **leukocytes** (LOO-kō-sīts). A main feature

Figure 34.1 The Composition of Whole Blood Whole blood is composed of a liquid portion, plasma, and a solid portion comprising three groups of blood cells collectively called the formed elements.

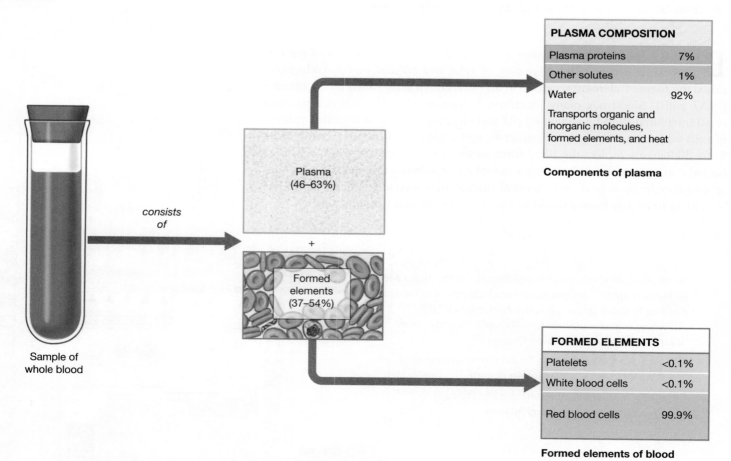

Sample of whole blood

consists of

Plasma (46–63%)

+

Formed elements (37–54%)

PLASMA COMPOSITION	
Plasma proteins	7%
Other solutes	1%
Water	92%
Transports organic and inorganic molecules, formed elements, and heat	

Components of plasma

FORMED ELEMENTS	
Platelets	<0.1%
White blood cells	<0.1%
Red blood cells	99.9%

Formed elements of blood

Figure 34.2 The Anatomy of Red Blood Cells

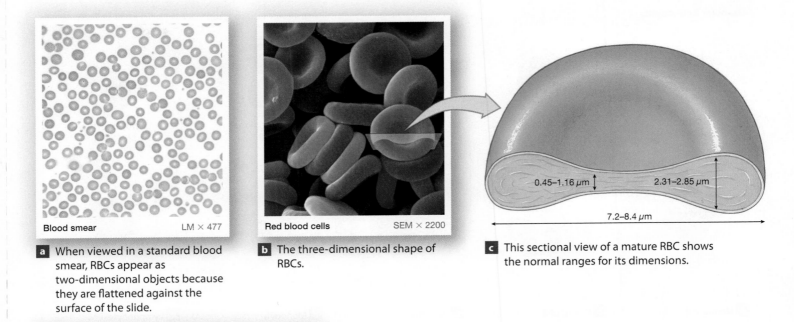

Blood smear   LM × 477

a When viewed in a standard blood smear, RBCs appear as two-dimensional objects because they are flattened against the surface of the slide.

Red blood cells   SEM × 2200

b The three-dimensional shape of RBCs.

0.45–1.16 µm   2.31–2.85 µm

7.2–8.4 µm

c This sectional view of a mature RBC shows the normal ranges for its dimensions.

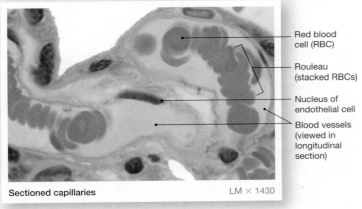

Red blood cell (RBC)

Rouleau (stacked RBCs)

Nucleus of endothelial cell

Blood vessels (viewed in longitudinal section)

Sectioned capillaries   LM × 1430

d When traveling through relatively narrow capillaries, RBCs may stack like dinner plates.

of WBCs is their nucleus, which takes a very dark stain and is often branched into two or more lobes (**Figure 34.3**). WBCs lack hemoglobin and therefore do not transport blood gases. They can pass between the endothelial cells of capillaries and enter the interstitial spaces of tissues. Most WBCs are **phagocytes,** scavenger cells that engulf foreign bodies and other unwanted materials circulating in the blood and destroy them, and are therefore part of the immune system.

There are two broad classes of WBCs: granular and agranular. The **granular leukocytes,** also called **granulocytes,** have granules in their cytoplasm and include the neutrophils, eosinophils, and basophils. **Agranular leukocytes,** which include the monocytes and lymphocytes, have few cytoplasmic granules.

Neutrophils (NOO-trō-filz) are the most common leukocytes and account for up to 70 percent of the WBC population. These granular leukocytes are also called **polymorphonuclear** (pol-ē-mōr-fō-NOO-klē-ar) **leukocytes** because the nuclei are complex and branch into two to five lobes. In addition to a dark-staining nucleus, neutrophils have many small cytoplasmic granules that stain pale purple, visible in Figure 34.3a.

Neutrophils are the first leukocytes to arrive at a wound site to begin infection control. They release cytotoxic chemicals and phagocytize (engulf and destroy) invading pathogens. They also release hormones called **cytokines** that attract other phagocytes, such as eosinophils and monocytes, to the site of injury. Neutrophils are short lived, surviving in the blood for

Figure 34.3 White Blood Cells

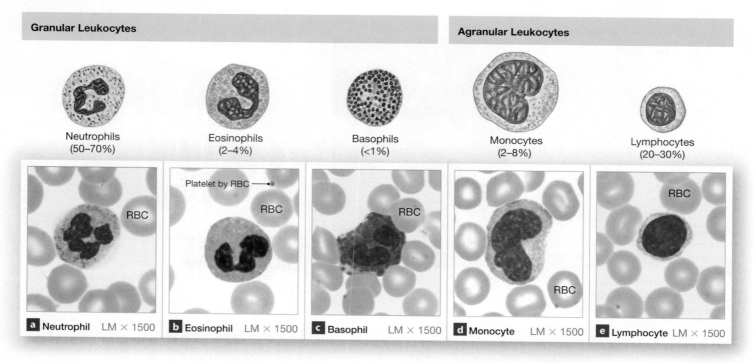

Granular Leukocytes			Agranular Leukocytes	
Neutrophils (50–70%)	Eosinophils (2–4%)	Basophils (<1%)	Monocytes (2–8%)	Lymphocytes (20–30%)
a Neutrophil LM × 1500	**b** Eosinophil LM × 1500	**c** Basophil LM × 1500	**d** Monocyte LM × 1500	**e** Lymphocyte LM × 1500

up to 10 hours. Active neutrophils in a wound may live only 30 minutes until they succumb to the toxins released by the pathogens they have ingested.

About the same size as neutrophils, the granular leukocytes known as **eosinophils** (ē-ō-SIN-ō-filz) are identified by the presence of medium-sized granules that stain orange-red, as shown in Figure 34.3b. The nucleus is conspicuously segmented into two lobes. Eosinophils are phagocytes that engulf bacteria and other microbes that the immune system has coated with antibodies. They also contribute to decreasing the inflammatory response at a wound or site of infection. Approximately 3 percent of the circulating WBCs are eosinophils.

Basophils (BĀ-sō-filz), the third type of granular leukocyte, constitute less than 1 percent of the circulating WBCs. They have large cytoplasmic granules that stain dark blue. The granules are so large and numerous that the nucleus is obscured, as illustrated in Figure 34.3c. Smaller than neutrophils and eosinophils, basophils are sometimes difficult to locate on a blood-smear slide because relatively few of them are present. They migrate to injured tissues and release histamines, which cause vasodilation, and heparin, which prevents blood from clotting. Mast cells in the tissue respond to these molecules and induce local inflammation.

Monocytes (MON-ō-sīts) are large agranular WBCs containing a dark-staining, kidney-shaped nucleus surrounded by a pale blue cytoplasm (Figure 34.3d). Approximately 2 to 8 percent of circulating WBCs are monocytes. On a blood-smear slide, monocytes appear roundish and may have small extensions, much like an amoeba. Even though monocytes are agranular leukocytes,

materials they ingest, such as phagocytized bacteria and debris, stain and may look like granules under the microscope.

Monocytes are wanderers. They leave the blood by squeezing between the capillary endothelium to patrol the body tissues in search of microbes and worn-out tissue cells. They are second to neutrophils in arriving at a wound site. When neutrophils die from phagocytizing bacteria, the monocytes phagocytize the neutrophils.

The agranular **lymphocytes** (LIM-fō-sīts) are the smallest of the WBCs and are approximately the size of a RBC (Figure 34.3e). The distinguishing feature of any lymphocyte is a large nucleus that occupies almost the entire cell, leaving room for only a small halo of pale blue cytoplasm around the edge of the cell. Lymphocytes are abundant in the blood and compose 20 to 30 percent of all circulating WBCs. Lymphocytes move freely between the blood and the tissues of the body. As their name suggests, they are the main cells populating lymph nodes, glands, and other lymphoid tissues.

Although several types of lymphocytes exist, they cannot be individually distinguished with a light microscope. Generally, lymphocytes provide immunity from microbes and defective cells by two methods. **T cells** attach to and destroy foreign cells in a cell-mediated response involving release of cytotoxic chemicals to kill the invaders. The second immunity method uses the lymphocytes known as **B cells,** which become sensitized to a specific antigen, then manufacture and pour antibodies into the blood. The antibodies attach to and help destroy foreign antigens.

Platelets (PLĀT-lets, Figure 34.3b), the third type of formed element, are small cellular pieces produced from the breakdown of **megakaryocytes,** which are large protein-producing cells located in the bone marrow. Platelets lack a nucleus and other organelles. They survive in the blood for a brief time and are involved in blood clotting.

QuickCheck Questions

1.1 What are the three types of formed elements in blood?

1.2 Which is the most abundant type of white blood cell?

In the Lab 1

Materials

☐ Compound microscope

☐ Immersion oil

☐ Human blood-smear slide (Wright's or Giemsa stained)

Procedures

1. Blood samples are thin and require careful focusing. Bring the sample into focus with the low-power objective. Then use the fine focus knob as you examine individual cells. Notice the abundance of red blood cells. The dark-stained cells are the various white blood cells.

2. Scan the slide at high-dry magnification and locate the different types of WBCs. Note the small platelets between the red and white cells.

3. Use the oil-immersion lens to observe the various blood cells. Place a small drop of immersion oil on the coverslip of the slide and gently rotate the oil-immersion objective lens so that the tip of the lens becomes covered with the oil. There should be oil, not air, between the lens and the slide. Use the mechanical stage and scan the slide slowly to avoid spreading the oil too thin. When you are finished, it is very important that you clean the oil off the lens and the slide correctly by using a sheet of microscope lens tissue to gently wipe the oil off the lens and slide. Then use a fresh sheet of tissue with a drop of lens cleaner and wipe the lens and slide clean of any remaining oil. To prevent damage to the lens, do not saturate the lens with the cleaner.

4. Complete the *Sketch to Learn* activity below.

 Sketch to Learn

The key to recognizing the various types of blood cells is to know how their nucleus is organized and characteristics of cytoplasmic granules and other unique features. Let's draw a neutrophil as seen at high magnification.

Sample Sketch

Step 1
• Draw a circle to represent a neutrophil.

Step 2
• Draw the nucleus with several lobes.
• Add nucleoli and fill in nucleus with small lines and dots.

Step 3
• Add small dots for cytoplasmic granules.

Your Sketch

5. ***Draw It!*** Sketch each blood cell in the space provided. ◼

Neutrophil	Eosinophil
Basophil	Lymphocyte
Monocyte	Platelet

Lab Activity 2 ABO and Rh Blood Groups

Your blood type is inherited from your parents' genes, and it does not change during your lifetime. Each blood type is a function of the presence or absence of specific antigen molecules on the surface of the red blood cells. (The antigens important in blood types are also called *agglutinogens* [a-gloo-TIN-ō-jenz], but we shall use the term *antigens*.) The antigens are like cellular nametags that inform your immune system that your red blood cells belong to "self" and are not "foreign."

Blood also contains specialized antibody molecules called *agglutinins* (a-GLOO-ti-ninz). The antibodies and antigens in an individual's blood do not interact with one another, but the antibodies do react with antigens of foreign red blood cells and cause the cells to burst, hence the need for blood type matching prior to a blood transfusion.

More than 50 surface antigens and blood groups occur in the human population. In this activity, you will study the two most common, the ABO group and the Rh group. Each blood group is controlled by a different gene, and your ABO blood type does not influence your Rh blood type. Table 34.1 shows the distribution of these two blood groups in the human population.

ABO Blood Group

There are four blood types in the **ABO blood group:** A, B, AB, and O (Figure 34.4). Two surface antigens, A and B, occur in different combinations that determine the blood type. Type A blood has the A surface antigen on its membrane, Type B blood has the B surface antigen, Type AB blood has both A and B surface antigens, and Type O blood has neither. Which antibodies are present in blood depends on type. Type A blood contains anti-B antibodies, which attack red blood cells carrying B surface antigens. Type B blood contains anti-A antibodies to defend against cells carrying A surface antigens. Type AB blood contains no antibodies, and Type O contains both anti-A and anti-B antibodies.

The anti-B antibodies in Type A blood do not react with the Type A surface antigens but do react with the B surface antigens present in blood Types B and AB. The same is true for Type B blood: The anti-A antibodies do not react with the B surface antigens but do destroy the cells carrying A surface antigens in

Table 34.1	Differences in ABO and Rh Blood Group Distribution				
	Percentage with Each Blood Type				
Population	**O**	**A**	**B**	**AB**	**Rh⁺**
U.S. (AVERAGE)	46	40	10	4	85
African-American	49	27	20	4	95
Caucasian	45	40	11	4	85
Chinese American	42	27	25	6	100
Filipino American	44	22	29	6	100
Hawaiian	46	46	5	3	100
Japanese American	31	39	21	10	100
Korean American	32	28	30	10	100
NATIVE NORTH AMERICAN	79	16	4	1	100
NATIVE SOUTH AMERICAN	100	0	0	0	100
AUSTRALIAN ABORIGINE	44	56	0	0	100

Figure 34.4 **Blood Typing and Cross-Reactions**

Type A	Type B	Type AB	Type O
Type A blood has RBCs with surface antigen A only.	**Type B** blood has RBCs with surface antigen B only.	**Type AB** blood has RBCs with both A and B surface antigens.	**Type O** blood has RBCs lacking both A and B surface antigens.

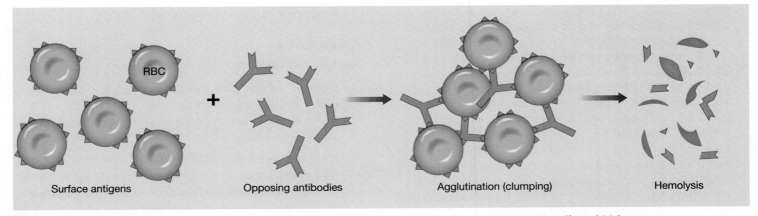

If you have Type A blood, your plasma contains anti-B antibodies, which will attack Type B surface antigens.

If you have Type B blood, your plasma contains anti-A antibodies, which will attack Type A surface antigens.

If you have Type AB blood, your plasma has neither anti-A nor anti-B antibodies.

If you have Type O blood, your plasma contains both anti-A and anti-B antibodies.

a Blood type depends on the presence of surface antigens (agglutinogens) on RBC surfaces. The plasma contains antibodies (agglutinins) that will react with foreign surface antigens.

RBC

Surface antigens + Opposing antibodies → Agglutination (clumping) → Hemolysis

b In a cross-reaction, antibodies react with their target antigens causing agglutination and hemolysis of the affected RBCs.

blood Types A and AB. Because AB blood contains neither anti-A antibodies nor anti-B antibodies, it does not react with blood of other types. People with AB blood are called *universal acceptors* because, lacking antibodies, they can accept blood of any type in a transfusion. Although surface antigens are absent in Type O blood, it has both anti-A and anti-B antibodies. With no surface antigens acting as nametags, Type O blood can be transfused to all blood types, and people with Type O blood are called *universal donors*.

To determine blood type, the presence of antigens is detected by adding to a blood sample drops of **antiserum** that contain either anti-A or anti-B antibodies. The antibodies in the antiserum react with the corresponding surface antigens

in the sample. The blood *agglutinates* (forms clumps of solid material that settle out from the plasma) as the antibodies react with the surface antigens.

Rh Blood Group

The **Rh blood group** has two blood types, **Rh positive** and **Rh negative.** Although this blood group is separate from the ABO group, the two are usually used together to identify blood type. For example, a blood sample may be A+ or A−. The Rh group has only one antigen, the Rh surface antigen (D antigen), plus a single Rh antibody designated anti-D. The D antigen is present only on RBCs that are Rh positive; Rh-negative blood cells lack the D antigen. (In case you are wondering, the Rh blood group is named after the Rhesus macaque, the animal in which this blood group was first discovered.)

Rh-positive blood has the Rh surface antigen and lacks the Rh antibody. Rh-negative blood does not have the Rh surface antigen and initially does not have the Rh antibody. However, if Rh-negative blood is exposed to Rh-positive blood, the Rh-negative person's immune system becomes sensitized to the Rh surface antigen and subsequently produces the anti-D antibody. This becomes clinically significant in cases of pregnancy with Rh incompatibilities between mother and fetus.

Make a Prediction

Predict the surface antigen on your own red blood cells and the antibodies in your plasma.

Clinical Application Rh Factor and Hemolytic Disease of the Newborn

If an expectant mother is Rh negative and her baby is Rh positive, a potentially life-threatening Rh incompatibility exists for the baby. Normally, fetal blood does not mix with maternal blood. Instead, the umbilical cord connects to the placenta, where fetal capillaries exchange gases, wastes, and nutrients with the mother's blood. If internal bleeding occurs, however, so that the mother is exposed to the D antigens in her baby's Rh-positive blood, she will produce anti-D antibodies. These antibodies cross the placental membrane and enter the fetal blood, where they hemolyze (rupture) the fetal blood cells of this fetus and those of any future Rh-positive fetuses. This Rh action is called either **hemolytic disease of the newborn** or **erythroblastosis fetalis** (e-rith-rō-blas-TŌ-is fe-TAL-sis). A dosage of anti-D antibodies, called **RhoGam**, may be given to the mother during pregnancy and after delivery to destroy any Rh-positive fetal cells in her blood. This treatment prevents her from developing anti-D antibodies. ■

⚠ Safety Alert: Handling Blood

1. Some infectious diseases are spread by contact with blood. Follow all instructions carefully and protect yourself by wearing gloves and working only with your own blood.
2. Materials contaminated with blood must be disposed of properly. Your instructor will inform you of methods for disposing of lancets, slides, prep pads, and toothpicks.

Your instructor may ask for a volunteer to "donate" blood in order to demonstrate how blood typing is done. Alternatively, many biological supply companies sell simulated blood-typing kits that contain a bloodlike solution and antisera. These kits contain no human or animal blood products and safely show the principles of typing human blood. ▲

QuickCheck Questions

2.1 What are the two major blood groups used to identify blood type?

2.2 What surface antigens does Type A blood have?

In the Lab 2

Materials

☐ Hand soap
☐ Paper towels
☐ Gloves
☐ Safety glasses
☐ Disposable sterile blood lancet
☐ Disposable sterile alcohol prep pad
☐ Disposable blood-typing plate or sterile microscope slide
☐ Wax pencil (if using microscope slide)
☐ Toothpicks
☐ Anti-A, anti-B, and anti-D blood-typing antisera
☐ Warming box
☐ Biohazardous waste container
☐ Bleach solution in spray bottle (optional)

Procedures
Sample Collection

1. If you are using a slide, use the wax pencil to draw three circles across the width of the slide. Label the circles "A," "B," and "D." If you are using a typing plate, label three of the depressions "A," "B," and "D."
2. Wash both hands thoroughly with soap, and then dry them with a clean paper towel. Obtain an additional paper towel to place blood-contaminated instruments on while collecting a blood sample. Wear gloves and safety glasses while collecting and examining blood. If collecting a sample from yourself, wear a glove on the hand used to hold the lancet.
3. Open a sterile alcohol prep pad, and clean the tip of the index finger from which the blood will be drawn. Be sure to thoroughly disinfect the entire fingertip, including the sides. Place the used prep pad on the paper towel.
4. Open a sterile blood lancet to expose only the sharp tip. Do not use an old lancet, even if it was used on one of your own fingers. Use the sterile tip *immediately* so that there is no time for it to inadvertently become contaminated.
5. With a swift motion, jab the point of the lancet into the lateral surface of the fingertip. Place the used lancet on the paper towel until it can be disposed of in a biohazard container.
6. Gently squeeze a drop of blood either into each depression on the blood-typing plate or into the circles

on the slide. If necessary, slowly "milk" the finger to work more blood out of the puncture site.

ABO and Rh Typing

1. Add a drop of anti-A antiserum to the sample labeled A, being very careful not to allow blood to touch (and thereby contaminate) the tip of the dropper. Repeat the process by adding a drop of anti-B antiserum to the B sample and a drop of anti-D antiserum to the D sample.

2. Immediately and gently mix each drop of antiserum into the blood with a clean toothpick. To prevent cross contamination, use a separate, clean toothpick for each sample. Place all used toothpicks on the paper towel until they can be disposed of in a biohazard container.

3. Place the slide or typing plate on the warming box and agitate the samples by rocking the box carefully back and forth for two minutes. *Note:* The anti-D agglutination reaction is often weaker and less easily observed than the anti-A and anti-B agglutination reactions. A microscope may help you observe the anti-D reaction.

4. Examine the drops for any agglutination visible with the unaided eye and compare your samples with Figure 34.5. Agglutination results when the antibodies in the antiserum react with the matching antigen on the red blood cells. For example, if blood agglutinates with the anti-A antiserum and the anti-D antiserum, the blood type is A positive.

5. Record your results in the first blank row of Table 34.2. In each cell of the table, indicate yes or no for the presence of agglutination.

6. Collect blood-typing data from three classmates to compare agglutination responses among blood types. How does the distribution of blood types in your four-person sample compare with the distribution of types given in Table 34.1?

Disposal of Materials and Disinfection of Work Space

1. Dispose of all blood-contaminated materials in the appropriate biohazard box. A box for sharp objects may

Figure 34.5 Blood Type Testing Test results for blood samples from four individuals. Drops are mixed with solutions containing antibodies to the surface antigens A, B, AB, and D (Rh). Clumping occurs when the sample contains the corresponding surface antigen(s). The individuals' blood type are shown at right.

Anti-A	Anti-B	Anti-D	Blood type
			A⁺
			B⁺
			AB⁺
			O⁻

be available to dispose of the lancets, toothpicks, and microscope slides.

2. Your instructor may ask you to disinfect your workstation with a bleach solution. If so, wear gloves and safety glasses while wiping the surfaces clean.

3. Lastly, remove your gloves and dispose of them in the biohazard box. Remember to wash your hands after disposing of all materials. ∎

Lab Activity 3 Hematocrit (Packed Red Cell Volume)

The **hematocrit** (he-MA-tō-krit), or packed cell volume (PCV), test measures the volume of packed formed elements in a given volume of blood. Because RBCs far outnumber all the other formed elements, the test mainly measures their

Table 34.2	Blood Typing Data				
Student	Anti-A Antiserum Reaction	Anti-B Antiserum Reaction	Anti-D Antiserum Reaction	Blood Type	Hematocrit Reading

volume. Hematocrit results provide information regarding the oxygen-carrying capacity of the blood. A low hematocrit value indicates that the blood has fewer RBCs to transport oxygen. Average hematocrit values range from 40 to 54 percent in males and from 37 to 47 percent in females.

Make a Prediction

Predict your hematocrit measurement.

QuickCheck Questions

3.1 What does a hematocrit test measure?

3.2 What is the average hematocrit range for males? For females?

In the Lab   3

Materials

- ☐ Hand soap
- ☐ Paper towels
- ☐ Gloves
- ☐ Safety glasses
- ☐ Disposable sterile blood lancet
- ☐ Disposable sterile alcohol prep pads
- ☐ Sterile heparinized capillary tubes
- ☐ Seal-easy clay
- ☐ Bleach solution in spray bottle
- ☐ Microcentrifuge
- ☐ Tube reader
- ☐ Biohazardous waste disposal container

Procedures

1. Review the Safety Alert in Lab Activity 2.

2. Follow steps 2 through 5 of Lab Activity 2, "Sample Collection," to obtain a blood sample.

3. *Gently* squeeze a drop of blood out of your finger. (Squeeze gently because excess pressure forces interstitial fluid into the blood, and the presence of this fluid may alter your hematocrit reading. If you are having difficulty obtaining a drop, use a clean, sterile lancet to lance your finger again in a different spot.)

4. Place a sterile heparinized capillary tube on the drop of blood. Orient the open end of the tube downward, as shown in **Figure 34.6**, to allow the blood to flow into the tube. Fill the tube at least two-thirds full with blood.

5. Carefully seal one end of the tube by dipping it into the seal-easy clay as shown in **Figure 34.7**. Do not force the delicate capillary tube into the clay, for it may break and cause you to jam glass into your hand. Instead, hold the tube the way you hold a pencil for writing, with

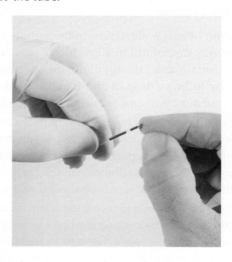

Figure 34.6 Filling a Capillary Tube with Blood A capillary tube is held slanting downward at the lance site to draw a drop of blood into the tube.

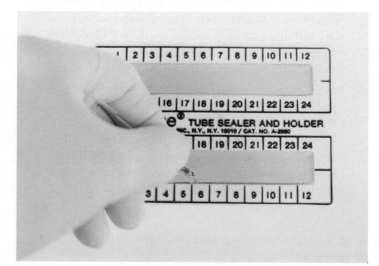

Figure 34.7 Plugging a Capillary Tube with Clay To avoid breaking the capillary tube, hold the tube at the end nearest the clay and gently press the tip into the clay.

your thumb and index finger close to the end where the blood has accumulated. Then gently turn the tube while pressing it into the clay. Leave the other end unplugged.

6. Clean any blood off the clay with the bleach solution and a paper towel.

7. Set the tube in the microcentrifuge with the clay end toward the outer margin of the chamber. Because the centrifuge spins at high speeds, the chamber must be balanced by placing tubes evenly in the chamber. Counterbalance your capillary tube by placing another sample directly across from yours. An empty tube sealed at one end with clay may be used if another student's sample is not available.

8. Screw the inner cover on with the centrifuge wrench. Do not overtighten the lid. Close the outer lid and push in the latch.

9. Set the timer to four to five minutes, and allow the centrifuge to spin. Do not attempt to open or stop the centrifuge while it is turning. Always keep loose hair and clothing away from the centrifuge.

10. After the centrifuge turns off and stops spinning, open the lid and the inner safety cover to remove the capillary tube. Your blood sample should have clear plasma at one end of the tube and packed RBCs at the other end.

11. Place the capillary tube in the tube reader. Because there are a variety of tube readers, your instructor will demonstrate how to use the reader in your laboratory.

12. Record your hematocrit measurement in Table 34.2. Is your hematocrit reading within the normal range?

13. Describe the appearance of your blood plasma:

_____

_____

_____

_____

_____

_____

14. Dispose of all used materials as described in Lab Activity 2, "Disposal of Materials and Disinfection of Work Space." ■

Lab Activity 4 Coagulation

Blood removed from the body and allowed to sit for three to four minutes changes from a liquid to a gel. This process is called either **coagulation** (cō-ag-ū-LĀ-shun) or **clotting,** and prevents excessive blood loss. Coagulation is a complex chemical chain reaction beyond the scope of this exercise. In brief, when you cut yourself, enzymes activate circulating proteins that ultimately convert the protein fibrinogen to an insoluble form called **fibrin.** The fibrin molecules join together in long threads that form a net to trap platelets and plug the wound.

In a coagulation test, you determine how fast these reactions occur in your blood. As noted in Lab Activity 1, heparin prevents blood from clotting, which means coagulation time is a measure of blood's heparin content. For a coagulation test, a nonheparinized capillary tube is used.

QuickCheck Questions

4.1 What is coagulation?

4.2 Why is a nonheparinized tube used for the coagulation time test?

In the Lab 4

Materials

- ☐ Hand soap
- ☐ Paper towels
- ☐ Gloves
- ☐ Safety glasses
- ☐ Disposable sterile blood lancet
- ☐ Disposable sterile alcohol prep pad
- ☐ Sterile nonheparinized capillary tube
- ☐ Stopwatch or clock with second hand
- ☐ Small metal file
- ☐ Bleach solution in spray bottle (optional)
- ☐ Biohazardous waste container

Procedures
Coagulation Time

1. Review the Safety Alert in Lab Activity 2.

2. Follow steps 2 through 5 of Lab Activity 2, "Sample Collection," to obtain a blood sample.

3. Gently squeeze a drop of blood out of your finger. If you have difficulty obtaining a drop, use a clean, sterile lancet to lance your finger again in a different spot.

4. Place the sterile nonheparinized capillary tube on the drop of blood. Orient the open end of the tube downward to allow the blood to flow into the tube, as shown in Figure 34.6. Fill the tube at least two-thirds full with blood. Once the tube is prepared, note the time on the stopwatch or clock.

5. Lay the tube on a paper towel, and after 30 seconds, break it as follows. (Make sure you are wearing safety glasses.) While holding one end down as the tube lies on the towel, gently scratch the glass with the edge of the metal file, making your mark about one-half inch from the free end. Place your thumbs and index fingers on either side of the scratch and break the tube by slowly bending it away from you.

6. Slowly separate the two broken ends of the tube, and look for a thin fibrin thread. If a thread is present, record 30 seconds as your coagulation time in Table 34.2.

7. If there is no fibrin, wait 30 seconds, and then break the tube again 1 cm from one end.

8. Repeat the sequence (wait 30 seconds, break the tube, look for a fibrin thread) until you see a thread. Record your coagulation time in Table 34.2.

9. Dispose of all used materials as described in Lab Activity 2, "Disposal of Materials and Disinfection of Work Space." ■

Name _____

Date _____

Section _____

Blood

A. Definition

Define or describe each of the following terms.

1. erythrocyte

2. polymorphonuclear cell

3. leukocyte

4. antibody

5. Type A blood

6. Rh-positive blood

7. red-orange stained blood cell

8. Type B blood

9. Rh-negative blood

10. antigen

B. Completion

Complete each typing slide in **Figure 34.8** by indicating with pencil dots where agglutination occurs.

Figure 34.8 **Simulated Blood-Typing Plates**

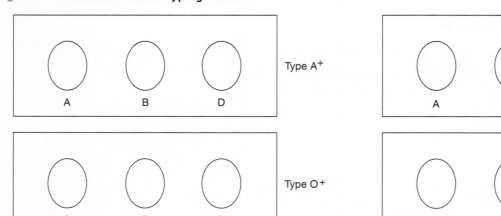

C. Labeling

Identify the blood cells in **Figure 34.9**.

Figure 34.9 **Blood Cells**

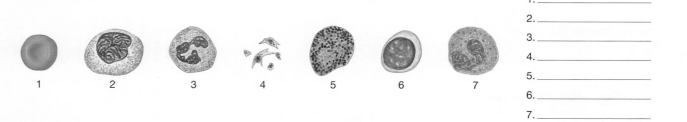

1. _____
2. _____
3. _____
4. _____
5. _____
6. _____
7. _____

D. Short-Answer Questions

1. What is the main function of RBCs?

2. List the five types of leukocytes, and describe the function of each.

3. What is the function of platelets?

E. Analysis and Application

1. Describe how to type blood to detect the ABO and Rh blood groups.

2. Describe how to test the coagulation time of a blood sample.

3. Describe how to do a hematocrit test. What are the average hematocrit values for males and females?

4. How does fibrin contribute to coagulation?

F. **Clinical Challenge**

1. How could you easily determine if two blood samples are compatible?

2. Describe what would happen if Type A blood were transfused into the bloodstream of someone with Type B blood.

3. What conditions are present when hemolytic disease of the newborn occurs?

Anatomy of the Heart

Learning Outcomes

On completion of this exercise, you should be able to:

1. Describe the gross external and internal anatomy of the heart.

2. Identify and discuss the function of the valves of the heart.

3. Identify the major blood vessels of the heart.

4. Trace a drop of blood through the pulmonary circuit and the systemic circuit.

5. Identify the vessels of coronary circulation.

6. List the components of the conduction system of the heart.

7. Describe the anatomy of a sheep heart.

T he cardiovascular system consists of blood; the heart, which pumps blood through the system; and all the blood vessels through which the blood flows. **Arteries** are the blood vessels that carry blood away from the heart, and **veins** are the blood vessels that return blood to the heart. In addition to arteries and veins, the cardiovascular system also contains small-diameter blood vessels called **capillaries.** It is across the walls of capillaries that gases, nutrients, and cellular waste products enter and exit the blood. The heart beats approximately 100,000 times daily to send blood flowing into thousands of miles of blood vessels, providing the body's cells with nutrients, regulating the amounts of substances and gases in the cells, and removing waste products from them. All organ systems of the body depend on the cardiovascular system. Damage to the heart often results in widespread disruption of homeostasis.

Need More Practice and Review?

Build your knowledge—and confidence!—in the Study Area of MasteringA&P® at **www.masteringaandp.com** with Pre-lab Quizzes, Post-lab Quizzes, Practice Anatomy Lab™ (PAL™) 3.0 virtual anatomy practice tool, PhysioEx™ 9.0 laboratory simulations, and A&P Flix™ with Quizzes.

PAL | practice anatomy lab   For this lab exercise, follow these navigation paths:
- PAL>Human Cadaver>Cardiovascular System>Heart
- PAL>Anatomical Models>Cardiovascular System>Heart
- PAL>Histology>Cardiovascular System

Your laboratory studies in this exercise include the histology of cardiac muscle tissue, external and internal heart anatomy, and circulation of blood through the pulmonary and systemic circuits of the cardiovascular system. The dissection of a sheep heart will reinforce your observations of the human heart.

Lab Activity 1 Heart Wall

The heart is located in the **mediastinum** (mē-dē-as-TĪ-num) of the thoracic cavity (**Figure 35.1**). Blood vessels join the heart at the **base,** positioned medially in the mediastinum. Because the left side of the heart has more muscle mass than the right side, the **apex** at the inferior tip of the heart is more on the left side of the thoracic cavity. (Note from Figure 35.1a that the heart's base and apex are "upside down" relative to what we usually mean by those words. The base is anterior to the apex.) Within the mediastinum, the heart is surrounded by the **pericardial** (per-i-KAR-dē-al) **cavity** formed by the **pericardium,** the serous membrane of the heart. The pericardial cavity contains **serous fluid** to reduce friction during muscular contraction. The superficial **parietal pericardium** attaches to the heart in the mediastinum, and the deep **visceral pericardium,** or **epicardium,** covers the heart surface and is considered the outermost layer of the cardiac wall.

The heart wall is organized into three layers: epicardium, myocardium, and endocardium (**Figure 35.2**). The epicardium is the same structure as the visceral pericardium, as just noted. The **myocardium** constitutes most of the heart wall and is composed of **cardiac muscle cells,** also called **cardiocytes.** Each cardiac muscle cell is **uninucleated** (containing a single nucleus) and branched. Cardiac muscle cells interconnect at their branches via junctions called **intercalated** (in-TER-ka-lā-ted) **discs.** Deep to the myocardium is the **endocardium,** a thin layer that lines the chambers of the heart. The endocardium is composed of endothelial tissue resting on a layer of areolar connective tissue.

Make a Prediction

Why would the myocardium be thicker in the left ventricle than in the right ventricle?

QuickCheck Questions

1.1 List the three layers of the heart wall, from superficial to deep.

1.2 How are cardiac muscle cells connected to one another?

In the Lab 1

Materials

☐ Heart model and specimens
☐ Compound microscope
☐ Prepared slide of cardiac muscle

Procedures

1. Review the heart anatomy in Figures 35.1 and 35.2.
2. Identify the layers of the heart wall on the heart model and specimens.
3. With the microscope at low power, examine the microscopic structure of cardiac muscle, using Figure 35.2c for reference. Increase the magnification to high and locate several cardiac muscle cells. Note the single nucleus in each cell and where each cell branches into two arms. Intercalated discs are dark-stained lines where cardiac muscle cells connect together.
4. *Draw It!* Sketch several cardiac muscle cells and intercalated discs in the space provided. ■

Cardiac muscle cells

Lab Activity 2 External and Internal Anatomy of the Heart

The heart is divided into right and left sides, with each side having an upper and a lower chamber (**Figure 35.3**). The upper chambers are the **right atrium** (Ā-trē-um; chamber) and the **left atrium,** and the lower chambers are the **right ventricle** (VEN-tri-kl; little belly) and the **left ventricle.** The atria are receiving chambers and fill with blood returning to the heart in veins. Blood in the atria flows into the ventricles, the pumping chambers, which squeeze their walls together to pressurize the blood and eject it into two large arteries for distribution to the lungs and body tissues. Most of the blood in the atria flows into the ventricles because of pressure and gravity. Before the ventricles contract, the atria contract and "top off" the ventricles.

For a drop of blood to complete one circuit through the body, it must be pumped by the heart twice—through the **pulmonary circuit,** which directs deoxygenated blood to the lungs; and through the **systemic circuit,** which takes oxygenated blood to the rest of the body (Figure 35.3). Each circuit delivers blood to a series of arteries, then capillaries, and finally veins that drain into the opposite side of the heart.

Figure 35.1 **The Location of the Heart in the Thoracic Cavity** The heart is situated in the anterior part of the mediastinum, immediately posterior to the sternum.

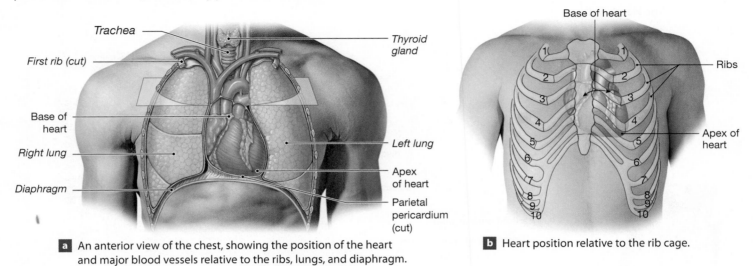

Trachea

Thyroid gland

First rib (cut)

Base of heart

Right lung

Left lung

Diaphragm

Apex of heart

Parietal pericardium (cut)

a An anterior view of the chest, showing the position of the heart and major blood vessels relative to the ribs, lungs, and diaphragm.

Base of heart

Ribs

Apex of heart

b Heart position relative to the rib cage.

Posterior mediastinum

Esophagus

Aorta (arch segment removed)

Right pleural cavity

Left pulmonary artery

Right lung

Left lung

Left pleural cavity

Bronchus of lung

Left pulmonary vein

Right pulmonary artery

Aortic arch

Pulmonary trunk

Right pulmonary vein

Left atrium

Superior vena cava

Left ventricle

Pericardial cavity

Right atrium

Epicardium

Right ventricle

Pericardial sac

Anterior mediastinum

c A superior view of the organs in the mediastinum; portions of the lungs have been removed to reveal blood vessels and airways. The heart is situated in the anterior part of the mediastinum, immediately posterior to the sternum.

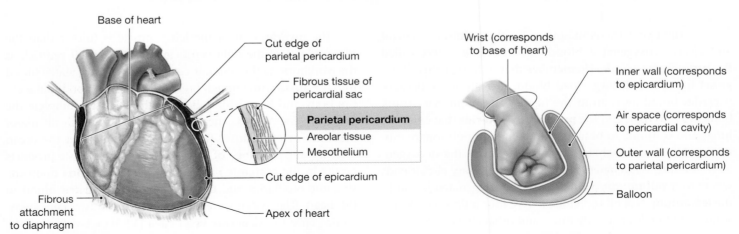

Base of heart

Cut edge of parietal pericardium

Fibrous tissue of pericardial sac

Wrist (corresponds to base of heart)

Inner wall (corresponds to epicardium)

Parietal pericardium
Areolar tissue
Mesothelium

Air space (corresponds to pericardial cavity)

Cut edge of epicardium

Outer wall (corresponds to parietal pericardium)

Fibrous attachment to diaphragm

Apex of heart

Balloon

d The relationship between the heart and the pericardial cavity; compare with the fist-and-balloon example.

493

Figure 35.2 The Heart Wall

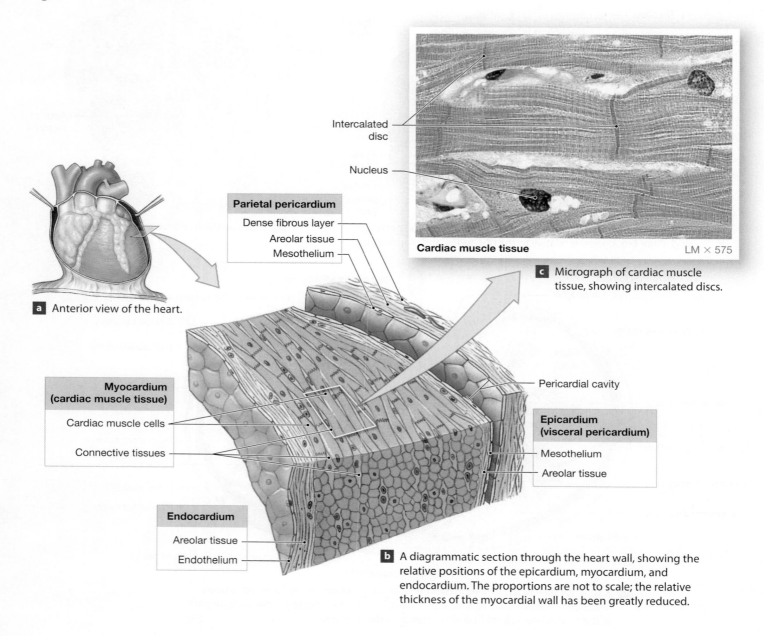

Intercalated disc

Nucleus

Cardiac muscle tissue   LM × 575

c Micrograph of cardiac muscle tissue, showing intercalated discs.

a Anterior view of the heart.

Parietal pericardium
Dense fibrous layer
Areolar tissue
Mesothelium

Myocardium (cardiac muscle tissue)
Cardiac muscle cells
Connective tissues

Pericardial cavity

Epicardium (visceral pericardium)
Mesothelium
Areolar tissue

Endocardium
Areolar tissue
Endothelium

b A diagrammatic section through the heart wall, showing the relative positions of the epicardium, myocardium, and endocardium. The proportions are not to scale; the relative thickness of the myocardial wall has been greatly reduced.

The right ventricle is the pump for the pulmonary circuit and ejects deoxygenated blood into the large artery called the **pulmonary trunk.** (Remember that although this blood vessel transports deoxygenated blood, it is an artery because it carries blood away from the heart.) The pulmonary trunk branches into right and left **pulmonary arteries** that enter the lungs and continue to branch ultimately into pulmonary capillaries, where gas exchange occurs to convert the deoxygenated blood to oxygenated blood. The pulmonary circuit ends where four **pulmonary veins** return the oxygenated blood to the left atrium. Not all individuals have four pulmonary veins; some individuals have only three, and others have five.

The myocardium of the left ventricle is thicker than the myocardium of the right ventricle. The thicker left ventricle is the workhorse of the systemic circuit; it ejects oxygenated blood into the **aorta** with enough pressure to deliver blood to the entire body and have it flow back to the heart to complete the pathway. The aorta is the main artery from which all major **systemic arteries** arise. The systemic arteries enter the organ systems, and exchange of gases, nutrients, and waste products occurs in the **systemic capillaries. Systemic veins** drain the systemic capillaries and transport the deoxygenated blood to the heart. The systemic veins merge into the two largest systemic veins: the **superior vena cava** (VĒ-na KĀ-vuh) and the

Figure 35.3 Generalized View of the Pulmonary and Systemic Circuits Blood flows through separate pulmonary and systemic circuits, driven by the pumping of the heart. Each circuit begins and ends at the heart and contains arteries, capillaries, and veins. Arrows indicate the direction of blood flow in each circuit.

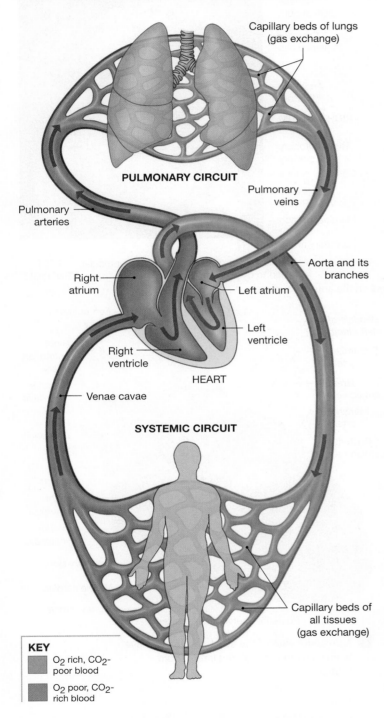

Capillary beds of lungs (gas exchange)

PULMONARY CIRCUIT

Pulmonary veins

Pulmonary arteries

Right atrium

Aorta and its branches

Left atrium

Left ventricle

Right ventricle

HEART

Venae cavae

SYSTEMIC CIRCUIT

Capillary beds of all tissues (gas exchange)

KEY

O$_2$ rich, CO$_2$-poor blood

O$_2$ poor, CO$_2$-rich blood

inferior vena cava, which empty the deoxygenated blood into the right atrium. The cycle of blood flow repeats as the deoxygenated blood enters the right ventricle and is pumped through the pulmonary circuit to the lungs to pick up oxygen for the next journey through the systemic circuit.

The external anatomy of the heart is detailed in **Figure 35.4.** The anterior surface of each atrium has an external flap called the **auricle** (AW-ri-kul; *auris*, ear), shown in Figure 35.4a. Adipose tissue and blood vessels occur along grooves in the heart wall. The **coronary sulcus** is a deep groove between the right atrium and right ventricle that extends to the posterior surface. The boundary between the right and left ventricles is marked anteriorly by the **anterior interventricular sulcus** and posteriorly by the **posterior interventricular sulcus.** Coronary blood vessels follow the sulci and branch to the myocardium. At the branch of the pulmonary trunk is the **ligamentum arteriosum,** a relic of a fetal vessel called the ductus arteriosus that joined the pulmonary trunk with the aorta. (Fetal circulation is discussed in Exercise 36.)

Figure 35.5 details the internal anatomy of the heart. Note how much thicker the myocardium is in the left ventricle, as mentioned previously. The wall between the atria is called the **interatrial septum,** and the ventricles are separated by the **interventricular septum.** In the right atrium, a depression called the **fossa ovalis** is located on the interatrial septum. This is a remnant of fetal circulation, where the foramen ovale allowed blood to bypass the fetal pulmonary circuit. Lining the inside of the right atrium are muscular ridges, the **pectinate** (*pectin*, comb) **muscles.** Folds of muscle tissue called **trabeculae carneae** (tra-BEK-ū-lē KAR-nē-ē; *carneus*, fleshy) occur on the inner surface of each ventricle. The **moderator band** is a ribbon of muscle that passes electrical signals from the interventricular septum to muscles in the right ventricle.

To control and direct blood flow, the heart has two **atrioventricular (AV) valves** and two **semilunar valves.** The two pairs generally work in opposition: When the AV valves are open, the semilunar valves are either closed or preparing to close; when the semilunar valves are open, the AV valves are either closed or preparing to close. The two atrioventricular valves prevent blood from reentering the atria when the ventricles contract. The **right atrioventricular valve,** which joins the right atrium and right ventricle, has three flaps, or cusps, and is also called the **tricuspid** (trī-KUS-pid; *tri*, three; *cuspid*, flap) **valve.** The **left atrioventricular valve** between the left atrium and left ventricle has two cusps and is called either the **bicuspid valve** or the **mitral** (MĪ-tral) **valve.** The cusps of each AV valve have small cords, the **chordae tendineae** (KOR-dē TEN-di-nē-ē; tendonlike cords), which are attached to **papillary** (PAP-i-ler-ē) **muscles** on the floor of the ventricles. When the ventricles contract, the AV valves are held closed by the papillary muscles pulling on the chordae tendineae.

The two semilunar valves are the **aortic valve** and **pulmonary valve,** each located at the base of its artery. These valves prevent backflow of blood into the ventricles when the ventricles are relaxed. Each semilunar valve has three small cusps that, when the ventricles relax, fill with blood and close the base of the artery.

Figure 35.4 **External Anatomy of the Heart**

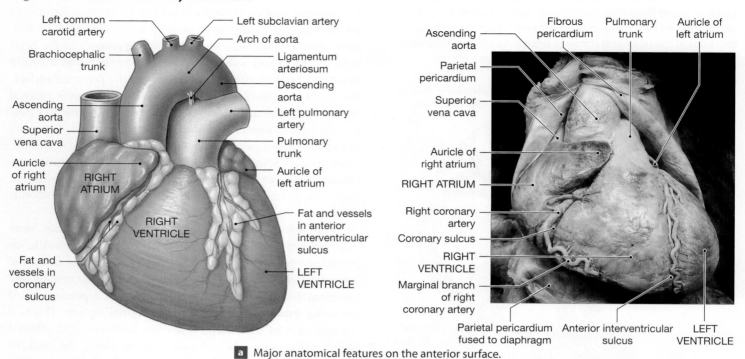

Left common carotid artery
Left subclavian artery
Brachiocephalic trunk
Arch of aorta
Ligamentum arteriosum
Ascending aorta
Descending aorta
Superior vena cava
Left pulmonary artery
Auricle of right atrium
Pulmonary trunk
RIGHT ATRIUM
Auricle of left atrium
RIGHT VENTRICLE
Fat and vessels in anterior interventricular sulcus
Fat and vessels in coronary sulcus
LEFT VENTRICLE

Ascending aorta
Fibrous pericardium
Pulmonary trunk
Auricle of left atrium
Parietal pericardium
Superior vena cava
Auricle of right atrium
RIGHT ATRIUM
Right coronary artery
Coronary sulcus
RIGHT VENTRICLE
Marginal branch of right coronary artery
Parietal pericardium fused to diaphragm
Anterior interventricular sulcus
LEFT VENTRICLE

a Major anatomical features on the anterior surface.

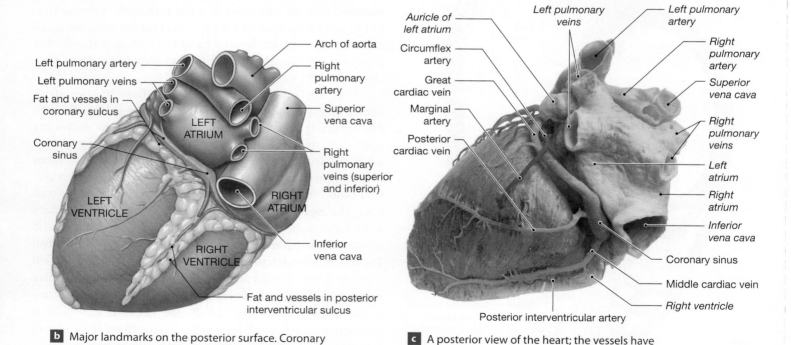

Left pulmonary artery
Left pulmonary veins
Fat and vessels in coronary sulcus
Arch of aorta
Right pulmonary artery
Superior vena cava
LEFT ATRIUM
Coronary sinus
Right pulmonary veins (superior and inferior)
LEFT VENTRICLE
RIGHT ATRIUM
RIGHT VENTRICLE
Inferior vena cava
Fat and vessels in posterior interventricular sulcus

Auricle of left atrium
Left pulmonary veins
Left pulmonary artery
Circumflex artery
Right pulmonary artery
Great cardiac vein
Superior vena cava
Marginal artery
Right pulmonary veins
Posterior cardiac vein
Left atrium
Right atrium
Inferior vena cava
Coronary sinus
Middle cardiac vein
Posterior interventricular artery
Right ventricle

b Major landmarks on the posterior surface. Coronary arteries (which supply the heart itself) are shown in red; coronary veins are shown in blue.

c A posterior view of the heart; the vessels have been injected with colored latex (liquid rubber).

Figure 35.5 Internal Anatomy of the Heart

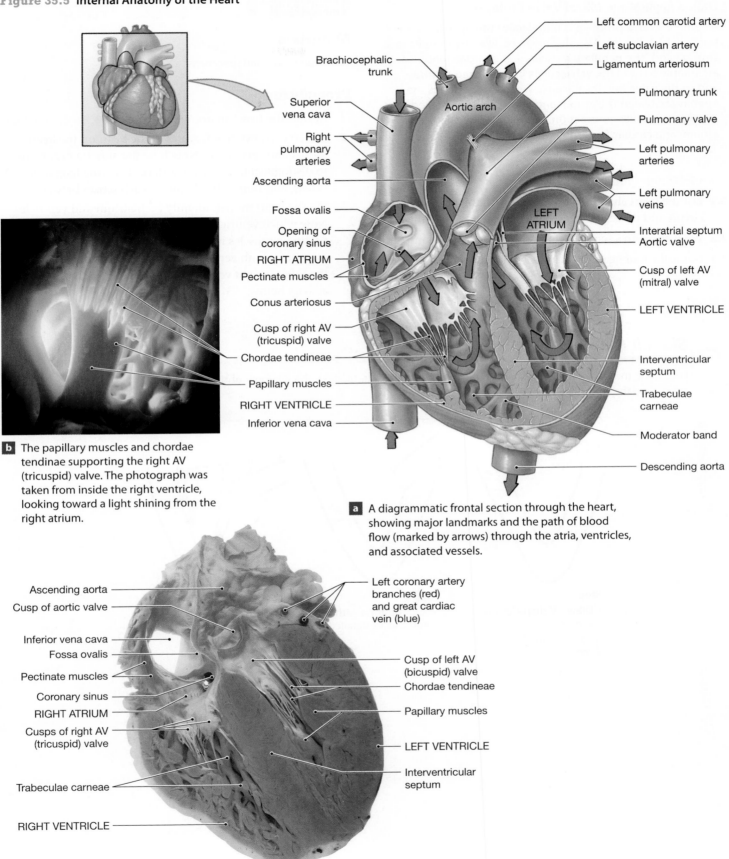

b   The papillary muscles and chordae tendinae supporting the right AV (tricuspid) valve. The photograph was taken from inside the right ventricle, looking toward a light shining from the right atrium.

Left common carotid artery
Left subclavian artery
Ligamentum arteriosum
Brachiocephalic trunk
Pulmonary trunk
Pulmonary valve
Superior vena cava
Aortic arch
Left pulmonary arteries
Right pulmonary arteries
Left pulmonary veins
Ascending aorta
LEFT ATRIUM
Interatrial septum
Aortic valve
Fossa ovalis
Opening of coronary sinus
RIGHT ATRIUM
Cusp of left AV (mitral) valve
Pectinate muscles
Conus arteriosus
LEFT VENTRICLE
Cusp of right AV (tricuspid) valve
Chordae tendineae
Interventricular septum
Papillary muscles
Trabeculae carneae
RIGHT VENTRICLE
Inferior vena cava
Moderator band
Descending aorta

a   A diagrammatic frontal section through the heart, showing major landmarks and the path of blood flow (marked by arrows) through the atria, ventricles, and associated vessels.

Ascending aorta
Cusp of aortic valve
Left coronary artery branches (red) and great cardiac vein (blue)
Inferior vena cava
Fossa ovalis
Pectinate muscles
Cusp of left AV (bicuspid) valve
Chordae tendineae
Coronary sinus
RIGHT ATRIUM
Papillary muscles
Cusps of right AV (tricuspid) valve
LEFT VENTRICLE
Interventricular septum
Trabeculae carneae
RIGHT VENTRICLE

c   A frontal section, anterior view.

Clinical Application Mitral Valve Prolapse

A common valve problem is **mitral valve prolapse,** a condition in which the left AV valve reverses, like an umbrella in a strong wind. The papillary muscles and chordae tendineae are unable to hold the valve cusps in the closed position, and so the valve inverts. Because when this happens the opening between the atrium and ventricle is not sealed shut during ventricular contraction, blood backflows into the left atrium, and cardiac function is diminished. ∎

QuickCheck Questions

2.1 List the heart chambers associated with the pulmonary circuit and those associated with the systemic circuit.

2.2 What structures separate the walls of the heart chambers?

2.3 Name the four heart valves and describe the function of each.

In the Lab 2

Materials

☐ Heart model and specimens

Procedures

1. Review the heart anatomy in Figures 35.3, 35.4, and 35.5.

2. Observe the external features of the heart on the heart model and specimens. Note how the auricles may be used to distinguish the anterior surface. Trace the length of each sulcus, and notice the chambers each passes between.

3. On the heart model, identify each atrium and ventricle. Note which ventricle has the thicker wall. Identify the pectinate muscles in the right atrium and the trabeculae carneae in both ventricles. Locate the moderator band in the inferior right ventricle. Complete the *Sketch to Learn* activity below.

 Sketch to Learn

Let's draw the left ventricle, isolated from the rest of the heart, to show the anatomy of the ventricle, valves, and the aorta.

Sample Sketch

Step 1
• Draw a V shape for the ventricle.
• Add an outer line to complete the heart wall.

Step 2
• Add the aorta with the aortic valve detailed.

Step 3
• Sketch in the bicuspid valve and associated anatomy.
• Label your sketch.

Your Sketch

4. Identify the two AV valves, their cusps, and the two semilunar valves.

5. Identify the major arteries and veins at the base of the heart.

6. Starting at the superior vena cava, trace a drop of blood though the heart model, and distinguish between the pulmonary and systemic circuits. ■

Lab Activity 3 Coronary Circulation

To produce the pressure required for blood to reach all through the cardiovascular system, the heart can never completely rest. The branch of the systemic circuit known as the **coronary circulation** supplies the myocardium with the oxygen necessary for muscle contraction (**Figure 35.6**). The right and left **coronary arteries** of the coronary circulation are the first vessels to branch off the base of the ascending aorta and penetrate the myocardium to the outer heart wall. As the right coronary artery (RCA) passes along the coronary sulcus, many **atrial arteries** supply blood to the right atrium and one or more **marginal arteries** arise to supply the right ventricle. The **posterior interventricular branch** off the RCA supplies adjacent posterior regions of the ventricles.

The left coronary artery (LCA) branches to supply blood to the left atrium, left ventricle, and interventricular septum.

The LCA divides into a **circumflex artery** and an **anterior interventricular artery.** The anterior interventricular branch supplies the left ventricle. The circumflex branch follows the left side of the heart, turns inferior, and passes along the left ventricle as the **marginal artery.** The posterior branches of the RCA and LCA often unite in the posterior coronary sulcus.

The **cardiac veins** of the coronary circulation collect deoxygenated blood from the myocardium (Figure 35.6). The **great cardiac vein** follows along the anterior interventricular sulcus and curves around the left side of the heart to drain the myocardium supplied by the anterior interventricular branch. The **posterior cardiac vein** drains the myocardium supplied by the LCA posterior ventricular branch. The **small cardiac vein** drains the superior right area of the heart. The **middle cardiac vein** drains the myocardium supplied by the posterior interventricular branch of the RCA. The cardiac veins merge as a large **coronary sinus** situated in the posterior region of the coronary sulcus. The coronary sinus empties deoxygenated blood from the myocardium into the right atrium. As noted previously, the right atrium also receives deoxygenated blood from the venae cavae.

QuickCheck Questions

3.1 Where do the right and left coronary arteries arise?

3.2 Where do the cardiac veins drain?

Figure 35.6 Coronary Circulation Coronary arteries and cardiac veins supply and drain the myocardium of blood. (These blood vessels are also shown in Figure 22.4b.)

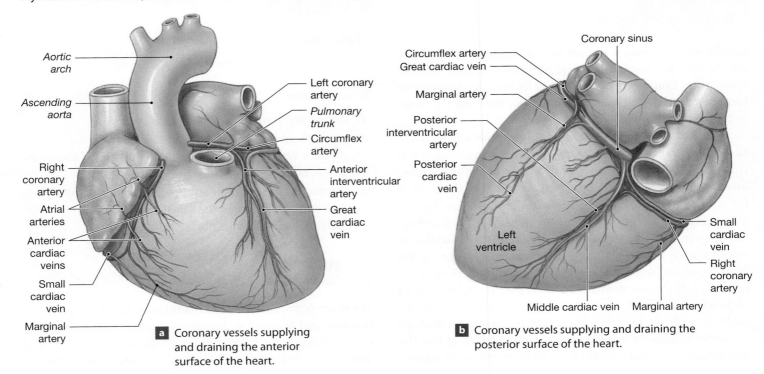

a Coronary vessels supplying and draining the anterior surface of the heart.

b Coronary vessels supplying and draining the posterior surface of the heart.

Clinical Application Anastomoses and Infarctions

The interventricular branches connect with one another, as do smaller arteries between the right coronary artery and the circumflex branch of the left coronary artery. These connections, called **anastomoses,** ensure that blood flow to the myocardium remains steady. In coronary artery disease, the arteries become narrower and narrower as fatty plaque is deposited on the interior walls of the vessels. As a result, blood flow is reduced. If enough plaque accumulates in critical areas, blood flow to that part of the heart becomes inadequate, and the heart muscle has an **infarction,** a heart attack. ■

In the Lab 3

Materials

☐ Heart model and specimens

Procedures

1. Review the blood vessels of the coronary circulation in Figure 35.6.
2. Follow the RCA and LCA on the heart model and identify their main branches.
3. Identify the cardiac veins and trace them into the coronary sinus. Identify where the coronary sinus drains. ■

Lab Activity 4   Conducting System of the Heart

Cardiac muscle tissue is unique in that it is *autorhythmic*, producing its own contraction and relaxation phases without stimulation from nerves. Nerves may increase or decrease the heart rate, but a living heart removed from the body continues to contract on its own.

Figure 35.7 details the **conducting system** of the heart. Special cells called **nodal cells** produce and conduct electrical currents to the myocardium, and it is these currents that coordinate the heart's contraction. The pacemaker of the heart is the **sinoatrial** (si-nō-Ā-trē-al) **node** (SA node), located where the superior vena cava empties into the upper right atrium. Nodal cells in the SA node self-excite faster than nodal cells in other areas of the heart and therefore set the pace for the heart's contraction. The **atrioventricular node** (AV node) is located on the lower medial floor of the right atrium. The SA node stimulates both the atria and the AV node, and the AV node then directs the impulse toward the ventricles through the **atrioventricular bundle,** also called the *bundle of His*. The atrioventricular bundle passes into the interventricular septum

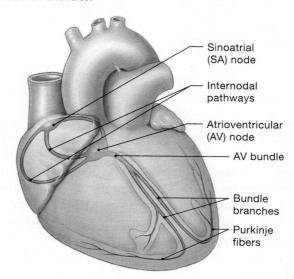

Figure 35.7 Conducting System of the Heart Components of the conducting system are specialized nodal cells that generate and distribute electrical signals, to coordinate the contraction of the atria and ventricles.

- Sinoatrial (SA) node
- Internodal pathways
- Atrioventricular (AV) node
- AV bundle
- Bundle branches
- Purkinje fibers

and branches into right and left **bundle branches.** The bundle branches divide into fine **Purkinje fibers,** which distribute the electrical impulses to the cardiocytes.

QuickCheck Questions

4.1 Where is the pacemaker of the heart located?

4.2 How is the AV node connected to the ventricles?

In the Lab 4

Materials

☐ Heart model and specimens

Procedures

1. Review the conducting system in Figure 35.7.
2. On the heart model, examine the sinus where the superior vena cava drains into the right atrium, and locate the SA node.
3. On the floor of the right atrium, locate the AV node. Trace the conducting path to the ventricles: AV bundle, bundle branches, and Purkinje fibers. ■

Lab Activity 5   Sheep Heart Dissection

The sheep heart, like all other mammalian hearts, is similar in structure and function to the human heart. One major difference is in where the great vessels join the heart. In four-legged animals, the inferior vena cava has a posterior connection to

⚠ Safety Alert: Dissecting the Heart

You *must* practice the highest level of laboratory safety while handling and dissecting the heart. Keep the following guidelines in mind during the dissection.

1. Wear gloves and safety glasses to protect yourself from the fixatives used to preserve the specimen.
2. Do not dispose of the fixative from your specimen. You will later store the specimen in the fixative to keep the specimen moist and to keep it from decaying.
3. Be extremely careful when using a scalpel or other sharp instrument. Always direct cutting and scissor motion away from you to prevent an accident if the instrument slips on moist tissue.
4. Before cutting a given tissue, make sure it is free from underlying and/or adjacent tissues so that they will not be accidentally severed.
5. Never discard tissue in the sink or trash. Your instructor will inform you of the proper disposal procedure. ▲

the heart instead of the inferior connection found in humans. Dissecting a sheep heart will enhance your studies of models and charts of the human heart. Take your time while dissecting and follow the directions carefully.

QuickCheck Questions

5.1 What type of safety equipment should you wear as you dissect the sheep heart?

5.2 How should you dispose of the sheep heart and scrap tissue?

In the Lab 5

Materials

☐ Gloves
☐ Safety glasses
☐ Dissecting tools
☐ Dissecting pan
☐ Fresh or preserved sheep heart

Procedures

1. Put on gloves and safety glasses, and clear your work space before obtaining your dissection specimen.
2. Wash the sheep heart with cold water to flush out preservatives and blood clots. Minimize your skin and mucous membrane exposure to the preservatives.
3. Carefully follow the instructions in this section. Cut into the heart only as instructed.

External Anatomy

1. Figure 35.8 details the external anatomy of the sheep heart. Examine the surface of the heart to see if the pericardium is present. (Often this serous membrane has been removed from preserved specimens.) Carefully scrape the outer heart muscle with a scalpel to loosen the epicardium.
2. Locate the anterior surface by orienting the heart so that the auricles face you. Under the auricles are the right and left atria. Note the base of the heart above the atria, where the large blood vessels occur. Squeeze gently just above the apex to locate the right and left ventricles. Locate the anterior interventricular sulcus, the fat-laden groove between the ventricles. Carefully remove some of the adipose tissue with the scalpel to uncover coronary blood vessels. Identify two grooves—the coronary sulcus between the right atrium and ventricle and the posterior interventricular sulcus between the ventricles on the posterior surface.
3. Identify the aorta and then the pulmonary trunk anterior to the aorta. If on your specimen the pulmonary trunk was cut long, you may be able to identify the right and left pulmonary arteries branching off the trunk. The brachiocephalic artery is the first major branch of the aorta and is often intact in preserved material.

Figure 35.8 **External Anatomy of the Sheep Heart**

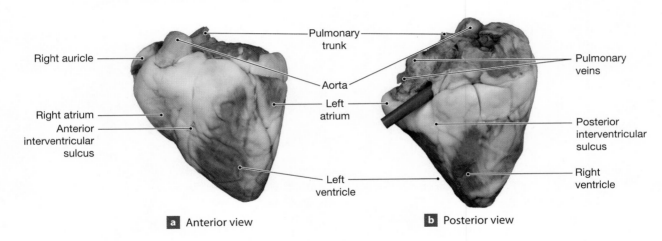

a Anterior view   b Posterior view

4. Follow along the inferior margin of the right auricle to the posterior surface. The prominent vessel at the termination of the auricle is the superior vena cava. At the base of this vessel is the inferior vena cava. Next, examine the posterior aspect of the left atrium and find the four pulmonary veins. You may need to carefully remove some of the adipose tissue around the superior region of the left atrium to locate these veins.

Internal Anatomy

1. Cut a frontal section passing through the aorta. Use Figure 35.9 as a reference to the internal anatomy.

2. Examine the two sides of the heart. Identify the right and left atria, right and left ventricles, and the interventricular septum. Compare the myocardium of the left ventricle with that of the right ventricle. Note the folds of trabeculae carneae along the inner ventricular walls. Examine the right atrium for the comblike pectinate muscles lining the inner wall.

3. Locate the tricuspid and bicuspid valves. Observe the papillary muscles with chordae tendineae attached.

4. Examine the wall of the left atrium for the openings of the four pulmonary veins.

5. At the entrance of the aorta, locate the small cusps of the aortic valve.

6. At the base of the pulmonary trunk, locate the pulmonary valve.

7. Locate the superior and inferior venae cavae, which drain into the right atrium.

8. Upon completion of the dissection, dispose of the sheep heart as directed by your instructor and wash your hands and dissecting instruments. ∎

Figure 35.9 Internal Anatomy of the Sheep Heart The major anatomical features of the sheep heart as shown in a frontal section.

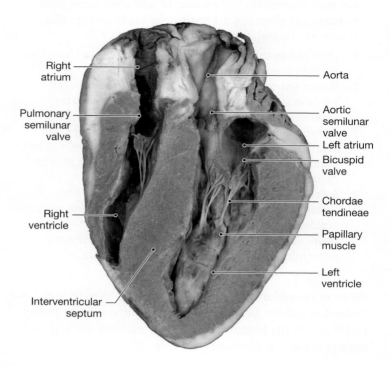

Name _____

Date _____

Section _____

Anatomy of the Heart

A. Descriptions

Write a description of each heart structure.

1. tricuspid valve

2. superior vena cava

3. interventricular septum

4. left ventricle

5. pulmonary veins

6. semilunar valve

7. bicuspid valve

8. pulmonary trunk

9. circumflex artery

10. trabeculae carneae

11. coronary sinus

12. epicardium

B. Short-Answer Questions

1. List the layers of the heart wall.

2. Describe how the AV valves function.

3. List the order in which an electrical impulse spreads through the conducting system.

C. Labeling

1. Label the anatomy of the heart in **Figure 35.10**.

Figure 35.10 Frontal Section Through the Heart

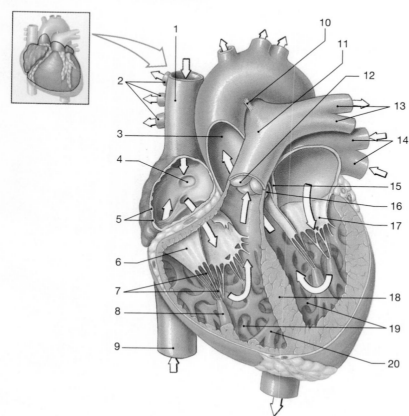

1. _____
2. _____
3. _____
4. _____
5. _____
6. _____
7. _____
8. _____
9. _____
10. _____
11. _____
12. _____
13. _____
14. _____
15. _____
16. _____
17. _____
18. _____
19. _____
20. _____

2. Label the major arteries and veins on the posterior of the heart in Figure 35.11.

Figure 35.11 **Posterior Surface of the Heart**

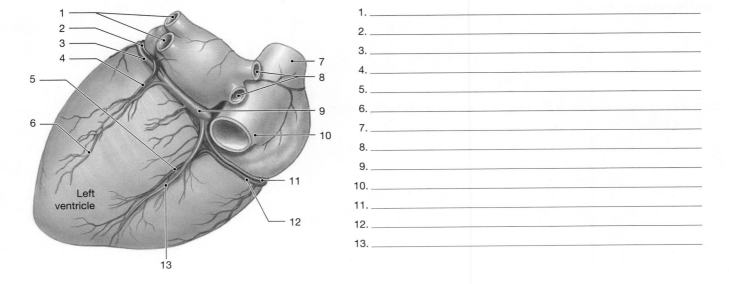

1. _____
2. _____
3. _____
4. _____
5. _____
6. _____
7. _____
8. _____
9. _____
10. _____
11. _____
12. _____
13. _____

D. Analysis and Application

1. Explain the difference between thickness of the myocardium in the right and left ventricles.

2. Does the pulmonary trunk transport oxygenated blood or deoxygenated blood? Why is it an artery rather than a vein?

3. Trace a drop of blood through the pulmonary and systemic circuits of the heart.

E. Clinical Challenge

1. Suppose a patient has a weakened bicuspid valve that does not close properly, a condition called mitral valve prolapse. How does this valve defect affect the flow of blood in the heart?

2. Coronary artery disease in the marginal arteries would affect which part of the myocardium?

Anatomy of the Systemic Circulation

Learning Outcomes

On completion of this exercise, you should be able to:

1. Compare the histology of an artery, a capillary, and a vein.

2. Describe the difference in the blood vessels serving the right and left arms.

3. Describe the anatomy and importance of the circle of Willis.

4. Trace a drop of blood from the ascending aorta into each abdominal organ and into the lower limbs.

5. Trace a drop of blood returning to the heart from the foot.

6. Discuss the unique features of the fetal circulation.

The body contains more than 60,000 miles of blood vessels to transport blood to the trillions of cells in the tissues. Arteries of the systemic circuit distribute oxygen and nutrient-rich blood to microscopic networks of thin-walled vessels called capillaries. At the capillaries, nutrients, gases, wastes, and cellular products diffuse either from blood to cells or from cells to blood. Veins drain deoxygenated blood from the systemic capillaries and direct it toward the heart, which then pumps it into the pulmonary circuit, to be carried to the lungs to pick up oxygen and release carbon dioxide.

In this exercise, you will study the major arteries and veins of the systemic circuit. (Refer to Exercise 35 and Figure 35.3 for a review of the pulmonary vessels.)

Need More Practice and Review?

Build your knowledge—and confidence!—in the Study Area of MasteringA&P® at www.masteringaandp.com with Pre-lab Quizzes, Post-lab Quizzes, Practice Anatomy Lab™ (PAL™) 3.0 virtual anatomy practice tool, PhysioEx™ 9.0 laboratory simulations, and A&P Flix™ with Quizzes.

PAL | practice anatomy lab   For this lab exercise, follow these navigation paths:

• PAL>Human Cadaver>Cardiovascular System>Blood Vessels
• PAL>Anatomical Models>Cardiovascular System>Veins
• PAL>Anatomical Models>Cardiovascular System>Arteries
• PAL>Histology> Cardiovascular System

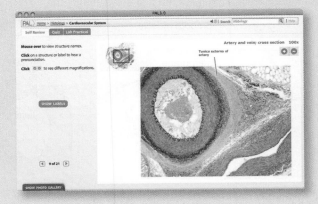

Lab Activity 1 Comparison of Arteries, Capillaries, and Veins

The walls of the body's blood vessels have three layers (Figure 36.1). The **tunica externa** is a layer of connective tissue that anchors the vessel to surrounding tissues. Collagen and elastic fibers give this layer strength and flexibility. The **tunica media** is a layer of smooth muscle tissue. In the tunica media of arteries are elastic fibers that allow the vessels to stretch and recoil in response to blood pressure changes. Veins have fewer elastic fibers; collagen fibers in the tunica media provide strength. Lining the inside of the vessels is the third layer, the **tunica intima,** a thin layer of simple squamous epithelium called **endothelium.** In arteries, the luminal surface of the endothelium has a thick, dark-staining **internal elastic membrane.**

Make a Prediction

Which vessels have greater pressure: arteries or veins, and which type of vessel has valves?

Because blood pressure is much higher in arteries than in veins and also because the pressure fluctuates more in arteries than in veins, the walls of arteries are thicker than those of veins. Notice how the artery cross section in the micrograph of Figure 36.1 is round and thick walled, whereas the adjacent vein is irregularly shaped and thin walled. In a slide preparation, the tunica intima of an artery may appear pleated because the vessel wall has recoiled due to a loss of pressure. In reality, the luminal surface is smooth and the vessel can expand and shrink to regulate blood flow.

A capillary consists of a single layer of endothelium that is continuous with the tunica intima of the artery and vein supplying and draining the capillary. Capillaries are so narrow that RBCs must line up in single file to squeeze through.

Veins have a thinner wall than arteries. The walls of a vein collapse if the vessel is emptied of blood. Blood pressure is low in veins; and to prevent backflow, the peripheral veins have valves that keep blood flowing in one direction, toward the heart.

QuickCheck Questions

1.1 Describe the three layers in the wall of an artery.

1.2 How do arterial walls differ from venous walls?

In the Lab 1

Materials

- ☐ Compound microscope
- ☐ Prepared slide of artery and vein
- ☐ Prepared slide of artery with plaque (atherosclerosis)

Procedures

1. Place the artery/vein slide on the microscope stage and locate the artery and vein at low magnification. Most slide preparations have one artery, an adjacent vein, and a nerve. The blood vessels are hollow and most likely have blood cells in the lumen. The nerve appears as a round, solid structure.

Figure 36.1 Comparison of the Structure of a Typical Artery and Vein Arteries have thicker walls and retain their shape compared to veins.

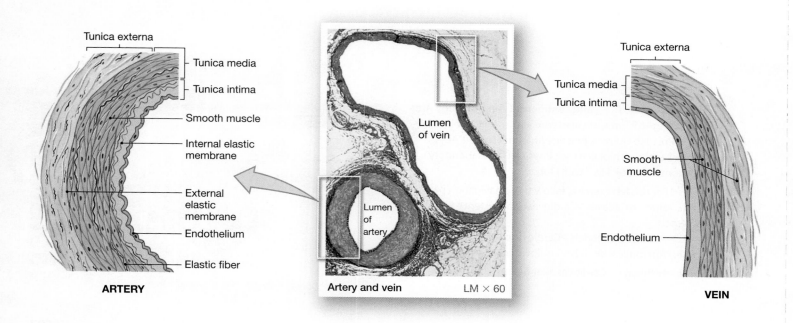

ARTERY   Artery and vein   LM × 60   VEIN

Clinical Application Arteriosclerosis

Arteriosclerosis (ar-tēr-ē-o-skler-Ō-sis) is the thickening and hardening of an artery. In cases of **focal calcification,** calcium salts gradually accumulate and damage smooth muscle tissue in the tunica media of the vessel wall. **Atherosclerosis** (ath-ĕr-o-skler-Ō-sis) is the buildup of lipid deposits in the tunica media (**Figure 36.2**). The deposits, called atherosclerotic **plaque,** eventually damage the vessel's endothelium and obstruct blood flow. Plaque accumulation in coronary arteries greatly increases the risk of heart attack and stroke. During balloon angioplasty, a catheter with an inflatable tip is used to push the plaque against the vessel wall to restore blood flow. A stent is frequently inserted into the narrowed region of the vessel to keep the vessel open. To reduce the risk of arteriosclerosis, lower the consumption of dietary fats such as saturated fats, trans fats, and cholesterol found in red meat, dairy cream, and egg yolks. Monitoring blood pressure and cholestrol levels and controlling weight are important for good vascular health. ■

Figure 36.2 A Plaque Within an Artery

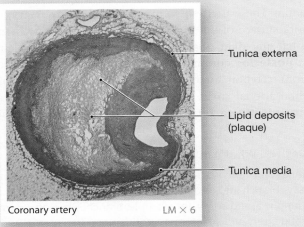

Tunica externa

Lipid deposits (plaque)

Tunica media

Coronary artery   LM × 6

a A cross-sectional view of a large plaque

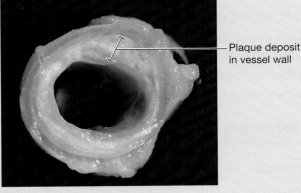

Plaque deposit in vessel wall

b A section of a coronary artery narrowed by plaque formation

2. Increase the magnification to high and compare each arterial layer with its venous counterpart.

3. ***Draw It!*** Draw and label a cross section of an artery and a vein in the space provided. Include enough detail in your drawings to show the anatomical differences between the vessels.

Artery cross section

Vein cross section

4. Examine the slide of an artery with atherosclerosis and note where the plaque has accumulated in the vessel. ■

Lab Activity 2 Arteries of the Head, Neck, and Upper Limb

Blood vessels are a continuous network of "pipes," and often there is little anatomical difference along the length of a given vessel as it passes from one region of the body to another. To facilitate identification and discussion, however, anatomists assign different names to a given vessel, depending on which part of the body the vessel is passing through. The subclavian artery becomes the axillary artery, for instance, and then the brachial artery. Each name is usually related to the name of a bone or organ adjacent to the vessel; therefore, because they often run parallel to each other, arteries and veins often have the same name.

The aorta receives oxygenated blood from the left ventricle of the heart and distributes the blood to the major arteries that arise from the aorta and supply the head, limbs, and trunk. The initial portion of the aorta is curved like an inverted letter *U*, and the various regions have different names. The **ascending aorta** exits the base of the heart, curves upward and to the left to form the **aortic arch,** and then as the **descending aorta** descends behind the heart (**Figure 36.3**). At the point where it passes through the diaphragm, the descending aorta becomes the **abdominal aorta.** Arteries that branch off the aortic arch serve the head, neck, and upper limb. Intercostal arteries stem from the thoracic aorta and supply the thoracic wall. Branches off the abdominal aorta serve the abdominal organs. The abdominal aorta enters the pelvic cavity and divides to send a branch into each lower limb.

Three Branches of the Aortic Arch

The first branch of the aortic arch, the **brachiocephalic** (brā-kē-ō-se-FAL-ik) **trunk,** or **innominate artery,** is short and divides into the **right common carotid artery** and the **right subclavian artery** (**Figure 36.4**). The right common carotid artery supplies blood to the right side of the head and neck; the right subclavian artery supplies blood to the right upper limb. The second and third branches of the aortic arch are the **left common carotid artery,** which supplies the left side of the head and neck, and **left subclavian artery,** which supplies the left upper limb as well as the shoulder and head. Note that only the right common carotid artery and right subclavian artery are derived from the brachiocephalic trunk. The left common carotid artery and left subclavian artery arise directly from the peak of the aortic arch. A **vertebral artery** branches off each subclavian artery and supplies blood to the brain and spinal cord.

Subclavian Arteries Supply the Upper Limb

The subclavian arteries supply blood to the upper limbs. Each subclavian artery passes under the clavicle, crosses the armpit as the **axillary artery,** and continues into the arm as the **brachial artery** (Figure 36.4). (Blood pressure is usually taken at the brachial artery.) At the antecubitis (elbow), the brachial artery divides into the lateral **radial artery** and the medial **ulnar artery,** each named after the bone it follows. In the palm of the hand, these arteries are interconnected by the **superficial** and **deep palmar arches,** which send small **digital arteries** to the fingers. Except in the vicinity of the heart, where the right arterial pathway has a brachiocephalic trunk that is absent from the left pathway, the arrangement of the arteries supplying the left and right upper limbs is symmetrical.

Carotid Arteries Supply the Head

Each common carotid artery ascends deep in the neck and divides at the larynx into an **external carotid artery** and an **internal carotid artery** (**Figure 36.5**). The base of the internal carotid swells as the **carotid sinus** and contains baroreceptors to monitor blood pressure. The external carotid artery branches to supply blood to the neck and face. The pulse in the external carotid artery can be felt by placing your fingers lateral to your thyroid cartilage (Adam's apple). The external carotid artery branches into the **facial artery, maxillary artery,** and **superficial temporal artery** to serve the external structures of the head. The internal carotid artery ascends to the base of the brain and divides into three arteries: the **ophthalmic artery,** which supplies the eyes, and the **anterior cerebral artery** and **middle cerebral artery,** both of which supply the brain.

Cerebral Arterial Circle

Because of its high metabolic rate, the brain has a voracious appetite for oxygen and nutrients. A reduction in blood flow to the brain may result in permanent damage to the affected area. To ensure that the brain receives a continuous supply of blood, branches of the internal carotid arteries and other arteries interconnect, or **anastomose,** as the **cerebral arterial circle,** also called the **circle of Willis** (Figure 36.5). The right and left vertebral arteries ascend in the transverse foramina of the cervical vertebrae and enter the skull at the foramen magnum. These arteries fuse into a single **basilar artery** on the inferior surface of the brain stem. The basilar artery branches into left and right **posterior cerebral arteries** and left and right **posterior communicating arteries.** The right and left anterior cerebral arteries form the anterior portion of the cerebral arterial circle. Between these arteries is the **anterior communicating artery,** which completes the anastomosis.

Study Tip   What's in a Name?

Arteries and veins with the term *common* as part of their name always branch into an external and an internal vessel. The common carotid artery, for example, branches into an external carotid artery and an internal carotid artery. The internal and external iliac veins join as the common iliac vein. ■

QuickCheck Questions

2.1 How does arterial branching in the left side of the neck differ from branching in the right side?

2.2 What is an anastomosis?

2.3 Which arteries in the brain anastomose with one another?

In the Lab   2

Materials

☐ Vascular system chart

☐ Torso model

☐ Head model

☐ Upper limb model

Figure 36.3 **Overview of the Major Systemic Arteries**

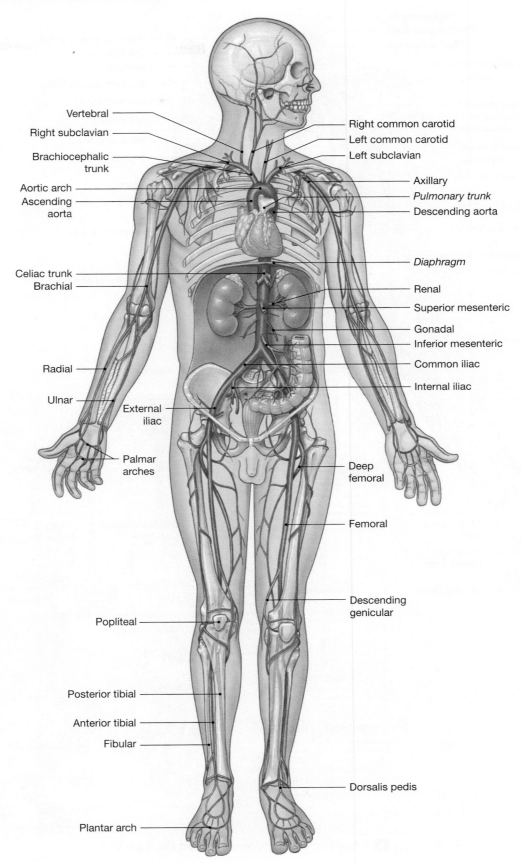

Figure 36.4 Arteries of the Chest and Upper Limb

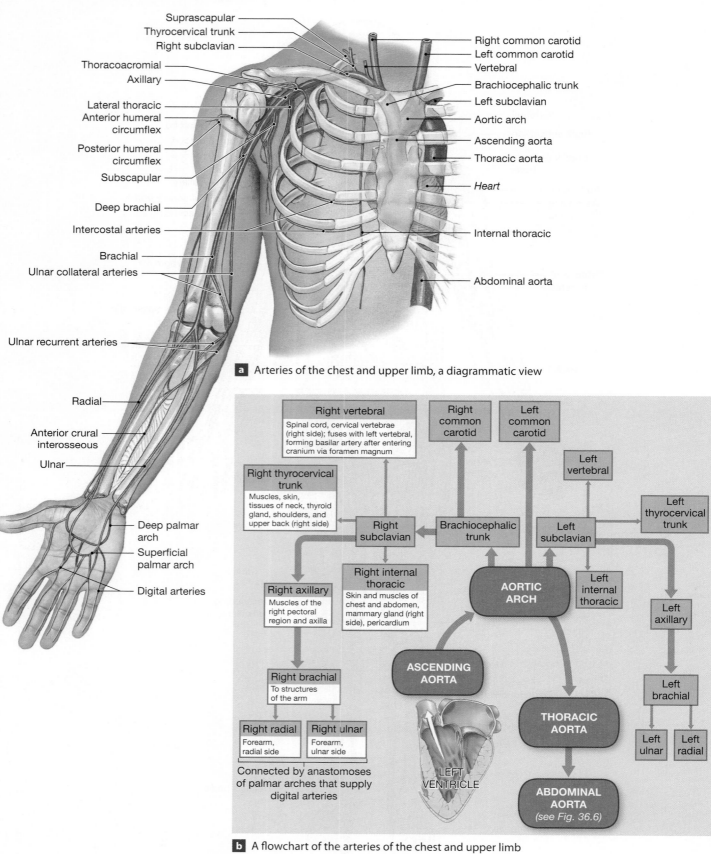

a Arteries of the chest and upper limb, a diagrammatic view

b A flowchart of the arteries of the chest and upper limb

Figure 36.5   Arteries of the Neck, Head, and Brain

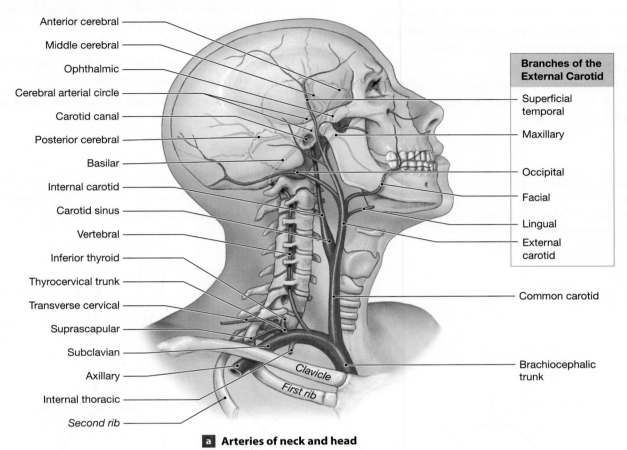

Anterior cerebral
Middle cerebral
Ophthalmic
Cerebral arterial circle
Carotid canal
Posterior cerebral
Basilar
Internal carotid
Carotid sinus
Vertebral
Inferior thyroid
Thyrocervical trunk
Transverse cervical
Suprascapular
Subclavian
Axillary
Internal thoracic
Second rib

Branches of the External Carotid

Superficial temporal
Maxillary
Occipital
Facial
Lingual
External carotid

Common carotid

Brachiocephalic trunk

Clavicle
First rib

a **Arteries of neck and head**

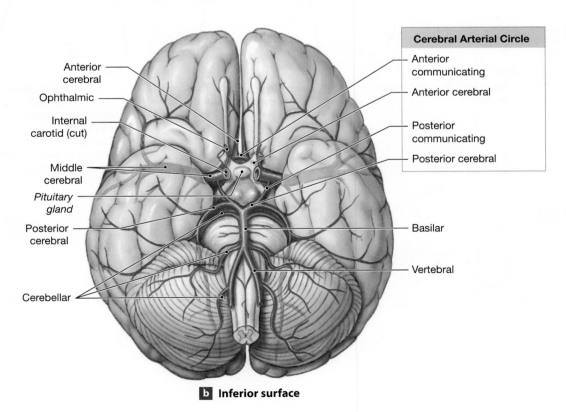

Anterior cerebral
Ophthalmic
Internal carotid (cut)
Middle cerebral
Pituitary gland
Posterior cerebral
Cerebellar

Cerebral Arterial Circle

Anterior communicating
Anterior cerebral
Posterior communicating
Posterior cerebral
Basilar
Vertebral

b **Inferior surface**

Procedures

1. Review the arteries in Figures 36.3, 36.4, and 36.5.

2. On the torso model, examine the aortic arch and identify the three branches arising from the superior margin of the arch.

3. On the torso model, identify the arteries of the shoulder and limb. Note the difference in origin of the right and left subclavian arteries.

4. On the head model, trace the arteries to the head and note the differences between the right and left common carotid arteries. Identify the arteries that converge at the cerebral arterial circle.

5. Using your index and middle fingers, locate the pulse in your radial and common carotid arteries.

For additional practice, complete the *Sketch to Learn* activity. ■

 Sketch to Learn!

An excellent way to remember blood vessels is to make simple "stick vessels." In this drawing we will show the arterial distribution branching from the aortic arch.

Sample Sketch

 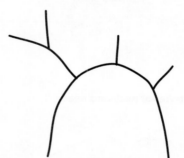 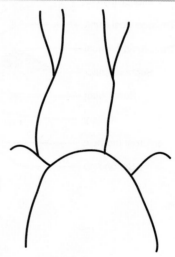

Step 1
- Draw an arch with 3 lines on top.

Step 2
- Branch the first line to show the right common carotid and subclavian arteries.

Step 3
- Sketch the left carotid and show branches of internal and external carotids.
- Extend subclavian arteries toward the "arms."
- Label your sketch.

Your Sketch

Lab Activity 3 Arteries of the Abdominopelvic Cavity and Lower Limb

The arteries stemming from the abdominal aorta are shown in Figure 36.3, as well as in Figures 36.6 and 36.7. An easy way to identify the branches of the abdominal aorta is to distinguish between paired arteries, which have right and left branches, and unpaired arteries. Also refer to the flowchart of arteries in Figure 36.6 for patterns and sequences of arteries as they arise from the abdominal aorta.

Celiac Trunk Has Three Branches

Three unpaired arteries arise from the abdominal aorta: celiac trunk, superior mesenteric artery, and inferior mesenteric artery. The short **celiac** (SĒ-lē-ak) **trunk** arises inferior to the diaphragm and splits into three arteries. The **common hepatic artery** divides to supply blood to the liver, gallbladder, and

part of the stomach. The **left gastric artery** supplies the stomach. The **splenic artery** supplies the spleen, stomach, and pancreas.

Mesenteric Arteries Supply the Intestines

Inferior to the celiac trunk is the next unpaired artery, the **superior mesenteric** (mez-en-TER-ik) **artery.** This vessel supplies blood to the large intestine, parts of the small intestine, and other abdominal organs. The third unpaired artery, the **inferior mesenteric artery,** originates before the abdominal aorta divides to enter the pelvic cavity and lower limbs. This artery supplies parts of the large intestine and the rectum.

Four major sets of paired arteries arise off the abdominal aorta. The right and left **adrenal arteries** arise near the superior mesenteric artery and branch into the adrenal glands, located on top of the kidneys. The right and left **renal arteries,** which supply the kidneys, stem off the abdominal aorta just inferior to the adrenal arteries. The right and left

Figure 36.6 **Arteries Supplying the Abdominopelvic Organs**

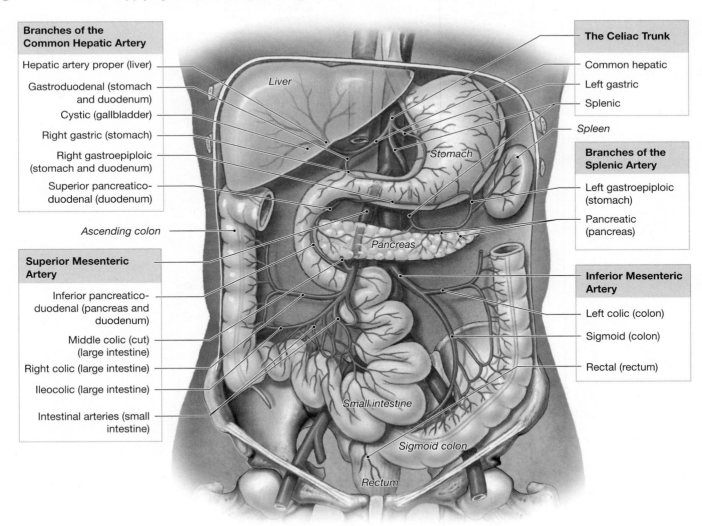

Branches of the Common Hepatic Artery
- Hepatic artery proper (liver)
- Gastroduodenal (stomach and duodenum)
- Cystic (gallbladder)
- Right gastric (stomach)
- Right gastroepiploic (stomach and duodenum)
- Superior pancreatico-duodenal (duodenum)

Ascending colon

Superior Mesenteric Artery
- Inferior pancreatico-duodenal (pancreas and duodenum)
- Middle colic (cut) (large intestine)
- Right colic (large intestine)
- Ileocolic (large intestine)
- Intestinal arteries (small intestine)

The Celiac Trunk
- Common hepatic
- Left gastric
- Splenic

Spleen

Branches of the Splenic Artery
- Left gastroepiploic (stomach)
- Pancreatic (pancreas)

Inferior Mesenteric Artery
- Left colic (colon)
- Sigmoid (colon)
- Rectal (rectum)

Liver

Stomach

Pancreas

Small intestine

Sigmoid colon

Rectum

Figure 36.7 Flowchart of the Major Arteries of the Trunk

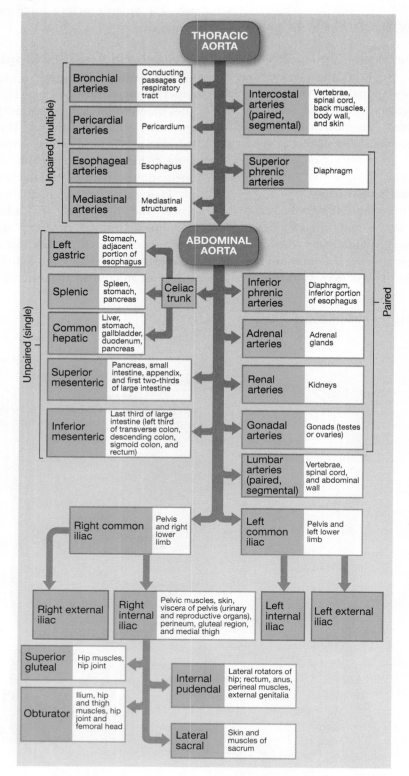

gonadal (gō-NAD-al) **arteries** arise near the inferior mesenteric artery and bring blood to the reproductive organs. The right and left **lumbar arteries** originate near the terminus of the abdominal aorta and service the lower body wall.

Iliac Arteries Branch to Supply the Pelvis and Lower Limb

At the level of the hips, the abdominal aorta divides into the right and left **common iliac** (IL-ē-ak) **arteries.** Each common iliac artery descends through the pelvic cavity and branches into an **external iliac artery,** which enters the lower limb, and an **internal iliac artery,** which supplies blood to the organs of the pelvic cavity. The external iliac artery pierces the abdominal wall and becomes the **femoral artery** of the thigh (Figure 36.8, p. 518). A **deep femoral artery** arising off the femoral artery supplies deep thigh muscles. The femoral artery passes through the posterior knee as the **popliteal** (pop-LIT-ē-al) **artery** and divides into the **posterior tibial artery** and the **anterior tibial artery,** each supplying blood to the leg. The **fibular artery,** also called the *peroneal artery,* stems laterally off the posterior tibial artery. The arteries of the leg branch into the foot and anastomose at the **dorsal arch** and the **plantar arch.**

QuickCheck Questions

3.1 What are the three branches of the celiac trunk?

3.2 Which arteries supply the intestines?

3.3 What does the external iliac artery become in the lower limb?

In the Lab 3

Materials

☐ Vascular system chart
☐ Torso model
☐ Lower limb model

Procedures

1. Review the arteries in Figures 36.5 through 36.8.
2. On the torso model, locate the celiac trunk and its three branches. Identify the superior and inferior mesenteric arteries and the four sets of paired arteries stemming from the abdominal aorta.
3. On the torso model, observe how the abdominal aorta branches into the left and right common iliac arteries.
4. On the lower limb model, locate the major arteries supplying the lower limb.
5. On your body, trace the location of your abdominal aorta, common iliac artery, external iliac artery, femoral artery, popliteal artery, and posterior tibial artery. ■

Lab Activity 4 Veins of the Head, Neck, and Upper Limb

Once you have learned the major systemic arteries, identifying the systemic veins is easy because most arteries have a corresponding vein (Figure 36.9). Unlike arteries, many veins are superficial and easily seen under the skin. Systemic veins are usually painted blue on vascular and torso models to indicate that they transport deoxygenated blood. When identifying veins, work in the direction of blood flow, from the periphery toward the heart.

Jugular Veins Drain the Head

Blood in the brain drains into large veins called *sinuses* (Figure 36.10). (Do not confuse this meaning of *sinus* with the more familiar meaning "cavity," as, for instance, the sinuses of the skull treated in Exercise 14.) Small-diameter veins deep inside the brain drain into progressively larger veins that empty into the **superior sagittal sinus** located in the falx cerebri separating the cerebral hemispheres. This large sinus drains into a **transverse sinus** on each side of the brain that, in turn, empties into a **sigmoid sinus.** The sigmoid sinus drains into the **internal jugular vein,** which exits the skull via the jugular foramen, descends the neck, and empties into the **brachiocephalic vein.** Superficial veins that drain the face and scalp empty into the **external jugular vein,** which descends the neck to join the **subclavian vein.** The **internal thoracic vein** joins the left brachiocephalic vein and drains the anterior thoracic wall. The right and left brachiocephalic veins merge at the **superior vena cava** and empty deoxygenated blood into the right atrium of the heart. The blood then enters the right ventricle, which contracts and pumps the blood to the lungs through the pulmonary circuit.

Study Tip Brachiocephalic Veins

One difference between the systemic arteries and the systemic veins is that the venous pathway has both a right and a left brachiocephalic vein, each formed by the merging of subclavian, vertebral, internal jugular, and external jugular veins. The arterial pathway has a single brachiocephalic trunk that branches into the right common carotid artery and right subclavian artery. On the left side of the body, the common carotid artery and subclavian artery originate directly off the aortic arch, as noted earlier. ■

Veins That Drain the Upper Limb

Figure 36.11 illustrates the venous drainage of the upper limb, chest, and abdomen. Figure 36.12 shows flowcharts of the venous circulation for the superior and inferior venae cavae. Small veins in the fingers drain into **digital veins** that

Figure 36.8 Major Arteries of the Lower Limb

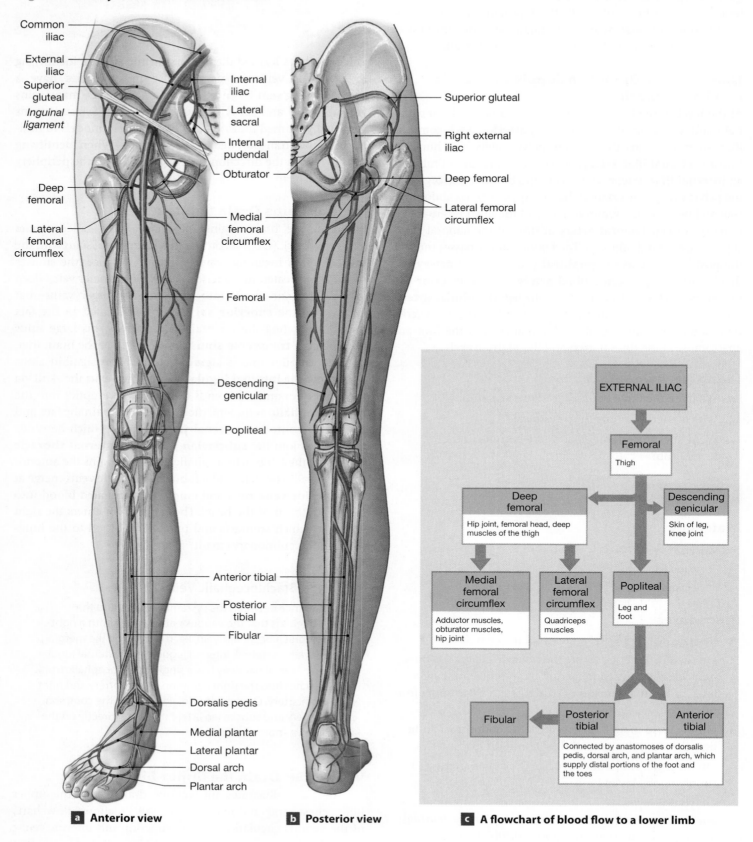

Common iliac
External iliac
Superior gluteal
Inguinal ligament
Deep femoral
Lateral femoral circumflex

Internal iliac
Lateral sacral
Internal pudendal
Obturator
Medial femoral circumflex

Femoral

Descending genicular

Popliteal

Anterior tibial
Posterior tibial
Fibular

Dorsalis pedis
Medial plantar
Lateral plantar
Dorsal arch
Plantar arch

a Anterior view

Superior gluteal
Right external iliac
Deep femoral
Lateral femoral circumflex

b Posterior view

EXTERNAL ILIAC

Femoral
Thigh

Deep femoral
Hip joint, femoral head, deep muscles of the thigh

Descending genicular
Skin of leg, knee joint

Medial femoral circumflex
Adductor muscles, obturator muscles, hip joint

Lateral femoral circumflex
Quadriceps muscles

Popliteal
Leg and foot

Fibular

Posterior tibial

Anterior tibial

Connected by anastomoses of dorsalis pedis, dorsal arch, and plantar arch, which supply distal portions of the foot and the toes

c A flowchart of blood flow to a lower limb

Figure 36.9 **Overview of the Major Systemic Veins**

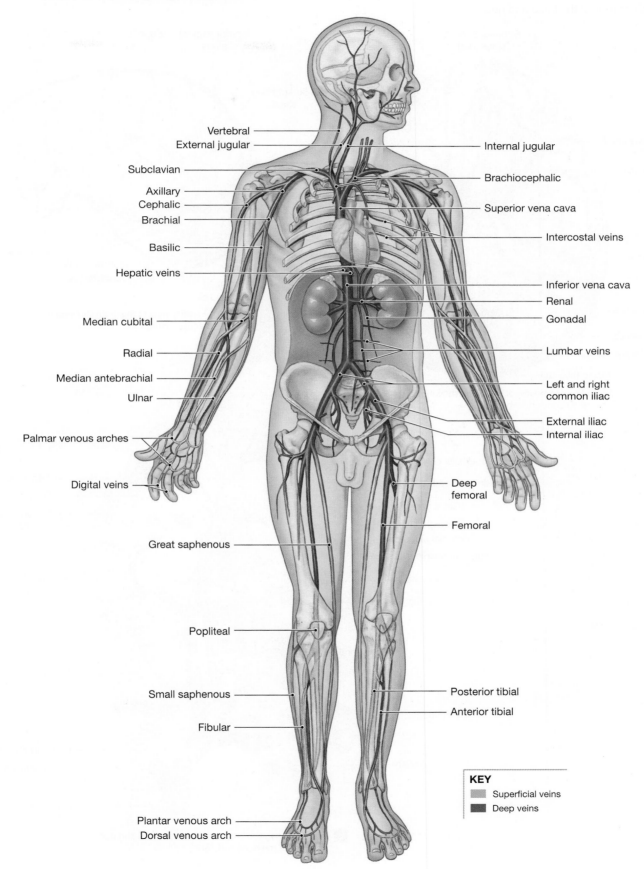

Vertebral
External jugular
Internal jugular
Subclavian
Brachiocephalic
Axillary
Cephalic
Superior vena cava
Brachial
Intercostal veins
Basilic
Hepatic veins
Inferior vena cava
Renal
Gonadal
Median cubital
Lumbar veins
Radial
Median antebrachial
Ulnar
Left and right common iliac
External iliac
Internal iliac
Palmar venous arches
Digital veins
Deep femoral
Femoral
Great saphenous
Popliteal
Small saphenous
Posterior tibial
Anterior tibial
Fibular
Plantar venous arch
Dorsal venous arch

KEY
Superficial veins
Deep veins

Figure 36.10 **Major Veins of the Head, Neck, and Brain** Veins draining the brain and the superficial and deep portions of the head and neck.

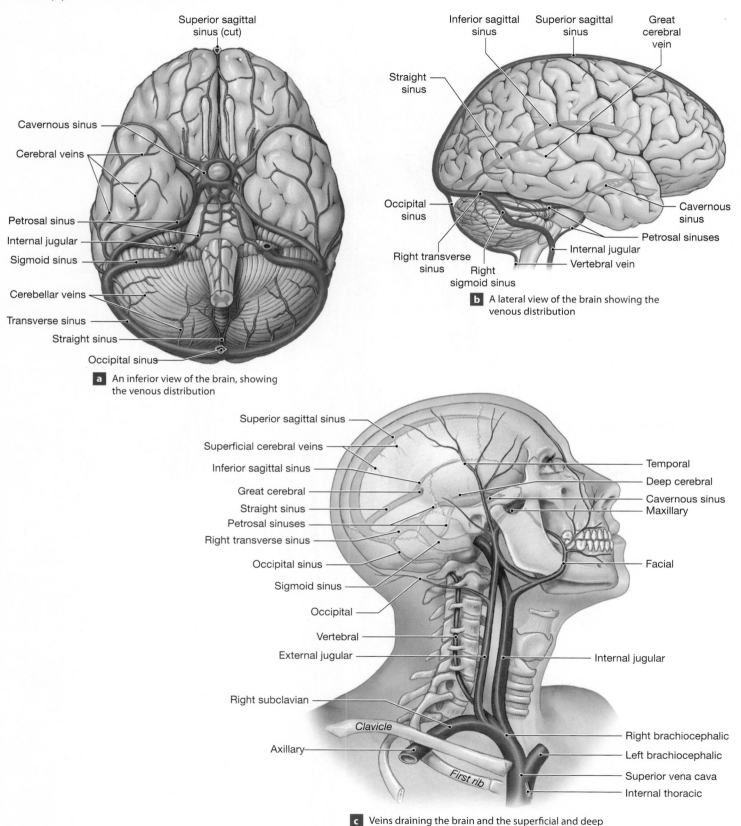

a An inferior view of the brain, showing the venous distribution

b A lateral view of the brain showing the venous distribution

c Veins draining the brain and the superficial and deep portions of the head and neck

Figure 36.11 Veins of the Upper Limb, Chest, and Abdomen The head, neck, and upper limb drain into the superior vena cava; the abdominopelvic organs and lower limb drain into the inferior vena cava.

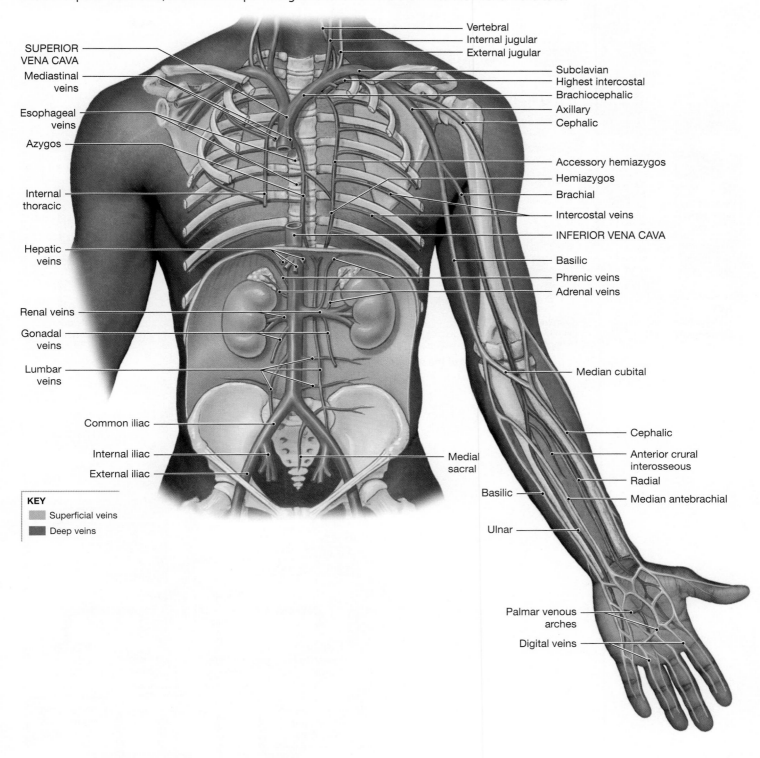

SUPERIOR VENA CAVA
Mediastinal veins
Esophageal veins
Azygos
Internal thoracic
Hepatic veins
Renal veins
Gonadal veins
Lumbar veins
Common iliac
Internal iliac
External iliac

Vertebral
Internal jugular
External jugular
Subclavian
Highest intercostal
Brachiocephalic
Axillary
Cephalic
Accessory hemiazygos
Hemiazygos
Brachial
Intercostal veins
INFERIOR VENA CAVA
Basilic
Phrenic veins
Adrenal veins
Median cubital
Cephalic
Anterior crural interosseous
Radial
Median antebrachial
Ulnar

Medial sacral
Basilic
Palmar venous arches
Digital veins

KEY
Superficial veins
Deep veins

Figure 36.12 Flowchart of Circulation to the Two Venae Cavae

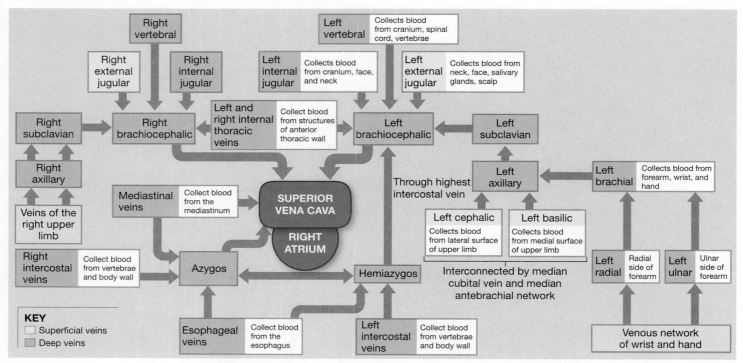

a Tributaries of the superior vena cava

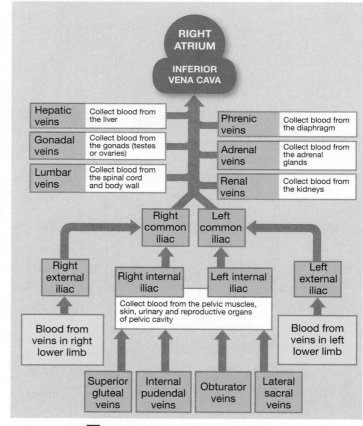

b Tributaries of the inferior vena cava

empty into a network of **palmar venous arches.** These vessels drain into the **cephalic vein,** which ascends along the lateral margin of the arm. The **median antebrachial vein** ascends to the elbow, is joined by the **median cubital vein** that crosses over from the cephalic vein, and becomes the **basilic vein.** The median cubital vein is often used to collect blood from an individual. Also in the forearm are the **radial** and **ulnar veins,** which fuse above the elbow into the **brachial vein.** The brachial and basilic veins meet at the armpit as the **axillary vein,** which joins the cephalic vein and becomes the subclavian vein. The subclavian vein plus veins from the neck and head drain into the brachiocephalic vein, which then empties into the superior vena cava, which empties into the right atrium. See Figure 36.12a for flowcharts of the venous circulation to the superior vena cava.

QuickCheck Questions

4.1 Which two veins combine to form the superior vena cava?

4.2 Where is the cephalic vein?

4.3 Where is the superior sagittal sinus?

In the Lab   4

Materials

- ☐ Vascular system chart
- ☐ Torso model
- ☐ Head model
- ☐ Upper limb model

Procedures

1. Review the head, neck, and upper limb veins in Figures 36.9 to 36.12.

2. On the head model, identify the superior sagittal sinus and other veins draining the head into the external and internal jugular veins.

3. Using the torso and upper limb models, start at one hand and name the veins draining the limb and shoulder. Notice how the right and left brachiocephalic veins join as the superior vena cava.

4. On your body, trace your cephalic vein, subclavian vein, brachiocephalic vein, and superior vena cava. ■

Lab Activity 5   Veins of the Abdominopelvic Cavity and Lower Limb

Veins that drain the lower limbs and abdominal organs empty into the **inferior vena cava,** the large vein that pierces the diaphragm and delivers deoxygenated blood to the right atrium

of the heart. Veins of the abdomen and lower limb are illustrated in Figures 36.13 and 36.14, as well as in Figures 36.9, 36.11, and 36.12.

Veins of the Abdomen

Six major veins from the abdominal organs drain blood into the inferior vena cava (Figure 36.11). The **lumbar veins** drain the muscles of the lower body wall and the spinal cord and empty into the inferior vena cava close to the common iliac veins. A pair of **gonadal veins** empty blood from the reproductive organs into the inferior vena cava above the lumbar veins. Pairs of **renal** and **adrenal veins** drain into the inferior vena cava next to their respective organs. Before entering the thoracic cavity to drain blood into the right atrium, the inferior vena cava collects blood from the **hepatic veins** draining the liver and the **phrenic veins** from the diaphragm. Figure 36.12b is a flowchart of the venous drainage into the inferior vena cava.

Hepatic Portal Vein

Veins leaving the digestive tract are diverted to the liver before continuing on to the heart. The **inferior** and **superior mesenteric veins** drain nutrient-rich blood from the digestive tract. These veins empty into the **hepatic portal vein** (Figure 36.13), which passes the blood through the liver, where blood sugar concentration is regulated. Phagocytic cells in the liver cleanse the blood of any microbes that may have entered it through the mucous membrane of the digestive system. Blood from the hepatic arteries and hepatic portal vein mixes in the liver and is returned to the inferior vena cava by the hepatic veins.

Veins of the Lower Limb

Figure 36.14 illustrates the venous drainage of the lower limb. Just like the hand, the foot contains digital veins, which in the foot drain into the **plantar venous arch** and the **dorsal venous arch,** which drain into the lateral **fibular vein** (also called *peroneal vein*) and the **anterior tibial vein,** located on the medial aspect of the anterior leg. These veins, along with the **posterior tibial vein,** merge and become the **popliteal vein** of the posterior knee. The **small saphenous** (sa-FĒ-nus) **vein,** which ascends from the ankle to the knee and drains blood from superficial veins, also empties into the popliteal vein. Superior to the knee, the popliteal vein becomes the **femoral vein,** which ascends along the femur to the inferior pelvic girdle, where it joins the **deep femoral vein** at the **external iliac vein.** The **great saphenous vein** ascends from the medial side of the ankle to the superior thigh and drains into the external iliac vein. In the pelvic cavity, the external iliac vein and the **internal iliac vein** fuse to form the **common iliac vein.** The right and left common iliac veins merge and drain into the inferior vena cava.

Figure 36.13 The Hepatic Portal System The hepatic portal vein receives blood from the superior and inferior mesenteric veins and passes the nutrient-rich blood into the liver for breakdown of toxins, removal of microbes, and regulation of blood sugar.

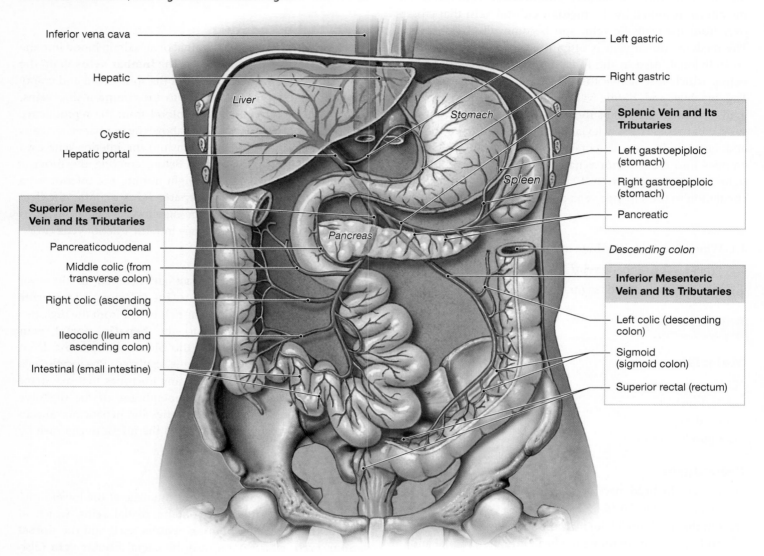

QuickCheck Questions

5.1 List the vessels that drain blood from the lower limb into the inferior vena cava.

5.2 Which veins drain into the hepatic portal vein?

In the Lab 5

Materials

☐ Vascular system chart
☐ Torso model
☐ Lower limb model

Procedures

1. Review the veins in Figures 36.11 through 36.14.

2. On the torso model, identify the veins draining the major abdominal organs. Locate where the superior and inferior mesenteric veins drain into the hepatic portal vein.

3. On the lower limb model, identify the veins that drain blood from the ankle to the inferior vena cava.

4. On your body, trace the location of the veins in your lower limb.

5. Although you have studied the arterial and venous divisions separately, they are anatomically connected to each other by capillaries. To reinforce this connectedness,

Figure 36.14 Veins of the Lower Limb

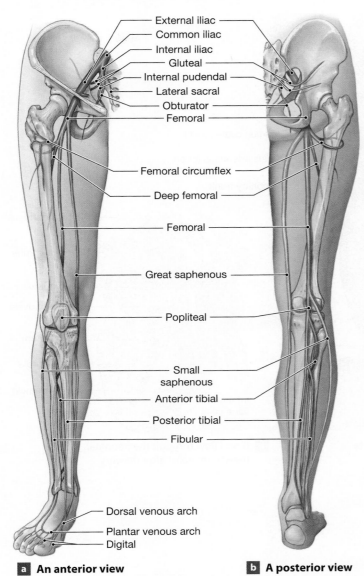

- External iliac
- Common iliac
- Internal iliac
- Gluteal
- Internal pudendal
- Lateral sacral
- Obturator
- Femoral
- Femoral circumflex
- Deep femoral
- Femoral
- Great saphenous
- Popliteal
- Small saphenous
- Anterior tibial
- Posterior tibial
- Fibular
- Dorsal venous arch
- Plantar venous arch
- Digital

a An anterior view **b A posterior view**

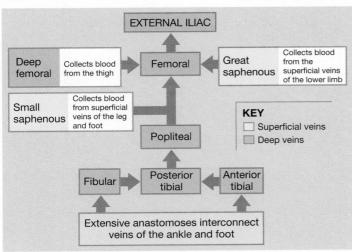

EXTERNAL ILIAC

| Deep femoral | Collects blood from the thigh | → | Femoral | ← | Great saphenous | Collects blood from the superficial veins of the lower limb |

Small saphenous — Collects blood from superficial veins of the leg and foot

KEY
☐ Superficial veins
▨ Deep veins

Popliteal

Fibular → Posterior tibial ← Anterior tibial

Extensive anastomoses interconnect veins of the ankle and foot

c A flowchart of venous circulation from a lower limb

practice identifying blood vessels while tracing the following systemic routes.

  a. From the heart through the left upper limb and back to the heart

  b. From the heart through the brain and back to the heart

  c. From the heart through the liver and back to the heart

  d. From the heart through the right lower limb and back to the heart ■

Lab Activity 6   Fetal Circulation

A fetus receives oxygen from the mother through the **placenta,** a vascular organ that connects the fetus to the wall of the mother's uterus. During development, the fetal lungs are filled with amniotic fluid, and for efficiency some of the blood is shunted away from the fetal pulmonary circuit by several structures (**Figure 36.15**). The **foramen ovale** is a hole in the interatrial wall. Much of the blood entering the right atrium from the inferior vena cava passes through the foramen ovale to the left atrium and avoids the right ventricle and the pulmonary circuit. Some of the blood that enters the pulmonary trunk may bypass the lungs through a connection with the aorta, the **ductus arteriosus.** At birth, the foramen ovale closes and becomes a depression on the interatrial wall, the **fossa ovalis.** The ductus arteriosus closes and becomes the **ligamentum arteriosum.**

QuickCheck Questions

6.1 Where is the foramen ovale?

6.2 Which two structures are connected by the ductus arteriosus?

In the Lab   6

Material
☐ Heart model

Procedures

1. Review the fetal structures in Figure 36.15.

2. Identify the fossa ovalis on the heart model. What was this structure in the fetus?

3. At the point where the pulmonary trunk branches, find the ligamentum arteriosum. What was this structure in the fetus, and what purpose did it serve?

4. On the heart model, trace a drop of blood through the fetal structures of the heart, starting at the right atrium. ■

Figure 36.15 Fetal Circulation

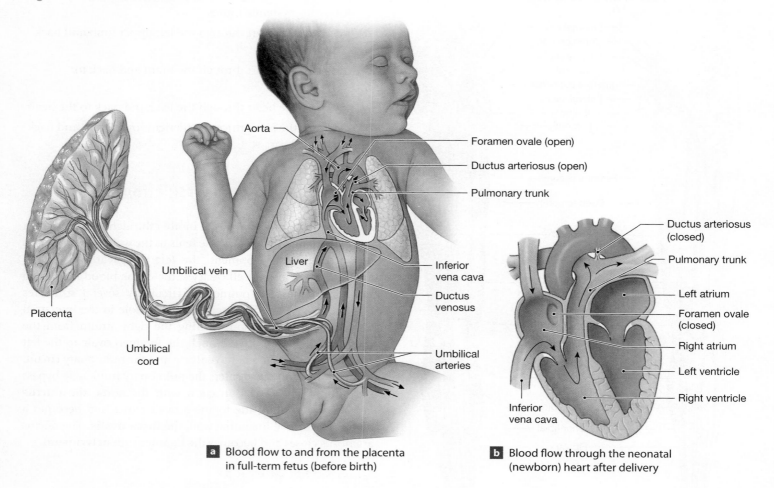

a Blood flow to and from the placenta in full-term fetus (before birth)

b Blood flow through the neonatal (newborn) heart after delivery

Review & Practice Sheet

EXERCISE 36

Anatomy of the Systemic Circulation

Name ______________________________

Date ______________________________

Section ______________________________

A. Matching

Match each term in the left column with its correct description from the right column.

________	**1.** artery in armpit	**A.**	subclavian
________	**2.** artery having three branches	**B.**	superior mesenteric
________	**3.** vein used for taking blood samples	**C.**	popliteal
________	**4.** artery on right side only	**D.**	cephalic
________	**5.** long vein of leg	**E.**	carotid
________	**6.** carries deoxygenated blood to liver	**F.**	gonadal
________	**7.** artery to large intestine	**G.**	valves
________	**8.** cerebral anastomosis	**H.**	axillary
________	**9.** long vein of arm	**I.**	great saphenous
________	**10.** vein in knee	**J.**	median cubital
________	**11.** artery to reproductive organ	**K.**	hepatic portal vein
________	**12.** major artery in neck	**L.**	circle of Willis
________	**13.** vein under clavicle	**M.**	celiac
________	**14.** found only in veins	**N.**	brachiocephalic artery

B. Drawing

Draw It! Draw and label a simple line sketch of the arteries of the right upper limb starting at the aorta.

C. Labeling

1. Label the arteries in **Figure 36.16**.

Figure 36.16 **Overview of the Major Systemic Arteries**

1. _____
2. _____
3. _____
4. _____
5. _____
6. _____
7. _____
8. _____
9. _____
10. _____
11. _____
12. _____
13. _____
14. _____
15. _____
16. _____
17. _____
18. _____
19. _____
20. _____
21. _____
22. _____
23. _____
24. _____
25. _____
26. _____

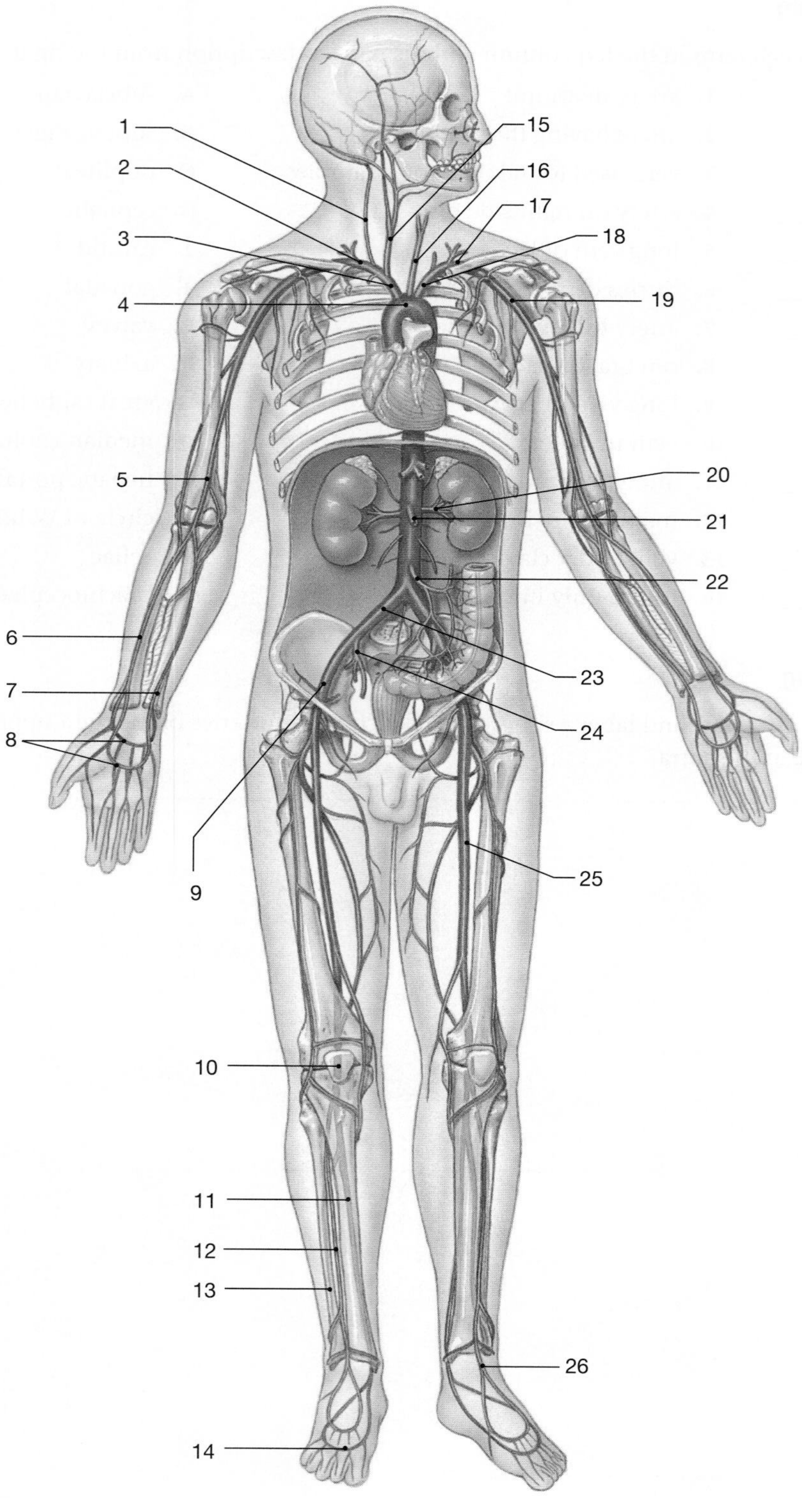

2. Label the veins in Figure 36.17.

Figure 36.17 **Overview of the Major Systemic Veins**

1. _____
2. _____
3. _____
4. _____
5. _____
6. _____
7. _____
8. _____
9. _____
10. _____
11. _____
12. _____
13. _____
14. _____
15. _____
16. _____
17. _____
18. _____
19. _____
20. _____
21. _____
22. _____
23. _____
24. _____
25. _____
26. _____

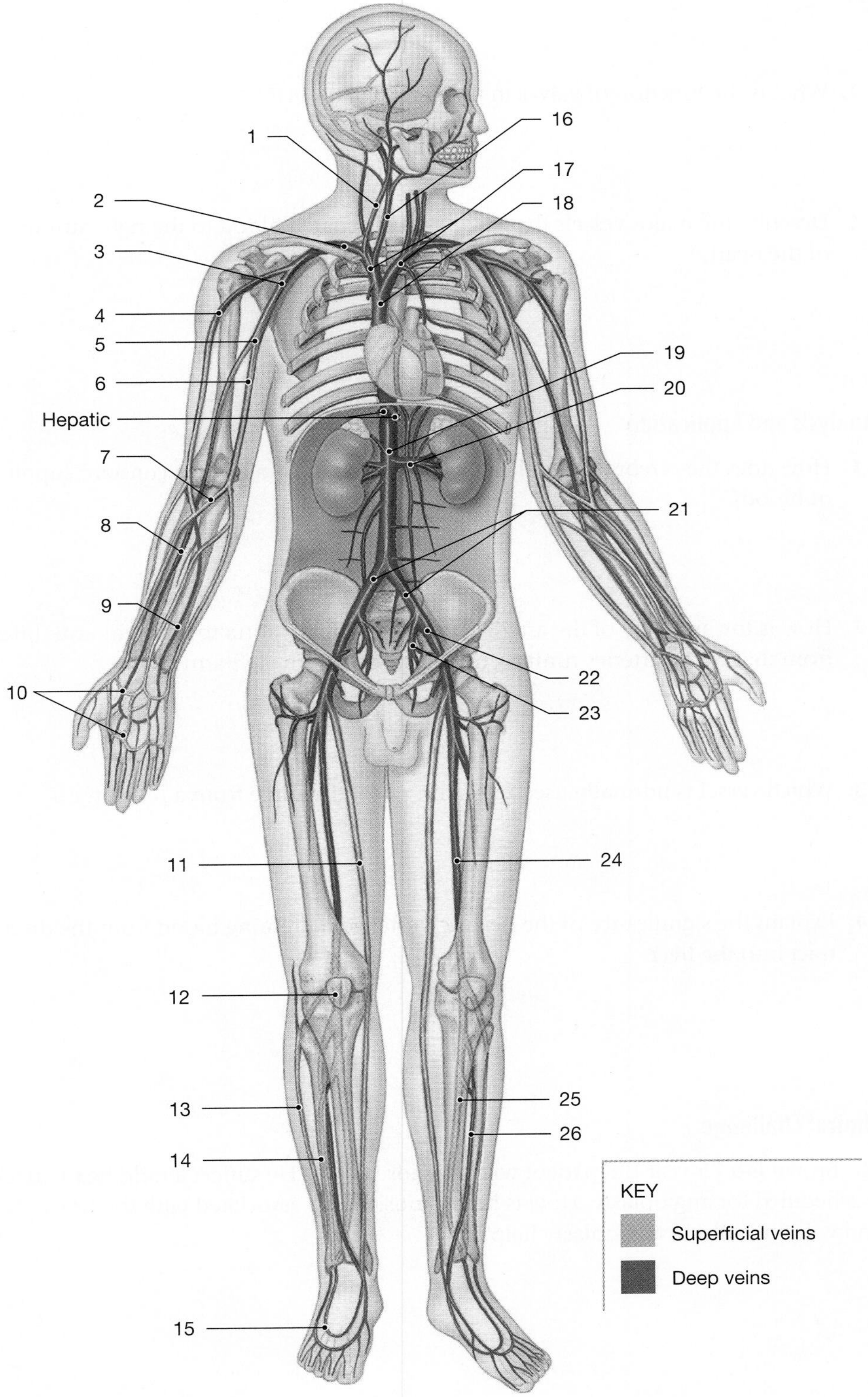

KEY

Superficial veins

Deep veins

D. Short-Answer Questions

1. List the vessels involved in supplying and draining blood from the small and large intestines.

2. What is the function of valves in the peripheral veins?

3. Describe the major vessels that return deoxygenated blood to the right atrium of the heart.

E. Analysis and Application

1. How does the cerebral arterial circle ensure that the brain has a constant supply of blood?

2. How is the anatomy of the arteries running from the aorta to the right arm different from that of the arteries running from the aorta to the left arm?

3. Which vessel is normally used to obtain a blood sample from a patient?

4. Explain the significance of the hepatic portal vein draining blood from the digestive tract into the liver.

F. Clinical Challenge

Mr. Brown is a 75-year-old patient with arteriosclerosis. He suffers a mild heart attack and is scheduled for angioplasty. How is his arteriosclerosis associated with the heart attack and will the balloon angioplasty help?

Cardiovascular Physiology

Learning Outcomes

On completion of this exercise, you should be able to:

1. Describe the pressure changes that occur during a cardiac cycle.

2. Demonstrate the steps involved in blood pressure determination.

3. Explain the differences in blood pressure caused by changes in body position.

4. Take a pulse rate at several locations on the body.

5. Read an electrocardiograph (ECG) and correlate electrical events as displayed on the ECG with the mechanical events of the cardiac cycle.

6. Observe ECG rate and rhythm changes associated with changes in body position and breathing.

7. Explain the principle of plethysmography and its usefulness in assessing changes in peripheral blood volume.

8. Observe and record changes in peripheral blood volume, pulse rate, and pulse strength under a variety of experimental and physiological conditions.

9. Determine the approximate speed of the pressure wave traveling between the heart and a finger.

To fully appreciate the cardiovascular experiments in this chapter, it is important that you understand the anatomical features of the heart and how blood circulates through its four chambers. (If necessary, review these concepts in Exercise 36 before proceeding.)

Need More Practice and Review?

Build your knowledge—and confidence!—in the Study Area of MasteringA&P® at www.masteringaandp.com with Pre-lab Quizzes, Post-lab Quizzes, Practice Anatomy Lab™ (PAL™) 3.0 virtual anatomy practice tool, PhysioEx™ 9.0 laboratory simulations, and A&P Flix™ with Quizzes.

PhysioEx™ 9.0 For this lab exercise, go to these topics:
- PhysioEx Exercise 5: Cardiovascular Dynamics
- PhysioEx Exercise 6: Cardiovascular Physiology

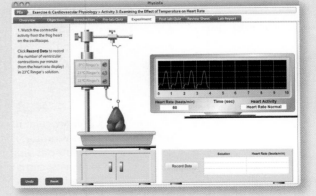

MasteringA&P®

One complete heartbeat is called a **cardiac cycle.** During a cardiac cycle, each atrium contracts and relaxes once and each ventricle contracts and relaxes once. The contraction phase of a chamber is called **systole** (SIS-tō-lē), and the relaxation phase is termed **diastole** (di-AS-tō-lē). The human heart averages 75 cardiac cycles, or heartbeats, per minute, with each cycle lasting 0.8 second(s). As **Figure 37.1** illustrates, a cycle begins with systole of the atria, lasting 0.1 s (100 milliseconds [msec]), to fill the relaxed ventricles. Next, the ventricles enter their systolic phase and contract for 0.3 s to pump blood out of the heart. For the remaining 0.4 s of the cycle, all four chambers are in diastole and fill with blood in preparation for the next heartbeat. Most blood enters the ventricles during this resting period. Atrial systole contributes only 30% of the blood volume in the ventricle prior to ventricular systole, with fluid pressure and gravity forcing the remaining 70% that flows into the ventricles.

Each cardiac cycle is marked by an increase and a decrease in blood pressure, both in the heart and in the arteries. When a heart chamber contracts, pressure increases as a result of the squeezing together of the chamber walls. The increase in blood pressure forces blood to move either from atrium to ventricle or from ventricle to outside the heart. As a chamber relaxes, its walls move apart and pressure decreases. This drop in pressure draws blood into the chamber and refills it for the next systole.

Figure 37.1 Phases of the Cardiac Cycle Thin black arrows indicate blood flow, and green arrows indicate contractions.

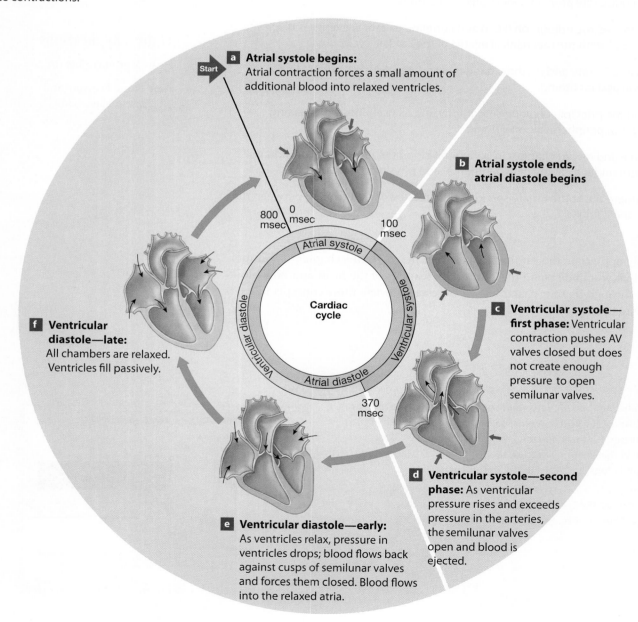

Start

a **Atrial systole begins:** Atrial contraction forces a small amount of additional blood into relaxed ventricles.

b **Atrial systole ends, atrial diastole begins**

c **Ventricular systole— first phase:** Ventricular contraction pushes AV valves closed but does not create enough pressure to open semilunar valves.

d **Ventricular systole—second phase:** As ventricular pressure rises and exceeds pressure in the arteries, the semilunar valves open and blood is ejected.

e **Ventricular diastole—early:** As ventricles relax, pressure in ventricles drops; blood flows back against cusps of semilunar valves and forces them closed. Blood flows into the relaxed atria.

f **Ventricular diastole—late:** All chambers are relaxed. Ventricles fill passively.

0 msec
800 msec
100 msec
370 msec

Atrial systole
Ventricular systole
Atrial diastole
Ventricular diastole

Cardiac cycle

When all four chambers are in diastole, the atrioventricular (AV) valves are open and blood flows from the atria into the ventricles. The semilunar (SL) valves are closed at this point to prevent backflow of blood from the aorta into the left ventricle and from the pulmonary artery into the right ventricle. When the left ventricle contracts, ventricular pressure increases to a point where it exceeds the pressure in the aorta that is holding the aortic SL valve shut. This difference in pressure forces blood through the valve into the aorta, and arterial blood pressure increases as a result of the increase in blood volume. When the ventricle relaxes, aortic and arterial pressures drop and the aortic SL valve closes. Similar events occur on the right side of the heart with the pulmonary SL valve.

In this exercise, you will investigate the physiology of blood pressure and the effect of posture on blood pressure. You will also listen to heart sounds and practice taking a subject's pulse.

Lab Activity 1 Listening to Heart Sounds

Listening to internal sounds of the body is called **auscultation.** A **stethoscope** is used to amplify the sounds to an audible level. The heart, lungs, and digestive tract are the most frequently auscultated systems. Auscultation provides the listener with valuable information concerning fluid accumulation in the lungs or blockages in the digestive tract. Auscultation of the heart is used as a diagnostic tool to evaluate valve function and detect the presence or absence of normal and abnormal heart sounds.

Four sounds are produced by the heart during a cardiac cycle (**Figure 37.2**). The first two are easily heard and are the familiar "lubb-dubb" of the heartbeat. The first heart sound (S_1), the **lubb,** is caused by the closure of the AV valves as the ventricles begin their contraction. The second sound (S_2), the **dubb,** occurs as the SL valves close at the beginning of ventricular diastole. The third and fourth sounds are faint and difficult to hear. The third sound (S_3) is produced by blood flowing into the ventricles, and the fourth (S_4) is generated by the contraction of the atria.

Figure 37.2a illustrates the landmarks for proper placement of the stethoscope **bell** (the flat metal disk that is placed against the patient's skin) to listen effectively to each heart valve. The bell has a delicate **diaphragm** that touches the skin and amplifies sounds. Notice in the figure that the sites for auscultation do not overlie the anatomical location of the heart valves. This is because the soft tissue and bone overlying the heart deflect the cardiac sound waves to locations lateral to the valves. Figure 37.2b is a graphical representation of one cardiac cycle.

Figure 37.2 Heart Sounds

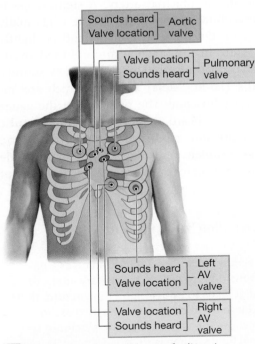

a Placements of a stethoscope for listening to the differrent sounds produced by individual valves

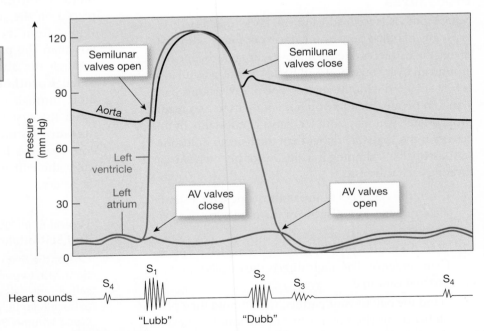

b The relationship between heart sounds and key events in the cardiac cycle

Clinical Application Heart Murmur

An unusual heart sound is called a **murmur.** Not all murmurs indicate an anatomical or functional anomaly of the heart. The sound may originate from turbulent flow in a heart chamber. Some murmurs, however, are diagnostic for certain heart defects. Septal defects are holes in a wall between two chambers. A murmur is heard as blood passes through the hole from one chamber to the other. Abnormal operation of a valve also produces a murmur. In mitral valve prolapse, the left AV valve does not seal completely during ventricular systole and blood regurgitates upward into the left atrium. ■

QuickCheck Questions

1.1 What is listening to body sounds called?

1.2 What event causes the lubb sound of the cardiac cycle?

1.3 What event causes the dubb sound of the cardiac cycle?

In the Lab 1

Materials

☐ Stethoscope

☐ Alcohol wipes

☐ Laboratory partner

Procedures

Note: From visits to the doctor, you know that a stethoscope is usually handled only by the doctor or nurse, who first places the earpieces and then moves the stethoscope bell around all over your chest and back. Cardiac auscultation is best achieved with the stethoscope placed against bare skin, but because neither you nor your partner is a medical professional, you may both feel most comfortable in this activity if the "patient" rather than the listener holds the stethoscope bell, slipping it inside his or her own shirt and locating each valve site.

1. Clean the earpieces of the stethoscope with a sterile alcohol wipe, and dispose of the used wipe in a trash can.

2. Wear the stethoscope by placing the angled earpieces facing anterior. The angle directs the earpiece into the external ear canal.

3. Hand the bell to your partner, who should then place it on her or his chest in any one of the four bell positions shown in Figure 37.2a. Have your partner then move the bell around so that you can auscultate the AV and SL valves. Can you discriminate between the lubb and the dubb sounds? ■

Lab Activity 2 Determining Blood Pressure

Blood pressure is a measure of the force the blood exerts on the walls of the systemic arteries (**Figure 37.3**). Arterial pressure increases when the left ventricle contracts and pumps blood into the aorta. When the left ventricle relaxes, less blood flows into the aorta and so arterial pressure decreases until the next ventricular systole. Two pressures are therefore used to express blood pressure, a systolic pressure and a lower diastolic pressure. Average blood pressure is considered to be 120/80 mm Hg (millimeters of mercury) for a typical male and closer to 110/70 mm Hg for most females. Do not be surprised when you take your blood pressure in the following exercise and discover it is not "average." Cardiovascular physiology is a dynamic mechanism, and pressures regularly change to adjust to the demands of the body.

Make a Prediction

Predict your blood pressure keeping in mind your overall health, weight, and most recent blood pressure measurement.

Blood pressure is measured using an inflatable cuff called a **sphygmomanometer** (sfig-mō-ma-NOM-e-ter). Figure 37.3b demonstrates proper placement of the cuff. It is wrapped around the arm just superior to the elbow and then inflated to approximately 160 mm Hg to compress and block blood flow in the brachial artery. A stethoscope is placed on the antecubital region, and pressure is gradually vented from the cuff. Once pressure in the cuff is slightly less than the pressure in the brachial artery, blood spurts through the artery and the turbulent flow makes sounds, called **Korotkoff's** (kō-ROT-kofs) **sounds,** which are audible through the stethoscope. The pressure on the gauge when the first sound is heard is recorded as the **systolic pressure.** As more pressure is relieved from the cuff, blood flow becomes less turbulent and quieter. The sounds fade when the cuff pressure matches the **diastolic pressure** of the artery.

Clinical Application High Blood Pressure and Salt Intake

One of the first recommendations a physician gives a patient with **hypertension** is to reduce the dietary intake of salt. A high-salt diet leads to saltier blood that, in turn, shifts extracellular fluid into the blood causing an increase in blood volume. Because the vascular system is a closed system, the additional fluid volume is "trapped" in the vessels, and blood pressure increases as a result. The next time you are in the grocery store, investigate the various salt substitutes currently available. ■

Figure 37.3 Checking the Pulse and Blood Pressure

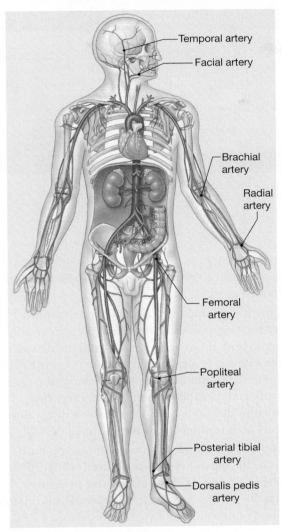

- Temporal artery
- Facial artery
- Brachial artery
- Radial artery
- Femoral artery
- Popliteal artery
- Posterial tibial artery
- Dorsalis pedis artery

a Pressure points used to check the presence and strength of the pulse

b Use of a sphygmomanometer to check aterial blood pressure

QuickCheck Questions

2.1 What is the name of the instrument used to measure blood pressure?

2.2 Which blood vessel is commonly used for measuring blood pressure?

In the Lab   2

Materials

- ☐ Sphygmomanometer
- ☐ Stethoscope
- ☐ Alcohol wipes
- ☐ Cot or laboratory table
- ☐ Laboratory partner

Procedures
I. Resting Blood Pressure

1. Have your partner sit comfortably and relax for several minutes. If your partner is wearing a long-sleeved shirt, roll up the right sleeve to expose the upper brachium. Clean the stethoscope earpieces with a sterile alcohol wipe and dispose of the used pad in the trash.

2. The sphygmomanometer consists of a **cuff** connected to a **pressure gauge** by rubber tubing and a **rubber bulb** used to inflate the cuff. A **valve** near the bulb closes or opens the cuff to hold or release the air. Force all air out of the sphygmomanometer by compressing the cuff against a flat surface. Loosely wrap the deflated cuff around your partner's right arm so that the lower edge of the cuff is just superior to the antecubital region of the elbow, as in Figure 37.3b.

3. If the cuff has **orientation arrows,** line up the arrows with the antecubitis; otherwise, position the rubber tubing over the antecubital region. Tighten the cuff so it is snug against the arm.

4. Gently close the valve on the cuff, and squeeze the rubber bulb to inflate the cuff to approximately 160 mm Hg. Do not leave the cuff inflated for more than one minute, because the inflated cuff prevents blood flow to the forearm, and the disruption in blood flow could lead to fainting.

5. Put the stethoscope earpieces in your ears, and place the bell below the cuff and over the brachial artery at the antecubitis.

6. Carefully open the valve to the sphygmomanometer, and slowly deflate the cuff while listening for Korotkoff's sounds. When the first sound is heard, note the systolic pressure reading on the gauge.

7. Continue to vent pressure from the cuff and to listen with the stethoscope. When you hear the last faint sound, note

Table 37.1	Blood Pressure Measurements		
	Resting	**Supine**	**Standing**
Measurement 1	_____	_____	_____
Measurement 2	_____	_____	_____

Table 37.2	Effect of Exercise on Blood Pressure					
	Start	**2 Min**	**4 Min**	**6 Min**	**8 Min**	**10 Min**
BP	_____	_____	_____	_____	_____	_____
MAP	_____	_____	_____	_____	_____	_____

the diastolic pressure on the gauge then open the pressure valve completely and quickly finish deflating the cuff.

8. Remove it from your partner's arm and record the pressure readings in the "Resting" column of **Table 37.1**.

II. Effect of Posture on Blood Pressure

Changing posture changes the way in which gravity influences blood pressure. This is readily apparent when standing on your head. In this section your partner will lie supine (on the back) for approximately five minutes to allow for cardiovascular adjustments. You will then determine blood pressure and compare your findings with the pressures obtained in procedure I.

1. Ask your partner to lie on the cot or laboratory table and relax for five minutes.

2. Wrap the sphygmomanometer around your partner's right arm, determine the supine blood pressure, and record it in the two lines of the "Supine" column of Table 37.1.

3. Next, ask your partner to stand up and remain still for five minutes. Take the blood pressure again, and record it in the "Standing" column of Table 37.1.

4. What was the effect of a supine posture on the blood pressure?

III. Effect of Exercise on Blood Pressure

In this section you will determine the effect of mild exercise on blood pressure. Be sure that you have determined the resting blood pressure of your subject beforehand, as outlined in procedure I. The pressure observed in procedure I will be used as a baseline.

1. Secure the sphygmomanometer cuff around your partner's arm loose enough that it is comfortable for exercise yet is in position for taking pressure readings.

2. Have your partner jog in place for five minutes, without stopping if possible. This is not a stress test; if the subject

becomes excessively winded or tired, he or she should stop immediately.

3. When the five minutes are up and your partner has stopped jogging, quickly take a blood pressure reading, and record both the systolic pressure and the diastolic pressure (write it as a fraction if you like) on the BP line of the "Start" column in **Table 37.2**. Have your partner stand still, and repeat the readings once every two minutes until the pressure returns to the resting values, recording each reading on the BP line of Table 37.2.

4. To compare your blood pressure readings, you must determine the **mean arterial pressure (MAP).** To calculate MAP, first determine the **pulse pressure,** which is the difference between the diastolic and systolic pressures (subtract diastolic pressure from systolic pressure). Next, add one-third of the pulse pressure to the diastolic pressure to get the MAP. Calculate the MAP for each pressure measurement, and record your data on the bottom row of Table 37.2. ■

Lab Activity 3 | Measuring the Pulse

Heart rate is usually determined by measuring the **pulse,** or **pressure wave,** in an artery. During ventricular systole, blood pressure increases and stretches the walls of arteries. When the ventricle is in diastole, blood pressure decreases and the arterial walls rebound to their relaxed diameter. This change in vessel diameter is felt as a throb—a pulse—at various **pressure points** on the body. The most commonly used pressure point is the radial artery on the lateral forearm just superior to the thumb (Figure 37.3a). Other pressure points include the common carotid artery in the neck and the popliteal artery of the posterior knee. The number of pulses in a given time interval indicates the number of cardiac cycles in that interval. As you will see in Laboratory Activity 5, arterial pulses are related to changes in the volume of blood passing a given point at a given time.

QuickCheck Questions

3.1 What does the pressure wave of a cardiac cycle represent?

3.2 Which events in the cardiac cycle cause the pulse you can feel in your anterior wrist?

In the Lab   3

Materials

☐ Watch or clock with second hand
☐ Laboratory partner

Procedures

1. Have your partner relax for several minutes.

2. Locate your partner's pulse in the right radial artery. Use either your index finger alone or your index and middle fingers to palpitate (feel) the pulse. Do not use your thumb for pulse measurements (because there is a pressure point in the thumb and you might not be able to distinguish your pulse from your partner's).

3. Apply light pressure to the pressure point, and count the pulse rate for 15 seconds. Multiply this number by 4 to obtain the rate per minute. Record your data in **Table 37.3**.

4. Repeat the pulse determination at the facial artery and the popliteal artery. Record your data in Table 37.3.

5. Is there any difference in the pulse strength at the various pressure points? ■

Lab Activity 4 — BIOPAC
Electrocardiography

During each cardiac cycle, a sequence of electrical impulses from pacemaker cells and nerves causes the heart muscle to produce electrical currents, or impulses, that result in contraction of the heart chambers. These impulses can be detected at the body surface with a series of electrodes. In this investigation, you will use the BIOPAC lead II configuration, which has a positive electrode on the left ankle, a negative electrode on the right wrist, and a ground electrode on the right ankle. A recording of the impulses is called an **electrocardiogram,** which is abbreviated either as **ECG** or as **EKG** (the latter abbreviation is an older one seldom used today).

The ECG is typically printed on a standard grid, with seconds on the *x* axis and either amplitude or intensity in millivolts (mV) on the *y* axis (**Figure 37.4**). In the grid shown in Figure 37.4, each small square represents 0.04 s in the horizontal direction and 0.1 mV in the vertical direction.

The basic components of an ECG are a straight baseline, the **isoelectric line,** and waves that indicate periods of depolarization and repolarization of the heart's chambers. During **depolarization,** positively charged sodium ions enter a cell, causing the cell membrane to reverse its internal charge from negative to positive and creating an electrical current. To return to the resting negative condition, the membrane **repolarizes** by allowing positively charged potassium ions to leave the cell. The **P wave** of an ECG occurs as the atria depolarize for contraction. Atrial systole occurs approximately 0.1 s after depolarization. The P wave is followed by a large spike called the **Q-R-S complex,** caused by depolarization of the ventricles. After ventricular systole, the **T wave** results from ventricular repolarization. The ECG returns to the baseline, and the next cardiac cycle soon occurs. Atrial repolarization occurs during the Q-R-S complex and is undetected.

Within the ECG are intervals that include a wave and the return to the baseline. The **P-R interval,** which occurs between the start of the P wave and the start of the Q-R-S complex, is the time required for an impulse to travel from the SA node to the ventricular muscle. The **Q-T interval** is the cycle of ventricular depolarization and repolarization. A **segment** on an ECG is the baseline recording between any two waves. The **P-R segment** represents the time for an impulse to travel from the AV node to the ventricles. The **S-T segment** measures the delay between ventricular depolarization and repolarization.

Table 37.3	Pulse Measurement
Pulse	**Pulse Rate (beats/min)**
Facial	_____
Radial	_____
Popliteal	_____

Figure 37.4 Components of the ECG

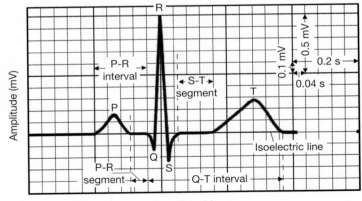

Sketch to Learn

Before running an ECG on your lab partner, let's sketch a sample ECG from one cardiac cycle and label each wave.

Sample Sketch

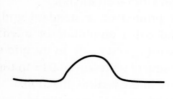 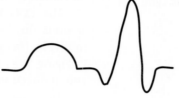 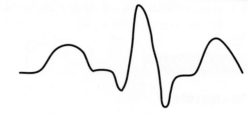

Step 1
• Start by drawing a dome shape to represent the P wave.

Step 2
• Add a small V shape then the tall, needle-like QRS wave and another V shape.

Step 3
• Draw the T wave as a dome.
• Label each wave and segment.

Your Sketch

If the electrical activity of the heart changes, then the ECG will reflect the changes. For example, damage to the AV node results in an extension of the P-R segment. Irregularities in the heartbeat are called **arrhythmias** and may indicate problems in cardiac function.

The ECG laboratory activity is organized into four major procedures. The setup section describes where to plug in the electrode leads and how to apply the skin electrodes. The calibration section adjusts the hardware so that it can collect accurate physiological data. Once the hardware has been calibrated, the data recording section describes taking ECG recordings with the subject in four situations. After the ECG data have been saved to a computer disk, the data analysis section instructs you how to use the software tools to interpret and evaluate the ECG.

Before beginning this lab, complete the *Sketch to Learn* activity above.

 Safety Alert: Read BIOPAC Safety Notices

Be sure to read the BIOPAC safety notices and carefully follow the procedures as outlined. *Under no circumstances should you deviate from the experimental procedures.* ▲

In the Lab 4

Materials

☐ BIOPAC acquisition unit (MP36/35/30)
☐ BIOPAC software: Biopac Student Lab (BSL) v3.7.5 or better
☐ BIOPAC electrode lead set (SS2L)
☐ BIOPAC electrode gel (GEL1)
☐ BIOPAC disposable vinyl electrodes (EL503), 3 electrodes per subject
☐ Computer: PC Windows 7, Vista, or XP; Mac OS X 10.4-10.6
☐ Cot or laboratory table
☐ Chair with armrests
☐ Skin cleanser or soap and water

Procedures

Section 1: Setup

1. Turn on your computer but keep the BIOPAC MP36/35/30 unit off.

2. Plug the electrode lead (SS2L) into CH 1 of the acquisition unit (**Figure 37.5**) and turn on the acquisition unit.

Figure 37.5 BIOPAC Acquisition Unit Setup

MP36/35/30 unit

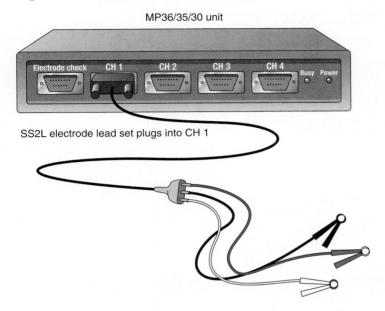

SS2L electrode lead set plugs into CH 1

Figure 37.6 Lead II Electrode Placement

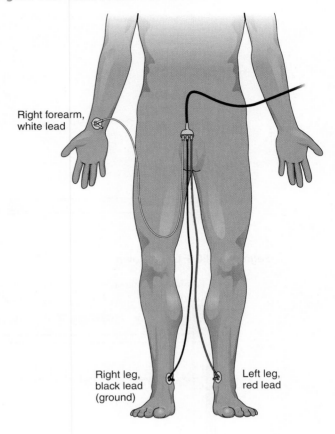

Right forearm, white lead

Right leg, black lead (ground)

Left leg, red lead

3. Have the subject remove all jewelry, especially rings, bracelets, and studs. Also, *be sure the subject is not in contact with any metal objects (faucets, pipes, and so forth)*.

4. Place the three EL503 electrodes on the subject as shown in **Figure 37.6**, using a small amount of gel (GEL1) on the skin where each electrode will be placed. *Note:* For optimal electrode adhesion, the electrodes should be placed on the skin at least five minutes before the start of the calibration procedure. Refer again to Figure 37.6 and connect the correct color pinch connector of the lead set (SS2L) to each electrode.

5. Have the subject lie down on the cot or table and relax. Position the SS2L leads so that they are not pulling on the electrodes. Connect the clip of the SS2L cable to a convenient location on the subject's clothes. This will relieve cable strain.

6. Start the Biopac Student Lab program and choose Lesson L05-ECG.1. Type in the subject's filename.

Section 2: Calibration

Calibration establishes the hardware's internal parameters (such as gain, offset, and scaling) and is critical for optimum performance. Pay close attention to the following steps.

1. Double check that the electrodes are adhered to the skin, and make sure the subject is relaxed and lying down. If the electrodes are peeling off the skin you will not get a good ECG signal. In addition, the electrocardiograph is very sensitive to small changes in voltage caused by contraction of skeletal muscles, and therefore the subject's arms and legs must be relaxed so that no muscle signal corrupts the ECG signal.

2. Click on the Calibrate button in the upper left corner of the Setup window. This will start the calibration. Wait for the calibration to stop, which will happen automatically after eight seconds.

3. At the end of the eight-second calibration recording, the screen should resemble **Figure 37.7**, a greatly reduced ECG waveform with a relatively flat baseline. If your recording is correct, proceed to Data Recording. If incorrect, click on Redo Calibration.

Section 3: Data Recording

You will make four ECG recordings of the subject: supine (lying down), immediately after sitting up, while sitting up and breathing deeply, and after exercise. A labeled marker is automatically inserted at the start of each recording segment. Heart Rate will be displayed on CH 40; heart rate is derived by finding each R-R interval in the ECG data and calculating the corresponding rate in beats per minute (BPM). To work efficiently, read this entire section so that you will know what to do for each recording segment. The subject should remain supine and relaxed while you review the lesson. Screen prompts are similar in the recent versions of the Biopac Student Lab (BSL) software. BSL 4 uses 'Continue' and 'Record' buttons to allow review/prep between segments; BSL 3.7.5-3.7.7 uses 'Record' and 'Resume' buttons.

Figure 37.7 **Calibration ECG**

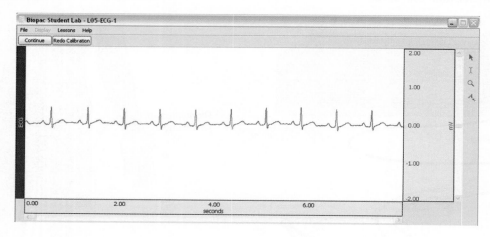

Figure 37.8 **Segment 1 ECG: Lying Down**

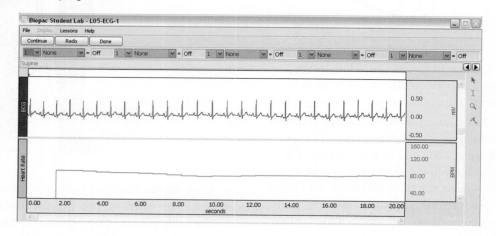

Hints for obtaining optimal data:

a. The subject should be relaxed and still, and should not talk or laugh during any recording segment.

b. When asked to sit up, the subject should do so in a chair, with arms relaxed on the armrest (if available).

Segment 1 Data Recording—Lying Down (Supine)

1. Click on Continue and when ready click on Record, and record with the subject lying down motionless for 20 seconds; then click on Suspend to stop the recording.

2. Review the data on the screen. If the screen resembles **Figure 37.8**, proceed to step 4. If the data are incorrect, erase the incorrect recording by clicking Redo, and repeat steps 1 and 2.

Segment 2 Data Recording—Seated

3. Click on Continue (if available). Do not click on Record while the subject is in the process of sitting up or you will

record a muscle artifact. Have the subject quickly sit up, and then immediately click the Record button.

4. Record for 20 seconds and then click on Suspend to stop the recording.

5. Proceed to Segment 3 or click Redo to repeat the recording.

Segment 3 Data Recording—Sitting and Deep Breathing

6. Click on Continue (if available) and prepare for the segment. Have the subject move from the cot or table to the chair and then sit without moving for about two minutes, arms supported comfortably on the armrests. In this segment, the Subject will inhale as deeply as possible and then exhale as deeply as possible, and you will click F4 and then click F5 to create corresponding "Inhale" and "Exhale" markers in the data. The subject will then start a series of slow, prolonged inhalations and exhalations.

Note: It is important that the Subject breathe with long, slow, deep breaths in order to minimize muscle artifacts in the recording.

7. Click on Record and then click F4 when the Subject starts to inhale as deeply as possible, then click F5 when the Subject starts to exhale as deeply as possible, and then have the Subject complete five prolonged breath cycles. Click on Suspend to stop the recording.

8. Proceed to Segment 4 or click Redo to repeat the recording.

   Note: The recording may have some baseline drift that is normal, and unless it is excessive, it does not necessitate repeating the recording.

Segment 4 Data Recording—After Exercise

9. Click on Continue (if available). Have the subject perform either pushups or jumping jacks for about 60 seconds to elevate the heart rate and then sit down in the chair.

   Note: You may remove the SS2L lead pinch connectors so that the subject can move about freely, but do not remove the EL503 electrodes. If you do remove the connectors, reattach them when the subject has finished exercising, following the color scheme of Figure 37.6. To capture the heart rate variation, it is important that you resume recording as quickly as possible after the subject has performed the exercise. However, it is also important that you do not click Record while the subject is exercising, for doing so will capture motion artifacts on the recording.

10. As soon as the subject has stopped exercising and is seated, click on Record, and record for 60 seconds. Click on Suspend to stop the recording. If necessary, click on Redo to repeat the recording.

    Note: The After Exercise recording may have some baseline drift, but unless the drift is excessive, do not redo the recording.

11. Click on Done, click on Yes, and choose from the pop-up options. Return to the Setup section if another student wishes to record.

12. Remove the lead pinch connectors from the EL503 electrodes, peel off the electrodes, and throw them away (BIOPAC electrodes are not reusable). Use soap and water to wash the electrode gel residue from the subject's skin. The electrodes may leave a slight ring on the skin for a few hours. This is normal and does not indicate anything wrong.

Section 4: Data Analysis

In this section, you will examine ECG components and measure amplitudes and durations of the ECG components. Interpreting ECGs is a skill; it requires practice to distinguish between normal variation and those arising from medical conditions. Do not be alarmed if your ECG is different from the examples shown or from the tables and figures.

1. Enter the Review Saved Data mode from the Lessons menu. The data window that comes up should look like Figure 37.9. The channel number (CH) designation on the left of the window are CH 1 ECG (Lead II) and CH 40 Heart Rate.

2. Set up your display window for viewing four successive beats from segment 1 (supine). The following tools will help you adjust the data window.

Autoscale horizontal	Horizontal (time) scroll bar
Autoscale waveforms	Vertical (amplitude) scroll bar
Zoom tool	Zoom previous

   Show Grid and Hide Grid buttons turn grids on and off. Use Adjust Baseline to position the waveform so that the baseline (isoelectric line) can be exactly zero. After Adjust Baseline is pressed, Up and Down buttons will be displayed—simply click on these to move the waveform in small increments. Baseline adjustment is not required to get accurate amplitude measurements, but you may want to make the adjustment before making a printout or when using grids.

Figure 37.9 Selection of ECG Waves

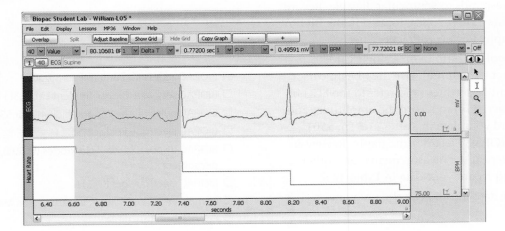

3. The measurement boxes are above the marker region in the data window. Each measurement has three sections: channel number, measurement type, and result. The first two sections are pull-down menus that are activated when you click on them. Set up the measurement boxes as follows:

Channel	Measurement
CH 40	Value (used to measure BPM, displays the amplitude at the point selected by I-beam tool). Note: CH 40 Heart Rate data is updated at the end of an R-R interval so it is constant within an R-R interval; therefore Value (BPM) is accurate from any point in the R-R interval.
CH 1	Delta T (delta time, difference in time between end and beginning of any area selected by I-beam tool)
CH 1	P-P (peak-to-peak, finds the max value in the selected area and subtracts the min value in the selected area)
CH 1	BPM (beats per minute, calculates delta T in seconds and converts this value to minutes; this measurement is only needed if CH 40 was not recorded)

4. Using the I-beam cursor, use your supine recording and select and measure the area from one R wave peak to the next R wave peak as precisely as possible. Figure 37.9 shows an example of the selected area. Record your Delta T and BPM data in **Table 37.4** (p. 549), located in the BIOPAC Electrocardiography Review & Practice Sheet. Take measurements at two other intervals in the supine recording, again recording your data in Table 37.4. Repeat the process with intervals from the other recordings as indicated in Table 37.4.

5. Use the I-beam cursor and measure the Q-T interval and measure from the end of a T wave to the next R wave. The Q-T interval corresponds to ventricular systole and the T to R wave corresponds to ventricular diastole. Record your data in **Table 37.5** (p. 550), located in the BIOPAC Electrocardiography Review & Practice Sheet.

6. Use the Zoom tool to zoom in on a single cardiac cycle from segment 1 (supine). Use the I-beam cursor and measurement-box values (and refer to the ECG in Figure 37.4 as necessary) to record the amplitudes and durations. Record your data in **Table 37.6** (p. 550), located in the BIOPAC Electrocardiography Review & Practice Sheet. Repeat the measurements for 2 other supine cycles and record your data in Table 37.6.

7. Save or print the data file and exit the program. ■

Lab Activity 5 — BIOPAC
Electrocardiography and Blood Volume

Each cardiac cycle pumps pressurized blood into the vascular system. This continual surge of blood from the heart creates a pressure wave, measured as the pulse, as you saw if you completed Laboratory Activity 3. A **plethysmogram** (PLĒ-thiz-mō-gram) is a recording of how the volume of blood at a given pressure point in the body changes as a pressure wave passes through that point. As the wave flows along the artery, it causes the vessel wall at any given point to first expand and then rebound to its original size.

In this activity, you will use a photoelectric transducer that passes light into the skin and measures how much light is reflected back. Blood absorbs light, and as the expansion part of a pressure wave passes through a given point along an artery, the increased blood volume at that point absorbs proportionally more light. The BIOPAC equipment converts the reflected light signals to electrical signals representing the pressure wave. Therefore, any change in the amplitude in the photoelectric transducer is directly proportional to the volume of blood in the pressure wave. In this activity you will collect ECG and blood-volume data from a subject under various physiological conditions.

The activity is organized into four major procedures. The setup section describes where to plug in the electrode leads, where and how to apply the skin electrodes, and how to wrap the transducer on the subject's finger. The calibration section adjusts the hardware so it can collect accurate physiological data. The data recording section describes taking ECG and blood-volume recordings with the subject in three positions. After the data have been saved to a computer disk, the data analysis section instructs you how to use the software tools to interpret and evaluate the data.

! **Safety Alert:** Read BIOPAC Safety Notices

Be sure to read the BIOPAC safety notices and carefully follow the procedures as outlined. *Under no circumstances should you deviate from the experimental procedures.* ▲

In the Lab 5

Materials

- ☐ BIOPAC acquisition unit (MP36/35/30)
- ☐ BIOPAC software: Biopac Student Lab (BSL) v3.7.5 or better
- ☐ BIOPAC pulse plethysmograph (SS4LA or SS4L)
- ☐ BIOPAC electrode lead set (SS2L)
- ☐ BIOPAC disposable vinyl electrodes (EL503), 3 electrodes per subject
- ☐ BIOPAC electrode gel (GEL1)
- ☐ Computer: PC Windows 7, Vista, or XP or; Mac OS X 10.4-10.6
- ☐ Cot or laboratory table

☐ Chair with armrests
☐ Ruler or measuring tape, calibrated in centimeters
☐ Ice water or warm water in *plastic* bucket (*not* metal bucket)
☐ Soft cloth
☐ Skin cleanser or soap and water

Procedures

Section 1: Setup

1. Turn on your computer but keep the BIOPAC MP36/35/30 unit off.

2. Plug the electrode lead (SS2L) into CH 1 and the pulse transducer (SS4LA or SS4L) into CH 2 (**Figure 37.10**). Turn on the acquisition unit.

3. Have the subject remove all jewelry, especially rings, bracelets, and studs. Also, *be sure the subject is not in contact with any metal objects (faucets, pipes, and so forth).*

4. Place the three EL503 electrodes on the subject as shown in Figure 37.6, using a small amount of gel (GEL1) on the skin where each electrode will be placed. *Note:* For optimal electrode adhesion, the electrodes should be placed on the skin at least five minutes before the start of the calibration procedure. Refer again to Figure 37.6 and connect the correct color pinch connector from the lead (SS2L) to each electrode.

5. Use the piece of cloth to clean the window of the pulse transducer sensor. Position the transducer so that the sensor is on the bottom of the fingertip (the part without the fingernail) of the index finger of one of the subject's hands, and wrap the tape around the finger so that

the transducer fits snugly but not so tightly that blood circulation is cut off (**Figure 37.11**).

6. With the ruler or measuring tape, measure two distances: from the fingertip where the sensor is attached to the subject's shoulder and from the shoulder to the middle of the sternum. Record these two distances in Section A, Data and Calculations, in the BIOPAC Electrocardiography and Blood Volume Review & Practice Sheet.

7. Have the subject lie down on the cot or table and relax. Position the SS2L leads so that they are not pulling on the electrodes. Connect the clip of the lead set cable to a convenient location on the subject's clothes. This will relieve cable strain.

8. Start the Biopac Student Lab program and choose Lesson L07ECG & Pulse. Type in the subject's filename, then click OK.

Section 2: Calibration

1. Double check that the electrodes are adhering to the skin, and make sure the subject is relaxed and lying down. If the electrodes are detaching from the skin, you will not

Figure 37.11 **Placement of Pulse Transducer**

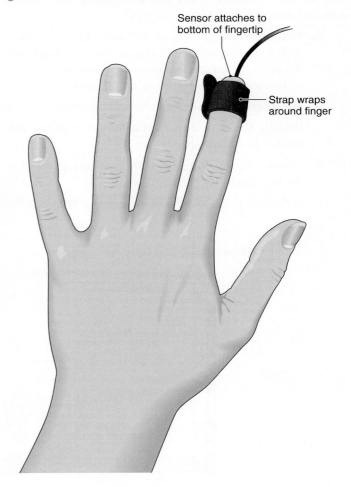

Sensor attaches to bottom of fingertip

Strap wraps around finger

Figure 37.10 **BIOPAC Acquisition Unit Setup**

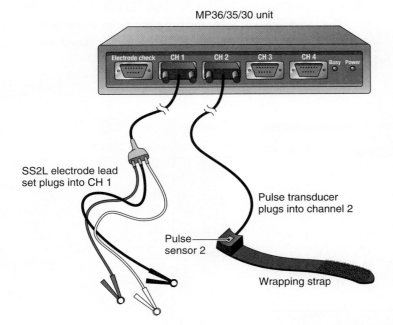

MP36/35/30 unit

SS2L electrode lead set plugs into CH 1

Pulse transducer plugs into channel 2

Pulse sensor 2

Wrapping strap

get a good ECG signal. In addition, the electrocardiograph is very sensitive to small changes in voltage caused by contraction of skeletal muscles, and therefore the subject's arms and legs must be relaxed so that no muscle signal corrupts the ECG signal.

2. Click on the Calibrate button. This will start the calibration. Wait for the calibration to stop, which will happen automatically after eight seconds.

3. At the end of the eight-second calibration recording, the screen should resemble **Figure 37.12** with a greatly reduced ECG waveform, and with a relatively flat baseline in the upper band and waveforms in the pulse (blood volume) band. Repeat the calibration if necessary.

Section 3: Data Recording

Have the subject sit in the chair and relax, with arms on the armrests. You will record ECG on CH 1 and changes in blood volume on CH 2 under three conditions: arm relaxed, hand in water, and arm up. (The blood volume changes will be measured indirectly as pressure-wave pulses in the subject's finger.) Screen prompts are similar in the recent versions of the BSL software. BSL 4 uses 'Continue' and 'Record' buttons to allow review/prep between segments; BSL 3.7.5-3.7.7 uses 'Record' and 'Resume' buttons. Hints for minimizing both baseline drift and muscle corruption of the ECG:

a. The subject should remain still and relaxed during each recording segment, because the recording from the pulse transducer is sensitive to motion and the ECG recording is sensitive to muscle artifacts.

b. The subject should be quiet for each recording segment.

c. Initially, the subject's forearms should be supported on the chair's armrests.

d. Always stop recording *before* the subject prepares for the next recording segment.

e. Make sure the electrodes do not peel up from the skin.

Segment 1—Seated with Arm Relaxed

1. After the subject has been sitting relaxed for several minutes, with arms on the chair armrests, click on Continue and when ready click on Record. Let the hardware collect data for 15 seconds.

2. Click on Suspend to stop the recording. Your screen should resemble **Figure 37.13**.

3. Proceed to Segment 2 or click Redo to repeat the recording.

Segment 2—Seated with Hand in Water

4. Have the subject remain seated and place the nonrecording hand in the warm or cold water.

⚠ **Safety Alert:** Avoid Metal Containers

The container for the water cannot be metal, as a metal container could bypass the electrical isolation of the system. ▲

5. Click on Continue and when ready click on Record. Record for 30 seconds. The recording will continue from the point where it last stopped, and a marker labeled "Seated, one hand in water" will automatically appear on the screen.

6. Click on Suspend to stop the recording.

7. Proceed to Segment 3 or click Redo to repeat the recording.

Segment 3—Seated with Arm Raised

8. Have the subject remain seated and raise the recording hand (with transducer) to extend the arm above the head and hold that position for the duration of the recording.

9. Click on Continue and when ready click on Record. The recording will continue from the point where it last stopped, and a marker labeled "Seated, arm raised above head" will automatically come up.

Figure 37.12 Calibration ECG and Pulse

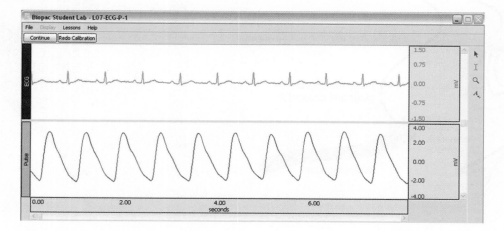

Figure 37.13 Segment 1: Arm Relaxed

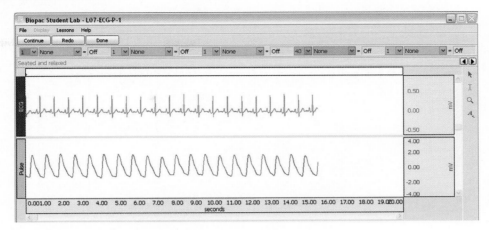

10. Record for 60 seconds then click on Suspend to stop the recording. If necessary, click on Redo to repeat the recording.

11. Click on Done, click on Yes, and choose from the pop-up options. Return to the Setup section if another student wishes to record.

12. Remove the transducer from the subject's finger. Disconnect the leads from their electrodes, peel off the electrodes, and throw them away. Use soap and water to wash the electrode gel residue from the subject's skin. The electrodes may leave a slight ring on the skin for a few hours. This is normal and does not indicate anything wrong.

Section 4: Data Analysis

1. Enter the Review Saved Data mode from the Lessons menu. The window that comes up should have ECG and Pulse Data. Note Channel Number (CH) designation:

Channel	Displays
CH 1	ECG
CH 40	Pulse

2. Set up your display window for optimal viewing of the entire recording. The following tools will help you adjust the window display.

Autoscale horizontal	Horizontal (time) scroll bar
Autoscale waveforms	Vertical (amplitude) scroll bar
Zoom tool	Zoom previous

Grids—click Show or Hide buttons, or choose Preferences from the File menu.

3. The measurement boxes are above the marker region in the data window. Each measurement has three sections: channel number, measurement type, and result. The first two sections are pull-down menus that are activated when you click on them. Set up the measurement boxes as follows:

Channel	Measurement
CH 1	Delta T (delta time, difference in time between end and beginning of any area selected by I-beam tool)
CH 1	BPM (beats per minute, calculates ΔT in seconds and converts this value to minutes)
CH 1	P-P (finds maximum value in selected area and subtracts minimum value found in selected area)
CH 40	P-P

4. Zoom in on a small section of the segment 1 data. Be sure to zoom in far enough that you can easily measure the intervals between peaks for approximately four cardiac cycles.

5. Using the I-beam cursor, select the area between two successive R waves (one cardiac cycle). Try to go from one R wave peak to the adjacent R wave peak as precisely as possible (**Figure 37.14**).

6. Measure Delta T and BPM for the selected area, and record your data in the "R-R interval" and "heart rate" portions of **Table 37.7** (p. 553), located in the BIOPAC Electrocardiography and Blood Volume Lab Review & Practice Sheet.

7. Using the I-beam cursor, select the area between two successive pulse peaks (one cardiac cycle). Measure Delta T and BPM for the selected area, and record your data in the "pulse interval" and "pulse rate" portions of Table 37.7.

8. Repeat the Delta T and BPM measurements for each data segment, and record your data in Table 37.7.

9. Select an individual pulse peak for each segment, and determine its amplitude, using the CH 40 P-P measurements. Record your data in **Table 37.8** (p. 553), located in the BIOPAC Electrocardiography and Blood Volume Lab Review & Practice Sheet.

Figure 37.14 **Measurement Between R Wave Peaks**

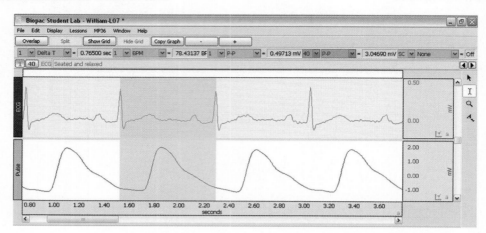

Important: Measure the first pulse peak after the recording is resumed. The body's homeostatic regulation of blood pressure and volume occurs quickly. The increase or decrease in your results will depend on the timing of your data relative to the speed of physiological adjustments.

10. Using the I-beam cursor, select the interval between one R wave and the adjacent pulse peak. Record the time interval (Delta T) between the two peaks in the Electrocardiography and Blood Volume Review & Practice Sheet, Section A, Data and Calculations.

11. Save or print the data file. You may save the data to a storage device, save notes that are in the journal, or print the data file.

12. Exit the program. ■

Name _____

Date _____

Section _____

Cardiovascular Physiology

A. Definition

Define or describe each term.

1. cardiac cycle

2. diastole

3. systole

4. auscultation

5. Korotkoff's sounds

6. first heart sound

7. second heart sound

8. murmur

B. Drawing

Draw It! Draw and label an example of a normal ECG recording for one cardiac cycle.

C. Short-Answer Questions

1. Briefly describe the events of a cardiac cycle.

2. Explain how blood pressure fluctuates in arteries.

D. Analysis and Application

1. Describe how a sphygmomanometer is used to determine blood pressure.

2. Describe how blood pressure changes during exercise and during rest.

3. Calculate the MAP for a blood pressure of 130/85 mm Hg.

E. Clinical Challenge

1. What is the rationale for reducing dietary salt for patients with high blood pressure?

2. An elderly patient complains of dizziness and lightheadedness when he stands up after lying down. Explain the cardiovascular response that may be involved in this case.

Name ______________________

Date ______________________

Section ______________________

BIOPAC: Electrocardiography

Review & Practice Sheet

EXERCISE 37

A. Data and Calculations

Subject Profile

Name ______________________ Height ______________

Age ______________________ Weight ______________

Sex: Male / Female

1. Record your data in Table 37.4.

Table 37.4 Measurement of R-R Interval for Cardiac Cycle Duration

Segment: Condition	Measurement	Cardiac Cycle: Cycle 1	Cycle 2	Cycle 3	Mean
1: Supine	Value [CH 40]	______	______	______	______
	BPM [CH 1]	______	______	______	______
2: Seated	Value [CH 40]	______	______	______	______
	BPM [CH 1]	______	______	______	______
3: Start of Inhale	Value [CH 40]	______	______	______	______
	BPM [CH 1]	______	______	______	______
3: Start of Exhale	Value [CH 40]	______	______	______	______
	BPM [CH 1]	______	______	______	______
4: After Exercise	Value [CH 40]	______	______	______	______
	BPM [CH 1]	______	______	______	______

2. Record your data in Table 37.5.

Table 37.5	Duration of Ventricular Systole and Ventricular Diastole	
	Delta T [CH 1]	
Segment: Condition	**Ventricular Systole**	**Ventricular Diastole**
1: Supine	_____	_____
	_____	_____
4: After Exercise	_____	_____
	_____	_____

3. Record your data in Table 37.6.

Table 37.6	ECG Duration and Amplitude Measurements									
ECG Component	**Normative Values Based on resting heart rate 75 bpm**		**Duration Delta T [CH1] of Segment 1 Cycle**			**Seg 1 Mean (calc.)**	**Amplitude (mV) P-P [CH 1] of Segment 1 Cycle**			**Seg 1 Mean (calc.)**
			1	2	3		1	2	3	
Waves	**Dur. (sec)**	**Amp. (mV)**								
P Wave	.07–.18	< .20	____	____	____	____	____	____	____	____
QRS Complex	.06–.12	.10–1.5	____	____	____	____	____	____	____	____
T wave	.10–.25	< .5	____	____	____	____	____	____	____	____
Intervals	**Duration (seconds)**									
P-R	.12–.20		____	____	____	____				
Q-T	.32–.36		____	____	____	____				
R-R	.80		____	____	____	____				
Segments	**Duration (seconds)**									
P-R	.02–.10		____	____	____	____				
S-T	< .20		____	____	____	____				
T-P	0–.40		____	____	____	____				

B. Data Summary and Questions

1. Is there always one P wave for every Q-R-S complex? Yes   No

2. Describe the shape of a P wave and of a T wave.

3. Do the wave durations and amplitudes for all subjects fall within the normal ranges listed in Table 37.4? Yes   No

4. Do the S-T segments mainly measure between –0.1 mV and 0.1 mV? Yes   No

5. Is there any baseline drift in the recording? Yes   No

C. Analysis and Application

1. Summarize the heart rate data in the space provided below. Explain the changes in heart rate as conditions change, and describe the physiological mechanisms causing these rate changes.

Condition	Mean Heart Rate (BPM)
Supine, regular breathing	_____
Sitting, regular breathing	_____
Seated, deep breathing, inhalation	_____
Seated, deep breathing, exhalation	_____
After exercise, start of recording	_____
After exercise, end of recording	_____

2. Duration (Delta T)

Condition	Mean Delta T
Supine, regular breathing	_____
Seated, regular breathing	_____
Seated, deep breathing	
Inhalation	_____
Exhalation	_____
After exercise	_____

a. Are there differences in the cardiac cycle with the respiratory cycle?

Condition	Mean QT Interval
Supine, regular breathing	_____
Ventricular systole	_____
Ventricular diastole	_____
After exercise	_____
Ventricular systole	_____
Ventricular diastole	_____

b. What changes do you observe between the duration of systole and diastole with the subject resting and the duration of systole and diastole after the subject has exercised?

Name _____

Date _____

Section _____

BIOPAC:
Electrocardiography
and Blood Volume

A. Data and Calculations

Subject Profile

Name _____   Height _____

Age _____   Weight _____

Sex: Male / Female

1. Record your data in Table 37.7.

Table 37.7							
Condition	**Measurement**			**Cycle 1**	**Cycle 2**	**Cycle 3**	**Mean**
Arm relaxed	R-R interval	**Delta T**	CH 1	_____	_____	_____	_____
Segment 1	Heart rate	**BPM**	CH 1	_____	_____	_____	_____
	Pulse interval	**Delta T**	CH 1	_____	_____	_____	_____
	Pulse rate	**BPM**	CH 1	_____	_____	_____	_____
Temp change	R-R interval	**Delta T**	CH 1	_____	_____	_____	_____
Segment 2	Heart rate	**BPM**	CH 1	_____	_____	_____	_____
	Pulse interval	**Delta T**	CH 1	_____	_____	_____	_____
	Pulse rate	**BPM**	CH 1	_____	_____	_____	_____
Arm up	R-R interval	**Delta T**	CH 1	_____	_____	_____	_____
Segment 3	Heart rate	**BPM**	CH 1	_____	_____	_____	_____
	Pulse interval	**Delta T**	CH 1	_____	_____	_____	_____
	Pulse rate	**BPM**	CH 1	_____	_____	_____	_____

2. Record your data in Table 37.8.

Table 37.8			
Measurement	**Arm Resting** Segment 1	**Temperature Change** Segment 2	**Arm Up** Segment 3
Q-R-S amplitude	_____	_____	_____
Pulse amplitude (mV)	_____	_____	_____

Calculation of Pulse Speed

1. Distance between subject's sternum and shoulder: _____ cm

2. Distance between subject's shoulder and fingertip: _____ cm

3. Total distance from sternum to fingertip: _____ cm

Segment 1 Data

1. Time between R wave and pulse peak: _____ s

2. Speed: _____ cm/s

Segment 3 Data

1. Time between R wave and pulse peak: _____ s

2. Speed: _____ cm/s

B. **Short-Answer Questions**

1. Are the values of heart rate and pulse rate given in Table 37.7 similar for each condition or different for each condition? Propose an explanation for any similarity or difference you observe.

2. Determine how much the Q-R-S amplitude values recorded in Table 37.8 changed as conditions changed:

   Segment 2 amplitude minus segment 1 amplitude: _____ mV

   Segment 3 amplitude minus segment 1 amplitude: _____ mV

3. Determine how much the pulse amplitude values recorded in Table 37.8 changed as arm position changed:

   Segment 2 amplitude minus segment 1 amplitude: _____ mV

   Segment 3 amplitude minus segment 1 amplitude: _____ mV

4. Does the amplitude of the Q-R-S complex change as the pulse amplitude changes? Why or why not?

5. Describe one mechanism that causes changes in blood volume to your fingertip.

6. In your calculation of pulse speed in part A of this Review & Practice Sheet, did you find a difference between the segment 1 speed and the segment 3 speed? If yes, explain the reason for the difference.

7. Which components of the cardiac cycle (atrial systole and diastole, ventricular systole and diastole) are discernible in the pulse tracing?

8. Would you expect the pressure-wave velocities measured in your own body to be very close to those measured in other students? Why or why not?

9. Explain any pressure-wave amplitude or frequency changes that occurred with changes in arm position.

Lymphatic System

Learning Outcomes

On completion of this exercise, you should be able to:

1. List the functions of the lymphatic system.

2. Describe the exchange of blood plasma, extracellular fluid, and lymph.

3. Describe the structure of a lymph node.

4. Explain how the lymphatic system drains into the vascular system.

5. Describe the gross anatomy and basic histology of the spleen.

The lymphatic system includes the lymphatic vessels, lymph nodes, tonsils, spleen, and thymus gland (**Figure 38.1**). **Lymphatic vessels** transport liquid called **lymph** from the extracellular spaces to the veins of the cardiovascular system. Scattered along each lymphatic vessel are **lymph nodes** containing lymphocytes (one type of white blood cell, see Exercise 34) and phagocytic macrophages. The macrophages remove invading microbes and other substances from the lymph before the lymph is returned to the blood. Although lymphocytes are classified as a formed element of the blood, they are the main cells of the lymphatic system and colonize dense populations in lymph nodes and the spleen. The antigens present in invading pathogens and other foreign substances cause the lymphocytes to produce antibodies to defend against the antigens. As macrophages capture the antigens in the lymph, lymphocytes exposed to the antigens activate the immune system to respond to the intruding cells. The lymphocytes called B cells produce antibodies that chemically combine with and destroy the antigens.

The thymus gland is involved in the development of the functional immune system in infants. In adults, this gland controls the maturation of lymphocytes.

Need More Practice and Review?

Build your knowledge—and confidence!—in the Study Area of MasteringA&P® at www.masteringaandp.com with Pre-lab Quizzes, Post-lab Quizzes, Practice Anatomy Lab™ (PAL™) 3.0 virtual anatomy practice tool, PhysioEx™ 9.0 laboratory simulations, and A&P Flix™ with Quizzes.

PAL | practice anatomy lab   For this lab exercise, follow these navigation paths:
- PAL>Human Cadaver>Lymphatic System
- PAL>Anatomical Models>Lymphatic System
- PAL>Histology>Lymphatic System

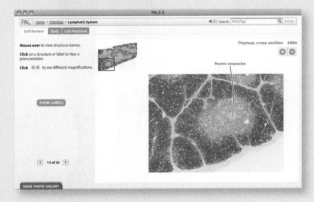

Figure 38.1 Lymphatic System An overview of the distribution of lymphatic vessels, lymph nodes, and the other organs of the lymphatic system.

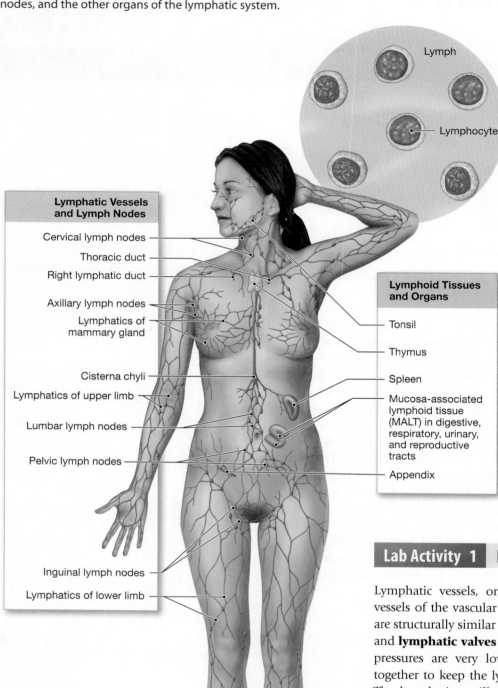

Lymphatic Vessels and Lymph Nodes

Cervical lymph nodes

Thoracic duct

Right lymphatic duct

Axillary lymph nodes

Lymphatics of mammary gland

Cisterna chyli

Lymphatics of upper limb

Lumbar lymph nodes

Pelvic lymph nodes

Inguinal lymph nodes

Lymphatics of lower limb

Lymph

Lymphocyte

Lymphoid Tissues and Organs

Tonsil

Thymus

Spleen

Mucosa-associated lymphoid tissue (MALT) in digestive, respiratory, urinary, and reproductive tracts

Appendix

(We covered the thymus gland with the endocrine system in Exercise 33.)

Pressure in the systemic capillaries forces liquids and solutes out of the capillaries and into the interstitial spaces. This constant renewal of extracellular fluid bathes the cells with nutrients, dissolved gases, hormones, and other materials. After this exchange, osmotic pressure forces most of the extracellular fluid back into the capillaries. Some extracellular fluid drains into lymphatic vessels and becomes lymph. The lymph travels through the lymphatic vessels to the lymph nodes, where macrophages remove abnormal cells and microbes from the lymph. Fluid buildup, called **edema,** can occur when injury or an increase in pressure on an area results in more fluid loss from capillaries than the lymphatic system can return to the blood. Extracellular fluid accumulates, often in the extremities, where swelling occurs.

Two **lymphatic ducts** join with veins near the heart and return the lymph to the blood. Approximately 3 L liquid per day are forced out of the capillaries and flow through lymphatic vessels as lymph.

Lab Activity 1   Lymphatic Vessels

Lymphatic vessels, or simply **lymphatics,** occur next to the vessels of the vascular system (**Figure 38.2**). Lymphatic vessels are structurally similar to veins. The vessel wall has similar layers and **lymphatic valves** to prevent backflow of liquid. Lymphatic pressures are very low, and the lymphatic valves are close together to keep the lymph circulating toward the body trunk. The lymphatic capillaries, which gradually expand to become the lymphatic vessels, are closed at the ends lying near the arterial blood capillaries. Lymph enters the lymphatic system at or near these closed ends and then moves into the lymphatic vessels, which conduct the lymph toward the body trunk and into the large-diameter lymphatics that empty into veins near the heart. Smaller lymphatic vessels combine into larger **lymphatic trunks** that eventually converge to empty lymph into the blood.

Two large lymphatic vessels, the **thoracic duct** and the **right lymphatic duct,** drain into the subclavian veins

Figure 38.2 Lymphatic Capillaries

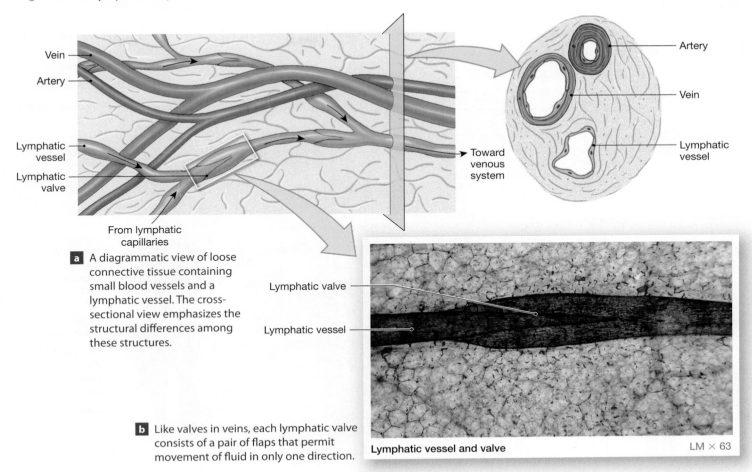

Vein

Artery

Lymphatic
vessel

Lymphatic
valve

From lymphatic
capillaries

Toward
venous
system

Artery

Vein

Lymphatic
vessel

a A diagrammatic view of loose
connective tissue containing
small blood vessels and a
lymphatic vessel. The cross-
sectional view emphasizes the
structural differences among
these structures.

Lymphatic valve

Lymphatic vessel

b Like valves in veins, each lymphatic valve
consists of a pair of flaps that permit
movement of fluid in only one direction.

Lymphatic vessel and valve   LM × 63

to return lymph to the blood (**Figure 38.3**). Most of the lymph is returned to the circulation by the thoracic duct, which commences at the level of the second lumbar vertebra on the posterior abdominal wall behind the abdominal aorta. Lymphatics from the lower limbs, pelvis, and abdomen drain into an inferior saclike portion of the thoracic duct called the **cisterna chyli** (KĪ-lī; *cistern*, storage well; *chyl*, juice). The thoracic duct ascends the abdomen and pierces the diaphragm. At the base of the heart, the thoracic duct joins with the left subclavian vein to return the lymph to the blood. The only lymph that does not drain into the thoracic duct is from lymphatic vessels in the right upper limb and the right side of the chest, neck, and head. These areas drain into the right lymphatic duct near the right clavicle. This duct empties lymph at or near the junction of the right internal jugular and the right subclavian veins near the base of the heart.

QuickCheck Questions

1.1 What is lymph?

1.2 Where is lymph returned to the vascular system?

1.3 What is the function of lymphocytes?

In the Lab   1

Materials

☐ Torso model or lymphatic system chart

☐ Compound microscope

☐ Prepared slide of lymphatic vessel

Procedures

1. Locate the thoracic duct and the cisterna chyli on the torso model or lymphatic system chart.

2. Which areas of the body drain lymph into the thoracic duct? Where does this lymphatic vessel return lymph to the blood?

3. Locate the right lymphatic duct on the torso model or lymphatic system chart. Where does this lymphatic vessel join the vascular system? Which regions of the body drain lymph into this vessel?

4. Observe the lymphatic vessel slide at low power and search for a lymphatic valve. Consider in which direction the valve allows lymph to flow. ■

Figure 38.3 Thoracic and Right Lymphatic Ducts Lymph is returned to the blood by lymphatic ducts near the heart.

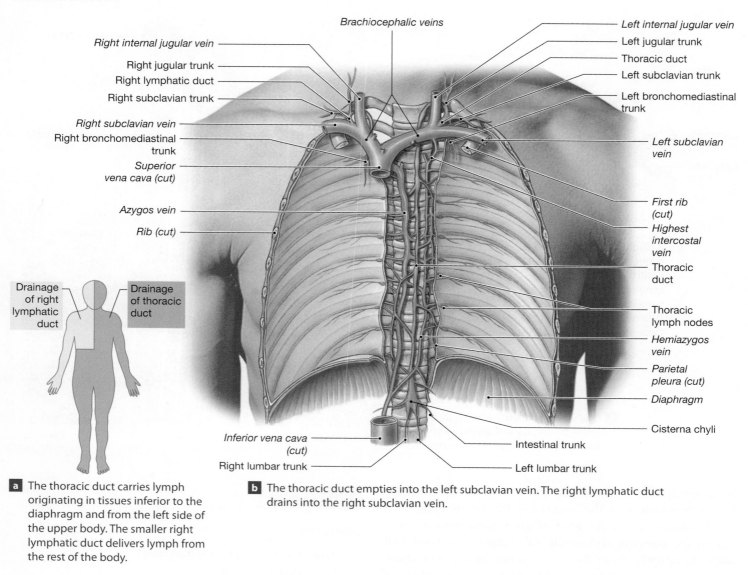

a The thoracic duct carries lymph originating in tissues inferior to the diaphragm and from the left side of the upper body. The smaller right lymphatic duct delivers lymph from the rest of the body.

b The thoracic duct empties into the left subclavian vein. The right lymphatic duct drains into the right subclavian vein.

Lab Activity 2 — Lymphatic Tissues and Lymph Nodes

Two major groups of lymphatic structures occur in connective tissues: **encapsulated lymph organs** and **diffuse lymphatic tissues.** The encapsulated lymph organs include lymph nodes, the thymus gland, and the spleen. Each encapsulated organ is separated from the surrounding connective tissue by a fibrous capsule. Diffuse lymphatic tissues do not have a defined boundary separating them from the connective tissue.

Each lymph node is an oval organ that functions like a filter cartridge. As lymph passes through a node, phagocytes remove microbes, debris, and other antigens from the lymph. Lymph nodes are scattered throughout the lymphatic system, as depicted in Figure 38.1. Lymphatic vessels from the lower limbs pass through a network of inguinal lymph nodes located in the groin region (*inguinal* means "pertaining to the groin"). Pelvic and lumbar nodes filter lymph from the pelvic and abdominal lymphatic vessels. Many lymph nodes occur in the upper limbs and in the axillary and cervical regions. The breasts in women also contain many lymphatic vessels and nodes. Often, infections occur in a lymph node before they spread systemically. A swollen or painful lymph node suggests an increase in lymphocyte abundance and general immunological activity in response to antigens in the lymph nodes.

Clinical Application Lymphatics and Breast Cancer

The female breast has milk-producing mammary glands embedded in a pectoral fat pad that lies against the pectoralis major muscle. An extensive network of lymphatic vessels and lymph nodes collects and filters lymph from the breast (Figure 38.4). The lymph nodes are of clinical importance in cases of breast and other cancers, because cancer cells can enter the lymphatic vessels and spread to other parts of the body *(metastasize)* when lymph is returned to the blood. Breast cancer is classified according to the extent to which metastasizing cancer cells have invaded the lymph nodes. Treatment typically begins with the removal of the tumor and a biopsy of the axillary lymph nodes. A *mastectomy* is removal of the breast. A *radical mastectomy* involves removal of the breast plus the regional lymphatic vessels, including the axillary lymph nodes. ∎

Figure 38.4 Lymphatic Vessels of the Female Breast
Superficial and deep lymphatic vessels and nodes in the female breast and chest.

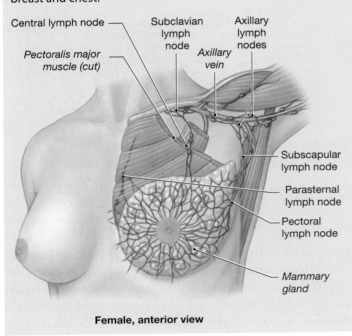

Central lymph node
Subclavian lymph node
Axillary lymph nodes
Axillary vein
Pectoralis major muscle (cut)
Subscapular lymph node
Parasternal lymph node
Pectoral lymph node
Mammary gland

Female, anterior view

Figure 38.5 details the anatomy of a lymph node. Each node is encased in a dense connective tissue **capsule.** Collagen fibers from the capsule extend as partitions called **trabeculae** into the interior of the node. The region immediately inside the capsule is the **subcapsular sinus.** Interspersed in the subcapsular sinus are regions of cortical tissue called **outer cortex,** rich in B cells. Each region of outer cortex surrounds a pale-staining **germinal center,** where lymphocytes are produced. Deep to the ring of germinal centers is the **deep cortex;** here lymphocytes carried into the node by the blood leave the blood and enter the node. The central region of the node is the **medulla,** where **medullary cords** made up of B cells and

plasma cells extend into a network of **medullary sinuses.** Lymph enters a node via several **afferent lymphatic vessels.** As the lymph flows through the subcapsular and medullary sinuses, macrophages phagocytize abnormal cells, pathogens, and debris. Draining the lymph node is a single **efferent lymphatic vessel,** which exits the node at a slit called the **hilum.**

Lymphatic nodules, which are diffuse lymphatic tissue, are found in connective tissue under the lining of the digestive, urinary, and respiratory systems (Figure 38.6). Microbes that penetrate the exposed epithelial surface pass into lymphatic nodules and into the lymph, where lymphocytes and macrophages destroy the foreign cells and remove them from the lymph. Some nodules have a germinal center where lymphocytes are produced by cell division.

Tonsils are lymphatic nodules in the mouth and pharynx. A pair of **lingual tonsils** sits at the posterior base of the tongue. The **palatine tonsils** are easily viewed hanging off the posterior arches of the oral cavity. A single **pharyngeal tonsil,** or **adenoid,** is located in the upper pharynx near the opening to the nasal cavity.

QuickCheck Questions

2.1 What are the names of the two types of lymphatic structures in the body?

2.2 Where are lymph nodes located?

Clinical Application Tonsillitis

The lymphatic system usually has the upper hand in the immunological battle against invading bacteria and viruses. Occasionally, however, microbes manage to populate a lymphatic nodule. When it is the tonsils that are infected, they swell and become irritated. This condition is called **tonsillitis** and is treated with antibiotics to control the infection. If the problem is recurrent, the tonsils are removed in a surgical procedure called a *tonsillectomy.* Usually, it is the palatine tonsils that are removed. If the pharyngeal tonsil is also infected or is abnormally large, it is also removed during the procedure. ∎

In the Lab 2

Materials

☐ Torso model
☐ Compound microscope
☐ Prepared slide of lymph node

Procedures

1. Review the structure of a lymph node in Figure 38.5 and tonsils in Figure 38.6.

2. On the torso model, locate the two pairs of tonsils and the single pharyngeal tonsil.

Figure 38.5 Structure of a Lymph Node Lymph nodes are covered by a capsule of dense, fibrous connective tissue. Lymphatic vessels and blood vessels penetrate the capsule to reach the lymphatic tissue within. Note that a lymph node has several afferent lymphatic vessels but only one efferent vessel.

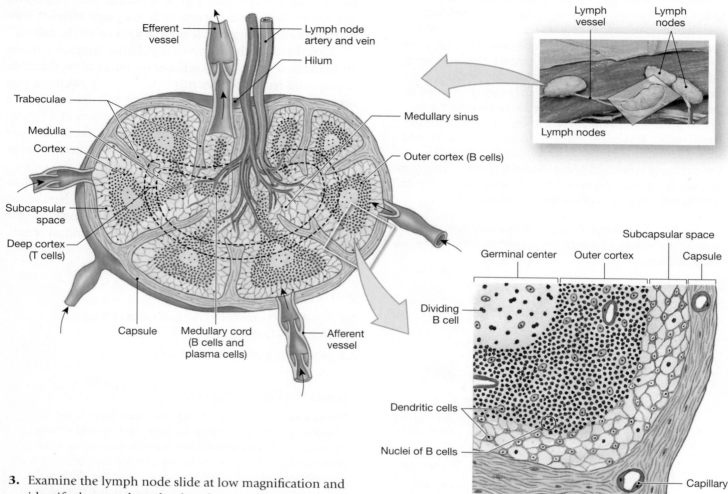

3. Examine the lymph node slide at low magnification and identify the capsule and trabeculae.

4. Change the microscope to high magnification and examine a germinal center inside the outer cortex. Identify the cells produced in the germinal center. ∎

Lab Activity 3 | The Spleen

The **spleen,** the largest lymphatic organ in the body, is located lateral to the stomach (Figure 38.7). A capsule surrounds the spleen and protects the underlying tissue of **red pulp** and **white pulp.** The color of the red pulp is due to the blood filtering through; white pulp appears blue because of the staining of the lymphocyte nuclei. Blood vessels and lymphatic vessels pass in and out of the spleen at the hilus. Branches of the splenic artery, called **trabecular arteries,** are distributed in the red pulp. **Central arteries** occur in the middle of white pulp. Capillaries of the trabecular arteries open into the red pulp. As blood flows through the red pulp, free and fixed phagocytes in the pulp

remove abnormal red blood cells and other antigens from the blood. Upon exposure to the antigens, the lymphocytes of the red pulp become sensitized to them and produce antibodies to counteract them. Blood drains from the sinuses of the red pulp into trabecular veins that eventually empty into the splenic vein.

QuickCheck Questions

3.1 Where is the spleen located?

3.2 What tissues are in the white pulp?

3.3 Which vessels open into the red pulp?

In the Lab 3

Materials

☐ Torso model or chart showing spleen
☐ Compound microscope
☐ Prepared slide of spleen

Figure 38.6 **Lymphoid Nodules**

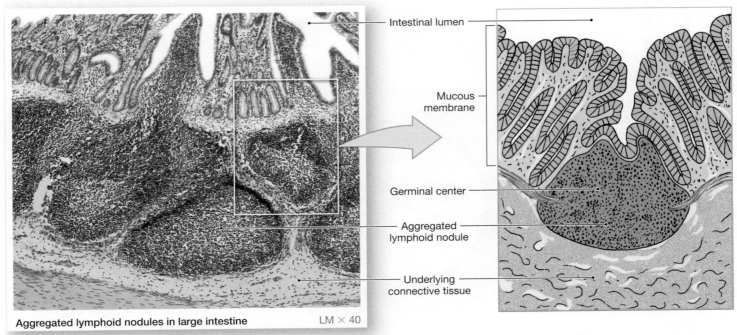

Intestinal lumen

Mucous membrane

Germinal center

Aggregated lymphoid nodule

Underlying connective tissue

Aggregated lymphoid nodules in large intestine   LM × 40

a Aggregated lymphoid nodules are shown in section.

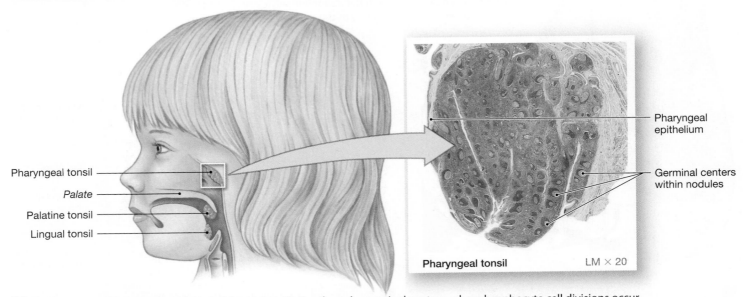

Pharyngeal tonsil

Palate

Palatine tonsil

Lingual tonsil

Pharyngeal epithelium

Germinal centers within nodules

Pharyngeal tonsil   LM × 20

b The positions of the tonsils and a tonsil in section. Notice the pale germinal centers, where lymphocyte cell divisions occur.

Procedures

1. Review the anatomy of the spleen in Figure 38.7.
2. Locate the spleen on the torso model or chart. Identify the hilus, splenic artery, and splenic vein. On the visceral surface, locate the gastric area of the spleen, which is in contact with the stomach, and the renal area, which is in contact with the kidneys.

3. Examine the spleen slide at low magnification and identify the dark-stained regions of white pulp and the lighter regions of red pulp. Examine several white pulp trabecular artery masses for the presence of an artery. ◼

Figure 38.7 **The Spleen**

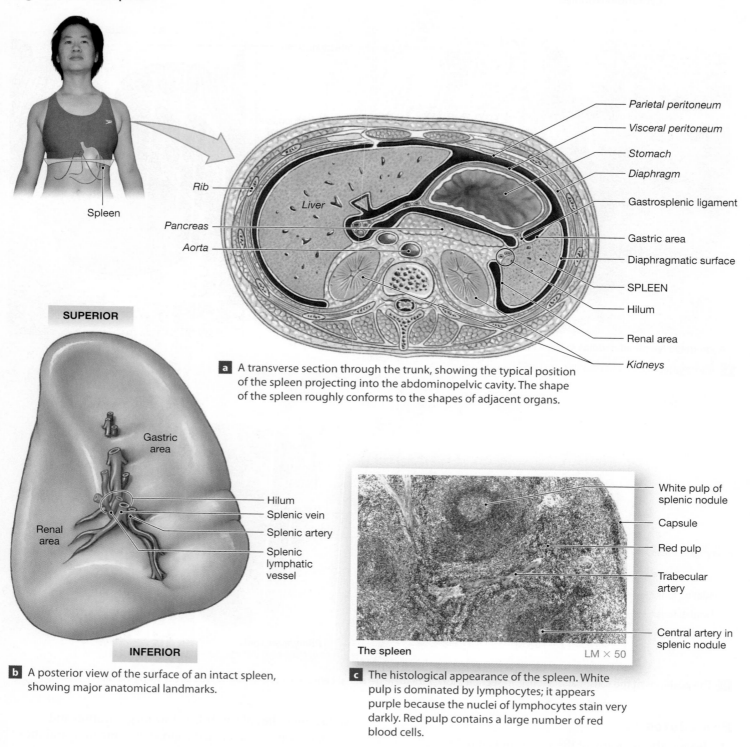

Spleen

a A transverse section through the trunk, showing the typical position of the spleen projecting into the abdominopelvic cavity. The shape of the spleen roughly conforms to the shapes of adjacent organs.

Rib

Liver

Pancreas

Aorta

Parietal peritoneum

Visceral peritoneum

Stomach

Diaphragm

Gastrosplenic ligament

Gastric area

Diaphragmatic surface

SPLEEN

Hilum

Renal area

Kidneys

SUPERIOR

Gastric area

Renal area

Hilum

Splenic vein

Splenic artery

Splenic lymphatic vessel

INFERIOR

b A posterior view of the surface of an intact spleen, showing major anatomical landmarks.

The spleen   LM × 50

White pulp of splenic nodule

Capsule

Red pulp

Trabecular artery

Central artery in splenic nodule

c The histological appearance of the spleen. White pulp is dominated by lymphocytes; it appears purple because the nuclei of lymphocytes stain very darkly. Red pulp contains a large number of red blood cells.

Name _____

Date _____

Section _____

Lymphatic System

A. Matching

Match each structure in the left column with its correct description in the right column.

_____ **1.** efferent vessel

_____ **2.** medullary cords

_____ **3.** cisterna chyli

_____ **4.** right lymphatic duct

_____ **5.** red pulp

_____ **6.** lymph node

_____ **7.** thoracic duct

_____ **8.** white pulp

_____ **9.** lymph

_____ **10.** afferent vessel

A. empties into right subclavian vein

B. empties into lymph node

C. splenic tissue containing red blood cells

D. fluid in lymphatic vessels

E. lymphocytes deep in node

F. empties into left subclavian vein

G. full of macrophages and lymphocytes

H. drains lymph node

I. lymphocytes surrounding trabecular artery

J. saclike region of thoracic duct

B. Short-Answer Questions

1. Describe the organization of a lymph node.

2. Discuss the major functions of the lymphatic system.

3. Explain how lymph is returned to the blood.

4. Describe the anatomy of the spleen.

C. Labeling

Label the structure of a lymph node in **Figure 38.8**.

Figure 38.8 Structure of a Lymph Node

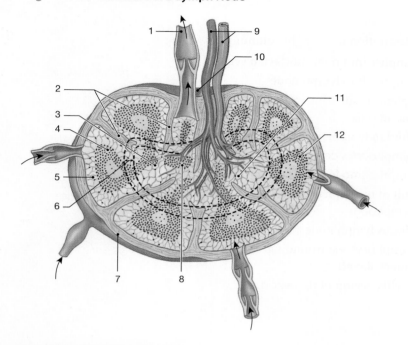

1. _____
2. _____
3. _____
4. _____
5. _____
6. _____
7. _____
8. _____
9. _____
10. _____
11. _____
12. _____

D. Analysis and Application

1. How are blood plasma, extracellular fluid, and lymph interrelated?

2. How does the way lymph drains from the right thoracic duct differ from the way it drains from the left thoracic duct?

E. Clinical Challenge

1. Explain the occurrence of edema in patients who are bedridden.

2. Explain the reasoning for sending a biopsy from a lump on a woman's breast to the pathology lab for microscopic analysis and evaluation.

Anatomy of the Respiratory System

Learning Outcomes

On completion of this exercise, you should be able to:

1. Identify and describe the structures of the nasal cavity.

2. Distinguish among the three regions of the pharynx.

3. Identify and describe the cartilages and ligaments of the larynx.

4. Identify the gross and microscopic structure of the trachea.

5. Identify and describe the gross and microscopic structure of the lungs.

6. Classify the branches of the bronchial tree.

All cells require a constant supply of oxygen (O_2) for the oxidative reactions of mitochondrial ATP production. A major by-product of these reactions is carbon dioxide (CO_2). The respiratory system exchanges these two gases between the atmosphere and the blood. Specialized organs of the airway filter, warm, and moisten the inhaled air before it enters the lungs. Once the air is in the lungs, the O_2 gas in the air diffuses into the surrounding capillaries to oxygenate the blood. As the blood takes up this oxygen, CO_2 gas in the blood diffuses into the lungs and is exhaled. Pulmonary veins return the oxygenated blood to the heart, where it is pumped into arteries of the systemic circulation.

The respiratory system, shown in **Figure 39.1**, consists of the nose, nasal cavity, sinuses, pharynx, larynx, trachea, bronchi, and lungs. The **upper respiratory system** includes the nose, nasal cavity, sinuses, and pharynx. These structures filter, warm, and moisten air before it enters the **lower respiratory system,** which

---

Need More Practice and Review?

Build your knowledge—and confidence!—in the Study Area of MasteringA&P® at www.masteringaandp.com with Pre-lab Quizzes, Post-lab Quizzes, Practice Anatomy Lab™ (PAL™) 3.0 virtual anatomy practice tool, PhysioEx™ 9.0 laboratory simulations, and A&P Flix™ with Quizzes.

PAL ⎸practice anatomy lab   For this lab exercise, follow these navigation paths:

- PAL>Human Cadaver>Respiratory System
- PAL>Anatomical Models>Respiratory System
- PAL>Histology>Respiratory System

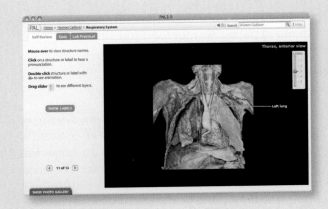

Figure 39.1 Structures of the Respiratory System Only the conducting portion of the respiratory system is shown; the smaller bronchioles and alveoli have been omitted.

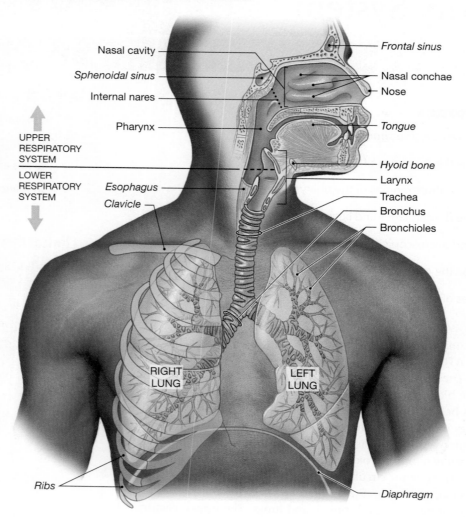

comprises the larynx, trachea, bronchi, and lungs. The larynx regulates the opening into the lower respiratory system and produces speech sounds. The trachea and bronchi maintain an open airway to the lungs where gas exchange occurs.

Make a Prediction

Predict which type of epithelial tissue lines the pharynx, which serves as a common passageway for food and air.

Lab Activity 1   Nose and Pharynx

The primary route for air entering the respiratory system is through two openings, the **external nares** (NA-rēz), or nostrils (Figure 39.2). Just inside each external naris is an expanded **nasal vestibule** (VES-ti-būl) containing coarse hairs. The hairs help to prevent large airborne materials such as dirt particles and insects from entering the respiratory system. The external portion

of the nose is composed of **nasal cartilages** that form the bridge and tip of the nose.

The **nasal cavity** is the airway from the external nares to the superior part of the pharynx. The perpendicular plate of the ethmoid and the vomer create the **nasal septum,** which divides the nasal cavity into right and left sides. The **superior, middle,** and **inferior nasal conchae** are bony shelves that project from the lateral walls of the nasal cavity. The distal edge of each nasal concha curls inferiorly and forms a tube, or **meatus,** that causes inhaled air to swirl in the nasal cavity. This turbulence moves the air across the sticky pseudostratified ciliated columnar epithelium lining, where dust and debris are removed. The floor of the nasal cavity is the superior portion of the **hard palate,** formed by the maxillae, palatine bones, and muscular **soft palate.** Hanging off the posterior edge of the soft palate is the conical **uvula** (Ū-vū-luh). The **internal nares** are the two posterior openings of the nasal cavity that connect with the superior portion of the pharynx.

Figure 39.2 The Nose, Nasal Cavity, and Pharynx

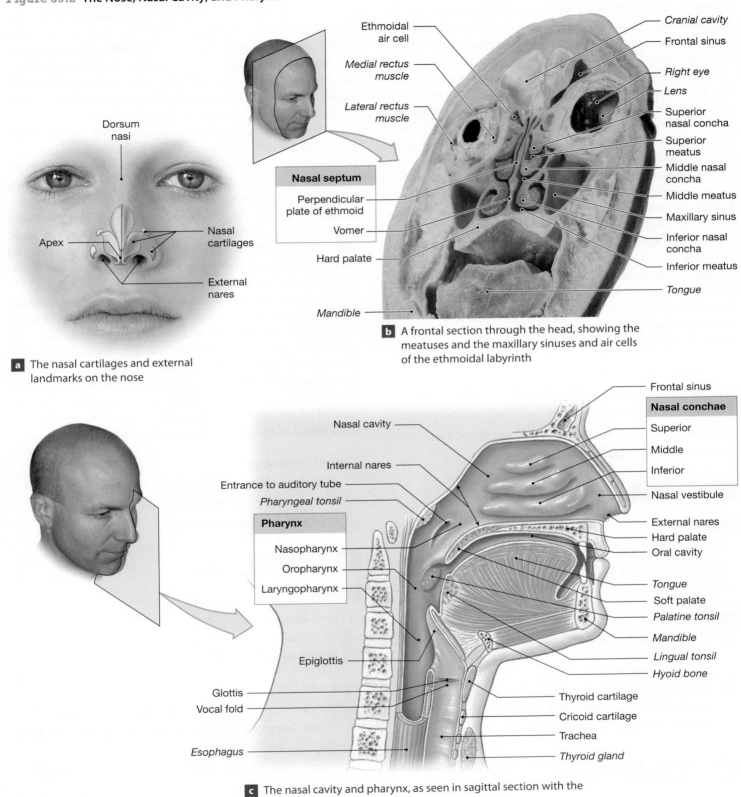

Dorsum nasi

Apex

Nasal cartilages

External nares

a The nasal cartilages and external landmarks on the nose

Ethmoidal air cell

Medial rectus muscle

Lateral rectus muscle

Nasal septum

Perpendicular plate of ethmoid

Vomer

Hard palate

Mandible

Cranial cavity

Frontal sinus

Right eye

Lens

Superior nasal concha

Superior meatus

Middle nasal concha

Middle meatus

Maxillary sinus

Inferior nasal concha

Inferior meatus

Tongue

b A frontal section through the head, showing the meatuses and the maxillary sinuses and air cells of the ethmoidal labyrinth

Nasal cavity

Internal nares

Entrance to auditory tube

Pharyngeal tonsil

Pharynx

Nasopharynx

Oropharynx

Laryngopharynx

Epiglottis

Glottis

Vocal fold

Esophagus

Frontal sinus

Nasal conchae

Superior

Middle

Inferior

Nasal vestibule

External nares

Hard palate

Oral cavity

Tongue

Soft palate

Palatine tonsil

Mandible

Lingual tonsil

Hyoid bone

Thyroid cartilage

Cricoid cartilage

Trachea

Thyroid gland

c The nasal cavity and pharynx, as seen in sagittal section with the nasal septum removed

The throat, or **pharynx** (FAR-inks), is divided into three regions: nasopharynx, oropharynx, and laryngopharynx. The **nasopharynx** (nā-zō-FAR-inks) is superior to the soft palate and serves as a passageway for airflow from the nasal cavity. Located on the posterior wall of the nasopharynx is the pharyngeal tonsil (Exercise 38). On the lateral walls are the openings of the auditory (pharyngotympanic) tubes (Exercise 32). The nasopharynx is lined with a pseudostratified ciliated columnar epithelium that functions to warm, moisten, and clean inhaled air. When a person is eating, food pushes past the uvula, and the soft palate raises to prevent the food from entering the nasopharynx.

The **oropharynx,** which extends inferiorly from the soft palate, is connected to the oral cavity at an opening called the **fauces** (FAW-sēz). The oropharynx contains the palatine and lingual tonsils (Exercise 38).

The **laryngopharynx** (la-rin-gō-FAR-inks) is located between the hyoid bone and the entrance to the esophagus, the muscular tube connecting the oral cavity with the stomach. (The esophagus is studied as part of the digestive system in Exercise 41.) The oropharynx and laryngopharynx have a stratified squamous epithelium to protect from abrasion by swallowed food passing through to the esophagus.

QuickCheck Questions

1.1 Name the components of the upper respiratory system.

1.2 Name the components of the lower respiratory system.

1.3 Describe the passageways into and out of the nasal cavity.

1.4 List the three regions of the pharynx.

In the Lab 1

Materials

☐ Head model
☐ Respiratory system chart
☐ Hand mirror

Procedures

1. Review the gross anatomy of the nose and pharynx in Figure 39.2. Locate these structures on the head model and respiratory system chart.

2. Using the hand mirror, examine the inside of your mouth. Locate your hard and soft palates, uvula, fauces, palatine tonsils, and oropharynx. ∎

Lab Activity 2   Larynx

The **larynx** (LAR-inks), or voice box, lies inferior to the laryngopharynx and anterior to cervical vertebrae C_4 through C_7. It consists of nine cartilages held together by **laryngeal ligaments.** The airway through the larynx is the **glottis** (Figure 39.2b).

Three large, unpaired cartilages form the body of the larynx (Figure 39.3). The first cartilage, the **epiglottis** (ep-i-GLOT-is), is the flap of elastic cartilage that lowers to cover the glottis during swallowing and helps direct the food to the esophagus. The **thyroid cartilage,** or Adam's apple, is composed of hyaline cartilage. It is visible under the skin on the anterior neck, especially in males. The **cricoid** (KRĪ-koyd) **cartilage** is a ring of hyaline cartilage forming the base of the larynx.

The larynx also has three pairs of smaller cartilages. The **arytenoid** (ar-i-TĒ-noyd) **cartilages** articulate with the superior border of the cricoid cartilage. **Corniculate** (kor-NIK-ū-lāt) **cartilages** articulate with the arytenoid cartilages and are involved in the opening and closing of the glottis and in the production of sound. The **cuneiform** (kū-NĒ-i-form) **cartilages** are club-shaped cartilages anterior to the corniculate cartilages. Spanning the glottis between the thyroid and arytenoid cartilages are two pairs of ligaments; the **superior vestibular ligaments** and the **inferior vocal ligaments**.

The vestibular and vocal ligaments are covered in epithelium that extend into the glottis as thick folds (Figure 39.4). The **vestibular folds** are inflexible and prevent foreign materials from entering the glottis. The vestibular folds also close the glottis during coughing and sneezing. Inferior to the vestibular ligaments are the elastic **vocal folds**, commonly called the *vocal cords*. These folds vibrate and produce speech and other sounds. Intrinsic muscles of the larynx move the arytenoid cartilages and change the tension on the vocal folds to produce different sounds.

QuickCheck Questions

2.1 How many pieces of cartilage are in the larynx?

2.2 What are the glottis and the epiglottis?

2.3 Describe the structures that produce speech.

In the Lab 2

Materials

☐ Larynx model
☐ Torso model
☐ Respiratory system chart

Procedures

1. Review the gross anatomy of the larynx in Figure 39.3.

2. On the larynx model, torso model, or respiratory system chart, do the following:

   ▪ Locate the thyroid cartilage. Is it continuous around the larynx?

   ▪ Locate the cricoid cartilage. Is it continuous around the larynx?

   ▪ Study the position of the epiglottis. How does it act like a chute to direct food into the esophagus?

Figure 39.3 The Larynx

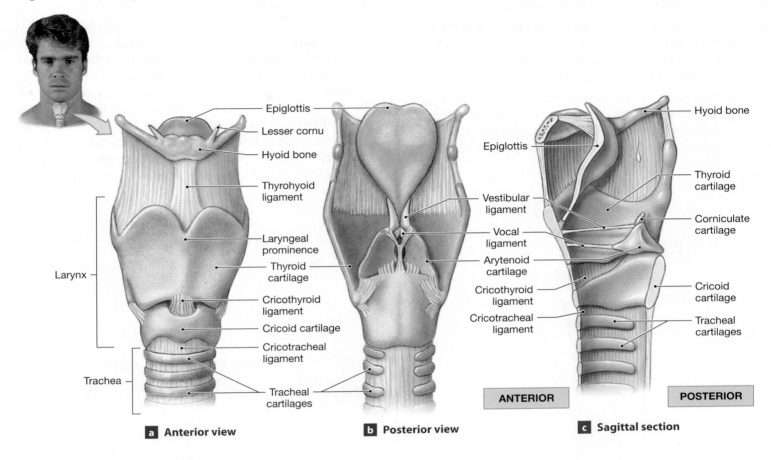

Epiglottis
Lesser cornu
Hyoid bone
Thyrohyoid ligament
Larynx
Laryngeal prominence
Thyroid cartilage
Cricothyroid ligament
Cricoid cartilage
Cricotracheal ligament
Trachea
Tracheal cartilages

a Anterior view

Vestibular ligament
Vocal ligament
Arytenoid cartilage

b Posterior view

Hyoid bone
Epiglottis
Thyroid cartilage
Corniculate cartilage
Cricothyroid ligament
Cricotracheal ligament
Cricoid cartilage
Tracheal cartilages

ANTERIOR   POSTERIOR

c Sagittal section

Figure 39.4 The Glottis

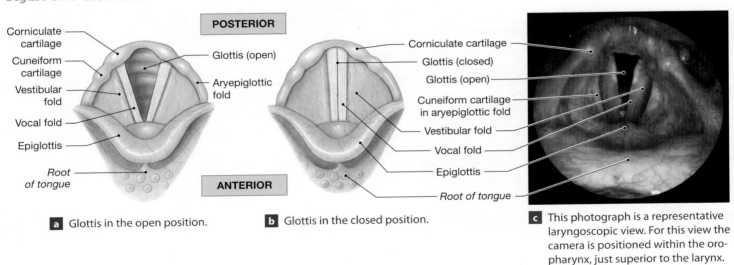

POSTERIOR

Corniculate cartilage
Cuneiform cartilage
Vestibular fold
Vocal fold
Epiglottis
Root of tongue
Glottis (open)
Aryepiglottic fold

ANTERIOR

a Glottis in the open position.

Corniculate cartilage
Glottis (closed)
Glottis (open)
Cuneiform cartilage in aryepiglottic fold
Vestibular fold
Vocal fold
Epiglottis
Root of tongue

b Glottis in the closed position.

c This photograph is a representative laryngoscopic view. For this view the camera is positioned within the oropharynx, just superior to the larynx.

■ Open the larynx model, and identify the arytenoid, corniculate, and cuneiform cartilages.

■ Locate the vestibular and vocal ligaments and folds.

3. Put your finger on your thyroid cartilage and swallow. How does the cartilage move when you swallow? Is it possible to swallow and make a sound simultaneously?

4. While holding your thyroid cartilage, first make a high-pitched sound and then make a low-pitched sound. Describe the tension in your throat muscles for each sound, and relate the muscle tension to the tension in the vocal folds. ■

Lab Activity 3 Trachea and Primary Bronchi

The **trachea** (TRĀ-kē-uh), or windpipe, is a tubular structure approximately 11 cm (4.25 in.) long and 2.5 cm (1 in.) in diameter (Figure 39.5). It lies anterior to the esophagus and can be felt on the front of the neck inferior to the thyroid cartilage of the larynx. Along the length of the trachea are 15 to 20 C-shaped pieces of hyaline cartilage called **tracheal cartilages** that keep the airway open. The **trachealis muscle** holds the two tips of each C-shaped tracheal cartilage together posteriorly. This

Figure 39.5 The Trachea and Primary Bronchi

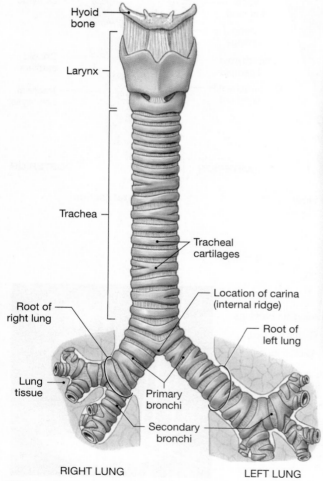

a This diagrammatic view shows the relationship of the trachea to the larynx and bronchi.

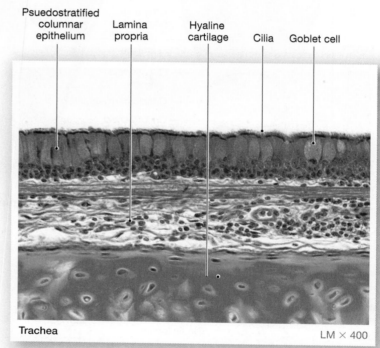

Trachea

LM × 400

b The lumen of the trachea is lined with a ciliated respiratory epithelium; the airway is kept open by hyaline cartilage formed into C-shaped rings called tracheal cartilages.

muscle allows the esophagus diameter to increase during swallowing so that the esophagus wall presses against the adjacent trachea wall and decreases the trachea diameter momentarily.

The trachea is lined with a pseudostratified ciliated columnar epithelium that is constantly sweeping the airway clean. Interspersed in the epithelium are goblet cells that secrete mucus to trap particles present in the inhaled air.

The trachea divides, at a ridge called the **carina,** into the left and right **primary bronchi** (BRONG-kī; singular *bronchus*). The right primary bronchus is wider and more vertical than the left primary bronchus. (For this reason, objects that are accidentally inhaled often enter the right primary bronchus.)

QuickCheck Questions

3.1 What is the lining epithelium of the trachea?

3.2 What is the connective tissue of a tracheal cartilage?

In the Lab   3

Materials

- ☐ Compound microscope
- ☐ Prepared slide of trachea
- ☐ Head model
- ☐ Torso model
- ☐ Lung model
- ☐ Respiratory system chart

Procedures

1. Review the gross anatomy of the trachea in Figure 39.5. Locate these structures on the head, torso, and lung models and the respiratory system chart. Palpate your trachea for the tracheal cartilages.

2. On the trachea slide, locate the structures labeled in Figure 39.5.

3. ***Draw It!*** Sketch a section of the trachea in the space provided.

Trachea section

4. Study the bronchial tree on the torso and/or lung models, and identify the right and left primary bronchi. ■

Clinical Application Asthma

Asthma (AZ-ma) is a condition that occurs when the smooth muscle encircling the delicate bronchioles contracts and reduces the diameter of the airway. The airway is further compromised by increased mucus production and inflammation of the epithelial lining. The individual has difficulty breathing, especially during exhalation, as the narrowed passageways collapse under normal respiratory pressures. An asthma attack can be triggered by a number of factors, including allergies, chemical sensitivities, air pollution, stress, and emotion.

Bronchodilator drugs are used to relax the smooth muscle and open the airway; other drugs reduce inflammation of the mucosa. *Albuterol* is an important bronchodilator, usually administered as an inhalant sprayed from a nebulizer. ■

Lab Activity 4   Lungs and Bronchial Tree

Each lung sits inside a pleural cavity located between the two layers of the pleura (Figure 39.6). The parietal pleura lies against the thoracic wall and the visceral pleura adheres to the surface of the lung. The pleural cavity between these layers contains pleural fluid that reduces friction on the lungs during breathing.

The lungs are a pair of cone-shaped organs lying in the thoracic cavity (Figure 39.7). The **apex** is the conical top of each lung, and the broad inferior portion is the **base.** The anterior, lateral, and posterior surfaces of each lung face the thoracic cage, and the medial surface faces the mediastinum. The heart lies on a medial concavity of the left lung called the **cardiac notch.** Each lung has a slitlike **hilum** on the medial surface where the bronchi, blood vessels, lymphatic vessels, and nerves access the lung.

Each lung is divided into lobes, two in the left lung and three in the right lung. Both lungs have an **oblique fissure** forming the lobes, and the right lung also has a **horizontal fissure.** The oblique fissure of the left lung separates the lung into its **superior** and **inferior lobes.** The oblique fissure of

Figure 39.6 **The Relationship Between the Lungs and the Heart** This transverse section was taken at the level of the cardia notch.

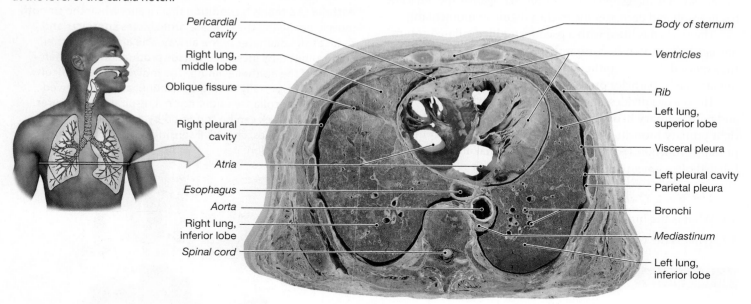

the right lung separates the **middle lobe** from the **inferior lobe,** and the horizontal fissure separates the middle lobe from the superior lobe.

The primary bronchi, called **extrapulmonary bronchi,** branch into increasingly smaller **intrapulmonary bronchi** to conduct air into the lungs (**Figure 39.8**). This branching pattern formed by the divisions of the bronchial structures is called the **bronchial tree.** At the superior terminus of the tree, the primary bronchi branch into as many **secondary bronchi** as there are lobes in each lung. The right lung has three lobes, and each lobe receives a secondary bronchus to supply it with air. The left lung has two lobes, and thus two secondary bronchi branch off the left primary bronchus. The secondary bronchi divide into **tertiary bronchi,** also called *segmental bronchi*. Smaller divisions called **bronchioles** branch into **terminal bronchioles.** The terminal bronchioles branch into **respiratory bronchioles,** which further divide into the narrowest passageways, the **alveolar ducts.**

As the bronchial tree branches from the primary bronchi to the respiratory bronchioles, cartilage is gradually replaced with smooth muscle tissue. The epithelial lining of the bronchial tree also changes from pseudostratified ciliated columnar at the superior end of the tree to simple squamous epithelium at the inferior end.

Inside a lobe, the region supplied by each tertiary bronchi is called a **bronchopulmonary segment** (**Figure 39.9**).

Subregions within each bronchopulmonary segment are called **lobules,** and each lobule is made up of numerous tiny air pockets called **alveoli** (al-VĒ-ō-lī; singular *alveolus*). Groups of alveoli clustered together are called **alveolar sacs.** Each lobule is served by a single terminal bronchiole. Inside a lobule, at the finest level of the bronchial tree, each alveolar duct serves a number of alveolar sacs.

The walls of the alveoli are constructed of simple squamous epithelium. Scattered throughout the simple squamous epithelium are **septal cells** that secrete an oily coating to prevent the alveoli from sticking together after exhalation. Also in the alveolar wall are macrophages that phagocytize debris. Pulmonary capillaries cover the exterior of the alveoli, and gas exchange occurs across the thin alveolar walls. Oxygen from inhaled air diffuses through the simple squamous epithelium of the alveolar wall, moves across the basal lamina membrane and the endothelium of the capillary, and enters the blood. The thickness of the combined alveolar wall and capillary wall is only about 0.5 mm, a size that permits rapid gas exchange between the alveoli and blood.

QuickCheck Questions

4.1 How many lobes does each lung have?

4.2 Which lung has the cardiac notch?

4.3 What is the bronchial tree?

Figure 39.7 Gross Anatomy of the Lungs

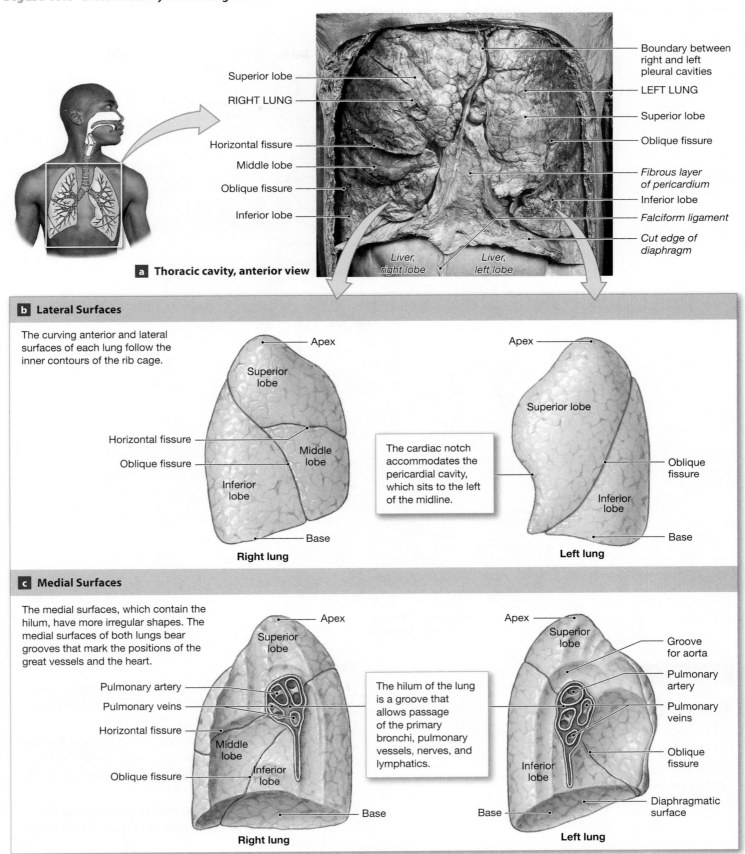

Superior lobe

RIGHT LUNG

Horizontal fissure

Middle lobe

Oblique fissure

Inferior lobe

Boundary between right and left pleural cavities

LEFT LUNG

Superior lobe

Oblique fissure

Fibrous layer of pericardium

Inferior lobe

Falciform ligament

Cut edge of diaphragm

Liver, right lobe

Liver, left lobe

a **Thoracic cavity, anterior view**

b **Lateral Surfaces**

The curving anterior and lateral surfaces of each lung follow the inner contours of the rib cage.

Apex

Superior lobe

Horizontal fissure

Oblique fissure

Middle lobe

Inferior lobe

Base

Right lung

Apex

Superior lobe

The cardiac notch accommodates the pericardial cavity, which sits to the left of the midline.

Oblique fissure

Inferior lobe

Base

Left lung

c **Medial Surfaces**

The medial surfaces, which contain the hilum, have more irregular shapes. The medial surfaces of both lungs bear grooves that mark the positions of the great vessels and the heart.

Apex

Superior lobe

Pulmonary artery

Pulmonary veins

Horizontal fissure

Middle lobe

Oblique fissure

Inferior lobe

Base

Right lung

The hilum of the lung is a groove that allows passage of the primary bronchi, pulmonary vessels, nerves, and lymphatics.

Apex

Superior lobe

Groove for aorta

Pulmonary artery

Pulmonary veins

Oblique fissure

Inferior lobe

Base

Diaphragmatic surface

Left lung

Figure 39.8 Bronchi and Bronchioles For clarity, the degree of branching has been reduced; an airway branches approximately 23 times before reaching the level of a lobule.

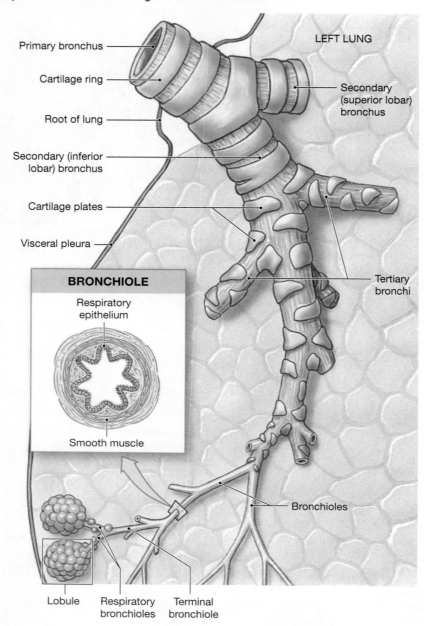

In the Lab   4

Materials

☐ Compound microscope
☐ Prepared slide of lung
☐ Torso model
☐ Respiratory system chart

Procedures

1. Review the gross anatomy of the lungs in Figure 39.6. Locate these structures on the torso models and on the respiratory system chart.

2. On the model, examine the right lung, and observe how the horizontal and oblique fissures divide it into three lobes. Note how the oblique fissure separates the left lung into two lobes.

3. Examine the model for the parietal pleura lining the thoracic wall. Where is the pleural cavity relative to the parietal pleura?

Figure 39.9 **Lobules and Alveoli of the Lung**

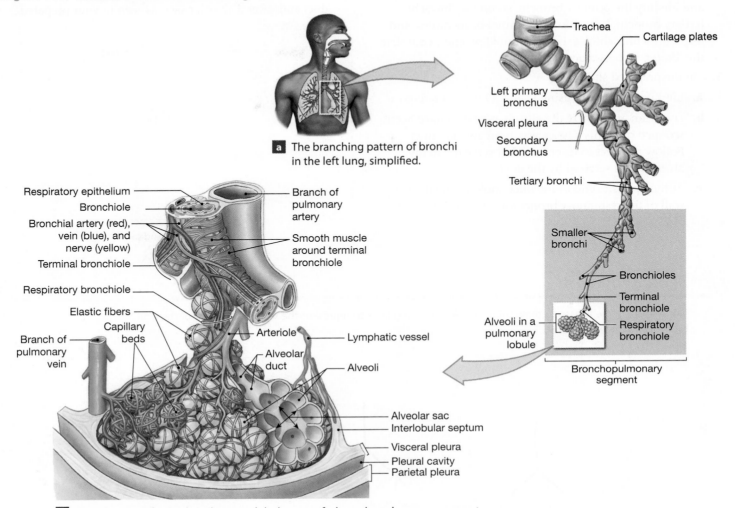

a The branching pattern of bronchi in the left lung, simplified.

Respiratory epithelium
Bronchiole
Bronchial artery (red), vein (blue), and nerve (yellow)
Terminal bronchiole
Respiratory bronchiole
Elastic fibers
Capillary beds
Branch of pulmonary vein

Branch of pulmonary artery
Smooth muscle around terminal bronchiole
Arteriole
Lymphatic vessel
Alveolar duct
Alveoli
Alveolar sac
Interlobular septum
Visceral pleura
Pleural cavity
Parietal pleura

Trachea
Cartilage plates
Left primary bronchus
Visceral pleura
Secondary bronchus
Tertiary bronchi
Smaller bronchi
Bronchioles
Terminal bronchiole
Respiratory bronchiole
Alveoli in a pulmonary lobule
Bronchopulmonary segment

b The structure of a single pulmonary lobule, part of a bronchopulmonary segment.

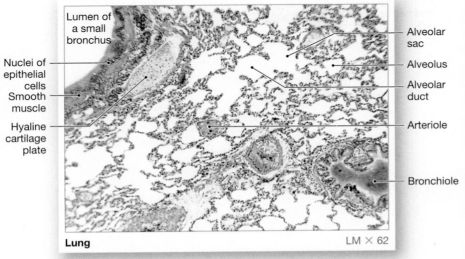

Lumen of a small bronchus
Nuclei of epithelial cells
Smooth muscle
Hyaline cartilage plate

Alveolar sac
Alveolus
Alveolar duct
Arteriole
Bronchiole

Lung   LM × 62

c This transverse section of lung shows a hyaline cartilage plate next to a small bronchus.

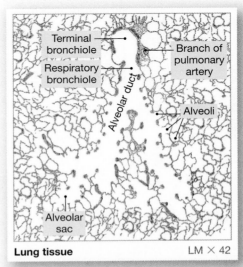

Terminal bronchiole
Respiratory bronchiole
Alveolar duct
Branch of pulmonary artery
Alveoli
Alveolar sac

Lung tissue   LM × 42

d This micrograph shows the distribution of a respiratory bronchiole that supplies a portion of a lobule.

4. Study the bronchial tree on the torso and/or lung models, and identify the primary bronchi, secondary bronchi, tertiary bronchi, bronchioles, terminal bronchioles, and respiratory bronchioles. For additional practice, complete the *Sketch to Learn* activity below.

5. On the prepared slide,

   a. Identify the alveoli, using Figure 39.9d as a reference.

   b. Locate an area where the alveoli appear to have been scooped out. This passageway is an alveolar duct. Follow the duct to its end, and observe the many alveolar sacs serviced by the duct.

   c. At the opposite end of the duct, look for the thicker wall of the respiratory bronchiole and blood vessels.

6. *Draw It!* In the space provided, sketch the alveolar duct and several alveolar sacs, as seen in your prepared slide. ■

Alveolar ducts and sacs

Sketch to Learn

The bronchial tree is easy to draw using simple branching lines to represent the various divisions of the bronchi and the bronchioles.

Sample Sketch

 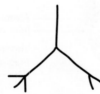 

Step 1
• Draw lines for the trachea and primary bronchi.

Step 2
• Add three secondary bronchi to the right primary bronchus and two secondaries on the left.

Step 3
• Draw branches off of each secondary bronchus.
• Add another set of branches for the bronchioles.
• Label your drawing

Your Sketch

Name _____

Date _____

Section _____

Anatomy of the Respiratory System

A. Matching

Match each structure in the left column with its correct description in the right column.

_____	**1.** C-shaped rings	**A.**	voice box
_____	**2.** internal nares	**B.**	elastic cartilage flap of larynx
_____	**3.** cricoid cartilage	**C.**	serous membrane of lungs
_____	**4.** pleurae	**D.**	left lung
_____	**5.** epiglottis	**E.**	connects nasal cavity with throat
_____	**6.** larynx	**F.**	tracheal cartilage
_____	**7.** vocal folds	**G.**	vocal cords
_____	**8.** cardiac notch	**H.**	protect vocal folds
_____	**9.** external nares	**I.**	nostrils
_____	**10.** three lobes	**J.**	base of larynx
_____	**11.** thyroid cartilage	**K.**	right lung
_____	**12.** vestibular folds	**L.**	Adam's apple

B. Short-Answer Questions

1. List the components of the upper and lower respiratory systems.

2. What are the functions of the superior, middle, and inferior conchae?

3. Where is the pharyngeal tonsil located?

4. Trace a breath of air from the external nares through the respiratory system to the alveolar sacs.

C. Labeling

Label the anatomy of the nose in **Figure 39.10**.

Figure 39.10 **The Nose, Nasal Cavity, and Pharynx**

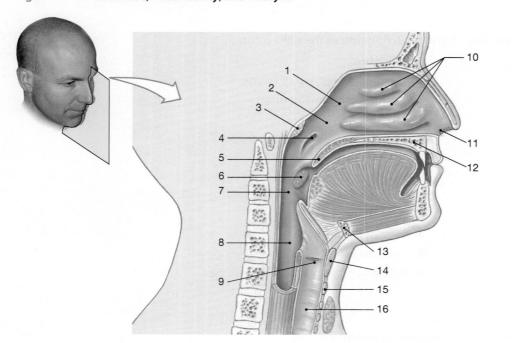

1. _____
2. _____
3. _____
4. _____
5. _____
6. _____
7. _____
8. _____
9. _____
10. _____
11. _____
12. _____
13. _____
14. _____
15. _____
16. _____

D. Analysis and Application

1. Where do goblet cells occur in the respiratory system, and what function do they serve?

2. What is the function of stratified squamous epithelium that lines the oropharynx and laryngopharynx?

E. Clinical Challenge

1. How does an asthma attack cause difficulty in breathing?

2. Emphysema from smoking and exposure to heavy pollution causes alveoli to expand and rupture. Describe how this would compromise respiratory function.

Physiology of the Respiratory System

Learning Outcomes

On completion of this exercise, you should be able to:

1. Discuss pulmonary ventilation, internal respiration, and external respiration.

2. Describe how the respiratory muscles move during inspiration and expiration.

3. Define the various lung capacities and explain how they are measured.

4. Show how to use a dry spirometer.

5. Observe, record, and/or calculate selected pulmonary volumes and capacities.

Respiration has three phases: pulmonary ventilation, external respiration, and internal respiration. Breathing, or **pulmonary ventilation,** is the movement of air into and out of the lungs. This movement requires coordinated contractions of the diaphragm, intercostal muscles, and abdominal muscles. **External respiration** is the exchange of gases between the lungs and the blood. Inhaled air is rich in oxygen, and this gas constantly diffuses through the alveolar wall of the lungs into the blood of the pulmonary capillaries. Simultaneously, carbon dioxide diffuses out of the blood and into the lungs, from where it is exhaled. The freshly oxygenated blood is pumped to the tissues to deliver the oxygen and take up carbon dioxide. **Internal respiration** is the exchange of gases between the blood and the tissues. To reinforce your understanding of the phases of respiration, complete the *Sketch to Learn* activity (on p. 583).

Pulmonary ventilation consists of inspiration and expiration. **Inspiration** is inhalation, the movement of oxygen-rich air into the lungs. **Expiration,** or exhalation, involves emptying the carbon dioxide–laden air from the lungs into the atmosphere. The average respiratory rate is approximately 12 breaths per minute.

Need More Practice and Review?

Build your knowledge—and confidence!—in the Study Area of MasteringA&P® at **www.masteringaandp.com** with Pre-lab Quizzes, Post-lab Quizzes, Practice Anatomy Lab™ (PAL™) 3.0 virtual anatomy practice tool, PhysioEx™ 9.0 laboratory simulations, and A&P Flix™ with Quizzes.

PhysioEx˙ 9.0   For this lab exercise, go to this topic:
  • PhysioEx Exercise 7: Respiratory System Mechanics

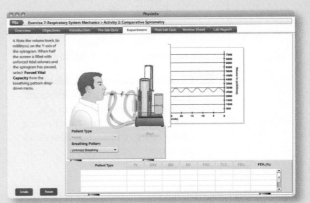

MasteringA&P®

This rate is modified by many factors, however, such as exercise and stress, which increase the rate, and sleep and depression, which decrease it.

For pulmonary ventilation to take place, the pressure in the thoracic cavity must be different from **atmospheric pressure,** which is the pressure of the air outside the body. Atmospheric pressure is normally 760 mm Hg, or approximately 15 pounds per square inch (psi). For inspiration to occur, the pressure in the thoracic cavity must be lower than atmospheric pressure, and for expiration the thoracic pressure must be higher than atmospheric pressure. **Pressure** is defined as the amount of force applied to a given surface area; and a relationship called **Boyle's law** explains how changing the size of the thoracic cavity (and thereby changing the volume of the lungs) creates the pressure gradient necessary for breathing.

The law states that the pressure of a gas in a closed container is inversely proportional to the volume of the container. Simply put, if the container is made smaller, the gas molecules exert the same amount of force on a smaller surface area and therefore the gas pressure increases.

Figure 40.1 illustrates the mechanisms of pulmonary ventilation. When the diaphragm is relaxed, it is dome shaped. As it contracts, it lowers and flattens the floor of the thoracic cavity. This results in an increase in thoracic volume and consequently a decrease in thoracic pressure. Simultaneously, the external intercostal muscles contract and elevate the rib cage, further increasing thoracic volume and decreasing the pressure. This decrease in thoracic pressure causes a concurrent expansion of the lungs and a decrease in the pressure of the air in the lungs, the **intrapulmonic** (in-tra-PUL-mah-nik) **pressure.**

Figure 40.1 **Mechanisms of Pulmonary Ventilation**

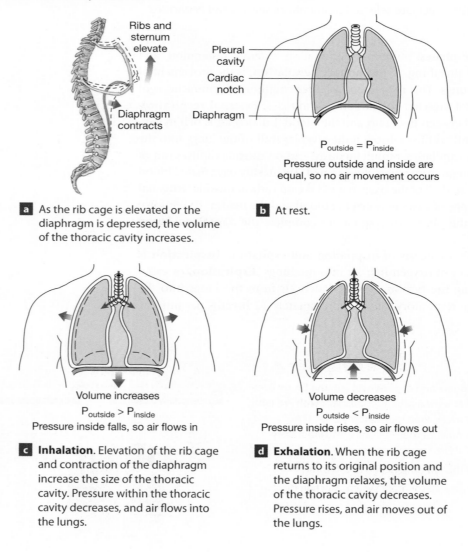

Ribs and sternum elevate

Diaphragm contracts

Pleural cavity

Cardiac notch

Diaphragm

$P_{outside} = P_{inside}$

Pressure outside and inside are equal, so no air movement occurs

a As the rib cage is elevated or the diaphragm is depressed, the volume of the thoracic cavity increases.

b At rest.

Volume increases

$P_{outside} > P_{inside}$

Pressure inside falls, so air flows in

Volume decreases

$P_{outside} < P_{inside}$

Pressure inside rises, so air flows out

c **Inhalation**. Elevation of the rib cage and contraction of the diaphragm increase the size of the thoracic cavity. Pressure within the thoracic cavity decreases, and air flows into the lungs.

d **Exhalation**. When the rib cage returns to its original position and the diaphragm relaxes, the volume of the thoracic cavity decreases. Pressure rises, and air moves out of the lungs.

Sketch to Learn

Let's do a process drawing this time and create a sketch that summarizes pulmonary ventilation, external respiration, and internal respiration.

Sample Sketch

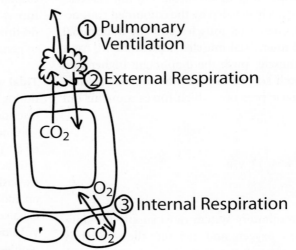

① Pulmonary Ventilation

② External Respiration

CO_2

O_2

③ Internal Respiration

O_2

CO_2

Step 1
- Draw an alveolar sac.
- Add a rectangle for blood vessels.
- Include a few cells at the bottom.

Step 2
- Draw a 2-headed arrow in the alveolar duct for pulmonary ventilation.
- Show which way CO_2 and O_2 move by drawing arrows.
- Label as shown.

Your Sketch

Once intrapulmonic pressure falls below atmospheric pressure, air flows into the lungs.

Inspiration is an **active process** because it requires the contraction of several muscles to change pulmonary volumes and pressures. Expiration is essentially a **passive process** that occurs when the muscles just used in inspiration relax and the thoracic wall and elastic lung tissue recoil. During exercise, however, air may be actively exhaled by the combined contractions of the internal intercostal muscles and the abdominal muscles. The internal intercostal muscles depress the rib cage, and the abdominal muscles push the diaphragm higher into the thoracic cavity. Both actions decrease the thoracic volume and increase the thoracic pressure, which forces more air out during exhalation.

Lab Activity 1 Lung Volumes and Capacities

During exercise, the respiratory system must supply the muscular system with more oxygen, and therefore the respiratory rate increases, as does the volume of air inhaled and exhaled. Pulmonary volumes and capacities are generally measured when assessing health of the respiratory system, because these values change with pulmonary disease. In this section you will measure a variety of respiratory volumes.

An instrument called a **spirometer** (spī-ROM-e-ter) is used to measure respiratory volumes. The wet spirometer is a bell-shaped container inside a chamber filled with water. As you exhale into the mouthpiece of the spirometer, the air displaces water in the container and the container rises and moves a gauge that indicates the volume of air exhaled. As you do this activity, keep in mind that lung volumes vary according to gender, height, age, and overall physical condition.

Figure 40.2 shows the volumes you will be measuring. **Tidal volume (TV)** is the amount of air one inspires and exhales during normal resting breathing. Tidal volume averages 500 mL, but additional air can be inhaled or exhaled beyond the tidal volume. The **inspiratory reserve volume (IRV)** is the amount of air that can be forcibly inspired above a normal inhalation. This volume averages 3300 mL; if you were to gulp in air while exercising strenuously, you would take in the 500 mL of TV plus an additional 3300 mL. The amount of air that can be forcefully exhaled after a normal exhalation is the **expiratory reserve volume (ERV).** The ERV averages 1000 mL, which means you would expel the 500 mL of TV plus another 1000 mL.

Vital capacity (VC) is the maximum amount of air that can be exhaled from the lungs after a maximum inhalation. This volume averages 4800 mL in men and 3100 mL in women

Figure 40.2 Respiratory Volumes and Capacities The graph diagrams the relationships between respiratory volumes and capacities.

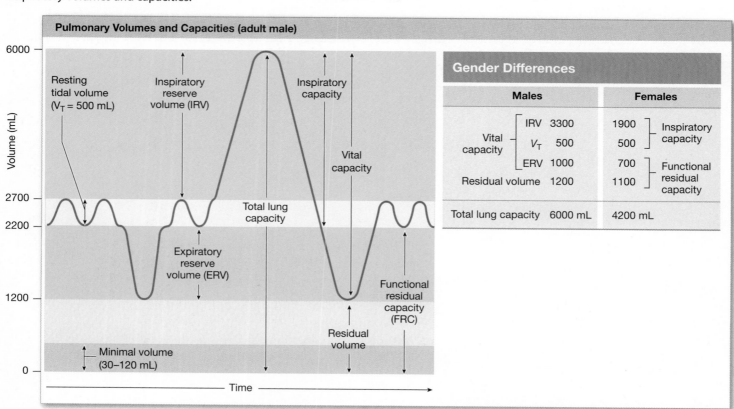

and includes the combined volumes of the IRV, TV, and ERV: VC = IRV + TV + ERV. The average vital capacity for individuals of your age, height, and gender is called the **predicted vital capacity (PVC).** The equation for determining your PVC will be provided later in this lab activity. Note that your vital capacity may differ from the average for numerous reasons. Genetics has an influence on potential lung capacities, for instance, and lung damage from smoking or air pollution decreases vital capacity. On the positive side, cardiovascular exercises such as swimming and jogging, increases lung volumes.

The respiratory system always contains some air. The **residual volume (RV)** is the amount of air that cannot be forcefully exhaled from the lungs. Surfactant produced by the septal cells of the alveoli prevents the alveoli from collapsing completely during exhalation. Because the alveoli are not allowed to empty completely, they always maintain a residual volume of air, which averages 1200 mL. **Minimal volume** is the amount of residual air—usually 30 to 120 mL—that stays in the lungs even if they are collapsed.

To calculate the **total lung capacity (TLC),** which averages 6000 mL, the vital capacity is added to the residual volume: TLC = VC + RV.

Respiratory rate (RR) is the number of breaths taken per minute. RR multiplied by tidal volume gives the **minute volume (MV),** defined as the amount of air exchanged between the lungs and the environment in one minute: MV = TV × RR.

Make a Prediction

Consider your overall health and respiratory fitness. How close do you think your vital capacity will be to the predicted vital capacity for your age, height, and gender?

QuickCheck Questions

1.1 What instrument measures respiratory volumes?

1.2 What is vital capacity?

1.3 How is respiratory rate calculated?

 Safety Alert: Spirometer Use

1. Do not use the wet spirometer if you have a cold or a communicable disease.
2. Always use a clean mouthpiece on the spirometer. Do not reuse a mouthpiece that has been removed from a spirometer.
3. A wet spirometer measures exhalations only. The instrument does not use an air filter, so *do not inhale through it. Only exhale into the instrument.*
4. Discard used mouthpieces in the designated biohazard box as indicated by your laboratory instructor. ▲

In the Lab 1

Materials

- ☐ Wet spirometer
- ☐ Disposable mouthpieces
- ☐ Noseclip (optional)
- ☐ Clock or watch with second hand
- ☐ Laboratory partner
- ☐ Biohazard box

Procedures

Setup

1. Insert a new, clean mouthpiece onto the breathing tube of the spirometer.
2. Remember: Only exhale into the spirometer; the instrument cannot measure inspiratory volumes. To obtain as accurate a reading as possible, use a noseclip or your fingers to pinch your nose closed while exhaling into the spirometer.
3. Set the dial face to zero by turning the silver ring surrounding the dial face until the zero point on the scale is aligned with the point of the needle, see **Figure 40.3**.

Figure 40.3 Wet Spirometer

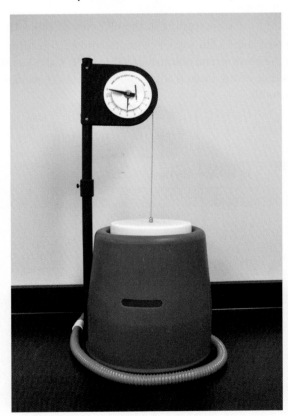

4. Following the steps listed next, measure each volume three times and then calculate an average.

Tidal Volume

1. Set the dial face so that the 1000 reading on the scale is aligned with the point of the needle. (You do this because tidal volume is small, and the scale on most dry spirometers is not graded before the 1000-mL setting.)

2. Take a normal breath, quickly place the mouthpiece in your mouth, pinch your nose closed with your fingers or a noseclip, and exhale normally into the spirometer. The exhalation should not be forcible; it should be more of a sigh.

3. Record the scale reading in the tidal-volume row of **Table 40.1**.

4. Repeat steps 1 and 2 twice. Record each reading in the appropriate column of Table 40.1. Calculate the average of the three readings and record that average in the rightmost column of the table.

Expiratory Reserve Volume

1. Set the dial face to zero.

2. Exhale normally into the air (not into the mouthpiece).

3. Stop breathing for a moment, place the mouthpiece in your mouth, pinch your nose closed with your fingers or a noseclip, and forcibly exhale all the remaining air from your lungs.

4. Record the scale reading in the ERV row of Table 40.1.

5. Repeat steps 1 through 4 twice. Record each reading in the appropriate column of Table 40.1. Calculate the average of the three readings and record that average in the table.

Vital Capacity

1. Set the dial face to zero.

2. Inhale maximally once, and then exhale maximally.

3. Inhale maximally, place the mouthpiece in your mouth, pinch your nose closed with your fingers or a noseclip, and forcibly exhale all the air from your lungs.

4. Record the scale reading in the vital capacity row of Table 40.1.

5. Repeat steps 1 through 4 twice. Record each reading in the appropriate column of Table 40.1. Calculate the average of the three readings and record that average in the table.

6. Use the equations provided in **Table 40.2** and calculate your predicted vital capacity. Then, compare your measured vital capacity (see Table 40.1) to your predicted vital capacity. Record these values in Table 40.2.

Respiratory Rate

1. Sit relaxed and read a textbook. Have your laboratory partner count the number of breaths you take in 20 seconds.

2. Multiply this number by 3 and record the value in the respiratory rate row of Table 40.1.

3. Repeat steps 1 and 2 twice. Record each reading in the appropriate column of Table 40.1. Calculate the average of the three readings and record that average in the table.

4. Calculate your average minute volume by multiplying your average tidal volume by your average respiratory rate. Enter this calculated value in Table 40.1.

Table 40.1	Spirometry Data			
Volume	**Reading 1**	**Reading 2**	**Reading 3**	**Average**
Tidal volume				
Expiratory reserve volume				
Vital capacity				
Respiratory rate				
Minute volume (calculated)				
Inspiratory reserve volume (calculated)				

Table 40.2	Comparison of Spirometry Data to Predicted Vital Capacity (PVC)				
Gender	PVC Equations	PVC	VC	Percent Difference between PVC and VC	
Male	PVC = 0.052H − 0.022A − 3.60				
Female	PVC = 0.041H − 0.018A − 2.69				

PVC = predicted vital capacity in liters (L)
H = height in centimeters (cm)
A = age in years

Inspiratory Reserve Volume

1. Use the average values for your vital capacity, tidal volume, and expiratory reserve volume to calculate your inspiratory reserve volume: IRV = VC − (TV + ERV).

2. Enter the calculated IRV in Table 40.1. ∎

Lab Activity 2 — BIOPAC
Volumes and Capacities

In this activity, you will use an **airflow transducer,** and a computer will convert airflow to volume. Although this is a quick method of obtaining lung capacity data, the disadvantage is that the recording procedure must be followed exactly for an accurate conversion from airflow to volume.

You will measure tidal volume, inspiratory reserve volume, and expiratory reserve volume and then use these data to calculate inspiratory capacity and vital capacity. The equations in Table 40.2 can be used to obtain predicted vital capacity based on gender, height, and age. For instance, the predicted vital capacity of a 19-year-old woman who is 167 cm tall (about 5.5 ft) is $0.041(167) − 0.018(19) − 2.69 = 3.8$ L.

QuickCheck Questions

2.1 What is the tidal volume of respiration?

2.2 What is the vital capacity volume of respiration?

In the Lab 2

Materials

☐ BIOPAC acquisition unit (MP36/35/30)

☐ BIOPAC software: Biopac Student Lab (BSL) v3.7.5 or better

☐ BIOPAC airflow transducer (SS11L or SS11LA)

☐ BIOPAC calibration syringe (AFT6)

☐ BIOPAC disposable bacteriological filters (AFT1), one per subject plus one for calibration

☐ BIOPAC mouthpiece, disposable (AFT2) or autoclavable (AFT8), one per subject

☐ BIOPAC noseclip (AFT3), one per subject

☐ Computer: PC Windows 7, Vista, or XP; Mac OS X 10.4-10.6

Procedures

This lesson has four sections: Setup, Calibration, Recording, and Data Analysis. Be sure to follow the setup instructions appropriate for your type of airflow transducer (SS11L or SS11LA). The calibration step is critical for getting accurate recordings. Four segments will be recorded and then analyzed. You may record the data analysis by hand or choose Edit > Journal > Paste Measurements to paste the data into your journal for future use.

Most markers and labels are automatically inserted into the data recordings. Markers appear at the top of the window as inverted triangles. This symbol indicates that you need to insert a marker and key in a marker label similar to the text in quotes. You can insert and label the marker during or after acquisition; on a Mac, press ESC; on a PC, press F9.

Section 1: Setup

1. Turn on your computer but keep the BIOPAC MP36/35/30 unit off.

2. Plug the airflow transducer (SS11L or SS11LA) into CH 1 as shown in **Figure 40.4** and turn the acquisition unit on.

3. Start the Biopac Student Lab program and choose Lesson 12 (L12-Pulmonary Function I). Type in the subject's filename.

Figure 40.4 Connecting the Airflow Transducer

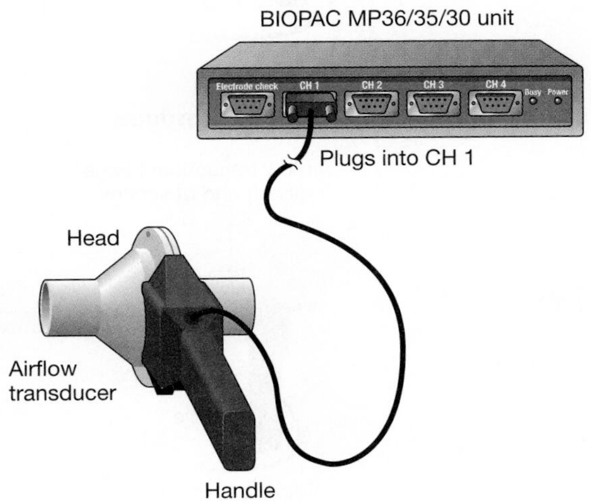

Section 2: Calibration

The calibration establishes the hardware's internal parameters and is critical for optimum performance. This exercise has two calibration stages: Stage 1 zeroes the baseline, which is critical for airflow to volume calculations, and is always required; Stage 2 sets the transducer amplitude and compensates for Standard temperature and Pressure (STP), and is only required once each time the BSL program is launched.

1. Hold the airflow transducer upright and still.
   Important: The transducer must be vertical to obtain a zero baseline and there must be no airflow through it.

2. Click on Calibrate. Two four-second recordings will be completed.

3. If prompted for stage 2, complete the calibration syringe assembly BEFORE clicking Calibrate or OK.

   a. Place a filter (AFT1) on the end of the calibration syringe (AFT6). The filter is required for calibration and recording because it forces the air to move smoothly through the transducer. This assembly can be left connected for future use. You need to replace the filter only if the paper inside the filter tears.

   b. Insert the syringe/filter assembly into the port of the transducer head (Figure 40.5). Inside the head is the sensor that measures airflow. If using the SS11L

transducer with nonremovable head, insert the assembly into the larger-diameter 120 port. If using the SS11LA transducer with removable, cleanable head, always insert the assembly on the transducer side labeled "Inlet" so that the transducer cable exits on the left, as shown in Figure 40.5.

   c. Pull the Calibration Syringe plunger all the way out. Hold the syringe horizontally and let the transducer hang upright off the end with no support.

   d. When you are ready to proceed, click on Calibrate or OK. The second calibration stage will begin and will run until you click on End Calibration.

4. Cycle the syringe plunger in and out completely five times (ten strokes), all the while holding the syringe with your hands placed as shown in Figure 40.6. Use a rhythm of about one second per stroke with a two-second rest between strokes: Take one second to push the plunger in completely, pause briefly and then take one second to pull the plunger out completely, then wait two seconds. Repeat this cycle four more times and then click on End Calibration. Check your calibration data, which should resemble Figure 40.7. If your screen shows five downward deflections and five upward deflections, proceed to Data Recording. If your screen shows any large spikes, click on Redo Calibration.

Figure 40.5 **Insertion of Calibration Syringe/Filter Assembly**

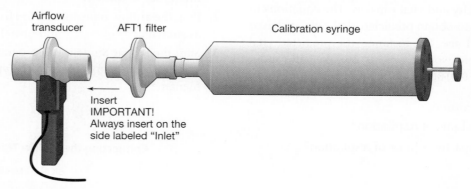

Figure 40.6 **Calibrating the Airflow Transducer**

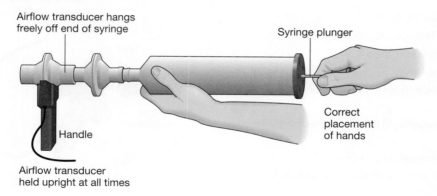

Figure 40.7 Calibration Data

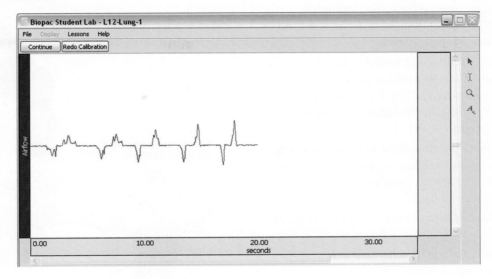

Section 3: Data Recording

To work efficiently, read this entire section now so that you will know what to do for each recording segment. You will be working with a partner, the subject, who should remain in a supine position and relaxed while you review the lesson. You will record airflow data for the subject for normal breathing, deep inhalation, deep exhalation, and return to normal breathing. The software will automatically calculate volumes based on the recorded airflow data.

Hints for obtaining optimal data:

a. Keep the airflow transducer upright at all times.

b. If you start the recording during an inhalation, try to end during an exhalation, and vice versa. This is not absolutely critical but does increase the accuracy of the calculations.

c. The subject should be facing away from the computer.

1. Find your transducer setup in **Figure 40.8**, and carefully follow the filter and mouthpiece instructions for that setup. *Important:* If your laboratory sterilizes the transducer heads after each use, make sure a clean head is installed now. Have the subject remove the filter and mouthpiece from the plastic packages. This mouthpiece will become the subject's personal one, and therefore the subject should write her or his name on the mouthpiece and filter with a permanent marker so they can be reused later.

Follow this procedure precisely to make sure the airflow transducer is sterile:

- If using the SS11L transducer with nonremovable head, insert a new filter and disposable mouthpiece (AFT1, AFT2) into the larger-diameter port on the transducer (Figure 40.8a).

Figure 40.8 Airflow Transducer Setups

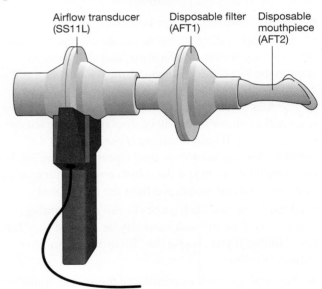

a SS11L (shown) or SS11LA with nonsterilized head

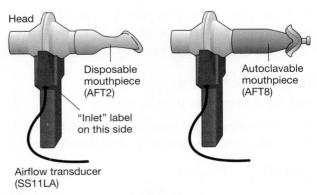

b S11LA with sterilized head

- If using the SS11LA transducer and *not sterilizing* the head after each use, insert a filter and disposable mouthpiece (AFT1, AFT2) into the transducer on the side labeled "Inlet" (Figure 40.8b).

- If the head will be sterilized in an autoclave after use, a filter is not required for the SS11LA transducer. Insert a disposable mouthpiece (ATF2) or an autoclavable mouthpiece (AFT8) into the transducer on the side labeled "Inlet" (Figure 40.8b).

2. Have the subject place the noseclip and begin breathing through the mouthpiece, holding the airflow transducer upright at all times and always breathing through the side labeled "Inlet" (**Figure 40.9**).

3. *After at least 20 seconds of normal breathing*, click on Continue and when ready click on Record and then have the subject:

   a. Breathe normally for five breaths. One breath is a complete inhale-exhale cycle.

   b. Inhale as deeply as possible.

   c. Exhale as deeply as possible.

   d. Breathe normally for five breaths.

4. Click on Stop, ending during an exhalation if you started the recording during an inhalation, and vice versa. As soon as the Stop button is pressed, the BSL software will automatically calculate volumes based on the recorded airflow data. At the end of the calculation, both an airflow wave and a volume wave will be displayed on the screen (**Figure 40.10**). If your recording is not similar to the volume waves on the screen, then repeat the recording. Your data would be incorrect if the subject coughed, for example, or if some exhaled air escaped from the mouthpiece.

5. Click on Done and click on Yes to exit the recording mode. Your data will automatically be saved in the "Data Files" folder. If you choose the "Record from another Subject" option:

   a. You will not need to recalibrate the airflow transducer. For this reason, all recordings should be completed before you proceed to data analysis.

   b. Remember to have each person use his or her own mouthpiece, bacterial filter, and noseclip.

   c. Repeat recording steps 1 through 4 for each new subject.

Section 4: Data Analysis

The first step is to evaluate the volume data.

1. Enter the Review Saved Data mode, and choose the correct file. Note the channel number designations:

Channel	Displays
CH 1	Airflow (hidden)
CH 2	Volume

Figure 40.9 Using the Airflow Transducer

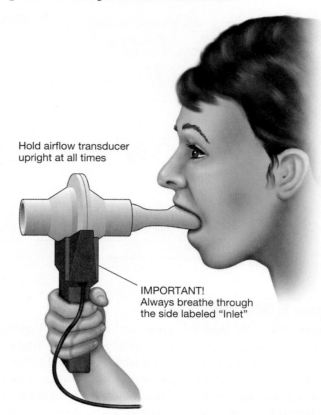

Hold airflow transducer upright at all times

IMPORTANT! Always breathe through the side labeled "Inlet"

Figure 40.10 Sample Recording CH 1 shows airflow. CH 2 shows volume.

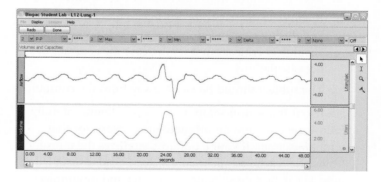

2. *Optional:* Airflow data do not have a lot of meaning for this lesson and may be a bit confusing at first glance, but they contain an interesting perspective on the recording. To review airflow data, enable CH 1 data display. PC: Alt-click or Ctrl-click the channel number box; Mac: Option-click the channel number box.

   Looking at the airflow waveform, note that the vertical scale is in liters per second and that the wave is centered on zero. Each upward-pointing region (called a *positive peak*) of the curve corresponds to inhalation, and each downward-pointing region (a *negative peak*) corresponds to an exhalation. The deeper the inhalation, the larger the positive peak will be; the more forceful an exhalation, the larger the negative peak.

3. The measurement boxes are above the marker region in the data window. Each box has three sections: channel number, measurement type (P-P, Max, Min, or Delta), and result. The first two sections are pull-down menus that are activated when you click on them. Set up the boxes as follows:

Channel	Displays
CH 2	P-P (finds maximum value in selected area and subtracts minimum value in selected area)
CH 2	Max (displays maximum value in selected area)
CH 2	Min (displays minimum value in selected area)
CH 2	Delta (difference in amplitude between last and first points of selected area)

The **selected area** is the part selected by the I-beam tool and includes the endpoints.

You can either record measurements in the data tables in the Lab Review and Practice Sheet or choose Edit > Journal > Paste Measurements to paste the data to your journal for future reference.

4. Measure observed Vital Capacity (VC): Use the I-beam cursor to select the area from the start of the forced inhalation to the peak of the forced exhalation and then record the P-P measurement in **Table 40.3** (p. 599) in the BIOPAC: Volumes and Capacities Review & Practice Sheet.

5. Take two measurements for an averaged Tidal Volume (TV) calculation: Zoom in to select the region of the first three breaths—from time 0 to the end of the third cycle Use the I-beam cursor to select the inhalation of cycle 3 (**Figure 40.11**) and record the P-P measurement in Table 40.3. Use the I-beam cursor to select the exhalation of cycle 3 and record the P-P measurement.

6. Use the I-beam cursor and measurements to determine all of the remaining values in Table 40.3.

For IRV, select from the third normal-inhalation peak to the peak of the forced inhalation and record the Delta measurement.

For ERV, locate the peaks for the series of three normal breaths taken after the forced deep inhalation but before the forced exhalation. Select from the third (downward-pointing) normal-exhalation peak to the (downward-pointing) peak of the forced exhalation and record the Delta measurement.

Figure 40.11 Selection of First Three Breaths

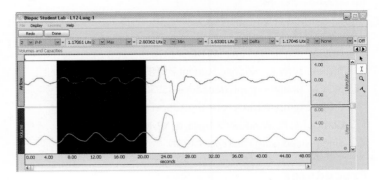

For Residual Volume (RV), select all data and record the Min measurement.

Finally, use the equations shown in **Table 40.4** (p. 599) in the BIOPAC: Volumes and Capacities Review & Practice Sheet to calculate inspiratory capacity (IC), expiratory capacity (EC), functional residual capacity (FRC), and total lung capacity (TLC).

7. Save or print the data and journal files and exit the program. ■

Lab Activity 3 BIOPAC
Respiratory Rate and Depth

In this activity, you will measure ventilation by recording the rate and depth of the breathing cycle using a **pneumograph transducer.** This transducer converts changes in chest expansion and contraction to changes in voltage, which will appear as a waveform. One respiratory cycle will then be recorded as an increasing voltage (ascending segment) during inspiration and a decreasing voltage (descending segment) during expiration.

You will also record the temperature of the air flowing in and out of one nostril with a **temperature probe.** The temperature of the air passing by the probe is inversely related to the expansion or contraction of the subject's chest. During inspiration (when the chest expands), the subject breathes in air that is cool relative to body temperature. This air is then warmed in the body. During expiration (when the chest contracts), the warmer air is compressed out of the lungs and out the respiratory passages.

QuickCheck Questions

3.1 What does a pneumograph transducer measure?

3.2 What is the temperature probe used for in this investigation?

In the Lab   3

Materials

☐ BIOPAC acquisition unit (MP36/35/30)

☐ BIOPAC software: Biopac Student Lab (BSL) v3.7.5 or better

☐ BIOPAC pneumograph transducer (SS5LB, SS5LA, or SS5L)

☐ BIOPAC temperature probe transducer (SS6L)

☐ Computer: PC Windows 7, Vista, or XP; Mac OS X 10.4-10.6

☐ Single-sided (surgical) tape (TAPE1)

Procedures

This lesson has four sections: setup, calibration, recording, and data analysis. Be sure to follow the setup instructions for your type of pneumograph transducer (SS5LB, SS5LA, or SS5L). The calibration step is critical for getting accurate recordings. Four segments will be recorded and then analyzed. You may record the data by hand or choose Edit > Journal > Paste Measurements to paste the data into your journal for future use.

Most markers and labels are automatically inserted into the data recordings. Markers appear at the top of the window as inverted triangles. This symbol indicates that you need to insert a marker and key in a marker label similar to the text in quotes. You can insert and label the marker during or after acquisition: on a Mac, press ESC; on a PC, press F9.

Section 1: Setup

1. Turn on your computer but keep the BIOPAC MP36/35/30 unit off.

2. Plug in the equipment as shown in **Figure 40.12**: the pneumograph transducer (SS5LB/LA/L) into CH 1 and the temperature probe (SS6L) into CH 2. (*Note:* Figure 40.14 shows the SS5LA model. Your laboratory might have the SS5LB or SS5L model, which both look a little different but work the same way.) If using the SS5LA transducer, be very careful not to pull or yank on the rubber bowtie portion that contains the sensor element.

   Note: The temperature probe is used to measure airflow. Each inhalation brings relatively cool air across the

Figure 40.12 Connecting the Respiratory and Temperature Transducers

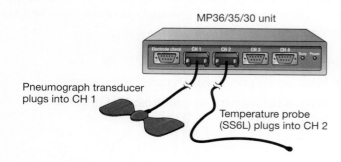

MP36/35/30 unit

Pneumograph transducer plugs into CH 1

Temperature probe (SS6L) plugs into CH 2

probe, and each exhalation blows warmer air across it. The probe records these temperature changes, which are proportional to the airflow output.

3. Turn on the acquisition unit. Attach the pneumograph transducer around the subject's chest below the armpits and above the nipples (**Figure 40.13**). The transducer can be worn over a shirt, but the correct tension is critical: slightly tight at the point of maximal expiration. If using the SS5LA model, attach the nylon belt by threading the nylon strap through the corresponding slots on the rubber bowtie such that the strap clamps into place when

Figure 40.13 Placement of Transducers

a Placement of respiratory transducer around chest

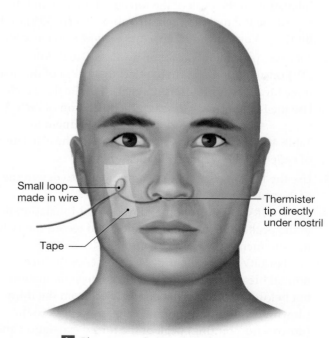

Small loop made in wire

Thermister tip directly under nostril

Tape

b Placement of temperature transducer

tightened. If using the SS5LB or SS5L model, attach the self-sticking ends together at the correct tension.

4. Attach the temperature probe (SS6L) to the subject's face. The probe should be firmly attached so that it does not move and should be positioned below the nostril and not touching the face. It is usually best to make a small loop in the cable about 2 inches from the tip and tape the loop to the subject's face, as shown in Figure 40.13.

5. **Start** the Biopac Student Lab Program and choose Lesson 8 (LO8-Respiration-1). Type in the subject's filename.

Section 2: Calibration

The calibration establishes the hardware's internal parameters (such as gain, offset, and scaling) and is critical for optimum performance.

1. Click the Calibrate button.

2. Instruct the subject to breathe normally until the calibration ends. The calibration will run for eight seconds and then stop automatically.

3. After the calibration has stopped, check your calibration data. Your screen should resemble **Figure 40.14**. The top channel displays data from the temperature probe and is labeled "airflow" because the temperature at the nostril is inversely proportional to airflow in and out of the nostril.

Both recording channels should show some fluctuation. If there is no fluctuation, it is possible that either the pneumograph transducer or the temperature probe is not connected properly, and you must redo the calibration by clicking on Redo Calibration and repeating the sequence.

Section 3: Recording Data

Have the subject sit down and relax. You will record four segments: normal breathing, hyperventilation, hypoventilation, and cough/reading. To work efficiently, read this entire section before proceeding so that you will know what to do for each segment. Screen prompts are similar in the recent versions of the BSL software. BSL 4 uses 'Continue' and 'Record' buttons to allow review/prep between segments; BSL 3.7.5–3.7.7 uses 'Record' and 'Resume' buttons.

Hints for obtaining optimal data:

a. Subject should stop hyperventilation or hypoventilation if dizziness develops.

b. The pneumograph transducer should fit snugly around the chest prior to inspiration.

c. The temperature probe should be firmly attached so that it does not move, positioned below the nostril and not touching the face.

d. The subject should be sitting for all segments.

e. The recording should be suspended after each segment so that the subject can prepare for the next segment.

Segment 1—Eupnea (Normal Breathing)

1. Click on Continue and when the subject is seated and breathing normally, click on Record.

2. After 15 seconds, click on Suspend to halt the recording. Review the data on the screen. If the recording is not similar to **Figure 40.15**, then adjust the placement of the transducer over the chest and repeat the calibration. The data would be incorrect if:

a. The pneumograph data has plateaus instead of waveforms.

b. The waveforms representing the temperature-probe (airflow) data are not offset from the respiration data.

c. The temperature probe moved and is no longer directly under the nostril.

d. The belt of the pneumograph transducer slipped.

e. The Suspend button was pressed prematurely.

f. Any of the channels have flat data, indicating no signal. In this case, be sure the cables are all securely in their respective ports.

Figure 40.14 Sample Calibration Data

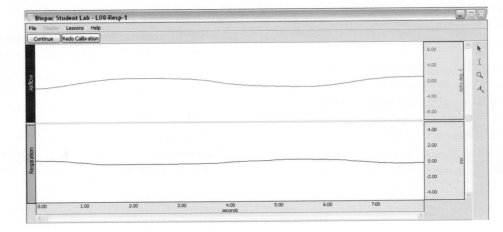

Figure 40.15 Segment 1: Normal Breathing

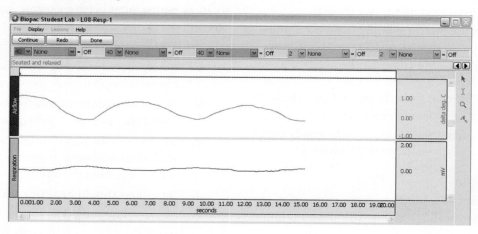

Segment 2—Hyperventilation and Recovery

3. Click on Continue and after subject knows how to breathe for the upcoming segment, click on Record. The recording will continue from the point where it last stopped, and a marker labeled "hyperventilation and recovery" will come up.

4. Have the subject *hyperventilate* by breathing rapidly and deeply through the mouth for 30 seconds (from the 15-s position on the screen to the 45-s position) and then *recover* by breathing through the nose for 30 seconds (45-s position to 75-s position). Record during hyperventilation and during recovery.

⚠ **Safety Alert:** Potential Dizziness

Stop the procedure immediately if the subject starts to feel sick or excessively dizzy. ▲

5. Click on Suspend. Review the data on the screen, using the horizontal scroll bar to look at different portions of the waveform. If your screen is not similar to Figure 40.15, click Redo and repeat the steps in Segment 2. The data would be incorrect for the same reasons listed in step 2.

Segment 3—Hypoventilation and Recovery

Important: You must not begin this segment until after the subject's breathing has returned to normal.

6. Click on Continue and after subject is breathing normally and knows how to breathe for the upcoming segment, click on Record. The recording will continue from the point where it last stopped, and a marker "hypoventilation and recovery" will be inserted.

7. Have the subject *hypoventilate* by breathing slowly and shallowly through the mouth for 30 seconds

(75-s position on the screen to 105-s position) and then *recover* by breathing through the nose for 30 seconds (105-s position to 135-s position). Record both during hypoventilation and during recovery.

8. Click on Suspend. Review the data on the screen, click Redo and repeat the recording if necessary.

Segment 4—Coughing Followed by Reading Aloud

9. Click on Continue and after subject is breathing normally and has reading material, click on Record. The recording will continue from the point where it last stopped, and a marker "cough, then read aloud" will come up.

10. Have the subject cough once and then begin reading aloud a passage from the laboratory manual.

11. After 60 seconds, click on Suspend. Review the data on the screen, click Redo and repeat the recording if necessary.

12. Click on Done and then click on Yes. Remove the pneumograph transducer and temperature probe from the subject. To record a different subject, refer back to the Setup to attach the pneumograph transducer and temperature probe.

Section 4: Data Analysis

1. Enter the Review Saved Data mode from the Lessons menu. Note the channel number designation:

Channel	Displays
CH 2	Airflow
CH 40	Respiration

2. The measurement boxes are above the marker region in the data window. Each box has three sections: channel number, measurement type (Delta T, BPM, or P-P), and result. The

Figure 40.16 **Selection of Inspiration Region**

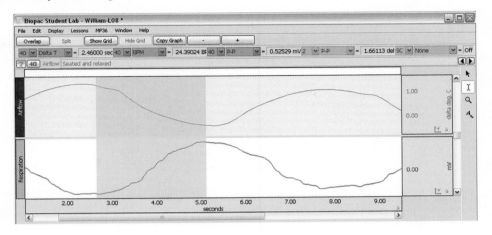

first two sections are pull-down menus that are activated when you click on them. Set up the boxes as follows:

Channel	Displays
CH 40	Delta T (delta time, difference in time between end and beginning of selected area)
CH 40	BPM (beats per minute; calculates time difference between end and beginning of selected area [same as ΔT] and converts difference from seconds to minutes; because BPM uses only time measurement of selected area, BPM value is not specific to a particular channel)
CH 40	P-P (finds maximum value in selected area and subtracts minimum value in selected area)
CH 2	P-P

The **selected area** is the part selected by the I-beam tool and includes the endpoints.

You can choose Edit > Journal > Paste Measurements to paste results into your computer journal for future reference to complete the Review & Practice Sheet.

3. **Zoom** in on a small section of the Segment 1 Eupnea (normal breathing) data, far enough to easily measure the intervals between peaks for approximately four cycles. The following tools help you adjust the data window:

Autoscale horizontal	Horizontal (time) scroll bar
Autoscale waveforms	Vertical (amplitude) scroll bar
Zoom tool	Zoom previous

4. Select the inspiration area, as shown in **Figure 40.16**. The Delta T measurement gives the duration of inspiration. Record the Delta T result in **Table 40.5** (p. 601) in the BIOPAC: Respiratory Rate and Depth Review & Practice Sheet or paste it into your journal.

5. Select the expiration area. Here the Delta T measurement gives the duration of expiration. Record the Delta T result in Table 40.5 or in your journal.

6. Repeat steps 4 and 5 for two additional cycles of the Segment 1 data. Record the data in the appropriate columns of Table 40.5 or in your journal.

7. Select a Segment 1 area from the beginning to the end of one breathing cycle (inspiration plus expiration). This time interval is called the **total duration.** Now the Delta T measurement is the total duration, and BPM is the breathing rate of the selected area. Record the measurement results in Table 40.5 or in your journal.

8. Repeat steps 3 through 7 for data segments 2, 3, and 4. Record the data either in **Table 40.6** (p. 601) in the BIOPAC: Respiratory Rate and Depth Review & Practice Sheet or in your journal. (The blacked-out cells in the Cough column of the table mean that only one cough measurement is required.)

9. Select three cycles in each of the four segments, and determine the respiration amplitude (maximum peak height) for each. The selected area should start at the middle of the descending wave in order to capture the minimum and maximum amplitudes. The P-P measurement will display the amplitude. Record the data either in **Table 40.7** (p. 602) in the BIOPAC: Respiratory Rate and Depth Review & Practice Sheet or in your journal. (Again, the blacked-out cells indicate that only one cough measurement is needed.)

10. Select the interval between maximum inspiration and maximum temperature change in each of the four data segments. Record the CH 2 P-P (temperature amplitude) data and the data for Delta T between the two peaks either in **Table 40.8** (p. 602) in the BIOPAC: Respiratory Rate and Depth Review & Practice Sheet or in your journal.

Name _____

Date _____

Section _____

Physiology of the Respiratory System

A. Matching

Match each term in the left column with its correct description in the right column.

_____ **1.** vital capacity

_____ **2.** tidal volume

_____ **3.** IRV

_____ **4.** ERV

_____ **5.** residual volume

_____ **6.** total lung capacity

_____ **7.** respiratory rate

_____ **8.** minute volume

_____ **9.** spirometer

_____ **10.** pulmonary ventilation

A. respiratory rate multiplied by tidal volume

B. volume of air that can be forcefully exhaled after normal exhalation

C. instrument used to measure respiratory volumes

D. amount of air normally inhaled or exhaled

E. inspiration and expiration

F. IRV + TV + ERV

G. number of breaths per minute

H. volume of air that can be forcefully inhaled after normal inhalation

I. vital capacity plus residual volume

J. volume of air that cannot be forcefully exhaled

B. Definitions

Define the following terms.

1. pulmonary ventilation

2. intrapulmonic pressure

3. inspiration

4. expiration

C. Short-Answer Questions

1. How do external and internal respiration differ from each other?

2. Which skeletal muscles contract during active exhalation?

D. Application and Analysis

1. Describe how skeletal muscles are used for inhalation.

2. Use Boyle's law to explain the process of pulmonary ventilation.

3. Describe how to calculate inspiratory reserve volume.

4. How did your vital capacity compare to your calculated predicted vital capacity? What health and lifestyle behaviors impact your vital capacity?

E. Clinical Challenge

Using your current height and gender, calculate your predicted vital capacity but add 5 years to your current age. Then, do the calculation again, but add 10 years to your current age. What changes are predicted in vital capacity as one ages? What do you think the reasons are for these age-related changes?

Name _____

Date _____

Section _____

BIOPAC:
Volumes and Capacities

A. Volume Measurements

Subject Profile

Name _____   Height _____

Age _____   Weight _____

Sex: Male/Female

1. Predicted Vital Capacity

Use the appropriate equation in Table 40.2 to calculate the subject's predicted vital capacity in liters.

2. Observed Volumes and Capacities

Use the data you entered into **Table 40.3** and the equations from the middle column of **Table 40.4** to calculate inspiratory, expiratory, functional residual, and total lung capacities. Enter your results in the rightmost column of Table 40.4.

Table 40.3	Respiratory Volume Measurements
Type of Volume	**Measurement (liters)**
Tidal volume (TV)	
Inspiratory reserve volume (IRV)	
Expiratory reserve volume (ERV)	
Vital capacity (VC)	

Residual volume (RV) used: _____ liters (Default is 1.2 L.)

Table 40.4	Calculated Respiratory Capacities	
Capacity	**Formula**	**Your Calculation**
Inspiratory (IC)	IC = TV + IRV	
Expiratory (EC)	EC = TV + ERV	
Functional residual (FRC)	FRC = ERV + RV	
Total lung (TLC)	TLC = IRV + TV + ERV + RV	

Compare the subject's lung volumes with the average volumes presented earlier in this exercise:

Volume	Average	Subject	
TV	500 mL	_____	mL
IRV	3300 mL	_____	mL
ERV	1000 mL	_____	mL

3. Observed Versus Predicted Vital Capacity

What is the subject's observed vital capacity as a percentage of the predicted vital capacity for her or his gender, age, and height (Table 40.2)?

_____ liters observed

_____ × 100 = _____ %

_____ liters predicted

Note: Vital capacities are dependent on other factors besides gender, age, and height. Therefore, 80% of predicted values are still considered normal.

B. Short-Answer Questions

1. Why does predicted vital capacity vary with height?

2. Explain how age and gender might affect lung capacity.

3. How would the volume measurements change if data were collected after vigorous exercise?

4. What is the difference between volume measurements and capacities?

5. Name the various types of pulmonary capacity studied in this exercise.

Name _____

Date _____

Section _____

BIOPAC:
Respiratory Rate and Depth

A. Data and Calculations

Subject Profile

Name _____   Height _____

Age _____   Weight _____

Sex: Male/Female

1. Normal Breathing (Segment 1)

Complete **Table 40.5** with values for each cycle and calculate the means.

Table 40.5	Segment 1: Eupnea Data					
Rate	Measurement	Channel	Cycle 1	Cycle 2	Cycle 3	Mean
Inspiration duration	Delta T	40				
Expiration duration	Delta T	40				
Total duration	Delta T	40				
Breathing rate	BPM	40				

2. Comparison of Ventilation Rates (Segments 2–4)

Complete **Table 40.6** with measurements from CH 40 for three cycles of each segment and calculate the means.

Table 40.6	Segments 2–4 Data							
Measurement	Hyperventilation (Segment 2)		Hypoventilation (Segment 3)		Cough (Segment 4)		Read Aloud (Segment 4)	
	Delta T	BPM	Delta T	BPM	Delta T	BPM	Delta T	BPM
Cycle 1								
Cycle 2								
Cycle 3								
Mean Ω								

Note: Delta T is cycle duration, BPM is breathing rate, and Cough has only one cycle.

3. Relative Ventilation Depths (Segments 1–4). Calculate the means in **Table 40.7**.

Table 40.7	Ventilation Depth Comparisons			
	P-P (CH 40)			
Depth	**Cycle 1**	**Cycle 2**	**Cycle 3**	**Mean**
Normal breathing (Segment 1)				
Hyperventilation (Segment 2)				
Hypoventilation (Segment 3)				
Cough (Segment 4)				

4. Relationship Between Respiratory Depth and Temperature (Segments 1–4). Record your data in **Table 40.8**.

Table 40.8	Respiratory Depth and Temperature (Segments 1–3)			
Measurement	**Channel**	**Normal Breathing (Segment 1)**	**Hyperventilation (Segment 2)**	**Hypoventilation (Segment 3)**
P-P (temperature amplitude)	2			
Delta T (between maximum inspiration and peak temperature amplitude)	40			

B. Short-Answer Questions

1. Suppose subjects in the Respiratory Rate and Depth Laboratory Activity were told to hold their breath immediately after hyperventilating and immediately after hypoventilating. Would a subject hold her or his breath longer after hyperventilating or after hypoventilating? Explain your answer.

2. What changes occur in the body as a person hypoventilates?

3. How does the body adjust respiratory rate and depth to counteract the effect of hypoventilation?

4. In which part of the respiratory cycle is the temperature of the air being breathed highest? In which part of the cycle is the temperature lowest?

5. Explain why temperature varies during the respiratory cycle.

6. What changes in breathing occur as a person coughs?

7. What changes in breathing occur as a person reads aloud?

8. Refer to Table 40.5 data. For each cycle and for the mean values, divide 60 seconds by the total-duration time (which is in seconds) to determine the breathing rate in breaths per minute. Do the rates you calculate this way match the breathing rates you entered for each cycle in the bottom row of Table 40.6 (the rate calculated by the computer)? If your calculated rates do not match the computer's, suggest a possible explanation for the difference.

9. In your Table 40.7 data, are there differences in the relative ventilation depths from one segment to another?

10. Refer to Table 40.8. How does ventilation depth influence the temperature of air being exhaled?

Anatomy of the Digestive System

Learning Outcomes

On completion of this exercise, you should be able to:

1. Identify the major layers and tissues of the digestive tract.

2. Identify all digestive anatomy on laboratory models and charts.

3. Describe the histological structure of the various digestive organs.

4. Trace the secretion of bile from the liver to the duodenum.

5. List the organs of the digestive tract and the accessory organs that empty into them.

T he five major processes of digestion are (1) ingestion of food into the mouth, (2) movement of food through the digestive tract, (3) mechanical and enzymatic digestion of food, (4) absorption of nutrients into the blood, and (5) formation and elimination of indigestible material and waste.

The **digestive tract** is a muscular tube extending from the mouth to the anus, a tube formed by the various hollow organs of the digestive system. Accessory organs outside the digestive tract plus the tract organs make up the **digestive system** (**Figure 41.1**). The accessory organs—salivary glands, teeth, liver, gallbladder, and pancreas—manufacture enzymes, hormones, and other compounds and secrete these substances onto the inner lining of the digestive tract. Food does not pass through the accessory organs.

The wet mucosal layer lining the mouth and the rest of the digestive tract is a mucous membrane. Glands drench the tissue surface with enzymes, mucus, hormones, pH buffers, and other compounds to orchestrate the step-by-step breakdown of food as it passes through the digestive tract.

Lab Activities

Clinical Application

Need More Practice and Review?

Build your knowledge—and confidence!—in the Study Area of MasteringA&P® at www.masteringaandp.com with Pre-lab Quizzes, Post-lab Quizzes, Practice Anatomy Lab™ (PAL™) 3.0 virtual anatomy practice tool, PhysioEx™ 9.0 laboratory simulations, and A&P Flix with Quizzes.

PAL | practice anatomy lab   For this lab exercise, follow these navigation paths:
- PAL>Human Cadaver>Digestive System
- PAL>Anatomical Models>Digestive System
- PAL>Histology>Digestive System

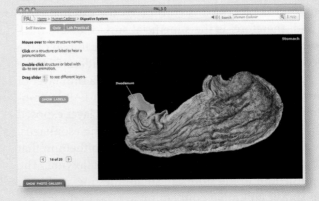

MasteringA&P®

Figure 41.1 Components of the Digestive System

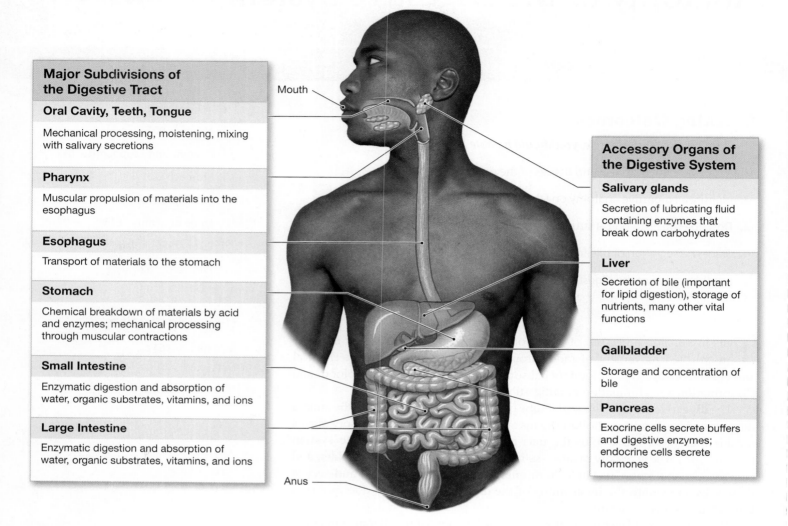

Major Subdivisions of the Digestive Tract

Oral Cavity, Teeth, Tongue

Mechanical processing, moistening, mixing with salivary secretions

Pharynx

Muscular propulsion of materials into the esophagus

Esophagus

Transport of materials to the stomach

Stomach

Chemical breakdown of materials by acid and enzymes; mechanical processing through muscular contractions

Small Intestine

Enzymatic digestion and absorption of water, organic substrates, vitamins, and ions

Large Intestine

Enzymatic digestion and absorption of water, organic substrates, vitamins, and ions

Mouth

Anus

Accessory Organs of the Digestive System

Salivary glands

Secretion of lubricating fluid containing enzymes that break down carbohydrates

Liver

Secretion of bile (important for lipid digestion), storage of nutrients, many other vital functions

Gallbladder

Storage and concentration of bile

Pancreas

Exocrine cells secrete buffers and digestive enzymes; endocrine cells secrete hormones

The histological organization of the digestive tract is similar throughout the length of the tract, and most of the tract consists of four major tissue layers: mucosa, submucosa, muscularis externa, and serosa (**Figure 41.2**). Each region of the digestive tract has anatomical specializations reflecting that region's role in digestion. Keep in mind that the inner surface where food is processed is considered the external environment and, therefore, is the superficial surface of the tract. The **mucosa** is the superficial layer exposed at the lumen of the tract. Three distinct layers in the mucosa can be identified: the **mucosal epithelium,** the **lamina propria,** and a thin layer of smooth muscle called the **muscularis** (mus-kū-LAR-is) **mucosae.** The mucosal epithelium is the superficial layer exposed to the lumen of the tract. From the mouth to the esophagus, the mucosal epithelium is stratified squamous epithelium that protects the mucosa from abrasion during swallowing. The mucosal epithelium in the stomach, small intestine, and large intestine is simple columnar epithelium, as food in these parts

of the tract is liquid and less abrasive. Deep to the mucosal epithelium is the lamina propria, a layer of connective tissue that attaches the epithelium and contains blood vessels, lymphatic vessels, and nerves. The muscularis mucosae is the deepest layer of the mucosa and in most organs has two layers, an inner circular layer that wraps around the tract and an outer longitudinal layer that extends along the length of the tract.

Deep to the mucosa is the **submucosa,** a loose connective-tissue layer containing blood vessels, lymphatic vessels, and nerves. Deep to the submucosa is a network of sensory and autonomic nerves, the **submucosal plexus,** that controls the tone of the muscularis mucosae.

Deep to the submucosal plexus is the **muscularis externa** layer, made up of two layers of smooth muscle tissue. Near the submucosa is a superficial circular muscle layer that wraps around the digestive tract; when this muscle contracts, the tract gets narrower. The deep layer is longitudinal muscle, with cells oriented parallel to the length of the tract. Contraction of

Figure 41.2 Structure of the Digestive Tract A diagrammatic view of a representative portion of the digestive tract. The features illustrated are typical of those of the small intestine.

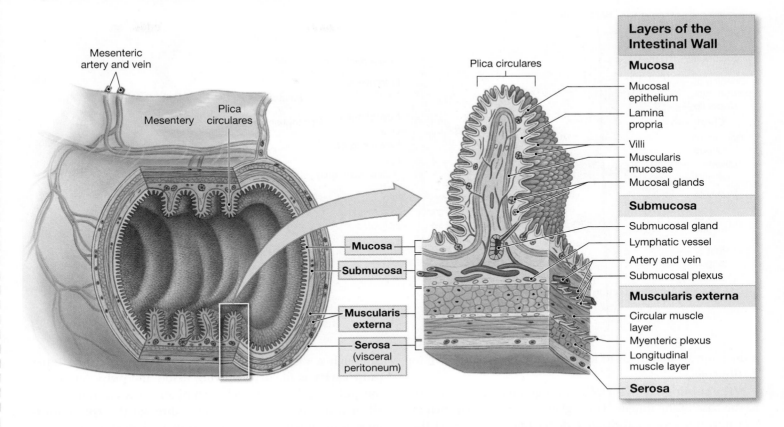

this muscle layer shortens the tract. The layers of the muscularis externa produce waves of contraction called **peristalsis** (per-i-STAL-sis), which move materials along the digestive tract. Between the circular and longitudinal muscle layers is the **myenteric** (mī-en-TER-ik) **plexus,** nerves that control the activity of the muscularis externa.

The deepest layer of the digestive tract is called the **adventitia** in the mouth, pharynx, esophagus, and inferior part of the large intestine and either the **serosa** or **visceral peritoneum** (Exercise 2) in the rest of the digestive tract. The adventitia is a network of collagen fibers, and the serosa is a serous membrane of loose connective tissue that attaches the digestive tract to the abdominal wall.

Lab Activity 1 Mouth

The mouth (**Figure 41.3**) is formally called either the **oral cavity** or the **buccal** (BUK-al) **cavity** and is defined by the space from the lips, or **labia,** posterior to the fauces (Exercise 39). The cone-shaped uvula is suspended from the cavity roof just anterior to the fauces. The lateral walls of the cavity are composed of the **cheeks,** and the roof is the hard palate and the soft palate. The **vestibule** is the region

between the teeth and the interior surface of the mouth; thus, the vestibule is bounded by the teeth and the cheeks laterally and by the teeth and the upper and lower lips anteriorly. The floor of the mouth is muscular, mostly because of the muscles of the **tongue.** A fold of tissue, the **lingual frenulum** (FREN-ū-lum), anchors the tongue yet allows free movement for food processing and speech. Between the posterior base of the tongue and the roof of the mouth is the **palatoglossal** (pal-a-tō-GLOS-al) **arch.** At the fauces is the **palatopharyngeal arch.**

The mouth contains two structures that act as digestive-system accessory organs: the salivary glands and the teeth. Three pairs of major salivary glands, illustrated in **Figure 41.4**, produce the majority of the saliva, enzymes, and mucus of the oral cavity. The largest, the **parotid** (pa-ROT-id) **glands,** are anterior to each ear between the skin and the masseter muscle. The **parotid ducts** (*Stensen's ducts*) pierce the masseter and enter the oral cavity to secrete saliva at the upper second molar.

The **submandibular glands** are medial to the mandible and extend from the mandibular arch posterior to the ramus. The **submandibular ducts** (*Wharton's ducts*) pass through the lingual frenulum and open at the swelling on the central margin of this tissue. Submandibular secretions are thicker than that of the parotid glands because of the presence of **mucin,**

Figure 41.3 The Oral Cavity

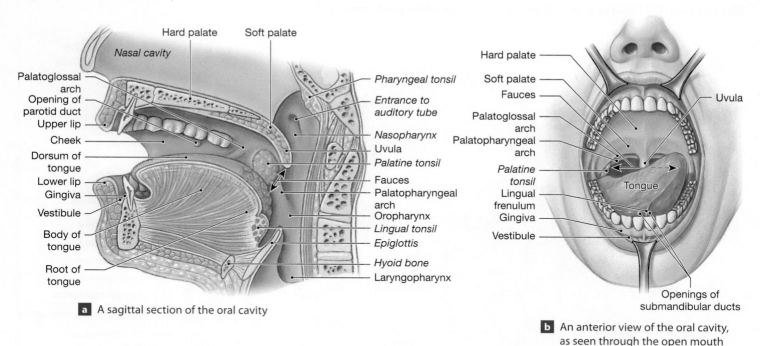

a A sagittal section of the oral cavity

b An anterior view of the oral cavity, as seen through the open mouth

a thick mucus that helps to keep food in a **bolus,** or ball, for swallowing.

The **sublingual** (sub-LING-gwal) **glands** are located deep to the base of the tongue. These glands secrete mucus-rich saliva into numerous **sublingual ducts** (*ducts of Rivinus*) that open along the base of the tongue.

Salivary glands predominately consist of two cell types; **serous cells** that produce a watery solution with enzymes and antibodies, and **mucous cells** that secrete the protein mucin for lubrication and sticking chewed food particles together for swallowing (Figure 41.4b). The cells have different responses to histological stains. The serous cells are *chromophilic* and pick up the dye and become dark-stained; the mucous cells are *chromophobic* and do not react with the stain so they appear much lighter than the chromophilic cells. The parotid glands are almost entirely serous cells and they secrete the bulk of the watery saliva. The submandibular gland consists of both serous and mucous cells and secretes saliva and mucin. Lingual glands are mostly mucous cells with few serous cells.

The teeth, as shown in **Figure 41.5**, are accessory digestive structures for chewing, or **mastication** (mas-tī-KĀ-shun). The **occlusal surface** is the superior area where food is ground, snipped, and torn by the tooth. Figure 41.5a details the anatomy of a typical adult tooth. The tooth is anchored in the alveolar bone of the jaw by a strong **periodontal ligament** that lines the embedded part of the tooth, the **root.** The **crown** is the portion of the tooth above the **gingiva** (JIN-ji-va), or gum. The crown and root meet at the **neck,** where the gingiva forms the **gingival sulcus,** a tight seal around the tooth.

Although a tooth has many distinct layers, only the inner **pulp cavity** is filled with living tissue, the **pulp.** Supplying the pulp are blood vessels, lymphatic vessels, and nerves, all of which enter the pulp cavity through the **apical foramen** at the inferior tip of the narrow U-shaped tunnel in the tooth root called the **root canal.** Surrounding the pulp cavity is **dentin** (DEN-tin), a hard, nonliving solid similar to bone matrix. Dentin makes up most of the structural mass of a tooth. In the root portion of the tooth, the dentin is covered by **cementum,** a material that provides attachment for the periodontal ligament. The crown is covered with **enamel,** the hardest substance produced by living organisms. Because of this hard enamel, which does not decompose, teeth are often used to identify accident victims and skeletal remains that have no other identifying features.

Humans have two sets of teeth during their lifetime. The first set, the **deciduous** (de-SID-ū-us; *decidua,* to shed) **dentition,** starts to appear at about the age of six months and is replaced by the **secondary dentition** (*permanent dentition*) starting at around the age of six years. The **deciduous teeth** (Figure 41.5d) are commonly called the *primary teeth, milk teeth,* or *baby teeth.* There are 20 of them, 5 in each jaw quadrant. (The mouth is divided into four quadrants: upper right, upper left, lower right, lower left.) Moving laterally from the midline of either jaw, the deciduous teeth are the **central incisor, lateral incisor, cuspid** (*canine*), **first molar,** and **second molar.** The secondary dentition consists of 32 adult teeth, each quadrant containing a central incisor, lateral incisor, cuspid, **first** and **second premolars** (*bicuspids*), and **first,**

Figure 41.4 **Salivary Glands**

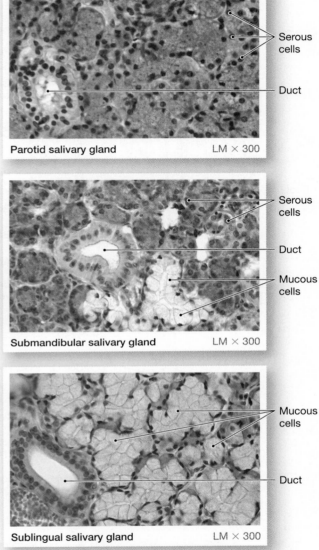

Parotid salivary gland — Serous cells — Duct — LM × 300

Submandibular salivary gland — Serous cells — Duct — Mucous cells — LM × 300

Sublingual salivary gland — Mucous cells — Duct — LM × 300

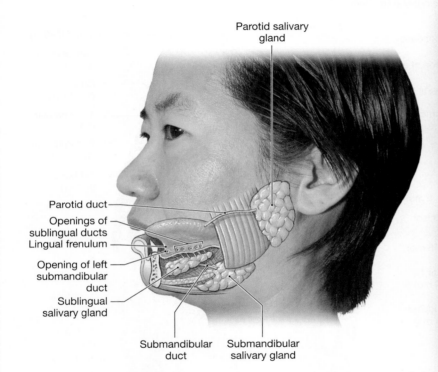

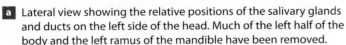

Parotid salivary gland

Parotid duct
Openings of sublingual ducts
Lingual frenulum
Opening of left submandibular duct
Sublingual salivary gland

Submandibular duct
Submandibular salivary gland

a Lateral view showing the relative positions of the salivary glands and ducts on the left side of the head. Much of the left half of the body and the left ramus of the mandible have been removed.

b Histological detail of the parotid, submandibular, and sublingual salivary glands. The parotid salivary gland produces saliva rich in enzymes. The gland is dominated by serous secretory cells. The submandibular salivary gland produces saliva containing enzymes and mucins, and it contains both serous and mucous secretory cells. The sublingual salivary gland produces saliva rich in mucins. This gland is dominated by mucous secretory cells.

second, and **third molars** (Figure 41.5c). The third molar is also called the *wisdom tooth*.

Each tooth is specialized for processing food. The incisors are used for snipping and biting off pieces of food. The cuspid is like a fang and is used to pierce and tear food. Premolars and molars are for grinding and processing food into smaller pieces for swallowing.

QuickCheck Questions

1.1 Which two mouth structures are digestive-system accessory organs?

1.2 Where is each salivary gland located?

1.3 Describe the main layers of a tooth.

Figure 41.5 Teeth

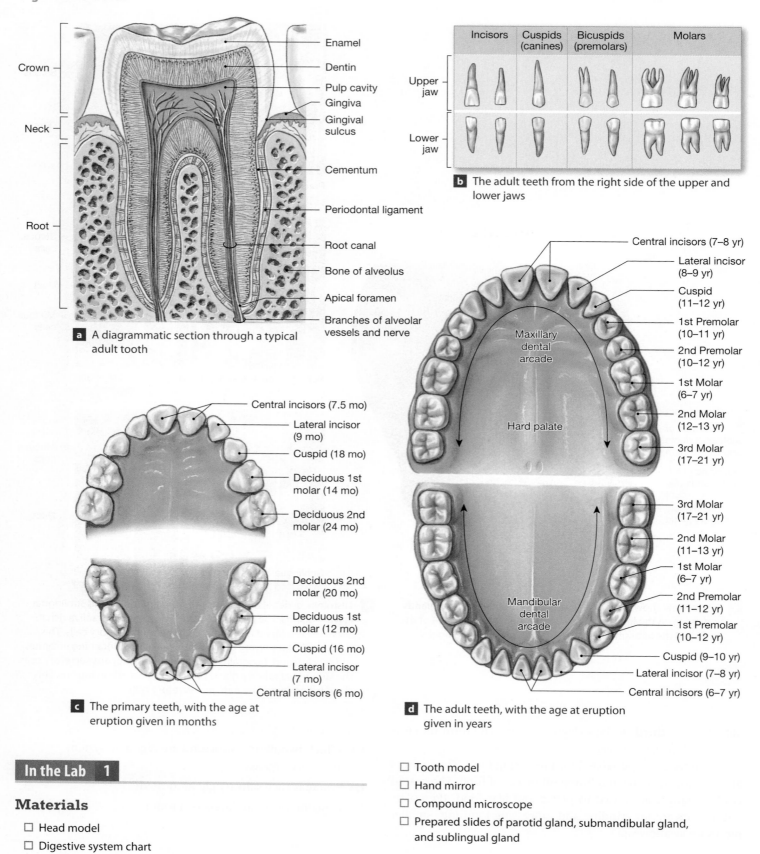

Enamel
Dentin
Pulp cavity
Gingiva
Gingival sulcus
Cementum
Periodontal ligament
Root canal
Bone of alveolus
Apical foramen
Branches of alveolar vessels and nerve

Crown
Neck
Root

a A diagrammatic section through a typical adult tooth

	Incisors	Cuspids (canines)	Bicuspids (premolars)	Molars		
Upper jaw						
Lower jaw						

b The adult teeth from the right side of the upper and lower jaws

Central incisors (7.5 mo)
Lateral incisor (9 mo)
Cuspid (18 mo)
Deciduous 1st molar (14 mo)
Deciduous 2nd molar (24 mo)
Deciduous 2nd molar (20 mo)
Deciduous 1st molar (12 mo)
Cuspid (16 mo)
Lateral incisor (7 mo)
Central incisors (6 mo)

c The primary teeth, with the age at eruption given in months

Central incisors (7–8 yr)
Lateral incisor (8–9 yr)
Cuspid (11–12 yr)
1st Premolar (10–11 yr)
2nd Premolar (10–12 yr)
1st Molar (6–7 yr)
2nd Molar (12–13 yr)
3rd Molar (17–21 yr)

Maxillary dental arcade
Hard palate

3rd Molar (17–21 yr)
2nd Molar (11–13 yr)
1st Molar (6–7 yr)
2nd Premolar (11–12 yr)
1st Premolar (10–12 yr)
Cuspid (9–10 yr)
Lateral incisor (7–8 yr)
Central incisors (6–7 yr)

Mandibular dental arcade

d The adult teeth, with the age at eruption given in years

In the Lab   1

Materials

☐ Head model
☐ Digestive system chart
☐ Tooth model
☐ Hand mirror
☐ Compound microscope
☐ Prepared slides of parotid gland, submandibular gland, and sublingual gland

Procedures

1. Review the mouth anatomy presented in Figure 41.3 and 41.4.

2. Identify the anatomy of the mouth on the head model and digestive system chart.

3. Use the hand mirror to locate your uvula, fauces, and palatoglossal arch. Lift your tongue and examine your submandibular duct.

4. Review the tooth anatomy in Figure 41.5.

5. Use the mirror to examine your teeth. Locate your incisors, cuspids, bicuspids, and molars. How many teeth do you have? Are you missing any because of extractions? Do you have any wisdom teeth?

6. Identify each salivary gland and duct on the head model and/or digestive system chart.

7. Examine the submandibular gland slide and distinguish between serous and mucous cells. Next view the parotid gland and note the abundance of serous cells. Observe the sublingual gland and locate the large clusters of mucous cells.

8. ***Draw It!*** Draw each salivary gland in the space provided. ■

Submanbidular gland

Parotid gland

Sublingual gland

Lab Activity 2   Pharynx and Esophagus

The pharynx is a passageway for both nutrients and air, and is divided into three anatomical regions—nasopharynx, oropharynx, and laryngopharynx (see Figure 41.5; Exercise 39). The nasopharynx is superior to the oropharynx, which is located directly posterior to the oral cavity. Muscles of the soft palate contract during swallowing and close the passageway to the nasopharynx to prevent food from entering the nasal cavity. When you swallow a bolus of food, it passes through the fauces into the oropharynx and then into the laryngopharynx. Toward the base of this area, the pharynx branches into the larynx of the respiratory system and the esophagus leading to the stomach. The epiglottis closes the larynx so that swallowed food enters only the esophagus and not the respiratory passageways. The lumen of the oropharynx and laryngopharynx is lined with stratified squamous epithelium to protect the walls from abrasion as swallowed food passes through this region of the digestive tract.

The food tube, or **esophagus,** connects the pharynx to the stomach. It is inferior to the pharynx and posterior to the trachea. The esophagus is approximately 25 cm (10 in.) long. It pierces the diaphragm at the **esophageal hiatus** (hī-Ā-tus) to connect with the stomach in the abdominal cavity. At the stomach, the esophagus terminates in a **lower esophageal sphincter,** a muscular valve that prevents stomach contents from backwashing into the esophagus. The four layers of the esophagus are shown in Figure 41.6, along with the three regions of the mucosa.

QuickCheck Questions

2.1 What are the three regions of the pharynx?

2.2 Which parts of the digestive tract does the esophagus connect?

2.3 Where is the esophageal hiatus?

Figure 41.6 Esophagus

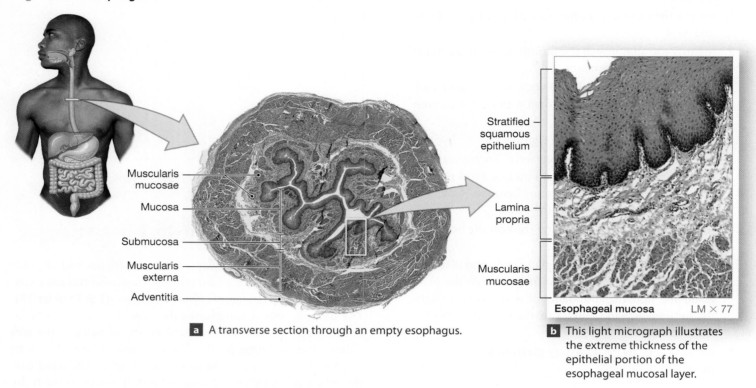

Muscularis mucosae

Mucosa

Submucosa

Muscularis externa

Adventitia

Stratified squamous epithelium

Lamina propria

Muscularis mucosae

Esophageal mucosa   LM × 77

a A transverse section through an empty esophagus.

b This light micrograph illustrates the extreme thickness of the epithelial portion of the esophageal mucosal layer.

Clinical Application   Acid Reflux

Acid reflux, also commonly called *heartburn,* occurs when stomach acid backflows into the esophagus and irritates the mucosal lining. The term *reflux* refers to a backflow, or regurgitation, of liquid—in this case, gastric juice, which is acidic. Some individuals have a weakened lower esophageal sphincter that allows the gastric juices to reflux during gastric mixing. Recent studies indicate that acid reflux is a major cause of esophageal and pharyngeal cancer. ■

In the Lab   2

Materials

☐ Head model
☐ Digestive system chart
☐ Torso model
☐ Hand mirror
☐ Compound microscope
☐ Prepared slide of esophagus

Procedures

1. Identify the anatomy of the pharynx and esophagus on the head model and digestive system chart.

2. Put your finger on your Adam's apple (thyroid cartilage of the larynx) and swallow. How does your larynx move, and what is the purpose of this movement?

3. Identify the anatomy of the esophagus on the torso model and digestive system chart.

4. Examine the esophagus slide at low magnification and observe the organization of the esophageal wall and identify the mucosa, submucosa, muscularis externa with its inner circular and outer longitudinal layers, and adventitia. Use Figure 41.6 for reference during your observations.

5. Increase the magnification and study the mucosa. Distinguish among the mucosal epithelium, which is stratified squamous epithelium, the lamina propria, and the muscularis mucosae. ■

Lab Activity   3   Stomach

The stomach is the J-shaped organ just inferior to the diaphragm (**Figure 41.7**). The four major regions of the stomach are the **cardia** (KAR-dē-uh), where the stomach connects with the esophagus; the **fundus** (FUN-dus), the superior rounded area; the **body,** the middle region; and the **pylorus** (pī-LOR-us), which joins the body at the **pyloric antrum** and moves into the **pyloric canal** at the distal end connected to the small intestine. The **pyloric sphincter** (also called the *pyloric valve*) controls movement of material from the stomach into the

Figure 41.7 Stomach

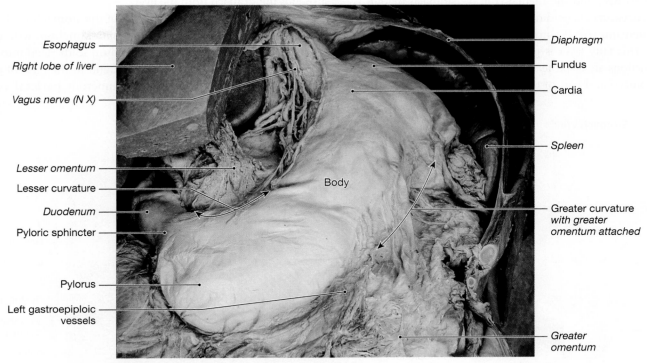

Esophagus

Right lobe of liver

Vagus nerve (N X)

Lesser omentum

Lesser curvature

Duodenum

Pyloric sphincter

Pylorus

Left gastroepiploic vessels

Body

Diaphragm

Fundus

Cardia

Spleen

Greater curvature *with greater omentum attached*

Greater omentum

a The position and external appearance of the stomach, showing superficial landmarks

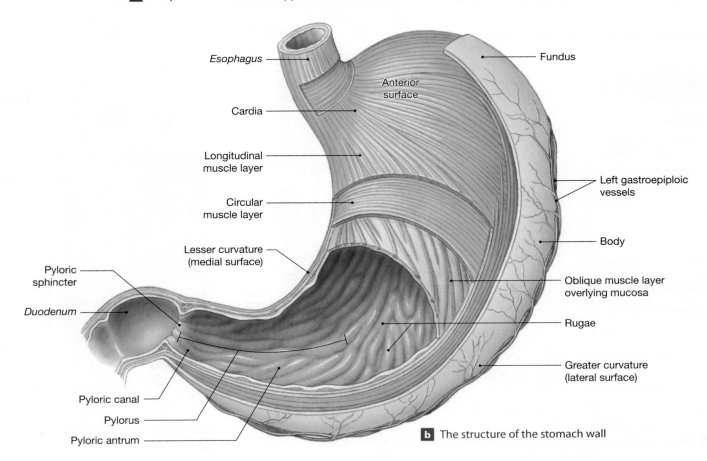

Esophagus

Cardia

Longitudinal muscle layer

Circular muscle layer

Lesser curvature (medial surface)

Pyloric sphincter

Duodenum

Pyloric canal

Pylorus

Pyloric antrum

Anterior surface

Fundus

Left gastroepiploic vessels

Body

Oblique muscle layer overlying mucosa

Rugae

Greater curvature (lateral surface)

b The structure of the stomach wall

small intestine. The lateral, convex border of the stomach is the **greater curvature,** and the medial concave stomach margin is the **lesser curvature.** Extending from the greater curvature is the **greater omentum** (ō-MEN-tum), commonly referred to as the *fatty apron.* This fatty layer is part of the serosa of the stomach wall. Its functions are to protect the abdominal organs and to attach the stomach and part of the large intestine to the posterior abdominal wall. The **lesser omentum,** also part of the serosa, suspends the stomach from the liver.

Figure 41.8 shows the histology of the stomach wall. The mucosal epithelium is simple columnar epithelium with cells called **mucus neck cells.** The epithelium folds deep into the lamina propria as **gastric pits** that extend to the base of **gastric glands.** The glands consist of numerous **parietal cells,**

Figure 41.8 Stomach Wall

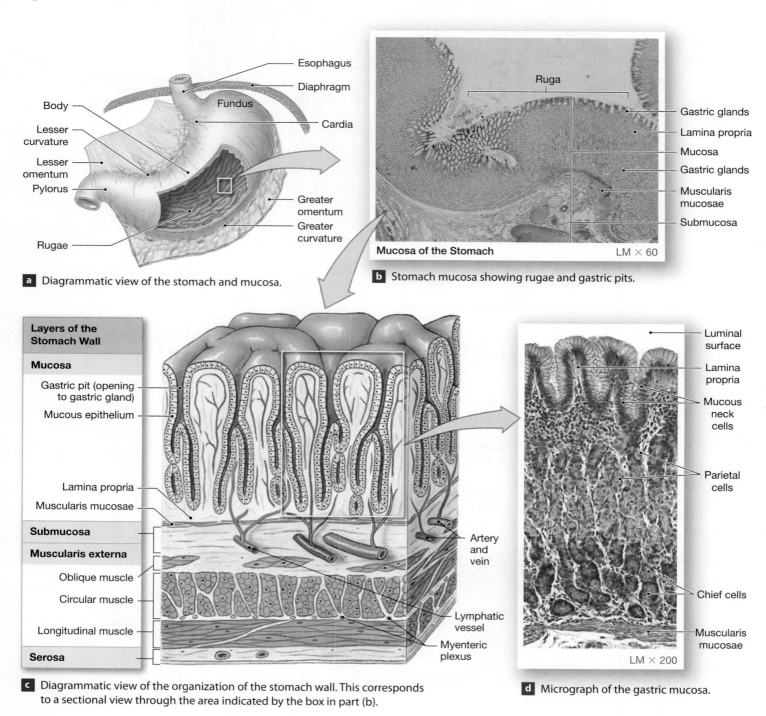

a Diagrammatic view of the stomach and mucosa.

Esophagus
Diaphragm
Fundus
Body
Cardia
Lesser curvature
Lesser omentum
Pylorus
Greater omentum
Rugae
Greater curvature

Ruga
Gastric glands
Lamina propria
Mucosa
Gastric glands
Muscularis mucosae
Submucosa
Mucosa of the Stomach   LM × 60

b Stomach mucosa showing rugae and gastric pits.

Layers of the Stomach Wall

Mucosa
Gastric pit (opening to gastric gland)
Mucous epithelium
Lamina propria
Muscularis mucosae
Submucosa
Muscularis externa
Oblique muscle
Circular muscle
Longitudinal muscle
Serosa
Artery and vein
Lymphatic vessel
Myenteric plexus

Luminal surface
Lamina propria
Mucous neck cells
Parietal cells
Chief cells
Muscularis mucosae
LM × 200

c Diagrammatic view of the organization of the stomach wall. This corresponds to a sectional view through the area indicated by the box in part (b).

d Micrograph of the gastric mucosa.

which secrete hydrochloric acid, and **chief cells,** which release an inactive protein-digesting enzyme called pepsinogen. The mucosa has **rugae** (ROO-gē), which are folds that enable the stomach to expand as it fills with food. Unlike what is found in other regions of the digestive tract, the muscularis externa of the stomach contains three layers of smooth muscle instead of two. The superficial layer (closest to the stomach lumen) is an **oblique layer,** surrounded by a **circular layer** and then a deep **longitudinal layer.** The three muscle layers contract and churn stomach contents, mixing gastric juice and liquefying the food into **chyme.** As mentioned previously, the serosa is expanded into the greater and lesser omenta.

QuickCheck Questions

3.1 What are the four major regions of the stomach?

3.2 How is the muscularis externa of the stomach unique?

3.3 Which structure of the stomach allows the organ to distend?

In the Lab 3

Materials

- ☐ Torso model
- ☐ Digestive system chart
- ☐ Preserved animal stomach (optional)
- ☐ Compound microscope
- ☐ Prepared slide of stomach

Procedures

1. Review the anatomy of the stomach in Figure 41.7.

2. Identify the gross anatomy of the stomach on the torso model and digestive system chart.

3. If specimens are available, examine the stomach of a cat or other animal. Locate the rugae, cardia, fundus, body, pylorus, greater and lesser omenta, lower esophageal sphincter, and pyloric sphincter.

4. Place the stomach slide on the microscope stage, focus at low magnification, and observe the rugae.

5. At medium magnification, identify the mucosa, submucosa, muscularis external, and serosa. Examine the muscularis externa and distinguish the three muscle layers.

6. Increase the magnification and, using Figure 41.8 as a guide, observe that the mucosal epithelium is simple columnar epithelium with mucous neck cells. Locate the numerous gastric pits, which appear as invaginations along the rugae. Within the pits, distinguish between parietal cells, which are more numerous in the upper areas, and chief cells, those that have nuclei at the basal region of the cells. ■

Lab Activity 4　Small Intestine

The small intestine (**Figure 41.9**) is approximately 6.4 m (21 ft) long and composed of three segments: duodenum, jejunum, and ileum. Sheets of serous membrane called the

Figure 41.9 **Segments of the Small Intestine**

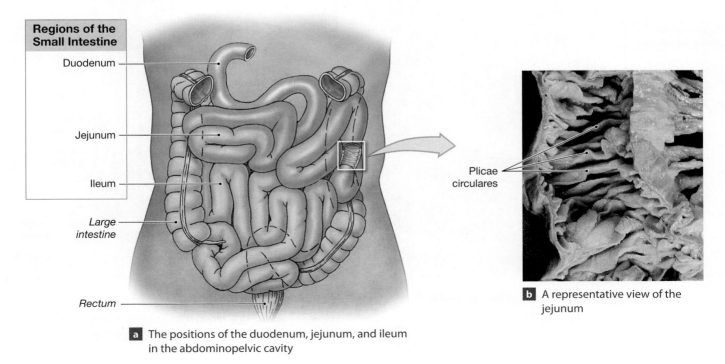

Regions of the Small Intestine

Duodenum

Jejunum

Ileum

Large intestine

Rectum

Plicae circulares

a The positions of the duodenum, jejunum, and ileum in the abdominopelvic cavity

b A representative view of the jejunum

mesenteries (MEZ-en-ter-ēz) **proper** extend from the serosa to support and attach the small intestine to the posterior abdominal wall. The first 25 cm (10 in.) is the **duodenum** (doo-ō-DĒ-num) and is attached to the distal region of the pylorus. Digestive secretions from the liver, gallbladder, and pancreas flow into ducts that merge and empty into the duodenum. This anatomy is described further in the upcoming section on the liver. The **jejunum** (je-JOO-num) is approximately 3.6 m (12 ft) long and is the site of most nutrient absorption. The last 2.6 m (8 ft) is the **ileum** (IL-ē-um), which terminates at the **ileocecal** (il-ē-ō-SĒ-kal) **valve** and empties into the large intestine.

The small intestine is the site of most digestive and absorptive activities and has specialized folds to increase the surface area for these functions (**Figure 41.10**; also see Figure 41.9). The submucosa and mucosa are creased together

Figure 41.10 Intestinal Wall

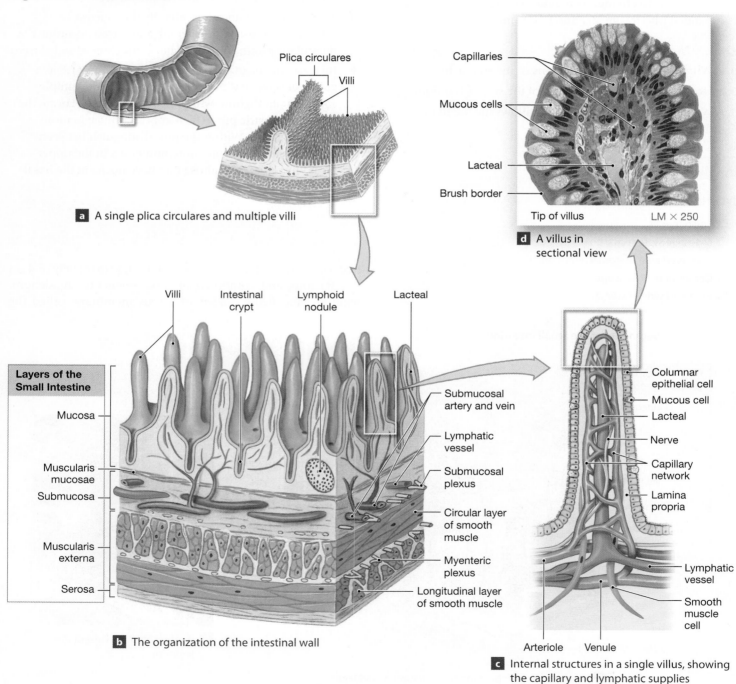

a A single plica circulares and multiple villi

d A villus in sectional view

b The organization of the intestinal wall

c Internal structures in a single villus, showing the capillary and lymphatic supplies

into large folds called **plicae** (PLĪ-sē) **circulars** (sir-kū-LAR-ēs). Along the plicae circulares, the lamina propria is pleated into small, fingerlike **villi** lined with simple columnar epithelium. The epithelial cells have a **brush border** of minute cell-membrane extensions or folds called **microvilli.**

At the base of the villi the epithelium forms pockets of cells called **intestinal glands** *(crypts of Lieberkuhn)* that secrete intestinal juice rich in enzymes and pH buffers to neutralize

stomach acid. Interspersed among the columnar cells are oval mucus-producing goblet cells. In the middle of each villus is a **lacteal** (LAK-tē-ul), a lymphatic vessel that absorbs fatty acids and monoglycerides from lipid digestion.

Each segment of the small intestine has unique histological features that reflect the specialized functions of the segment (**Figure 41.11**). In the duodenal submucosa are scattered **submucosal** (Brunner's) **glands** that secrete an alkaline mucin to

Figure 41.11 Regional Specialization of the Small Intestine

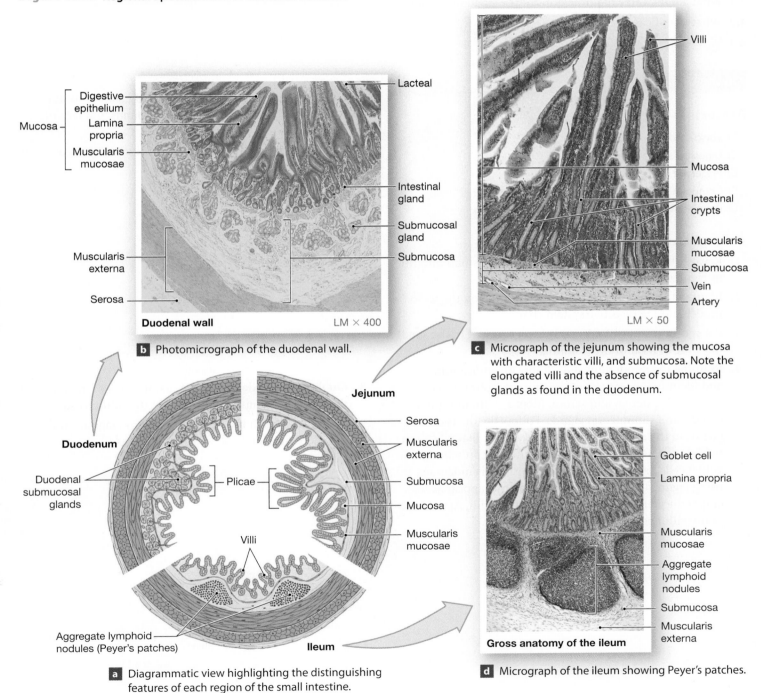

Duodenal wall LM × 400

b Photomicrograph of the duodenal wall.

LM × 50

c Micrograph of the jejunum showing the mucosa with characteristic villi, and submucosa. Note the elongated villi and the absence of submucosal glands as found in the duodenum.

Gross anatomy of the ileum

d Micrograph of the ileum showing Peyer's patches.

a Diagrammatic view highlighting the distinguishing features of each region of the small intestine.

protect the intestinal lining from the harsh acidic chyme arriving from the stomach. The jejunum has many intestinal crypts to manufacture enzymes for chemical digestion and elongated villi to increase surface area for nutrient absorption. The ileum has fewer plicae and the submucosa has **aggregate lymphoid nodules,** also called **Peyer's patches,** which are large lymphatic nodules that prevent bacteria from the colon entering the blood.

QuickCheck Questions

4.1 What are the three major regions of the small intestine?

4.2 What is a plica?

4.3 Where are the intestinal glands located?

In the Lab 4

Materials

- ☐ Torso model
- ☐ Digestive system chart
- ☐ Preserved animal intestines (optional)
- ☐ Compound microscope
- ☐ Prepared slides of duodenum, jejunum, and ileum

Procedures

1. Review the regions and organization of the small intestine in Figures 41.9 and 41.10.
2. Identify the anatomy of the small intestine on the torso model and the digestive system chart.
3. If a specimen is available, examine a segment of the small intestine of a cat or other animal.
4. Examine the duodenum slide at low magnification, and identify the features of the mucosa, submucosa, muscularis externa, and serosa. Identify villi, intestinal glands, and submucosal glands. Follow the ducts of the glands to the mucosal surface. Increase the magnification to high and identify the simple columnar epithelium, goblet cells, lamina propria, and muscularis mucosae, using Figure 41.11 as a guide. The lacteals appear as empty ducts in the lamina propria of the villi. At the base of the villi, locate the intestinal glands.

5. *Draw It!* Draw the duodenum at medium magnification in the space provided. For additional practice, complete the *Sketch to Learn* activity (on p. 619).

Duodenum

6. Observe the jejunum slide and identify features of the wall. Note the numerous intestinal glands and lack of submucosal glands.
7. *Draw It!* Draw the jejunum at medium magnification in the space provided.

Jejunum

8. On the ileum slide, locate the major layers of the wall and the aggregate lymphoid nodules in the submucosa.
9. *Draw It!* Sketch the ileum at medium magnification in the space provided. ■

Ileum

Sketch to Learn

Drawing the unique features of the duodenal wall is an excellent technique to reinforce your lab studies of the small intestine. Take your time drawing and labeling this sketch and you will have a great study guide to the small intestine!

Sample Sketch

Your Sketch

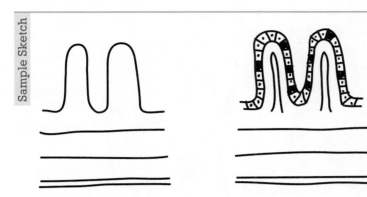

Step 1
- Draw two cones for villi.
- Add lines for the other layers of the intestinal wall.

Step 2
- Add a line over the villi and draw the cells of the digestive epithelium.
- Draw ovals in the epithelium for goblet cells.
- Add a space in each villus for a lacteal.

Step 3
- Add dots in the lamina propria and draw small ovals at base between villi for crypts.
- Add detail for muscularis mucosae and submucosal glands.

Lab Activity 5 Large Intestine

The large intestine is the site of electrolyte and water absorption and waste compaction. It is approximately 1.5 m (5 ft) long and divided into two regions: the **colon** (KŌ-lin), which makes up most of the intestine, and the **rectum** (**Figure 41.12**). The ileocecal valve regulates what enters the colon from the ileum. The first part of the colon, a pouchlike **cecum** (Sē-kum), is located in the right lumbar region. At the medial floor of the cecum is the wormlike **appendix.** Distal to the cecum, the **ascending colon** travels up the right side of the abdomen, bends left at the **right colic (hepatic) flexure,** and crosses the abdomen inferior to the stomach as the **transverse colon.** The **left colic (splenic) flexure** turns the colon inferiorly to become the **descending colon.**

The S-shaped **sigmoid** (SIG-moyd) **colon** passes through the pelvic cavity to join the **rectum,** which is the last 15 cm (6 in.) of the large intestine and the end of the digestive tract. The opening of the rectum, the **anus,** is controlled by an

internal anal sphincter of smooth muscle and an **external anal sphincter** of skeletal muscle. Longitudinal folds called **anal columns** occur in the rectum where the digestive epithelium changes from simple columnar to stratified squamous.

In the colon, the longitudinal layer of the muscularis externa is modified into three bands of muscle collectively called the **taenia coli** (TĒ-neē-a KŌ-lī). The muscle tone of the taenia coli constricts the colon wall into pouches called **haustra** (HAWS-truh, singular *haustrum*), which permit the colon wall to expand and stretch.

The wall of the colon lacks plicae and villi (**Figure 41.13**). It is thinner than the wall of the small intestine and contains more glands. The mucosal epithelium is simple columnar epithelium that folds into intestinal glands lined by goblet cells.

QuickCheck Questions

5.1 What are the major regions of the colon?

5.2 Where is the appendix located?

Figure 41.12 Large Intestine

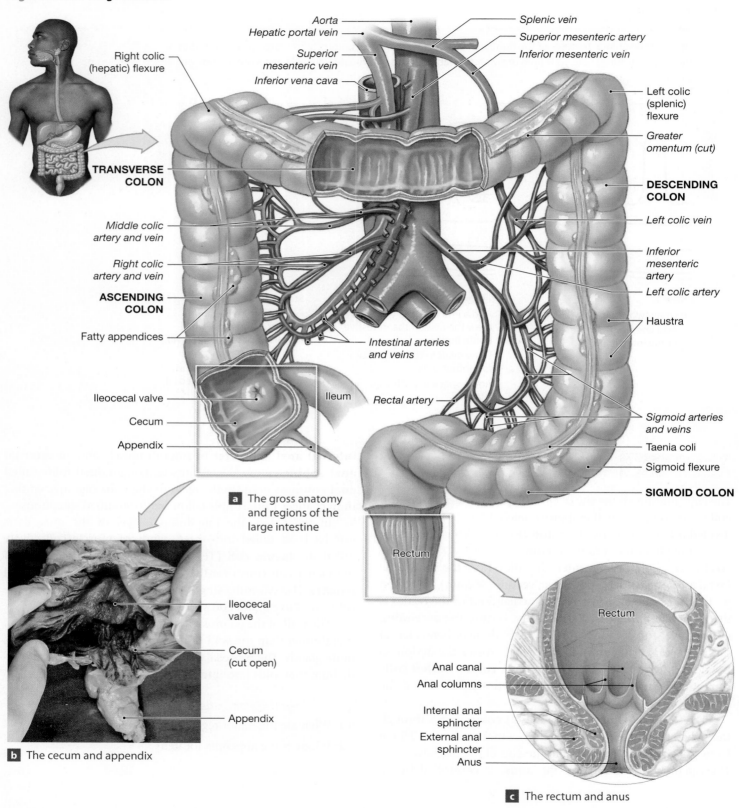

Aorta

Hepatic portal vein

Superior mesenteric vein

Inferior vena cava

Splenic vein

Superior mesenteric artery

Inferior mesenteric vein

Right colic (hepatic) flexure

TRANSVERSE COLON

Middle colic artery and vein

Right colic artery and vein

ASCENDING COLON

Fatty appendices

Ileocecal valve

Cecum

Appendix

Ileum

Left colic (splenic) flexure

Greater omentum (cut)

DESCENDING COLON

Left colic vein

Inferior mesenteric artery

Left colic artery

Haustra

Intestinal arteries and veins

Rectal artery

Sigmoid arteries and veins

Taenia coli

Sigmoid flexure

SIGMOID COLON

a The gross anatomy and regions of the large intestine

Ileocecal valve

Cecum (cut open)

Appendix

b The cecum and appendix

Rectum

Rectum

Anal canal

Anal columns

Internal anal sphincter

External anal sphincter

Anus

c The rectum and anus

Figure 41.13 **Wall of the Colon**

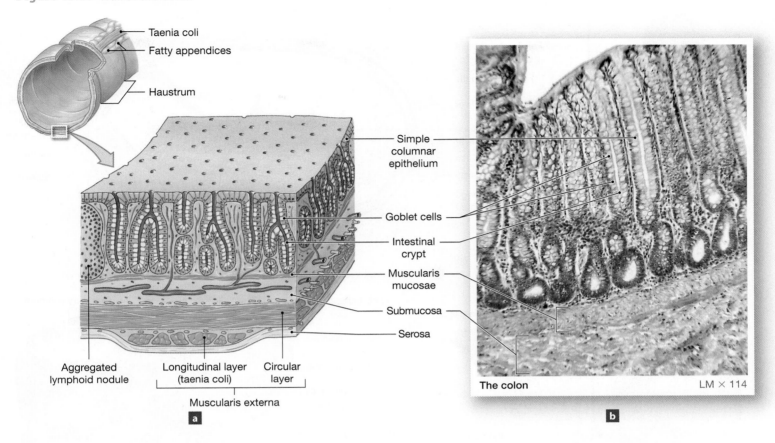

- Taenia coli
- Fatty appendices
- Haustrum
- Simple columnar epithelium
- Goblet cells
- Intestinal crypt
- Muscularis mucosae
- Submucosa
- Serosa
- Aggregated lymphoid nodule
- Longitudinal layer (taenia coli)
- Circular layer
- Muscularis externa

a

The colon   LM × 114

b

In the Lab 5

Materials

- ☐ Torso model
- ☐ Digestive system chart
- ☐ Preserved animal intestines (optional)
- ☐ Compound microscope
- ☐ Prepared slide of large intestine

Procedures

1. Review the anatomy of the large intestine in Figures 41.12 and 41.13.

2. Identify the gross anatomy of the large intestine on the torso model and the digestive system chart.

3. If a specimen is available, examine the colon of a cat or other animal. Locate each region of the colon, the left and right colic flexures, the taenia coli, and the haustra.

4. View the microscope slide of the large intestine and, referring to Figure 41.13, locate the intestinal glands. Distinguish between the simple columnar cells and goblet cells. ◼

Lab Activity 6 Liver and Gallbladder

The liver is located in the right upper quadrant of the abdomen and is suspended from the inferior of the diaphragm by the **coronary ligament** (Figure 41.14). Historically the liver has been divided into four lobes visible in gross observation. Current medical and surgical classification of the liver is based on vascular supply to individual segments; however, the blood vessels are apparent only in dissection. For gross observations, we shall use the four-lobe description. The **right** and **left lobes** are separated by the **falciform ligament,** which attaches the lobes to the abdominal wall. Within the falciform ligament is the **round ligament,** where the fetal umbilical vein passed. The square **quadrate lobe** is located on the inferior surface of the right lobe, and the **caudate lobe** is posterior, near the site of the inferior vena cava.

Each lobe is organized into approximately 100,000 smaller lobules (Figure 41.15). In the lobules, cells called **hepatocytes** (he-PAT-ō-sīts) secrete **bile,** a watery substance that acts like dish soap and breaks down the fat in ingested food. The bile is released into small ducts called **bile canaliculi,** which empty into **bile ductules** (DUK-tūlz) surrounding

Figure 41.14 Anatomy of the Liver

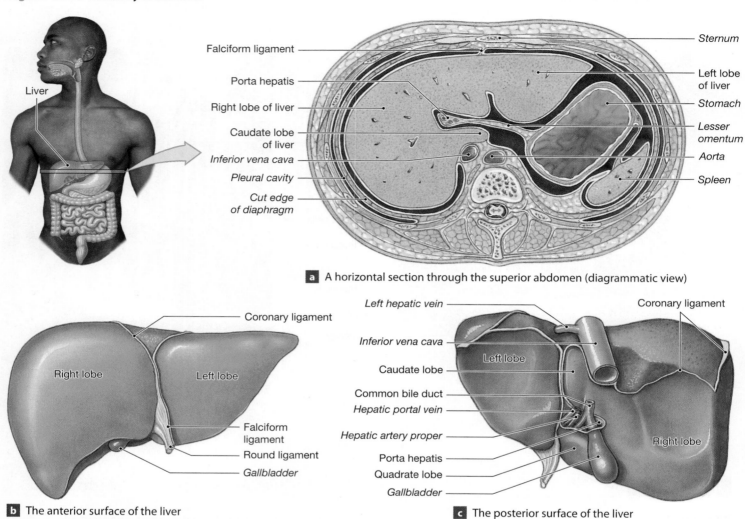

Liver

Falciform ligament
Porta hepatis
Right lobe of liver
Caudate lobe of liver
Inferior vena cava
Pleural cavity
Cut edge of diaphragm

Sternum
Left lobe of liver
Stomach
Lesser omentum
Aorta
Spleen

a A horizontal section through the superior abdomen (diagrammatic view)

Coronary ligament

Right lobe
Left lobe

Falciform ligament
Round ligament
Gallbladder

b The anterior surface of the liver

Left hepatic vein
Coronary ligament

Inferior vena cava
Left lobe
Caudate lobe
Common bile duct
Hepatic portal vein
Hepatic artery proper
Porta hepatis
Quadrate lobe
Gallbladder
Right lobe

c The posterior surface of the liver

Figure 41.15 Histology of the Liver

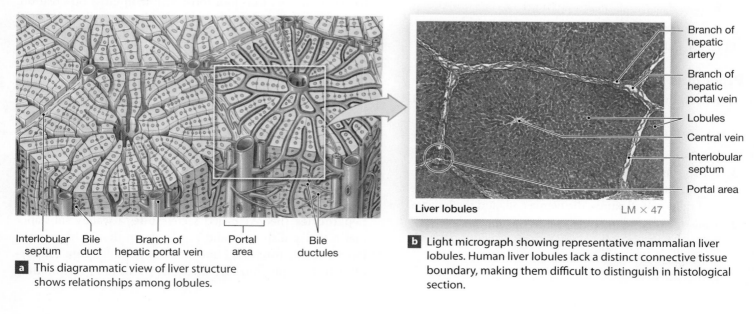

Interlobular septum
Bile duct
Branch of hepatic portal vein
Portal area
Bile ductules

Branch of hepatic artery
Branch of hepatic portal vein
Lobules
Central vein
Interlobular septum
Portal area

Liver lobules   LM × 47

a This diagrammatic view of liver structure shows relationships among lobules.

b Light micrograph showing representative mammalian liver lobules. Human liver lobules lack a distinct connective tissue boundary, making them difficult to distinguish in histological section.

each lobule. Progressively larger ducts drain bile into the **right** and **left hepatic ducts,** which then join a **common hepatic duct.** Blood flows through spaces called **sinusoids,** with each sinusoid receiving blood from a branch of either the hepatic artery or the hepatic portal vein. The sinusoids empty into a **central vein** in the middle of each lobule. Hepatocytes lining the sinusoids phagocytize wornout blood cells and reprocess the hemoglobin pigments for new blood cells.

The **gallbladder** is a small, muscular sac that stores and concentrates bile salts used in the digestion of lipids. It is located inferior to the right lobe of the liver (**Figure 41.16**). The wall of the gallbladder consists of three layers and does not include a muscularis mucosae or a submucosa., The mucosa is simple columnar epithelium that is folded and pinched into **mucosal crypts.** A **lamina propria** of connective tissue underlies the epithelium. The **muscularis externis** forms

the outer wall and is organized into an inner (superficial to lumen) longitudinal and an outer (deep) circular layer.

The liver and gallbladder are connected with ducts to transport bile. The common hepatic duct from the liver meets the **cystic duct** of the gallbladder to form the **common bile duct.** This duct passes through the lesser omentum and continues on to a junction called the **duodenal ampulla** (am-PUL-luh). The ampulla projects into the lumen of the duodenum at the **duodenal papilla.** A band of muscle called the **hepatopancreatic sphincter** *(sphincter of Oddi)* regulates the flow of bile and other secretions into the duodenum.

QuickCheck Questions

6.1 What are the four visible lobes of the liver?

6.2 How does bile enter the small intestine?

Figure 41.16 Gallbladder and Bile Ducts

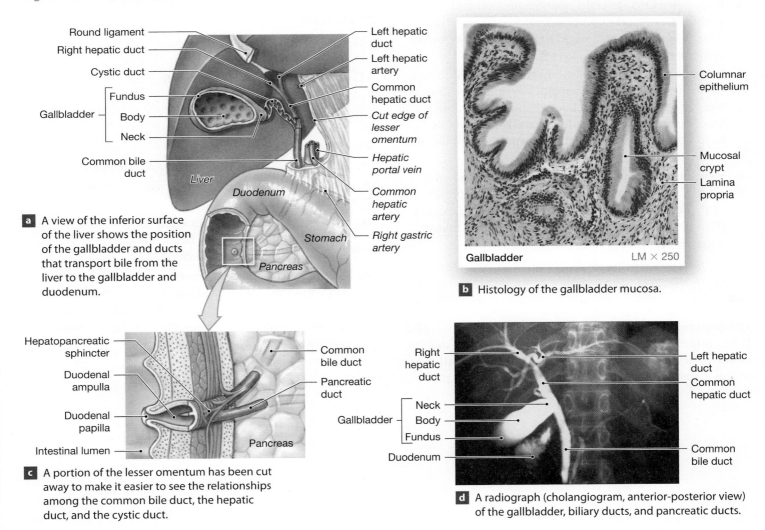

a A view of the inferior surface of the liver shows the position of the gallbladder and ducts that transport bile from the liver to the gallbladder and duodenum.

b Histology of the gallbladder mucosa.

c A portion of the lesser omentum has been cut away to make it easier to see the relationships among the common bile duct, the hepatic duct, and the cystic duct.

d A radiograph (cholangiogram, anterior-posterior view) of the gallbladder, biliary ducts, and pancreatic ducts.

In the Lab 6

Materials

☐ Torso model
☐ Digestive system chart
☐ Liver model
☐ Preserved animal liver and gallbladder (optional)
☐ Compound microscope
☐ Prepared slide of liver
☐ Prepared slide of gallbladder

Procedures

1. Review the anatomy of the liver in Figures 41.14 and 41.15.
2. Review the anatomy of the gallbladder in Figure 41.16.
3. Identify the gross anatomy of the liver and gallbladder on the torso model, liver model, and digestive system chart. Trace the ducts that transport bile from the liver and gallbladder into the small intestine.
4. If specimens are available, examine the liver and gallbladder of a cat or other animal. Locate each liver lobe, the falciform and round ligaments, and the hepatic, cystic, and common bile ducts.
5. Examine the liver slide at low magnification and identify the many lobules. Notice hepatocytes lining the sinusoids and the central vein of each lobule. In humans, the lobules are not well defined. Pigs and other animals have a connective tissue septum around each lobule; this septum can be seen in Figure 41.15b.
6. View the gallbladder slide at low magnification and identify the three components of the wall: columnar epithelium, lamina propria, and the muscularis externa (Figure 41.16). Observe how the epithelium is folded into pockets of mucosal crypts. ■

Lab Activity 7 Pancreas

The **pancreas** is a gland located posterior to the stomach. It has three main regions: the **head** is adjacent to the duodenum, the **body** is the central region, and the **tail** tapers to the distal end of the gland (**Figure 41.17**). The pancreas is characterized as a *double gland*, which means it has both endocrine and exocrine functions. The endocrine cells occur in **pancreatic islets** and secrete hormones for sugar metabolism (Exercise 33). Most of the glandular epithelium of the pancreas has an exocrine function. These exocrine cells, called **acini** (AS-i-nī) **cells,** secrete pancreatic juice into small ducts called **acini** located in the pancreatic glands. The acini drain into progressively larger ducts that merge as the **pancreatic duct** and, in some individuals, an **accessory pancreatic duct.** The pancreatic duct joins the common bile duct at the duodenal ampulla (see Figure 41.16).

QuickCheck Questions

7.1 What are the exocrine and endocrine functions of the pancreas?

7.2 Where does the pancreatic duct connect to the duodenum?

In the Lab 7

Materials

☐ Torso model
☐ Digestive system chart
☐ Preserved animal pancreas (optional)
☐ Compound microscope
☐ Prepared slide of pancreas

Procedures

1. Review the anatomy of the pancreas in Figure 41.17.
2. Identify the anatomy of the pancreas on the torso model and the digestive system chart.
3. If a specimen is available, examine the pancreas of a cat or other animal. Locate the head, body, and tail of the organ and the pancreatic duct.
4. On the pancreas slide, observe the numerous oval pancreatic ducts at low and medium magnifications. The exocrine cells are the dark-stained acini cells that surround groups of endocrine cells, the light-stained pancreatic islets. ■

Figure 41.17 Pancreas

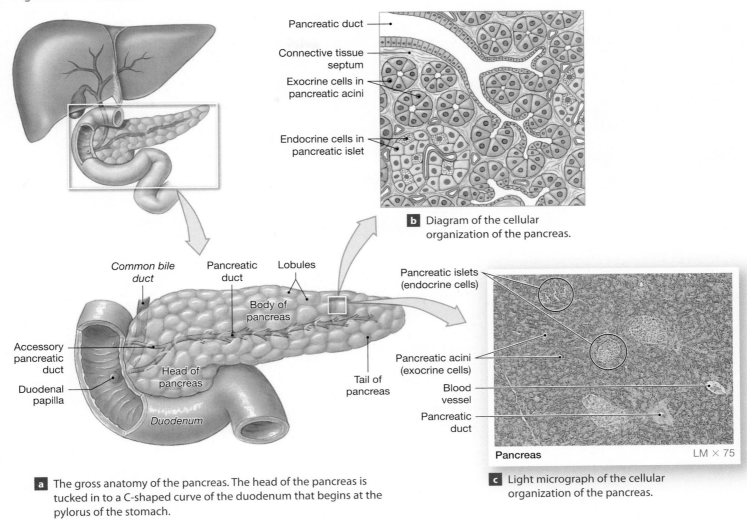

Pancreatic duct

Connective tissue
septum

Exocrine cells in
pancreatic acini

Endocrine cells in
pancreatic islet

b Diagram of the cellular
organization of the pancreas.

*Common bile
duct*

Pancreatic
duct

Lobules

Body of
pancreas

Accessory
pancreatic
duct

Head of
pancreas

Tail of
pancreas

Duodenal
papilla

Duodenum

Pancreatic islets
(endocrine cells)

Pancreatic acini
(exocrine cells)

Blood
vessel

Pancreatic
duct

Pancreas   LM × 75

a The gross anatomy of the pancreas. The head of the pancreas is
tucked in to a C-shaped curve of the duodenum that begins at the
pylorus of the stomach.

c Light micrograph of the cellular
organization of the pancreas.

Name ______________________________

Date ______________________________

Section ______________________________

Anatomy of the Digestive System

Review & Practice Sheet

EXERCISE 41

A. Definition

Define or state an anatomical description for each of the following structures.

1. pyloric sphincter
2. greater omentum
3. incisor
4. haustra
5. muscularis mucosae
6. serosa
7. gingiva
8. enamel
9. rugae
10. plicae circulares
11. molar
12. common bile duct

B. Drawing

Draw It! Draw a transverse section of the stomach wall showing the four major layers and the unique regional specializations such as gastric pits.

C. Labeling

1. Label the structures of the digestive tract in **Figure 41.18**.

Figure 41.18 Small Intestine

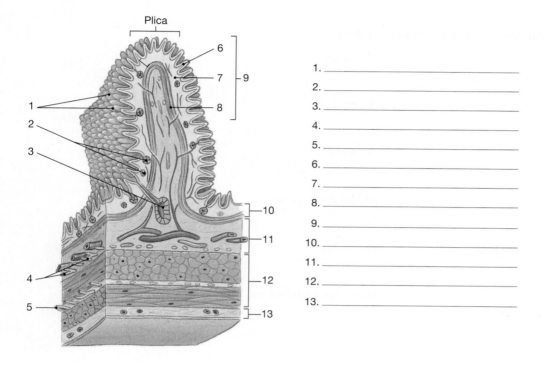

Plica

1. _____
2. _____
3. _____
4. _____
5. _____
6. _____
7. _____
8. _____
9. _____
10. _____
11. _____
12. _____
13. _____

2. Label the features of a typical tooth in **Figure 41.19**.

Figure 41.19 A Typical Tooth

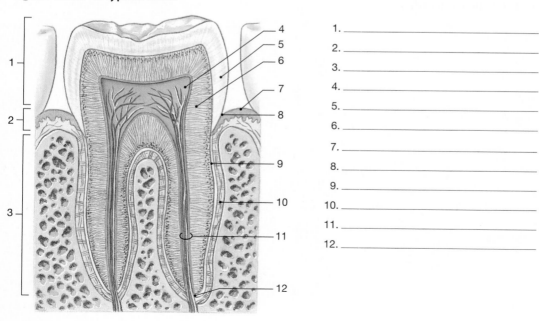

1. _____
2. _____
3. _____
4. _____
5. _____
6. _____
7. _____
8. _____
9. _____
10. _____
11. _____
12. _____

3. Label the anatomy of the liver in **Figure 41.20**.

Figure 41.20 **Liver**

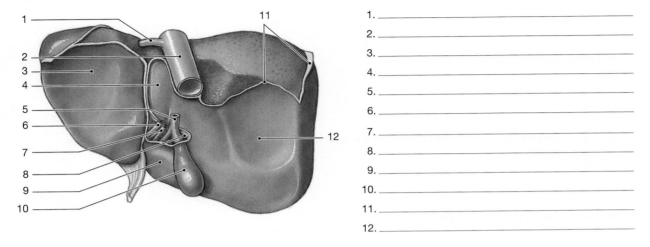

1. _____
2. _____
3. _____
4. _____
5. _____
6. _____
7. _____
8. _____
9. _____
10. _____
11. _____
12. _____

4. Label the structures of the stomach in **Figure 41.21**.

Figure 41.21 **Stomach**

1. _____
2. _____
3. _____
4. _____
5. _____
6. _____
7. _____
8. _____
9. _____
10. _____
11. _____
12. _____
13. _____
14. _____
15. _____
16. _____

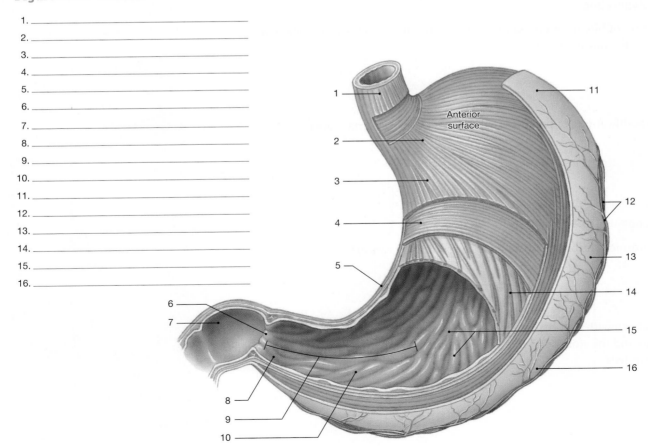

Anterior surface

D. Short-Answer Questions

1. List the three major pairs of salivary glands and the type of saliva each gland secretes.

2. List the accessory organs of the digestive system.

3. How is the wall of the stomach different from the wall of the esophagus?

4. Describe the gross anatomy of the large intestine.

E. Analysis and Application

1. Trace a drop of bile from the point where it is produced to the point where it is released into the intestinal lumen.

2. List the modifications of the intestinal wall that increase surface area.

F. Clinical Challenge

1. How is chronic heartburn associated with esophageal cancer?

2. A baby is born with esophageal atresia, an incomplete connection between the esophagus and the stomach. What will most likely happen to the infant if this defect is not corrected?

Digestive Physiology

Learning Outcomes

On completion of this exercise, you should be able to:

1. Explain why enzymes are used to digest food.

2. Describe a dehydration synthesis and a hydrolysis reaction.

3. Describe the chemical composition of carbohydrates, lipids, and proteins.

4. Discuss the function of a control group in scientific experiments.

Lab Activities

Clinical Application

Before nutrients can be converted to energy that is usable by the body's cells, the large organic **macromolecules** in food must be broken down into **monomers,** the building blocks of macromolecules. This chemical breakdown of food is called **catabolism** and is accomplished by a variety of enzymes secreted by the digestive system. **Enzymes** are protein **catalysts** that lower **activation energy,** which is the energy required for a chemical reaction (**Figure 42.1**). Without enzymes, the body would have to heat up to dangerous temperatures to provide the activation energy necessary to decompose ingested food. **Figure 42.2** highlights chemical digestion of carbohydrates, lipids, and proteins. In this exercise you will perform experiments that use enzymes to break down each of these main groups of nutrients.

An enzymatic reaction involves reactants, called **substrates,** and results in a **product.** Each enzyme molecule has one or more **active sites** where substrates bind. Enzymes have **specificity** because only substrates that are compatible with the active sites are metabolized by the enzyme. When the enzymatic reaction is complete, the product is released from the enzyme molecule, and the enzyme, unaltered in the reaction, can bind to other substrates and repeat the reaction. The function of each enzyme is related to the shape of the enzyme molecules, much as

Need More Practice and Review?

Build your knowledge—and confidence!—in the Study Area of MasteringA&P® at www.masteringaandp.com with Pre-lab Quizzes, Post-lab Quizzes, Practice Anatomy Lab™ (PAL™) 3.0 virtual anatomy practice tool, PhysioEx™ 9.0 laboratory simulations, and A&P Flix™ with Quizzes.

PhysioEx™ 9.0 For this lab exercise, go to this topic:
- PhysioEx Exercise 8: Physical and Chemical Processes of Digestion

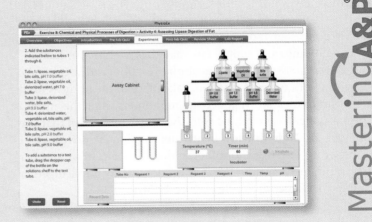

Figure 42.1 Enzymes and Activation Energy Enzymes lower the activation energy requirements, so a reaction can occur readily, in order from 1 to 4, under conditions in the body.

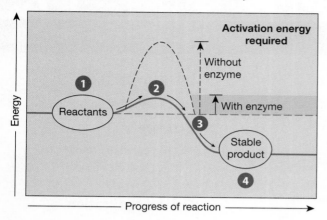

the shape of a key determines which lock it fits. Anything that changes the molecular shape, a process called **denaturation,** renders the enzyme nonfunctional. Heat denatures some enzymes, for instance, and destroys them. This is what you do every time you cook an egg—the heat you apply denatures the protein in the egg.

Clinical Application Lactose Intolerance

The digestive system requires a complex sequence of enzymes to catabolize food. If one enzyme in the sequence is absent or secreted in an insufficient quantity, the substrate cannot be digested. For example, individuals who are **lactose intolerant** do not produce the enzyme lactase, which digests lactose, commonly called milk sugar. If a lactose-intolerant individual consumes dairy products, which are high in lactose, the sugar remains in the digestive tract and is slowly digested by bacteria. This results in gas, intestinal cramps, and diarrhea. ■

Enzymes as a group are involved in both catabolic (decomposition) and anabolic (synthesis) reactions. A specific enzyme, however, functions in only one type of reaction. Digestive enzymes generally cause catabolic reactions, which metabolize ingested food into molecules small enough to cross cell membranes and supply raw materials for ATP production.

Figure 42.3 details an anabolic reaction and a catabolic reaction. Both types of reactions take place in the body as food macromolecules are digested to monomers, and the monomers are then reassembled into the larger molecules that the body needs. **Dehydration synthesis** reactions remove an OH group from one free monomer and an H atom from another free monomer, causing the two monomers to bond together and form a larger molecule (Figure 42.3b).

Table 42.1 summarizes the time and materials required for each activity. Use this table to help manage your laboratory time.

 Safety Alert: Digestion Tests

- Read through each activity from beginning to end before starting the activity.
- Wear gloves and safety glasses while pouring reagents and working near water baths.
- Report all spills and broken glass to your instructor. ▲

Lab Activity 1   Digestion of Carbohydrate

Starch and sugar molecules are classified as **carbohydrates** (kar-bō-HĪ-drātz). These molecules are primary energy sources for cellular production of ATP. Dietary sources of carbohydrates are fruits, grains, and vegetables. Snack food and soft drinks are loaded with carbohydrates and, considering today's serving "proportion distortion," the calories of these foods can account for weight gain.

Carbohydrates are composed of smaller **saccharide** molecules. A **monosaccharide** (mon-ō-SAK-uh-rīd) is a simple sugar with one saccharide. Glucose and fructose are examples of monosaccharides. Monosaccharides are the monomers used to build larger sugar and starch molecules. A **disaccharide** (dī-SAK-uh-rīd) is formed when two monosaccharides bond by a dehydration synthesis reaction, as in Figure 42.3a. Table sugar, sucrose, is a common disaccharide. Lactose, mentioned previously, is also a disaccharide.

Complex carbohydrates are **polysaccharides** (pol-ē-SAK-uh-rīdz) and consist of long chains of monosaccharides (**Figure 42.4**). Cells store polysaccharides as future energy sources. Plant cells store them as **starch,** the molecules found in potatoes and grains; animal cells store polysaccharides as **glycogen** (GLĪ-kō-jen).

In this activity, you will use the carbohydrate-digesting enzyme called *amylase* to digest the polysaccharide macromolecules in a solution of starch. Although amylase is easily obtained from your saliva, your instructor may provide amylase from a nonhuman source. **Figure 42.5** summarizes the activity.

QuickCheck Questions

1.1 What is the general name for the monomers used to make carbohydrates?

1.2 What is the name of the enzyme used in the carbohydrate-digesting activity?

1.3 What does starch break down to?

Figure 42.2 Chemical Events in Digestion A typical meal contains carbohydrates, proteins, lipids, water, minerals (electrolytes), and vitamins. The digestive system handles each component differently. Large organic molecules must be broken down by digestion before they can be absorbed. Water, minerals, and vitamins can be absorbed without processing, but they may require special transport mechanisms.

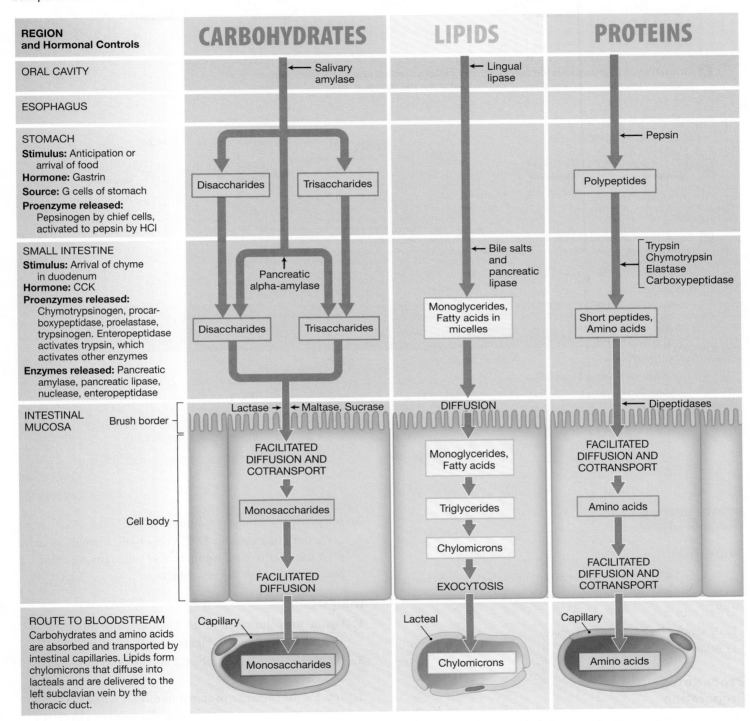

Figure 42.3 Formation and Breakdown of Complex Sugars

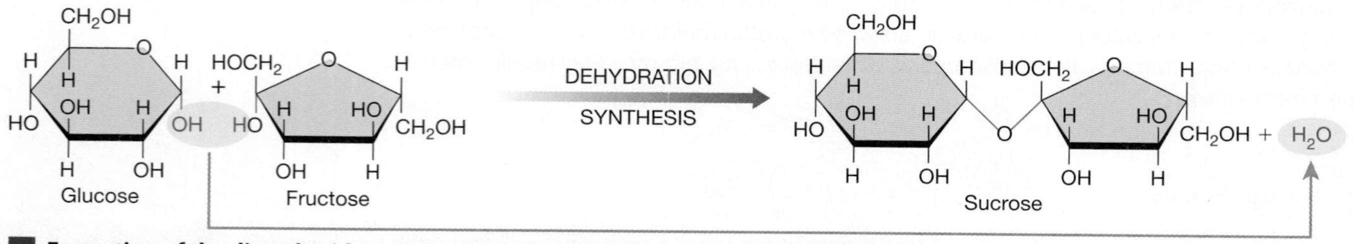

a **Formation of the disaccharide sucrose through dehydration synthesis.** During dehydration synthesis, two molecules are joined by the removal of a water molecule.

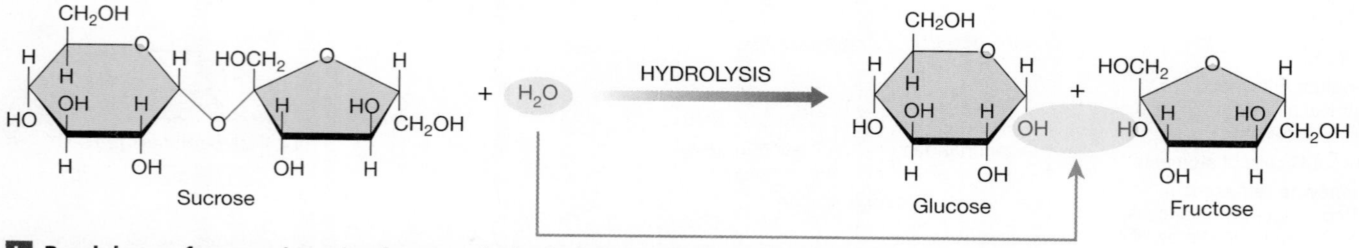

b **Breakdown of sucrose into simple sugars by hydrolysis.** Hydrolysis reverses the steps of dehydration synthesis; a complex molecule is broken down by the addition of a water molecule.

Table 42.1 **Summary of Enzyme Experiments**

	Carbohydrate	Lipid	Protein
Incubation time	0.5 hr	1 hr	1 hr
Number of test tubes required	6	2	2
Solutions required	Starch solution	Litmus cream	Protein solution
	Amylase	Pancreatic lipase	Pepsinogen
	Lugol's solution		0.5 M hydrochloric acid
	Benedict's reagent		Biuret's reagent

In the Lab 1

Materials

- ☐ Refer to Table 42.1 for number of test tubes and solutions needed
- ☐ Wax pencil
- ☐ Test tube rack
- ☐ 37°C water bath (body temperature)
- ☐ Boiling water bath

Procedures
Preparation

1. Number the six test tubes C1 to C6 with the wax pencil.
2. Add 20 mL of the starch solution to tube C1 and another 20 mL to tube C2. Be sure to shake the solution before pouring it into the tubes.

3. Add 20 mL of amylase solution to tube C1.
4. Place both tubes in the 37°C water bath, leave them there for a minimum of 30 minutes, and then remove them and place them in the test tube rack.

Analysis

1. Divide the solution in tube C1 equally into tubes C3 and C4, as indicated in Figure 42.5.
2. Divide the solution in tube C2 equally into tubes C5 and C6.
3. Test for the presence of starch in tubes C3 and C5 by placing one or two drops of **Lugol's solution** in each tube. A dark blue color indicates that starch is present, a light brown is a negative test indicating no starch.
4. Record your results in **Table 42.2**.

Figure 42.4 Polysaccharides Liver and muscle cells store glucose in glycogen molecules.

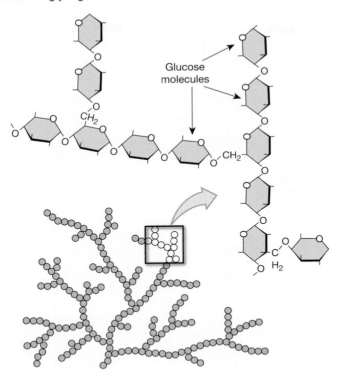

Glucose molecules

Figure 42.5 Summary of Carbohydrate Digestion Procedures

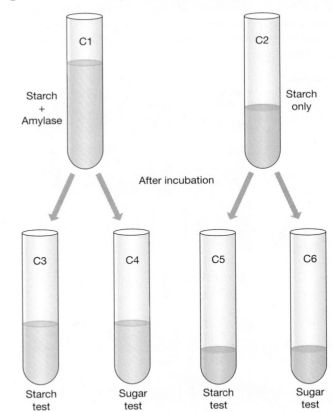

Table 42.2	Digestion of Starch by Amylase		
Lugol's Test for Starch		**Benedict's Test for Sugar**	
Tube C3	_____	Tube C4	_____
Tube C5	_____	Tube C6	_____
Conclusion	_____	Conclusion	_____

5. Test for the presence of monosaccharides in tubes C4 and C6 by placing 10 mL of **Benedict's reagent** in each tube. Mix by gently swirling the tubes, and then set them in the boiling water bath for five minutes. Do not point the tubes toward anyone because the solution could splatter and cause a burn. A color change indicates the presence of monosaccharides: from light olive to dark orange, depending on how much monosaccharide is present. A (+) and (−) system is often used to indicate the amount of sugar present:

blue (−) negative Benedict's test, no sugar present

green (+) some sugar

yellow (++) more sugar

orange (+++) high sugar concentration

red (++++) saturated sugar solution

6. Record your results in Table 42.2. ∎

Lab Activity 2 Digestion of Lipid

Lipids are oils, fats, and waxes. Fats are solid at room temperature while oils and waxes are usually liquid. Most dietary lipids are **monoglycerides** and **triglycerides.** These lipids are constructed of one or more **fatty acid** molecules bonded to a

Figure 42.6 Triglyceride Formation The formation of a triglyceride involves the attachment of fatty acids to the carbons of a glycerol molecule. In this example, a triglyceride is formed by the attachment of one unsaturated and two saturated fatty acids to a glycerol molecule.

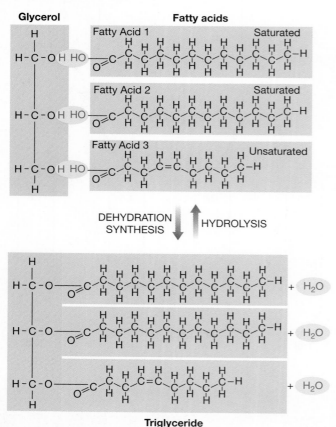

Figure 42.7 Summary of Lipid Digestion Procedures

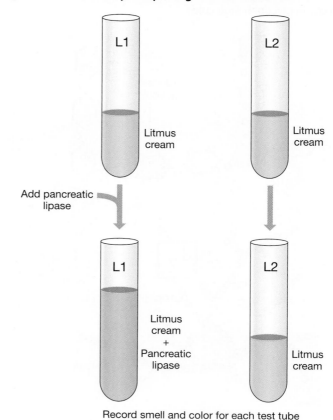

Record smell and color for each test tube

glycerol molecule. A **saturated fat** is a fatty acid with only single covalent bonds between the carbon atoms resulting in more hydrogen atoms in the fatty acid, hence the term *saturated*. An **unsaturated fat** has at least one double bond between carbon atoms and therefore has less hydrogen atoms. **Figure 42.6** shows the formation (dehydration synthesis) and decomposition (hydrolysis) of a triglyceride.

Pancreatic juice contains **pancreatic lipase,** a lipid-digesting enzyme that hydrolyzes triglycerides to monoglycerides and free fatty acids. In this activity, the released fatty acids will cause a pH change in the test tube, an indication that lipid digestion has occurred. The lipid substrate you will use is whipping cream to which a pH indicator, **litmus,** has been added. Litmus is blue in basic (alkaline) conditions and pink in acidic conditions. **Figure 42.7** summarizes the lipid digestion activity.

QuickCheck Questions

2.1 What are the monomer units of lipids called?

2.2 What is the enzyme used in the lipid-digesting activity?

In the Lab 2

Materials

☐ Refer to Table 42.1 for number of test tubes and solutions needed

☐ Wax pencil

☐ 37°C water bath (body temperature)

☐ Test tube rack

Table 42.3	Digestion of Lipid by Pancreatic Lipase			
	Tube L1		**Tube L2**	
	LITMUS CREAM + ENZYME		**LITMUS CREAM ONLY**	
	Start	End	Start	End
Smell	_____	_____	_____	_____
Color	_____	_____	_____	_____
Conclusion	_____	_____	_____	_____

Procedures

Preparation

1. Number the two test tubes L1 and L2. Fill each tube one-fourth full with litmus cream.

2. Add to tube L1 the same amount of pancreatic lipase as there is litmus cream in the tube.

3. Record the smell and color of the solution in each tube in the "Start" columns of **Table 42.3**.

4. Incubate the tubes for one hour in the 37°C water bath.

Analysis

1. After one hour, remove the tubes from the water bath and place them in the test tube rack.

2. Carefully smell the solutions by wafting fumes from each tube up to your nose. Record the smells in the "End" columns of Table 42.3.

3. Record the color of each solution in the "End" columns of Table 42.3. ■

Lab Activity 3 Digestion of Protein

Proteins are key molecules in the body with structural roles as in the cytoskeleton and functional activities such as the enzymatic catabolism of nutrients for cellular absorption. Dietary sources of proteins are meat and vegetables. Proteins are composed of long chains of **amino acids** bonded together by **peptide bonds** (**Figure 42.8**). There are 20 different types of amino acids, and their position in a protein molecule determines the structure and function of the protein, much like how the location of teeth on a key determines how the key works. Ten of the amino acids are only obtained in the diet and are called essential amino acids. The other amino acids are also ingested and can be manufactured by cells.

Figure 42.8 Protein Structure

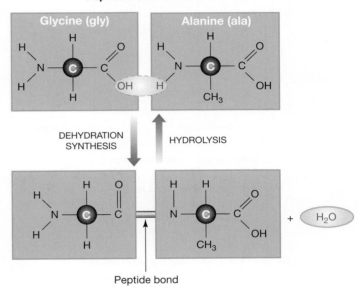

Peptide Bond Formation

DEHYDRATION SYNTHESIS

HYDROLYSIS

Peptide bond

In this activity you will use the proenzyme pepsinogen. The proenzyme alone cannot cause digestion; it must be activated by certain chemical conditions. By doing this activity, you will not only learn about protein digestion, but also discover the importance of the environment in which the enzyme operates. The protein substrate you will use is albumin, protein found in egg whites and in blood plasma. **Figure 42.9** summarizes the protein digestion activity.

Make a Prediction

In this experiment you will use the enzyme pepsin to digest protein similar to egg white. In what pH is this enzyme most active?

QuickCheck Questions

3.1 What are the monomer units of proteins called?

3.2 What is the enzyme used in the protein-digesting activity?

Figure 42.9 Summary of Protein-Digestion Procedures

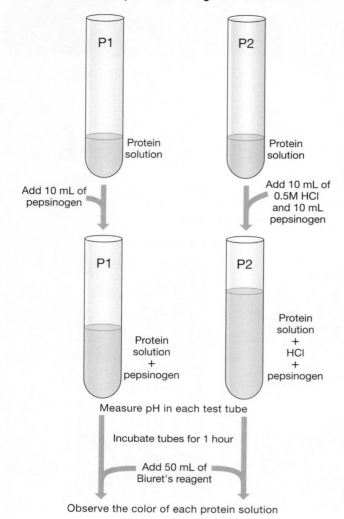

Materials

☐ Refer to Table 42.1 for number of test tubes and solutions needed
☐ Wax pencil
☐ 37°C water bath (body temperature)
☐ Test tube rack
☐ pH paper

Procedures
Preparation

1. Number two clean test tubes P1 and P2, and then add 10 mL of protein solution to each tube.
2. To tube P1, add 10 mL of the pepsinogen solution.
3. To tube P2, add 10 mL of the pepsinogen solution and 10 mL of 0.5 M hydrochloric acid.
4. Use the pH paper to measure the pH in each tube, and record your data in **Table 42.4**.
5. Incubate both tubes for one hour in the 37°C water bath.

Analysis

1. After one hour, remove the tubes from the water bath and place them in the test tube rack.
2. Test for protein digestion by adding 50 mL of **Biuret reagent** to each tube. In the presence of amino acids, a Biuret test turns the solution a lavender to light pink color, undigested protein is indicated by a dark purple color. If the color is too subtle to allow you to distinguish between pink and purple, add 20 mL more of Biuret's reagent.
3. Record the color in each tube in Table 42.4. ■

Table 42.4	Digestion of Protein by Pepsinogen	
	Tube P1 PROTEIN + ENZYME	**Tube P2** PROTEIN + ENZYME + HCl
pH of solution	_____	_____
Color after Biuret's test	_____	_____
Conclusion	_____	_____

Name ______________________________

Date ______________________________

Section ______________________________

Digestive Physiology

Review & Practice Sheet

EXERCISE 42

A. Definition

Define each term.

1. substrate

2. product

3. amylase

4. lipase

5. pepsinogen

6. hydrochloric acid

7. Benedict's reagent

8. pepsin

9. Lugol's solution

10. Biuret's reagent

B. **Short-Answer Questions**

    1. In each activity, you used a control tube that had no enzyme or other reagent added to it. What was the function of the control tubes?

    2. Discuss the difference between dehydration synthesis reactions and hydrolysis reactions.

    3. Describe the general chemical composition of carbohydrates, lipids, and proteins.

    4. How do enzymes initiate chemical reactions?

C. **Analysis and Application**

    1. Why were test tubes C1 and C2 in the carbohydrate laboratory activity tested for both starch *and* sugar?

    2. Why did the solution in tube L1 in the lipid-digestion activity turn pink after the incubation?

    3. Why was a warm-water bath used to incubate all the solutions in all three activities?

    4. What chemical conditions were necessary for protein digestion?

D. **Clinical Challenge**

    1. List the three primary groups of nutrients and give two examples of each group.

    2. How does lactose intolerance affect digestion?

Anatomy of the Urinary System

Learning Outcomes

On completion of this exercise, you should be able to:

1. Identify and describe the basic anatomy of the urinary system.

2. Trace the blood flow through the kidney.

3. Explain the function of the kidney.

4. Identify the basic components of the nephron.

5. Describe the differences between the male and female urinary tracts.

The primary function of the urinary system is to control the composition, volume, and pressure of the blood. The system exerts this control by adjusting both the volume of the liquid portion of the blood (the *plasma*) and the concentration of solutes in the blood as they pass through the kidneys. Any excess water and solutes that accumulate in the blood and waste products are eliminated from the body via the urinary system. These eliminated products are collectively called *urine*. The urinary system, highlighted in **Figure 43.1**, comprises a pair of kidneys, a pair of ureters, a urinary bladder, and a urethra.

Lab Activity 1   Kidney

The kidneys lie on the posterior surface of the abdomen on either side of the vertebral column between vertebrae T_{12} and L_3. The right kidney is typically lower than the left kidney because of the position of the liver. The kidneys are *retroperitoneal*,

Need More Practice and Review?

Build your knowledge—and confidence!—in the Study Area of MasteringA&P® at **www.masteringaandp.com** with Pre-lab Quizzes, Post-lab Quizzes, Practice Anatomy Lab™ (PAL™) 3.0 virtual anatomy practice tool, PhysioEx™ 9.0 laboratory simulations, and A&P Flix with Quizzes.

PAL practice anatomy lab   For this lab exercise, follow these navigation paths:
- PAL>Human Cadaver>Urinary System
- PAL>Anatomical Models>Urinary System
- PAL>Histology>Urinary System

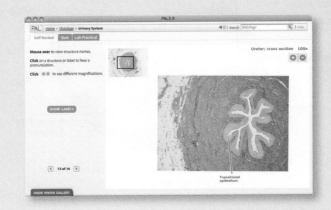

Figure 43.1 An Introduction of the Urinary System An anterior view of the urinary system, showing the positions of its components.

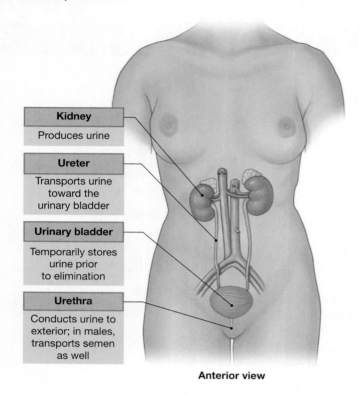

| **Kidney** |
| Produces urine |

| **Ureter** |
| Transports urine toward the urinary bladder |

| **Urinary bladder** |
| Temporarily stores urine prior to elimination |

| **Urethra** |
| Conducts urine to exterior; in males, transports semen as well |

Anterior view

meaning they are located outside of the peritoneal cavity, behind the parietal peritoneum. Each kidney is secured in the abdominal cavity by three layers of tissue: renal fascia, adipose capsule, and renal capsule. Superficially, the **renal fascia** anchors the kidney to the abdominal wall. The **perinephric fat capsule** is a mass of adipose tissue that envelopes the kidney, protects it from trauma, and helps to anchor it to the abdominal wall. Deep to the adipose capsule, on the surface of the kidney, the fibrous tissue of the **fibrous capsule** protects from trauma and infection.

A kidney is about 13 cm (5 in.) long and 2.5 cm (1 in.) thick. The medial aspect contains a **hilum** through which blood vessels, nerves, and other structures enter and exit the kidney (**Figure 43.2**). The hilum also leads to a cavity in the kidney called the **renal sinus.** The **cortex** is the outer, light red layer of the kidney, located just deep to the renal capsule. Deep to the cortex is a region called the **medulla,** which consists of triangular **renal pyramids** projecting toward the kidney center. Areas of the cortex extending between the renal pyramids are **renal columns.** A **renal lobe** is a renal pyramid and accompanying cortex, and the adjacent renal columns. At the apex of each renal pyramid is a **renal papilla** that empties urine into a small cuplike space called the **minor calyx** (KĀ-liks). Several minor calyces (KĀL-i-sēz) empty into a common space, the **major calyx.** These larger calyces merge to form the **renal pelvis.**

Figure 43.2 Structure of the Kidney

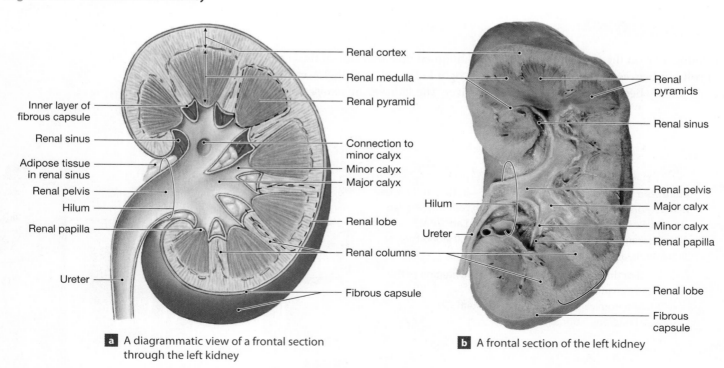

a A diagrammatic view of a frontal section through the left kidney

Renal cortex
Renal medulla
Renal pyramid
Inner layer of fibrous capsule
Renal sinus
Adipose tissue in renal sinus
Renal pelvis
Hilum
Renal papilla
Ureter
Connection to minor calyx
Minor calyx
Major calyx
Renal lobe
Renal columns
Fibrous capsule

b A frontal section of the left kidney

Renal pyramids
Renal sinus
Hilum
Ureter
Renal pelvis
Major calyx
Minor calyx
Renal papilla
Renal lobe
Fibrous capsule

QuickCheck Questions

1.1 What is the hilum?

1.2 Where are the renal pyramids located?

1.3 Where are the renal columns located?

In the Lab 1

Materials

☐ Kidney model

☐ Kidney chart

Procedures

1. Review the anatomy of the kidney in Figure 43.2.

2. Locate each structure shown in Figure 43.2 on the kidney model and/or chart. ■

Lab Activity 2 Nephron

Each kidney contains more than 1 million microscopic tubules called **nephrons** (NEF-ronz) that produce urine. As blood circulates through the blood vessels of the kidney, blood pressure forces materials such as water, excess ions, and waste products out of the blood and into the nephrons. This aqueous solution, called **filtrate,** circulates through the nephrons. As this circulation takes place, any substances in the filtrate still needed by the body move back into the blood. The remaining filtrate is excreted as urine.

Approximately 85% of the nephrons are **cortical nephrons,** which are found in the cortex and barely penetrate into the medulla (Figure 43.3). The remaining 15% are **juxtamedullary** (juks-ta-MED-ū-lar-ē) **nephrons,** located primarily at the junction of the cortex and the medulla and extending deep into the medulla before turning back toward the cortex. These longer nephrons produce a urine that is more concentrated than that produced by the cortical nephrons.

Each nephron, whether cortical or juxtamedullary, consists of two regions: a renal corpuscle and a renal tubule (Figure 43.3). The **renal corpuscle** is where blood is filtered. It consists of a **glomerular capsule,** also called **Bowman's capsule,** that houses a capillary called the **glomerulus** (glo-MER-ū-lus). As filtration occurs, materials are forced out of the blood that is in the glomerulus and into the **capsular space** in the glomerular capsule.

The renal corpuscle empties filtrate into the **renal tubule,** which consists of twisted and straight ducts of primarily cuboidal epithelium. The first segment of the renal tubule, coming right after the glomerular capsule, is a twisted segment called the **proximal convoluted tubule (PCT)** (Figure 43.3). The **nephron loop,** also called the **loop of Henle** (HEN-lē), is a straight portion that begins where the proximal convoluted tu-

bule turns toward the medulla. The nephron loop has both thick portions near the cortex and thin portions extending into the medulla. The **descending limb** is mostly a thin tubule that turns back toward the cortex as the **ascending limb.** The ascending limb leads to a second twisted segment, the **distal convoluted tubule (DCT).** The nephron ends where the distal convoluted tubule empties into a **connecting tubule,** which drains into a **collecting duct.** Adjacent nephrons join the same collecting duct that, in turn, joins other collecting ducts and collectively open into a common **papillary duct** that empties urine into a minor calyx. There are between 25 and 35 papillary ducts per renal pyramid.

At its superior end, the ascending limb of the nephron loop twists back toward the renal corpuscle and comes into contact with the blood vessel that supplies its glomerulus. This point of contact is called the **juxtaglomerular apparatus** (Figure 43.4). Here the cells of the renal tubule become tall and crowded together and form the **macula densa** (MAK-ū-la DEN-sa), which monitors NaCl concentrations in this area of the renal tubule.

The renal corpuscle is specialized for filtering blood, the physiological process that forces water, ions, nutrients, and wastes out of the blood and into the capsular space. The glomerular capsule has a superficial layer called the **parietal epithelium** (*capsular epithelium*) and a deep **visceral epithelium** (*glomerular epithelium*), the latter wrapping around the surface of the glomerulus. Between these two layers is the capsular space. The visceral epithelium consists of specialized cells called **podocytes** (PŌ-dō-sīts). These cells wrap extensions called **pedicels** around the endothelium of the glomerulus. Small gaps between the pedicels are pores called **filtration slits.** To be filtered out of the blood passing through the glomerulus, a substance must be small enough to pass through the capillary endothelium and its basement membrane and squeeze through the filtration slits to enter the capsular space. Any substance that can pass through these layers is removed from the blood as part of the filtrate. The filtrate therefore contains both essential materials and wastes.

QuickCheck Questions

2.1 What are the two main regions of a nephron?

2.2 What are the two kinds of nephrons?

In the Lab 2

Materials

☐ Kidney model

☐ Nephron model

☐ Compound microscope

☐ Prepared slide of kidney

Figure 43.3 **Cortical and Juxtamedullary Nephrons**

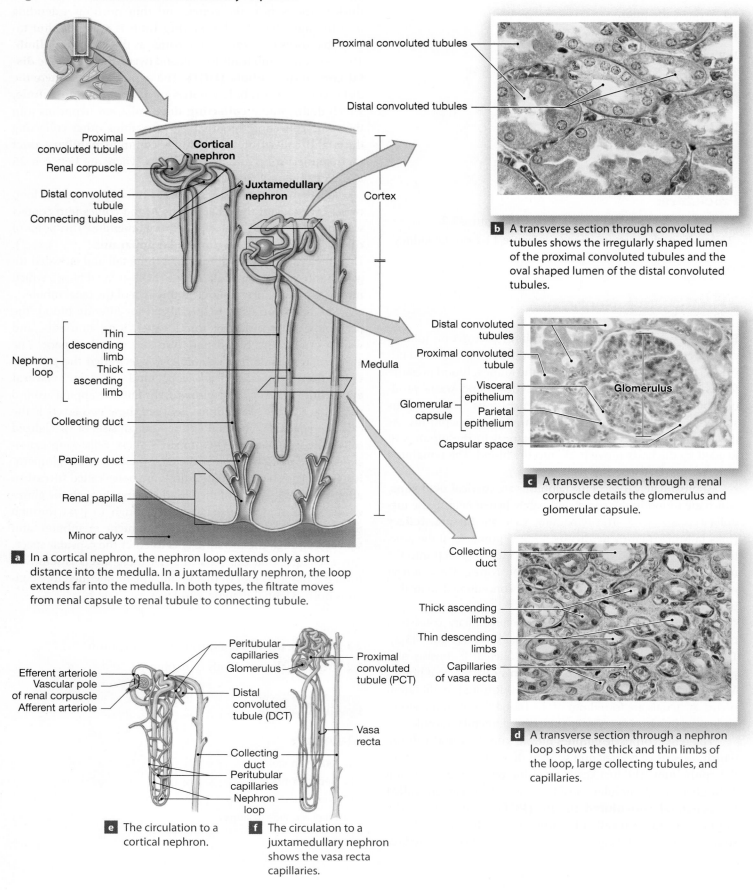

b A transverse section through convoluted tubules shows the irregularly shaped lumen of the proximal convoluted tubules and the oval shaped lumen of the distal convoluted tubules.

c A transverse section through a renal corpuscle details the glomerulus and glomerular capsule.

d A transverse section through a nephron loop shows the thick and thin limbs of the loop, large collecting tubules, and capillaries.

a In a cortical nephron, the nephron loop extends only a short distance into the medulla. In a juxtamedullary nephron, the loop extends far into the medulla. In both types, the filtrate moves from renal capsule to renal tubule to connecting tubule.

e The circulation to a cortical nephron.

f The circulation to a juxtamedullary nephron shows the vasa recta capillaries.

Figure 43.4 The Renal Corpuscle This diagrammatic view shows the important structural features of a renal corpuscle.

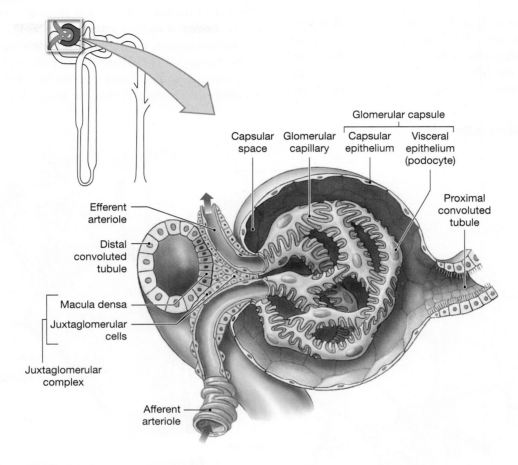

Procedures

1. Review the nephron anatomy in Figures 43.3 and 43.4.
2. Identify each structure of a kidney on the kidney model.
3. Identify each structure of a nephron on the nephron model.
4. Observe the kidney slide at low and medium magnifications and determine whether the renal capsule is present on the specimen. Increase the magnification to high and identify the renal cortex and, if present, the renal medulla. The medulla is usually not present on most kidney slides.
5. Examine several renal tubules, visible as ovals on the slide. Use the micrographs in Figure 43.3 as a guide. Cells of the PCT have microvilli facing the lumen of the tubule that make the lining appear fuzzy. The tubular cells of the DCT do not have microvilli and the lumen appears smoother as compared to the PCT.
6. Locate a renal corpuscle, which appears as a small knot in the cortex. Distinguish among the parietal and visceral epithelia of the glomerular capsule, the capsular space, and the glomerulus. The visceral epithelium is visible as the cells covering the glomerulus.
7. ***Draw It!*** Draw a section of the kidney slide in the space provided. Label the cortex, a renal corpuscle, and a renal tubule. ■

Cross section of kidney

Lab Activity 3   Blood Supply to the Kidney

Each minute, approximately 25% of the total blood volume travels through the kidneys. This blood is delivered to a kidney by the **renal artery,** which branches off the abdominal aorta. Once it enters the hilum, the renal artery divides into five **segmental arteries,** which then branch into **interlobar arteries,** which pass through the renal columns (**Figure 43.5**). The interlobar arteries divide into **arcuate** (AR-kū-āt) **arteries,** which cross the bases of the renal pyramids and enter the renal cortex as **cortical radiate arteries** *(interlobular arteries).*

In the nephron, an **afferent arteriole** branches off from one of the cortical radiate arteries serving the nephron, passes into the glomerular capsule, and supplies blood to the glomerulus. An **efferent arteriole** drains the blood from the glomerulus

Figure 43.5 **Blood Supply to the Kidney**

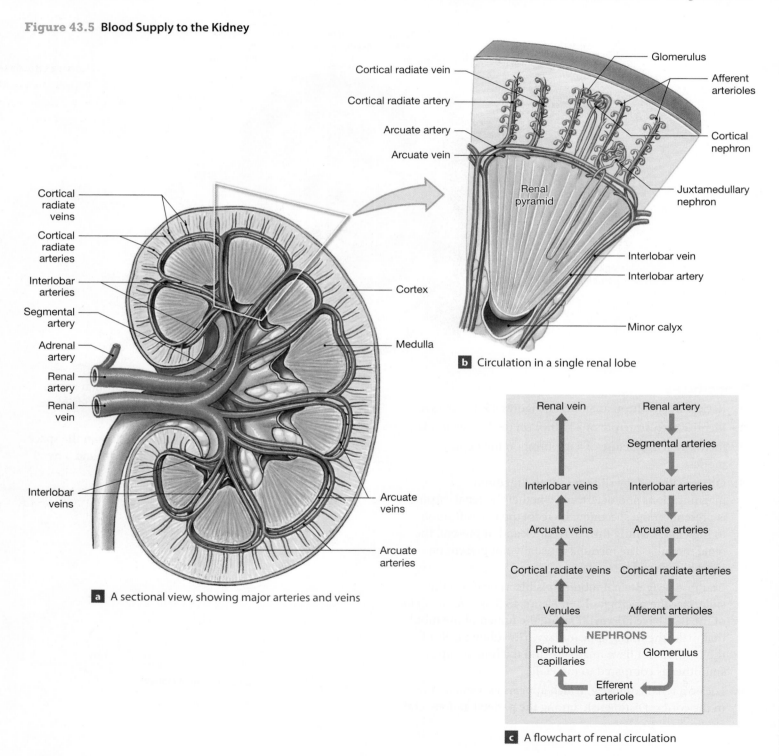

a A sectional view, showing major arteries and veins

b Circulation in a single renal lobe

c A flowchart of renal circulation

and branches into capillaries that surround the nephrons and reabsorb water, nutrients, and ions from the filtrate in the renal tubule. **Peritubular capillaries** in the cortex surround cortical nephrons and parts of juxtamedullary nephrons. The nephron loops of juxtamedullary nephrons have thin vessels collectively called the **vasa recta** (see Figure 43.3e, f). Both the peritubular capillaries and the vasa recta are involved in reabsorbing materials from the filtrate of the renal tubules back into the blood. Both networks drain into **cortical radiate veins,** which then drain into **arcuate veins** along the base of the renal pyramids. **Interlobar veins** pass through the renal columns and join the **renal vein,** which drains into the inferior vena cava. Although there are segmental arteries, there are no segmental veins.

QuickCheck Questions

3.1 Which vessel branches from the abdominal aorta to supply blood to the kidney?

3.2 Where are the interlobar and cortical radiate arteries located?

 In the Lab **3**

Materials

☐ Kidney model
☐ Nephron model

Procedures

1. Review the blood vessels depicted in Figure 43.5.

2. On the kidney and nephron models, identify the blood vessels that supply and drain the kidneys. Start with the renal artery and follow the blood supply to a renal corpuscle, the capillary beds, and the venous drainage toward the renal vein. For additional practice, complete the *Sketch to Learn* activity below. ■

Sketch to Learn

This activity is helpful for reviewing gross anatomy of the kidney while highlighting the blood supply to this vital organ. First let's sketch the outline of a kidney, then we'll add internal features.

Sample Sketch

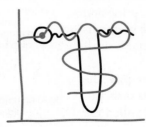

Step 1
- Draw a simple nephron as shown.
- In red, add a horizontal line for the arcuate artery and a vertical line for the cortical radiate artery.

Step 2
- Add the afferent arteriole entering the renal corpuscle and a glomerulus.
- Draw the efferent arteriole and the capillary beds that surround the nephron.

Step 3
- Use blue to draw the veins.
- Draw a vertical line for the cortical radiate (CR) vein.
- Connect the capillaries to this vein with a blue line.
- Make a horizontal line at the bottom of the CR vein for the arcuate vein.
- Label the vessels in your drawing.

Your Sketch

Lab Activity 4 Ureter, Urinary Bladder, and Urethra

Each kidney has a single **ureter** (ū-RĒ-ter), a muscular tube that transports urine from the renal pelvis to the **urinary bladder,** a hollow, muscular organ that stores urine temporarily (**Figure 43.6**). The two ureters conduct urine from kidney to bladder by means of gravity and peristalsis. Folds in the mucosa of the urinary bladder called **rugae** allow the bladder wall to expand and shrink as it fills with urine and then empties. The submucosa is deep to the mucosa. Deep to the submucosa, the muscular wall of the bladder is known as the **detrusor** (de-TROO-sor) **muscle.** In males (Figure 43.6a), the urinary bladder lies between the pubic symphysis and the rectum. In females (Figure 43.6b), the urinary bladder is posterior to the pubic symphysis, inferior to the uterus, and superior to the vagina.

A single duct, the **urethra** (ū-RĒ-thra), drains urine from the bladder out of the body (Figure 43.6). Around the opening to the urethra are two sphincter muscles that control the voiding of urine from the bladder, the **internal urethral sphincter** and the **external urethral sphincter.** In males, the urethra passes through the penis and opens at the distal tip of the penis. The **prostatic urethra** is the portion of the male urethra that passes through the prostate gland, located inferior to the bladder. The urethra in males transports urine and semen, each at the appropriate time. In females, the urethra is separate from the reproductive organs and opens anteriosuperior to the vaginal opening.

The point where the urethra exits the bladder plus the two points where the ureters enter the bladder on its posterior surface define a triangular area of the bladder wall called the **trigone** (TRĪ-gōn). In this region, the lumenal bladder wall is smooth rather than folded into rugae.

Make a Prediction

The urinary bladder expands and recoils as it fills and empties with urine. What type of lining epithelium does the urinary bladder have to facilitate this change in shape?

The ureter is lined with a mucosa consisting of a layer of **transitional epithelium** covering a **lamina propria** (**Figure 43.7**). The function of this mucus-producing covering is to protect the ureteral walls from the acidic urine. The ureters enter the bladder low on the posterior bladder surface.

Histological details of the bladder are shown in Figure 43.7b. Because the bladder is a passageway to the external environment,

Figure 43.6 Organs for Conducting and Storing Urine

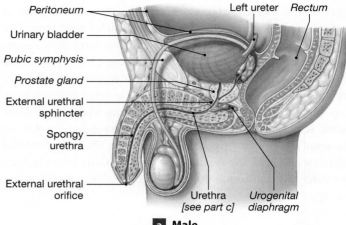

a Male

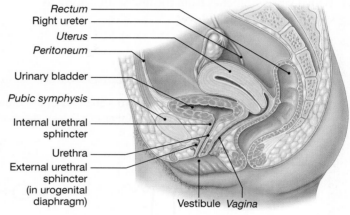

b Female

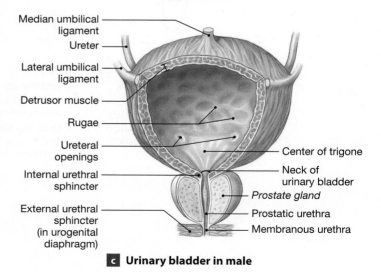

c Urinary bladder in male

Figure 43.7 Histology of the Organs That Collect and Transport Urine

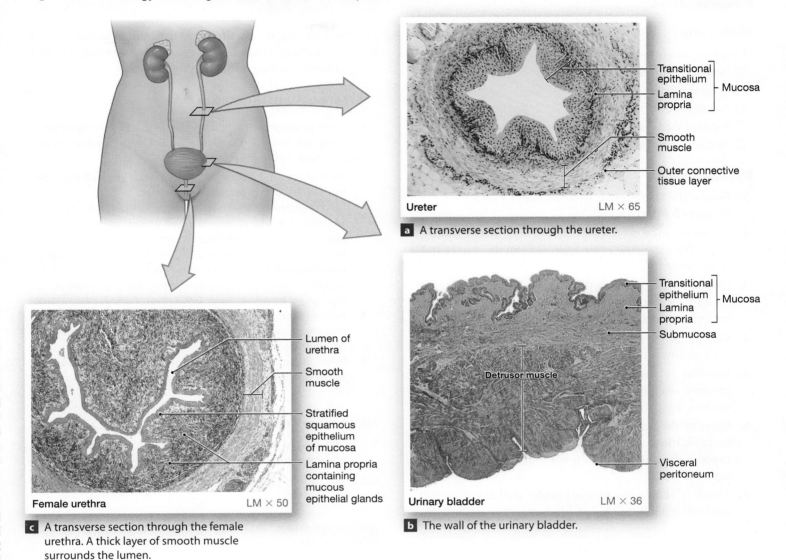

Ureter   LM × 65

a A transverse section through the ureter.

- Transitional epithelium
- Lamina propria
- Mucosa
- Smooth muscle
- Outer connective tissue layer

Female urethra   LM × 50

- Lumen of urethra
- Smooth muscle
- Stratified squamous epithelium of mucosa
- Lamina propria containing mucous epithelial glands

c A transverse section through the female urethra. A thick layer of smooth muscle surrounds the lumen.

Urinary bladder   LM × 36

- Transitional epithelium
- Lamina propria
- Mucosa
- Submucosa
- Detrusor muscle
- Visceral peritoneum

b The wall of the urinary bladder.

the mucosal lining facing the lumen is the superficial layer. This **mucosa** is made up of **transitional epithelium** overlying a **lamina propria.** The transitional epithelium consists of many different cell shapes to facilitate the stretching and recoiling of the bladder wall. Deep to the mucosa is the connective tissue of the **submucosa.** The detrusor muscle is deep to the submucosa.

Unlike the ureters and urinary bladder, the urethral mucosa is lined with stratified squamous epithelium rather than transitional epithelium. Deep to the epithelium is the lamina propria with mucous glands that secrete mucus to protect the urethral wall from the acidity of urine.

Clinical Application Floating Kidneys

Nephroptosis, or "floating kidneys," is the condition that results when the integrity of either the adipose capsule or the renal fascia is jeopardized, often because of excessive weight loss. There is less adipose tissue available to secure the kidneys around the renal fascia. This lack of support can result in the pinching or kinking of one or both ureters, preventing the normal flow of urine to the urinary bladder. ∎

QuickCheck Questions

4.1 Where do the ureters join the urinary bladder?

4.2 What is the trigone?

In the Lab 4

Materials

☐ Urinary system model
☐ Urinary system chart
☐ Compound microscope
☐ Prepared slide of the ureter
☐ Prepared slide of the urinary bladder
☐ Prepared slide of the urethra

Procedures

1. Review the anatomy of the lower urinary tract in Figures 43.6 and 43.7.

2. Locate the ureters on the urinary system model and/or chart. Trace the path urine follows from the renal papilla to the ureter.

3. On the model, examine the wall of the urinary bladder. Identify the trigone and the rugae. Which structures control emptying of the bladder?

4. On the model, examine the urethra. Note how the male urethra differs from the female urethra.

5. Examine the ureter slide with the microscope at low and medium magnifications. Refer to Figure 43.7 and identify the major layers of the ureteral wall.

6. Examine the urinary bladder slide at different magnifications. Observe transitional epithelium and rugae of the mucosa and the smooth muscle tissue of the detrusor muscle.

7. **Draw It!** Draw the urinary bladder wall as you view it at medium magnification.

Urinary bladder wall

8. Observe the urethral slide and note the lining epithelium in the mucosa and the mucous epithelia glands. ■

Lab Activity 5 Sheep Kidney Dissection

The sheep kidney is very similar to the human kidney in both size and anatomy. Dissection of a sheep kidney reinforces your observations of kidney models in the laboratory.

⚠ **Safety Alert:** Dissecting a Kidney

You *must* practice the highest level of laboratory safety while handling and dissecting the kidney. Keep the following guidelines in mind during the dissection.

1. Wear gloves and safety glasses to protect yourself from the fixatives used to preserve the specimen.

2. Do not dispose of the fixative from your specimen. You will later store the specimen in the fixative to keep the specimen moist and to keep it from decaying.

3. Be extremely careful when using a scalpel or other sharp instrument. Always direct cutting and scissor motions away from you to prevent an accident if the instrument slips on moist tissue.

4. Before cutting a given tissue, make sure it is free from underlying and/or adjacent tissues so that they will not be accidentally severed.

5. Never discard tissue in the sink or trash. Your instructor will inform you of the proper disposal procedure. ▲

QuickCheck Questions

5.1 What type of safety equipment should you wear during the sheep kidney dissection?

5.2 How should you dispose of the sheep kidney and scrap tissue?

In the Lab 5

Materials

☐ Gloves
☐ Safety glasses
☐ Dissecting tools
☐ Dissecting pan
☐ Preserved sheep kidney

Procedures

1. Put on gloves and safety glasses, and clear your workspace before obtaining your dissection specimen.

2. Rinse the kidney with water to remove excess preservative. Minimize your skin and mucous membrane exposure to the preservatives.

3. Examine the external features of the kidney. Using Figure 43.8 as a guide, locate the hilum. Locate the renal capsule and gently lift it by teasing with a needle. Below this capsule is the light pink cortex.

4. With a scalpel, make a longitudinal cut to divide the kidney into anterior and posterior portions. A single long, smooth cut is less damaging to the internal anatomy than a sawing motion.

5. Distinguish between the cortex and the darker medulla, which is organized into many triangular renal pyramids. The base of each pyramid faces the cortex, and the tip narrows into a renal papilla.

6. The renal pelvis is the large, expanded end of the ureter. Extending from this area are the major calyces and then the smaller minor calyces into which the renal papillae project.

7. Upon completion of the dissection, dispose of the sheep kidney as directed by your instructor; then wash your hands and dissecting instruments. ■

Figure 43.8 Gross Anatomy of Sheep Kidney A frontal section of a sheep kidney that has been injected with latex dye to highlight arteries (red), veins (blue), and urinary passageways (yellow).

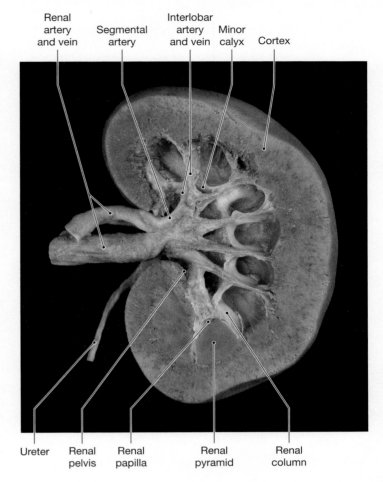

Name _____

Date _____

Section _____

Anatomy of the Urinary System

A. Definition

Define each term.

1. renal papilla

2. cortex

3. glomerular capsule

4. nephron loop

5. renal sinus

6. nephron

7. efferent arteriole

8. hilum

9. ureter

10. arcuate artery

11. renal capsule

12. renal column

B. Short-Answer Questions

1. Describe the components of the renal corpuscle.

2. What are two differences between cortical and juxtamedullary nephrons?

3. How does the urethra differ between males and females?

4. Where are the internal and external urethral sphincters located?

C. Labeling

Label the gross anatomy of the kidney in Figure 43.9.

Figure 43.9 **Gross Anatomy of the Kidney**

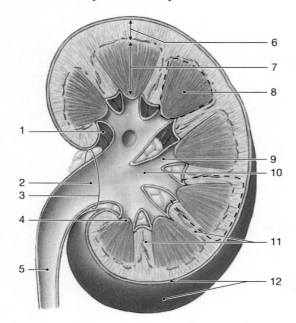

1. _____
2. _____
3. _____
4. _____
5. _____
6. _____
7. _____
8. _____
9. _____
10. _____
11. _____
12. _____

D. Analysis and Application

1. List the layers in the renal corpuscle through which filtrate must pass to enter the capsular space.

2. Trace a drop of blood from the abdominal aorta, through a kidney, and into the inferior vena cava.

3. Trace a drop of urine from a minor calyx to the urinary bladder.

E. Clinical Challenge

1. A patient has lost much weight while ill and now has difficulty urinating. Describe how a diagnosis of nephroptosis affects the urinary system.

2. Describe the effect of an enlarged prostate gland on the urinary function of a male.

Physiology of the Urinary System

Learning Outcomes

On completion of this exercise, you should be able to:

1. Define glomerular filtration, tubular reabsorption, and tubular secretion.

2. Describe the physical characteristics of normal urine.

3. Recognize normal and abnormal urine constituents.

4. Conduct a urinalysis test.

The kidneys maintain the chemical balance of body fluids by removing metabolic wastes, excess water, and electrolytes from the blood plasma. Three physiological processes occur in the nephrons to produce urine: filtration, reabsorption, and secretion (Figure 44.1). **Filtration** occurs in the renal corpuscle as blood pressure in the glomerulus forces water, ions, and other solutes small enough to pass through the filtration slits surrounding the glomerulus out of the blood and into the capsular space. Because size is the only thing that determines what passes through the filtration slits and becomes part of the filtrate, both wastes and essential solutes are removed from the blood in the glomerulus. Any solutes and water still needed by the body reenter the blood during **reabsorption,** as cells in the renal tubule reclaim the needed materials. Movement is in both directions along the length of the renal tubule, however, and in the process called **secretion,** any unneeded blood materials that did not leave the blood in the glomerulus leave it now and become part of the filtrate. The tubular cells actively transport ions in both directions (from blood to filtrate and from filtrate to blood), often using countertransport mechanisms that result in the reabsorption of necessary ions and the secretion of unneeded ones. As the filtrate passes through the entire length of

Need More Practice and Review?

Build your knowledge—and confidence!—in the Study Area of MasteringA&P® at www.masteringaandp.com with Pre-lab Quizzes, Post-lab Quizzes, Practice Anatomy Lab™ (PAL™) 3.0 virtual anatomy practice tool, PhysioEx™ 9.0 laboratory simulations, and A&P Flix™ with Quizzes.

PhysioEx™ 9.0 For this lab exercise, go to these topics:
- PhysioEx Exercise 9: Renal System Physiology
- PhysioEx Exercise 10: Acid-Base Balance

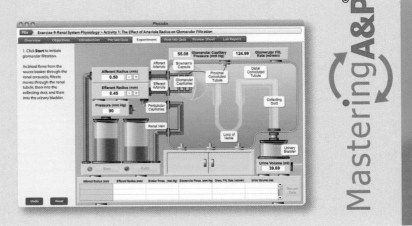

MasteringA&P®

Figure 44.1 A Summary of Renal Function A summary of the major processes of urine production.

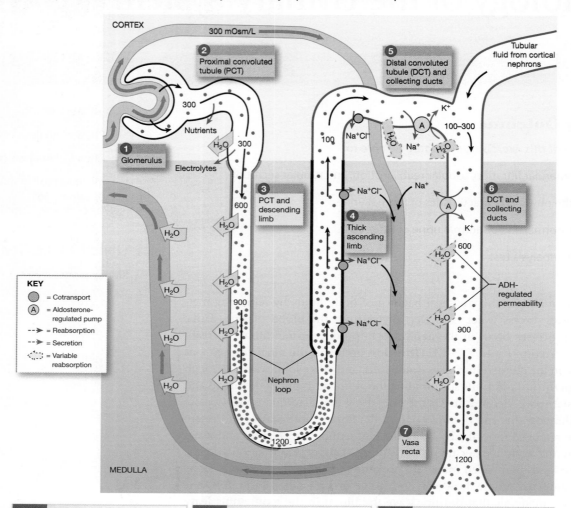

1	**Glomerulus**	**2**	**Proximal convoluted tubule (PCT)**	**3**	**PCT and descending limb**
The filtrate produced at the renal corpuscle has the same osmotic concentration as plasma—about 300 mOsm/L. It has the same composition as plasma without the plasma proteins.		In the proximal convoluted tubule (PCT), the active removal of ions and organic substrates produces a continuous osmotic flow of water out of the tubular fluid. This reduces the volume of filtrate but keeps the solutions inside and outside the tubule isotonic.		In the PCT and descending limb of the nephron loop, water moves into the surrounding peritubular fluids, leaving a small volume of highly concentrated tubular fluid. This reduction occurs by obligatory water reabsorption.	

4	**Thick ascending limb**	**5**	**DCT and collecting ducts**	**6**	**DCT and collecting ducts**	**7**	**Vasa recta**
The thick ascending limb is impermeable to water and solutes. The tubular cells actively transport Na+ and Cl− out of the tubule, thereby lowering the osmotic concentration of the tubular fluid. Because just Na+ and Cl− are removed, urea accounts for a higher proportion of the total osmotic concentration at the end of the loop.		The final adjustments in the composition of the tubular fluid occur in the DCT and the collecting system. The osmotic concentration of the tubular fluid can be adjusted through active transport (reabsorption or secretion).		The final adjustments in the volume and osmotic concentration of the tubular fluid are made by controlling the water permeabilities of the distal portions of the DCT and the collecting system. The level of exposure to ADH determines the final urine concentration.		The vasa recta absorbs the solutes and water reabsorbed by the nephron loop and the collecting ducts. By removing these solutes and water into the main circulatory system, the vasa recta maintains the concentration gradient of the medulla.	

the renal tubule of the nephron, reabsorption and secretion occur over and over. Once out of the renal tubule, the filtrate is processed into urine, which drips out of the renal papillae into the minor calyxes.

As the filtrate moves through the proximal convoluted tubule (PCT), 60% to 70% of the water and ions and 100% of the organic nutrients, such as glucose and amino acids, are reabsorbed into the blood. The simple cuboidal epithelium in this part of the nephron has microvilli to increase the surface area for reabsorption. The loop of Henle conserves water and salt while concentrating the filtrate for modification by the distal convoluted tubule (DCT). Reabsorption in the DCT is controlled by two hormones, aldosterone and antidiuretic hormone (ADH). Most of the secretion that takes place occurs in the DCT.

The kidneys filter 25% of the body's blood each minute, producing on average 125 mL/min of filtrate. About 180 L of filtrate are formed by the glomerulus per day, which eventually results in the production of an average daily output of 1.8 L of urine. The composition of urine can change on a daily basis depending on one's metabolic rate and urinary output. Water accounts for about 95% of the volume of urine. The other 5% contains excess water-soluble vitamins, drugs, electrolytes, and nitrogenous wastes. Abnormal substances in urine can usually be detected by **urinalysis,** an analysis of the chemical and physical properties of urine.

To reinforce your understanding of urine production, complete the *Sketch to Learn* activity (on p. 658).

Lab Activity 1   Physical Analysis of Urine

Normal constituents of urine include water, urea, creatinine, uric acid, many electrolytes, and possibly small amounts of hormones, pigments, carbohydrates, fatty acids, mucin, and enzymes.

The average pH of urine is 6, and a normal range is 4.5 to 8. Urine pH is greatly affected by diet. Diets high in vegetable fiber result in an alkaline pH value (above 7), and high-protein diets yield an acidic pH value (below 7).

Specific gravity is the ratio of the weight of a volume of a substance to the weight of an equal volume of distilled water. The specific gravity of water is therefore 1.000. The average specific gravity for a normal urine sample is between 1.003 and 1.030. Urine contains solutes and solids that affect its specific gravity. The amount of fluids ingested affects the volume of urine excreted and therefore the amount of solutes and solids per given volume. Drinking a lot of liquids results in more frequent urination of a dilute urine that contains few solutes and solids per given volume and therefore has a low specific gravity. Drinking very little liquid results in less frequent urination of a concentrated urine that has a high specific gravity. Excessively concentrated urine results in the crystallization of solutes, usually salts, into insoluble kidney stones.

During this urinalysis you will examine *only your own urine* and will study its volume, color, cloudiness, odor, and specific gravity. Alternatively, your laboratory instructor may provide your class with a mock urine sample for analysis. This artificial sample will probably include several abnormal urine constituents for instructional purposes.

QuickCheck Questions

1.1 What is the normal range of pH of urine?

1.2 What is the definition of specific gravity?

 Safety Alert: Handling a Urine Sample

Collect, handle, and test *only your own urine*. Dispose of all urine-contaminated materials in the biohazard disposal container as described by your instructor. If you spill some urine, wear gloves as you clean your workspace with a mild bleach solution. ▲

In the Lab   1

Materials

☐ 8-oz disposable cup

☐ Bleach solution

☐ Biohazard disposal container

Procedures

1. Sample Collection: Before collecting, void a small volume of urine from your bladder. By not collecting the first few milliliters, you avoid contaminating the sample with substances such as bacteria and pus from the urethra or menstrual blood. Then void into the disposable cup until it is about one-half full.

2. Observe the **physical characteristics** of the sample and record your observations in **Table 44.1**.

   a. The color of urine varies from colorless to amber. **Urochrome** is a by-product of the breakdown of hemoglobin that gives urine its yellowish color. Color also varies because of ingested food. Vitamin supplements, certain drugs, and the amount of solutes also influence urine color. A dark red or brown color indicates blood in the urine.

   b. Turbidity (cloudiness) is related to the amount of solids in the urine. Contributing factors include bacteria, mucus, cell casts, crystals, and epithelial cells. Observe and describe the turbidity of the urine sample. Use descriptive words such as *clear, clouded,* and *hazy*.

   c. To smell the sample, place it approximately 12 in. from your face and wave your hand over the sample toward your nose. Normally, freshly voided urine has no odor, and therefore odor serves as a diagnostic tool for fresh urine. Starvation causes the body to break

Sketch to Learn

A great way to understand how urine is made is to draw a nephron and show where each step of the process occurs.

Sample Sketch

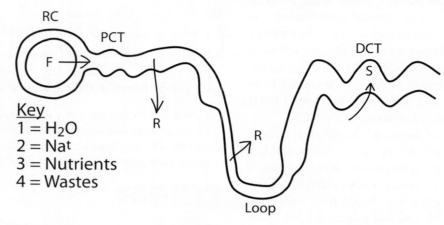

Key
1 = H_2O
2 = Na^t
3 = Nutrients
4 = Wastes

Step 1
- Draw a nephron as shown. Take your time and plan this part of the sketch.
- Add a circle in the renal corpuscle to show the glomerulus.

Step 2
- Place a letter "F" in the glomerulus and an arrow showing filtration.
- Add numbers 1, 2, 3, 4 inside PCT to indicate what was filtered.
- Make a key showing what each number represents.

Step 3
- Add arrow and "R" for reabsorption and list of materials reclaimed.
- Show reabsorption of water (1) in loop.
- Add "S" in DCT to show secretion of wastes (4).

Your Sketch

Table 44.1	Physical Observations of Urine Sample
Characteristic	**Observation**
Volume	_____
Color	_____
Turbidity	_____
Odor	_____
Specific gravity	_____

Figure 44.2 Using a Urinometer A hydrometer is floated in urine to measure the specific gravity of the sample.

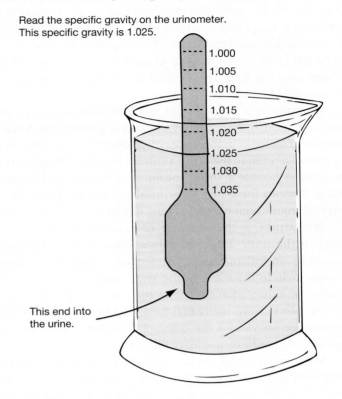

Read the specific gravity on the urinometer. This specific gravity is 1.025.

1.000
1.005
1.010
1.015
1.020
1.025
1.030
1.035

This end into the urine.

down fats and produce ketones, which give urine a fruity or acetone-like smell. Individuals with diabetes mellitus often produce sweet-smelling urine. (The characteristic odor associated with a urine sample that is not fresh is the result of the chemical breakdown of substances in the urine; the most characteristic odor is that of ammonia.)

d. Use a **urinometer,** shown in **Figure 44.2,** to determine the specific gravity of your sample. A urinometer consists of a small glass cylinder and a urine hydrometer. The cylinder is used to hold the urine sample being tested. The hydrometer is a float that has been calibrated against water. Along the stem of the hydrometer is a scale used to determine the specific gravity of the sample.

- Swirl the sample in the collection cup to suspend any materials that may have settled after collection.
- Fill the glass cylinder of the urinometer at least two-thirds full with the urine sample.
- Carefully lower the urine hydrometer into the cylinder of urine. If the hydrometer does not float, add more urine to the cylinder until the hydrometer does float.
- The urine will adhere to the walls of the glass cylinder and form a trough called a **meniscus.** The scale on the hydrometer stem is read at the bottom of the meniscus, where the meniscus intersects a line on the scale. Record the specific gravity of the sample in Table 44.1.

3. Wash the glass cylinder and the hydrometer with soap and water and dry with clean paper towels. Dispose of the urine cup in the biohazard box as indicated by your instructor.

4. Wash and dry your hands before leaving the laboratory. ∎

Lab Activity 2 Chemical Analysis of Urine

Certain materials in the urine suggest renal disease or injury. Excessive consumption of a substance may cause the substance to saturate the filtrate and overload the transport mechanisms of reabsorption. Because the overworked renal tubular cells cannot reclaim all of the substance, it appears in the urine. Hormones, enzymes, carbohydrates, fatty acids, pigments, and mucin all typically occur in small quantities in the urine.

Ketones in the urine, a condition called **ketosis** (Kē-TŌ-sis), may be the result of starvation, diabetes mellitus, or a diet very low in carbohydrates. When the blood carbohydrate concentration is low, cells begin to catabolize fats. The products of fat catabolism are glycerol and fatty acids. Liver cells convert the fatty acids to ketones, which then diffuse out of the liver and into the blood, where they are filtered by the kidneys. In diabetes mellitus, commonly called sugar diabetes, not enough glucose enters the cells of the body. As a result, the cells use fatty acids to produce ATP. This increase in fatty acid catabolism results in the appearance of ketones in the urine.

Glucosuria is glucose in the urine, which usually indicates diabetes mellitus. Because diabetic individuals do not have the normal cellular intake of glucose, the blood glucose concentration is abnormally high. The amount of glucose filtered out of the blood is greater than the amount the tubular cells of the nephrons can reabsorb back into the blood, and the glucose that is not reabsorbed appears in the urine. Glucosuria may also be the result of a very-high-carbohydrate meal that produces a temporary overload of glucose. Another cause of glucosuria is stress. Production of epinephrine in response to stress results in the conversion of glycogen to glucose and its release from the liver. The elevated levels of glucose may then be secreted into the urine.

Albumin is a large protein molecule that normally cannot pass through the filtration slits of the glomerulus. A trace amount of albumin in the urine is considered normal. However, excessive albumin in the urine, a condition called **albuminuria,** suggests an increase in the permeability of the glomerular membrane. Reasons for increased permeability can be the result of physical injury, high blood pressure, disease, or bacterial toxins.

Hematuria is the presence of erythrocytes (red blood cells) in the urine and is usually an indication of bleeding caused by an inflammation or infection of the urinary tract. Causes include irritation of the renal tubules from the formation of kidney stones, trauma such as a hard blow to the kidney, blood from menstrual flow, and possible tumor formation. Leukocytes in the urine, a condition called **pyuria** (pī-Ū-rē-uh), indicate a urinary tract infection.

When erythrocytes are hemolytized in the blood, the hemoglobin molecules break down into two chains that are filtered by the kidneys and excreted in the urine. If a large number of erythrocytes are being broken down in the circulation, the urine develops a dark brown to reddish color. The presence of hemoglobin in the urine is called **hemoglobinuria.**

Bilirubin in large amounts in the urine, the condition known as **bilirubinuria,** is a result of the breakdown of hemoglobin from old red blood cells being removed from the circulatory system by phagocytic cells in the liver. When red blood cells are removed from the blood by the liver, the globin portion of the hemoglobin molecule is split off the molecule and the heme portion is converted to biliverdin. The biliverdin is then converted to bilirubin, which is a major pigment in bile.

Urobilinogen in the urine is called **urobilinogenuria.** Small amounts of urobilinogen in the urine are normal. It is a product of the breakdown of bilirubin by the intestines and is responsible for the normal brown color of feces. Greater than trace levels in the urine may be due to infectious hepatitis, cirrhosis, congestive heart failure, or a variety of other diseases.

Urea is produced during **deamination** (dē-am-i-NĀ-shun) reactions that remove ammonia (NH_3) from amino acids. The ammonia combines with CO_2 and forms urea (CH_4ON_2). About 4600 mg of urea are produced daily, accounting for approximately 80% of the nitrogen waste in urine. **Creatinine** is formed from the breakdown of creatine phosphate, an energy-producing molecule found in muscle tissue. Uric acid is produced from the breakdown of the nucleic acids DNA and RNA, two molecules obtained either from foods or when body cells are destroyed.

Nitrites in the urine indicate a possible urinary tract infection.

Several inorganic ions and molecules are found in urine. Their presence is a reflection of diet and general health. Na+ and Cl– are ions from sodium chloride, the principal salt of the body. The amount of **Na+** and **Cl– ions** present in the urine varies with how much table salt is consumed in the diet. **Ammonium** (NH_4+) **ion** is a product of protein catabolism and must be removed from the blood before it reaches toxic concentrations. Many types of ions bind with sodium and form a buffer in the blood and urine to stabilize pH.

Dip sticks are a fast, inexpensive method of determining the chemical composition of urine. These sticks hold from one to nine testing pads containing reagents that react with certain substances found in urine. Single-test sticks are primarily used to determine whether glucose or ketones are present in the urine, and multiple-test sticks are used to give a more informative evaluation of the urine's chemical content.

QuickCheck Questions

2.1 What might cause glucose to be present in the urine of a person who does not have diabetes mellitus?

2.2 What are ketones, and when might they appear in the urine?

⚠ Safety Alert: Handling a Urine Sample

Collect, handle, and test *only your own urine.* Dispose of all urine-contaminated materials in the biohazard disposal container as described by your instructor. If you spill some urine, wear gloves as you clean your workspace with a mild bleach solution. ▲

In the Lab 2

Materials

- ☐ 8-oz disposable cup
- ☐ Bleach solution
- ☐ Urine dip sticks
- ☐ Paper towels
- ☐ Biohazard disposal container

Procedures

1. Sample Collection: You may use the urine collected in Lab Activity 1. If you do not use that sample, before collecting, void a small volume of urine from your bladder. By not collecting the first few milliliters, you avoid contaminating the sample with substances such as bacteria and pus from the urethra or menstrual blood. Then void into the disposable cup until it is about one-half full.

Table 44.2	Chemical Evaluation of Urine Sample
	Remark
pH	_____
Specific gravity	_____
Glucose	_____
Ketone	_____
Protein	_____
Erythrocytes	_____
Bilirubin	_____
Other	_____

2. Review the color chart on the dip stick bottle, and note which test pads require reading at specific times.

3. Swirl your sample of urine before placing a dip stick into the urine.

4. Holding a dip stick by the end that does not have any test pads, immerse the stick in the urine so that all the test pads are wetted and then withdraw the stick. Lay it on a clean, dry paper towel to absorb any excess urine.

5. How long to wait after removing a stick from the urine varies from immediately to two minutes. Use the color chart on the side of the dip stick bottle to determine how long you must wait.

6. Record your data from the dip stick in **Table 44.2**.

7. Dispose of all used dip sticks and paper towels (and gloves if you used them) in the biohazard disposal container. Dispose of the urine cup in the biohazard box as indicated by your instructor.

8. Wash and dry your hands before leaving the laboratory. ■

Lab Activity 3 Microscopic Examination of Urine

Examination of the sediment of a centrifuged urine specimen reveals the solid components of the sample. This can be a valuable test to determine or confirm the presence of abnormal contents in the urine. A wide variety of solids can be in urine, including cells, crystals, and mucus (**Figure 44.3**).

Figure 44.3 Sediments Found in the Urine Various cells, casts, crystals, and sediment occur in urine.

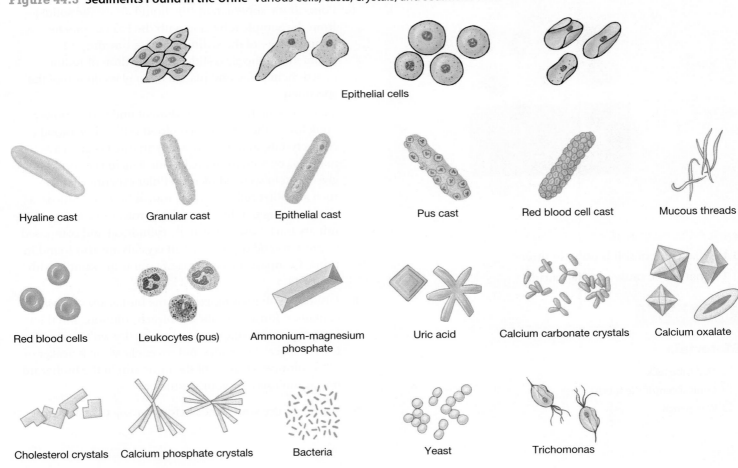

Epithelial cells

Hyaline cast   Granular cast   Epithelial cast   Pus cast   Red blood cell cast   Mucous threads

Red blood cells   Leukocytes (pus)   Ammonium-magnesium phosphate   Uric acid   Calcium carbonate crystals   Calcium oxalate

Cholesterol crystals   Calcium phosphate crystals   Bacteria   Yeast   Trichomonas

Urine is generally sterile, but **microbes** can be present in a sample for several reasons, and microbes present in large numbers usually indicate infection. Microbes may contaminate a urine sample when they are present at the urethral opening and in the urethra.

Clinical Application Kidney Stones

Crystals can form in the urinary tract and are usually voided with the urine. This sediment consists of **casts,** which usually are small clots of blood, tissue, or crystals of mineral salts. Complete blockage of the urinary tract can occur as a result of the formation of kidney stones, or **renal calculi,** which are solid pebbles of urinary salts containing calcium, magnesium, or uric acid (Figure 44.4). Calculi can occur in the kidneys, ureters, bladder, or urethra, and may cause severe pain. If a stone completely blocks the urinary tract and will not pass out of the body, it must be removed surgically. *Lithotripsy* is a nonsurgical procedure that uses sound waves to shatter the stone into pieces small enough to pass through the tract. The patient is immersed in water, and the sound energy is directed to the area overlying the stone to destroy the calculi. ∎

Figure 44.4 Kidney Stones

QuickCheck Questions

3.1 What types of solids occur in urine?

3.2 What is one cause of renal calculi?

In the Lab   3

Materials

☐ Test tube rack
☐ Conical centrifuge tubes
☐ Wax pencil
☐ Centrifuge
☐ Pasteur pipette with bulb
☐ Iodine or sediment stain
☐ Glass slides
☐ Cover glasses
☐ Compound microscope
☐ Biohazard disposal container
☐ Bleach solution
☐ 10% chlorine solution

Procedures

1. Use the urine sample collected in the previous two lab activities. Swirl the sample to suspend any solids that have settled.

2. With the wax pencil, mark two centrifuge tubes with a horizontal line two-thirds up from the bottom. Then label one tube "Sample" and the other "Blank."

3. Fill the Sample tube to the mark with urine, and fill the Blank tube to the mark with tap water.

4. Place the two tubes in the centrifuge opposite each other so that the centrifuge remains balanced. This step is very important to keep from damaging the centrifuge.

5. Centrifuge the sample for eight to ten minutes.

6. Pour off the supernatant (the liquid above the solids) from the Sample tube, and with the Pasteur pipette remove some of the sediment. Place one drop of sediment on the glass slide, add one drop of iodine or sediment stain, and place a cover glass on top of the specimen.

7. Begin viewing the stained sediment under 10× power. Look for epithelial cells, red blood cells, white blood cells, crystals, and microbes. Mucin threads and casts may also be seen in the sediment. Mucin is a complex glycoprotein secreted by unicellular exocrine glands, such as goblet cells. In water, mucin becomes mucus, a slimy coating that lubricates and protects the lining of the urinary tract. Casts are usually cylindrical and composed of proteins and dead cells. Salt crystals are also found in urine. Compare the contents of your urine sample with Figure 44.3.

8. Dispose of all used pipettes in the biohazard disposal container. Rinse test tubes in bleach solution, and then wash them with soap and water and dry with clean paper towels. Place glass slides and cover glasses in a beaker of 10% chlorine. Dispose of the urine cup in the biohazard box as indicated by your instructor.

9. Wash and dry your hands before leaving the laboratory. ∎

Review & Practice Sheet

EXERCISE 44

Name ____________________

Date ____________________

Section ____________________

Physiology of the Urinary System

A. Definition

Define each of the following terms.

1. pyuria

2. hematuria

3. glucosuria

4. secretion

5. albuminuria

6. bilirubin

7. urochrome

8. deamination

9. reabsorption

10. ketone

11. filtrate

12. urobilinogen

B. Short-Answer Questions

1. What is the normal pH range of urine?

2. What is the specific gravity of a normal sample of urine?

3. List five abnormal components of urine.

4. What substances in the urine might indicate that a person has diabetes?

5. Describe the three physiological processes of urine production.

C. Analysis and Application

1. A diabetic woman has been dieting for several months and has lost more than 25 lb. At her annual medical checkup, a urinalysis is performed. What would you expect to find in her urine?

2. What factors might affect the odor, color, and pH of a sample of urine?

3. Mike and Fred have been hiking in the desert all afternoon. While on the trail, Fred drinks much more water than Mike. If urine samples were collected from both men, what differences in specific gravity of the samples would you expect to measure?

D. Clinical Challenge

A patient with a history of renal calculi is scheduled for lithotripsy. Describe her condition and the procedure she will have.

Anatomy of the Reproductive System

Learning Outcomes

On completion of this exercise, you should be able to:

1. Identify the male testes, ducts, and accessory glands.

2. Describe the composition of semen.

3. Identify the three regions of the male urethra.

4. Identify the structures of the penis.

5. Identify the female ovaries, ligaments, uterine tubes, and uterus.

6. Describe and recognize the three main layers of the uterine wall.

7. Identify the vagina and the features of the vulva.

8. Identify the structures of the mammary glands.

9. Compare the formation of gametes in males and females.

Whereas all the other systems of the body function to support the continued life of the organism, the reproductive system functions to ensure continuation of the species. The primary sex organs, or **gonads** (GŌ-nads), of the male and female are the **testes** (TES-tēz; singular *testis*) and **ovaries,** respectively. The testes produce the male sex cells, **spermatozoa** (sper-ma-tō-ZŌ-uh; singular **spermatozoon;** also called *sperm cell*), and the ovaries produce the female sex cells, **ova** (singular **ovum**). These reproductive cells, collectively called **gametes** (GAM-ēts),

Need More Practice and Review?

Build your knowledge—and confidence!—in the Study Area of MasteringA&P® at www.masteringaandp.com with Pre-lab Quizzes, Post-lab Quizzes, Practice Anatomy Lab™ (PAL™) 3.0 virtual anatomy practice tool, PhysioEx™ 9.0 laboratory simulations, and A&P Flix with Quizzes.

PAL | practice anatomy lab   For this lab exercise, follow these navigation paths:
- PAL>Human Cadaver>Reproductive System
- PAL>Anatomical Models>Reproductive System
- PAL>Histology>Reproductive System

PhysioEx™ 9.0 For this lab exercise, go to this topic:
- PhysioEx Exercise 12: Serological Testing

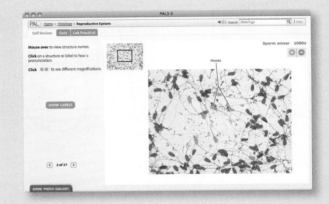

MasteringA&P®

are the parental cells that combine and become a new life. The gonads have important endocrine functions and secrete hormones that support maintenance of the male and female sex characteristics. The gametes are stored and transported in ducts, and several accessory glands in the reproductive system secrete products to protect and support the gametes.

Lab Activity 1 **Male: Testes, Epididymis, and Ductus Deferens**

In addition to the pair of testes, the male reproductive system consists of ducts, glands, and the penis (**Figure 45.1**). The testes are located outside the pelvic cavity, and the ducts transport the spermatozoa produced in the testes to inside the pelvic cavity, where glands add secretions to form a mixture

called **semen** (SĒ-men), the liquid that is ejaculated. The testes are located in the **scrotum** (SKRŌ-tum), a pouch of skin hanging from the pubis region. The pouch is divided into two compartments by the notch in the scrotum called the **raphe** (RĀ-fē) and by the **scrotal septum** (**Figure 45.2**). The **dartos** (DAR-tōs) **muscle** in the dermis also contributes to the septum separating the testes and is responsible for the wrinkling of the scrotum skin. Deep to the scrotal skin is the superficial scrotal fascia that encases the testes and the **spermatic cord** from which the testes are suspended. Other structures in the spermatic cord include blood and lymphatic vessels, nerves, and the **cremaster** (krē-MAS-ter) **muscle,** which encases the testes and raises or lowers them to maintain an optimum temperature for spermatozoa production.

Each testis is about 5 cm (2 in.) long and 2.5 cm (1 in.) in diameter. The scrotum and testes are lined with a serous

Figure 45.1 Male Reproductive System in Midsagittal View A sagittal section of the male reproductive organs.

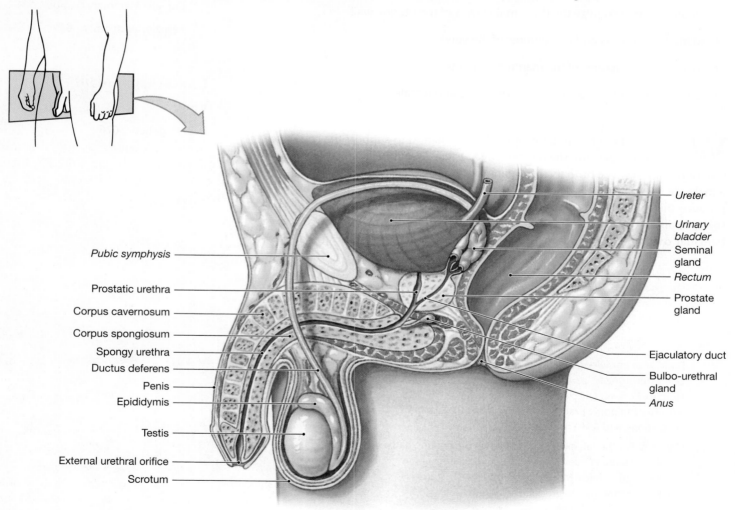

Pubic symphysis

Prostatic urethra

Corpus cavernosum

Corpus spongiosum

Spongy urethra

Ductus deferens

Penis

Epididymis

Testis

External urethral orifice

Scrotum

Ureter

Urinary bladder

Seminal gland

Rectum

Prostate gland

Ejaculatory duct

Bulbo-urethral gland

Anus

Figure 45.2 Male Reproductive System in Anterior View The frontal section of the scrotum to show the cremaster muscle and the spermatic cord.

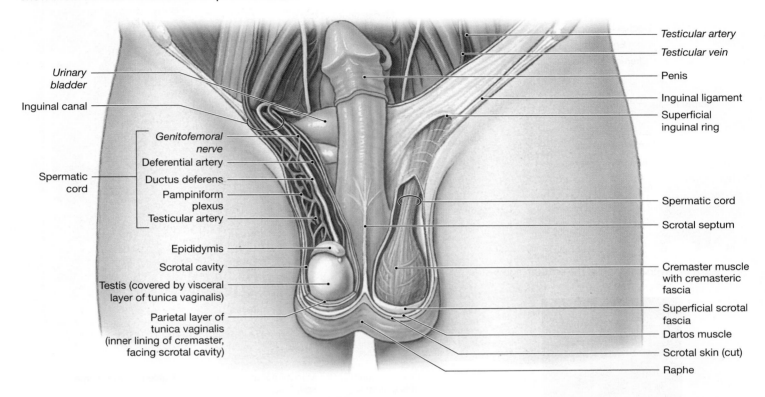

membrane called the **tunica vaginalis** that allows for movement of the tests within the scrotum (**Figure 45.3**). Deep to the tunica vaginalis is the **tunica albuginea** (al-bū-JEN-ē-uh) that branches into septa that divide the testis into sections called **lobules** that contain highly coiled **seminiferous** (se-mi-NIF-e-rus) **tubules** where spermatozoa are produced. Between the seminiferous tubules are small clusters of cells called **interstitial cells** that secrete **testosterone,** the male sex hormone. Testosterone is responsible for the male sex drive and for development and maintenance of the male secondary sex characteristics, such as facial hair and increased muscle and bone development.

Spermatozoa are flagellated cells each with a **head** containing a nucleus with the male's genetic contribution to an offspring. The **acrosome** (ak-rō-SŌM) on the head contains enzymes to break down the outer layer of the egg so fertilization may occur. The **neck** of the spermatozoon contains centrioles; the **middle piece** has mitochondria to generate ATP that is required to whip the tail-like **flagellum** around to move the sperm.

After spermatozoa are produced in the seminiferous tubules, they pass through a series of tubules called the **rete testis** and enter the **epididymis** (ep-i-DID-i-mus), a highly coiled tubule

located on the posterior side of the testis (**Figure 45.4**). The wall of the epididymis consists mainly of smooth muscle tissue; the lumen is lined with pseudostratified columnar epithelium that has stereocilia to help transport the spermatozoa out of the epididymis during ejaculation. The spermatozoa mature in the epididymis and are stored until ejaculation out of the male reproductive system. Peristalsis of the smooth muscle of the epididymis and surface transport by the stereocilia propels the spermatozoa into the **ductus deferens** (DUK-tus DEF-e-renz), or *vas deferens*, the duct that empties into the urethra. The ductus deferens is 46 to 50 cm (18 to 20 in.) long and is lined with pseudostratified columnar epithelium. Peristaltic waves propel spermatozoa toward the urethra.

Within the scrotum, the ductus deferens ascends into the pelvic cavity as part of the spermatic cord. The ductus deferens passes through the **inguinal** (ING-gwi-nal) **canal** in the lower abdominal wall to enter the body cavity. This canal is a weak area and is frequently injured. An **inguinal hernia** occurs when portions of intestine protrude through the canal and slide into the scrotum. The ductus deferens continues around the posterior of the urinary bladder and widens into the **ampulla** (am-PŪL-uh) before joining the seminal vesicle at the ejaculatory duct.

Figure 45.3 Scrotum and Testes

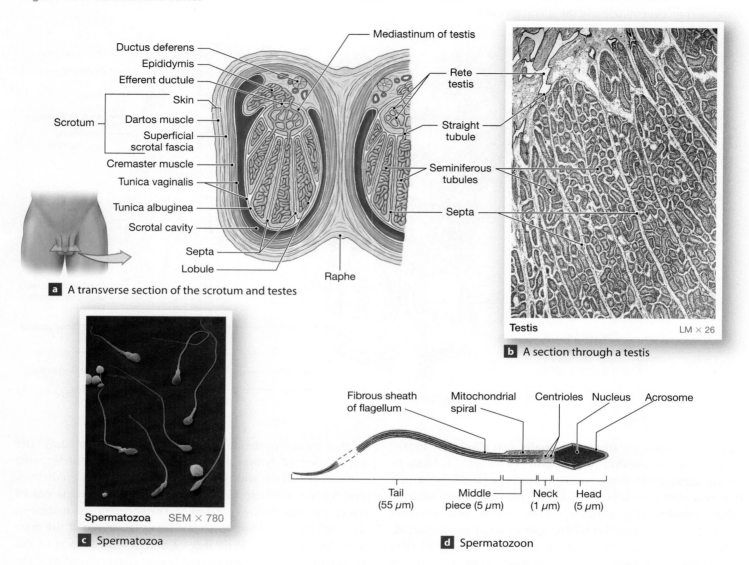

a A transverse section of the scrotum and testes

Ductus deferens
Epididymis
Efferent ductule
Skin
Dartos muscle
Superficial scrotal fascia
Scrotum
Cremaster muscle
Tunica vaginalis
Tunica albuginea
Scrotal cavity
Septa
Lobule
Raphe

Mediastinum of testis
Rete testis
Straight tubule
Seminiferous tubules
Septa

Testis LM × 26

b A section through a testis

Spermatozoa SEM × 780

c Spermatozoa

Fibrous sheath of flagellum
Mitochondrial spiral
Centrioles
Nucleus
Acrosome

Tail (55 μm)
Middle piece (5 μm)
Neck (1 μm)
Head (5 μm)

d Spermatozoon

Clinical Application Vasectomy

A common method of birth control for men is a procedure called **vasectomy** (vaz-EK-tō-mē). Two small incisions are made in the scrotum, and a small segment of the ductus deferens on each side is removed. A vasectomized man still produces spermatozoa, but because the duct that transports them from the epididymis to the urethra is removed, the semen that is ejaculated contains no spermatozoa. As a result, no female ovum can be fertilized. Men who have had a vasectomy still produce testosterone and have a normal sex drive. They have orgasms, and the ejaculate is approximately the same volume as in men who have not been vasectomized. ∎

QuickCheck Questions

1.1 Where are the testes located?

1.2 Where are spermatozoa stored?

1.3 Where does the ductus deferens enter the abdominal cavity?

In the Lab 1

Materials

☐ Male urogenital model and chart
☐ Compound microscope
☐ Prepared slide of testis
☐ Prepared slide of epididymis

Figure 45.4 The Epididymis

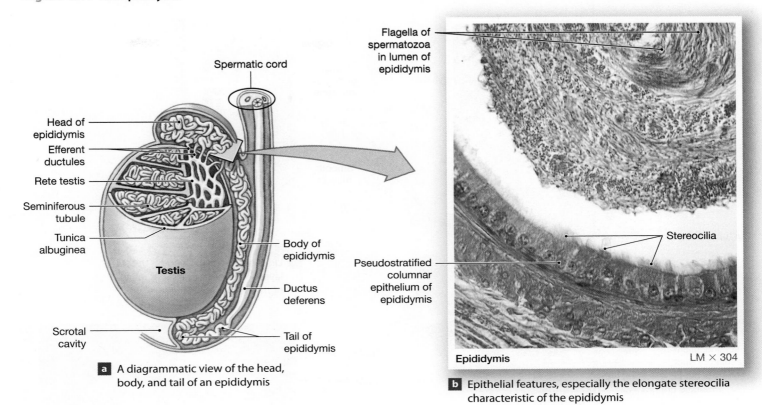

Head of epididymis
Efferent ductules
Rete testis
Seminiferous tubule
Tunica albuginea
Testis
Scrotal cavity
Spermatic cord
Body of epididymis
Ductus deferens
Tail of epididymis

a A diagrammatic view of the head, body, and tail of an epididymis

Flagella of spermatozoa in lumen of epididymis
Stereocilia
Pseudostratified columnar epithelium of epididymis
Epididymis   LM × 304

b Epithelial features, especially the elongate stereocilia characteristic of the epididymis

Procedures

1. Review the male anatomy in Figures 45.1 and 45.2. Locate the scrotum, testes, epididymis, and ductus deferens on the urogenital model and chart.

2. Locate the spermatic cord, the cremaster muscle, and the inguinal canal on the model and chart.

3. Examine the testis slide at different magnifications and identify the rete testis, seminiferous tubules, and interstitial cells. At high magnification, look carefully in the lumen of the tubules for spermatozoa.

4. On the epididymis slide, examine the epithelium and observe the stereocilia extending into the lumen. Observe the layers of smooth muscle tissue of the wall. ■

Lab Activity 2   Male: Accessory Glands

Three accessory glands—seminal vesicles, prostate gland, and bulbo-urethral glands—produce fluids that nourish, protect, and support the spermatozoa (**Figure 45.5**). The spermatozoa and fluids from these glands mix together as semen. The average number of spermatozoa per milliliter of semen is between 50 million and 150 million, and the average volume of ejaculate is between 2 and 5 mL.

The **seminal** (SEM-i-nal) **vesicles** are a pair of glands posterior and lateral to the urinary bladder. Each gland is approximately 15 cm (6 in.) long and merges with the ductus deferens into an **ejaculatory duct.** The seminal vesicles contribute about 60% of the total volume of semen. They secrete a viscous, alkaline **seminal fluid** containing the sugar fructose. The alkaline nature of this liquid neutralizes the acidity of the male urethra and the female vagina. The fructose provides the energy needed by each spermatozoon for beating its flagellum tail to propel the cell on its way to an ovum. Seminal fluid also contains fibrinogen, which causes the semen to temporarily clot after ejaculation.

The **prostate** (PROS-tāt) **gland** is a single gland just inferior to the urinary bladder. The ejaculatory duct passes into the prostate gland and empties into the first segment of the urethra, the **prostatic urethra.** The prostate gland secretes a milky white, slightly acidic liquid that contains clotting enzymes to coagulate the semen. These secretions contribute about 20% to 30% of the semen volume.

The prostatic urethra exits the prostate gland and passes through the floor of the pelvis, the urogenital diaphragm, as the **membranous urethra.** A pair of **bulbo-urethral** (bul-bō-ū-RĒ-thral) **glands,** also called *Cowper's glands*, occur on either side of the membranous urethra and add an alkaline

Figure 45.5 **Ductus Deferens and Accessory Glands**

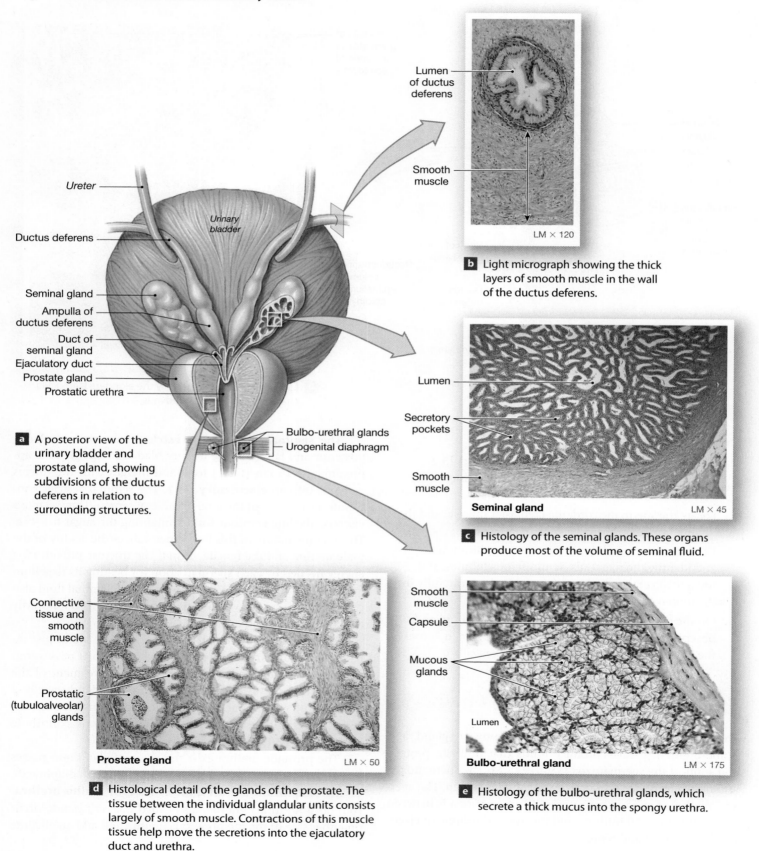

Lumen of ductus deferens

Smooth muscle

LM × 120

b Light micrograph showing the thick layers of smooth muscle in the wall of the ductus deferens.

Ureter

Urinary bladder

Ductus deferens

Seminal gland

Ampulla of ductus deferens

Duct of seminal gland

Ejaculatory duct

Prostate gland

Prostatic urethra

a A posterior view of the urinary bladder and prostate gland, showing subdivisions of the ductus deferens in relation to surrounding structures.

Bulbo-urethral glands

Urogenital diaphragm

Lumen

Secretory pockets

Smooth muscle

Seminal gland LM × 45

c Histology of the seminal glands. These organs produce most of the volume of seminal fluid.

Connective tissue and smooth muscle

Prostatic (tubuloalveolar) glands

Prostate gland LM × 50

d Histological detail of the glands of the prostate. The tissue between the individual glandular units consists largely of smooth muscle. Contractions of this muscle tissue help move the secretions into the ejaculatory duct and urethra.

Smooth muscle

Capsule

Mucous glands

Lumen

Bulbo-urethral gland LM × 175

e Histology of the bulbo-urethral glands, which secrete a thick mucus into the spongy urethra.

mucus to the semen. Before ejaculation, the bulbo-urethral secretions neutralize the acidity of the urethra and lubricate the end of the penis for sexual intercourse. These glands contribute about 5% of the volume of semen.

QuickCheck Questions

2.1 What are the three accessory glands that contribute to the formation of semen?

2.2 Where is the membranous urethra located?

In the Lab 2

Materials

☐ Male urogenital model and chart

Procedures

1. Review the anatomy in Figure 45.5.
2. On the model and/or chart, trace each ductus deferens through the inguinal canal, behind the urinary bladder, to where each unites with a seminal vesicle. Identify the enlarged ampulla of the ductus deferens.
3. Identify the prostate gland, and note the ejaculatory duct that drains the ductus deferens and the seminal vesicle on each side of the body. Identify the prostatic urethra passing from the urinary bladder through the prostate gland.
4. Find the membranous urethra in the muscular pelvic floor. Identify the small bulbo-urethral glands on either side of the urethra. ∎

Lab Activity 3 Male: Penis

The **penis,** detailed in Figure 45.6, is the male copulatory organ that delivers semen into the vagina of the female. The penis is cylindrical and has an enlarged, acorn-shaped head called the **glans.** Around the base of the glans is a margin called the **corona** (crown). On an uncircumcised penis, the glans is covered with a loose-fitting skin called the **prepuce** (PRĒ-pūs) or *foreskin.* **Circumcision** is surgical removal of the prepuce. The **spongy urethra** transports both semen and urine through the penis and ends at the **external urethral orifice** in the tip of the glans. The **root** of the penis anchors the penis to the pelvis. The **body** consists of three cylinders of erectile tissue: a pair of dorsal **corpora cavernosa** (KOR-po-ruh ka-ver-NŌ-suh), and a single ventral **corpus spongiosum** (spon-jē-Ō-sum). During sexual arousal, the three erectile tissues become engorged with blood and cause the penis to stiffen into an erection.

QuickCheck Questions

3.1 What is the enlarged structure at the tip of the penis?

3.2 Which structures fill with blood during erection?

3.3 What duct transports urine and semen in the penis?

In the Lab 3

Materials

☐ Male urogenital model and chart

Procedures

1. Review the anatomy of the penis in Figure 45.6.
2. Identify the glans, corona, body, and root of the penis on the model and/or chart.
3. On the model, identify the corpora cavernosa and the corpus spongiosum. ∎

Lab Activity 4 Male: Spermatogenesis

Millions of spermatozoa are produced each day by the seminiferous tubules, in a process called **spermatogenesis** (sper-ma-tō-JEN-e-sis), shown in Figure 45.7. During this process, cells go through a series of cell divisions, called **meiosis** (mī-Ō-sis), that ultimately reduce the number of chromosomes in each cell to one-half the initial number. Cells containing this lower number of chromosomes are called **haploid** (HAP-loyd) cells. In females, a similar process (called *oogenesis* and discussed in Lab Activity 8) occurs in an ovary to produce a haploid ovum. When a haploid spermatozoon with its 23 chromosomes joins a haploid ovum with its 23 chromosomes, the resulting fertilized ovum has all 46 chromosomes and is **diploid** (DIP-loyd). From this first new diploid cell, called the **zygote,** an incomprehensible number of divisions ultimately shape a new human.

The term **somatic cells** refers to all the cells in the body except the cells that produce gametes. **Mitosis** (mī-TŌ-sis) is cell division in somatic cells, where one parent cell divides to produce two identical diploid daughter cells. Meiosis, as just noted, is cell division of cells in the testes and ovaries that produces haploid gametes. Meiosis occurs in two cycles, meiosis I and II, and in many ways is similar to mitosis. For simplicity, Figure 45.7 illustrates meiosis in a diploid cell containing 3 chromosome pairs (6 individual chromosomes) instead of the 23 pairs found in humans.

When a male reaches puberty, hormones stimulate the testes to begin spermatogenesis (Figure 45.8). Cells called **spermatogonia** (sper-ma-tō-GŌ-nē-uh) located in the outer wall of the seminiferous tubules divide by mitosis and produce, in addition to new (haploid) spermatogonia, some

Figure 45.6 The Penis

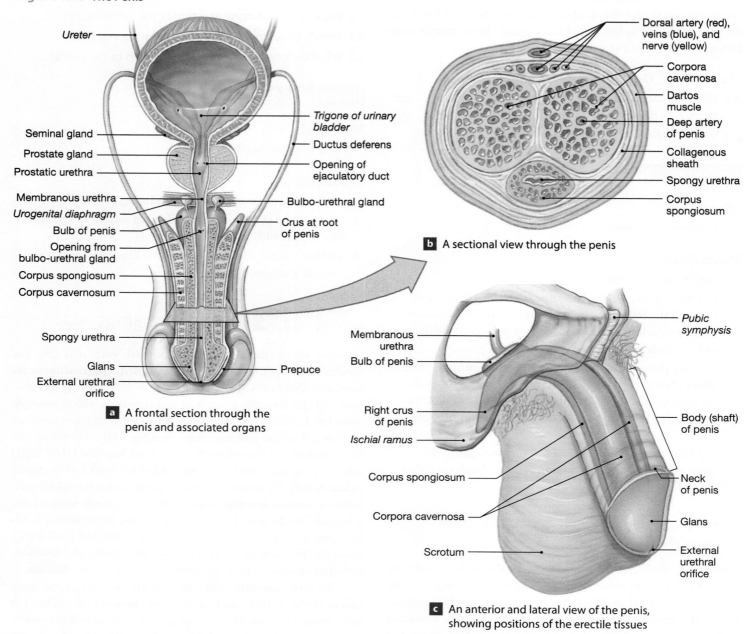

a A frontal section through the penis and associated organs

b A sectional view through the penis

c An anterior and lateral view of the penis, showing positions of the erectile tissues

diploid **primary spermatocytes** (sper-MA-tō-sīts). A primary spermatocyte prepares for meiosis by duplicating its genetic material. After replication, each chromosome is double stranded and consists of two **chromatids.** Thus each original pair of chromosomes, which are called **homologous chromosomes,** now consists of four chromatids. The primary spermatocyte is now ready to proceed into meiosis.

Meiosis I begins as the nuclear membrane of the primary spermatocyte dissolves and the chromatids condense into chromosomes. The homologous chromosomes match into pairs in a process called **synapsis,** and the four chromatids of the pair are collectively called a **tetrad.** Because each chromatid in a tetrad belongs to the same chromosome pair, genetic information may be exchanged between chromatids. This **crossing over,** or mixing, of the genes contained in the chromatids increases the genetic variation within the population.

Next the tetrads line up in the middle of the cell, and the critical step of reducing the chromosome number to haploid occurs. The tetrads separate, and the double-stranded chromosomes move to opposite sides of the cell. This separation step is called the **reduction division** of meiosis because haploid

Figure 45.7 Seminiferous Tubules and Meiosis

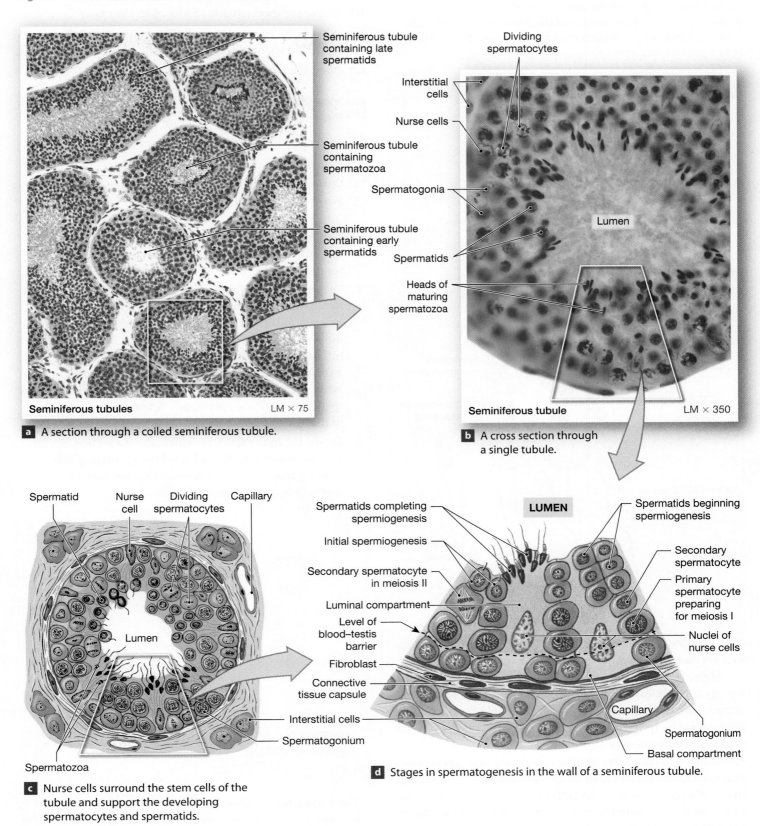

Seminiferous tubule containing late spermatids

Seminiferous tubule containing spermatozoa

Seminiferous tubule containing early spermatids

Seminiferous tubules   LM × 75

a A section through a coiled seminiferous tubule.

Dividing spermatocytes

Interstitial cells

Nurse cells

Spermatogonia

Lumen

Spermatids

Heads of maturing spermatozoa

Seminiferous tubule   LM × 350

b A cross section through a single tubule.

Spermatid   Nurse cell   Dividing spermatocytes   Capillary

Lumen

Spermatozoa

c Nurse cells surround the stem cells of the tubule and support the developing spermatocytes and spermatids.

Spermatids completing spermiogenesis

Initial spermiogenesis

Secondary spermatocyte in meiosis II

Luminal compartment

Level of blood–testis barrier

Fibroblast

Connective tissue capsule

Interstitial cells

Spermatogonium

LUMEN

Spermatids beginning spermiogenesis

Secondary spermatocyte

Primary spermatocyte preparing for meiosis I

Nuclei of nurse cells

Capillary

Spermatogonium

Basal compartment

d Stages in spermatogenesis in the wall of a seminiferous tubule.

Figure 45.8 Spermatogenesis Stem cells in the wall of the seminiferous tubule undergo meiosis, cell division that results in gametes with half of the number of chromosomes. Human cells contain 23 pairs of chromosomes in diploid stages, but for clarity only 3 pairs are illustrated here.

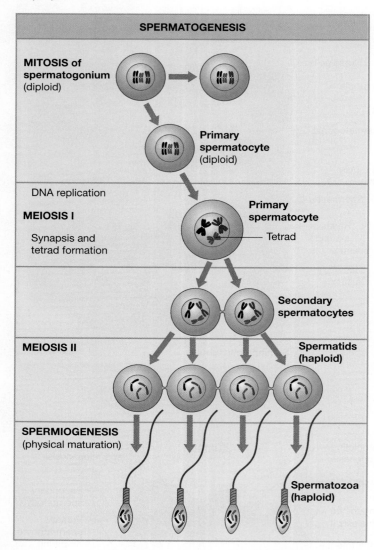

SPERMATOGENESIS

MITOSIS of spermatogonium (diploid)

Primary spermatocyte (diploid)

DNA replication

MEIOSIS I

Synapsis and tetrad formation

Primary spermatocyte

Tetrad

Secondary spermatocytes

MEIOSIS II

Spermatids (haploid)

SPERMIOGENESIS (physical maturation)

Spermatozoa (haploid)

QuickCheck Questions

4.1 Where are spermatozoa produced in the male?

4.2 What is the name of the cell that divides to produce a primary spermatocyte?

4.3 What is a tetrad?

In the Lab 4

Materials

☐ Meiosis models
☐ Compound microscope
☐ Prepared slide of testis

Procedures

1. Identify the different cell types shown on the meiosis models.

2. Examine the testis slide, using the micrographs in Figure 45.8 for reference. Scan the slide at low magnification and observe the many seminiferous tubules. Increase the magnification and locate the interstitial cells between the tubules. At high power, pick a seminiferous tubule that has distinct cells within the walls. Identify the spermatogonia, primary and secondary spermatocytes, and spermatids. Spermatozoa are visible in the lumen of the tubule.

3. *Draw It!* In the space provided, draw a section of a seminiferous tubule, and label the spermatogonia, primary spermatocytes, secondary spermatocytes, spermatids, and spermatozoa. ■

Seminiferous tubule

cells are produced. Next the cell pinches apart into two haploid **secondary spermatocytes.**

Meiosis II is necessary because, although the secondary spermatocytes are haploid, they have double-stranded chromosomes that must be reduced to single-stranded chromosomes. The process is similar to mitosis, with the double-stranded chromosomes lining up and separating. The two secondary spermatocytes produce four haploid **spermatids** that contain single-stranded chromosomes. In approximately five weeks the spermatids develop into spermatozoa, enter the lumen of the seminiferous tubules, and are transported to the epididymis where they undergo several weeks of maturation into a mature, active spermatozoa.

Lab Activity 5 Female: Ovaries, Uterine Tubes, and Uterus

The female reproductive system, highlighted in Figure 45.9, includes two ovaries, two uterine tubes, the uterus, the vagina, external genitalia, and two mammary glands. **Gynecology** is

Figure 45.9 Female Reproductive System in Sagittal Section A midsagittal section of the female pelvis showing the anatomical location of the reproductive organs.

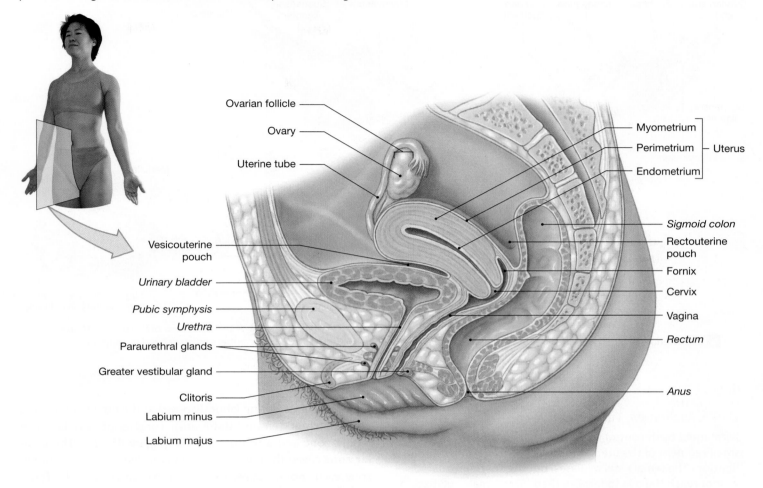

the branch of medicine that deals with the care and treatment of the female reproductive system.

The ovaries are paired structures approximately the size and shape of an almond and are located along the lateral walls of the pelvic cavity. A double-layered fold of peritoneum called the **mesovarium** (mes-ō-VAR-ē-um) holds the ovaries to the **broad ligament** of the uterus (**Figure 45.10**). The **suspensory ligaments** hold the ovaries to the wall of the pelvis, and the **ovarian ligaments** hold the ovaries to the uterus. The **round ligaments** extend laterally from the ovaries and provide posterior support.

Like around the testis, a layer of dense connective tissue called the **tunica albuginea** surrounds the ovary. The **stroma** or interior of the ovary has a central **medulla** and outer **cortex** where the ova are produced. The process of oogenesis, egg production, begins before birth, therefore, the cortex of a mature ovary is full of **egg nests** of immature eggs called **oocytes** (ō-ō-sīts) that can develop into **mature follicles** that can ovulate ova for fertilization. (See Lab Activity 8 for the study of oogenesis.)

Upon ovulation, an ovum is released from ovary and transported to the uterus by one of two **uterine tubes,** commonly called *fallopian tubes* (**Figure 45.11**). At the tip of the uterine tubes are fingerlike projections called **fimbriae** (FIM-brē-ē). These projections sweep over the surface of the ovary to capture the released ovum and draw it into the expanded **infundibulum** region of the uterine tube. The lumen of the uterine tube is lined with **ciliated simple columnar epithelium** with an underlying bed of connective tissue called the **lamina propria** (Figure 45.11b). Deep to the lamina propria is smooth muscle. Once the ovum is inside the uterine tube, movements of the cilia and peristaltic waves of muscle contraction transport the ovum toward the uterus. The tube widens midway along its length in the **ampulla** and then narrows at the **isthmus** (IS-mus) to enter the uterus. Fertilization of the ovum usually occurs between the infundibulum and the ampulla of the uterine tube.

Figure 45.10 **Ovaries and Their Relationships to the Uterine Tubes and Uterus**

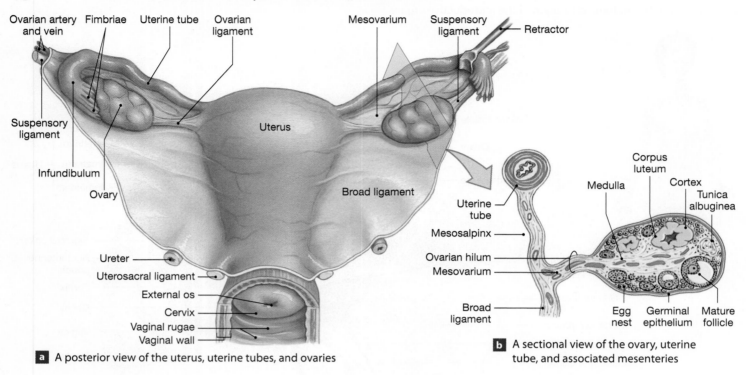

a A posterior view of the uterus, uterine tubes, and ovaries

b A sectional view of the ovary, uterine tube, and associated mesenteries

Clinical Application Tubal Ligation

Permanent birth control for females involves removing a small segment of the uterine tubes in a process called **tubal ligation.** The female still ovulates, but the spermatozoa cannot reach the ova to fertilize them. The female still has a monthly menstrual period. ■

The **uterus,** the pear-shaped muscular organ located between the urinary bladder and the rectum, is the site where a fertilized ovum is implanted and where the fetus develops during pregnancy. The uterus consists of three major regions: fundus, body, and cervix. The superior, dome-shaped portion of the uterus is the **fundus,** and the inferior, narrow portion is the **cervix** (SER-viks). The rest of the uterus is called the **body.** Within the uterus is a space called the **uterine cavity** that narrows at the cervix as the **cervical canal.**

The uterine wall consists of three main layers: perimetrium, myometrium, and endometrium. The **perimetrium** is the outer covering of the uterus. It is an extension of the visceral peritoneum and is therefore also called the *serosa.* The thick middle layer, the **myometrium** (mī-ō-MĒ-trē-um), is composed of three layers of smooth muscle and is responsible for the power-

ful contractions during labor. Exposed at the uterine cavity, the **endometrium** (en-dō-MĒ-trē-um), consists of two layers, a basilar zone and a functional zone (Figure 45.11c). The **basilar zone** covers the myometrium and produces a new functional zone each month. Superficial to the basilar zone is the **functional zone.** This layer is very glandular and is highly vascularized to support an implanted embryo. The functional zone is the endometrial layer that is shed each cycle during menstruation.

As a woman's monthly cycle progresses, the histology of the endometrium changes and the uterus prepares for the possibility of pregnancy (**Figure 45.12**). The cycle starts with bleeding, called **menses,** and is characterized by the breaking down of the functional zone. After menses, the functional zone is rebuilt during the **proliferative phase** and small uterine glands appear. Toward the end of the cycle, the **secretory phase** is distinguished by a thick endometrium with many elongated uterine glands. If pregnancy does not occur, the cycle repeats as menses occurs.

QuickCheck Questions

5.1 What structure transports an ovum from the ovary to the uterus?

5.2 What are the three layers of the uterine wall?

5.3 Which layer of the uterine wall is shed during menses?

Figure 45.11 **Uterine Tubes and Uterus**

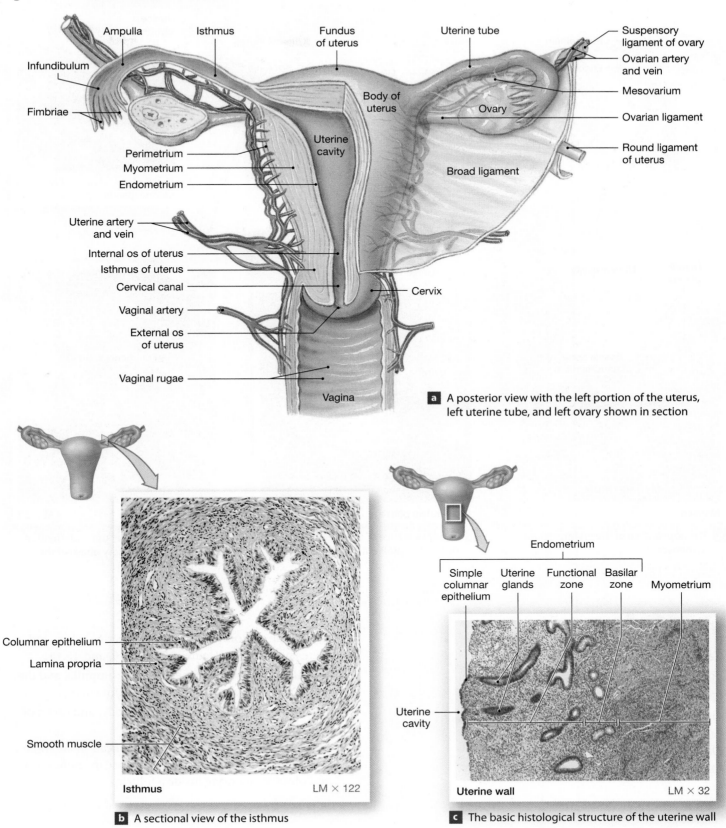

Ampulla

Isthmus

Fundus of uterus

Uterine tube

Suspensory ligament of ovary

Ovarian artery and vein

Infundibulum

Mesovarium

Body of uterus

Ovary

Ovarian ligament

Fimbriae

Perimetrium

Uterine cavity

Round ligament of uterus

Myometrium

Endometrium

Broad ligament

Uterine artery and vein

Internal os of uterus

Isthmus of uterus

Cervical canal

Cervix

Vaginal artery

External os of uterus

Vaginal rugae

Vagina

a A posterior view with the left portion of the uterus, left uterine tube, and left ovary shown in section

Columnar epithelium

Lamina propria

Smooth muscle

Isthmus

LM × 122

b A sectional view of the isthmus

Endometrium

Simple columnar epithelium

Uterine glands

Functional zone

Basilar zone

Myometrium

Uterine cavity

Uterine wall

LM × 32

c The basic histological structure of the uterine wall

Figure 45.12 Appearance of the Endometrium During the Uterine Cycle

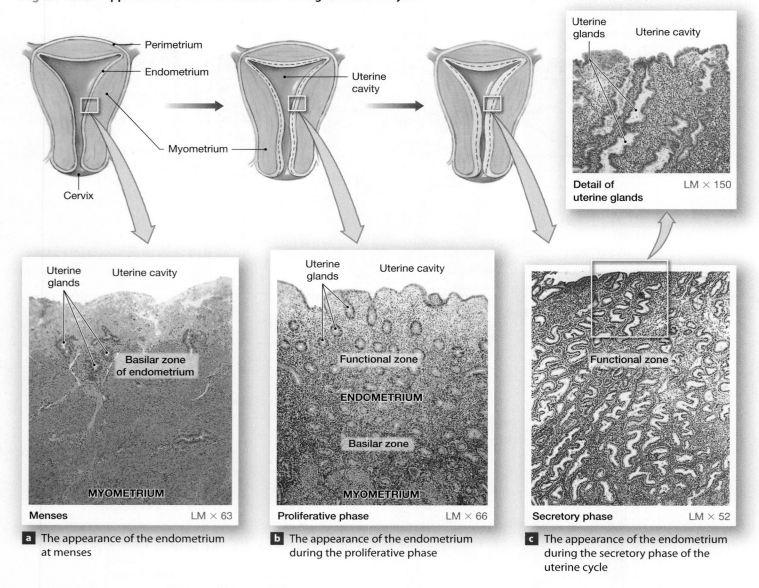

a The appearance of the endometrium at menses

b The appearance of the endometrium during the proliferative phase

c The appearance of the endometrium during the secretory phase of the uterine cycle

In the Lab 5

Materials

- ☐ Female reproductive system model and chart
- ☐ Compound microscope
- ☐ Prepared slide of ovary
- ☐ Prepared slide of uterine tube
- ☐ Prepared slide of uterus
- ☐ Prepared slides of endometrium series

Procedures

1. Review the anatomy of the ovaries, uterine tubes, and uterus presented in Figures 45.9, 45.10, and 45.11.

2. Identify the ovaries, uterine tubes, the ampulla, and the isthmus on the laboratory model and/or chart.

3. On the model, identify the fundus, body, and cervix of the uterus.

4. Examine the ovary slide at low magnification and note the cortex with many egg nests. Increase magnification and observe an oocyte inside a follicle.

5. *Draw It!* Draw some follicles in the space provided.

Follicles of the ovary

6. Scan the uterine tube slide at low and medium magnifications. Observe the lining epithelium and smooth muscle tissue.

7. *Draw It!* Draw a section of the uterine tube in the space provided.

Uterine tube

8. Observe the uterus slide and locate the perimetrium and the thick myometrium composed of smooth muscle tissue. Identify the endometrium.

9. *Draw It!* Draw a section of the uterine wall in the space provided.

Uterine wall

10. Using Figure 45.12 for reference, examine the endometrium slide set and compare the functional zone and uterine glands during the menses, proliferative, and secretory phases. ■

Lab Activity 6 Female: Vagina and Vulva

The **vagina** is a muscular tube approximately 10 cm (4 in.) long (**Figure 45.13**). It is lined with stratified squamous epithelium and is the female copulatory organ, the pathway for menstrual flow, and the lower birth canal. The **fornix** is the pouch formed where the uterus protrudes into the vagina. The **vaginal orifice** is the external opening of the vagina. This opening may be partially or totally occluded by a thin fold of vascularized mucous membrane called the **hymen** (HĪ-men). On either side of the vaginal orifice are openings of the **greater vestibular glands,** glands that produce a mucous secretion that lubricates the vaginal entrance for sexual intercourse. These glands are similar to the bulbourethral glands of the male.

The **vulva** (VUL-vuh), which is the collective name for the female **external genitalia** (jen-i-TĀ-lē-uh), includes the following structures (**Figure 45.14**).

- The **mons pubis** is a pad of adipose over the pubic symphysis. The mons is covered with skin and pubic hair and serves as a cushion for the pubic symphysis.

- The **labia** (LĀ-bē-uh) **majora** are two fatty folds of skin extending from the mons pubis and continuing posteriorly. They are homologous to the scrotum of the male. They usually have pubic hair and contain many sudoriferous (sweat) and sebaceous (oil) glands.

- The **labia minora** (mi-NOR-uh) are two smaller parallel folds of skin containing many sebaceous glands. This pair of labia lacks hair.

- The **clitoris** (KLIT-ō-ris) is a small, cylindrical mass of erectile tissue analogous to the penis. Like the penis, the clitoris contains a small fold of covering skin called the prepuce. The exposed portion of the clitoris is called the **glans.**

- The **vestibule** is the area between the labia minora that contains the vaginal orifice, hymen, and external urethral orifice.

- **Paraurethral glands** (*Skene's glands*) surround the urethra.

- The **perineum** is the area between the legs from the clitoris to the anus. This area is of clinical significance because of the tremendous pressure exerted on it during childbirth. If the vagina is too narrow during childbirth, an **episiotomy** (e-pēz-ē-OT-uh-mē) is performed by making a small incision at the base of the vaginal opening toward the anus to expand the vaginal opening.

QuickCheck Questions

6.1 Where is the mons pubis located?

6.2 The vestibule is between what two sets of folds?

6.3 Which female organ has a glans?

Figure 45.13 The Vagina

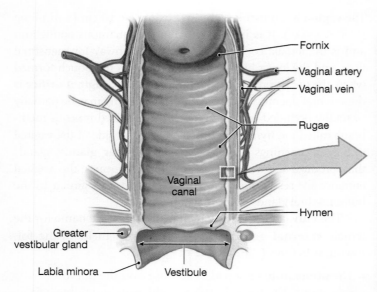

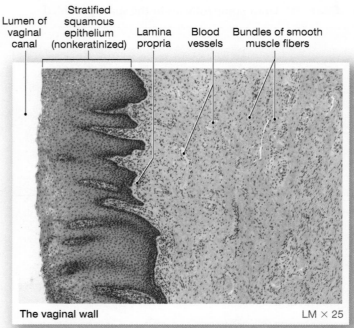

The vaginal wall   LM × 25

Figure 45.14 Female External Genitalia The external anatomy of the female is collectively called the vulva.

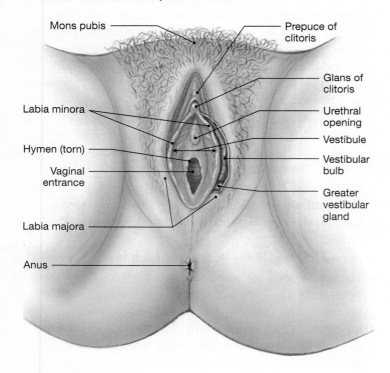

In the Lab   6

Materials

☐ Female reproductive system model and chart
☐ Compound microscope
☐ Prepared slide of vagina

Procedures

1. Review the anatomy of the vulva in Figure 45.14.
2. Locate the vagina and vaginal orifice on the laboratory model and/or chart. Examine the fornix, which is the point where the cervix and vagina connect.
3. Observe the vagina slide and study the histology of the wall. Identify the stratified squamous epithelium, lamina propria, and smooth muscle.
4. ***Draw It!*** Draw and label the wall in the space provided.

Vaginal wall

5. Locate each component of the vulva. Note the positions of the clitoris, urethra, and vagina. ◼

Figure 45.15 Mammary Glands

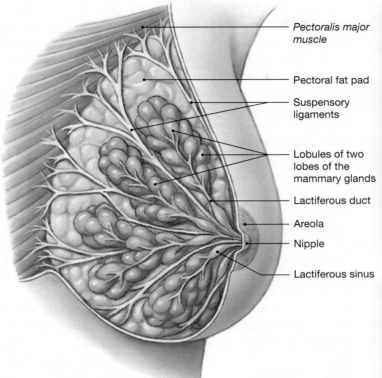

- Pectoralis major muscle
- Pectoral fat pad
- Suspensory ligaments
- Lobules of two lobes of the mammary glands
- Lactiferous duct
- Areola
- Nipple
- Lactiferous sinus

a The mammary glands of the left breast

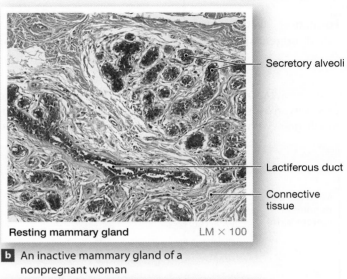

- Secretory alveoli
- Lactiferous duct
- Connective tissue

Resting mammary gland   LM × 100

b An inactive mammary gland of a nonpregnant woman

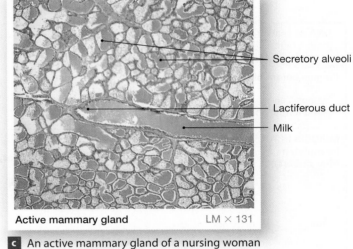

- Secretory alveoli
- Lactiferous duct
- Milk

Active mammary gland   LM × 131

c An active mammary gland of a nursing woman

Lab Activity 7 Mammary Glands

The **mammary glands** (**Figure 45.15**) are modified sweat glands that, in the process called **lactation** (lak-TĀ-shun), produce milk to nourish a newborn infant. At puberty, the release of estrogens stimulates an increase in the size of these glands. Fat deposition is the major contributor to the size of the breast, and size does not influence the amount of milk produced. Each gland consists of 15 to 20 lobes separated by fat and connective tissue. Each lobe contains smaller lobules that contain milk-secreting cells called **alveoli. Lactiferous** (lak-TIF-e-rus) **ducts** drain milk from the lobules toward the **lactiferous sinuses.** These sinuses empty the milk at the raised portion of the breast called the *nipple*. A circular pigmented area called the **areola** (a-RĒ-ō-luh) surrounds the nipple.

QuickCheck Questions

7.1 What are the milk-producing cells of the breast called?

7.2 What is the areola?

In the Lab 7

Materials

☐ Breast model

☐ Female reproductive system model and chart

Procedures

1. Review the anatomy of the breast presented in Figure 45.15.

2. On the model and chart, trace the pathway of milk from a lobule to the surface of the nipple. ∎

Lab Activity 8   Female: Oogenesis

Formation of the female gamete, the ovum (or *egg*), is called **oogenesis** (ō-ō-JEN-e-sis) and occurs in the ovaries (Figure 45.16). In a female fetus, meiosis I begins when cells

Figure 45.16 Oogenesis In oogenesis, a single primary oocyte produces an ovum and two nonfunctional polar bodies. Compare this diagram with Figure 45.3 which summarizes spermatogenesis.

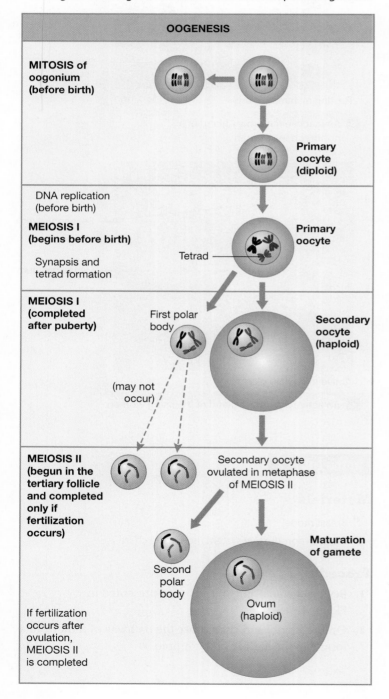

OOGENESIS

MITOSIS of oogonium (before birth)

Primary oocyte (diploid)

DNA replication (before birth)

MEIOSIS I (begins before birth)

Synapsis and tetrad formation

Tetrad

Primary oocyte

MEIOSIS I (completed after puberty)

First polar body

Secondary oocyte (haploid)

(may not occur)

MEIOSIS II (begun in the tertiary follicle and completed only if fertilization occurs)

Secondary oocyte ovulated in metaphase of MEIOSIS II

Maturation of gamete

Second polar body

Ovum (haploid)

If fertilization occurs after ovulation, MEIOSIS II is completed

called **oogonia** (ō-ō-GŌ-nē-uh, singular *oogonium*) divide by mitosis and produce **primary oocytes** (ō-ō-sīts), which remain suspended in this stage until the child reaches puberty. At puberty, each month, a primary oocyte divides into two **secondary oocytes.** One of the secondary oocytes is much smaller than its sister cell and is a nonfunctional cell called the **first polar body** (Figure 45.8). The other secondary oocyte remains suspended in meiosis II until it is ovulated. If fertilization occurs, the secondary oocyte completes meiosis II and divides into another polar body, called the **second polar body,** and an ovum. The haploid ovum and haploid spermatozoon combine their haploid chromosomes and become the first cell of the offspring, the diploid zygote.

Note from Figure 45.16 that females produce only a single ovum by oogenesis, whereas in males, spermatogenesis results in four spermatozoa (Figure 45.8).

Each ovary contains 100,000 to 200,000 oocytes clustered in groupings called **egg nests.** Within the nests are **primordial follicles,** which are primary oocytes surrounded by follicular cells. Figure 45.17 details the monthly ovarian cycle, during which hormones stimulate the follicular cells of the primordial follicles to proliferate and produce several **primary follicles,** each one a primary oocyte surrounded by follicular cells. These follicles increase in size, and a few become **secondary follicles** containing primary oocytes. Eventually, one secondary follicle develops into a **tertiary follicle,** also called a *mature Graafian* (GRAF-ē-an) *follicle*. By now the oocyte has completed meiosis I and is now a secondary oocyte starting meiosis II. The tertiary follicle fills with liquid and ruptures, casting out the secondary oocyte during ovulation. This follicle secretes **estrogen,** the hormone that stimulates rebuilding of the spongy lining of the uterus. After ovulation, the follicular cells of the tertiary follicle become the **corpus luteum** (LOO-tē-um) and secrete primarily the hormone **progesterone** (prō-JES-ter-ōn), which prepares the uterus for pregnancy. If the secondary oocyte is not fertilized, the corpus luteum degenerates into the **corpus albicans** (AL-bi-kanz), and most of the rebuilt lining of the uterus is shed as the menstrual flow.

QuickCheck Questions

8.1 Where are ova produced in the female?

8.2 Which structure ruptures during ovulation to release an ovum?

8.3 What are polar bodies?

Figure 45.17 The Ovarian Cycle Ovaries contain oocytes which become surrounded by follicle cells. Ovulation occurs when the tertiary follicle ruptures and releases the secondary oocyte from the ovary. The torn follicle develops into the corpus luteum and produces progesterone.

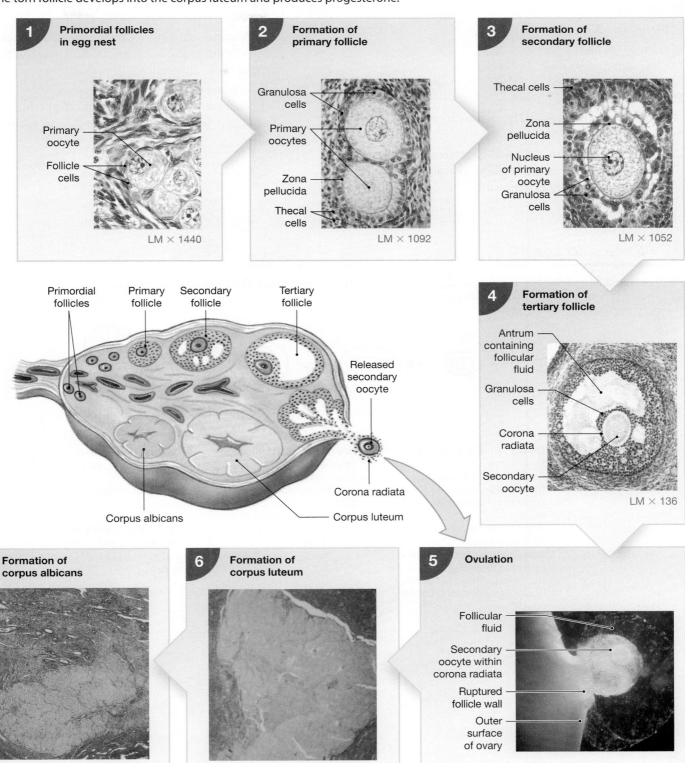

In the Lab   8

Material

☐ Meiosis models

☐ Compound microscope

☐ Prepared slide of ovary

Procedures

1. Identify the different cell types shown on the meiosis models.

2. Using Figures 45.16 and 45.17 as references, scan the ovary slide at low magnification, and locate an egg nest along the periphery of the ovary.

3. Identify the primary follicles, which are larger than the primordial follicles in the nests. In the primary-follicle stage, the oocyte has increased in size and is surrounded by follicular cells.

4. Identify some secondary follicles, which are larger than primary follicles and have a separation between the outer and inner follicular cells.

5. Identify some tertiary follicles, which are easily distinguished by the large, liquid-filled space they contain.

For additional practice, complete the *Sketch to Learn* activity below. ■

 Sketch to Learn

In this drawing let's show the various stages of follicle development before and after ovulation of an oocyte.

Sample Sketch

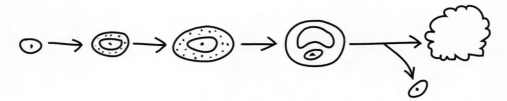

Step 1
- Draw a series of three ovals for oocytes.
- Add a circle close around the second oocyte and a larger circle around the third.
- Add dots in the outer circles for follicular cells.
- Add arrows between follicles.
- Label the three follicles in sequence; primordial follicle, primary follicle, and secondary follicle.

Step 2
Draw the Graafian follicle:
- Draw an oocyte.
- Draw a large circle around the oocyte with oocyte at bottom.
- Add an inner circle that wraps around the top of the oocyte.

Step 3
- Add a branched arrow.
- Show the oocyte out of follicle.
- Draw a large corpus luteum.
- Label the Graafian follicle in Step 2 and the corpus luteum and ovulated oocyte in Step 3.

Your Sketch

Name _____

Date _____

Section _____

Anatomy of the
Reproductive System

A. Description

Write a description of each of the following structures.

1. epididymis

2. ductus deferens

3. bulbo-urethral gland

4. corpora cavernosa

5. prostatic urethra

6. seminiferous tubule

7. labia minora

8. myometrium

9. fundus

10. infundibulum

11. vulva

12. cervix

B. Labeling

1. Label the anatomy of the male in **Figure 45.18**.

Figure 45.18 **Male Reproductive System**

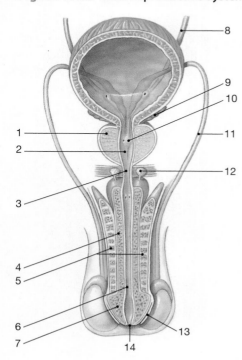

1. _____
2. _____
3. _____
4. _____
5. _____
6. _____
7. _____
8. _____
9. _____
10. _____
11. _____
12. _____
13. _____
14. _____

2. Label the anatomy of the vulva in **Figure 45.19**.

Figure 45.19 **The Vulva**

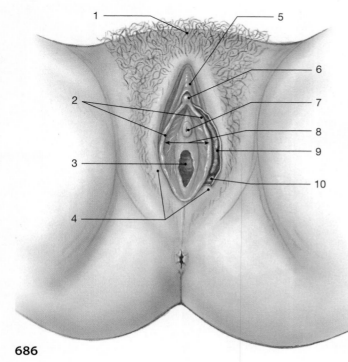

1. _____
2. _____
3. _____
4. _____
5. _____
6. _____
7. _____
8. _____
9. _____
10. _____

3. Label the anatomy of the female in **Figure 45.20**.

Figure 45.20 Female Reproductive System

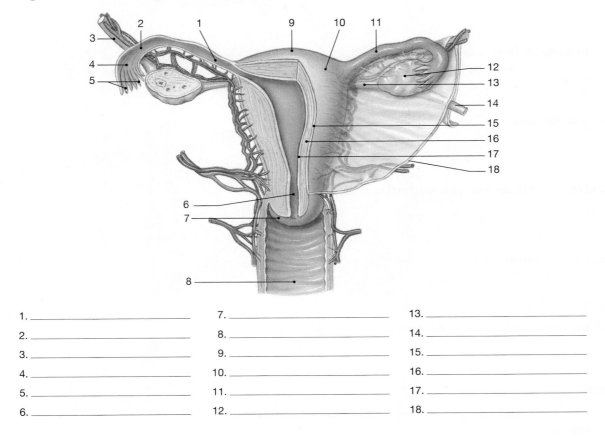

1. _____	7. _____	13. _____
2. _____	8. _____	14. _____
3. _____	9. _____	15. _____
4. _____	10. _____	16. _____
5. _____	11. _____	17. _____
6. _____	12. _____	18. _____

4. Label the anatomy of the seminiferous tubule in **Figure 45.21**.

Figure 45.21 Seminiferous Tubule

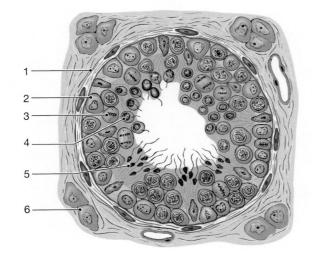

1. _____

2. _____

3. _____

4. _____

5. _____

6. _____

C. Short-Answer Questions

1. List the three layers of the uterus, from superficial to deep.

2. Describe the gross anatomy of the female breast.

3. List the components of the vulva.

4. How is temperature regulated in the testes for maximal spermatozoa production?

5. Name the three regions of the male urethra.

6. What are the three accessory glands of the male reproductive system?

D. Analysis and Application

1. Explain the division sequence that leads to four spermatids in male meiosis but only one ovum in female meiosis.

2. How are the clitoris and the penis similar to each other?

E. Clinical Challenge

1. How does a vasectomy or a tubal ligation sterilize an individual?

2. Do castration and vasectomy have the same effects on endocrine and reproductive functions?

Development

Learning Outcomes

On completion of this exercise, you should be able to:

1. Describe the process of fertilization and early cleavage to the blastocyst stage.

2. Describe the process of implantation and placenta formation.

3. List the three germ layers and the embryonic fate of each.

4. List the four extraembryonic membranes and the function of each.

5. Describe the general developmental events of the first, second, and third trimesters.

6. List the three stages of labor.

The cell theory of biology states that cells come from preexisting cells. In animals, gametes from the parents unite and form a new cell, the zygote, which has inherited the parental genetic material. The zygote quickly develops into an **embryo,** the name given to the organism for approximately the first two months after fertilization. By the end of the second month, most organ systems have started to form, and the embryo is then called a **fetus.**

In humans, the prenatal period of development occurs over a nine-month **gestation** (jes-TĀ-shun) that is divided into three-month trimesters. During the first trimester, the embryo develops cell layers that are precursors to organ systems. The second trimester is characterized by growth in length, mass gain, and the appearance of functional organ systems. In the third trimester, increases in length and mass occur, and all organ systems either become functional or are prepared to become

Lab Activities

1. First Trimester: Fertilization, Cleavage, and Blastocyst Formation   690

2. First Trimester: Implantation and Gastrulation   692

3. First Trimester: Extraembryonic Membranes and the Placenta   695

4. Second and Third Trimesters and Birth   698

Clinical Application

Ectopic Pregnancy   690

Need More Practice and Review?

Build your knowledge—and confidence!—in the Study Area of MasteringA&P® at www.masteringaandp.com with Pre-lab Quizzes, Post-lab Quizzes, Practice Anatomy Lab™ (PAL™) 3.0 virtual anatomy practice tool, PhysioEx™ 9.0 laboratory simulations, and A&P Flix™ with Quizzes.

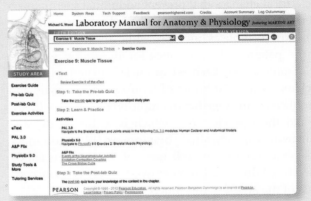

functional at birth. After 38 weeks' gestation, the uterus begins to rhythmically contract to deliver the fetus into the world. Although maternal changes occur during the gestation period, this exercise focuses on the development of the fetus.

Morphogenesis (mor-fō-JEN-uh-sis) is the general term for all the processes involved in the specialization of cells in the developing fetus and the migration of those cells to sproduce anatomical form and function.

<hr>

Lab Activity 1 First Trimester: Fertilization, Cleavage, and Blastocyst Formation

Fertilization is the act of the spermatozoon and ovum joining their haploid nuclei to produce a diploid zygote, the genetically unique cell that develops into an individual. The male ejaculates approximately 300 million spermatozoa into the female's reproductive tract during intercourse. Once exposed to the female's reproductive tract, the spermatozoa complete a process called **capacitation** (ka-pas-i-TĀ-shun), during which they increase their motility and become capable of fertilizing an ovum. Most spermatozoa do not survive the journey through the vagina and uterus, and only an estimated 100 of them reach the ampulla. Normally, only a single ovum is released from a single ovary during one ovulation cycle. Fertilization of the ovum by a spermatozoon typically occurs in the upper third of the uterine tube.

Figure 46.1 illustrates fertilization. Ovulation releases a secondary oocyte from the ovary, and the oocyte begins moving along the uterine tube. Layers of follicular cells still encase the ovulated oocyte and now constitute a layer called the **corona radiate** (koō-RŌ-nuh rā-dē-A-tuh). Spermatozoa reaching the oocyte in the uterine tube must pass through the corona radiata to reach the cell membrane of the oocyte. Spermatozoa swarm around the oocyte and release from their acrosome an enzyme called *hyaluronidase*. The combined action of the hyaluronidase contributed by all the spermatozoa eventually creates a gap between some coronal cells, and a single spermatozoon slips into the oocyte. The membrane of the oocyte instantly undergoes chemical and electrical changes that prevent additional spermatozoa from entering the cell. The oocyte, suspended in meiosis II since ovulation, now completes meiosis, while the spermatozoon prepares the paternal chromosomes for the union with the maternal chromosomes. Each set of nuclear material is called a **pronucleus.** Within 30 hours of fertilization, the male and female pronuclei come together in **amphimixis** (am-fi-MIK-sis) and undergo the first **cleavage,** which is a mitotic division resulting in two cells, each called a **blastomere** (BLAS-toō-měr). During cleavage, the existing cell mass of the ovum is subdivided by each cell division. (In other words, there is no increase in the mass of the zygote at this time.)

As the zygote slowly descends in the uterine tube toward the uterus, cleavages occur approximately every 12 hours. By the third day, the blastomeres are organized into a solid ball of nearly identical cells called a **morula** (MOR-ū-la), shown in **Figure 46.2**. Around day 6, the morula has entered the uterus and changed into a **blastocyst** (BLAS-tō-sist), a hollow ball of cells with an internal cavity called the **blastocoele** (BLAS-tō-sēl). Now the process of **differentiation,** or specialization, begins. The blastomeres making up the blastocyst are now of various sizes and have migrated into two regions. Cells on the outside compose the **trophoblast** (TRŌ-fō-blast), which will burrow into the uterine lining and eventually form part of the placenta. Cells clustered inside the blastocoele form the **inner cell mass,** which will develop into the embryo.

Clinical Application Ectopic Pregnancy

Implantation of an embryo normally occurs in the endometrium of the uterus. An **ectopic pregnancy** is when implantation is not in the uterus. In most cases, the ectopic implantation is in the uterine tube, but occasionally it is in the pelvic cavity. Placental development may be impaired in ectopic pregnancies and most of these embryos do not survive past the first trimester. Implantation in the uterine tube leads to expansion and eventual rupturing of the tube as the embryo grows. Massive bleeding of the placental blood is life threatening to the embryo, and in many instances, to the mother, too. If gestation is successful, a natural vaginal birth is not possible because of the extrauterine implantation, and surgery is necessary to remove the fetus from the mother. ∎

QuickCheck Questions

1.1 Where does fertilization normally occur?

1.2 What is a morula?

1.3 What is a blastocyst?

In the Lab 1

Materials

☐ Fertilization model

☐ 6-, 10-, and 12-day embryo models

Procedures

1. Review the steps of fertilization in Figure 46.1 and those of cleavage in Figure 46.2.
2. On the fertilization model, note how the male and female pronuclei join to create the diploid zygote.
3. On the embryo models, identify some blastomere cells and the morula.
4. On the embryo models, identify the blastocoele and the trophoblast. ∎

Figure 46.1 Fertilization

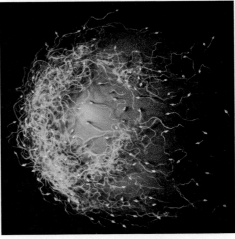

a A secondary oocyte and numerous sperm at the time of fertilization. Notice the difference in size between the gametes.

Oocyte at Ovulation

Ovulation releases a secondary oocyte and the first polar body; both are surrounded by the corona radiata. The oocyte is suspended in metaphase of meiosis II.

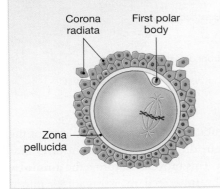

Corona radiata   First polar body

Zona pellucida

1 Fertilization and Oocyte Activation

Acrosomal enzymes from multiple sperm create gaps in the corona radiata. A single sperm then makes contact with the oocyte membrane, and membrane fusion occurs, triggering oocyte activation and completion of meiosis.

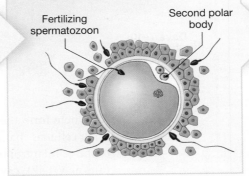

Fertilizing spermatozoon

Second polar body

2 Pronucleus Formation Begins

The sperm is absorbed into the cytoplasm, and the female pronucleus develops.

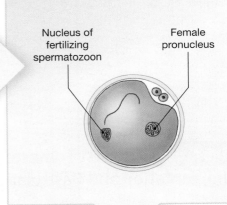

Nucleus of fertilizing spermatozoon

Female pronucleus

5 Cleavage Begins

The first cleavage division nears completion roughly 30 hours after fertilization.

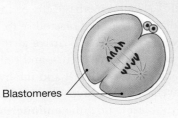

Blastomeres

4 Amphimixis Occurs and Cleavage Begins

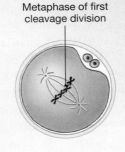

Metaphase of first cleavage division

3 Spindle Formation and Cleavage Preparation

The male pronucleus develops, and spindle fibers appear in preparation for the first cleavage division.

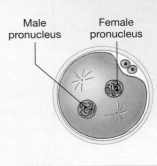

Male pronucleus

Female pronucleus

b Fertilization and the preparations for cleavage.

Figure 46.2 **Cleavage and Blastocyst Formation** Fertilization occurs in ampulla. It takes approximately six days for the embryo, now a hollow ball of cells called the blastocyst, to pass into the uterus.

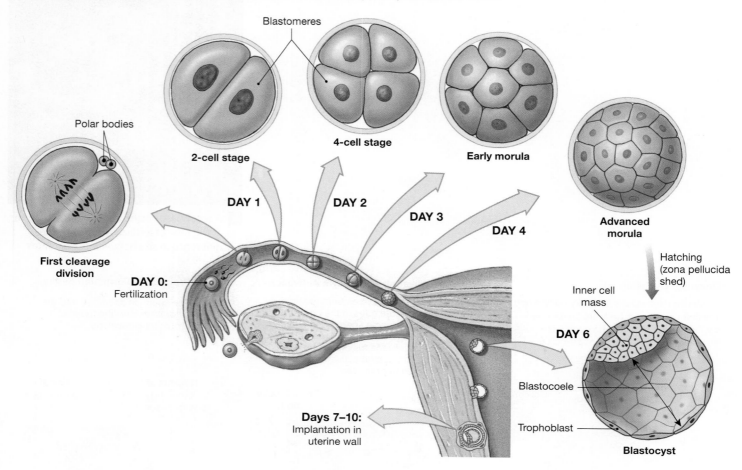

Lab Activity 2 First Trimester: Implantation and Gastrulation

Implantation begins on day 7 or 8, when the blastocyst touches the spongy uterine lining (**Figure 46.3**). The tropho-blast layer of the blastocyst burrows into the functional zone of the endometrium. The cell membranes of the trophoblast cells dissolve, and the cells mass together as a cytoplasmic layer of multiple nuclei called the **syncytial** (sin-SISH-al) **trophoblast.** The cells secrete hyaluronidase to erode a path for implanta-tion, which continues until the embryo is completely covered by the functional zone of the endometrium, about day 14. To establish a diffusional link with the maternal circulation, the syncytial trophoblast sprouts villi that erode into the endome-trium and create spaces in the endometrium called **lacunae.** Maternal blood from the endometrium seeps into the lacunae and bathes the villi with nutrients and oxygen. These materials diffuse into the blastocyst to support the inner cell mass. Deep to the syncytial layer is the **cellular trophoblast,** which will

soon help form the placenta. By day 9, the middle layer of the inner cell mass has gradually dropped away from the layer next to the cellular trophoblast. This movement forms the **amniotic** (am-nē-OT-ik) **cavity.** The inner cell mass organizes into a **blastodisc** (BLAS-tō-disk) made up of two cell layers: the **superficial layer (epiblast)** (EP-i-blast) and the **deep layer (hypoblast)** (HĪ-pō-blast) facing the blastocoele.

Once the blastodisc has developed, rapid changes in the embryo take place (**Figure 46.4**). By day 10, the amniotic membrane forms around the amniotic cavity and the yolk sac appears to supply the embryo with nutrients. Within the next few days, cells begin to migrate in the process called **gastrulation** (gas-troo-LĀ-shun) (Figure 46.4b). Cells of the epiblast move toward the medial plane of the blastodisc to a region known as the **primitive streak.** As cells arrive at the primitive streak, infolding, or **invagination,** occurs, and cells are liberated into the region between the epiblast and the hy-poblast, producing three cell layers in the embryo. The epiblast becomes the **ectoderm,** the hypoblast is now the **endoderm,** and the cells proliferating between the two layers form the

Figure 46.3 Stages of Implantation The syncytial trophoblast of the embryo secretes enzymes that allow the trophoblast to implant into the uterine wall for gestation.

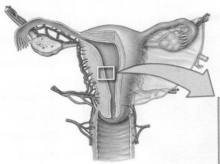

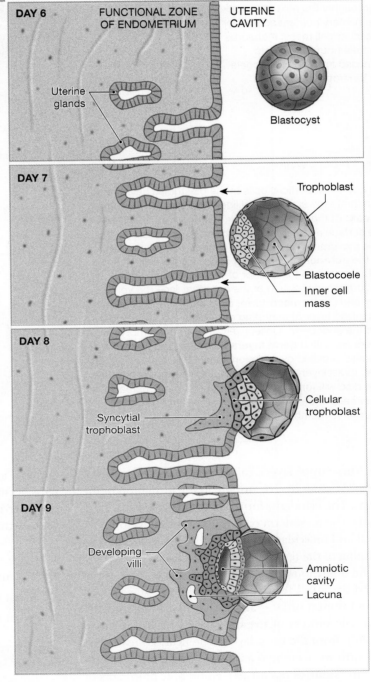

DAY 6

FUNCTIONAL ZONE OF ENDOMETRIUM

UTERINE CAVITY

Uterine glands

Blastocyst

DAY 7

Trophoblast

Blastocoele

Inner cell mass

DAY 8

Syncytial trophoblast

Cellular trophoblast

DAY 9

Developing villi

Amniotic cavity

Lacuna

Figure 46.4 **Blastodisc Organization and Gastrulation**

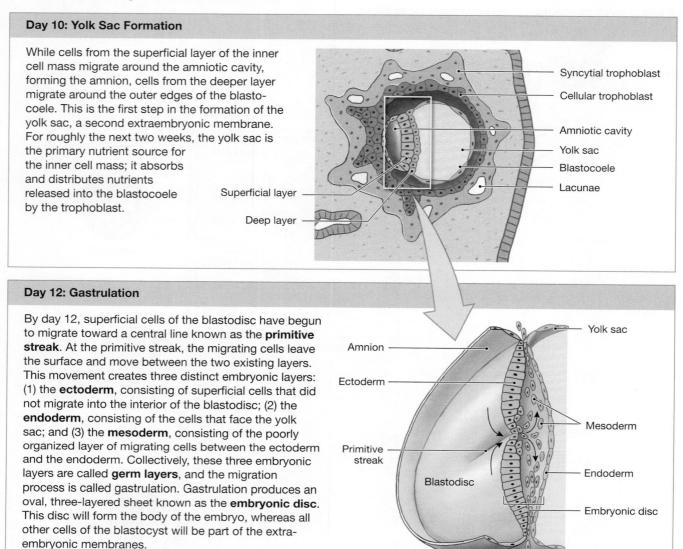

Day 10: Yolk Sac Formation

While cells from the superficial layer of the inner cell mass migrate around the amniotic cavity, forming the amnion, cells from the deeper layer migrate around the outer edges of the blastocoele. This is the first step in the formation of the yolk sac, a second extraembryonic membrane. For roughly the next two weeks, the yolk sac is the primary nutrient source for the inner cell mass; it absorbs and distributes nutrients released into the blastocoele by the trophoblast.

Syncytial trophoblast
Cellular trophoblast
Amniotic cavity
Yolk sac
Blastocoele
Lacunae
Superficial layer
Deep layer

Day 12: Gastrulation

By day 12, superficial cells of the blastodisc have begun to migrate toward a central line known as the **primitive streak**. At the primitive streak, the migrating cells leave the surface and move between the two existing layers. This movement creates three distinct embryonic layers: (1) the **ectoderm**, consisting of superficial cells that did not migrate into the interior of the blastodisc; (2) the **endoderm**, consisting of the cells that face the yolk sac; and (3) the **mesoderm**, consisting of the poorly organized layer of migrating cells between the ectoderm and the endoderm. Collectively, these three embryonic layers are called **germ layers**, and the migration process is called gastrulation. Gastrulation produces an oval, three-layered sheet known as the **embryonic disc**. This disc will form the body of the embryo, whereas all other cells of the blastocyst will be part of the extra-embryonic membranes.

Yolk sac
Amnion
Ectoderm
Mesoderm
Primitive streak
Endoderm
Blastodisc
Embryonic disc

mesoderm. These three layers, called **germ layers,** each produce specialized tissues that contribute to the formation of the organ systems. The ectoderm forms the nervous system, skin, hair, and nails. The mesoderm contributes to the development of the skeletal and muscular systems, and the endoderm forms part of the lining of the respiratory and digestive systems.

By the end of the fourth week of development, the embryo is distinct and has a **tail fold** and a **head fold** (Figure 46.5). The dorsal and ventral surfaces and the right and left sides are well defined. The process of **organogenesis** begins as organ systems develop from the germ layers. The heart is clearly visible in the fourth week embryo and it has beat since the third week of growth. **Somites** (so-MĪ-tis), embryonic precursors of skeletal muscles, appear. Elements of the nervous system are also developing. Buds for the upper and lower limbs and small discs for the eyes and ears are also present. By week 8, fingers

and toes are present, and the embryo is now usually called the fetus, as noted earlier. At the end of the third month, the first trimester is completed, and every organ system has appeared in the fetus.

QuickCheck Questions

2.1 Where does implantation normally occur?

2.2 What is the syncytial trophoblast?

2.3 What are the two cellular layers of the blastodisc?

In the Lab 2

Materials

☐ 6-, 10-, and 12-day embryo models and charts

Figure 46.5 **First Trimester**

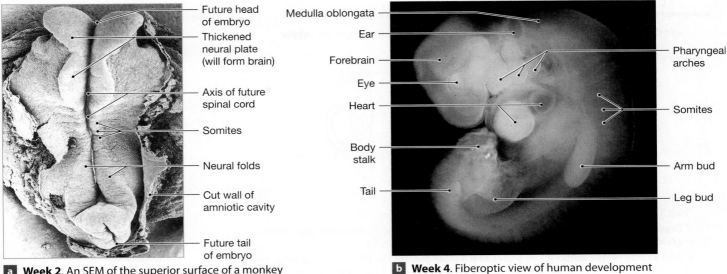

Future head of embryo
Thickened neural plate (will form brain)
Axis of future spinal cord
Somites
Neural folds
Cut wall of amniotic cavity
Future tail of embryo

a **Week 2**. An SEM of the superior surface of a monkey embryo at 2 weeks of development. A human embryo at this stage would look essentially the same.

Medulla oblongata
Ear
Forebrain
Eye
Heart
Body stalk
Tail
Pharyngeal arches
Somites
Arm bud
Leg bud

b **Week 4**. Fiberoptic view of human development at week 4.

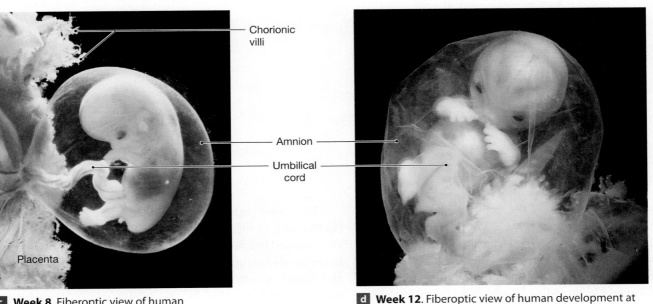

Chorionic villi
Amnion
Umbilical cord
Placenta

c **Week 8**. Fiberoptic view of human development at week 8.

d **Week 12**. Fiberoptic view of human development at week 12.

Procedures

1. Review the anatomy of the blastocyst during implantation in Figure 46.3.

2. On the six-day model or chart, locate the cellular and syncytial trophoblasts. The model may show the development of villi where the syncytial trophoblast has dissolved the endometrium for implantation.

3. Review the anatomy of the blastodisc in Figure 46.4a.

4. On the 10-day model or chart, examine the blastodisc and identify the epiblast and the hypoblast.

5. On the 12-day model or chart, identify the ectoderm, mesoderm, and endoderm.

For additional practice, complete the *Sketch to Learn* activity (on p. 696). ■

Lab Activity 3 First Trimester: Extraembryonic Membranes and the Placenta

Four extraembryonic membranes develop from the germ layers: the yolk sac, amnion, chorion, and allantois (**Figure 46.6**). These membranes lie outside the blastodisc and provide

Sketch to Learn

Let's draw an eight-day-old embryo as it begins to implant into the endometrium.

Sample Sketch

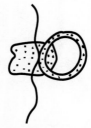

Step 1
- Draw a wavy line as the endometrium.
- Add a circle for the embryo.

Step 2
- Add an inner circle and fill between circles with dots for trophoblast cells.
- Draw a small oval on the left side of the inner circle. Fill this oval with dots to show the inner cell mass.

Step 3
- Add an irregular shape to the left of the embryo for the syncytial trophoblast.
- Label your drawing.

Your Sketch

protection and nourishment for the embryo/fetus. The **yolk sac** is the first membrane to appear, around the ten-day stage. Initially, cells from the hypoblast form a pouch under the blastodisc. Mesoderm reinforces the yolk sac, and blood vessels appear. As the syncytial trophoblast develops more villi, the yolk sac's importance in providing nourishment for the embryo diminishes.

While the yolk sac is forming, cells in the epiblast portion of the blastodisc migrate to line the inner surface of the amniotic cavity with a membrane called the **amnion** (AM-nē-on), the "water bag." As with the yolk sac, the amnion is soon reinforced with mesoderm. Embryonic growth continues, and by week 10, the amnion has mushroomed and envelops the embryo in a protective environment of **amniotic fluid** (Figure 46.6).

The **allantois** (a-LAN-tō-is) develops from the endoderm and mesoderm near the base of the yolk sac. The allantois forms part of the embryonic urinary bladder and contributes to the **body stalk,** the tissue between the embryo and the developing chorion. Blood vessels pass through the body stalk and into the villi protruding into the lacunae of the endometrium.

The outer extraembryonic membrane is the **chorion** (KOR-ē-on), formed by the cellular trophoblast and mesoderm. The chorion completely encases the embryo and the blastocoele. In the third week of growth, the chorion extends **chorionic villi** and blood vessels into the endometrial lacunae to establish the structural framework for the development of the **placenta** (pla-SENT-uh), the temporary organ through which nutrients, blood gases, and wastes are exchanged between the mother and the embryo. The embryo is connected to the placenta by the body stalk. The **yolk stalk,** where the yolk sac attaches to the endoderm of the embryo, and the body stalk together form the **umbilical cord.** Inside the umbilical cord are two **umbilical arteries,** which transport deoxygenated blood to the placenta, and a single **umbilical vein,** which returns oxygenated blood to the embryo.

By the fifth week of development, the chorionic villi have enlarged only where they face the uterine wall, and villi that face the uterine cavity become insignificant (Figure 46.6). Only the part of the chorion where the villi develop becomes the placenta. The rest of this membrane remains chorion, as the week 10 part of Figure 46.6 indicates. Thus the placenta

Figure 46.6 **Extraembryonic Membranes and Placenta Formation** Four extraembryonic membranes protect and support the embryo and fetus: the yolk sac, chorion, placenta, and allantois.

1 Week 2

Migration of mesoderm around the inner surface of the trophoblast creates the chorion. Mesodermal migration around the outside of the amniotic cavity, between the ectodermal cells and the trophoblast, forms the amnion. Mesodermal migration around the endodermal pouch creates the yolk sac.

- Amnion
- Syncytial trophoblast
- Cellular trophoblast ⎫
- Mesoderm   ⎬ Chorion
- Yolk sac
- Blastocoele

2 Week 3

The embryonic disc bulges into the amniotic cavity at the head fold. The allantois, an endodermal extension surrounded by mesoderm, extends toward the trophoblast.

Yolk sac

- Amniotic cavity (containing amniotic fluid)
- Allantois
- Head fold of embryo
- Chorion
- Syncytial trophoblast
- Chorionic villi of placenta

4 Week 5

The developing embryo and extraembryonic membranes bulge into the uterine cavity. The trophoblast pushing out into the uterine lumen remains covered by endometrium but no longer participates in nutrient absorption and embryo support. The embryo moves away from the placenta, and the body stalk and yolk stalk fuse to form an umbilical stalk.

- Uterus
- Myometrium
- Decidua basalis
- Umbilical stalk
- Placenta
- Yolk sac
- Chorionic villi of placenta
- Decidua capsularis
- Decidua parietalis
- Uterine lumen

3 Week 4

The embryo now has a head fold and a tail fold. Constriction of the connections between the embryo and the surrounding trophoblast narrows the yolk stalk and body stalk.

- Tail fold
- Body stalk
- Yolk stalk
- Yolk sac
- Embryonic gut
- Embryonic head fold

5 Week 10

The amnion has expanded greatly, filling the uterine cavity. The fetus is connected to the placenta by an elongated umbilical cord that contains a portion of the allantois, blood vessels, and the remnants of the yolk stalk.

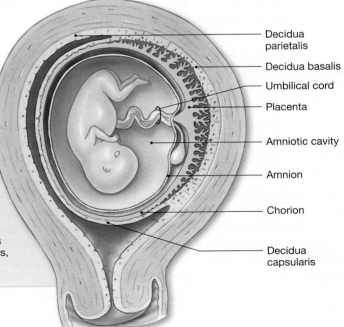

- Decidua parietalis
- Decidua basalis
- Umbilical cord
- Placenta
- Amniotic cavity
- Amnion
- Chorion
- Decidua capsularis

does not completely surround the embryo. The placenta is in contact with the area of the endometrium called the **decidua basalis** (dē-SID-ū-uh bā-SĀ-lis). The rest of the endometrium, where villi are absent, isolates the embryo from the uterine cavity and is called the **decidua capsularis** (kap-sū-LA-ris). The endometrium on the wall opposite the embryo is called the **decidua parietalis.**

QuickCheck Questions

3.1 List the four extraembryonic membranes.

3.2 Which membrane gives rise to the placenta?

In the Lab 3

Materials

- ☐ 3-, 5-, and 10-week embryo models and charts
- ☐ Placenta model or biomount

Procedures

1. Review the extraembryonic membranes in Figure 46.6.

2. On the embryology models or charts, locate the yolk sac and the amnion. How does each of these membranes form, and what is the function of each?

3. On the embryology models or charts, locate the allantois and the chorion. How does each of these membranes form? Describe the chorionic villi and their significance to the embryo.

4. On the placenta model or biomount, note the appearance of the various placental surfaces. Are there any differences in appearance from one surface to another? Is the amniotic membrane attached to the placenta?

5. Examine the umbilical cord attached to the placenta, and describe the vascular anatomy in the cord. ■

Lab Activity 4 Second and Third Trimesters and Birth

By the start of the second trimester, all major organ systems have started to form. Growth during the second trimester is fast, and the fetus doubles in size and increases its mass by 50 times. As the fetus grows, the uterus expands and displaces the other maternal abdominal organs (**Figure 46.7**). The fetus begins to move as its muscular system becomes functional, and articulations begin to form in the skeleton. The nervous system organizes the neural tissue that developed in the first trimester, and many sensory organs complete their formation. During the third trimester, all organ systems complete their development and become functional, and the fetus responds to sensory stimuli such as a hand rubbing across the mother's abdomen.

Birth, or **parturition** (par-tū-RISH-un), involves muscular contractions of the uterine wall to expel the fetus. Delivering the fetus is much like pulling on a turtleneck sweater. Muscle contractions must stretch the cervix over the fetal head, pulling the uterine wall thinner as the fetus passes into the vagina. Once true labor contractions begin, positive feedback mechanisms increase the frequency and force of uterine contractions.

Labor is divided into three stages: dilation, expulsion, and placental (**Figure 46.8**). The **dilation stage** begins at the onset of true labor contractions. The cervix dilates, and the fetus moves down the cervical canal. To be maximally effective at dilation, the contractions must be less than ten minutes apart. Each contraction lasts approximately one minute and spreads from the upper cervix downward to *efface*, or thin, the cervix for delivery. Contractions usually rupture the amnion, and amniotic fluid flows out of the uterus and the vagina.

The **expulsion stage** occurs when the cervix is dilated completely, usually to 10 cm, and the fetus passes through the cervix and the vagina. This stage usually lasts less than two hours and results in birth. Once the baby is breathing independently, the umbilical cord is cut, and the baby must now rely on its own organ systems to survive.

During the **placental stage,** uterine contractions break the placenta free of the endometrium and deliver it out of the body as the **afterbirth.** This stage is usually short, and many women deliver the afterbirth within five to ten minutes after the birth of the fetus.

QuickCheck Questions

4.1 What are the three stages of labor?

4.2 What is the afterbirth?

In the Lab 4

Materials

- ☐ Second-trimester model
- ☐ Third-trimester model
- ☐ Parturition model

Procedures

1. Review the anatomical changes during pregnancy in Figure 46.7.

2. Describe how the fetus is positioned in the uterus in the second-trimester model. If shown in the model, describe the location of the amnion and the placenta.

3. Describe how the fetus is positioned in the uterus in the third-trimester model. If shown in the model, describe the location of the amnion and the placenta.

4. Using the parturition model as an aid, describe the contractions that force the fetus out of the uterus. ■

Figure 46.7 Growth of the Uterus and Fetus

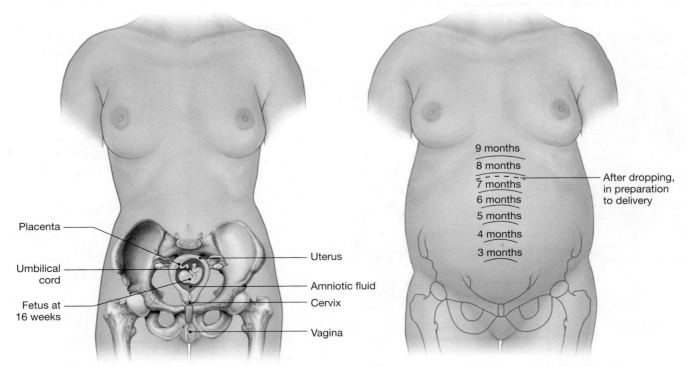

Placenta

Umbilical cord

Fetus at 16 weeks

Uterus

Amniotic fluid

Cervix

Vagina

a Pregnancy at 16 weeks, showing the positions of the uterus, fetus, and placenta.

9 months
8 months
7 months
6 months
5 months
4 months
3 months

After dropping, in preparation to delivery

b Pregnancy at three months to nine months (full term), showing the superior-most position of the uterus within the abdomen.

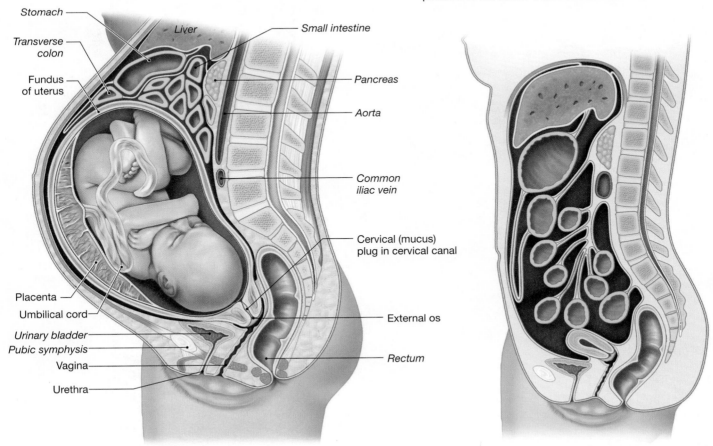

Stomach

Liver

Transverse colon

Fundus of uterus

Placenta

Umbilical cord

Urinary bladder

Pubic symphysis

Vagina

Urethra

Small intestine

Pancreas

Aorta

Common iliac vein

Cervical (mucus) plug in cervical canal

External os

Rectum

c Pregnancy at full term. Note the positions of the uterus and full-term fetus within the abdomen, and the displacement of abdominal organs.

d A sectional view through the abdominopelvic cavity of a woman who is not pregnant.

Figure 46.8 Stages of Labor At birth, the cervix dilates and the myometrium contracts to deliver the fetus. After the baby is born, the placenta is expelled.

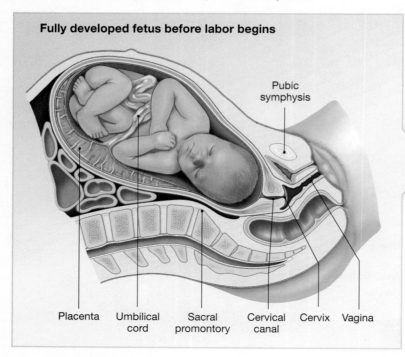

Fully developed fetus before labor begins

Pubic symphysis

Placenta Umbilical cord Sacral promontory Cervical canal Cervix Vagina

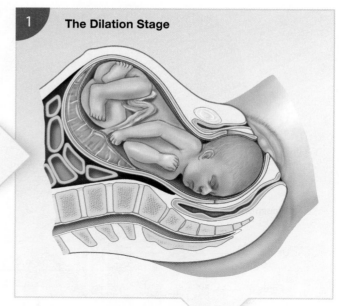

1 The Dilation Stage

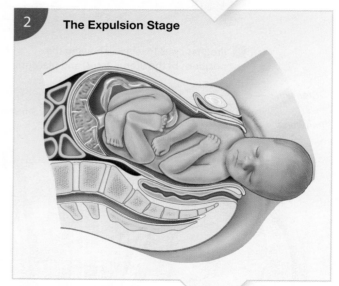

2 The Expulsion Stage

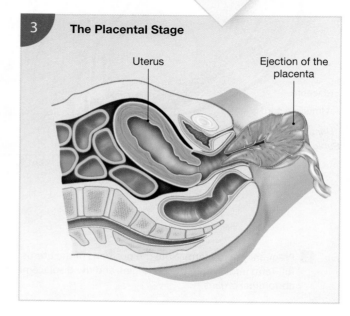

3 The Placental Stage

Uterus

Ejection of the placenta

Name _____

Date _____

Section _____

Development

A. Description

Describe each structure.

1. blastomere

2. amnion

3. allantois

4. morula

5. blastocoele

6. syncytial trophoblast

7. chorion

8. mesoderm

9. endoderm

10. epiblast

11. parturition

12. gastrulation

B. Short-Answer Questions

1. Describe the process of fertilization.

2. Discuss the formation of the three germ layers.

3. Describe the structure of the blastocyst.

4. List the three stages of labor and explain what takes place in each stage.

C. Labeling

Label the anatomy of the embryo and fetus in **Figure 46.9**.

Figure 46.9 **Embryonic and Fetal Development**

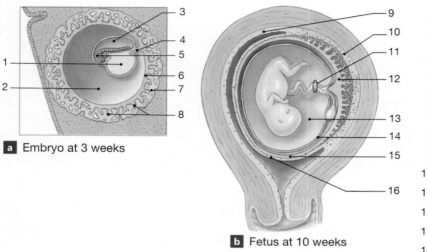

a Embryo at 3 weeks

b Fetus at 10 weeks

1. _____
2. _____
3. _____
4. _____
5. _____
6. _____
7. _____
8. _____
9. _____
10. _____
11. _____
12. _____
13. _____
14. _____
15. _____
16. _____

D. Analysis and Application

1. How does the amniotic cavity form?

2. What is the function of each of the four extraembryonic membranes?

3. How does a fetus obtain nutrients and gases from the maternal blood?

E. Clinical Challenge

A patient who is five weeks' pregnant starts to experience abdominal pain and a bloody discharge from her vagina. An ultrasound procedure indicates damage to the left uterine tube. How does this condition affect her pregnancy?

Surface Anatomy

Learning Outcomes

On completion of this exercise, you should be able to:

1. Describe the surface anatomy of the head, neck, and trunk.

2. Describe the surface anatomy of the shoulder and upper limb.

3. Describe the surface anatomy of the pelvis and lower limb.

4. Identify the major surface features on your body or on the body of a partner.

Lab Activities

S urface anatomy is the study of anatomical landmarks that can be identified on the body surface. Most of the features are either skeletal structures or muscles and tendons. A regional approach to surface anatomy is presented in this exercise. Because the models in the photographs are muscular and have little body fat, all the anatomical landmarks discussed in this exercise are easily seen. Depending on your body type, it may be difficult to precisely identify a structure on yourself.

Lab Activity 1 Head, Neck, and Trunk

The head is a complex region where many body systems are integrated for such vital functions as breathing, eating, and speech production (**Figure 47.1**). Main surface features include the eyes and eyebrows, the nose, mouth, and ears. The zygomatic bone is the prominent cheek bone (Figure 47.1a). The surface anatomy of the head is divided into regions corresponding to the underlying bones.

The sternocleidomastoid muscle divides the neck into an **anterior cervical triangle** and a **posterior cervical triangle** (Figure 47.1b). The anterior cervical

Need More Practice and Review?

Build your knowledge—and confidence!—in the Study Area of MasteringA&P® at www.masteringaandp.com with Pre-lab Quizzes, Post-lab Quizzes, Practice Anatomy Lab™ (PAL™) 3.0 virtual anatomy practice tool, PhysioEx™ 9.0 laboratory simulations, and A&P Flix™ with Quizzes.

MasteringA&P®

Figure 47.1 Surface Anatomy of Head and Neck

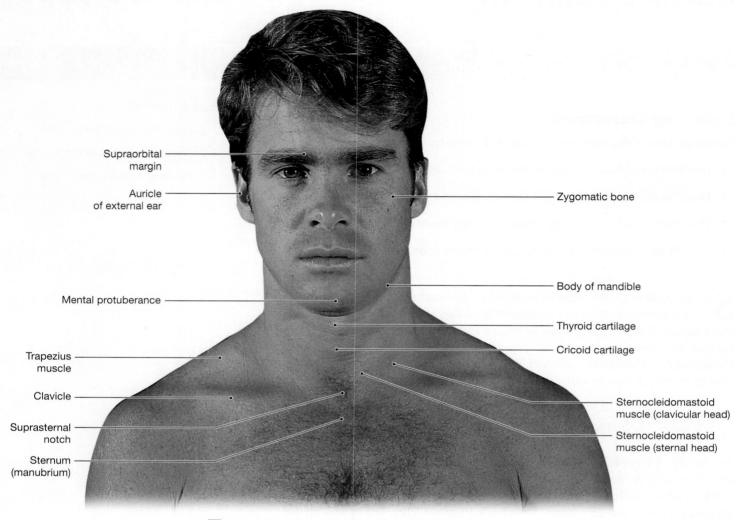

Supraorbital margin

Auricle of external ear

Zygomatic bone

Body of mandible

Mental protuberance

Thyroid cartilage

Cricoid cartilage

Trapezius muscle

Clavicle

Sternocleidomastoid muscle (clavicular head)

Suprasternal notch

Sternocleidomastoid muscle (sternal head)

Sternum (manubrium)

a Anterior view of head, neck, and upper trunk

triangle lies inferior to the mandible and anterior to the sternocleidomastoid muscle. It is subdivided into four smaller triangles, as shown in Figure 47.1c. The **suprahyoid triangle** is the superior region of the anterior neck. Inferior is the **submandibular triangle.** The **superior carotid triangle** is at the midpoint of the neck and surrounds the thyroid cartilage of the larynx. The pulse of the carotid artery is often palpated within this region. The base of the neck is the **inferior carotid triangle.**

The trunk comprises the **thorax,** or chest, the **abdominopelvic region,** and the **back** (Figures 47.2 and 47.3). The jugular notch marks the boundary between the neck and the thorax. The pectoralis major muscles, nipples, and umbilicus are prominent on the thorax. The jugular notch of the sternum can be palpated at the base of the neck. The inferior sternum is the xiphoid process. When CPR is being performed, it is critical that the xiphoid process not be pushed on during chest

compressions. Muscles of the abdomen are difficult to palpate on most individuals because of the presence of a layer of adipose tissue along the waistline. The commonly called "six pack" is the rectus abdominis muscle and the linea alba.

QuickCheck Questions

1.1 What are the two main regional divisions of the neck?

1.2 In what region of the neck is the pulse easily felt?

1.3 What are the main surface features of the thorax?

In the Lab 1

Materials

☐ Lab partner

☐ Mirror

Figure 47.1 Surface Anatomy of Head and Neck (*continued*)

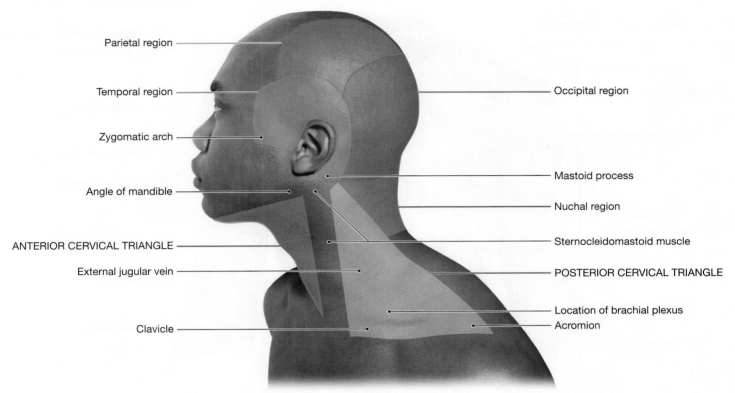

Parietal region

Temporal region

Zygomatic arch

Angle of mandible

ANTERIOR CERVICAL TRIANGLE

External jugular vein

Clavicle

Occipital region

Mastoid process

Nuchal region

Sternocleidomastoid muscle

POSTERIOR CERVICAL TRIANGLE

Location of brachial plexus

Acromion

b The posterior cervical triangles and the larger regions of the head and neck

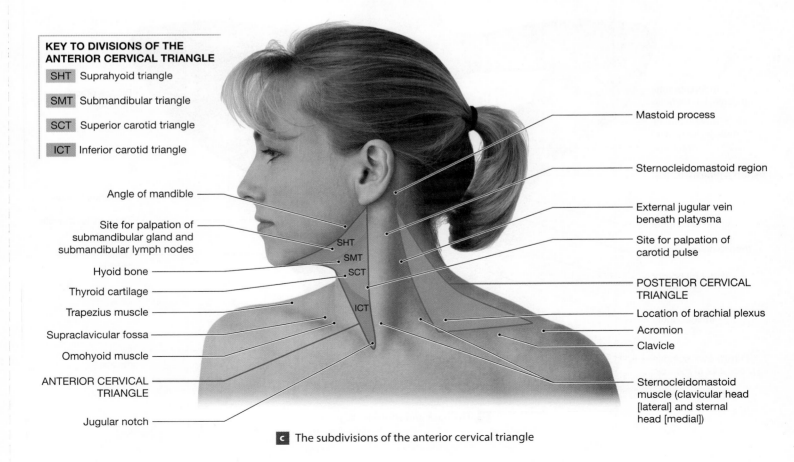

**KEY TO DIVISIONS OF THE
ANTERIOR CERVICAL TRIANGLE**

SHT	Suprahyoid triangle
SMT	Submandibular triangle
SCT	Superior carotid triangle
ICT	Inferior carotid triangle

Angle of mandible

Site for palpation of
submandibular gland and
submandibular lymph nodes

Hyoid bone

Thyroid cartilage

Trapezius muscle

Supraclavicular fossa

Omohyoid muscle

ANTERIOR CERVICAL
TRIANGLE

Jugular notch

Mastoid process

Sternocleidomastoid region

External jugular vein
beneath platysma

Site for palpation of
carotid pulse

POSTERIOR CERVICAL
TRIANGLE

Location of brachial plexus

Acromion

Clavicle

Sternocleidomastoid
muscle (clavicular head
[lateral] and sternal
head [medial])

c The subdivisions of the anterior cervical triangle

Figure 47.2 Thorax

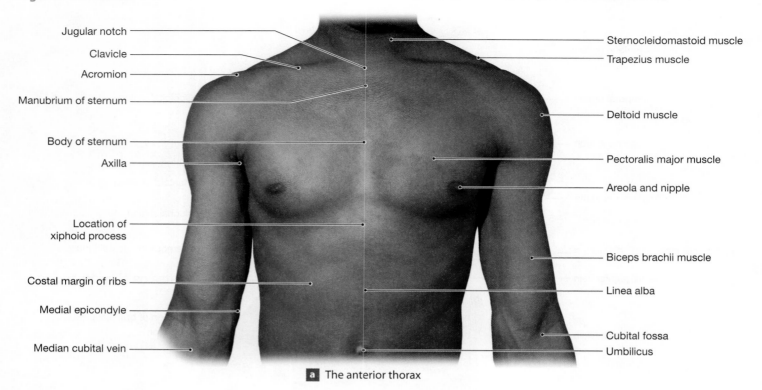

Jugular notch

Clavicle

Acromion

Manubrium of sternum

Body of sternum

Axilla

Location of
xiphoid process

Costal margin of ribs

Medial epicondyle

Median cubital vein

Sternocleidomastoid muscle

Trapezius muscle

Deltoid muscle

Pectoralis major muscle

Areola and nipple

Biceps brachii muscle

Linea alba

Cubital fossa

Umbilicus

a The anterior thorax

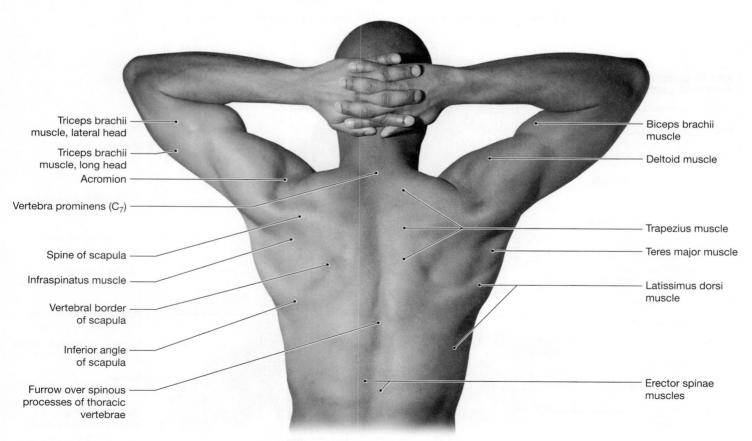

Triceps brachii
muscle, lateral head

Triceps brachii
muscle, long head

Acromion

Vertebra prominens (C$_7$)

Spine of scapula

Infraspinatus muscle

Vertebral border
of scapula

Inferior angle
of scapula

Furrow over spinous
processes of thoracic
vertebrae

Biceps brachii
muscle

Deltoid muscle

Trapezius muscle

Teres major muscle

Latissimus dorsi
muscle

Erector spinae
muscles

b The back and shoulder regions

Figure 47.3 **Abdominal Wall**

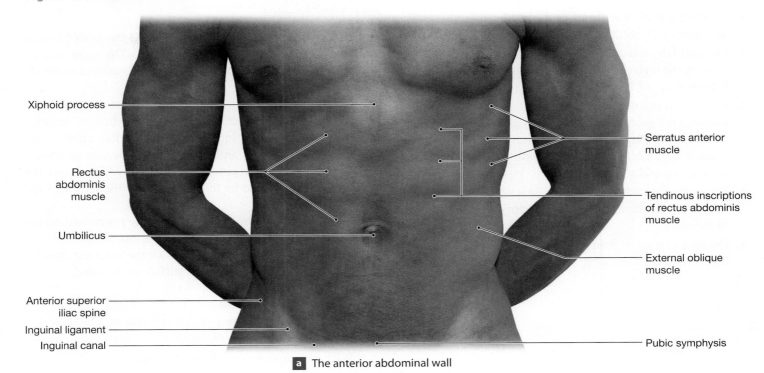

Xiphoid process

Rectus abdominis muscle

Umbilicus

Anterior superior iliac spine

Inguinal ligament

Inguinal canal

Serratus anterior muscle

Tendinous inscriptions of rectus abdominis muscle

External oblique muscle

Pubic symphysis

a The anterior abdominal wall

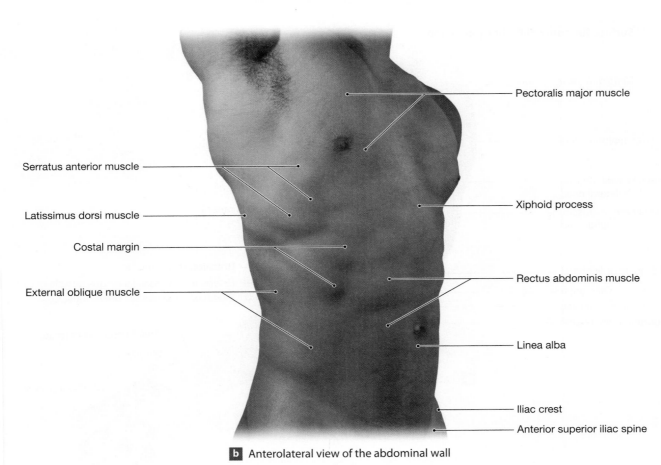

Serratus anterior muscle

Latissimus dorsi muscle

Costal margin

External oblique muscle

Pectoralis major muscle

Xiphoid process

Rectus abdominis muscle

Linea alba

Iliac crest

Anterior superior iliac spine

b Anterolateral view of the abdominal wall

Procedures

1. Review the surface anatomy of the head, neck, and trunk in Figures 47.1, 47.2, and 47.3.

2. On your lab partner or on yourself in a mirror, identify the regions of the head and neck identified in Figures 47.1, 47.2, and 47.3.

3. Palpate around the superior part of the neck and attempt to detect the submandibular salivary gland or a cervical lymph gland. Palpate midway up your neck and find the pulse in the carotid artery. This region of the neck is the superior carotid triangle.

4. Locate the thyroid cartilage and use it as reference in identifying the divisions of the anterior cervical triangle of the neck.

5. On your partner or yourself, locate the rectus abdominis muscle and the linea alba. ∎

Lab Activity 2 Shoulder and Upper Limb

The surface anatomy of the shoulder includes the deltoid muscle and bony features of the clavicle and scapula. The scapular spine and acromion are easy to palpate, as is the sternal end and body of the clavicle.

Anatomy visible on the surface of the arm includes the biceps brachii, brachialis, and triceps brachii muscles (**Figure 47.4**). In some individuals, the median cubital vein is clearly visible in the anterior of the elbow, the *antecubitis* (**Figure 47.5**). The olecranon of the ulna, which forms the point of the elbow, can be felt on the posterior surface of the elbow (Figure 47.4).

Muscles that flex the wrist and hand are positioned on the anterior forearm (Figure 47.5). The extensors are located posteriorly (Figure 47.4b). The tendons to the extensor muscles are clearly visible on the posterior surface of the hand.

QuickCheck Questions

2.1 Where are the flexor muscles of the wrist located on the forearm?

2.2 What are the three muscles of the arm?

Figure 47.4 Surface Anatomy of Right Upper Limb

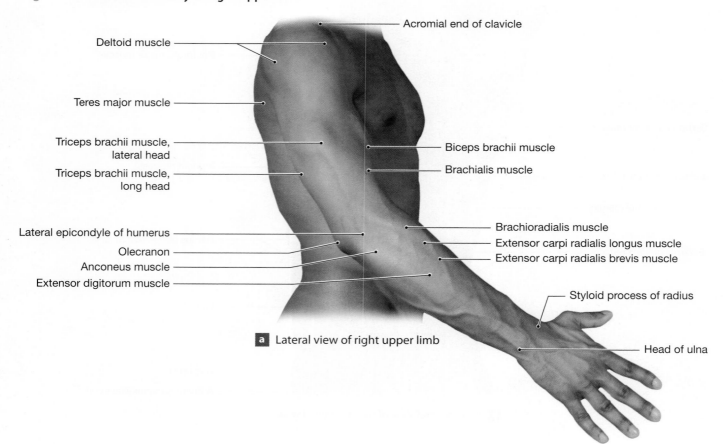

Acromial end of clavicle

Deltoid muscle

Teres major muscle

Triceps brachii muscle, lateral head

Triceps brachii muscle, long head

Biceps brachii muscle

Brachialis muscle

Lateral epicondyle of humerus

Olecranon

Anconeus muscle

Extensor digitorum muscle

Brachioradialis muscle

Extensor carpi radialis longus muscle

Extensor carpi radialis brevis muscle

Styloid process of radius

a Lateral view of right upper limb

Head of ulna

Figure 47.4 **Surface Anatomy of Right Upper Limb** (*continued*)

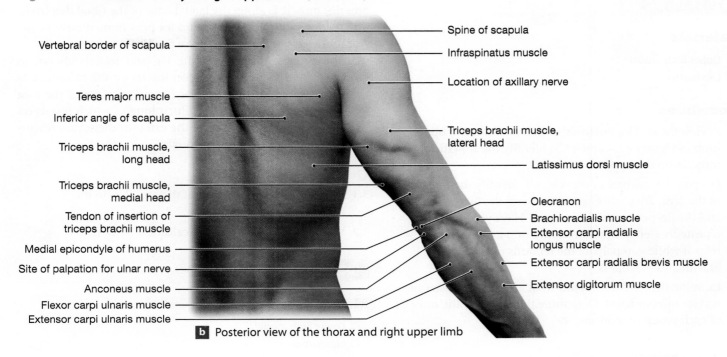

Vertebral border of scapula

Teres major muscle

Inferior angle of scapula

Triceps brachii muscle, long head

Triceps brachii muscle, medial head

Tendon of insertion of triceps brachii muscle

Medial epicondyle of humerus

Site of palpation for ulnar nerve

Anconeus muscle

Flexor carpi ulnaris muscle

Extensor carpi ulnaris muscle

Spine of scapula

Infraspinatus muscle

Location of axillary nerve

Triceps brachii muscle, lateral head

Latissimus dorsi muscle

Olecranon

Brachioradialis muscle

Extensor carpi radialis longus muscle

Extensor carpi radialis brevis muscle

Extensor digitorum muscle

b Posterior view of the thorax and right upper limb

Figure 47.5 **Arm, Forearm, and Wrist** Anterior view of left arm, forearm, and wrist.

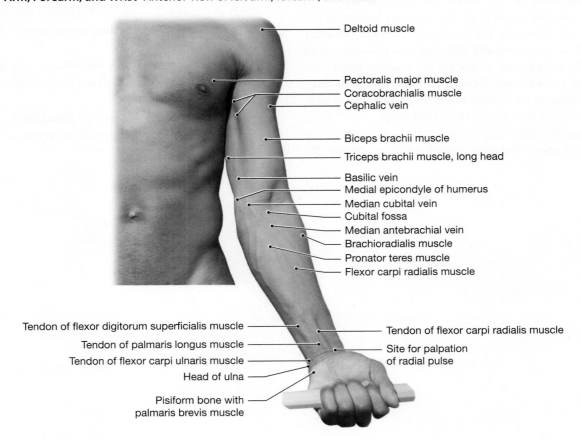

Deltoid muscle

Pectoralis major muscle

Coracobrachialis muscle

Cephalic vein

Biceps brachii muscle

Triceps brachii muscle, long head

Basilic vein

Medial epicondyle of humerus

Median cubital vein

Cubital fossa

Median antebrachial vein

Brachioradialis muscle

Pronator teres muscle

Flexor carpi radialis muscle

Tendon of flexor digitorum superficialis muscle

Tendon of palmaris longus muscle

Tendon of flexor carpi ulnaris muscle

Head of ulna

Pisiform bone with palmaris brevis muscle

Tendon of flexor carpi radialis muscle

Site for palpation of radial pulse

In the Lab 2

Materials

☐ Upper limb model
☐ Lab partner

Procedures

1. Review the surface anatomy of the shoulder and upper limb in Figures 47.4 and 47.5. Identify the superficial muscles on the upper limb model.

2. On your lab partner or on yourself, identify the muscles of the arm. Try to distinguish between the biceps brachii and the deeper brachialis muscles. Have your partner repeatedly clench her or his hand into a fist and then relax it while you palpate the tendons of the biceps brachii and brachialis.

3. Examine the tendons at your wrist and on the posterior surface of your hand. Determine the action of the muscles of each group of tendons. ■

Lab Activity 3 Pelvis and Lower Limb

The surface anatomy of the pelvis and thigh is shown in Figure 47.6. The iliac crests mark the superior border of the hips. The gluteus maximus is easily located on the posterior of the pelvis. The rectus femoris, sartorius, vastus lateralis, and vastus medialis muscles are visible on the anterior thigh, and the hamstrings are seen in the posterior view. Adductor muscles are positioned on the medial thigh; the tensor fasciae latae muscle and iliotibial tract are on the lateral thigh.

At the posterior knee, the popliteal fossa is a palpation site for the popliteal artery. The gastrocnemius and soleus muscles of the leg are well defined on many individuals. The tibialis anterior muscle is along the lateral edge of the tibial diaphysis. The tendons of the fibularis muscles pass immediately posterior to the lateral malleolus of the fibula.

The surface anatomy of the leg and foot is shown in Figure 47.7. The calcaneal tendon inserts on the calcaneus of the foot. Visible on the superior (dorsal) surface of the foot is the tendon of the extensor hallucis longus, which inserts on the big toe, and the tendons of the extensor digitorum longus muscle that insert on toes 2 through 5.

QuickCheck Questions

3.1 What is the general action of the muscles on the medial thigh?

3.2 What muscles insert on the calcaneal tendon?

In the Lab 3

Materials

☐ Lower limb model
☐ Lab partner

Procedures

1. Review the surface anatomy of the pelvis and lower limb in Figures 47.6 and 47.7. Identify the tibialis anterior, gastrocnemius, and soleus muscles on the lower limb model.

2. On your lab partner or on yourself, identify the muscles of the leg. Try to distinguish between the gastrocnemius and the soleus muscles.

3. Examine the tendons at your heel and on the anterior surface of your foot. Determine the action of the muscles of each group of tendons. ■

Figure 47.6 Pelvis and Lower Limb

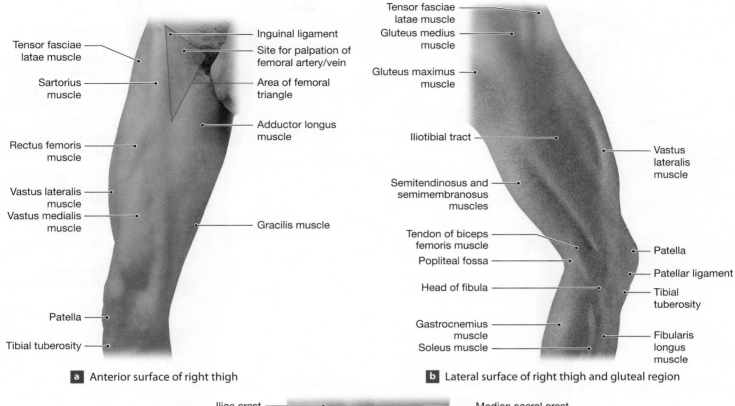

Tensor fasciae latae muscle

Sartorius muscle

Rectus femoris muscle

Vastus lateralis muscle

Vastus medialis muscle

Patella

Tibial tuberosity

Inguinal ligament

Site for palpation of femoral artery/vein

Area of femoral triangle

Adductor longus muscle

Gracilis muscle

a Anterior surface of right thigh

Tensor fasciae latae muscle

Gluteus medius muscle

Gluteus maximus muscle

Iliotibial tract

Semitendinosus and semimembranosus muscles

Tendon of biceps femoris muscle

Popliteal fossa

Head of fibula

Gastrocnemius muscle

Soleus muscle

Vastus lateralis muscle

Patella

Patellar ligament

Tibial tuberosity

Fibularis longus muscle

b Lateral surface of right thigh and gluteal region

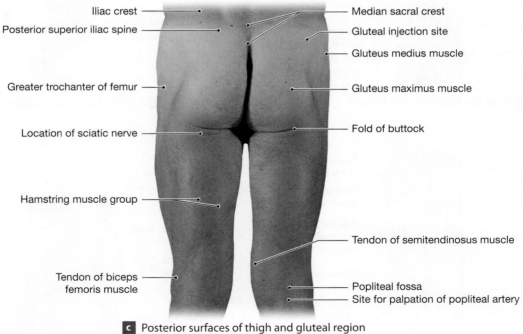

Iliac crest

Posterior superior iliac spine

Greater trochanter of femur

Location of sciatic nerve

Hamstring muscle group

Tendon of biceps femoris muscle

Median sacral crest

Gluteal injection site

Gluteus medius muscle

Gluteus maximus muscle

Fold of buttock

Tendon of semitendinosus muscle

Popliteal fossa

Site for palpation of popliteal artery

c Posterior surfaces of thigh and gluteal region

Figure 47.7 Lower Limb

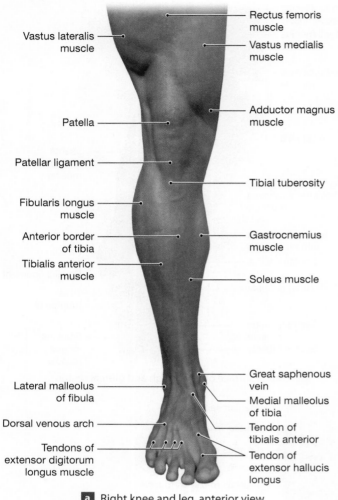

Vastus lateralis muscle
Rectus femoris muscle
Vastus medialis muscle
Patella
Adductor magnus muscle
Patellar ligament
Tibial tuberosity
Fibularis longus muscle
Anterior border of tibia
Gastrocnemius muscle
Tibialis anterior muscle
Soleus muscle
Lateral malleolus of fibula
Great saphenous vein
Medial malleolus of tibia
Dorsal venous arch
Tendon of tibialis anterior
Tendons of extensor digitorum longus muscle
Tendon of extensor hallucis longus

a Right knee and leg, anterior view

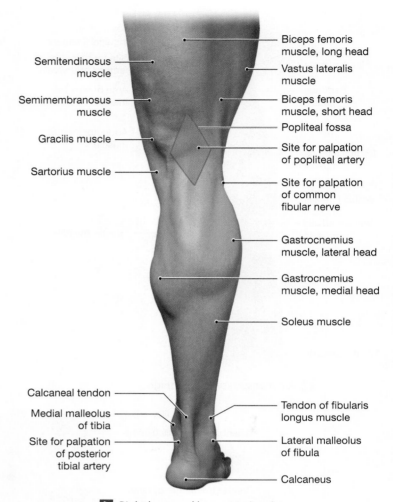

Semitendinosus muscle
Biceps femoris muscle, long head
Vastus lateralis muscle
Semimembranosus muscle
Biceps femoris muscle, short head
Gracilis muscle
Popliteal fossa
Site for palpation of popliteal artery
Sartorius muscle
Site for palpation of common fibular nerve
Gastrocnemius muscle, lateral head
Gastrocnemius muscle, medial head
Soleus muscle
Calcaneal tendon
Medial malleolus of tibia
Tendon of fibularis longus muscle
Site for palpation of posterior tibial artery
Lateral malleolus of fibula
Calcaneus

b Right knee and leg, posterior view

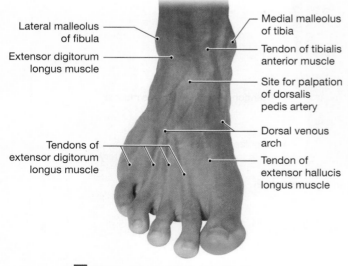

Lateral malleolus of fibula
Medial malleolus of tibia
Extensor digitorum longus muscle
Tendon of tibialis anterior muscle
Site for palpation of dorsalis pedis artery
Dorsal venous arch
Tendons of extensor digitorum longus muscle
Tendon of extensor hallucis longus muscle

c Right ankle and foot, anterior view

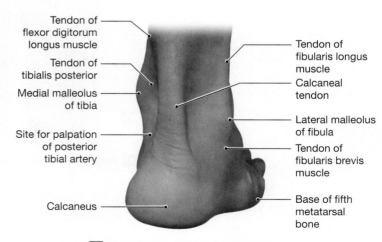

Tendon of flexor digitorum longus muscle
Tendon of fibularis longus muscle
Tendon of tibialis posterior
Calcaneal tendon
Medial malleolus of tibia
Lateral malleolus of fibula
Site for palpation of posterior tibial artery
Tendon of fibularis brevis muscle
Calcaneus
Base of fifth metatarsal bone

d Right ankle and foot, posterior view

Name _____

Date _____

Section _____

Surface Anatomy

A. Matching

Match each structure in the left column with its correct anatomical region in the right column.

_____ 1. muscle dividing neck into anterior and posterior cervical triangles

_____ 2. neck region immediately inferior to body of mandible

_____ 3. neck region immediately inferior to suprahyoid triangle

_____ 4. base of neck

_____ 5. neck region containing thyroid cartilage

_____ 6. olecranon process

_____ 7. xiphoid process

_____ 8. sartorius muscle

_____ 9. gastrocnemius muscle

_____ 10. brachialis

A. elbow

B. thigh

C. superior carotid triangle

D. sternum

E. arm

F. leg

G. suprahyoid triangle

H. submandibular triangle

I. inferior carotid triangle

J. sternocleidomastoid muscle

B. Short-Answer Questions

1. Name at least five surface features of the upper limb.

2. What tendon is visible at the posterior ankle, and which muscles insert on this tendon?

C. Labeling

Label the surface anatomy of the thorax in **Figure 47.8**.

Figure 47.8 **Anterior View of the Thorax**

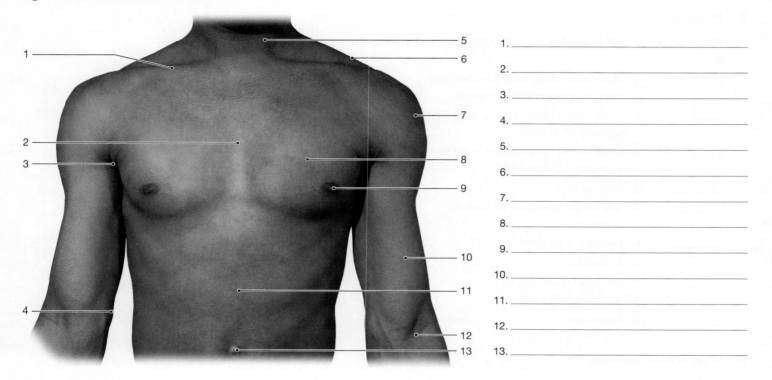

1. _____
2. _____
3. _____
4. _____
5. _____
6. _____
7. _____
8. _____
9. _____
10. _____
11. _____
12. _____
13. _____

D. Analysis and Application

1. Sam is injured on the anterior surface of his arm. What muscles may be involved in his injury?

2. An accident victim is not breathing and needs CPR. What structure of the thorax should you first palpate to properly position your hands for chest compressions? How should you then position your hands in relation to this structure?

Appendix A

Table 1	The U.S. System of Measurement		
Physical Property	**Unit**	**Relationship to Other U.S. Units**	**Relationship to Household Units**
Length	inch (in.)	1 in. = 0.083 ft	
	foot (ft)	1 ft = 12 in. = 0.33 yd	
	yard (yd)	1 yd = 36 in. = 3 ft	
	mile (mi)	1 mi = 5,280 ft = 1,760 yd	
Volume	fluidram (fl dr)	1 fl dr = 0.125 fl oz	
	fluid ounce (fl oz)	1 fl oz = 8 fl dr = 0.0625 pt	= 6 teaspoons (tsp) = 2 tablespoons (tbsp)
	pint (pt)	1 pt = 128 fl dr = 16 fl oz = 0.5 qt	= 32 tbsp = 2 cups (c)
	quart (qt)	1 qt = 256 fl dr = 32 fl oz = 2 pt = 0.25 gal	= 4 c
	gallon (gal)	1 gal = 128 fl oz = 8 pt = 4 qt	
Mass	grain (gr)	1 gr = 0.002 oz	
	dram (dr)	1 dr = 27.3 gr = 0.063 oz	
	ounce (oz)	1 oz = 437.5 gr = 16 dr	
	pound (lb)	1 lb = 7,000 gr = 256 dr = 16 oz	
	ton (t)	1 t = 2,000 lb	

Table 2	The Metric System of Measurement			
Physical Property	**Unit**	**Relationship to Standard Metric Units**	**Conversion to U.S. Units**	
Length	nanometer (nm)	1 nm = 0.000000001 m (10^{-9})	= 3.94×10^{-8} in.	25,400,000 nm = 1 in.
	micrometer (µm)	1 µm = 0.000001 m (10^{-6})	= 3.94×10^{-5} in.	25,400 mm = 1 in.
	millimeter (mm)	1 mm = 0.001 m (10^{-3})	= 0.0394 in.	25.4 mm = 1 in.
	centimeter (cm)	1 cm = 0.01 m (10^{-2})	= 0.394 in.	2.54 cm = 1 in.
	decimeter (dm)	1 dm = 0.1 m (10^{-1})	= 3.94 in.	0.25 dm = 1 in.
	meter (m)	standard unit of length	= 39.4 in.	0.0254 m = 1 in.
			= 3.28 ft	0.3048 m = 1 ft
			= 1.093 yd	0.914 m = 1 yd
	kilometer (km)	1 km = 1,000 m	= 3,280 ft	
			= 1,093 yd	
			= 0.62 mi	1.609 km = 1 mi
Volume	microliter (µl)	1 µl = 0.000001 l (10^{-6}) = 1 cubic millimeter (mm^3)		
	milliliter (mL)	1 mL = 0.001 l (10^{-3}) = 1 cubic centimeter (cm^3 or cc)	= 0.0338 fl oz	5 mL = 1 tsp
				15 mL = 1 tbsp
				30 mL = 1 fl oz
	centiliter (cL)	1 cl = 0.01 l (10^{-2})	= 0.338 fl oz	2.95 cl = 1 fl oz
	deciliter (dL)	1 dl = 0.1 l (10^{-1})	= 3.38 fl oz	0.295 dl = 1 fl oz
	liter (L)	standard unit of volume	= 33.8 fl oz	0.0295 l = 1 fl oz
			= 2.11 pt	0.473 l = 1 pt
			= 1.06 qt	0.946 l = 1 qt
Mass	picogram (pg)	1 pg = 0.000000000001 g (10^{-12})		
	nanogram (ng)	1 ng = 0.000000001 g (10^{-9})	= 0.000000015 gr	66,666,666 mg = 1 gr
	microgram (µg)	1 µg = 0.000001 g (10^{-6})	= 0.000015 gr	66,666 mg = 1 gr
	milligram (mg)	1 mg = 0.001 g (10^{-3})	= 0.015 gr	66.7 mg = 1 gr
	centigram (cg)	1 cg = 0.01 g (10^{-2})	= 0.15 gr	6.67 cg = 1 gr
	decigram (dg)	1 dg = 0.1 g (10^{-1})	= 1.5 gr	0.667 dg = 1 gr
	gram (g)	standard unit of mass	= 0.035 oz	28.4 g = 1 oz
			= 0.0022 lb	454 g = 1 lb
	dekagram (dag)	1 dag = 10 g		
	hectogram (hg)	1 hg = 100 g		
	kilogram (kg)	1 kg = 1,000 g	= 2.2 lb	0.454 kg = 1 lb
	metric ton (kt)	1 mt = 1,000 kg	= 1.1 t	
			= 2,205 lb	0.907 kt = 1 t

Temperature	Centigrade	Fahrenheit
Freezing point of pure water	0°	32°
Normal body temperature	36.8°	98.6°
Boiling point of pure water	100°	212°
Conversion	°C → °F: °F = (1.8 × °C) + 32	°F → °C: °C = (°F − 32) × 0.56

Appendix B

Eponym	Equivalent Term	Individual Referenced
THE CELLULAR LEVEL OF ORGANIZATION *(EXERCISES 5–6)*		
Golgi apparatus		Camillo Golgi (1844–1926), Italian histologist; shared Nobel Prize in 1906
Krebs cycle	Citric acid cycle, TCA cycle, or tricarboxylic acid cycle	Hans Adolph Krebs (1900–1981), British biochemist; shared Nobel Prize in 1953
THE SKELETAL SYSTEM *(EXERCISES 13–17)*		
Colles fracture		Abraham Colles (1773–1843), Irish surgeon
Haversian canals	Central canals	Clopton Havers (1650–1702), English anatomist and microscopist
Haversian systems	Osteons	Clopton Havers
Pott's fracture		Percivall Pott (1713–1788), English surgeon
Sharpey's fibers	Perforating fibers	William Sharpey (1802–1880), Scottish histologist and physiologist
Volkmann's canals	Perforating canals	Alfred Wilhelm Volkmann (1800–1877), German surgeon
Wormian bones	Sutural bones	Olas Worm (1588–1654), Danish anatomist
THE MUSCULAR SYSTEM *(EXERCISES 17–22)*		
Achilles tendon	Calcaneal tendon	Achilles, hero of Greek mythology
Cori cycle		Carl Ferdinand Cori (1896–1984) and Gerty Theresa Cori (1896–1957), American biochemists; shared Nobel Prize in 1947
THE NERVOUS SYSTEM *(EXERCISES 23–26)*		
Broca's area	Speech center	Pierre Paul Broca (1824–1880), French surgeon
Foramen of Lushka	Lateral foramina	Hubert von Lushka (1820–1875), German anatomist
Meissner's corpuscles	Tactile corpuscles	Georg Meissner (1829–1905), German physiologist
Merkel discs	Tactile discs	Friedrich Siegismund Merkel (1845–1919), German anatomist
Foramen of Munro	Interventricular foramen	John Cummings Munro (1858–1910), American surgeon
Nissl bodies		Franz Nissl (1860–1919), German neurologist
Pacinian corpuscles	Lamellated corpuscles	Fillippo Pacini (1812–1883), Italian anatomist
Purkinje cells		Johannes E. Purkinje (1787–1869), Bohemian anatomist and physiologist
Nodes of Ranvier	Nodes	Louis Antoine Ranvier (1835–1922), French physiologist
Island of Reil	Insula	Johann Christian Reil (1759–1813), German anatomist
Fissure of Rolando	Central sulcus	Luigi Rolando (1773–1831), Italian anatomist
Ruffini corpuscles		Angelo Ruffini (1864–1929), Italian anatomist
Schwann cells	Neurolemmocytes	Theodor Schwann (1810–1882), German anatomist
Aqueduct of Sylvius	Cerebral aqueduct, aqueduct of the midbrain, or mesencephalic aqueduct	Jacobus Sylvius (Jacques Dubois, 1478–1555), French anatomist
Sylvian fissure	Lateral sulcus	Franciscus Sylvius (Franz de le Boë, 1614–1672), Dutch anatomist
Pons varolii	Pons	Costanzo Varolio (1543–1575), Italian anatomist

(continued)

Eponym	Equivalent Term	Individual Referenced
SENSORY FUNCTION (EXERCISES 27–28)		
Organ of Corti	Spiral organ	Alfonso Corti (1822–1888), Italian anatomist
Eustachian tube	Auditory tube	Bartolomeo Eustachio (1520–1574), Italian anatomist
Golgi tendon organs	Tendon organs	Camillo Golgi (1844–1926), Italian histologist; shared Nobel Prize in 1906
Hertz (Hz)		Heinrich Hertz (1857–1894), German physicist
Meibomian glands	Tarsal glands	Heinrich Meibom (1638–1700), German anatomist
Canal of Schlemm	Scleral venous sinus	Friedrich S. Schlemm (1795–1858), German anatomist
THE ENDOCRINE SYSTEM (EXERCISE 33)		
Islets of Langerhans	Pancreatic islets	Paul Langerhans (1847–1888), German pathologist
Interstitial cells of Leydig	Interstitial Cells	Franz von Leydig (1821–1908), German anatomist
THE CARDIOVASCULAR SYSTEM (EXERCISES 34–37)		
Bundle of His	AV Bundle	Wilhelm His (1863–1934), German physician
Purkinje fibers		Johannes E. Purkinje (1787–1869), Bohemian anatomist and physiologist
Frank-Starling principle (Starling's law)		Otto Frank (1865–1944), German physiologist, and Ernest Henry Starling (1866–1927), English physiologist
Circle of Willis	Cerebral arterial circle	Thomas Willis (1621–1675), English physician
THE LYMPHATIC SYSTEM (EXERCISE 38)		
Hassall's corpuscles	Thymic corpuscles	Arthur Hill Hassall (1817–1894), English physician
Kupffer cells	Stellate reticuloendothelial cells	Karl Wilhelm Kupffer (1829–1902), German anatomist
Langerhans cells	Dendritic cells	Paul Langerhans (1847–1888), German pathologist
Peyer's patches	Aggregated lymphoid nodules	Johann Conrad Peyer (1653–1712), Swiss anatomist
THE RESPIRATORY SYSTEM (EXERCISES 39–40)		
Bohr effect		Christian Bohr (1855–1911), Danish physiologist
Boyle's law		Robert Boyle (1621–1691), English physicist
Charles' law		Jacques Alexandre César Charles (1746–1823), French physicist
Dalton's law		John Dalton (1766–1844), English physicist
Henry's law		William Henry (1775–1837), English chemist
THE DIGESTIVE SYSTEM (EXERCISES 41–42)		
Plexus of Auerbach	Myenteric plexus	Leopold Auerbach (1827–1897), German anatomist
Brunner's glands	Duodenal glands	Johann Conrad Brunner (1653–1727), Swiss anatomist
Kupffer cells	Stellate reticuloendothelial cells	Karl Wilhelm Kupffer (1829–1902), German anatomist
Crypts of Lieberkühn	Intestinal glands	Johann Nathaniel Lieberkühn (1711–1756), German anatomist
Plexus of Meissner	Submucosal plexus	Georg Meissner (1829–1905), German physiologist
Sphincter of Oddi	Hepatopancreatic sphincter	Ruggero Oddi (1864–1913), Italian physician
Peyer's patches	Aggregated lymphoid nodules	Johann Conrad Peyer (1653–1712), Swiss anatomist
Duct of Santorini	Accessory pancreatic duct	Giovanni Domenico Santorini (1681–1737), Italian anatomist
Stensen duct	Parotid duct	Niels Stensen (1638–1686), Danish physician/priest
Ampulla of Vater	Duodenal ampulla	Abraham Vater (1684–1751), German anatomist
Wharton duct	Submandibular duct	Thomas Wharton (1614–1673), English physician
Duct of Wirsung	Pancreatic duct	Johann Georg Wirsung (1600–1643), German physician

(continued)

Eponym	Equivalent Term	Individual Referenced
THE URINARY SYSTEM *(EXERCISES 43–44)*		
Bowman's capsule	Glomerular capsule	Sir William Bowman (1816–1892), English physician
Loop of Henle	Nephron loop	Friedrich Gustav Jakob Henle (1809–1885), German histologist
THE REPRODUCTIVE SYSTEM *(EXERCISES 45–46)*		
Bartholin's glands	Greater vestibular glands	Casper Bartholin, Jr. (1655–1738), Danish anatomist
Cowper's glands	Bulbo-urethral glands	William Cowper (1666–1709), English surgeon
Fallopian tube	Uterine tube/oviduct	Gabriele Fallopio (1523–1562), Italian anatomist
Graafian follicle	Tertiary follicle	Reijnier de Graaf (1641–1673), Dutch physician
Interstitial cells of Leydig	Interstitial cells	Franz von Leydig (1821–1908), German anatomist
Glands of Littré	Lesser vestibular glands	Alexis Littré (1658–1726), French surgeon
Sertoli cells	Nurse cells, sustentacular cells	Enrico Sertoli (1842–1910), Italian histologist

Photo Credits

Index